Transition States of Biochemical Processes

Transition States of Biochemical Processes

Edited by

Richard D. Gandour
Louisiana State University

and

Richard L. Schowen
University of Kansas

Plenum Press · New York and London

Library of Congress Cataloging in Publication Data

Main entry under title:

Transition states of biochemical processes.

Includes bibliographical references and index.
1. Biological chemistry. 2. Chemical reaction, Conditions and laws of. I. Gandour,
Richard D. II. Schowen, Richard L. [DNLM: 1. Biochemistry. QU4.3 T772]
QP514.2.T7   574.1'92   78-6659
ISBN 0-306-31092-9

© 1978 Plenum Press, New York
A Division of Plenum Publishing Corporation
227 West 17th Street, New York, N. Y. 10011

Printed in the United States of America

Preface

The transition-state theory has been, from the point of its inception, the most influential principle in the development of our knowledge of reaction mechanisms in solution. It is natural that as the field of biochemical dynamics has achieved new levels of refinement its students have increasingly adopted the concepts and methods of transition-state theory. Indeed, every dynamical problem of biochemistry finds its most elegant and economical statement in the terms of this theory. Enzyme catalytic power, for example, derives from the interaction of enzyme and substrate structures in the transition state, so that an understanding of this power must grow from a knowledge of these structures and interactions. Similarly, transition-state interactions, and the way in which they change as protein structure is altered, constitute the pivotal feature upon which molecular evolution must turn. The complete, coupled dynamical system of the organism, incorporating the transport of matter and energy as well as local chemical processes, will eventually have to yield to a description of its component transition-state structures and their energetic response characteristics, even if the form of the description goes beyond present-day transition-state theory. Finally, the importance of biochemical effectors in medicine and agriculture carries the subject into the world of practical affairs, in the use of transition-state information for the construction of ultrapotent biological agents.

Because there has been no general treatment of transition-state approaches to mechanistic biochemistry, we thought it worthwhile to compile this volume, in which the principles and some of the findings and prospects are set forth by leading scientists in the field. We hope the volume will serve several purposes. It ought to introduce the advantages and disadvantages of these techniques to mechanistic biochemists who are not now using them. It should be a useful textbook for students and for researchers in other fields who may be planning work in mechanistic biochemistry. We are particularly optimistic that it may encourage biochemists to use methods traditionally associated with physical organic chemistry while at the same time alerting other mechanisms scientists to the difficult and exciting unsolved problems of mechanistic biochemistry.

The book has four parts. In the first, the basic principles of transition-state theory are reviewed and set in a biochemical context. In the second part there are fairly detailed presentations of the major techniques in current use. In the third part, transition-state information about several important processes is discussed. Finally a brief consideration is given to the use of transition-state concepts in inhibitor and drug design.

The authors have aimed in all cases to write chapters accessible to scientists inexperienced in the subject, capable of bringing the reader up to the level of current research, but nevertheless containing much of interest to the specialist as well. We believe they have all succeeded. Our principle of editing was to persuade, in each case, a scientist whose knowledge and ideas we knew to be thorough and valuable to write about his subject. Thereafter, we took a *laissez-faire* attitude. The reader will therefore detect a certain amount of repetition, as the same matters are taken up by several authors. A Russian proverb says, "Repetition is the mother of learning," in agreement with which we take the view that many will find successive encounters with the same subject, seen from various angles, as helpful and welcome. Indeed, in a not inconsiderable number of cases, direct conflicts between the views of different authors will be found. This is quite in accord with the mode of operation of science and gives an accurate picture of opinion in the field. We hope the reader will consider this feature an advantage.

Naturally the selection of subjects to be treated reveals various idiosyncracies of the editors. We have tried to be as broad as possible in the selection, without duplicating extensively material easily available elsewhere. Because of the importance of enzyme mechanisms as a field for the application of transition-state theory to biochemistry at the current time, there is a certain concentration on methods especially adaptable to enzyme systems. Introductory notes are supplied at the beginning of each of the four parts of the book to show how the chapters fit together with each other and with the general shape of thought in the field.

The task of persuading overworked scientists to write contributions for this volume and the companion task of extracting the latest chapter from its author before the earliest one received had become thoroughly outdated have had their challenging aspects. Nevertheless, it is hard to imagine a more cooperative and patient group than our colleagues in this enterprise, whom we offer our cordial thanks.

Richard D. Gandour
Richard L. Schowen

Contributors

Stephen J. Benkovic, Department of Chemistry, Pennsylvania State University, University Park, Pennsylvania 16802

Herbert G. Bull, Department of Chemistry, Indiana University, Bloomington, Indiana 47401

Ralph E. Christoffersen, Department of Chemistry, University of Kansas, Lawrence, Kansas 66045

Eugene H. Cordes, Department of Chemistry, Indiana University, Bloomington, Indiana 47401

James K. Coward, Department of Pharmacology, Yale University School of Medicine, New Haven, Connecticut 06510

Richard D. Gandour, Department of Chemistry, Louisiana State University, Baton Rouge, Louisiana 70803

Mohamed F. Hegazi, Department of Chemistry, University of Kansas, Lawrence, Kansas 66045

John L. Hogg, Department of Chemistry, Texas A & M University, College Station, Texas 77843

Judith P. Klinman, The Institute for Cancer Research, Fox Chase Cancer Center, Philadelphia, Pennsylvania 19111

Gerald M. Maggiora, Department of Biochemistry, University of Kansas, Lawrence, Kansas 66045

Albert S. Mildvan, The Institute for Cancer Research, Fox Chase Cancer Center, Philadelphia, Pennsylvania 19111

Marion H. O'Leary, Department of Chemistry, University of Wisconsin, Madison, Wisconsin 53706

Ralph M. Pollack, Department of Chemistry, University of Maryland Baltimore County, Baltimore, Maryland 21228

Daniel M. Quinn, Department of Chemistry, University of Kansas, Lawrence, Kansas 66045

Katharine B. J. Schowen, Department of Chemistry, Baker University, Baldwin City, Kansas 66006

Richard L. Schowen, Department of Chemistry, University of Kansas, Lawrence, Kansas 66045

Keith J. Schray, Department of Chemistry, Lehigh University, Bethlehem. Pennsylvania 18015

Eli Shefter, Department of Pharmaceutics, School of Pharmacy, State University of New York, Amherst, New York 14260

Edward R. Thornton, Department of Chemistry, University of Pennsylvania, Philadelphia, Pennsylvania 19174

Elizabeth K. Thornton, Department of Chemistry, Widener College, Chester, Pennsylvania 19013

Richard Wolfenden, Department of Biochemistry, University of North Carolina, Chapel Hill, North Carolina 27514

Contents

4 • Primary Hydrogen Isotope Effects

Judith P. Klinman

5 • Secondary Hydrogen Isotope Effects

John L. Hogg

6 • Solvent Hydrogen Isotope Effects

Katharine B. J. Schowen

7 • Heavy-Atom Isotope Effects in Enzyme-Catalyzed Reactions

Marion H. O'Leary

8 • Magnetic-Resonance Approaches to Transition-State Structure
Albert S. Mildvan

9 • Mapping Reaction Pathways from Crystallographic Data
Eli Shefter

Part III
Studies of Transition-State Properties in Enzymic and Related Reactions

10 • Transition-State Properties in Acyl and Methyl Transfer
Mohamed F. Hegazi, Daniel M. Quinn, and Richard L. Schowen

Part IV
Applications of Biochemical Transition-State
Information 553

15 • Transition-State Affinity as a Basis for the Design of Enzyme Inhibitors
Richard Wolfenden

16 • Transition-State Theory and Reaction Mechanism in Drug Action and Drug Design
James K. Coward

Transition States of Biochemical Processes

PART I

The Role of the Transition State in Chemical and Biological Catalysis

In this initial section the applications of transition-state theory that are important to biochemistry are explored. Aspects that focus on the "chemical" side of the question (that is, problems which already arise in small-molecule chemistry) are the substance of Chapter 1, by Thornton and Thornton. Problems that come into view at the "biological" level, particularly in enzyme catalysis, are treated in Chapter 2.

The Thorntons discuss what transition states are "in reality," what can be gained by using the idea of a transition state and what precautions one must exercise. Since a transition state is defined, in one sense, by the absolute rate theory, they examine this theory, its assumptions, its place in theoretical chemical kinetics, and the implications for practical use of the concept in mechanisms research. Because transition states are, in a different sense, defined by experiments, the Thorntons show what experiments can be performed to discover the properties of transition states. Finally they address the question, obviously critical if the findings in simplified systems are to be generalized to a biological context, of how transition states are likely to respond to changes in structure and environment.

In Chapter 2, Schowen relates the prevailing or "canonical" descriptions of enzyme catalytic action (the proximity effect, the destabilization of reactants through geometrical distortion or desolvation, etc.) to a "Fundamentalist" view of catalysis as originating solely in the stabilization of the transition state through binding to the catalyst. He shows that a simple translation of the two formulations is always possible. The relationship of transition-state interpretations of experimental findings to the customary parameters of enzyme kinetics is described. He discusses transition-state structure for situations in which more than one step determines the rate and develops the idea of a "virtual transition-state structure."

These two chapters should provide the reader with a firm basis for the critical appreciation of transition-state concepts applied to biochemical problems and for proceeding to Part II on methods of determining transition-state structures.

Scope and Limitations of the Concept of the Transition State

Elizabeth K. Thornton and Edward R. Thornton

1. INTRODUCTION

The study of biological reactions in ever-increasing numbers has raised intricate questions whose answers can best be sought through theories and experimental techniques which, until recently, were applied primarily to nonbiological (organic) systems. One such theory is that of absolute reaction rates (ART), commonly called transition-state theory (TST). TST has been evoked innumerable times in the treatment of experimental data and has provided the underlying concept for most mechanistic studies. Central to the theory is the transition state.

It is tempting to say that the scope of the concept of the transition state is very wide and the limitations are very few. The transition state as a physical entity is well defined, and it plays an obviously important role in every thermal rate process, since it is defined as the top of the energy barrier separating reactants from products. This optimistic view must be qualified by explaining that the quantitative relationship between the properties of the transition state and the rate of reaction, which is the crucial theoretical problem, can only be approximated for reactions of any significant complexity. Therefore, while the concept is sound, we are a very long way from being able to use it to calculate the rate of, say, an enzyme-catalyzed reaction.

We shall spend half of this chapter defining and qualifying, because the application of the concept of the transition state to experimental study is

Elizabeth K. Thornton • Department of Chemistry, Widener College, Chester, Pennsylvania. *Edward R. Thornton* • Department of Chemistry, University of Pennsylvania, Philadelphia, Pennsylvania.

highly dependent on knowing what the inherent approximations are for each kind of experiment. In the second half we shall be concerned with experimental approaches to transition-state structure. We hope to have shown that mapping the transition state for a reaction is a fascinating structural problem and that the connection between transition-state structure and rate is direct and conceptually useful, even though in most cases it is not quantitative.

2. NATURE OF THE TRANSITION STATE

2.1. Definition of the Transition State: Transition-State Theory

A measure of the usefulness and strength of transition-state theory (or theory of absolute reaction rates) is its success in predicting, with acceptable accuracy, the rate of a chemical reaction. Since it is not feasible at this time to use the Schrödinger equation to calculate rates for any but the simplest of systems, we sacrifice a certain degree of rigor in return for a prediction which will come close to experimental observation.

A classical mechanical expression for the rate constant of a reaction* derived by Johnston[4] and based on the transition-state theory of reaction rates (or ART) is given by

$$k = (f_{\ddagger}^{3N-1}/f_A^{\alpha} f_B^{\beta} f_C^{\gamma})(\mathbf{k}T/\mathbf{h}) \exp(-\Delta\varepsilon_0^{\ddagger}/\mathbf{k}T)\langle 1 + \phi \rangle_{6N-6}^{T} \langle \kappa \rangle_{6N-8}^{B} \qquad (1)$$

for the reaction $\alpha A + \beta B + \gamma C \rightarrow [\text{T. S.}]^{\ddagger} \rightarrow$ products, where f_A, f_B, and f_C are partition functions per unit volume for reactants; $f_{\ddagger}^{3N-1}$ is the partition function of the activated complex (as defined by the theory[5] and commonly referred to as the transition state) for $3N - 1$ degrees of freedom, N being the number of nuclei in the structure, i.e., excluding the reaction coordinate motion; and $\langle 1 + \phi \rangle_{6N-6}^{T}$ is an average value of the ratio of actual concentration to equilibrium concentration of activated complexes for those complexes that react, $6N - 6$ referring to the fact that only $3N - 3$ coordinates and $3N - 3$ momenta of the activated complex are included since it must be assumed that the three translations of the center of mass of the reactants and activated complex are at equilibrium in order to have a defined temperature T. The average reaction probability for complexes which cross the energy barrier separating reactants from products but which may have, in principle, some probability of returning to the reactant side of the energy barrier without becoming equilibrated product molecules is $\langle \kappa \rangle_{6N-8}^{B}$. The $6N - 8$ refers to the fact that only $3N - 4$ coordinates and $3N - 4$ momenta of the activated complex are included since the three translations of the center of mass and also

---

* For a complete derivation of this expression, as well as others frequently used in absolute rate theory (ART), see References 1, 2, and 3.

the reaction coordinate motion are omitted (the translations are omitted because κ cannot depend on the position in space or momentum of the center of mass, and the reaction coordinate motion is omitted because classically κ for the reaction coordinate motion must be either zero or unity, depending on the energy in the reaction coordinate, which is exactly taken into account in the factor $\mathbf{k}T/\mathbf{h}$). The terms $\langle 1 + \phi \rangle^T_{6N-6}$ and $\langle \kappa \rangle^B_{6N-8}$ are approximate corrections for the nonflat potential energy surface between reactants and activated complexes and activated complexes and products, respectively[6]; their omission simply gives the ordinary classical form of transition-state theory.

Since, for real systems at ordinary temperatures, the vibrations (at least) have to be treated quantum mechanically, Eq. (1) can be replaced by

$$k = (f_{\ddagger}^{3N-1} \Gamma^* / f_A^\alpha f_B^\beta f_C^\gamma)(\mathbf{k}T/\mathbf{h}) \exp(-\Delta \varepsilon_0^{\ddagger}/\mathbf{k}T) \langle 1 + \phi \rangle^T_{6N-6} \langle \kappa \rangle^B_{6N-8} \qquad (2)$$

where the f's are quantum-mechanical partition functions and Γ^* is the barrier penetration correction for the reaction coordinate motion.

It is commonly assumed that $\phi = 0$ in Eqs. (1) and (2), i.e., that there is thermal (Boltzmann) equilibrium between activated complex and reactants. The validity of this assumption has been questioned in light of possible nonequilibrium solvation effects in proton transfers.[7] Although "nonequilibrium proton transfer" has been demonstrated,[7,8] a violation of the thermal equilibrium assumption has not been demonstrated in liquid solution where collisions are extremely frequent. A distinction should be made between those systems "not at equilibrium" with respect to products (and, potentially, to intermediates) in a rate process not yet at equilibrium and systems in which "non-Boltzmann solvation" and the like are involved. In the former, an equilibrium situation between transition state and immediate precursor will exist, and unstable intermediates will be present in steady-state (but not necessarily in equilibrium) concentration, while in the latter, a lack of Boltzmann equilibrium of the transition state with its immediate precursor is implied. "Nonequilibrium proton transfers"[8] and many nonequilibrium solvation effects are examples of the former case. That solvent molecules may not be capable of "following" fast reaction coordinate motions such as proton transfers is a demonstration of uncoupled motions of light and heavy nuclei[9] in the normal vibrations of a particular equilibrium nuclear geometry (ENG), which, in this case, happens to be a transition state.

Tests of the transition-state theory (ART), carried out on simple three-atom collinear reaction systems,[10-13] have indicated its validity within the limits set by the various approximations (corrections) and variety of potential energy surfaces used. Attempts at refinement of the theory through elimination of approximations[14,15] and redefinition of the transition state[16] have resulted in treatments which, while possibly enabling a more satisfactory *a priori* calculation of reaction rates, still retain the concept of a transition state; questions of the validity of transition-state theory would appear to be questions of how precisely the measured rate constants reflect the true structure of the transition state.

2.2. Existence and Structure of the Transition State

The accuracy of the calculation of an absolute reaction rate within the framework of transition-state theory (TST) is certainly limited by the various necessary assumptions as well as by the lack of complete knowledge of the detailed shape of the potential surface of the reacting system[10-12] (although, with the aid of the computer, surfaces for some very simple systems have been plotted). TST is therefore approximate with respect to predictions of absolute rates, but it is also true that TST is conceptually precise. It defines a reaction mechanism based on a transition state with a well-defined equilibrium nuclear geometry (ENG). Despite the fact that, based on this ENG, it may not be possible to calculate a reaction rate with absolute precision because this depends on an average quantum-mechanical transmission coefficient, averaged over each (∞) energy level of reactant (and its population) and its probability of going to each of the (∞) energy levels of the products (i.e., the "number of effective reaction paths"), nevertheless, each reaction must have one or more well-defined transition-state ENG's (with zero forces on the nuclei) which interconnect any two ENG-stable structures.

To be sure, a transition state has a *metastable ENG* (rather than the *stable ENG* of an ordinary molecule) due to its single coordinate of negative curvature and may be best thought of more as a *construct* which has certain associated properties and a definite ENG than as an isolable "species" of chemical "reality," although if thought of as the latter, then exceedingly short-lived. However, to make *use* of the transition state does not necessitate its population and lifetime as a thermodynamic species. It should obey rules such that it is present at "equilibrium concentration" (from which approximate predictions of reaction rates can be made, using this concentration and its "frequency of decomposition"); it vibrates with $3N^{\ddagger} - 7$ normal coordinate motions ($3N^{\ddagger} - 6$ if linear) and has well-defined thermodynamic properties. Vibrations must be restricted to the $3N^{\ddagger} - 7$ (or $3N^{\ddagger} - 6$) normal coordinates, in that vibration along the negatively curved "reaction coordinate" results in a distorting (rather than restoring) force and a destruction of the transition-state ENG.

Despite the "elusiveness" of transition states (their nonisolability) and the difficulties involved in structure proof, they are, *in principle*, observable[17] (although direct observation is prohibited by experimental limitations). In fact, their elucidation is a fine structural goal, the achievement of which would provide the information necessary to detail the sequence of intermediate states between reactants and products, i.e., the reaction mechanism, as well as to calculate such rate-related quantities as Hammett ρ values, Brønsted α and β values, and isotope effects. (Of course, at present it is the latter experimental quantities which are used to shed light on transition-state structure.)

Several early attempts to define a transition-state structure using only steric parameters and to test the validity of the structure by calculating rates and thermodynamic quantities based on the structure have been reasonably successful. For certain types of reactions, namely those involving interconversion between isomers controlled primarily by steric factors, one can easily envision a transition-state structure for which a reasonably good molecular model (of a "stable" molecule) exists. Providing pertinent data are available, quantitative (or semiquantitative) calculations can be made which, when compared with experimental parameters, give information about the validity of the model chosen. A pioneering and elegant *tour de force* by Westheimer and co-workers[18-21] demonstrates this approach. With parameters from "real" molecules (a molecular mechanics approach using force constants for bending and stretching various bonds and van der Waals potentials for nonbonded interactions between pertinent atoms or groups of atoms) and their application to planar biphenyls (models for the transition states), calculations of the activation energies for racemization of 2,2'-dibromo-4,4'-dicarboxybiphenyl, for 2,2'-diiodo-5,5'-dicarboxybiphenyl, and for 2,2',3,3'-tetraiodo-5,5'-dicarboxybiphenyl were carried out (an admirable undertaking, considering the lack of computerized help at the time). Since the racemization of these *ortho*-substituted biphenyls will occur via a molecular configuration which leads to the smallest value of the activation energy, the latter was minimized. Despite the approximations and uncertainties inherent in the calculations, very good agreement with experimental values was obtained. The small differences (the worst case was 5.8 kcal/mol) can easily be ascribed to the approximations used. The results give a very clear picture of bond and angle displacements and their corresponding energy requirements for the planar biphenyl transition states. Since the onset of the computer age, this basic method of Westheimer has been further developed and extended to calculations on a wide variety of molecules.[22-25]

Ingold and co-workers[26-28] have calculated transition-state geometries for some S_N2 displacement reactions, again using only steric parameters. In these cases, however, the situation is more complicated than for those reactions in which no bonds are being made or broken. Many simplifying assumptions were made, and calculations were performed for reactions leading only to symmetrical transition states. As only energy differences between reactions were considered, certain energy terms could be omitted (since they ultimately cancelled). Comparison between calculated and observed relative rate parameters for S_N2 displacements on methyl, ethyl, *i*-propyl, *t*-butyl, *n*-propyl, *i*-butyl, and neopentyl bromides (by bromide ion) gave remarkably good agreement, considering the approximations which were made.

Another isomeric interconversion, although one which is clearly not controlled *primarily* by steric factors, is that between staggered ethane structures. The rotation about the C—C bond requires passage through a well-defined, again easily envisioned transition state, i.e., the eclipsed, higher-energy species. In fact, by providing the appropriate restraining forces to overcome the inherent instability in a transition state, one could, in principle, "freeze" it into a stable molecule. For example, if the "real" molecule bicyclo[2.1.1]hexane were perfectly rigid and unable to distort (even slightly) to relieve some of the unfavorable (repulsive) forces between nonbonded atoms, it would contain within its structure the "frozen" transition state for interconversion of the staggered ethane conformers. There are obviously strong structural similarities between "real" molecules having stable ENG's and transition states having metastable ENG's; these will be examined in greater detail in a later section.

2.3. Relationship of the Transition State and Intermediates to Reaction Mechanism

If one is interested in reaction *mechanisms*, consideration must be given to both the structures of molecules and the energy requirements for distorting them. A valuable analogy pictured in Fig. 1 relates chemical reactions to the wanderings of molecules among mountains, where height corresponds to energy and where the paths most easily taken are the ones involving the least climbing and therefore proceed through mountain passes rather than across the summits.

Fig. 1. Molecular meanderings in the mountains. [Reprinted with permission from G. C. Pimentel and R. D. Spratley, *Understanding Chemical Thermodynamics*, Holden-Day, Inc., San Francisco (1970), p. 149, copyright G. C. Pimentel and R. D. Spratley.]

It happens that, under the control of heat, molecular structures are determined by the valleys—the lower the valley floor and the wider the valley, the more stable the structure. Also, reaction rates are determined by the mountain passes—the lower the top of the pass and the wider, the faster the reaction. The terrain between valley floor and pass is thermally irrelevant, for all routes from the one to the other can and will be taken. Nobody knows much about the regions between valley bottoms and passes for molecules larger than diatomic.

It is recognized that when one talks about the effect of structure upon reaction rates, the structural changes investigated do not constitute an inert probe of the rate process. Rather, changing structure changes not only the *rate* (height and width of the pass) but also the actual *geometry*—bond lengths and bond angles—characteristic of that pass, or *transition state*.

The transition state, conceptually the most important geometry in characterizing the molecular distortions which lead to chemical reactions, can now be defined as the saddle point or mountain pass on such an energy surface. It is the most *unfavorable* geometry (a metastable ENG) of a reaction complex along the most favorable (energetically) reaction path leading to products. It is a *true equilibrium geometry* in the sense that it is a geometric arrangement of nuclei in which no forces are exerted on those nuclei.

For any chemical reaction, then, there must be *at least* one transition state involved in the interconversion of one stable ENG into another.* Since many reactions occur in more than one elemental step, one must consider transition states for *each* of these steps. If one of these steps is considerably slower than the others, such that the overall rate for the reaction is determined essentially by the rate of conversion of the species involved in this one step, then it becomes the *rate-determining step*, and characterizing the transition state for this step would become a prime objective in elucidating the reaction pathway (mechanism). Commonly, that transition state with the highest energy is called the *rate-determining transition state*, although there are cases in which two or more transition states of nearly equal energy can each partially control the overall reaction rate. In such cases, one would want to know as much as possible about the transition-state geometry for each of the steps which contribute to the overall rate.

For these multistep cases, one recognizes that there must exist at least one stable ENG between reactants and products; this is called a reaction intermediate. Two classes of intermediates occurring prior to some rate-determining transition state, as well as two classes of transition states, i.e., energy barriers, for formation of such intermediates can be defined:

Intermediates:
1. Low energy: close in energy to reactants.
2. High energy: significantly higher in energy than reactants.

---

* There may, in principle, be more than one; e.g., see References 29 and 30.

Transition states:
1. Low barrier: significantly lower in energy than the rate-determining transition state.
2. High barrier: close in energy to the rate-determining transition state.

We know that *if* energy levels exist (between reactants and products) and *are accessible,* that is, the energy barrier to their population is lower than the energy of the transition state, then they must be populated in thermal processes.* Therefore, if a reaction pathway includes *accessible intermediates* ("Boltzmann holes"?), some of the molecules on the way to becoming transition states must fall into these holes; they cannot precariously skirt around the rim. The necessity of population of such an accessible intermediate is easily visualized in physical terms if one remembers that the Boltzmann hole is really a region in which there are forces acting on the nuclei, pulling them toward the ENG. Thermal equilibration requires that such forces essentially always succeed in pulling any nuclear geometry upon which they act into the Boltzmann hole, thus populating that intermediate. The population of a low-energy, low-barrier intermediate will be significant and therefore usually detectable by ordinary spectroscopic techniques, but a high-energy, low-barrier intermediate will be hardly detectable because of its extremely low population. Both of these low-barrier intermediate types will be present in essentially Boltzmann equilibrium populations. On the other hand, high-barrier, high-energy intermediates will be present in nonequilibrium, steady-state concentrations. The presence of a high-energy intermediate may affect the concentration dependence of the overall reaction rate if the concentration dependence of the rate of conversion of the intermediate back to reactants is different from the concentration dependence of the rate of its conversion forward into products. If so, changing these concentrations will alter the population of the intermediate, thus demonstrating its existence by bringing about a predictable change in rate or rate-determining step, even though its concentration is very low. Hence, any intermediate which can form at a rate faster than the rate of formation of products can be said to be *obligatory,* in the sense that it *will* form as the reaction proceeds, because thermodynamically it *must.*[32] In contrast to obligatory (i.e., accessible) intermediates, intermediates which are formed more slowly than the rate-determining transition state for a particular reaction mechanism will be *inaccessible,* or *nonobligatory,* for that particular mechanism.

Inherent in TST is the principle of *microscopic reversibility,*[33,34] which states that a reaction and its kinetically equivalent reverse will occur via the same path. That this should be true is supported by the argument that the

---

* This is true only for systems in which collisional deactivation can occur and thus does not necessarily apply to reactions in the gas phase, where the existence of a potential well between reactant and product may provide an opportunity for intermediate formation but certainly does not require it.[31]

path of lowest energy over a barrier involves formation of a lowest-energy transition state and the energy of this route and of this transition state will be the same whether starting out from the reactant or from the product side of the reaction. Because changes in concentration can cause changes of mechanism, it is possible that a reaction and its reverse may be studied under different experimental conditions such that they may proceed by predominantly different pathways.[35] Therefore, the preceding considerations involving intermediates are applicable to both a reaction (conversion of reactants into products) and its *kinetically equivalent* reverse (conversion of products into reactants). Two important mechanistic consequences of the principle of microscopic reversibility are worth noting: (1) If a particular mechanistic pathway is postulated for a given reaction, then one must avoid postulating a *different* mechanism for the kinetically equivalent reverse of the reaction, and (2) in cases where one wishes to study a reaction which is experimentally difficult to effect, one may, instead, study its kinetically equivalent reverse.

It may be noticed that, up to this point, we have not mentioned the words enzyme or catalysis even once. Since the ultimate concern of most readers of this book will undoubtedly be with biochemical systems, it is probably worth remarking that since enzymes are very large molecules, they are subject to all the preceding generalizations and accompanying problems, though more profusely than are smaller molecules.

3. MORE PRECISELY DEFINING THE TRANSITION STATE

3.1. Potential Energy Surfaces and Energy Barriers

The previous comparison of chemical reactions to "molecular meanderings among mountains" (Fig. 1) really corresponds exactly to ART; the theory can be examined more precisely by discussing the nature of potential energy surfaces (PES's).

In the early 1930's Polanyi and Eyring developed a relationship between chemical reactions and energy surfaces.[36] They studied the very simple exchange reaction $H + H_2(ortho) \rightarrow H_2(para) + H$. Applying theoretical concepts developed by London,[37] they calculated the potential energy for this triatomic system [Eq. (3)] as a function of the two internal coordinates r_{AB} and r_{BC}, where r_{AB} = the distance of nucleus H_A from nucleus H_B and r_{BC} = the distance of nucleus H_B from nucleus H_C:

$$H_A \text{----------} H_B \text{--------} H_C \qquad (3)$$
$$\quad r_{AB} \qquad\qquad r_{BC}$$

Figure 2 illustrates the results of these calculations, assuming the angle H_A—H_B—H_C to be fixed at 180°, as a *contour diagram* (or *map*), showing the height of the surface in terms of equal-elevation contours. The numbers associated with the contour lines indicate only *relative* energies (usually expressed in kilocalories per mole). The upper valley represents the reactant

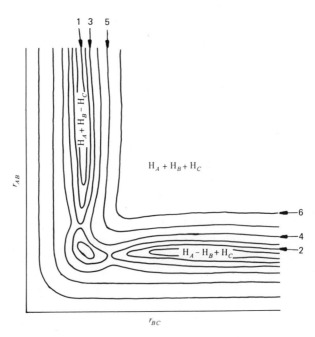

Fig. 2. Energy contour diagram for the H + H_2 system.

valley ($H_A + H_B$—H_C) and the lower one the product valley (H_A—$H_B + H_C$). The "well at the top of the pass" (lower left-hand corner) is an artifact of the particular function used and does disappear when other, more accurate quantum-mechanical calculations are used. One then gets the now-familiar kind of diagram for a typical, one-step displacement reaction (e.g., S_N2). Such a contour diagram allows for simple representation of a three-dimensional potential energy surface in two dimensions. Of course, all such diagrams are really abstractions of more complete ones; in this particular case the coordinate representing the H_A—H_B—H_C angle is missing. For more complex systems (more than three atoms), even more coordinates, the inclusion of which would require more than three dimensions, would be missing from the contour map. One can certainly imagine a complete description by a hypersurface involving more than three dimensions, but, of course, one cannot draw it. However, for any given pair of values of the coordinates specified on the map there will be a set of values for all the other coordinates which will give the lowest possible energy for the system. It is usually assumed that the energy plotted on a contour map has been calculated using those sets of missing coordinates for which the energy is a minimum.

A PES represents energies resulting from *electronic* motion; therefore, the contour diagram of such a surface is a plot of *electronic energy* (both kinetic and potential) and *nuclear potential energy* as a function of different *nuclear geometries*. That it is valid to do this is supported by the Born–Oppenheimer approximation, which allows for the separability of electronic and nuclear motions, the former being so much faster than the latter. Only zero-point electronic energies are taken into account since the molecular distortions defined by the surface are assumed to occur *adiabatically* (that is, equilibrium is assumed between all possible nuclear configurations defined by the surface) and to be independent of temperature. All nuclear kinetic energy, zero point and otherwise, is excluded. Therefore, because forces on nuclei are determined by changes in the plotted energy with changes in nuclear geometry, the distortion of molecules can be defined by a surface. For any point (i.e., geometry) on the surface with a zero slope, the forces on the nuclei will be zero in that direction; hence, any point for which the slope is zero in every direction is an ENG. At any point where the slope is nonzero, there are forces on the nuclei. The PES is a construct representing the distortion of *real* molecules (i.e., chemical reactions), but, because of the considerations mentioned previously, it does not represent the *entire* energy of any real system, since it excludes the kinetic energy of the nuclei. Because it is a function of the electron–electron repulsions, the electron–nucleus attractions, the nucleus–nucleus repulsions, and the kinetic energy of the rapidly moving electrons, it really represents the potential energy for nuclear motion.

The difficulties involved in calculating, as well as in drawing, potential surfaces have been somewhat lessened by the advent of the computer, in conjunction with the CAL-COMP plotter.[38] Work done by Schowen and co-workers[39] on coupled proton motion in systems of the type ^+A—H $\cdots$ BH $\cdots$ A, where A is H_2O or NH_3 and BH is H_2O or HF, presents a nice illustration of computer-calculated potential surfaces and contour maps. Figure 3 shows a typical result. Of course these as well as other calculations do not take into account the effect of solvent molecules on the potential surface, an obviously difficult factor to include. There is some work being done in this area[40] which will undoubtedly be expanded in the future.

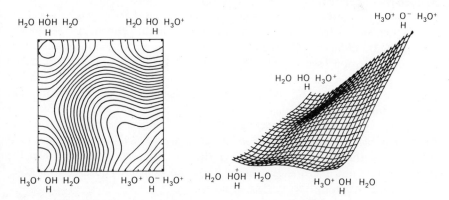

Fig. 3. Contour map (at right, resolution 5 kcal/mol) and transect diagram (at left, graphical representation) of the PES for $H_3O^+ + H_2O + H_2O \rightarrow H_2O + H_2O + H_3O^+$. (Reprinted with permission from Reference 39, copyright the American Chemical Society.)

If, as we have said, a potential energy surface represents molecular distortions, or chemical reactions, then the question arises as to which part or parts of the surface the molecules must "meander" over in order to effect the reaction. Transition-state theory has provided a widely accepted definition of the path or paths over which the reactants travel in order to become products—the *reaction path* (frequently referred to as the reaction coordinate). It is that locus of points on a PES which defines the *minimum-energy* path between reactants and products. Both reactants and products, being stable ENG's, lie in valleys or basins on the surface. Since during conversion of reactants into products the molecules must increase their energy, there must exist between the two basins at least one point (nuclear geometry) on the surface and lying along the reaction path which is of higher energy than either reactants or products. That particular nuclear geometry which is the *highest* energy point on the reaction pathway is called the *transition state* and occurs at a *saddle point* or mountain col.

A somewhat different perspective of the reaction pathway or coordinate is that of the path of *steepest descent*. Operationally, it is defined by starting from the saddle point and moving toward either reactant or product valley in small steps each of which is chosen to correspond to the steepest descent.[41] It is of importance to note that a minimum-energy definition of a reaction path has been shown to give rise to a discontinuous derivative or *kink* at the saddle point. A steepest-descent definition results always in a *kinkless* reaction path,[42] thus satisfying the requirement of ART.

Although the terms reaction path, reaction pathway, and reaction coordinate are commonly used interchangeably, it should be recognized that the term reaction coordinate is frequently used to refer to the specific normal coordinate of the transition state which has a negative force constant. This unique feature requires that the reaction coordinate motion of a transition state, if treated as a "vibration," have an imaginary frequency, since a vibration frequency is proportional to the square root of the force constant.

Passage of molecules over the potential energy barrier (i.e., attainment of sufficient energy to become a transition-state nuclear geometry) is dependent on, among other factors, Γ^*, the barrier penetration correction for the reaction coordinate motion [see Eq. (2)]. This factor is commonly referred to as the *quantum-mechanical tunnel effect, factor,* or *correction*[43] and is a type of *transmission coefficient*.[44] Many extensive discussions of the theory and applications of the tunnel effect are available in the literature.[43-47] The most important consequence of tunneling is its ability to affect the rate of a reaction. Obviously, if a significant fraction of reactant molecules becomes product molecules without having passed over the transition-state barrier, then the predicted rate of a thermal reaction, without appropriate consideration given to tunneling, will fall short of the observed rate.

Since the tunnel effect is dependent on the mass of the particle which is tunneling, the effect is not important except possibly for reactions involving motion of light particles, e.g., H, D, or T nuclei. The question is not whether tunneling occurs for a given system but whether or not its effects are large enough to be detected. The effects of tunneling, if present, for a reaction involving transfer of light nuclei are usually detected through measurement of isotope effects on reaction rates or by determining the temperature dependence of the rate constant.[48,49] There are many experimental results the magnitudes of which have been ascribed to the tunnel effect.[50]* A recent example is work by Saunders *et al.* involving a study of both carbon and hydrogen isotope effects for the same molecule.[51]

The *calculated* effects of tunneling depend on the height and shape of the potential energy barrier, particularly on the curvature at the top (usually assumed to be parabolic), as well as on other assumptions.[43] Since detailed information on the shape of the barrier is usually not available, effects cannot be predicted with accuracy for any but the simplest of systems, e.g., $H + H_2$.[52] However, recent attempts at predicting limiting values of such rate-related factors as kinetic isotope effects, activation energies, and Arrhenius parameters that have taken tunneling into account have given results that so far are consistent with experimental observations.[53] Thus, there are quite a few cases where a significant tunneling correction seems to be required in order to give good agreement with experimental data, although there is perhaps not yet universal acceptance of such a requirement. In any case, it does not appear to be possible to specify the scope or the limits of possible tunnel factors at present.

3.2. Reacting Species in More Detail: Potential Energy Surfaces and Reaction Profiles

Regardless of whether molecules tunnel through or pass over the transition-state barrier, the latter is still that particular nuclear geometry separating the "reactants" from the "products" in a reaction. According to transition-state theory, the rate constant is independent of the path taken by reactants in getting to the transition state. However, the observed rate constant is the rate constant for crossing the rate-determining saddle point (transition state) of "all reactants," where "all reactants" includes everything on the reactant side of the saddle point (on the potential energy hypersurface). Thus, "all reactants" *may* include vast numbers of intermediates and *does* include all *accessible* geometries. Intermediates of low energy will be significantly populated, while those of high energy will be only sparsely populated. It is only by convention (or for convenience) that we generally identify "reactant" as that particular ENG which has the lowest molar free energy. It is convenient since "major" covalency changes usually do change the molar free energy substantially, and thus we can *usually* identify one ENG as being practically the "only" one present at equilibrium. It is really the *average* (molar free energy) of all "reactants" that is being compared with the transition state; the average,

---

* A large number of them have been collected and commented upon in Reference 44.

however, is composed *primarily* of only this one "reactant" ENG to several significant figures.

In the case where the "reactants" A and B react to give an "intermediate," C, whose free energy is *lower* than the total free energy of the A + B system (and which is formed *prior* to the transition state for the reaction), the "reactant" would, by convention, be C, not A + B. A more accurate description of this system can be given by employing a reaction coordinate or energy profile diagram in which the potential energy of a system is plotted as a function of distance along the reaction coordinate (DARP), meaning here the reaction pathway. Figure 4 illustrates such a profile. The "reactant" is the equilibrium *average* of the entire shaded area, which indicates all accessible ENG's, "excited" (vibrationally, rotationally, translationally, but not electronically) as well as "ground" (lowest energy level) states. This equilibrium average is determined by the Boltzmann distribution of energies; therefore the population of higher energy levels decreases exponentially, proportional to $\exp(-\varepsilon/\mathbf{k}T)$, where ε is the energy, $\mathbf{k}$ the Boltzmann constant, and T the absolute temperature (all per molecule). In fact, the presence of partition functions for "reactants" in the transition-state formulation [see Eq. (2)] precisely takes this into account. However, if one wants to use the correct partition function, the formulation is determined by identifying the correct, lowest energy level of "reactants" and measuring all possible excited states from there. This is also why the observed rate constants are dependent on the stability of any intermediate but only slightly dependent on the stability of unstable (high-energy) intermediates: the reactant average will include unstable intermediates, but since they are only sparsely populated, their effect will be very small and is frequently negligible.

One can construct energy profiles similar to the one depicted in Fig. 4 for the three common types of intermediates formed prior to the transition state, as previously defined in this chapter: low energy, low barrier; high energy, low barrier; and high energy, high barrier. These cases are illustrated

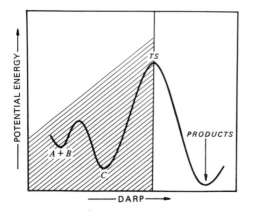

Fig. 4. Energy profile for a reaction in which a very low-energy "intermediate," C, is formed from "reactants" A + B prior to the transition state. DARP = distance along the reaction pathway.

in Figs. 5(a), (b), and (c), respectively. There will be essentially no contribution of C to the reaction rate in Fig. 5(b) since its population (i.e., concentration) is exceedingly small. Therefore, it does not significantly affect the energy of the reactant, and the experimentally determined rate constant will be that for conversion of A + B to products. In contrast, the situations in Fig. 5(a) with significant population of C, and in Fig. 5(c), with the transition state for formation of C partly rate determining, are such that the experimentally determined rate constant is not simply that for conversion of A + B to products. In Fig. 5(a), the rate is influenced by an effect on the *reactants*, while in Fig. 5(c), the rate is influenced by an effect on the *transition states*. These factors also illustrate an aspect of the very important fact that rates are determined by *differences* between transition state and reactant properties.

The situation may be summarized for the mathematically simple case of a single "reactant" A and an "intermediate" B:

$$A \xrightleftharpoons{K} B \xrightarrow{k_2} \text{products} \tag{4}$$

The rate of conversion of B into products is $k_2(B)$; however, the concentration of B is equal to the total concentration of reactant species (A + B) multiplied by the fraction present as B. This fraction is readily shown to be $K/(1 + K)$. The experimental rate constant, k_{obsd}, is therefore $k_2K/(1 + K)$, and this rate constant will be found whether one measures the rate of formation of products, or the rate of disappearance of B, or the rate of disappearance of A, or the rate of disappearance of the sum of (A) + (B). If $K \ll 1$, $k_{obsd} =$

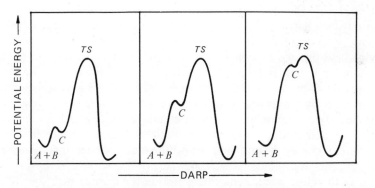

Fig. 5. Energy profiles for reactions in which (a) a low-energy, low-barrier intermediate is formed from reactants and is less than 2 or 3 kcal/mol greater in energy than A + B, while the barrier for its formation is more than 2 or 3 kcal/mol lower in energy than TS; (b) a high-energy, low-barrier intermediate is formed from reactants and is more than 2 or 3 kcal/mol greater in energy than A + B, while the barrier is more than 2 or 3 kcal/mol lower in energy than TS; (c) a high-energy, high-barrier intermediate is formed from reactants and is more than 2 or 3 kcal/mol greater in energy than A + B, while the barrier is less than 2 or 3 kcal/mol lower in energy than TS. The figure of 2 or 3 kcal/mol arises from the fact that around room temperature a difference of 1.4 kcal/mol corresponds to a factor of 10 in equilibrium constant or rate constant.

$k_2 K = k_A$, where k_A is the rate constant for conversion of A into products; this corresponds to Fig. 5(b).* If $K \gg 1$, $k_{obsd} = k_2$; this corresponds to Fig. 4. If K is near unity, no simplification of the expression exists, as in Fig. 5(a). Finally, under the conditions of Fig. 5(c), A and B are not at equilibrium, so Eq. (4) does not apply, and the observed rate constant can be expressed using the steady-state assumption and

$$A \underset{k_{-1}}{\overset{k_1}{\rightleftharpoons}} B \overset{k_2}{\longrightarrow} \text{products} \tag{5}$$

as $k_{obsd} = k_1 k_2 / (k_{-1} + k_2)$, which reduces to the expression for Fig. 5(b) if $k_{-1} \gg k_2$, since $k_1 / k_{-1} = K$.

The question of which species actually constitute the "reactants" and which the "intermediates" also raises the question of the relationship between intermediate and transition state for a given reaction. Population of an accessible intermediate (in a thermal reaction) is inevitable; however, the transition state through which a particular intermediate must pass in order to become products may not be the same transition state through which reactant molecules must pass in order to undergo the same reaction. In fact, an intermediate may end up in a "blind alley" in that its direct conversion to products through one transition state may be energetically unfavorable as compared to its conversion to products through a different transition state, namely, the one through which *reactant* molecules must pass.

Consider the reaction scheme

$$X \rightleftharpoons Y \xrightarrow[\text{TS(1)}]{\text{slow}} Z \tag{6}$$

where Y represents the original reactants, X some accessible intermediate, and Z the products. The slow conversion of Y into Z involves passage over the transition-state barrier TS(1). The intermediate X can, of course, pass into products by passing over the same transition-state barrier that Y passes over, in which case one would consider X a "blind alley" intermediate in that its formation, while unavoidable (since it is *accessible*), is certainly unnecessary for formation of products. On the other hand, X may be able to be converted into products by traversing a *different* transition-state barrier [e.g., TS(2)]. This is clear if one remembers that the potential energy surface for the reaction in question is a multidimensional one on which X, Y, and Z occupy three distinct and different energy basins. The transition-state barrier for conversion of Y into Z occurs at a saddle point somewhere on the surface. A *different* saddle point may therefore occur somewhere else on the surface such that X can be converted into Z directly by following a reaction pathway which does not pass through the energy basin (well) "containing" Y. If this saddle point [TS(2)] is of higher energy than TS(1), then it is *not* the transition state for the

---

* In this paragraph we refer to Figs. 4 and 5 because these curves and those that *could* be drawn for Eqs. (4) and (5) are similar; it should be noted, however, that the "reactants" in Figs. 4 and 5 are (A + B) and the "intermediate" C, whereas the "reactant" referred to in this paragraph [Eqs. (4) and (5)] is A and the "intermediate" B.

reaction involving conversion of Y into Z, and X is indeed a "blind alley" species. If, however, TS(2) is of lower energy than TS(1), then it [TS(2)] will be the lower-energy saddle point and will define the reaction pathway for this reaction. In this case, then, direct conversion of Y to Z through TS(1) would be unfavorable relative to its conversion through TS(2). Therefore, it is not necessary to be able to follow the reaction pathway in detail to determine whether X is on the reaction pathway or is a "blind alley" intermediate; it is only necessary to distinguish which one of the two possible transition states [here TS (2) and TS (1), respectively] is rate determining.

For a system in which A and B [Eq. (5)] or X and Y [Eq. (6)] represent rapidly equilibrating conformational isomers, e.g., equatorial and axial cyclohexanes, such that $k_{-1} \gg k_2$ [Eq. (5)], the rate constant can be expressed as

$$k_{\text{obsd}} = f_e k_e + f_a k_a \tag{7}$$

where f_e and f_a represent the mole fractions of the equatorial and axial conformers, respectively, and k_e and k_a what are frequently called their rate constants.[54]

However, since the two conformers exist as an equilibrium (reactant) mixture, it is more accurate to associate k_e and k_a with the two *transition states*, "equatorial" and "axial," surmounted by the two conformers, equatorial and axial. Picturing these two transition states on a PES, we can imagine molecules of *either* conformer arriving at *either* transition state since rapid equilibrium occurs between axial and equatorial species. Hence, it is not possible to determine the *original* conformation of a molecule arriving at a particular transition state. Equation (7) is, nonetheless, correct. *If* we could somehow "freeze" the system in the totally equatorial conformation, we could measure k_e directly; similarly, if it could be "frozen" in the totally axial conformation, we could measure k_a directly. Once k_{obsd}, k_e, and k_a were obtained, then f_e and f_a could be calculated for a given mixture, which is why Eq. (7) is so useful in conformational analysis. It is not necessarily easy to "freeze" a complex system, but good results have been obtained with substituted cyclohexanes by use of the t-butyl group at the remote, 4-position. The bulky t-butyl group apparently "freezes" the conformation of the ring such that it occupies the preferred equatorial position; thus *cis*- and *trans*-1-substituted-4-t-butyl-cyclohexanes will have the 1-substituent axial and equatorial, respectively. The observed rate constant for a 1-substituted cyclohexane will be the sum of the rate constants for the equilibrium mixture of reactants going to equatorial and to axial transition states. Referring back then to Eq. (4), we can calculate the rate constant for B going to products (k_2) or for A going to products $(k_2 K)$ if we measure k_{obsd} and K. Therefore, it is an arbitrary decision (made because of our desire to compare k_{obsd} with rate constants for "frozen" systems) to express Eq. (7) in terms of rate constants k_e and k_a for conversion of the equatorial conformer to the equatorial transition state and the axial conformer to the axial transition state. We could equally well calculate rate constants for an axial conformer going to an equatorial transition state, and so on, from the same data, but these rate constants would not be desirable for comparison with frozen systems. Thus, by definition, the equatorial transition state is "that transition state (or transition states) reached by the 'frozen' equatorial isomer"; the axial transition state is analogously defined.

If isomeric products result from a reaction involving a mixture of rapidly equilibrating reactant conformers, the ratio of the isomers formed is independent of the conformations of the reactant and depends solely on the relative energies of the transition states from which the products are formed—provided that the energy of activation for product formation is large compared to the activation energy for the interconversion of the reactant isomers. This statement of the Curtin–Hammett principle[55,56] dramatically emphasizes the importance of transition-state structure (energy) on the eventual outcome of a reaction.

The *distinction* between "reactants" and "products" on a PES is clear at low temperatures, where vibrational partition functions are close to unity and therefore most systems are in their lowest vibrational energy level. Energy *basins* (stable ENG's) will be highly populated (the extent being governed by the Boltzmann distribution) relative to hillsides and saddle points (metastable ENG's). Hence, "reactants" and "products" will be obviously separated by extensive areas that are virtually unpopulated.[5] One can then discuss the rate of a reaction in terms of the flux of "reactant" molecules through a surface which traverses the PES anywhere between reactants and products. This dividing surface (DS) is then represented as a line drawn in the plane of the two variable coordinates plotted on a contour diagram (Figs. 2 and 3). ART evaluates the rate using that DS which passes through (or very near to) the lowest-energy saddle point (between reactants and products) and is perpendicular to the unstable normal coordinate (i.e., the reaction coordinate); the DS is therefore very specifically defined and located. One can, however, envisage other possible DS's, within the framework of ART, still passing through the saddle point (i.e., that point which is the transition-state metastable ENG) but with varying orientations. One such orientation is perpendicular to the path of steepest descent from the transition state, and this path may not lie parallel to the reaction coordinate.

Some work on reaction rates based on the flux of molecules through DS's of variable orientations through the transition state has been done recently.[15] Results for a number of triatomic systems for which the transition states were all symmetrical or very close to symmetrical indicate little difference between rates arrived at using a DS which is perpendicular to the reaction coordinate and those arrived at using a DS perpendicular to the path of steepest descent. However, for those triatomic systems having very reactant-like or very product-like transition states (i.e., unsymmetrical), large rate differences were noted. Further investigation of this situation is obviously of interest.

It is generally accepted, within the framework of ART, that the transition state for a chemical reaction, being the lowest barrier separating reactants and products on a PES, is a critical point on this surface. Murrell and Laidler[57] have pointed out that the special characteristic of this critical point is that it have one and only one normal coordinate of negative curvature (i.e., its force constant matrix must have one and only one negative eigenvalue). If there were *no* coordinates with negative curvature, the point would obviously be a stable ENG; if there were *more* than one negatively curved coordinate, the point would resemble a *mountain peak* rather than a *mountain pass*. A mountain *peak* can never be a transition state, since there is always a *pass* somewhere below (at lower energy) through which the molecules can go. Proof of these concepts has been demonstrated by McIver and Stanton,[58-60] who apply group theory to develop symmetry rules for transition states. They also demonstrate that the existence of a "monkey saddle," that is, a surface on which three valleys meet at a single transition state, is exceedingly improbable, since this type of phenomenon would require the transition state (meeting point) to be at a point of zero slope (zero force) and zero curvature (zero eigenvalue), i.e., a point where the change of energy with geometry is dependent only on the third and higher powers of the change in geometry.[60]

3.3. Potential Energy Surfaces and Reaction Types

We have, so far, tried to establish general, qualitative concepts of, and relationships between, reaction species in terms of a PES. In addition, there are, of course, specific factors which will influence the nature of a particular PES for a specific reaction and thereby affect the energy of the transition state relative to reactants and products. One can group these factors into three major classes:

1. The strengths of bonds being made and/or broken during the reaction [including possible compensation of energy needed to break bond(s) partially by the energy of partially forming new bonds].

2. The exo- or endothermicity of the reaction; this is related to the Hammond postulate which implies that exothermic reactions tend to have rather low activation energies (i.e., the structure of the transition state resembles closely that of the reactant) and that endothermic reactions (being the microscopic reverse of exothermic reactions) tend to have high activation energies.

3. The special effect associated with *orbital phases*, or *orbital symmetry*, which arises primarily from geometric restrictions on approaches of reacting species to each other; this may control the stereochemical course of a reaction and even make certain reactions essentially "forbidden" by making the energy barrier prohibitively high (perhaps 20 kcal/mol higher than it might otherwise be). In other words, it turns out that certain otherwise reasonable reaction types may have enforced, unavoidable antibonding interactions. There is extensive comment on this factor in the literature.[61-63]

Fig. 6. Schematic PES (omitting energy contours) showing a *concerted* reaction pathway, proceeding directly from R to P (as a dashed line), and two *stepwise* pathways, proceeding from R to intermediate A (or intermediate B) first and then to P (as two dashed lines). For this figure and for all others like it, it is understood that the transition state for a given conversion between two species (always at the corners of the figure) lies somewhere along a dashed line and exists as a maximum along this line. On either side of the transition state leading away from it, the surface *descends* into valleys at the corners. At every point along a dashed line and perpendicular to it, the surface curves *upward* on either side.

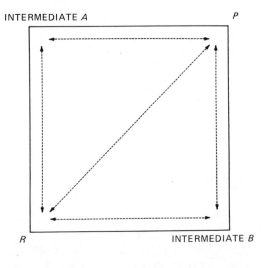

INTERMEDIATE *A*   *P*

R   INTERMEDIATE *B*

In addition to the above factors, one must reconsider, in a more quantitative way, the previously mentioned considerations of concerted and stepwise reactions, the latter involving the formation of intermediates, in terms of a PES. Figure 6 illustrates both the "typical" concerted and stepwise processes for a general reaction type. Generally, a concerted reaction is considered to be a one-step process in which there are no potential energy minima between reactant and product wells,* while stepwise pathways involve potential wells somewhere between reactant and product. Of course, *each* step of a stepwise process can be likened to a concerted process in that both require the existence of a saddle point between potential wells, be they those of reactant and intermediate or intermediate and product. In Fig. 6, the species which occupy the corners of the diagram are stable ENG's (i.e., exist in potential energy wells), whereas any transition states (metastable ENG's) can be found elsewhere on the surface.

Figure 7 illustrates (in the manner of Fig. 6) three alternative reaction pathways for the substitution reaction (S_N)

$$RX + Y^- \xrightarrow{\;\;\;b\;\;\;} [YRX^-]^{\ddagger} \longrightarrow YR + X^- \tag{8}$$

with branches labeled a to $[YRX^-]$ and c to $[R^+X^-, Y^-]$.

Pathways a and c [Fig. 7(a)] involve population of reaction intermediates. Pathway c is traditionally known as the S_N1 mechanism, whereas pathway b is the well-known S_N2 mechanism. The upper triangular portion of the sur-

---

* See the footnote on p. 10.

face, involving pathway a, is rarely discussed for displacements on carbon, as there is negligible evidence for the formation of a pentavalent carbon intermediate. However, some type of intermediate involving association of a nucleophile and substrate appears to be involved in certain nucleophilic displacement reactions in benzyl systems.[64]

The pentavalent pathway is a viable one for displacements on phosphorus and silicon; its significance in biological processes is obvious. Pathway b can offer a spectrum of transition-state structures, depending on the relative importance of bond making and bond breaking in the transition state for the reaction. Curved Hammett plots, observed in the nucleophilic displacement reactions of some substituted benzyl halides, have been interpreted as reflecting a gradation of transition states.[65] As one moves on the PES from a rather centrally located transition state (traditional S_N2) where bond making and bond breaking are of approximately equal importance to a transition state in which bond making exceeds bond breaking, one approaches the upper-left-hand corner of Fig. 7(a), that is, formation of a pentavalent intermediate. Analogously, as one alters the importance of bond breaking relative to bond making such that the former predominates at the transition state, one approaches the lower right-hand corner of Fig. 7(a), that is, formation of a positively charged intermediate. If the reaction involves displacement on carbon, this intermediate is, of course, a carbonium ion. In this way, a continuum of reaction mechanisms (e.g., between S_N2 and S_N1) can be easily envisioned [Figs. 7(b) and (c)]. With manipulation of experimental variables, it becomes possible for a given reaction to produce a gradation of S_N2-type transition states until a mechanistic changeover from S_N2 to S_N1 is observed. This amounts to stabilizing a "corner" of the PES (i.e., an intermediate) such

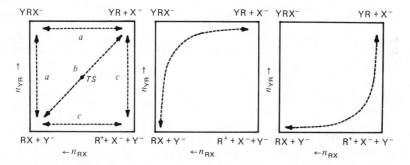

Fig. 7. Schematic PES's for (a) an S_N2 reaction (pathway b) where bond making and bond breaking are of approximately equal importance, an S_N1 reaction (pathway c) involving carbonium-ion formation for displacement on carbon, and a pentavalent species-forming reaction (pathway a) involving displacement on P or Si; (b) an S_N2 reaction where bond making is more advanced at the transition state than bond breaking; (c) an S_N2 reaction where bond breaking is more advanced at the transition state than bond making. Since, for all of these mechanisms where one coordinate is increasing while the other is decreasing, plotting bond distance leads to a curved reaction coordinate even for a central transition state, bond order (n) is used instead ($n = 0$ for no bond; $n = 1$ for a fully formed bond).

that it becomes significantly populated, thereby causing a change in mechanism.

Base-promoted elimination reactions involving loss of a proton can be depicted using diagrams comparable to those of Fig. 7. In fact, one of the steps in a stepwise (E1cB or E1) elimination mechanism [Fig. 8(a)] is a displacement on H by a base. A concerted E2 mechanism is shown in Fig. 8(b). Increasing the stability of the incipient carbanion in the lower right-hand corner by decreasing its energy will result in an *anti-Hammond* effect on the E2 transition-state structure, moving it closer to the carbanion [i.e., a "carbanion (E1cB)-like" transition state]. This occurs when the effect of increasing the stability of a species is transmitted to the transition state from a direction *perpendicular* to the reaction coordinate and thus acts across the energy *minimum* of the latter, causing a shift in its structure *toward* the direction of increased stability (i.e., the *perpendicular effect*); on the other hand, when an increase in the stability of a species is transmitted *along* the reaction coordinate and acts across the energy *maximum* of the transition state, there is a shift in its structure *away* from the species of increased stability (i.e., the *parallel effect*).[66-67] Stabilizing either incipient carbonium ion (HC^+) or carbanion (XC^-) sufficiently may decrease the barrier to formation of either to the degree that it becomes lower than the barrier to formation of the E2 transition state, thereby shifting the mechanism from E2 to either E1 or E1cB. Using this model and an appropriate PES, one can analyze the effects of changes in substrate and reaction conditions upon the structure of the transition state in E2 eliminations. If effects can be expressed in terms of a change in energy of the stable species—reactants, products, carbonium ions, or carbanions—they can be correlated with a change in the location of the E2 transition state on the PES. This has been done elegantly for base-promoted β elimination of 9-fluorenylmethyl derivatives by More O'Ferrall,[68] who examined in detail the relationship between E2 and E1cB mechanisms.

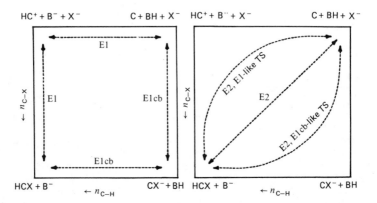

Fig. 8. Schematic PES's for a base-promoted elimination reaction: (a) stepwise (E1 or E1cB) pathways and (b) concerted pathways (E2; E2, E1-like transition state; E2, E1cB-like transition state).

Reactions involving the reversible addition of a nucleophilic reagent to a carbonyl group serve as models for a large number of biochemical reactions and, if for no other reason, have been and still are the subject of extensive mechanistic investigation. Being subject to general acid–general base catalysis, these reactions are, in the forward direction (i.e., addition), the reverse of the base-promoted elimination reactions discussed in the previous paragraph, except, of course, the proton being added (in the forward reaction) and removed (in the reverse, elimination reaction) in the case of the carbonyl reactions is bonded to an oxygen rather than to a carbon. While the energetics of the reaction may be altered by this substitution, the possible alternative mechanisms and the conclusions to be drawn from them are not.[69a] Figure 9 illustrates both stepwise (paths 1 and 2) and concerted (path 3) pathways involving general acid catalysis for the addition of a nucleophile to a carbonyl group. If addition of the nucleophile to the carbonyl occurs prior to transfer of the proton from B—H to oxygen (as in path 1) and if this is the slow (rate-determining) step, then the addition will *not* be general-acid-catalyzed at all. If transfer of H from B—H to oxygen is rate determining, the reaction will be general-acid-catalyzed. Rate-determining proton transfer in the first step of pathway 2 is also subject to general acid catalysis, whereas a rate-determining attack by nucleophile on the protonated carbonyl center will be *specific*-acid-catalyzed. Specific acid catalysis refers to kinetic dependence on the conjugate acid of the solvent (commonly H_3O^+) and may also be called "apparent hydronium ion catalysis." If the second step of pathway 2 is rate determining, then the protonated carbonyl must be formed rapidly and reversibly; therefore, its concentration must be determined by pH and not affected by B—H concentration. The concerted mechanism depicted by pathway 3 is, of course, general-acid-catalyzed.

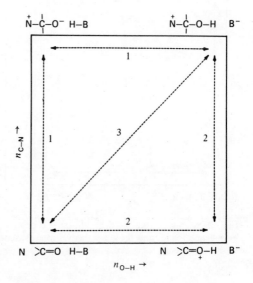

Fig. 9. Schematic PES for stepwise (pathways 1 and 2) and concerted (pathway 3) nucleophilic addition to a carbonyl center, subject to general acid catalysis.

It is interesting to note that each of the proton-transfer steps indicated along the top and bottom horizontal borders of Fig. 9, although shown as a concerted process, may in fact be stepwise, as illustrated in Figs. 10(a) and (b), which are analogous to Fig. 7(a). Although displacement on H involving a bare proton (pathway 2) seems highly improbable for general acid catalysis, it is certainly more probable that an intermediate of the type shown in the upper left-hand corners of Figs. 10(a) and (b) may be formed. Such an intermediate would be similar to the well-known HF_2^- species.

General-base-catalyzed nucleophilic addition to a carbonyl center, illustrated in Fig. 11, is analogous to Fig. 9 and subject to the same analysis.

Construction of the appropriate diagram, comparable to Figs. 7–11 (with or without contours), provides a relatively simple way to decide whether or not an unambiguous effect, due to a structural change in reactants (e.g., a substituent effect), on the stability of an ionic "corner" species can be predicted; if it can, then, using perpendicular and parallel effects, one can predict the resulting structural adjustment made by the transition state in response to the change. Jencks[69a] has done this for a substantial number of general-acid/general-base-catalyzed additions to carbonyls.

Three-dimensional drawings of energy surfaces for various reaction types, including nucleophilic addition to an unsaturated center[69] and nucleophilic substitution,[69b] are available elsewhere.

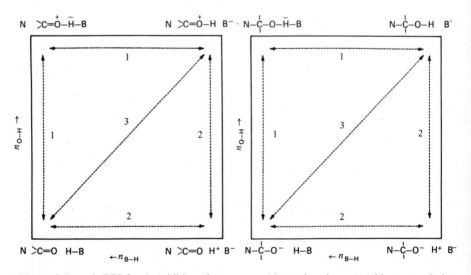

Fig. 10. Schematic PES for the addition of a proton to (a) a carbonyl center, subject to catalysis by a general acid, and (b) a tetrahedral intermediate formed by addition of a nucleophile to a carbonyl center.

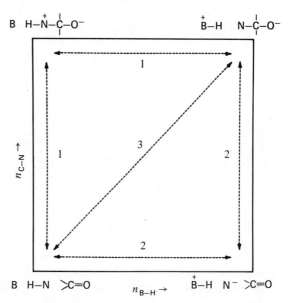

B H–Ṅ–Ċ–O⁻

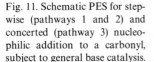

Fig. 11. Schematic PES for stepwise (pathways 1 and 2) and
concerted (pathway 3) nucleophilic addition to a carbonyl,
subject to general base catalysis.

4. PROPERTIES OF THE TRANSITION STATE

In this section we shall discuss the experimentally accessible quantities which can be used as probes of transition-state structure.

4.1. Thermodynamic Properties

As we have seen, according to ART, one can calculate a rate constant from a PES (if it is known), since the latter provides one with all the information necessary to utilize Eq. (1) or (2). It can be seen from inspection of either equation that the partition functions for reactant(s) and transition state, along with the exponential term, which arises from the difference between the zero-point energy (ZPE) of transition state and reactant(s), define an equilibrium constant for the formation of a transition state from reactant(s):

$$K^{\ddagger} = (f_{\ddagger}^{3N-1}/f_A^{\alpha} f_B^{\beta} f_C^{\gamma}) \exp(-\Delta\varepsilon_0/\mathbf{k}T) \tag{9}$$

For a bimolecular reaction between A and B, on a molar basis, we then have

$$K^{\ddagger} = (f_{\ddagger}^{3N-1}/f_A f_B) \exp(-\Delta E_0^{\ddagger}/\mathbf{R}T) \tag{10}$$

The constant $K^{\ddagger}$ is generally referred to as a "quasi" equilibrium constant since one vibrational degree of freedom, the reaction coordinate, is missing from the partition function for the transition state. However, it certainly does closely resemble a true equilibrium constant and *is* the equilibrium constant for a hypothetical transition-state species, a "pencil balanced on its point," which has only $3N - 1$ degrees of freedom of nuclear motion, i.e., which is not moving in the reaction coordinate.

Since for an equilibrium between "ordinary" (non-transition-state) molecules

$$\Delta G_0 = -\mathbf{R}T \ln K \tag{11}$$

and

$$\Delta G_0 = \Delta H_0 - T\,\Delta S_0 \tag{12}$$

then, for transition states, we can write

$$\Delta G_0^{\ddagger} = -\mathbf{R}T \ln K^{\ddagger} \tag{13}$$

and

$$\Delta G_0^{\ddagger} = \Delta H_0^{\ddagger} - T\,\Delta S_0^{\ddagger} \tag{14}$$

where ΔG_0 is the (Gibbs) free-energy difference between species involved in the equilibrium defined by K, ΔH_0 is the enthalpy difference, and ΔS_0 is the entropy difference. ΔG_0, ΔH_0, and ΔS_0 are *average* quantities for each substance *in its standard state*. $\Delta G_0^{\ddagger}$, $\Delta H_0^{\ddagger}$, and $\Delta S_0^{\ddagger}$ are then the free-energy, enthalpy, and entropy differences between the *hypothetical* "pencil balanced on its point" and the "real" reactants. By eliminating the approximate correction factors for a nonflat PES and for tunneling in Eq. (2) so as to give the ordinary classical form of the equation, the rate constant may be expressed as

$$k = (\mathbf{k}T/\mathbf{h})\,K^{\ddagger} \tag{15}$$

The quantity $\Delta E_0^{\ddagger}$ [Eq. (10)] is the activation energy (from the PES) at $0°\text{K}$; hence at $0°\text{K}$

$$\Delta E_0^{\ddagger} = \Delta H_0^{\ddagger} \qquad \text{(enthalpy of activation)} \tag{16}$$

We can rewrite Eq. (15), using Eqs. (13) and (14), as

$$k = (\mathbf{k}T/\mathbf{h})\,e^{-\Delta H_0^{\ddagger}/\mathbf{R}T}e^{\Delta S_0^{\ddagger}/\mathbf{R}} \tag{17}$$

Therefore, using an ART equation and rate constants experimentally determined under varying reaction conditions (temperatures and pressures), one can calculate not only the free-energy change between reactant(s) and transition state but other thermodynamic quantities usually associated with ordinary molecules.

Those thermodynamic functions which have been evaluated most frequently and appear best suited to supplying information concerning transition-state structure are the standard free energy ($\Delta G_0^{\ddagger}$), enthalpy ($\Delta H_0^{\ddagger}$), entropy ($\Delta S_0^{\ddagger}$), and volume ($\Delta V_0^{\ddagger}$) of activation and heat capacity ($\Delta C_p^{\ddagger}$, at constant pressure). Just as with "ordinary" molecules, the enthalpy change between reactants and transition state is related to changes in bond energies at the temperature in question. Since bond energies are not completely temperature independent, a *heat capacity term*, $\Delta C_p^{\ddagger}$, which measures the dependence of the enthalpy on the temperature, is included as a correction factor in the enthalpy measured at other than $0°K$. It should be recognized, however, that the heat capacity term is often so small as to be experimentally undetectable over the limited temperature range through which it is usually possible to conduct experiments. The entropy of activation reflects the difference in the translational, rotational, and vibrational degrees of freedom between transition state and reactants, while the volume of activation reflects the change in volume accompanying the conversion of 1 mol of starting material to 1 mol of transition state.

Evaluation of $\Delta G_0^{\ddagger}$ from an experimentally determined rate constant is carried out using the relationship

$$\Delta G_0^{\ddagger} = 2.303RT \log(kT/hk_{obsd}) \tag{18}$$

(assuming all activity coefficients are unity).[70,71] Assuming, then, that $\Delta H_0^{\ddagger}$ is independent of temperature over the range being studied, one has

$$\log(k_{obsd}/T) = (-\Delta H_0^{\ddagger}/2.303R)(1/T) + \text{constant} \tag{19}$$

and measurement of k_{obsd} at various temperatures will give $-\Delta H_0^{\ddagger}/2.303R$ as the slope of a plot of $\log(k_{obsd}/T)$ vs. $1/T$. From $\Delta G_0^{\ddagger}$ and $\Delta H_0^{\ddagger}$, $\Delta S_0^{\ddagger}$ can easily be calculated. A better assumption, however, is that $\Delta H_0^{\ddagger}$ is *not* temperature independent, in which case

$$\Delta H_0^{\ddagger} = \Delta H_0^{0^{\ddagger}} + \Delta C_p^{\ddagger} T \tag{20}$$

and

$$\Delta S_0^{\ddagger} = \Delta S_0^{0^{\ddagger}} + \Delta C_p^{\ddagger} \ln T \tag{21}$$

where $\Delta H_0^{0^{\ddagger}}$ and $\Delta S_0^{0^{\ddagger}}$ are the enthalpy and entropy changes at $0°K$ and $\Delta C_p^{\ddagger}$ is the heat capacity (at constant pressure) difference between reactants and transition state. If experimental data can be fitted to an equation of the form

$$\log k = A/T + B \log T + C \tag{22}$$

then the thermodynamic quantities can be calculated[72]:

$$A = -\Delta H_0^{0^{\ddagger}}/2.303R \tag{23}$$

$$B = (\Delta C_p^{\ddagger}/R) + 1 \tag{24}$$

$$C = \log(k/h) + [(\Delta S_0^{0^{\ddagger}} - \Delta C_p^{\ddagger})/2.303R] \tag{25}$$

When the data are over only a relatively short temperature range (within $100°$ or so) and/or when $\Delta C_p^{\ddagger}$ is close to zero, the nonlinear equation (22) assumes the linear form of Eq. (19). The volume of activation may be determined experimentally by studying the effect of pressure on the rate constant, since

$$d(\ln k)/dp = -(\Delta V_0^{\ddagger}/RT) \tag{26}$$

For reactions in solution, the presence of solvent molecules may have a significant effect on this quantity, since the volume change includes not only reactant and transition-state species but their solvation shells as well.

The quantity we would like to measure, but cannot do so experimentally, is the molecular property $\Delta E_0^{\ddagger}$ [Eq. (10)]. The experimental parameters $\Delta G_0^{\ddagger}$, $\Delta H_0^{\ddagger}$, and $\Delta S_0^{\ddagger}$ are average properties. The question of which experimentally determinable parameter, $\Delta G_0^{\ddagger}$ or $\Delta H_0^{\ddagger}$, is a better measure of $\Delta E_0^{\ddagger}$ remains unresolved.

Although not a thermodynamic quantity, the Arrhenius activation energy, also called the apparent activation energy, E_A, is an often-calculated, useful property which can be directly related to $\Delta H_0^{\ddagger}$[73]:

$$E_A = \Delta H_0^{\ddagger} - P\,\Delta V_0^{\ddagger} + \mathbf{R}T \tag{27}$$

which arises when data are measured and used in the form of the Arrhenius relationship:

$$k = Ae^{-E_A/\mathbf{R}T} \tag{28}$$

For a typical bimolecular reaction in solution, $\Delta V_0^{\ddagger}$ turns out to be of the order of $-20\ \mathrm{cm^3/mol}$, which corresponds to ca. $-0.5\ \mathrm{cal/mol}$ at 1 atm pressure. Hence, the $P\,\Delta V_0^{\ddagger}$ may be reasonably ignored for reactions run in solution at atmospheric pressure and

$$E_A = \Delta H_0^{\ddagger} + \mathbf{R}T \tag{29}$$

At absolute zero then

$$E_A = \Delta H_0^{\ddagger} = \Delta E_0^{\ddagger} \tag{30}$$

all of which measure the difference between zero-point levels in reactants and transition state, whereas at any other temperature E_A and $\Delta H_0^{\ddagger}$ measure the difference between the average energies of reactants and transition state.

More recently Kurz[17] has developed an approach to the experimental determination of transition-state structure using a thermodynamic-related quantity known as $pK_a^{\ddagger}$. The latter may be obtained from observed rate constants for two closely related reactions in the following manner: For a system in which S is the reactant, one can write

$$S + X + \cdots \xrightarrow{\ k_S\ } [S^{\ddagger}] \longrightarrow \text{products} \tag{31}$$

For an analogous reaction in which SH^+, the conjugate acid of S, is the reactant, we have

$$SH^+ + X + \cdots \xrightarrow{\ k_{SH}\ } [SH^{\ddagger}] \longrightarrow \text{products} \tag{32}$$

The two rate constants, k_S and k_{SH}, are experimentally observable. For a system containing both S and SH^+ in kinetically significant concentrations, the equilibrium

$$SH^{\ddagger} + H_2O \xrightarrow{\ K_a^{\ddagger}\ } S^{\ddagger} + H_3O^+ \tag{33}$$

is indirectly established. Since the lifetime of a transition state is not sufficiently long to permit it to engage in direct dynamic equilibrium, $K_a^{\ddagger}$ is a virtual equilibrium constant. It may be related to observable quantities via the thermodynamic cycle:

$$H^+ + S + X + \cdots \; \overset{K_S^{\ddagger}}{\rightleftharpoons} \; [S^{\ddagger} + H^+] \; \rightleftharpoons \; \cdots \tag{34}$$

$$\Big\updownarrow K_a(SH^+) \qquad\qquad\qquad \Big\updownarrow K_a^{\ddagger}$$

$$SH^+ + X + \cdots \; \overset{K_{SH}^{\ddagger}}{\rightleftharpoons} \; [SH^{\ddagger}] \; \rightleftharpoons \; \cdots \tag{35}$$

$K_S^{\ddagger}$ and $K_{SH}^{\ddagger}$ are equilibrium constants for conversion of S and SH^+ into their respective transition states, while $K_a(SH^+)$ is the acid dissociation constant for SH^+. Since $K_S^{\ddagger}$ and $K_{SH}^{\ddagger}$ are related to k_S and k_{SH},

$$k_S = (\mathbf{k}T/\mathbf{h}) \, K_S^{\ddagger} \tag{36}$$

$$k_{SH} = (\mathbf{k}T/\mathbf{h}) \, K_{SH}^{\ddagger} \tag{37}$$

(ignoring the tunneling factors and transmission coefficients and using all activity coefficients = 1), we have

$$K_a^{\ddagger}/K_a(SH^+) = K_S^{\ddagger}/K_{SH}^{\ddagger} = k_S/k_{SH} \tag{38}$$

Therefore

$$pK_a^{\ddagger} = pK_a(SH^+) + \log(k_{SH}/k_S) \tag{39}$$

Equation (39) is applicable, then, to investigating transition states which differ from each other by a proton. The "acidic proton" of the transition state serves as a sensing device for factors affecting its acidity [as measured by Eq. (39)], just as the acidity of the carboxyl group in benzoic acid serves as a probe for the electronic effects of various substituents. Factors which determine pK_a in all sorts of "ordinary," non-transition-state molecules have been extensively studied, appear to be well understood, and are readily interpreted. Since Kurz's approach nicely lends itself to interpretation of pK_a data in terms of transition-state structure by analogy, it may be expected to permit reaction mechanism differentiation.

In a manner analogous to Kurz's, it is possible to investigate transition states which differ from each other only by virtue of their surrounding solvent molecules. It is possible to measure the free energy of solvation of the same set of reactants in two different solvents and the free energy of activation for the reaction in these two different solvents and, by difference, to calculate the thermodynamic parameters associated with the transfer of the transition state for the reaction from one solvent to another, i.e., transition-state solvation.[74]

The application of thermodynamic data to elucidation of transition-state structure can be best illustrated by selected examples from among the very large number which have been reported. The selection is arbitrary; no attempt will be made to review the literature on this subject.

Although the interpretation of entropies of activation for reactions other than simple unimolecular processes is less than straightforward, the magnitude of the data can still be of value in providing clues to both reaction mechanism and transition-state structure. Schaleger and Long[75] have reviewed a number of interesting reactions for which data relating structural change in a nonreacting part of the molecule to changes in entropy and enthalpy of activation are available.

For example, the differences between the rates of alkaline hydrolysis of formic acid esters and the corresponding acetic acid esters have been shown to be entirely entropy controlled, as can be seen from the data in Table I.[76] Humphreys and Hammett have estimated the entropy of acetic acid or its derivative to be about 4–6 eu greater than the entropy of formic acid or its corresponding derivative, due to the internal freedom of the methyl group. From the entropy changes of reactions (40),

$$HCOO^- + CH_3COOH \rightleftharpoons CH_3COO^- + HCOOH \qquad (40)$$

$$\Delta S_0 = -4.94 \text{ eu}$$

and (41), the acid-catalyzed esterification of formic and acetic acids in CH_3OH,

$$\left[\begin{array}{c} OH \\ | \\ H-C-OH \\ | \\ +OCH_3 \\ H \end{array} \right]^{\ddagger} + CH_3COOH \rightleftharpoons HCOOH + \left[\begin{array}{c} OH \\ | \\ CH_3-C-OH \\ | \\ +OCH_3 \\ H \end{array} \right]^{\ddagger} \qquad (41)$$

$$\Delta S_0^{\ddagger} = -5 \text{ eu}$$

they conclude that the entropy of the ionic species in each of the reactions must be about the same, presumably because the internal motion of the methyl group is frozen out in these species. It can then be reasoned that the entropy of the transition state for acetate hydrolysis (Table I) is more negative than that for formate hydrolysis by about 5 eu.[75] This may be due to a greater degree of solvent immobilization in the acetate transition state as compared with the formate transition state. In addition, it is consistent with the qualitative rule proposed by Price and Hammett[77]: for a reaction with a highly polar transition state, the more entropy contributed to a reactant molecule by the substituent group, the more entropy is frozen out in the transition state. Although this generalization has been found to hold in many kinetic studies, caution must be exercised in its application, since there are known cases in which it fails.[75]

Table I[76]

Alkaline Hydrolysis of Formic and Acetic Acid Esters

Solvent	Compound	$\Delta H^{\ddagger}$ (kcal/mol)	$\Delta S^{\ddagger}$ (eu)
85% C_2H_5OH	Ethyl formate	13.86	−9.83
85% C_2H_5OH	Ethyl acetate	14.1	−21.1
H_2O	Methyl formate	9.81	−18.41
H_2O	Methyl acetate	9.62	−30.1

For many acid-catalyzed reactions, it is of interest to know whether or not there is a water molecule, in addition to the proton, firmly bound to the substrate in the transition state. Kinetic data are inadequate in obtaining this information since water is usually present in great excess. Solvent isotope effects in H_2O–D_2O mixtures can be used, as has been demonstrated by Swain and Thornton.[78] The entropy of activation, also an extremely sensitive probe for acquiring this information, is of great value in elucidating this important aspect of transition-state structure, one which is certainly of considerable interest in enzyme-catalyzed reactions. Furthermore, comparisons between entropies of activation for similar transition states provide insight into solvent reorganization and specific interactions between solvent and transition state during a reaction.

Long et al.,[79] amplifying a suggestion of Taft and co-workers,[80] have shown that there is a substantial negative entropy of activation (usually a decrease of more than 20 eu) for many reactions in which a water molecule is bound to the transition state. On the other hand, the entropy of activation appears to be either positive or negative and numerically small for those reactions in which the transition state consists of simple substrate plus proton. Long and co-workers[79] suggest a decrease of 20–30 eu on binding a water molecule to a protonated (organic) substrate. Kreevoy[81] and Schaleger and Long[75] have tabulated a series of reactions for which this generalization is nicely borne out. Included are acid-catalyzed reactions such as epoxide hydrolysis, ring openings of substituted ethyleneimines, the hydrolysis of simple acetals and ketals, and the enolization of 1,2-cyclohexanedione.

The variations in $\Delta S_0^{\ddagger}$, as well as in $\Delta H_0^{\ddagger}$, with solvent composition are complex and have long been the subject of extensive investigation. Recent work on variations in $\Delta G_0^{\ddagger}$, $\Delta H_0^{\ddagger}$, and $\Delta S_0^{\ddagger}$ with solvent composition has shown that glycine is a suitable transition-state model for the solvolysis of *t*-butyl chloride in H_2O–C_2H_5OH mixtures.[82] The use of such stable molecules provides a potentially valuable method for investigating specific structural features of postulated transition states.

In an attempt to learn more about the nature and magnitude of the maximum increase in rate that may be expected by bringing together two properly oriented reactants in the active site of an enzyme without recourse to strain or desolvation, Page and Jencks[83] have calculated values for the entropy of reactions occurring in an organic solvent in which two molecules react to give one molecule of product and no charge is produced in the transition state. They conclude that translational and rotational entropy provides the important entropic driving force for enzymic and intramolecular rate accelerations and support the view that enzymes perform a large part of their extraordinary rate accelerations by utilizing substrate-binding forces to overcome unfavorable entropy "cost" in bringing together substrate(s) and enzyme.

Since enzymic reactions occur in water, one ought to compare their rates with the uncatalyzed reaction in water. Thus, Larsen[84] has extended the treatment of Page and Jencks to the solvent water. He finds that the loss of translational degrees of freedom due to the combination of two reactants (an unfavorable entropy contribution) is accompanied by the destruction of one structural solvent cage as the two reactants form a single species (a favorable entropy contribution) and hence that these effects cancel. The loss of one structural solvent cage is expected to lead to a decrease in hydrogen bonding (an unfavorable enthalpy effect). Thus, bringing together two molecules to form a single species in an aqueous environment has an overall unfavorable effect just as it does in a nonaqueous environment. However, this appears in the $\Delta H_0^{\ddagger}$ rather than in the $\Delta S_0^{\ddagger}$ term.

Measurement of $\Delta G_0^{\ddagger}$ to test transition-state models assumes a sensitivity of this parameter to structural changes, which, in some cases, may be unwarranted. The hydrolyses of methyl, ethyl, isopropyl, and t-butyl trifluoroacetates are mechanistically ambiguous with respect to order of reactivity at various temperatures and overall range of reactivity.[85] However, the entropies of activation and the corresponding enthalpies of activation (along with the solvent isotope effects) allowed for a more favorable mechanistic diagnosis, demonstrating that in some instances these parameters may provide useful mechanistic criteria when the free energy of activation proves to be relatively insensitive to structural changes.

Although ideally it ought to be possible to separate the contributions of concentration and orientation effects (which should appear as a more favorable $\Delta S_0^{\ddagger}$) from those which result from reaction (which should appear as a more favorable $\Delta H_0^{\ddagger}$), it is not always possible to do so. For example, while it is the difference in the entropy term which accounts for the increased rate of the intramolecular aminolysis of phenyl esters over the intermolecular reaction,[86] it is the decrease in the *enthalpy* of activation for the hydrolysis of tetramethylsuccinanilic acid as compared to the hydrolysis of the unsubstituted succinanilic acid which appears to account for the rate acceleration.[87] In fact, there is an *unfavorable* entropy effect accompanying this rate increase.

Mutually compensatory changes in enthalpy and entropy were observed by Cane and Wetlaufer[88] in the rates of deacylation of the acylchymotrypsins $CH_3(CH_2)_n CO$–enzyme, where an increase in $\Delta H_0^{\ddagger}$ is accompanied by an increase in the $\Delta S_0^{\ddagger}$. Since these two increases are large, they undoubtedly involve a change in the enzyme conformation, probably accompanied by changes in solvent structure. The rates of the deacylation step for certain specific and nonspecific substrates are also associated with a large increase in the $\Delta S_0^{\ddagger}$, which could result from a change in the enzyme conformation.[89] However, there is no apparent compensating enthalpy effect in these cases. This suggests that the enzyme conformational change is programmed to compensate for $\Delta H_0^{\ddagger}$ while increasing $\Delta S_0^{\ddagger}$ for specific substrates.

The reaction of n-butylamine with p-nitrophenyl trifluoroacetate is accompanied by substantial *negative* activation energies when carried out in both chlorobenzene and 1,2-dichloroethane solvents.[90] Activation parameters for the third-order aminolysis (first order in ester and second order in amine) are as follows: $\Delta H_0^{\ddagger}$ (kcal/mol) $= -11.5$, $\Delta G_0^{\ddagger}$ (kcal/mol) $= 9.62$, and $\Delta S_0^{\ddagger}$ (eu) $=$

-70.9 in 1,2-dichloroethane; $\Delta H_0^{\ddagger} = -10.5$, $\Delta G_0^{\ddagger} = 10.02$, and $\Delta S_0^{\ddagger} = -68.9$ in chlorobenzene. The mechanism is believed to be general-base(amine)-catalyzed attack of a second amine molecule on the ester. The $\Delta G_0^{\ddagger}$ is positive, even though the $\Delta H_0^{\ddagger}$ is negative, because of the very negative $\Delta S_0^{\ddagger}$ associated with bringing three molecules together into one transition state. A highly exothermic addition of amine to the ester carbonyl followed by expulsion of leaving group with a small, positive $\Delta H_0^{\ddagger}$ accounts for the overall negative $\Delta H_0^{\ddagger}$; the negative entropy change upon amine addition prevents accumulation of the tetrahedral intermediate. The authors suggest that changes in activation parameters between "hydrophobic" and "hydrophilic" environments may provide information applicable to a model for temperature regulatory action in certain enzyme processes.

The relationship between heat capacities of activation, $\Delta C_p^{\ddagger}$ (at constant pressure), and other activation parameters, reaction mechanisms, and transition-state structures has been reviewed by Kohnstam.[91] The value of $\Delta C_p^{\ddagger}$, when nonzero, becomes important in order to draw valid mechanistic conclusions based on kinetic parameters. For example, $\Delta H_0^{\ddagger}$ and $\Delta S_0^{\ddagger}$ will vary as the temperature is altered. Therefore, comparisons of activation parameters of different reactions should refer to the same temperature. The large majority of studies of $\Delta C_p^{\ddagger}$ have involved the solvolysis of a neutral substrate via a highly polar transition state, and it is primarily for these processes that measurements have provided information about reaction mechanisms.

A large number of S_N1 reactions in aqueous organic solvents, mainly acetone–water mixtures, show a $\Delta C_p^{\ddagger}/\Delta S_0^{\ddagger}$ ratio which is independent of the nature of the substrate. Since these results are applicable to a wide variety of structures, leaving groups, and reactivities, it would appear that $\Delta C_p^{\ddagger}$ and $\Delta S_0^{\ddagger}$ in S_N1 solvolyses reflect changes associated with the activation process in the same manner.[91] This is consistent with Kohnstam's proposal that for these systems the solvent–solvent and solvent–solute interactions are identical in the initial and transition states, except for solvent–solute interactions arising from the increased solvation of the transition state. The ratio $\Delta C_p^{\ddagger}/\Delta S_0^{\ddagger}$ is noticeably smaller for S_N2 solvolyses in aqueous organic solvents than for the S_N1 reactions under the same conditions, and these ratios (for S_N1 and S_N2 solvolyses) can be useful in providing mechanistic information. That the ratio $\Delta C_p^{\ddagger}/\Delta S_0^{\ddagger}$ is *not* independent of the nature of the substrate for a large number of solvolysis reactions in *water* indicates that the solvation processes here are more complicated than those in the aqueous organic solvents, which led to Kohnstam's simple solvation model. The $\Delta C_p^{\ddagger}$ usually gives a good indication of the mechanism for the solvolyses of halides in water $[\Delta C_p^{\ddagger}(S_N1) < -70$ eu, $\Delta C_p^{\ddagger}(S_N2) \sim -40$ to -60 eu$]$. This generalization, however, cannot be extended to other, similar systems.

Recent work on the hydrolysis of a series of p-substituted benzyl chlorides was aimed at correlating changes in charge development in the transition state with possible changes in $\Delta C_p^{\ddagger}$. Values of $\Delta C_p^{\ddagger}$ showed a distinct dichotomy, the p-methyl compound giving -79 eu, close to that for t-butyl chloride, and more electron-withdrawing substituents giving near -50 eu, close to that for methyl chloride.[92]

The volume of activation, $\Delta V_0^{\ddagger}$, can be very useful in elucidation of reaction mechanism and transition-state structure. Interpretation of $\Delta V_0^{\ddagger}$ values may be complicated by ionic solutes in that their activity coefficients can be highly pressure dependent, especially in solvents of low dielectric constant.[93] Extrapolation of $\Delta V_0^{\ddagger}$ to 1 atm pressure is also difficult since pressure has only a small effect on rates and therefore measurements must be made at several thousand atmospheres.

The effect of pressure on kinetics of reactions in solution has been reviewed.[94] More recently, $\Delta V_0^{\ddagger}$ for the solvolysis of several γ-amino alcohol derivatives (bromides, chlorides, and tosylates) and their carbon analogs has been measured to test the relationship between concertedness of reaction and $\Delta V_0^{\ddagger}$.[95] It was found that the average value for the simple solvolysis reactions [S_N1, E1, F1 (two-step fragmentation)] in 80% aqueous ethanol at 40.0°C is -21.5 ± 1.8 cm^3/mol, while for concerted fragmentation the average value is -13.3 ± 2.0 cm^3/mol. These results lend further support to the idea that activation volumes are quite uniform within a given reaction type. The data also indicate that any kind of nitrogen assistance (as evidenced by rate data) has a pronounced effect on $\Delta V_0^{\ddagger}$, increasing it relative to the carbon analog by 5–10 cm^3/mol.

The usefulness of thermodynamic data in testing a hypothesized transition-state structure relies considerably on comparisons between experimentally obtained values and calculated, or estimated, values based on models. Thermochemical data[96] and extensive rules and additivity schemes for the estimation of thermochemical properties and rate parameters[97] are available for a wide variety of molecules and reactions. The success of these data for predictions involving stable (ENG) molecules suggests the development of similar data for transition states. Benson has done some work on this.[96] The effect of solvation is, of course, a main impediment.

Although the Arrhenius equation does not have quite the same temperature dependence as the ART equation, it has been used extensively in the treatment of rate data [see Eqs. (27)–(30)]. Arrhenius parameters are available for a large number of gas phase reactions[96,98] and for some reactions in solution.[43,44,99] The combination of kinetic isotope effects (k_H/k_D) and the Arrhenius preexponential factor [Eq. (28), A_H/A_D] is frequently useful in obtaining information about possible tunneling effects in hydrogen-transfer reactions[45,48,49,50,53] as well as about reaction mechanism.

The thermodynamics-related quantity pK_a [Eq. (39)] is of particular interest with respect to acid- and base-catalyzed reactions. The presence of an acidic site in a nucleophile allows measurement of a $pK_a^{\ddagger}$ which is sensitive to the extent of nucleophile–carbon bond making in the transition state. The magnitude of the ratio of rate constants, k_1/k_2, for the neutral and lyate-ion-promoted hydrolyses for alkyl halides has commonly been considered diagnostic of whether k_1 refers to an S_N1-like or S_N2-like process.[17] Since this ratio (k_1/k_2) is proportional to the factor by which the acidity of a solvent molecule is changed upon transfer from bulk solvent into the transition state for the neutral reaction [Eq. (38)], one can draw conclusions about bonding and charge distribution in the transition states for both neutral and lyate-ion-promoted processes. It is the vast amount of data available on the correlation between structure (particularly internal charge distribution) and acidity for stable (ENG) molecules that is used to reach these conclusions.

Data are available[17] which measure the increase of acidity of a water molecule when it is transferred from bulk solvent to the transition state for hydrolysis of various monoalkyl sulfate ions. Of the five alkyl sulfates involved, none appears to have more than about 20% bond making in the transition state. As alkyl substitution at the central carbon increases, the acidity of the water in the transition state decreases and finally falls into the range expected for *no* bond formation in the transition state. This is what would be predicted based on other theories describing S_N2 transition-state structure.[66,100]

Methods comparable to those which have been successful in predicting pK_a's for stable molecules can be used to calculate substituent effects on the $pK_a^{\ddagger}$ for a given geometry and charge distribution in the transition state, although it is more accurate to choose a stable acid which resembles the transition state as closely as possible and measure substituent effects on its pK_a. These values can then be used to estimate corrections to the values predicted by calculations. For example, the values obtained for the $pK_a^{\ddagger}$'s for the ionization of the $—CO_2H$ group in a series of haloacetic acids undergoing (1) neutral and (2) base-promoted hydrolyses indicate essentially no difference in internal charge distribution between reactant and transition state.[100] Measurements of the temperature dependencies of the neutral and base-promoted hydrolyses of methyl tosylate have enabled calculation of $pK_a^{\ddagger}$, $\Delta H_0^{\ddagger}$, $\Delta S_0^{\ddagger}$, and $\Delta C_p^{\ddagger}$ for the virtual equilibrium between the transition states for these two reactions. Comparisons between these values and corresponding values for model acids were used to yield information about transition-state structure.[101]

Use of transition-state acid dissociation constants has been shown to be valuable in the analysis of possible enzyme reaction schemes. A rate equation can be derived by inspection which both fits any experimental $\log k_{obsd}$ vs. pH profile accurately and contains just that number of parameters which can be determined from experimental data.[102]

4.2. Symmetry Numbers and Isomeric Structures

The problem of how to determine the symmetry numbers (σ) of those species whose rotational partition functions contribute to the rate of reaction as calculated within the ART framework [Eq. (1)] has been the subject of considerable discussion for many years. Symmetry numbers are, of course, directly related to transition-state structure and constitute a unique feature of the rotational partition function, specifically influencing the entropic contribution to the reaction rate. One approach to this problem has been to omit symmetry numbers entirely and to employ statistical factors, defined in a particular way.[57] The development of inconsistencies using this procedure has led to a reworking of the definition of "statistical factors."[103] Laidler[104] discusses this approach and its difficulties at length. However, it ought to be possible, using correct symmetry numbers, to arrive at chemically correct rate data. Consider the following proton transfers from lyonium ion to substrate:

$$H_3O^+ + S \longrightarrow [S\text{----------}H_3O^+]^\ddagger \longrightarrow product \qquad (42)$$

$$H_2DO^+ + S \longrightarrow [S\text{----------}H_2DO^+]^\ddagger \longrightarrow product \qquad (43)$$

The reactant H_3O^+ has $\sigma = 3$, while H_2DO^+ has $\sigma = 1$. One might, on first thought, ascribe a $\sigma = 1$ to transition states for both reactions (42) and (43), thereby necessitating the conclusion that reaction (42) is favored over reaction (43) by a factor of 3 simply from symmetry considerations alone, since the reactant H_2DO^+ is favored over H_3O^+ by a factor of 3. However, if one examines the situation more closely, one finds that there are *three* possible transition states for reaction (43) [and only one for reaction (42)]:

$$(44)$$

$$(45)$$

$$(46)$$

$$(47)$$

The transition states in reactions (46) and (47) are enantiomeric forms, and hence each must be taken into account. Thus, if symmetry numbers are correctly treated, taking into consideration all *possible* transition states including enantiomeric forms, a correct rate analysis will result. It is particularly important to do this in H_2O-D_2O mixtures where a transition state involving a water molecule may exist in several different forms, some of which may be enantiomers. A lucid and comprehensive discussion of solvent isotope effects has been presented by Schowen[105]; Swain and Thornton[78] discuss the structures of the various transition states which may exist in H_2O-D_2O mixtures.

The structural relationship between the transition states in reactions (45) and (46) indicates another aspect of transition-state structure, namely, the isomeric forms possible. While kinetic data give essentially unambiguous answers about the atomic composition, they do not directly permit selection of the correct isomeric structure of a transition state. However, the application of other criteria, in conjunction with kinetic data, can allow the selection of a transition-state structure which is unique, at least in some cases, or permit a significant narrowing down of possible transition states. The proposed transition-state structure must be consistent with (1) the products formed in the reaction; (2) the stereochemistry of the reaction; (3) the results of isotopic or other kinds of labeling experiments; (4) any intermediate(s) detected by chemical or physical methods; (5) the rate at which a compound, suspected of being an intermediate, reacts to form the correct products; (6) the effects on the nature of the product(s) and the rate of a reaction produced by changes in reaction conditions such as medium, temperature, and catalysts; and (7) the relative reactivities of related compounds, with well-defined differences, in the same reaction.[106,107] It is obvious that the use of these criteria in proposing a unique transition-state structure is entirely analogous to the structural characterization of stable (ENG) molecules, particularly prior to the development of spectroscopic methods.

Transition-state structures for reactions involving proton transfer are difficult to characterize. Studies involving the simultaneous use of hydrogen and heavy-atom isotope effects can provide information about the extent of coupling between proton motion and heavy-atom reorganization (HAR) in the transition state. Particularly in general acid–base-catalyzed reactions, energy requirements for HAR may determine the specific isomeric form for the proton transfer (as well as the degree of coupling).[105] In many cases it is expected that HAR and proton movement will not be coupled but will occur consecutively either through intermediate formation or possibly through consecutive transition states.[29,30,108]

It is interesting, in connection with enzyme catalysis, that certain reactions contain a catalytic rate term involving both a general acid and a general base. It has recently been shown that the enolization of oxalacetic acid contains such a catalytic term and proceeds by a stepwise mechanism involving nucleophile and general base,[108] while enolization of acetone, which also contains such a catalytic term, seems to involve concerted proton transfers.[109] In the former case, HAR and proton transfer appear to be *uncoupled*; in the latter case, two proton transfers appear to be *coupled*.

4.3. Free-Energy Diagrams

With a given PES, structures, including the transition state, can be determined. Furthermore, partition functions and, from these, standard free-energy changes (ΔG_0) and the standard free energies of activation for all the steps in the reaction may be calculated. Finally, a standard free-energy diagram, analogous to an energy profile, may then be drawn. However, since the standard free energy is an average quantity of a collection of molecules and therefore can only be defined for a specific ENG, the diagram is drawn to represent standard free energies of stable structures and transition states by short, horizontal lines, not as a continuous curve, as is the case with an energy profile abstracted from the PES.

The ΔG_0 values for many kinds of reactions are dependent on the concentration units or standard state. This dependency exists wherever the stoichiometric number of molecules of reactants differs from that of products, since then the equilibrium constant is not dimensionless [see Eq. (11)]. In the most common concentration units, M, ΔG_0 is the free-energy difference between products and reactants which would exist if all species were present at 1 M concentration. However, concentrations are usually far different from 1 M, especially in biochemical systems. For example, in the single enzyme–substrate association

$$E + S \rightleftharpoons ES \tag{48}$$

the Michaelis–Menten constant K_m is the dissociation constant, K_d, for the complex ES. From

$$K_d = [E][S]/[ES] \tag{49}$$

it can be seen that $[E] = [ES]$, i.e., the enzyme is half-saturated, when $[S] = K_d$. In many biochemical systems, $[S]$ is comparable with K_d, which is frequently on the order of 10^{-3} M, so that ΔG_0 does not reflect the fact that $[E]$ is similar to $[ES]$. For this reason, use of a standard state more nearly comparable with natural concentrations can make the standard free-energy diagram more informative. The procedure is simply to express K in the desired units. For example, a standard state of 40 μM was used for the enzyme-catalyzed interconversion of dihydroxyacetone phosphate and D-glyceraldehyde-3-phosphate.[110] In pH-dependent processes, it is common to use a standard state of 10^{-7} M (i.e., pH 7) for the concentration of hydronium ion.

A related method has been used by Klotz, who discusses enzyme-catalyzed reactions by means of "concentration profiles."[111]

Many sequential biochemical reactions exist with steady-state, but not equilibrium, concentrations of reactants and products, because these species are continually being produced or destroyed by yet other reactions. Under these conditions, free-energy relationships between reactants and products, rather than the *standard* free-energy relationships we have been discussing up to this point, are more appropriate quantities for consideration. When one studies a single reaction *in vitro*, equilibrium will be approached as the reaction proceeds, but *in vivo*, the steady state may obtain over times that are long compared with reaction rates. For the simple reaction

$$A + B \longrightarrow C \tag{50}$$

the free-energy difference between product and reactants, ΔG, is given by

$$\Delta G = \Delta G_0 + \mathbf{R}T(\ln c_C/c_A c_B) \tag{51}$$

where ΔG_0 is the standard free-energy change and c is the concentration (e.g., in M, if this is the standard state for ΔG_0).[112] Figure 12 illustrates the basic difference between ΔG and ΔG_0 for reaction (50),[113] where concentrations are such that $\Delta G < 0$, although $\Delta G_0 > 0$. The significance of ΔG is that a reaction will proceed spontaneously only in the direction in which $\Delta G < 0$; under conditions of Fig. 12, $A + B \rightarrow C$, not $C \rightarrow A + B$. This analysis has been carried out by Hamori[113] for the glycolytic conversion of glucose to lactate in red blood cells, as shown in Fig. 13, in which one sees that all reactions proceed spontaneously (or are at equilibrium), since $\Delta G < 0$, whereas ΔG_0 values, some being > 0, do not directly demonstrate the spontaneity of the processes at all.

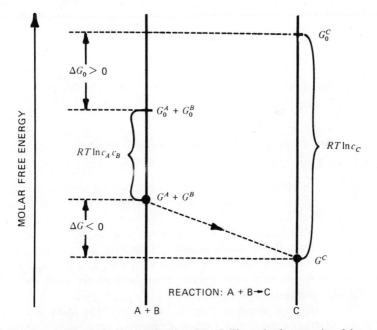

Fig. 12. Free-energy diagram for the reaction $A + B \rightarrow C$. The molar free energies of the system in states $A + B$ and C are represented on vertical bars. The short horizontal lines denote standard free energies (at unit concentrations), and the dark circles denote free energies corresponding to concentrations c_A, c_B, and c_C.

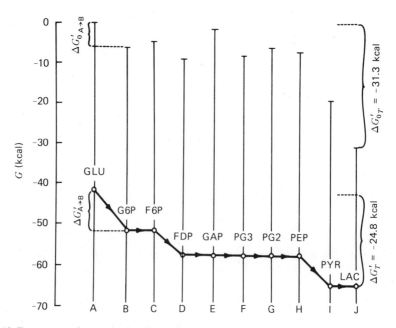

Fig. 13. Free-energy changes in the glycolytic conversion of glucose to lactate. The vertical bars represent the free energies of the major chemical states of glycolysis. For a complete definition of these states, the meaning of the symbols, and the experimental data used in the construction of the diagram, see Reference 113.

5. EXPERIMENTAL APPROACHES TO DETERMINING TRANSITION-STATE STRUCTURE

5.1. Reaction Rates and Correlation with Transition-State Structure

The transition state, being a well-defined molecular entity possessing the usual properties associated with molecules, is subject to structural studies just as are other molecular species. However, the experimental approaches employed in elucidating transition-state structure are, by necessity, more indirect and inferential than those conventional methods (primarily spectroscopic) used on stable ENG's, because of its transient nature as a metastable ENG. This feature obviously limits its concentration, thereby reducing the number of feasible approaches to its study.

Of course, once the general structural features of the transition state for a particular reaction have been established, one may then study a stable (relative to the transition state) molecule which can serve as a reasonably good model. For example, alkyl halide–halide association ions, $[RX_2]^-$, observed in the negative chemical ionization mass spectra of alkyl halides, appear to be related to the corresponding S_N2 transition states in solution.[114] Bicyclo[2.1.1.]hexane, mentioned previously (Section 2.2), is a stable molecular model for the transition state (i.e., eclipsed conformation) between two staggered conformations of ethane. A systematic crystal-structure analysis of a series of compounds containing recurring molecular fragments has led to correlation curves the shape of which is reminiscent of the structural changes occurring along the pathways of several different types of chemical reactions, including S_N1, S_N2, and nucleophilic addition to carbonyl groups.[115] Comparison between reaction pathways determined from structural correlations (using crystallographic data) with those obtained from models of approximate surfaces (from spectroscopic, force field, and quantum-mechanical calculations) shows fair agreement.

For the most part, however, transition-state structures are characterized through mechanistic pathways, which, in turn, are elucidated through the interpretation of various kinds of experimental data obtained from "dynamic" reactions, rather than from "static" molecules.

5.1.1. Structural, Solvent, and Catalytic Effects

Since the rate of a reaction is inextricably related to its transition state, it is obvious that rate data are a prime source of information about the transition state. However, to clarify transition-state structure using rates, one needs data from not just one but a series of closely related reactions such that any changes in rate observed can be related, systematically, and hopefully unambiguously, to changes in transition-state structure. Ideally, one would like to change one, and only one, variable, holding all others constant, for a set of (closely related) reactions and then be able to draw general conclusions about the nature of the transition state for that set.

One obvious "variable" to change is an atom or group of atoms, close enough to the reaction center (that part of the molecule which actually undergoes changes in bonding on forming a transition state) to exert an influence on it and hence on the rate. If the atom, or group of atoms, is part of the reacting molecule, it is generally referred to as a "substituent." Since the reaction rate is reflective of a *difference* in (free) energy between reactant(s) and transition state, the effect of the substituent on both must be considered. The influence of a particular substituent on the reaction center is usually categorized as electronic and/or steric; electronic effects can be further divided into (1) inductive, (2) field, (3) polarizability, (4) resonance, and (5) hyperconjugative effects. These general categories have been reviewed elsewhere.[107,116,117] A modification suggested by Pople and Gordon,[118] classifies substituents in terms of two characteristic features:

1. Those which either withdraw electrons from or donate them to a hydrocarbon fragment (saturated or unsaturated) as a whole and are therefore described as inductive $-I$ and $+I$ types, respectively, and

2. Those which polarize the remaining electrons of the hydrocarbon fragment such that electrons are drawn to or from the site of substitution. The effect of these substituents is denoted by $-$ and $+$ superscripts, respectively. These superscripts correspond to the signs usually used to denote the resonance effect of the substituent.

This classification results in four types of substituents ($-I^-$, $-I^+$, $+I^-$, and $+I^+$), provides a consistent approach to the effect of a large number of substituents, and successfully predicts many details of electric dipole moment data. The role of field and inductive effects—in fact, the question of their separability—has been investigated by many workers and recently reviewed by Hirsch.[116]

Since an isotopic replacement of an atom or group of atoms in a molecule can exert an influence on a reaction rate, isotopes do qualify as "substituents." However, because they are unique, nonperturbing structural probes quite different from ordinary substituents, isotopic substituents and their effects will be treated in a later section.

For reactions carried out in solution rather than in the gas phase, the influence of solvent on the reaction rate can provide useful information about a reaction mechanism. A continuum of interactions, ranging from very weak and nonspecific to relatively strong and specific, is exhibited between solvent and solute molecules. Furthermore, to relate the effect of solvent to the transition-state structure, solvent interactions with both reactant and transition state must be analyzed. If coupling between the internal motions of the solute (reactant and/or transition state) and those of the surrounding solvent molecules, to the extent that it occurs at all, is very weak, then the solvent will exert only a gross effect, called solvation. Interactions between solute and surrounding solvent molecules are such that the latter are not significantly altered from molecules in the bulk solvent. Of course, if the solvent molecules are far enough away from the solute, their environment is essentially that of bulk solvent, whereas if they are close to the solute, they may be perturbed, albeit weakly. On the other hand, if coupling between the internal motions of solute and those of surrounding solvent molecules is strong, the latter may be significantly altered from molecules in the bulk solvent. Interactions of this kind may be so specific that solvent itself becomes a reactant and is wholly or in part covalently bound to a fragment of the solute during the reaction. Hence, there is certainly a continuum of interactions possible, one extreme being complete covalency changes occurring as a result of interaction between solvent and solute and the other being complete noninteractions. For the case in which solute–solvent interactions are strong, solvent effects are not unlike ordinary "substituent effects," albeit the "substituent" is not initially a part of the reacting species. Finally, the degree or character of solute–solvent interactions, weak or strong, may change on progressing from reactant to transition state. If the reactant has solvation properties quite different from the transition state, changing the solvent may affect the reactivity or even the reaction pathway.

The prediction of the effect of solvent polarity on reaction rates was first made by Hughes and Ingold.[119]

The combination of solvent with isotope effects has been applied to water and other protic solvents. Solvent isotope effects which come about when a hydroxylic protium is replaced by a deuterium may be the result of (1) solvent becoming (wholly or in part) covalently bound to reactant (solute), (2) positions in the reactant becoming labeled by rapid exchange with solvent so that solutes differ isotopically in the two solvents, and/or (3) changes in the degree or character of weak solute–solvent interactions on going from reactant to transition state.[105] Since factors (1) and (2) result from changes in the frequencies of internal vibrations on formation of the transition state and factor (3) from solvent changes, the isotope effects, depending on their origin, can provide information about internal structural features of the transition state as well as about its solvational environment.

There is a relationship between solvent effects and catalysis. For protic solvents and for water in particular, favorable interactions may stabilize the transition state for a reaction to the extent that "solvation catalysis"[105] based on the "solvation rule" of Swain *et al.*[120] may be possible. This catalytic mechanism presumably involves no covalency changes of the solvent prior to, or during, formation of the transition state for the reaction. It will be recognized that the functions of solvation can be performed by a molecule of some *solute*, not the solvent, present in a reaction solution. Hence, one can envision the extreme of catalysis by a nonsolvent molecule ("solvation catalysis") for catalysts such as an acid, a base, or an enzyme. In fact, interactions between solute (substrate) and solvating species (e.g., acid, base, enzyme) may become so strongly coupled that covalency changes may then take place—frequently involving rapid proton transfers—which speed up the reaction. Regeneration of the solvating species at the end of the reaction designates it as a catalyst molecule (by definition). There can be (and undoubtedly is) a continuum of types of interactions between solute (substrate) and solvating species (solutes) acting as catalysts.

The quintessential catalyst is, of course, an enzyme, a molecule which can be viewed as a "different kind of environment" (or solvent), one in which multiple solvent molecules (of the "ordinary" variety) are replaced by a *single* molecule—the enzyme. This unique molecule can, at the region of its active site, contribute both solvation and multiple catalytic groups simultaneously. In addition, the active site may be so different (e.g., less polar) from the aqueous solution in which enzyme and substrate are located that processes entirely different from those occurring in aqueous solution may be favored. For instance, it has been found that reducing the solvent polarity can greatly accelerate the rate of general-base-catalyzed hydrolysis of Schiff bases.[121] The synergistic effect observed between general base catalysis and reduced solvent polarity in this type of reaction, certainly potentially available to enzymes, may, in fact, be the basis of significant rate accelerations in enzymic reactions.

The dramatic rate accelerations brought about by enzymes have stimu-lated a great many investigators to speculate about the nature of these ex-traordinary catalytic effects. Many excellent reviews on the subject have appeared.[122,123] Although some have suggested that altogether new phenom-ena must be considered to explain enzymic catalysis, others have reached the conclusion that it can be explained according to well-known principles of physics and chemistry which are used to explain nonenzymic catalysis. Ac-cording to ART, a catalyst (including an enzyme) increases the rate of a re-action by reducing the free energy of activation. This decrease (in $\Delta G_0^{\ddagger}$) can result from a decrease in the enthalpy ($\Delta H_0^{\ddagger}$) or/and an increase in the entropy ($\Delta S_0^{\ddagger}$) of activation for a reaction. There are various ways in which a non-enzymic catalyst can bring about the decrease in $\Delta G_0^{\ddagger}$; it is recognized that the same effects are operative in enzymic catalysis.[122,123] One apparent means whereby an enzyme can effect rate enhancement of a bimolecular reaction is by bringing the reacting molecules together at the active site of the enzyme by use of weak binding forces. This is known as the *proximity* or *approximation* effect. As a consequence of this effect (or, according to some, identical to it), there will be a second effect, that of increasing the effective local concentration of reactants. A further rate acceleration might result from favorable *orientation* or rotamer distribution of the reactants if they are fixed in a reactive position (i.e., one resembling the transition state).[124] In addition, the term "orbital steering" has been coined by Storm and Koshland[125,126] to describe orienta-tion effects which include potential energy effects that may be involved in strict steric and angular requirements but not gross orientation of substrate and catalytic groups on the enzyme surface or the juxtaposition of reacting atoms. Extensive discussion has ensued since the introduction of the term, and the exact nature of the effect itself and the significance of its contribution, if any, to rate enhancement are, as yet, unresolved.[83,127,128] At present, it is difficult to see how this effect could be distinct from, and add onto, the orienta-tion effect. Strain effects, such as the induction of strain or distortion in the substrate, the enzyme, or both, have also been hypothesized as responsible, at least in part, for the rate accelerations observed in enzymic reactions.[129]

If a decrease in the translational and overall rotational degrees of freedom (i.e., entropy) of the reactants occurs prior to reaction, it will no longer be necessary for them to lose this entropy in order to react; thus a rate enhance-ment, due to increasing the ground-state free energy from loss of entropy, should be observed.[83,130] An approach toward relating interpretations of rate increases attributed to loss of entropy in the ground state, favorable orientation and rotamer distribution effects in the ground state, and strain effects, or some combination of these effects, is that of Delisi and Crothers, who attempt to formulate and assess these effects in terms of molecular struc-ture parameters.[131]

Other often-cited catalytic effects applicable to enzyme reactions include formation of covalently bonded intermediates; general acid–base catalysis, where catalytic groups are already present and favorably positioned (for reaction) on the enzyme; solvent effects; formation of hydrogen bonds; electrostatic interactions; and hydrophobic forces. All of these effects have been comprehensively reviewed elsewhere.[123]

We can envision what features an enzyme-catalyzed reaction has which tend to accelerate its rate. However, it is not quite so easy to estimate the rate accelerations which would be brought about by such features. It has been found that enzyme reaction rates are so large that the problem at hand is whether all the features which can be mustered are sufficient to account for these rates.

Pauling demonstrated that the binding of transition state by enzyme is the single feature which determines enzyme catalytic efficiency, if we take as the basis of comparison the uncatalyzed reaction.[132] To be a catalyst, the enzyme must interact with the transition state more favorably than it interacts with the substrate. Put another way, the dissociation constant of the transition state from the enzyme, $K_S^{\ddagger}$, must be smaller than that for substrate, K_S,

$$ES^{\ddagger} \underset{}{\overset{K_S^{\ddagger}}{\rightleftharpoons}} E + S^{\ddagger}$$
$$ES \underset{}{\overset{K_s}{\rightleftharpoons}} E + S \tag{52}$$

where $S^{\ddagger}$ is the transition state for the uncatalyzed reaction. Of course, this statement applies to any kind of catalyst, not just an enzyme, but we normally do not think of substrate binding by small-molecule catalysts, because K_S is generally expected to be quite large. If a catalyst is fairly large, then it is possible that it interacts significantly with parts of the substrate which are removed from the actual catalytic reaction center. This leads to the property of "molecular recognition" or specificity for certain substrates and the property of "entropy saving," where such substrate binding compensates for the large amount of entropy lost upon bringing two molecules, e.g., E + S, together into a single transition state. Since recognition can take place at parts of the substrate which do not undergo chemical change, such a catalyst binds both the substrate and the transition state in similar ways by this means. Therefore, insofar as catalytic efficiency is concerned, binding of substrate is a sort of incidental consequence of the specific binding of the transition state.

A maximum rate will be observed when the enzyme is saturated with substrate (and reactant is ES), but at sufficiently low substrate concentrations the rate will decrease as enzyme becomes less than completely saturated and the reactant species become E + S rather than ES (see Section 3.2). The distinction as to which is the reactant is not very important, however, because the main feature of enzymes is that they interact with the substrate strongly enough so that, at physiological concentrations, the enzyme is either saturated with substrate or, if not (as is probably the usual case)[122a], it costs very *little* to bind substrate to enzyme. There is one exception to this statement that may have biochemical importance. Since the enzyme specifically recognizes the transition state, it may be that when it binds substrate, the binding causes distortion of the substrate at the site of chemical change, thus destabilizing substrate relative to transition state and increasing reactivity of the ES complex. This strain effect will not increase catalytic efficiency of the enzyme for the substrate alone, but if reaction occurs under conditions where the enzyme is saturated so that ES is the reactant, then this effect will increase the rate of the catalytic step, ES $\rightarrow$ ES‡. A strain effect on substrate has been found in the case of the enzyme chymotrypsin.[133-135] The translational entropy loss upon bringing two species, E + S‡, together is compensated for by interactions at parts of the transition state which are not undergoing chemical change. Thus, specific interactions of catalytic groups with the site of chemical change do not also have to compensate for the entropy loss, as they must do in the case of a small-molecule catalyst. This entropy-saving binding can contribute on the order of a factor of 10^6 to the rate, since the energetics of the catalytic mechanism can contribute entirely to catalysis itself if it does not have to compensate the entropy loss.

Since the active site of an enzyme is not analogous to, say, an aqueous solution, it can be asked why the aqueous solution is considered to be an appropriate "uncatalyzed reaction." The answer is that we must take into account *all* differences between the "uncatalyzed" and the catalyzed reaction; therefore, the standard of comparison is not an absolute. The gas phase would be a truly "uncatalyzed" system, but in solution there would be solvent catalysis. However, most reactions of interest will not proceed in the gas phase or cannot be studied there for experimental reasons. In fact, many enzyme reactions do not proceed even in aqueous or other solutions in the absence of enzyme, and much research has been directed toward finding uncatalyzed model reactions. An aqueous solution simply provides a convenient basis for comparison, where all molecules interacting with the substrate are the same and are small. From this standard, comparisons can be made not only with the enzyme itself but also with other, small-molecule catalysts, in order to dissect further the effects contributing to the overall catalytic efficiency of the enzyme.

Two types of effects have already been mentioned: the effects of catalytic groups and the translational entropy-saving feature. Effects of catalytic groups can be studied with models; for example, general acid, general base, metal-ion, and nucleophilic catalyses are features of certain enzymes and have been studied extensively in solution. The translational entropy-saving feature can be studied by combining substrate and catalyst groups together into a single molecule. In some cases, large rate enhancements are observed.[127,128] In an enzyme or an intramolecular model reaction, one can also expect additional rate enhancements from orientation effects, in which the reacting groups are properly aligned for reaction. This alignment is an entropy-saving effect similar to the translational effect we have discussed, but it is additionally based on the loss of rotational entropy upon binding. If an enzyme can orient the groups for reaction, then the stabilizing effects of the catalytic groups will not have to be used to compensate for the entropy loss in attaining proper orientation in the ES‡ complex. A factor of 10^2–10^4 can be expected from orientational (rotational) entropy saving.[130]

In addition, an enzyme may bring together several different functional groups which can simultaneously catalyze a given reaction step. To bring several small-molecule catalysts into association with substrate would involve an enormous entropy cost, on the order of the factor of 10^6 for each, not counting orientation. The enzyme avoids this entropy cost by means of the loss of entropy involved in the biosynthesis of the enzyme from its amino acid components, which is paid for by metabolic energy. Once biosynthesized, the enzyme can act with multiple catalytic groups, without extra entropy cost in the catalytic reaction. Additional entropy savings are realized if the enzyme's catalytic groups are already properly oriented by biosynthesis, or even by the binding of substrate.[136] Since the enzyme acts as the "solvent" for the transition state, groups within the enzyme which specifically solvate the transition state may be classed as catalytic groups and therefore are all subject to entropy savings through covalent bonding and orientation effects.[135] Even groups within the peptide backbone may be involved in specific hydrogen bonding. Binding of substrate can be thought of simply as a way of saving entropy where covalent bonds are not involved, while the functional groups within the enzyme save entropy through prior biosynthesis involving covalent bonds. It is one of the wonders of nature that the substrate binding phenomenon leads not only to entropy saving but also to specific molecular recognition and to possibilities for control of enzyme rates through saturation with substrate, product, or inhibitors as well as allosteric effectors.

With the combination of specific substrate binding, multiple catalytic groups, and alignment of substrate and catalytic groups, there is at present every reason to believe that the catalytic efficiency of enzymes compared with uncatalyzed and model reactions will be fully explained. In fact, these are essentially all the effects which could possibly be involved. One more possibility exists, which can be called "vertical catalysis and force compensation." We shall define it in a later section.

If the transition state is very favorably bound by enzyme [see Eq. (52)], then it follows that stable molecules resembling the transition state more than they resemble the substrate might bind more tightly to the enzyme than the actual substrate.[137-140] This approach has great potential for the study of enzyme–transition-state complexes.[141] A similar approach, using conformationally restricted substrate analogs, has been employed to study enzyme–substrate complexes and, in those cases where the analog actually reacts, the enzyme–transition-state complex.[142]

5.1.2. Linear Free-Energy Relationships

A great deal is known about linear free-energy relationships (LFER) and the limitations thereon; several of these relationships have found particular use in the discussion of enzymic as well as nonenzymic reactions.[116,117,143,144] Interesting results may be derived from the combination of certain LFER's[145,146]—primarily the Hammett, Brønsted, and Swain–Scott equations—and we have chosen to center our discussion around these.

An LFER simply means that the free-energy changes for one series of reactions are linearly related to the free-energy changes for another series of reactions, i.e.,

$$\Delta G_{0(i)} = x \, \Delta G'_{0(i)} + b \tag{53}$$

where the $\Delta G_{0(i)}$ are a series of free-energy differences for similar reactions and the $\Delta G'_{0(i)}$ are a series of free-energy differences for another series of similar reactions in which the same structural changes are made as in the first series. When several series of reactions are correlated well with a single series of $\Delta G'_{0(i)}$ (with different x and b for each series of $\Delta G_{0(i)}$), the reactions probably are responding to a common interaction mechanism.[143]

As early as 1935,[147] Hammett recognized that when two series containing logarithms of rate or equilibrium constants are both linearly related to a third series, they are also linearly related to each other [see Eqs. (11), (13), and (15)]. Thus, various series could be related to a single "reference" series. The now well-known Hammett equation, which describes the effect of a *meta*- or *para*-substituent on the rate or equilibrium constant of an aromatic side-chain reaction, takes the form

$$\log(k_i/k_0) = \rho\sigma_i \tag{54}$$

where ρ is a proportionality constant characteristic of the reaction and conditions but independent of substituents and σ is a constant characteristic of the substituent but independent of the reaction and conditions. Hammett's σ was originally defined by the ionization constants of substituted benzoic acids in water at $25°$:

$$C_6H_5CO_2H \overset{K'_0}{\rightleftharpoons} C_6H_5CO_2^- + H_3O^+ \tag{55}$$

$$m\text{- or } p\text{-}XC_6H_4CO_2H \overset{K_i}{\rightleftharpoons} m\text{- or } p\text{-}XC_6H_4CO_2^- + H_3O^+ \tag{56}$$

$$\sigma_i \equiv \log(K'_i/K'_0)_{H_2O, 25°} \tag{57}$$

Hence, the standard or "reference" series is the ionization of the benzoic acids (with $\rho = 1$ and $\sigma = 0$ for benzoic acid itself), and the $\rho\sigma$ relationship might be expected to work best for equilibria similar to this. In fact, the Hammett relationship holds for a very large number and wide variety of rates and equilibria.

Because σ constants are essentially empirical proportionality constants, and because deviations from linearity are observed in some cases, there is a question of whether the same σ constant for a given substituent (on a benzene or other aromatic ring) should apply to all kinds of reactions. Since a substituent will interact differently with the reaction site depending on its position, two sets of σ values are required (σ_m and σ_p), both of which may include electrostatic and resonance effects. Sigma values based on measured thermodynamic ionization of the substituted benzoic acids (in water at $25°$) are called *primary* values, while those obtained by comparison with another set of compounds or reaction conditions are called *secondary*. Since the success of the Hammett equation depends on the relatively constant relationship between electrostatic and resonance effects in reactions occurring on aromatic side chains, σ values *different* from those involved in the original Hammett equation may be required to give an LFER with few (if any) deviations for reactions which are different from the ionization of benzoic acids. Hence, a number of different sigma scales have been tabulated, for example,[148] (1) σ^0 and σ^n, called normal substituent constants, characterizing substituents which are not involved in direct resonance and (whose effects) are unaffected by changes in solvation; (2) σ^+, electrophilic substituent constants, most useful for electrophilic aromatic substitution reactions; and (3) σ^-, nucleophilic substituent constants, most useful for nucleophilic aromatic substitution reactions. In fact, a direct relation of σ constants (specifically $\sigma_R^{0[149]}$) to an energy scale has been proposed.[150] Using the energy barrier to rotation of a substituent about the ring–substituent bond in a substituted benzene as equivalent to the difference in the energy of resonance interaction between the position of maximum energy (usually the orthogonal position of 90° twist) and the position of lowest energy (usually at or near coplanarity of substituent and ring) *less* the corresponding difference in strain energy, Grindley and Katritzky[150] calculate a (rounded) figure of 33 kcal/mol per σ_R^0 unit as the standard figure to obtain values of ring–substituent resonance interaction energies in monosubstituted benzenes.

Various attempts have been made to separate substituent effects into electrostatic (σ') and resonance (σ_R) components (in the absence of steric effects). Swain and Lupton,[151] in an attempt to reduce the large number of σ scales, have calculated field and resonance constants for 42 substituents using Hammett σ_m and σ_p values and find that, from these two parameters, the substituent effects on a large body of data analyzed in terms of the Hammett equation can be predicted. Each set of experimental substituent constants, e.g., σ, σ', σ^+, etc., is expressed as

$$\text{``}\sigma\text{''} = f\mathscr{F} + r\mathscr{R} \tag{58}$$

in which f and r express the sensitivity of the experimental or analytical technique to $\mathscr{F}$ and $\mathscr{R}$, the electrostatic (field or inductive) and resonance (delocalization) constants, respectively. Although the degree of separation of field and resonance effects achieved by the Swain–Lupton approach has been extensively debated, their technique or analogous techniques provide a means of estimating for each reaction series the mix of field and resonance contributions for that reaction. This mix is a very significant kind of information; it can be derived for transition states through rate correlations which determine the values of f and r that best fit the data. Extensive analysis of substituent effects in the benzene series has led Ehrenson et al.[149] to conclude that better fit of data is obtained using four different types of resonance parameters (σ_R^0, σ_R, σ_R^+, and σ_R^-) for different classes of reactions or effects. They use a generalized dual substituent parameter treatment which attributes substituent effects to an additive blend of polar (I) and resonance (pi delocalization, R) effects, each of which may be represented as a $\sigma\rho$ product:

$$\log(k/k_0)_i, \text{ or other substituent property, } P^i = I^i + R^i = \sigma_I\rho_I^i + \sigma_R\rho_R^i \quad (59)$$

The coefficients, ρ_I^i and ρ_R^i depend on the substituent position (indicated by the index, i) with respect to the reaction (or detector) center, the nature of the measurement at this center, and the conditions of solvent and temperature. Sjöström and Wold, also using a statistical approach, find that a unified sigma scale, based on reactions of σ^0, σ, and σ^- type, gives as good a data fit as the fit obtained using each of the different sigma scales separately.[152]

That the Hammett equation is a free energy comparison between two series of reactions can be demonstrated as follows (for equilibria, but equally valid for rates):

$$\log(K_i/K_{ref})^{\text{benzoic acids (benz)}} = \rho\sigma \quad (60)$$

and

$$\Delta G_0 = -2.303\mathbf{R}T\log K \quad (61)$$

Therefore

$$\delta\Delta G_0^{\text{benz}} = \Delta G_{0(i)}^{\text{benz}} - \Delta G_{0(ref)}^{\text{benz}} \quad (62)$$

Also,

$$\log(K_i/K_{ref})^{\text{other series}} = \rho\sigma \quad (63)$$

Since

$$\sigma \equiv \log(K_i/K_{ref})_{\text{benzoic acids, }H_2O, 25°C} \quad (64)$$

then

$$\log(K_i/K_{ref})^{\text{other}} = \rho\log(K_i/K_{ref})^{\text{benz}} \quad (65)$$

and

$$\delta\Delta G_0^{\text{other}} = \rho\,\delta\Delta G_0^{\text{benz}} \quad (66)$$

Since the quantity (ΔG_0) being compared is thermodynamic, yet the relationship is not part of the formal structure of thermodynamics, the Hammett equation is called an *extrathermodynamic relationship*. The validity of the equation over a range of temperature requires that $\delta\Delta H_0$ (or $\delta\Delta H_0^{\ddagger}$) and $\delta\Delta S_0$ (or $\delta\Delta S_0^{\ddagger}$) be proportional to σ and therefore that $\delta\Delta H_0$ be proportional to $\delta\Delta S_0$, i.e.,[71,153-157] that

$$\delta\Delta H_0 = \beta\,\delta\Delta S_0 \quad (67)$$

Such a relationship is termed *isergonic*, and frequently the behavior of the quantities involved is compensatory [see Eq. (14)], as denoted by $\beta > 0$.[156]

The idea that reaction rates should tend to parallel equilibrium constants was tested many years ago by Brønsted and Pedersen, who found a correlation between the rate constants, k_B, for a series of base-catalyzed decomposition reactions and the basicity constants, K_B, of the catalysts.[158] It was suggested, and subsequently borne out, that similar correlations might be found for other proton-transfer reactions. These correlations, called Brønsted relations (law, equations),

$$k_A = G_A(K_A)^\alpha \qquad \text{for general acid catalysis} \qquad (68)$$

$$k_B = G_B(K_B)^\beta = G'_B(K_A)^{-\beta} \qquad \text{for general base catalysis} \qquad (69)$$

were the first-known LFER's.[159-161] An energy profile for a proton-transfer reaction (zero-point energies have been omitted) can be drawn as in Fig. 14[162a,b], where curves I and II represent the potential energies of the reactants SH (substrate) and B (basic catalyst) and reaction products, S^- and HB^+, respectively, as a function of the distances S—H and H—B^+, assuming that the centers S and B remain stationary at a fixed distance apart (a reasonable assumption as both are always considerably heavier than H). If we assume that free energy *differences* can here be approximated by potential energy differences, and if the relative positions of the two curves are appropriately chosen, the vertical distance between the two minima is equal to the free-energy change (ΔG_0) in the reaction, and the distance between the "reactant" minimum and the point of intersection of the two curves is equal (approximately) to the

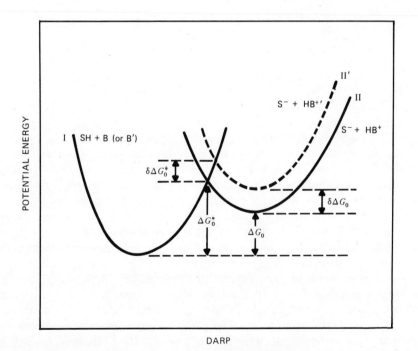

DARP

Fig. 14. Energy profile for a proton-transfer reaction from substrate (curve I) to bases of different strengths (product curves II and II'). See the text for the assumptions involved.

free energy of activation ($\Delta G_0^{\ddagger}$) for the proton transfer.[159] If the basic catalyst is modified (perhaps by alteration of a substituent) in some way so that it becomes weaker, curve II will be displaced upward in the direction of higher energy (curve II′), with consequent changes of $\delta \Delta G_0$ and $\delta \Delta G_0^{\ddagger}$ as indicated (see Fig. 14). If s_1 and s_2 are the slopes of curves I and II at their point of intersection (both taken as positive), then

$$\delta(\Delta G_0^{\ddagger})_{SH} = [s_1/(s_1 + s_2)] \, \delta(\Delta G_0)_{HB^+} = \beta \, \delta(\Delta G_0)_{HB^+} \qquad (70)$$

which is an LFER relating the standard free energy of activation and the standard free-energy change for the acid–base pair HB^+–B and is equivalent to the logarithmic form of Eq. (69).

Since the Brønsted relationship compares a system in which proton transfer is incomplete (the catalysis *rate*) with one in which it is complete (*equilibrium constant* of the acid or base), a rate–equilibrium relationship is defined in which values of α (or β) are expected to range from 0 to 1 and to reflect the amount of proton transfer in the transition state for the catalytic process. The more completely the transition state resembles the products (i.e., the more completely transferred the proton), the closer the slope (α or β) will be to 1, while the less shifted the proton in the transition state, the closer the value of α or β to 0.[160,161] Figure 14 suggests that the β for curves I and II′ should be larger than the β for curves I and II, since s_1 for curve I at the point of intersection with curve II′ is *larger* than s_1 for curve I at the point of intersection with curve II, while s_2 for curve II′ at its point of intersection with curve I is *smaller* than s_2 for curve II at its point of intersection with curve I. However, for real molecular systems, the parabolic curves I, II, and II′ would be replaced by Morse-type curves. The intersection of these curves (I and II or I and II′), or any other comparable curves representing species in a reacting system for which there is a "reasonable" energy of activation (possibly 10–30 kcal/mol, although the limits are certainly unknown at this time), occurs along the side of the Morse curve where the slope remains *essentially* constant, and therefore the β's for the two sets of curves are essentially the same. Constant β's are frequently observed, but it is expected that if the base or the acid (catalyst) is made strong enough or weak enough, the slope should approach 0 or 1; i.e., plots (of $\log k$ vs. $\log K$) should be curved if studied over a wide enough range of acidity or basicity. Some such trends have been observed experimentally.[161] Brønsted relationships ought to be observed for various degrees of proton transfer, all the way from rate-limiting diffusion-controlled proton transfer, where the approach of the acid (or base) is rate determining and α is expected to be 0, to prior (to the rate-determining step) proton transfer, where α is expected to be 1. Extremes ($\alpha = 0$, $\alpha = 1$) are not usually observed; however, an $\alpha = 0.03$ (essentially 0) has been reported for general acid protonation of an unstable intermediate. The pathway was reasonably interpreted as involving rate-determining diffusion-controlled proton transfer to this intermediate on the basis of an observed change of mechanism which appears to require proton transfer to be rate determining.[163] This small α can be observed because water (the solvent) has a reactivity much lower than diffusion-controlled reactivity.

Slopes outside the range of expected values ($0 \leq$ slope ≤ 1) have been reported.[161,164] The "abnormal" Brønsted slopes (<0 or >1) have been explained by invoking extra interaction mechanisms in the transition state.[160,161] However, in view of the wide range of applicability of normal Brønsted slopes, one probably should first look for a more complex reaction mechanism (as, for example, the two-step, "two-anion" mechanism recently proposed for nitroalkane deprotonations[165]), and only if such a mechanism can be ruled out, should one accept the idea of a true, abnormal Brønsted slope.

In the study of proton transfers, a question of major concern is why some are fast while others are slow. The fact that proton transfer between electronegative atoms such as oxygen or nitrogen is very fast while that involving carbon is usually slow would appear to be related to the localization of the electron pair receiving the proton on a single atom in the former case and the usual delocalization of the electron pair away from the atom of the carbon base receiving the proton in the latter. A study of the rate and equilibrium for deprotonation of phenylacetylene and chloroform shows that the rate constants for the reverse reactions are essentially diffusion controlled ($k \approx 10^{10}\ M^{-1}\ \mathrm{sec}^{-1}$); i.e., proton transfer to these localized carbanions is very fast.[160]

A Hammett-type LFER correlating reaction rates with "nucleophilicity" is the Swain–Scott equation[166]

$$\log(k/k_0) = sn \tag{71}$$

where s measures substrate discrimination (sensitivity) to various nucleophiles and n is characteristic of the nucleophile itself.[167,168] It has been shown that combining the Hammett [Eq. (54)] and Brønsted [Eq. (68)] equations (in their logarithmic forms) for the reactions between a series of i substrates and j general acid catalysts, one derives the relationship (for any two arbitrary acids, here denoted as 1 and 2)[146]

$$(pK_{a2} - pK_{a1})/(\rho_2 - \rho_1) = \sigma_i/(\alpha_0 - \alpha_i) = c_1 \tag{72}$$

where the symbols have their usual meanings. Since the left-hand side of Eq. (72) depends only on the nature of the general acid catalysts, while the right-hand side depends only on the nature of the substrates, the two sides are independently variable and therefore equal to a constant, c_1. Analogously, it has been shown[145,146] that a combination of the Brønsted and Swain–Scott equations (in their logarithmic forms) for the reactions of a series of k nucleophilic reagents (each characterized by n_k) gives

$$(pK_{a2} - pK_{a1})/(s_2 - s_1) = n_k/(\alpha_0 - \alpha_k) = c_2 \tag{73}$$

If we use Eq. (69) and a series of general bases instead of general acids, we can derive analogous relationships [to Eqs. (72) and (73)] for general base catalysis:

$$(pK_{a2} - pK_{a1})/(\rho_1 - \rho_2) = \sigma_i/(\beta_0 - \beta_i) = c_4 \qquad (74)$$

$$(pK_{a2} - pK_{a1})/(s_1 - s_2) = n_k/(\beta_0 - \beta_k) = c_5 \qquad (75)$$

Utilizing the Hammett and Swain–Scott equations for a series of i substrates and k nucleophiles, then for two arbitrary nucleophiles we get

$$\sigma_i/(s_0 - s_i) = (n_1 - n_2)/(\rho_2 - \rho_1) = c_3 \qquad (76)$$

Equation (72) states that (1) as substrate reactivity is varied by a change of substituents, the Brønsted α value for general acid catalysis will vary linearly with σ and (2) as the acid is varied, the variation in the ρ value for the reaction will be linearly related to the variation in the pK_a of the catalyzing acid. If we assume that c_1 is positive, then as a substrate is made more reactive by catalysis with a stronger acid (pK_{a2} decreases), the sensitivity to substrate substituent changes, ρ, decreases, and, conversely, as the substituent changes to make the substrate more electrophilic (σ_i increases), the sensitivity to the strength of the acid catalyst decreases (α_i decreases). Similarly, if we assume the constants c in Eqs. (73)–(76) are positive, then Eq. (73) states that as a substrate is made more reactive by catalysis with a stronger acid (pK_{a2} decreases), its sensitivity toward the attacking nucleophile decreases (s_2 decreases), and, conversely, as the reactivity of the attacking nucleophile increases (n_k increases), the sensitivity to the strength of the acid catalyst decreases (α_k decreases). For general base catalysis, Eq. (74) states that as the catalyzing base becomes stronger (pK_{a2} increases), the sensitivity of the rate to changes in substrate substituents decreases (ρ_2 decreases), and as the substituent is changed to make the substrate more electrophilic (σ_i increases), the sensitivity to general base catalysis (of attack by the nucleophile) decreases (β_i decreases). Equation (75) states that as the catalyzing base becomes stronger (pK_{a2} increases), the sensitivity of the substrate toward the nucleophile will decrease (s_2 decreases), while if the reactivity of the nucleophile increases (n_k increases), the sensitivity of the reaction to general base catalysis will decrease (β_k decreases). Equation (76) states that as the reactivity of the substrate is increased (σ_i increases), the sensitivity toward the attacking nucleophile will decrease (s_i decreases), while if the nucleophilicity of the attacking nucleophile is increased (n_2 increases), the ρ value for the reaction will decrease (ρ_2 decreases).

It can be seen, assuming positive values of the c's, that Eqs. (72)–(76) are essentially quantitative statements of the Hammond postulate[169] (also known as the Hammond–Leffler postulate, HLP, or the reactivity–selectivity principle, RSP), which implies that the greater the rate at which a reaction proceeds, the more the transition state for the reaction will resemble the reactants; i.e., the faster and thermodynamically more favored reaction (in a comparison of reactions involving structurally similar reactants) will have its transition state occur earlier along the reaction coordinate. This means less bond breakage and bond formation at the transition state in the faster reaction. The implications of this postulate will be discussed in Section 6.

Equations (72)–(75) apply only to true general acid or general base catalysis. If instead the mechanisms are the kinetically indistinguishable specific acid–general base or general acid–specific base types, respectively, then the c's will be negative. This change of sign follows from the fact that any two kinetically indistinguishable mechanisms are related by $\alpha = 1 - \beta$, which can be derived from Eqs. (68) and (69) and which is directly related to the fact that a stronger acid has by definition a weaker conjugate base. If, for example, one calculates α values from an experimental general acid catalytic rate term, Eq. (73) will give a positive c_2 if the mechanism is true general acid catalysis, but with β values (calculated from the observed α as $\beta = 1 - \alpha$), Eq. (75) will give a negative c_5. Conversely, Eq. (73) gives a negative c_2 and Eq. (75) gives a positive c_5 if the mechanism is specific acid–general base.[145] Therefore, the sign of the constant c_1, c_2, c_4, or c_5, through its connection with the Hammond postulate, provides a powerful way of distinguishing between kinetically equivalent catalytic mechanisms.

5.2. Nature of Isotope Effects as a Tool for Probing Transition-State Structure: Origin, Interpretation, and Types*

The extraordinary and simplifying feature of an isotope effect results from the fact that both unsubstituted and isotopically substituted species have virtually identical electronic distributions and hence move on the same electronic PES (see Section 3.2), whereas changing substituents in the customary way (see Section 5.1.1) completely changes some nuclei, thereby changing the PES with each substituent. The isotope effect, therefore, provides a probe of the nature of a *single transition state*—in contrast with ordinary substituent effects.[3] With ART and Eq. (1) as a starting point, Bigeleisen's complete formulation[170c] (expressing the ratio of rate constants, i.e., the isotope effect, in terms of vibrational frequencies of reactants and transition states) can be developed.[171] From this expression it can be seen that an isotope effect (e.g., k_H/k_D) arises almost entirely from changes in vibrational force constants on going from reactants to transition state. An important corollary of this conclusion is that isotope effects are *not* determined by real effects of the isotopic atom on the structure or properties of the rest of the molecule, except through coupling of vibrational motions of the isotopic atom with the motions of other atoms. Instead, the isotope effect is determined by the effect of the rest of the molecule on the vibrational motions of the isotopic atom. A further simplification in the interpretation of isotope effects is that for large molecules (e.g., enzymes) and for hydrogen isotopes, where vibrational frequencies and, therefore, zero-point energies are large, the zero-point vibrational energy

---

* In view of the fact that there are numerous excellent, rigorous reviews on the origin of isotope effects (see References 2–4, 26, 70, and 170), our remarks here will be brief and essentially qualitative in nature.

effects are the only major contributors. Although it will not be proven here, it can be established that "heavier isotopes tend to accumulate in those positions where they are most closely confined by potential barriers,"[172] where force constants tend to be larger. This generalization is helpful in predicting isotope effects, although one must base the prediction on the total amount of confinement. Only the true vibrations contribute to confinement of nuclear motion; therefore, the reaction coordinate motion of a transition state does not confine at all, making nuclei moving in it less confined in the transition state than in the reactant, where they engage in true vibrations. For example, if the motion of the isotope is predicted to be less confined in the transition state than in the reactants, a normal isotope effect is expected ($k_{light}/k_{heavy} > 1$); if the isotope is predicted to be more confined in the transition state than in the reactants, an *inverse* isotope effect ($k_{light}/k_{heavy} < 1$) is expected.

Kinetic isotope effects can be divided into two types: *primary* and *secondary*. Although these categories are not entirely distinct, a useful definition of each can be made: (1) A *primary* effect is one resulting from the making and/or breaking of bonds to the isotopically substituted atom in the course of the reaction *and* the motion of this atom in the *reaction coordinate*; (2) a *secondary* effect is one which is not primary. Maximum isotope effects for various elements can be roughly estimated[170c]; the rather wide range between the maximum estimated secondary effects and those estimated for primary effects enables one to conclude that if an effect is larger than the estimated maximum secondary effect then it is probably primary. However, a small isotope effect may be primary or secondary. Inverse isotope effects near the estimated minima are probably secondary, while primary (inverse) isotope effects are unlikely to be much below unity.

One may utilize secondary hydrogen isotope effects in determining Brønsted α's (or β's). For example, by measuring the rate of a general-acid-catalyzed reaction using first one acid catalyst and then a second acid catalyst differing from the first only in being isotopically substituted at a position close to the site of proton transfer, giving rise to a secondary isotope effect upon the pK_a, one can calculate an α value, provided the pK_a values for the two acids are known. Assuming that the acid catalyst used is "typical" and would fall on the line from which an α value could be calculated using a series of different general acid catalysts for this reaction, the α value calculated from the isotope effect would be expected to be the same as that calculated from the series of general acids. However, the isotopic α value would be for reaction on a single PES, in contrast with a Brønsted plot, which involves different acids, each moving on a different PES; therefore, it should be possible to utilize the isotopic α value to study anomalous acids or to investigate possible variations in α for different acids (i.e., as a direct probe of nonlinear Brønsted plots). Entirely analogous considerations apply to determination of isotopic β values for general base catalysis.

Both solvent and heavy-atom isotope effects can provide information about transition-state structure; both are treated elsewhere.[173]

6. EFFECTS OF STRUCTURAL CHANGES ON TRANSITION-STATE GEOMETRY

6.1. Selectivity and Transition-State Structure

Referring back to the definitions of α and β, both of which measure the proportionality of $\delta\Delta G_0^{\ddagger}$ with $\delta\Delta G_0$, and to Eq. (70), we see that while the limits of α and β are clear, one cannot assume that the value of α or β is a direct measure of the extent of proton transfer in the transition state. If the effect of structural change on the reaction pathway is primarily on the PES, then "curve crossing" (e.g., Fig. 14) is a good model. There is a wide portion over which *Morse-type* curves for bond stretching are approximately linear; moreover, many single bonds have similar strengths and force constants. Therefore, intersection of PE curves in these regions will result in $\beta \simeq 0.5$, since then $s_1 \simeq s_2$ [see Eq. (70) and Fig. 14], and a variation of β with the percent of proton transfer as shown in Fig. 15 can be predicted.[162b,c]* In general, this argument suggests that the selectivity constants ($\alpha, \beta, \rho, s, et\ al.$) for LFER's should be similar for a wide range of transition-state structures, that is, selectivity should be independent of reactivity, and this has been found experimentally.[144] It also suggests that LFER's may easily turn out to be *linear* over wide ranges of reactivity, with little curvature except if the transition

---

* Complex, nonlinear relationships between α and ΔG_0, using BEBO and intersecting parabola models, have been found. The order of the forming bond (n) is also a nonlinear function of ΔG_0 (BEBO model), and, interestingly, in many cases, n follows α quite closely (A. J. Kresge, private communication, 1977).

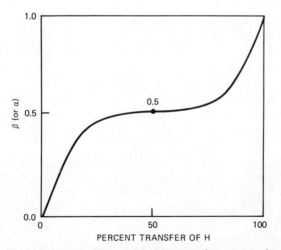

Fig. 15. Variation of Brønsted β with extent of proton transfer.

state becomes very reactant-like or product-like; again this has been observed experimentally.[144] However, this is really a problem of sensitivity, not inaccuracy, in that we should see expected changes in selectivity constants provided a wide enough range of reactivities is investigated.

Similarly, primary isotope effects, anticipated to be maximal at half-transfer,[170b] may be expected to be large and approximately constant except if the transition state is very reactant-like or product-like, since linearity of PE curves implies that forces acting on the proton being transferred are approximately constant (and approximately equal for S--------H and H--------B) over a wide range of transition-state structures. This also suggests that near-maximal primary isotope effects may be influenced by bending or other factors more strongly than by the actual degree of proton transfer. This situation tends to bear out Bordwell and Boyle's[174] widely quoted conclusion that "the hope, which at one time seemed bright, for a simple general correlation of Brønsted coefficients, kinetic isotope effects, and solvent isotope effects with the extent of proton transfer in the transition state has proved vain." However, if the above explanation reasonably describes the cause for near-constant selectivities and isotope effects, then there is some hope, since:

1. Secondary isotope effects are not subject to this "constant slope" (they do not involve bonds being made or broken) and there is, at present, good reason to believe that they are reliable indicators of transition-state structure; for example, β values calculated from isotope effects involving DO^- vs. HO^- should be reasonable measures of actual degree of proton transfer.
2. In reactions which do have highly reactant-like or product-like transition states, variation of selectivity and isotope effects and even the expected curvature in LFE plots should be observable.

It appears that such variations (of selectivity and isotope effects) should be good measures of relative transition-state structures but not of absolute structures. For example, a constant, but large primary deuterium isotope effect, e.g., > 5, is consistent with variable degrees of proton transfer, while a small k_H/k_D is expected to vary with transition-state structure. If it does not, then a constant degree of proton transfer is probable.[30] It would therefore seem valuable to explore methods other than those involving selectivities related to reacting bonds for information on transition-state structural variations. Steric effects, for example, may be particularly useful, along with secondary isotope effects.

Despite the large number of reactions which obey LFER's, there are situations which do not conform to expectations, reactions where "kinetically controlled" and "equilibrium-controlled" conditions lead to different products.[175] These reactions are characterized by rapid formation of a less stable product, followed by slower rearrangement (of this initially formed product) to the more stable one. This situation is incompatible with a Bronsted-like relationship between rate and equilibrium constant, since that relationship predicts a larger rate constant of formation for the more stable product. Because the Brønsted relation is expected to hold only for a single step, an observed incompatibility can, in many cases, be taken as evidence of a multi-step reaction, as in electrophilic aromatic substitution.

There are, however, some reactions which appear to be one-step reactions and yet lead initially to the less stable product. They frequently involve reaction of an unstable species to form two isomeric products, as in reactions of un-symmetrical allyl cations, protonation of conjugated carbanions, and other ambient reactions.[176] The energy effects involved are usually quite small but important, since they actually control the nature of the product formed. It seems probable that many of these effects result from the fact that the nature of the reaction tends to cancel out the large energy effects of a Brønsted relationship (Fig. 14). Figure 16 shows (a) what is expected for the reaction $S^-(S^{-\prime}) + HA \rightarrow SH(S'H) + A^-$ for two different products SH and S'H, based on Fig. 14; (b) what is expected for the case where SH and SH' are isomers, both formed from the *same* S^- by protonation at two different sites (for example, formation of keto and enol forms from enolate anion); and (c) what might happen if distortion of S^- were actually more favorable toward the less stable product SH', prior to its formation. Since the curves for disso-ciation of SH and SH' must coincide asymptotically if both form from the same S^-, then, assuming the transition state resembles S^-, one expects prac-tically no rate difference for the two reactions (b). However, in the absence of a large effect (a), it is possible that a small effect of the opposite kind (c) can arise. Distortion of S^- toward the product it more nearly resembles in energy may well be more favorable, in that it would require less electronic reorganization. (This, in turn, would result in less nuclear reorganization.) Similar arguments have been applied to explain the "principle of least motion" (PLM), which relates faster reaction rates to distortions of reactants requiring less energy (which, in turn, usually means smaller distortions of nuclear geometry).[176-178] Figure 16 may therefore provide an explanation of the working of the PLM, at least in some cases. Presumably, these reactant-distortion, or PLM, effects exist in all kinds of reactions but are masked, in many cases, by much larger selectivity effects [e.g., Figs. 14 and 16(a)].

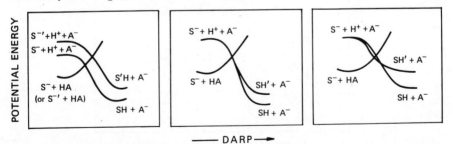

Fig. 16. Energy profiles of proton-transfer reactions for (a) two different products, SH and S'H; (b) isomeric products formed from the same S^-; and (c) isomeric products formed from S^- which distorts toward SH and SH' along noncoincident curves.

6.2. Vertical and Adiabatic Effects

The question of the relationship of selectivities to transition-state geometries returns us to the entire problem of transition-state structure. The fact that structural changes in reactants change not only the energy of the transition state but also its geometry has long been recognized.[169,179,180] Figure 14 indicates that a higher activation energy is expected to give a more product-like transition-state geometry; Fig. 17 illustrates the general principle.

Structural changes in reactants may change the transition-state structure other than along the reaction pathway, since the entire PES is altered. The effect on transition-state structure can be thought of as the superposition of effects on each independent normal mode of nuclear motion, including the reaction coordinate.[66] Effects such as the Hammond postulate, then, result from the difference in such structural changes between the transition state and the reactants. The *major* effects are expected to be on the relatively weak and polarizable bonds of the transition state which are actually being made or broken during the course of the reaction ("reacting bonds"). Strong, covalent bonds such as all the bonds of the reactants and all except the reacting bonds of the transition state are less likely to be perturbed by the addition of substituents; therefore, effects on reacting bonds in the transition state are frequently sufficient to characterize the main geometric effects, thus greatly simplifying the analysis.

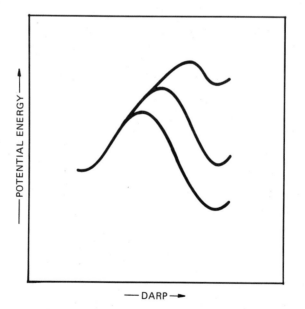

Fig. 17. Energy profiles illustrating how higher activation energies are correlated with more product-like transition-state geometries: the Hammond postulate.

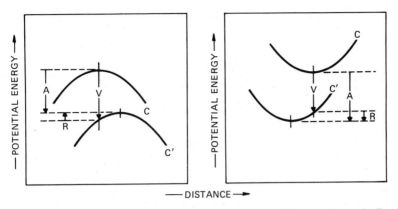

Fig. 18. Effect of structural change on (a) a reaction coordinate (RC) and (b) a real vibration: C = unperturbed RC or real vibration; C' = perturbed RC or real vibration; A = adiabatic energy effect; V = vertical energy effect; R = relaxation energy effect.

The effect of a substituent or other structural change on the PES can be divided into two parts: (1) a *vertical* effect, the effect upon the energy of the unperturbed (without the substituent) ENG of the transition state, and (2) a *relaxation* of the geometry to that characteristic of the perturbed system. If we plot both of these effects for a reaction coordinate, the result is as schematically shown in Fig. 18(a); this illustrates that the vertical energy effect may be quite different from the relaxed (adiabatic) one if the ENG of the transition state (maximum) is significantly shifted. It can be seen from Fig. 18(a) and shown algebraically[66] that if the perturbed ENG is different from the unperturbed ENG along the reaction coordinate, the relaxation effect of the reaction coordinate is always destabilizing, i.e., no matter what the vertical effect is (positive, zero, or negative), the reaction coordinate relaxation effect is always positive and therefore works in the direction of making the adiabatic energy effect less favorable (or more unfavorable) than the vertical effect.

The same situation exists for a real vibration [Fig. 18(b)], except that the relaxation effect is always stabilizing, i.e., negative, and works in the direction of making the adiabatic energy effect more favorable (or less unfavorable) than the vertical effect. Note that a real molecule has only one vertical effect (not one for each normal coordinate); therefore, the total energy effect is the sum of V (for the real molecule or transition state in question) plus the individual R effects for each normal mode.

If we subtract curve C from curve C' in Figs. 18(a) and (b), the result is a straight line of positive slope in both cases. It is straight, however, only if parabolas C and C' have exactly the same curvature, but this is expected to be a good first-order approximation for real molecules and probably transition states.[66] Since force is, by definition, the negative slope of energy vs. distance, in both cases we have pictured, a constant negative force is being added onto the unperturbed forces of curve C to give C'. It must therefore be generally true that for negative curvature a constant force shifts the ENG in the direction opposite to the force, and for positive curvature the ENG shifts in the same direction as the force.[66,67]

A macroscopic analogy for this effect consists of picturing a "champagne bottle model," in which it will be obvious that if we place a weight atop the cork, a downward force will be exerted, pushing the cork downward into the bottle a finite amount. That is, the stable ENG shifts in the direction of the force. On the other hand, the transition state for cork popping (the point at which the holding forces are overcome; i.e., the top of the potential energy barrier has been reached and the cork is pushed outward) will be reached farther upwards in the neck of the bottle when the weight is atop the cork. That is, the metastable ENG shifts in the direction opposite to the force.

The effects of structural changes upon transition-state geometries can then be summarized in a "force formulation" in which it is recognized that geometric changes in ENG's, whether transition states or stable species, result entirely from the perturbations upon nuclei induced by forces arising from structural change. It is possible to make predictions of such forces (which cause the nuclei to relax to a new ENG) based on well-studied stereoelectronic considerations, which will usually be agreed to be unambiguous. Such predictions are easily made since they are for a given model structure which is well defined; the question to be asked is whether observed data fit the predictions and, therefore, fit the model.

The "force formulation" states that:

1. A structural change will exert a force on the nuclei and thus displace the ENG.
2. For each normal mode of true vibration, the ENG will be shifted in the direction of the net force on the nuclei closest to the site of structural change (or most affected by the structural change).
3. For a reaction-coordinate normal mode, the ENG will be shifted in the direction *opposite* to the net force on the nuclei closest to the site of structural change (or most affected by the structural change).
4. The shift in ENG is larger when the effective force constant for the normal mode is nearer zero. In a transition state, normal modes involving mainly motions of nuclei which are in reacting (partial) bonds [i.e., the reaction coordinate and "antireaction coordinate(s)"] will have effective force constants near zero. Therefore, these nuclei and these normal modes will be among the ones which are most affected by the structural change. The net force can be made zero or near zero by opposing force effects within a single normal mode, as when more than one nucleus is affected, but in opposite directions.
5. The shift in ENG is the *sum* of the separate effects on the normal modes.

One should include torsions and other low-frequency vibrations in both reactants and transition states among the ones which may be most affected by structural changes; however, unless these vibrations involve reacting bonds, such effects will usually not be mechanistically important. An exception might be a hyperconjugative effect, in which a methyl group attached to a developing carbonium ion center could have much more restricted torsional motion in the transition state than in the reactant.

6.2.1. Parallel vs. Perpendicular Effects

Various quantitative models for these effects have been proposed.[100,181,182] An important point made is that effects along the reaction coordinate (*parallel effects*) are not expected always to dominate effects on the real vibrations (*perpendicular effects*), especially in the case of the "anti-reaction coordinates," which are real vibrations involving the same atoms as the reaction coordinate and thus involving partial bonds also. Dominance of parallel over perpendicular effects was originally suggested,[66] but it was recognized[67] that the accumulated data on E2 elimination reactions fit much better to a model which considered both effects to be of approximately equal importance.

In fact, the existence of important perpendicular effects of the same type as suggested in the E2 case may very well prove to explain the abnormal Brønsted slopes and ρ values for deprotonation of nitroalkanes which have been interpreted in terms of a two-anion model.[165] The transition state is quite analogous to that for E2,

$$\text{B:}\overset{\curvearrowright}{\text{H}}\text{—}\overset{|}{\underset{|}{\text{C}}}\text{—}\overset{+}{\text{N}}\overset{\diagup\text{O}^-}{\diagdown} \longrightarrow \text{BH}^+ \quad \overset{\diagdown}{\underset{\diagup}{\text{C}}}\text{=}\overset{+}{\text{N}}\overset{\diagup\text{O}^-}{\diagdown}_{\text{O}^-} \tag{77}$$

where Ar = substituted phenyl, except the "leaving group" is effectively a π bond to one of the oxygen atoms of the nitro group. Large ρ values for strong bases are at least consistent with an "E1cB-like" transition state as predicted for E2[67] which has considerable carbanion character but little C=N character. In contrast, the product has little carbanion character and instead has most of the negative charge localized on oxygen and hence a smaller ρ.

Quantitative evaluations of expectations for transition states are difficult to make; however, the force formulation provides a way to make predictions of the *directions* of numerous possible structural changes in any given transition-state model. The predictions for parallel effects agree with extrapolations of the Hammond postulate. If *trends* can be experimentally determined, and if sufficient structural changes are investigated, it is possible that only one class of transition-state model will be found to agree with all observed trends. Therefore, this approach suggests the possibility of "transition-state mapping," which can provide a good deal of structural detail about transition states without relying on difficultly accessible quantitative interpretations.[67] To understand the structural features of a transition state, one needs to bring in both the parallel (Hammond) effect which measures reactant-like or product-like features (RP character) and the perpendicular effects which generally measure the tightness or looseness (TL character) of the transition state.[183]

The force formulation can be directly applied to structural effects on reaction pathways using schematic PES's (Figs. 6–11).

Normally, discussion of substituent effects is in terms of whether a given substituent stabilizes or destabilizes the transition state. This corresponds to the vertical effect and is a zero-order approximation to the true situation, which is undoubtedly adequate in many cases. The geometric changes associated with the relaxation of forces to a new ENG are the next higher approximation. In addition to selectivity effects and isotope effects, recent work on $\Delta V_0^{\ddagger}$ for nucleophilic displacements provides another kind of evidence for the significance of nonvertical, geometric effects consistent with the Hammond postulate.[184]

We can analyze catalysis in terms of forces, considering a catalyst as a "substituent" or structural change and examining its effect on transition-state energy. The most important effect is that the catalyst must stabilize the transition state, probably by a vertical stabilization mechanism. However, if the interactions leading to such stabilization are very strong, one may expect a strong force effect, particularly on the reaction coordinate and the antireaction coordinate(s). If we wish to maximize catalysis, then the force effects should be such that they are strong on real vibrations, since this adds to the stabilizing influence according to the discussion above [see Fig. 18(b)]. The force effect on the reaction coordinate should be as weak as possible, however, since it is always destabilizing [see Fig. 18(a)]. In other words, the forces exerted along the reaction coordinate by stabilizing groups may stabilize the unperturbed transition state but in so doing may shift the transition state geometry to a new, less stabilized ENG. Therefore, maximal catalysis would be expected for a catalyst which could interact with a substrate at several points such that the forces along the reaction coordinate could be neutralized and the geometry along the reaction coordinate left unchanged. This situation might be called "force compensation leading to vertical catalysis," and since a large catalyst such as an enzyme might be ideally suited to produce this effect, it is possible that enzyme catalysis is enhanced relative to small-molecule catalysis by this effect as well as by the others discussed in Section 5.1.1.

6.2.2. Solvation and Anthropomorphic Rules

Application of the concepts of ART to the experimental investigation of transition-state structure has led to significant knowledge of many kinds of mechanisms. Along with the characterization of individual reactions, a search for generalizations or rules which might explain broad classes of reactions has been undertaken. For example, in reaction types such as nucleophilic displacement or bimolecular elimination, specific structural models of transition states have developed, involving Walden inversion and antiperiplanar elimination, respectively. Catalysis mechanisms have been classified as nucleophilic, specific acid, specific base, general acid, general base, and combinations of these.

The categories of catalysis do not define mechanisms thoroughly in many cases, particularly because rapid proton transfers can take place along the reaction pathway and thus lead to several different types of transition states for each category. Intuitively, one expects a catalyst to be located in a transition state where it can "assist" the process taking place. However, it was pointed out[120] that any process by which a catalyst could stabilize the transition state would be a catalytic mechanism. Analysis of experimental information led to the "solvation rule,"[120] which postulated that HAR would always be rate determining in catalysis involving proton transfer between electronegative atoms such as oxygen, leading to "spectator" or solvation catalysis.[9] This rule was questioned,[185] and a so-called "anthropomorphic rule" was proposed, which has since been further defined as a "catalysis rule"[186a]:

> Concerted general acid–base catalysis of complex reactions in aqueous solution can occur only (a) at sites that undergo a large change in pK in the course of the reaction, and (b) when this change in pK converts an unfavorable to a favorable proton transfer with respect to the catalyst, i.e., the pK of the catalyst is intermediate between initial and final pK values of the substrate site.

This idea has been illustrated by use of PES's.[69a] It has also been shown that there are some reactions in which catalysis *must* occur because of certain properties of unstable intermediates.[186b] Also, solvation mechanisms have been shown to be "anthropomorphic" after all.[187] Possibilities for coupling of proton transfer with HAR have been analyzed.[188]

Thus it is that ART, experimental approaches, and empirical generalizations such as the "catalysis rule" (which really predicts features of PES's) are being combined to explain the basis of catalysis. Application of these new discoveries can be expected to lead to detailed understanding of enzyme mechanisms.

7. CONCLUSION

From what has been said, we feel it is possible to conclude that there are unifying concepts in the study of reaction mechanisms, all of which can be understood on the basis of transition-state structure. The time-tested tool of comparing the changes in rate caused by structural changes in the reactants has proven to be so useful that it is hardly even disturbing to realize that we cannot calculate rates from transition-state structures with real accuracy. Calculations actually giving correct orders of magnitude for rates have reinforced the feeling that ART is conceptually accurate, though not exact.

This theory has led to mechanistic ideas and approximate calculations which now strongly suggest that enzyme catalysis does not differ in detail from small-molecule catalysis. Enzymes are simply taking advantage of most, if not all, of the possible ways by which a specially designed (by evolution) and constructed large molecule could have very high catalytic efficiency. This concept, its elaboration, and presumably its eventual proof will, it seems, prove to be a historic chapter in the human effort to understand nature.

The future may provide new tools and new theoretical models both for PES's and for rates, but even now it appears that, with the right choice of experimental approaches, it will be possible to detail enzyme mechanisms. Certainly it will be possible to learn much about transition-state structures; at present it seems entirely possible to map the reaction coordinate as well.

If this much can be discovered, we shall have learned most of what nature has to tell about how she carries out her remarkable reactions—for there is always thermal averaging, so that the ultimate details of molecular motions are averaged in determining rates. Rates must be slow enough, on the average, to give selectivity, yet fast enough to permit motion and self-replication. This thermal agitation and accompanying reactions may even supply the (small) degree of uncertainty which we call "free will."

REFERENCES

1. (a) S.Glasstone, K. J. Laidler, and H. Eyring, *The Theory of Rate Processes*, McGraw-Hill, New York (1941); (b) K. J. Laidler, *Theories of Chemical Reaction Rates*, McGraw-Hill, New York (1969).
2. L. Melander, *Isotope Effects on Reaction Rates*, Ronald, New York (1960).
3. E. K. Thornton and E. R. Thornton, in: *Isotope Effects in Chemical Reactions* (C. J. Collins and N. S. Bowman, eds.), Chap. 4, pp. 213–285, Van Nostrand Reinhold, New York (1970).
4. H. S. Johnston, *Gas Phase Reaction Rate Theory*, Ronald, New York (1966), Chap. 8, p. 131.
5. H. S. Johnston, Reference 4, pp. 118–119.
6. H. S. Johnston, Reference 4, pp. 129–130.
7. J. Kurz and L. Kurz, On the mechanism of proton transfer in solution. Factors determining whether the activated complex has an equilibrated environment, *J. Am. Chem. Soc.* **94**, 4451–4461 (1972).
8. R. Wolfenden and W. P. Jencks, Acetyl transfer reactions of 1-acetyl-3-methylimidazolium chloride, *J. Am. Chem. Soc.* **83**, 4390–4393 (1961).
9. R. L. Schowen, in: *Progress in Physical Organic Chemistry* (A. Streitwieser, Jr., and R. W. Taft, eds.), Vol. 9, pp. 275–332, Wiley-Interscience, New York (1972).
10. D. G. Truhlar and A. Kuppermann, A test of transition state theory against exact quantum mechanical calculations, *Chem. Phys. Lett.* **9**, 269–272 (1971).
11. A. Persky and M. Baer, Exact quantum mechanical study of kinetic isotope effects in the collinear reaction $Cl + H_2 \rightarrow HCl + H$. The H_2/D_2 and the H_2/T_2 isotope effects, *J. Chem. Phys.* **60**, 133–136 (1974).
12. G. W. Koeppl, Comparison of quantum mechanical and transition state theory reaction probabilities for the reaction $O + HBr \rightarrow OH + Br$, *J. Chem. Phys.* **59**, 2168–2169 (1973).
13. G. W. Koeppl, Best *ab initio* surface transition state theory rate constants for the $D + H_2$ and $H + D_2$ reactions, *J. Chem. Phys.* **59**, 3425–3426 (1973).
14. W. H. Miller, Quantum mechanical transition state theory and a new semiclassical model for reaction rate constants, *J. Chem. Phys.* **61**, 1823–1834 (1974).
15. G. W. Koeppl, Alternate locations for the dividing surface of transition state theory. Implications for application of the theory, *J. Am. Chem. Soc.* **96**, 6539–6548 (1974).
16. R. P. Bell, Derivation of the fundamental expression of transition state theory, *Trans. Faraday Soc.* **66**, 2770–2771 (1970).

17. J. L. Kurz, Transition states as acids and bases, *Acc. Chem. Res.* **5**, 1–9 (1972).
18. F. H. Westheimer, in: *Steric Effects in Organic Chemistry* (M. S. Newman, ed.), Chap. 12, pp. 542–554, Wiley, New York (1956).
19. F. H. Westheimer and J. E. Mayer, The theory of the racemization of optically active derivatives of diphenyl, *J. Chem. Phys.* **14**, 733–738 (1946).
20. F. H. Westheimer, A calculation of the energy of activation for the racemization of 2,2′-dibromo-4,4′-dicarboxydiphenyl, *J. Chem. Phys.* **15**, 252–260 (1947).
21. M. Rieger and F. H. Westheimer, The calculation and determination of the buttressing effect for the racemization of 2,2′,3,3′-tetraiodo-5,5′-dicarboxybiphenyl, *J. Am. Chem. Soc.* **72**, 19–28 (1950).
22. J. B. Hendrickson, Molecular geometry. I. Machine computation of the common rings, *J. Am. Chem. Soc.* **83**, 4537–4547 (1961).
23. K. B. Wiberg, A scheme for strain energy minimization. Application to the cycloalkanes, *J. Am. Chem. Soc.* **87**, 1070–1078 (1965).
24. N. L. Allinger, M. T. Tribble, M. A. Miller, and D. W. Wertz, Conformational analysis. LXIX. An improved force field for the calculation of the structures and energies of hydrocarbons, *J. Am. Chem. Soc.* **93**, 1637–1648 (1971). See also the previous papers in this series.
25. E. M. Engler, J. D. Andose, and P. von R. Schleyer, Critical evaluation of molecular mechanics, *J. Am. Chem. Soc.* **95**, 8005–8025 (1973).
26. K. B. Wiberg, *Physical Organic Chemistry*, Wiley, New York (1964), pp. 370–373.
27. (a) P. B. D. de la Mare, L. Fowden, E. D. Hughes, C. K. Ingold, and J. D. H. MacKie, Mechanism of substitution at a saturated carbon atom. Part XLIX. Analysis of steric and polar effects of alkyl groups in bimolecular nucleophilic substitution, with special reference to halogen exchange, *J. Chem. Soc.* **1955**, 3200–3236; (b) I. Dostrovsky, E. D. Hughes, and C. K. Ingold, The role of steric hindrance. XXXII. Magnitude of steric effects, range of occurrence of steric and polar effects, and place of the Wagner rearrangement in nucleophilic substitution and elimination, *J. Chem. Soc.* **1946**, 173–194.
28. C. K. Ingold, Quantitative study of steric hindrance, *Q. Rev. (London)* **11**, 1–14 (1957).
29. J. N. Murrell and G. L. Pratt, Statistical factors and the symmetry of transition states, *Trans. Faraday Soc.* **66**, 1680–1684 (1970).
30. M. Choi and E. R. Thornton, A kinetic study of the hydrolysis of substituted *N*-benzoylimidazoles and *N*-benzoyl-*N*′-methylimidazolium ions in light and heavy water. Hydrogen bridging without rate-determining proton transfer as a mechanism of general base catalyzed hydrolysis and a model for enzymic charge-relay, *J. Am. Chem. Soc.* **96**, 1428–1436 (1974).
31. A. H. Andrist, Concertedness: A function of dynamics or nature of the potential energy surface?, *J. Org. Chem.* **38**, 1772–1773 (1973).
32. E. K. Thornton and E. R. Thornton, Reference 3, p. 260.
33. B. W. Morrissey, Microscopic reversibility and detailed balance: An overview, *J. Chem. Educ.* **52**, 296–298 (1975).
34. B. H. Mahan, Microscopic reversibility and detailed balance: An analysis, *J. Chem. Educ.* **52**, 299–302 (1975).
35. R. M. Krupka, H. Kaplan, and K. J. Laidler, Kinetic consequences of the principle of microscopic reversibility, *Trans. Faraday Soc.* **62**, 2754–2759 (1966).
36. H. Eyring and M. Polanyi, On simple gas reactions, *Z. Phys. Chem. Abt. B* **12**, 279–311 (1931).
37. F. London, Quantum-mechanical explanation of activation, *Z. Elektrochem.* **35**, 552–555 (1929).
38. J. E. Hulse, R. A. Jackson, and J. S. Wright, Energy surfaces, trajectories, and the reaction coordinate, *J. Chem. Educ.* **51**, 78–82 (1974).
39. R. D. Gandour, G. M. Maggiora, and R. L. Schowen, Coupling of proton motions in catalytic activated complexes. Model potential-energy surfaces for hydrogen-bond chains, *J. Am. Chem. Soc.* **96**, 6967–6979 (1974).
40. O. Sinanoğlu, The *C*-potential surface for predicting conformations of molecules in solution, *Theor. Chim. Acta* **33**, 279–284 (1974).

41. D. M. Silver, Character of the least-energy trajectory near the saddle-point on H_3 potential surface, *J. Chem. Phys.* **57**, 586–587 (1972).

42. E. A. McCullough, Jr., and D. M. Silver, Reaction path properties at critical points on potential surfaces, *J. Chem. Phys.* **62**, 4050–4052 (1975).

43. R. P. Bell, *The Proton in Chemistry*, 2nd ed., Cornell University Press, Ithaca, N.Y. (1973), Chap. 12.

44. R. P. Bell, Recent advances in the study of kinetic hydrogen isotope effects, *Chem. Soc. Rev.* **3**, 513–544 (1974).

45. M. D. Harmony, Quantum mechanical tunnelling in chemistry, *Chem. Soc. Rev.* **1**, 211–228 (1972).

46. H. S. Johnston, Reference 4, pp. 190–197.

47. E. F. Caldin, Tunneling in proton-transfer reactions in solution, *Chem. Rev.* **69**, 135–156 (1969).

48. M. J. Stern and R. E. Weston, Jr., Phenomenological manifestations of quantum-mechanical tunneling. I. Curvature in Arrhenius plots, *J. Chem. Phys.* **60**, 2803–2807 (1974).

49. M. J. Stern and R. E. Weston, Jr., Phenomenological manifestations of quantum-mechanical tunneling. II. Effect on Arrhenius preexponential factors for primary hydrogen kinetic isotope effects, *J. Chem. Phys.* **60**, 2808–2814 (1974).

50. E. F. Caldin and S. Mateo, Kinetic isotope effects and tunnelling in the proton-transfer reaction between 4-nitrophenylnitromethane and tetramethylguanidine in various aprotic solvents, *J. Chem. Soc. Faraday Trans. 1*, **71**, 1876–1904 (1975).

51. J. Banger, A. Jaffe, A. Lin, and W. H. Saunders, Jr., Carbon isotope effects on proton transfers from carbon, and the question of hydrogen tunneling, *J. Am. Chem. Soc.* **97**, 7177–7178 (1975).

52. D. G. Truhlar and A. Kuppermann, Exact tunneling calculations, *J. Am. Chem. Soc.* **93**, 1840–1851 (1971).

53. M. Simonyi and I. Mayer, Barrier width: A powerful parameter for hydrogen transfer reactions, *J. Chem. Soc. Chem. Commun.* **1975**, 695–696.

54. J. E. Leffler and E. Grunwald, *Rates and Equilibria of Organic Reactions*, Wiley, New York (1963), p. 119.

55. D. Y. Curtin, Stereochemical control of organic reactions. Differences in behavior of diastereomers. I. Ethane derivatives. The cis effect, *Rec. Chem. Prog.* **15**, 111–128 (1954).

56. E. L. Eliel, N. L. Allinger, S. J. Angyal, and G. A. Morrison, *Conformational Analysis*, Wiley, New York (1967), p. 28.

57. J. H. Murrell and K. J. Laidler, Symmetries of activated complexes, *Trans. Faraday Soc.* **64**, 371–377 (1968).

58. J. W. McIver, Jr., The structure of transition states: Are they symmetric, *Acc. Chem. Res.* **7**, 72–77 (1974).

59. J. W. McIver, Jr., and R. E. Stanton, Symmetry selection rules for transition states, *J. Am. Chem. Soc.* **94**, 8618–8620 (1972).

60. (a) R. E. Stanton and J. W. McIver, Group theoretical selection rules for the transition states of chemical reactions, *J. Am. Chem. Soc.* **97**, 3632–3646 (1975); (b) P. Pechukas, On simple saddle points of a potential surface, the conservation of nuclear symmetry along paths of steepest descent, and the symmetry of transition states, *J. Chem. Phys.* **64**, 1516–1521 (1976).

61. (a) R. G. Pearson, Symmetry rules for predicting the course of chemical reactions, *Theor. Chim. Acta* **16**, 107–110 (1970); (b) R. G. Pearson, Symmetry rules for chemical reactions, *Acc. Chem. Res.* **4**, 152–160 (1971); (c) R. G. Pearson, Orbital symmetry rules for unimolecular reactions, *J. Am. Chem. Soc.* **94**, 8287–8293 (1972).

62. R. B. Woodward and R. Hoffmann, *The Conservation of Orbital Symmetry*, Academic Press, New York (1971).

63. R. F. W. Bader, Vibrationally induced perturbations in molecular electron distributions, *Can. J. Chem.* **40**, 1164–1175 (1962).

64. J. Hayami, N. Tanaka, N. Hihara, and A. Kaji, Nucleophile–substrate complex in solution.

Detection of chloride–organic chloride association and the potential role of the complexes in the S_N2 reaction, *Tetrahedron Lett.* **1973**, 385–388.

65. C. G. Swain and W. P. Langsdorf, Jr., Concerted displacement reactions. VI. *m-* and *p*-Substituent effects as evidence for a unity of mechanism in organic halide reactions, *J. Am. Chem. Soc.* **73**, 2813–2819 (1951).

66. E. R. Thornton, A simple theory for predicting the effects of substituent changes on transition-state geometry, *J. Am. Chem. Soc.* **89**, 2915–2927 (1967).

67. D. A. Winey and E. R. Thornton, Elimination mechanisms. Deuteroxide/hydroxide isotope effects as a measure of proton transfer in the transition states for E2 elimination of 2-(*p*-trimethylammoniophenyl) ethyl 'onium ions and halides. Mapping of the reaction-coordinate motion, *J. Am. Chem. Soc.* **97**, 3102–3108 (1975).

68. R. A. More O'Ferrall, Relationships between E2 and E1cB mechanisms of β-elimination, *J. Chem. Soc. B* **1970**, 274–277.

69. (a) W. P. Jencks, General acid–base catalysis of complex reactions in water, *Chem. Rev.* **72**, 705–718 (1972); (b) T. H. Lowry and K. S. Richardson, *Mechanism and Theory in Organic Chemistry*, Harper & Row, New York (1976), pp. 247, 409.

70. E. R. Thornton, *Solvolysis Mechanisms*, Ronald, New York (1964), pp. 42–44.

71. J. E. Leffler and E. Grunwald, Reference 54, Chap. 2.

72. E. R. Thornton, Reference 70, p. 183.

73. K. J. Laidler, Reference 1(b), Chap. 3.

74. (a) A. J. Parker, Protic–dipolar aprotic solvent effects on rates of bimolecular reactions, *Chem. Rev.* **69**, 1–32 (1969); (b) M. H. Abraham, in: *Progress in Physical Organic Chemistry* (A. Streitwieser, Jr., and R. W. Taft, eds.), Vol. 11, pp. 1–87, Wiley-Interscience, New York (1974).

75. L. L. Schaleger and F. A. Long, in: *Advances in Physical Organic Chemistry* (V. Gold, ed.), Vol. 2, pp. 1–33, Academic Press, New York (1963).

76. H. M. Humphreys and L. P. Hammett, Rate measurements on fast reactions in the stirred flow reactor; the alkaline hydrolysis of methyl and ethyl formate, *J. Am. Chem. Soc.* **78**, 521–524 (1956).

77. F. P. Price, Jr., and L. P. Hammett, Effect of structure on reactivity of carbonyl compounds; temperature coefficients of rate of formation of several semicarbazones, *J. Am. Chem. Soc.* **63**, 2387–2393 (1941).

78. (a) C. G. Swain and E. R. Thornton, Calculated isotope effects for reactions of lyonium ion in mixtures of light and heavy water, *J. Am. Chem. Soc.* **83**, 3884–3889 (1961); (b) C. G. Swain and E. R. Thornton, Calculated isotope effects for reactions of lyoxide ion or water in mixtures of light and heavy water, *J. Am. Chem. Soc.* **83**, 3890–3896 (1961).

79. F. A. Long, J. G. Pritchard, and F. E. Stafford, Entropies of activation and mechanism for the acid-catalyzed hydrolysis of ethylene oxide and its derivatives, *J. Am. Chem. Soc.* **79**, 2362–2364 (1957).

80. (a) R. W. Taft, Jr., E. L. Purlee, P. Riesz, and C. A. DeFazio, π-Complex and carbonium ion intermediates in olefin hydration and E1 elimination from *t*-carbinols. II. Trimethylene, methylenecyclobutane, triptene, and the effect of acidity on their hydration rate, *J. Am. Chem. Soc.* **77**, 1584–1590 (1955); (b) R. H. Boyd, R. W. Taft, Jr., A. P. Wolf, and D. R. Christman, Studies on the mechanism of olefin–alcohol interconversion. The effect of acidity on the O^{18} exchange and dehydration of *t*-alcohols, *J. Am. Chem. Soc.* **82**, 4729–4736 (1960).

81. M. M. Kreevoy, in: *Investigation of Rates and Mechanisms of Reactions*, 2nd ed. (S. L. Friess, E. S. Lewis, and A. Weissberger, eds.), Vol. VIII, Part II, pp. 1361–1406, Wiley-Interscience, New York (1963).

82. M. H. Abraham, D. H. Buisson, and R. A. Schulz, Activation parameters for the solvolysis of *t*-butyl chloride in water–ethanol mixtures. Glycine as a transition state model, *J. Chem. Soc. Chem. Commun.* **1975**, 693–694.

83. M. I. Page and W. P. Jencks, Entropic contributions to rate accelerations in enzymic and intramolecular reactions and the chelate effect, *Proc. Nat. Acad. Sci. USA* **68**, 1678–1683 (1971).

84. J. W. Larsen, Entropy contributions to rate accelerations of intramolecular reactions in water vs. non-structured solvents, *Biochem. Biophys. Res. Commun.* **50**, 839–845 (1973).
85. J. G. Martin and J. M. W. Scott, Thermodynamic parameters and solvent isotope effects as mechanistic criteria in the neutral hydrolysis of some alkyl trifluoroacetates in water and deuterium oxide, *Chem. Ind. (London)* **1967**, 665.
86. T. C. Bruice and S. J. Benkovic, A comparison of the bimolecular and intramolecular nucleophilic catalysis of the hydrolysis of substituted phenyl acylates by the dimethylamino group, *J. Am. Chem. Soc.* **85**, 1–8 (1963).
87. T. Higuchi, L. Eberson, and A. K. Herd, The intramolecular facilitated hydrolytic rates of methyl substituted succinanilic acids, *J. Am. Chem. Soc.* **88**, 3805–3808 (1966).
88. W. P. Cane and D. Wetlaufer, *Abstr. Am. Chem. Soc. 152d Ann. Meeting* **1966**, 110C; W. P. Cane, Homologousacyl Chymotrypsins, *Diss. Abstr. B.* **28**, 4038 (1968).
89. W. P. Jencks, *Catalysis in Chemistry and Enzymology*, McGraw-Hill, New York (1969), Chap. 5, p. 298.
90. T. D. Singh and R. W. Taft, Novel activation parameters and catalytic constants in the aminolysis and methanolysis of *p*-nitrophenyl trifluoroacetate, *J. Am. Chem. Soc.* **97**, 3867–3869 (1975).
91. G. Kohnstam, in: *Advances in Physical Organic Chemistry* (V. Gold, ed.), Vol. 5, pp. 121–172, Academic Press, New York (1967).
92. K. M. Koshy, R. E. Robertson, and W. M. J. Strachan, Pseudo-thermodynamic parameters and isotope effects for hydrolysis of a series of benzyl chlorides in water, *Can. J. Chem.* **51**, 2958–2962 (1973).
93. E. R. Thornton, Reference 70, p. 192.
94. W. J. le Noble, in: *Progress in Physical Organic Chemistry* (A. Streitwieser, Jr., and R. W. Taft, eds.), Vol. 5, pp. 207–330, Wiley-Interscience, New York (1967).
95. W. J. le Noble, H. Guggisberg, T. Asano, L. Cho, and C. A. Grob, Pressure effects in solvolysis and solvolytic fragmentation. A correlation of activation volume with concertedness, *J. Am. Chem. Soc.* **84**, 920–924 (1976).
96. S. W. Benson, *Thermochemical Kinetics*, 2nd ed., Wiley-Interscience, New York (1976).
97. S. W. Benson, F. R. Cruickshank, D. M. Golden, G. R. Haugen, H. E. O'Neal, A. S. Rodgers, R. Shaw, and R. Walsh, Additivity rules for the estimation of thermochemical properties, *Chem. Rev.* **69**, 279–324 (1969).
98. D. R. Herschbach, H. S. Johnston, K. S. Pitzer, and R. E. Powell, Theoretical pre-exponential factors for twelve bimolecular reactions, *J. Chem. Phys.* **25**, 736–741 (1956).
99. V. J. Shiner, Jr., in: Reference 3, Chap. 2.
100. J. C. Harris and J. L. Kurz, A direct approach to the prediction of substituent effects on transition state structure, *J. Am. Chem. Soc.* **92**, 349–355 (1970).
101. J. L. Kurz and Y.-N. Lee, The acidity of water in the transition state for methyl tosylate hydrolysis, *J. Am. Chem. Soc.* **97**, 3841–3842 (1975).
102. H. B. Dunford, J. E. Critchlow, R. J. Maguire, and R. Roman, The advantages of transition state and group acid dissociation constants for pH-dependent enzyme kinetics, *J. Theor. Biol.* **48**, 283–298 (1974).
103. D. M. Bishop and K. J. Laidler, Statistical factors for chemical reactions, *Trans. Faraday Soc.* **66**, 1685–1687 (1970).
104. K. J. Laidler, Reference 1(b), Chap. 4.
105. R. L. Schowen, Reference 9, pp. 317–321.
106. R. Breslow, *Organic Reaction Mechanisms*, 2nd ed., Benjamin, Reading, Mass. (1969), pp. 38–40.
107. E. Gould, *Mechanism and Structure in Organic Chemistry*, Holt, Rinehart and Winston, New York (1959).
108. P. Y. Bruice and T. C. Bruice, The lack of concertedness in the general acid–base catalysis of the enolization of oxalacetic acid. A case for stepwise nucleophilic–general base catalysis, *J. Am. Chem. Soc.* **98**, 844–845 (1976).

109. A. F. Hegarty and W. P. Jencks, Bifunctional catalysis of the enolization of acetone, *J. Am. Chem. Soc.* **97**, 7188–7189 (1975).

110. A. Hall and J. R. Knowles, The uncatalyzed rates of enolization of dihydroxyacetone phosphate and of glyceraldehyde 3-phosphate in neutral aqueous solution. The quantitative assessment of the effectiveness of an enzyme catalyst, *Biochemistry* **14**, 4348–4352 (1975).

111. I. M. Klotz, Free energy diagrams and concentration profiles for enzyme-catalyzed reactions, *J. Chem. Educ.* **53**, 159–160 (1976).

112. J. E. Leffler and E. Grunwald, Reference 54, p. 19.

113. E. Hamori, Illustration of free energy changes in chemical reactions, *J. Chem. Educ.* **52**, 370–373 (1975).

114. R. C. Dougherty, J. Dalton, and J. D. Roberts, S_N2 reactions in the gas phase. Structure of the transition state, *Org. Mass Spectrom.* **8**, 77–79 (1974).

115. (a) H. B. Bürgi, J. D. Dunitz, and E. Shefter, Geometrical reaction coordinates. II. Nucleophilic addition to a carbonyl group, *J. Am. Chem. Soc.* **95**, 5065–5067 (1973); (b) H. Bürgi, Stereochemistry of reaction paths as determined from crystal structure data—a relationship between structure and energy, *Angew. Chem. Int. Ed. Engl.* **14**, 460–473 (1975).

116. J. A. Hirsch, *Concepts in Theoretical Organic Chemistry*, Allyn and Bacon, Boston (1974), Chap. 4.

117. L. P. Hammett, *Physical Organic Chemistry*, 2nd ed., McGraw-Hill, New York (1970), Chap. 11.

118. J. A. Pople and M. Gordon, Molecular orbital theory of the electronic structure of organic compounds. I. Substituent effects and dipole moments, *J. Am. Chem. Soc.* **89**, 4253–4261 (1967).

119. E. D. Hughes and C. K. Ingold, Mechanism of substitution at a saturated carbon atom. IV. Discussion of constitutional and solvent effects on the mechanism, kinetics, velocity, and orientation of substitution, *J. Chem. Soc.* **1935**, 244–255.

120. C. G. Swain, D. A. Kuhn, and R. L. Schowen, Effect of structural changes in reactants on the position of hydrogen-bonding hydrogens and solvating molecules in transition states. The mechanism of tetrahydrofuran formation from 4-chlorobutanol, *J. Am. Chem. Soc.* **87**, 1553–1561 (1965).

121. R. M. Pollack and M. Brault, Synergism of the effect of solvent and of general base catalysis in the hydrolysis of a Schiff base, *J. Am. Chem. Soc.* **98**, 247–248 (1976).

122. (a) W. P. Jencks, in: *Advances in Enzymology* (A. Meister, ed.), Vol. 43, pp. 219–410, Wiley, New York (1975); (b) W. P. Jencks, Approximation, chelation, and enzymic catalysis, *Paabs Revista* **2**, 235–243 (1973), and references therein.

123. W. P. Jencks, Reference 89, Chaps. 1–9.

124. T. C. Bruice and U. K. Pandit, Intramolecular models depicting the kinetic importance of fit in enzymic catalysis, *Proc. Nat. Acad. Sci. USA* **46**, 402–404 (1960).

125. D. R. Storm and D. E. Koshland, A source for the special catalytic power of enzymes: Orbital steering, *Proc. Nat. Acad. Sci. USA* **66**, 445–452 (1970).

126. A. Dafforn and D. E. Koshland, Proximity, entropy and orbital steering, *Biochem. Biophys. Res. Commun.* **52**, 779–785 (1973).

127. T. C. Bruice, Is "orbital steering" a new concept?, *Nature (London)* **237**, 335 (1972).

128. T. C. Bruice, in: *Annual Review of Biochemistry* (E. S. Snell, ed.), Vol. 45, pp. 331–373, Annual Reviews, Inc., Palo Alto, Calif. (1976).

129. W. P. Jencks, Reference 89, Chap. 5.

130. W. P. Jencks and M. I. Page, "Orbital steering," entropy, and rate accelerations, *Biochem. Biophys. Res. Commun.* **57**, 887–891 (1974).

131. C. Delisi and D. M. Crothers, The contribution of proximity and orientation to catalytic reaction rates, *Biopolymers* **12**, 1689–1704 (1973).

132. L. Pauling, Molecular architecture and biological reactions, *Chem. Eng. News* **24**, 1375–1377 (1946).

133. J. C. Powers, B. L. Baker, J. Brown, and B. K. Chelm, Inhibition of chymotrypsin A_{α} with *N*-acyl- and *N*-peptidyl-2-phenylethylamines. Subsite binding free energies, *J. Am. Chem. Soc.* **96**, 238–243 (1974).

134. D. W. Ingles and J. R. Knowles, The stereospecificity of α-chymotrypsin, *Biochem. J.* **108**, 561–569 (1968).

135. A. R. Fersht, Catalysis, binding and enzyme-substrate complementarity, *Proc. R. Soc. London Ser. B* **187**, 397–407 (1974).

136. R. Breslow, Reference 106, pp. 64–65.

137. R. Wolfenden, Transition state analogs for enzyme catalysis, *Nature (London)* **223**, 704–705 (1969).

138. R. Wolfenden, Analog approaches to the structure of the transition state in enzyme reactions, *Acc. Chem. Res.* **5**, 10–18 (1972).

139. G. E. Lienhard, Enzymatic catalysis and transition-state theory, *Science* **180**, 149–154 (1973).

140. K. Schray and J. P. Klinman, The magnitude of enzyme transition state analog binding constants, *Biochem. Biophys. Res. Commun.* **57**, 641–648 (1974).

141. R. Wolfenden, Chapter 15 in this volume.

142. G. L. Kenyon and J. A. Fee, in: *Progress in Physical Organic Chemistry* (A. Streitwieser and R. W. Taft, eds.), Vol. 10, pp. 381–410, Wiley-Interscience, New York (1973).

143. J. E. Leffler and E. Grunwald, Reference 54, Chaps. 6 and 7.

144. C. D. Johnson, Linear free energy relationships and the reactivity–selectivity principle, *Chem. Rev.* **75**, 755–765 (1975), and references therein.

145. W. P. Jencks, Reference 89, pp. 195–197.

146. E. H. Cordes and W. P. Jencks, General acid catalysis of semicarbazone formation, *J. Am. Chem. Soc.* **84**, 4319–4328 (1962).

147. L. P. Hammett and H. L. Pfluger, The rate of addition of methyl esters to trimethylamine, *J. Am. Chem. Soc.* **55**, 4079–4089 (1933); L. P. Hammett, Some relations between reaction rates and equilibrium constants, *Chem. Rev.* **17**, 125–136 (1935).

148. J. A. Hirsch, Reference 116, Chap. 6.

149. S. Ehrenson, R. T. C. Brownlee, and R. W. Taft, in: *Progress in Physical Organic Chemistry* (A. Streitwieser, Jr., and R. W. Taft, eds.), Vol. 10, pp. 1–80, Wiley-Interscience, New York (1973).

150. T. B. Grindley and A. R. Katritzky, A direct relation of sigma constants to the energy scale, *Tetrahedron Lett.* **1972**, 2643–2646.

151. C. G. Swain and E. C. Lupton, Jr., Field and resonance components of substituent effects, *J. Am. Chem. Soc.* **90**, 4328–4337 (1968).

152. M. Sjöström and S. Wold, Statistical analysis of the Hammett equation, *Chem. Scr.* **6**, 114–121 (1974).

153. L. P. Hammett, Reference 117, Chaps. 11 and 12.

154. O. Exner, in: *Progress in Physical Organic Chemistry* (A. Streitwieser, Jr., and R. W. Taft, eds.), Vol. 10, pp. 411–482, Wiley-Interscience, New York (1973).

155. J. E. Leffler, The interpretation of enthalpy and entropy data, *J. Org. Chem.* **31**, 533–537 (1966).

156. R. L. Schowen, Isergonic relations and their significance for catalysis, *J. Pharm. Sci.* **56**, 931–943 (1967).

157. L. G. Hepler, Thermodynamic analysis of the Hammett equation, the temperature dependence of ρ, and the isoequilibrium (isokinetic) relationship, *Can. J. Chem.* **49**, 2803–2807 (1971).

158. J. N. Brønsted and K. Pedersen, The catalytic decomposition of nitramide and its physicochemical applications, *Z. Phys. Chem.* **108**, 185–235 (1924).

159. R. P. Bell, Reference 43, Chap. 10.

160. A. J. Kresge, What makes proton transfer fast?, *Acc. Chem. Res.* **8**, 354–360 (1975).

161. A. J. Kresge, The Brønsted relation—recent developments, *Chem. Soc. Rev.* **2**, 475–503 (1973); M. Eigen, Proton transfer, acid–base catalysis, and enzymatic hydrolysis, Part I:

Elementary processes, *Angew. Chem. Int. Ed. Engl.* **3**, 1–19 (1964); R. A. Marcus, Theoretical relations among rate constants, barriers, and Brønsted slopes of chemical reactions, *J. Phys. Chem.* **72**, 891–899 (1968); J. R. Murdoch, Rate–equilibria relationships and proton-transfer reactions, *J. Am. Chem. Soc.* **94**, 4410–4418 (1972).

162. (a) R. P. Bell, Reference 43, p. 207; (b) R. P. Bell, The development of ideas about proton transfer reactions, *Faraday Symp. Chem. Soc.* **10**, 7–19 (1975); (c) R. P. Bell, Potential energy curves and Brønsted exponents in proton-transfer reactions, *J. Chem. Soc. Faraday Trans. 2* **72**, 2088–2094 (1976).

163. R. Barnett and W. P. Jencks, Rate-limiting diffusion-controlled proton transfer in an acetyl transfer reaction, *J. Am. Chem. Soc.* **90**, 4199–4200 (1968).

164. F. G. Bordwell, W. J. Boyle, Jr., J. A. Hautala, and K. C. Yee, Brønsted coefficients larger than 1 and less than 0 for proton removal from carbon acids, *J. Am. Chem. Soc.* **91**, 4002–4003 (1969).

165. F. G. Bordwell and W. J. Boyle, Jr., Kinetic isotope effects for nitroalkanes and their relationship to transition-state structure in proton-transfer reactions, *J. Am. Chem. Soc.* **97**, 3447–3452 (1975).

166. C. G. Swain and C. B. Scott, Quantitative correlation of relative rates. Comparison of hydroxide ion with other nucleophilic reagents toward alkyl halides, esters, epoxides, and acyl halides, *J. Am. Chem. Soc.* **75**, 141–147 (1953).

167. J. A. Hirsch, Reference 116, Chap. 8.

168. T. H. Lowry and K. S. Richardson, Reference 69(b), pp. 185–190.

169. G. S. Hammond, A correlation of reaction rates, *J. Am. Chem. Soc.* **77**, 334–338 (1955).

170. (a) M. Wolfsberg, Theoretical evaluation of experimentally observed isotope effects, *Acc. Chem. Res.* **5**, 225–233 (1972); (b) F. H. Westheimer, The magnitude of the primary kinetic isotope effect for compounds of hydrogen and deuterium, *Chem. Rev.* **61**, 265–273 (1961); (c) J. Bigeleisen and M. Wolfsberg, in: *Advances in Chemical Physics* (I. Prigogine, ed.), Vol. 1, pp. 15–76, Wiley-Interscience, New York (1958).

171. J. Klinman, Chapter 4 in this volume.

172. M. M. Kreevoy, The exposition of isotope effects on rates and equilibria, *J. Chem. Educ.* **41**, 636–638 (1964).

173. K. B. J. Schowen and M. H. O'Leary, Chapters 6 and 7, respectively, in this volume.

174. F. G. Bordwell and W. J. Boyle, Jr., Kinetic isotope effects as guides to transition-state structures in deprotonation reactions, *J. Am. Chem. Soc.* **93**, 512–514 (1971).

175. J. Hine, *Physical Organic Chemistry*, 2nd ed., McGraw-Hill, New York (1962), pp. 69, 380–382.

176. J. Hine, The principle of least motion. Application to reactions of resonance-stabilized species, *J. Org. Chem.* **31**, 1236–1244 (1966).

177. O. S. Tee, J. A. Altmann, and K. Yates, Application of the principle of least motion to organic reactions. III. Eliminations, enolizations, and homoenolizations, *J. Am. Chem. Soc.* **96**, 3141–3146 (1974).

178. S. Ehrenson, Application of analytic least motion forms to organic reactivities, *J. Am. Chem. Soc.* **96**, 3784–3793 (1974).

179. J. E. Leffler, Parameters for the description of transition states, *Science* **117**, 340–341 (1953).

180. C. G. Swain and E. R. Thornton, Effect of structural changes in reactants on the structure of transition states, *J. Am. Chem. Soc.* **84**, 817–821 (1962).

181. J. E. Critchlow, Prediction of transition state configuration in concerted reactions from the energy requirements of the separate processes, *J. Chem. Soc. Faraday Trans. 1* **68**, 1774–1792 (1972).

182. D. J. McLennan, Effect of substituents on the geometry of transition states for slow proton transfer reactions, *J. Chem. Soc. Faraday Trans. 1* **71**, 1516–1527 (1975).

183. G. J. Frisone and E. R. Thornton, Solvolysis mechanisms. β-Deuterium isotope effects for *t*-butyl chloride solvolysis at constant ionizing power and effect of structure of reactant on S_N1 transition-state geometry, *J. Am. Chem. Soc.* **90**, 1211–1215 (1968).

184. W. J. le Noble and T. Asano, Special effect of pressure on highly hindered reactions as a

possible manifestation of the Hammond postulate, *J. Am. Chem. Soc.* **97**, 1778–1782 (1975).

185. J. E. Reimann and W. P. Jencks, The mechanism of nitrone formation. A defense of anthropomorphic electrons, *J. Am. Chem. Soc.* **88**, 3973–3982 (1966).

186. (a) W. P. Jencks, Requirements for general acid–base catalysis of complex reactions, *J. Am. Chem. Soc.* **94**, 4731–4732 (1972); (b) W. P. Jencks, Enforced general acid–base catalysis of complex reactions and its limitations, *Acc. Chem. Res.* **9**, 425–432 (1976).

187. E. K. Thornton and E. R. Thornton, Reference 3, pp. 261–263.

188. R. L. Schowen, Reference 9, pp. 309–329.

2

Catalytic Power and Transition-State Stabilization

Richard L. Schowen

1. INTRODUCTION

This book is about the use of the transition-state concept, and of ideas about transition-state structure and the stabilization of transition states in catalysis, in understanding the rate processes of biochemistry. Essentially all the reactions considered are enzyme-catalyzed reactions or reactions studied in order to illuminate enzyme catalysis. In view of our thesis in this volume—that the transition state and its structure and interactions are the central feature of catalysis—it is interesting to note that in much of the current writing on enzyme catalysis, the transition state is mentioned rather more rarely than might have been anticipated and its stabilization is sometimes relegated to the position of a single item in a long list of potential contributions to catalytic power. This is true in the pioneering volumes of Bruice and Benkovic,[1] in the monographs of Jencks[2] and Bender,[3] and in the recent reviews of Jencks[4] and Bruice.[5] It is likewise true of many other valuable books and articles, of which well over 5000 are cited in the sources just mentioned. Obvious exceptions to this custom are, of course, the authors interested in transition-state analogs, such as Lienhard[6] and Wolfenden,[7] but the general tendency is clear. There is certainly no agreement that all we need to know in order to understand enzyme catalysis is the structure of the transition state and the manner in which it is stabilized.

It is nevertheless a physical fact that the transition state enjoys the primacy claimed for it by its advocates. Its apparent neglect is a matter of linguistics. In this chapter, I will show that every one of the customary sources of catalytic

Richard L. Schowen • Department of Chemistry, University of Kansas, Lawrence, Kansas.

power may be reformulated in the language of transition-state stabilization and that the two formulations differ only in language and in the depth of the model (i.e., in the degree of physical detail which is proposed). Otherwise, they are equivalent.

To cast the comparison in high relief, I assume a Fundamentalist position:

> ... that the entire and sole source of catalytic power is the stabilization of the transition state; that reactant-state interactions are by nature inhibitory and only waste catalytic power.

The motivation for making this comparison and translation is to clarify the equivalency of the two approaches (it makes no sense, for example, to attempt to "distinguish" them experimentally) and, in addition, to certify the correctness of transition-state stabilization descriptions so that they may replace the customary descriptions wherever they have an expository or pedagogical superiority.

A second, closely related, purpose of this chapter is to establish the relationship of transition-state structure and stabilization to the observable parameters of enzyme kinetics. A number of researchers are accustomed to the use of rate constants and combinations of rate constants in discussing enzymic accelerations but are unaccustomed to using transition-state structural concepts. Rate-constant language, of course, has a straightforward relationship to transition-state language. The relationship is outlined for the convenience of all concerned. The discussion also deals with the natural question of the meaning of transition-state structures and stabilization under conditions where more than one transition state contributes to determining the rate of reaction.

2. ISSUES IN THE DESCRIPTION OF ENZYME CATALYTIC POWER

2.1. The Canon of Enzyme Catalysis

It will be convenient to have a name for the usual descriptions of enzyme catalysis that do not seem to accord any position of great importance to transition-state structure or stabilization. I propose to refer to these descriptions as "canonical formulations" of enzyme catalysis. It is tempting to define some corresponding body of literature as the *canon of enzyme catalysis*; the major works cited above[1-5] might make a natural choice. I intend to resist that temptation, in part because I lack the courage to appoint myself the Athanasius of enzyme catalysis and then to face the authors whose writings I would have omitted from the canon. It may be remembered that the other Athanasius worked at a remove of several centuries from the authors among whom he selected and that they were all saints. Thus while the references already given and others cited below form a useful source of examples of canonical formulations, they should not be presumed to exhaust the canon. Rather, a *canonical* formulation of catalysis is simply one that omits or de-emphasizes the transition state but that is correct in principle and is therefore transformable to transition-state language. A *Fundamentalist* formulation is, by contrast, one which dwells on transition-state language to the exclusion of other descriptive apparatus.

2.2. The Transition State as the Focus of Catalytic Action

Figure 1 shows a familiar kind of free-energy diagram for a simple enzyme-catalyzed reaction and the corresponding uncatalyzed reaction under two sets of limiting conditions that are of practical enzymological interest: (a) substrate concentration low relative to K_m such that the enzyme is mostly free and the enzyme–substrate complex is a steady-state intermediate of negligible concentration, and (b) substrate concentration high relative to K_m such that the enzyme exists wholly as complex, which is thus the effective reactant species for the catalyzed reaction. As usual, the enzyme concentration is considered very small relative to substrate concentration. Under both sets of conditions, we can define a quantity $\Delta G_{cat} = \Delta G_U^* - \Delta G_C^*$, the difference in the free energies of activation for the uncatalyzed reaction (subscript U) and the catalyzed reaction (subscript C). This *catalytic free energy* is a positive number which will be the larger as the enzyme is more effective as a catalyst.

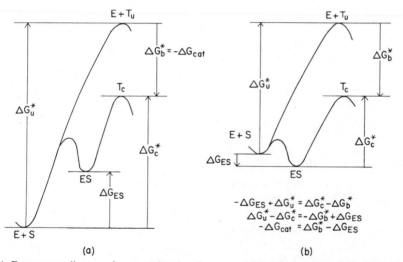

$$-\Delta G_{ES} + \Delta G_u^* = \Delta G_c^* - \Delta G_b^*$$
$$\Delta G_u^* - \Delta G_c^* = -\Delta G_b^* + \Delta G_{ES}$$
$$-\Delta G_{cat} = \Delta G_b^* - \Delta G_{ES}$$

(a)　　　　　　　　　　　　(b)

Fig. 1. Free-energy diagrams for uncatalyzed and enzyme-catalyzed reactions of the substrate S at different standard-state concentrations of S. (a) Concentration of S low compared to K_m, so that no substantial concentration of enzyme-substrate complex is present. The entire reduction in free energy of activation effected by the enzyme originates in stabilization of the uncatalyzed transition state T_U with formation of the catalyzed transition state T_C. The whole binding or stabilization energy appears as catalytic free energy. (b) Concentration of S high compared to K_m so that the enzyme is mainly complexed. The inhibitory complexation of the substrate by the enzyme countervails against the transition-state stabilization and the substrate stabilization energy must be subtracted from the transition-state stabilization energy to obtain the catalytic free energy.

The situation is simpler when the substrate concentration is low [Fig. 1(a)]. Then, as the figure illustrates, the catalytic free energy is given directly by the free energy released when the enzyme (in formal terms) binds the uncatalyzed transition state to form the catalytic transition state. Therefore the catalysis arises wholly from the transition-state stabilization effected by the enzyme, as measured by the free energy of transition-state binding ΔG_b^*, and this entire binding energy is realized as catalytic free energy:

$$\Delta G_{cat} = \Delta G_U^* - \Delta G_C^* = -\Delta G_b^* \qquad (1)$$

The most straightforward approach to understanding the catalytic power of the enzyme is to investigate this transition-state binding and the structural and energetic changes which accompany it. Under these circumstances all other factors (in particular, events in the enzyme–substrate complex) are totally irrelevant.

When the substrate concentration is made high enough to induce the large-scale formation of enzyme–substrate complex [Fig. 1(b)], the enzyme is permitted to reveal its capacities as an inhibitor as well as a catalyst. The free energy released by the binding of the transition state remains, of course, the same as at low substrate concentration (ΔG_b^*), but it no longer appears completely as catalytic free energy. Instead, the free energy expended by the enzyme in stabilizing the reactant state (ΔG_{ES}) must now be subtracted from the transition-state stabilization energy to yield the *net* stabilization which is applied to catalysis:

$$\Delta G_{cat} = -\Delta G_b^* + \Delta G_{ES} \qquad (2)$$

Here again, the entire catalytic potential of the enzyme lies in its capacity for stabilization of the transition state. Its inhibitory potential, as expressed in reactant-state stabilization, now appears and prevents complete realization of the transition-state stabilization as catalysis.

Enzymes are, of course, only one member of the class of multifunctional, multivalent macromolecules having the capacity to bind smaller molecules. Other members of this class, such as certain transport proteins and antibodies, have a highly developed capacity for binding reactant-state species but no special affinity for transition states. They function, therefore, as inhibitors: Antibodies inhibit reactions of their ligands (which may themselves be of quite large size) deleterious to the host, and transport proteins may inhibit enzymic or nonenzymic degradation of the species bound to them. Enzymes, in having a strong transition-state affinity, function essentially as catalysts. Their reactant-state affinities, which are anticatalytic, may in some cases be coincidental and in other cases may be connected with biological factors such as specificity or regulation.

A complete understanding of enzyme catalysis for such a simple case as that illustrated therefore resolves into a characterization of two binding processes: that for the transition state, which yields a model for catalysis, and that for the reactant state, which yields a model for the inhibitory effects which arise in the substrate-saturation region of concentration. More complex systems require the characterization of further binding processes, but the principle is always the same. The binding of transition states promotes catalysis; the binding of other species produces inhibition. The differential stabilization of the transition state (total stabilization of the transition state minus stabilization of reactant species) always gives the catalytic acceleration.

2.3. The Choice of Standard Reaction and of Standard States

To discuss quantitatively the catalytic power of an enzyme, we must select a set of standard reaction conditions. This selection corresponds to a choice of thermodynamic standard states for the free energies treated in the previous section. Such a choice is a feature common to the discussion of all reaction rates and equilibria. The particular subject of catalysis, on the other hand, demands a further choice. A catalyst must have a reaction to catalyze; catalysis implies an "uncatalyzed" or *standard reaction.*

What appears not to have been sufficiently clear is that the choice of standard reaction is just as arbitrary as the choice of standard states. To remain true to the definition of catalysis, a standard reaction should have the same stoichiometry as the catalyzed reaction, but there are no further requirements. An enzyme catalyzes *all* reactions of the same stoichiometry as its own reaction as long as they are slower. Faster reactions may be inhibited. There is no such thing as a universal "catalytic acceleration" produced by a particular enzyme; the catalytic acceleration must always be stated with respect to a chosen standard reaction. In general, each such standard reaction will produce a different value of the catalytic acceleration. The situation is equivalent to that with free energies of reaction: They can have different values, depending on the choice of standard states.

It is very important to realize that there is no requirement whatever for the standard-reaction transition state to resemble in any way the structure of the substrate-derived part of the enzymic transition state. Such a resemblance may in certain discussions be convenient, but there is nothing which makes it a necessity. It is perfectly proper to envision internal structural changes in both substrate-derived and enzymic parts as the standard-reaction transition state binds to the enzyme, liberating ΔG_b^*. It is further true that even the *composition* of the standard-reaction transition state need not be the same as that of the substrate-derived part of the enzymic transition state. The binding equilibrium to form the enzymic transition state, if these compositions are not the same, must then involve other reactants and products to balance the chemical equation. Thus the use of acid-catalyzed or base-catalyzed standard reactions will require the expulsion of a proton or a hydroxide ion as enzymic binding of the standard transition state occurs. This is the kind of process which Lienhard[6] has called "transition-state interchange."

The choice of standard states should be made so as to coincide with the conditions under which catalysis is to be considered. It is therefore desirable to take enzyme standard-state concentrations as being low (say, 10^{-6} to 10^{-9} M), corresponding to the actual levels in the biological or experimental environment. Substrate concentrations may be taken as low (below K_m) if it is desired to consider catalysis under conditions where the enzyme is largely free or as high (above K_m) if the discussion pertains to catalysis under saturating conditions. These points have been very clearly discussed by Jencks.[4] It should always be remembered that standard states are *implied* if one calculates free-energy changes directly from rate or equilibrium constants which are expressed in concentration units. Such calculations require in principle that the activities of the species, rather than their concentrations, be used. In effect, of one expresses the rate or equilibrium constants in molar units, one has implicitly divided all concentrations by a standard-state concentration of 1 M in order to produce (dimensionless) activities. Thus if one calculates $\Delta G_{ES} = -RT \ln(1/K_m)$ (for binding of S to E, for which $K_{eq} = 1/K_m$), the standard states depend on the units of K_m. Expression of this constant in M units implies a standard state of 1 M, mM implies standard states of 1 mM, etc. A change from one set of standard states to another can readily be effected by

$$\Delta G_{new} = \Delta G_{old} + 2.303RT \{\log[([C]_{new}/[C]_{old})$$
$$\times ([D]_{new}/[D]_{old}) \cdots] - \log[([A]_{new}/[A]_{old})$$
$$\times ([B]_{new}/[B]_{old} \cdots)]\} \tag{3}$$

assuming the reaction of

$$A + B + \cdots \longrightarrow C + D + \cdots \tag{4}$$

For a free energy of activation, one merely allows one of the product species to represent the transition state. Some free-energy changes (corresponding to rate or equilibrium constants which have no concentration units, such as first-order rate constants or equilibrium constants for reactions having the same number of product species as reactant species) are independent of standard state as long as the same state is chosen for all species.

Equivalent considerations apply to the choice of temperature, pressure, pH, and other variables for standard states. Realistic choices should be made. If values are known for a single standard state and the appropriate quantities are available (i.e., the derivatives of the free energy with respect to each variable), transformation to any desired set of standard states is possible.

The most effective way of avoiding confusion through the choice of standard states in treating questions of catalysis is to concentrate attention on the relative rates of the catalyzed and uncatalyzed reactions under conditions of interest, instead of employing rate constants themselves, which are simply the rates at unit concentration. Thus we can use equations such as

$$v_U = k_U S \tag{5}$$

$$v_C = (k_{ES}E_T S)/(K_m + S) \tag{6}$$

to represent the rates of a typical standard reaction [Eq. (5) for a first-order reaction of the substrate S] and of a typical enzymic reaction [Eq. (6), the Michaelis–Menten law for the initial rate of a simple one-substrate reaction with k_{ES} the rate constant for reaction from the enzyme–substrate complex and K_m the substrate concentration at half-maximal velocity]. Formulating the reciprocal of the enzymic reaction velocity as in

$$v_C^{-1} = (k_E E_T S)^{-1} + (k_{ES} E_T)^{-1} \tag{7}$$

(familiar from the Lineweaver–Burk plot; $k_E = k_{ES}/K_m$), we can obtain the catalytic acceleration v_C/v_U, given as its reciprocal in

$$(v_C/v_U)^{-1} = [(k_E/k_U) E_T]^{-1} + [(k_{ES}/k_U) E_T]^{-1} S^{-1} \tag{8}$$

or its equivalent in free-energy terms as in

$$v_C/v_U = \exp\{\Delta G_{cat}/RT\} \tag{9}$$

The reciprocal form v_U/v_C is used in Eq. (8) because it separates the catalytic acceleration into two terms corresponding to the high and low substrate-concentration domains. Equation (8) shows that a higher total *enzyme* concentration will always favor the catalyzed reaction (increasing v_C/v_U by its effect in both terms), while a higher *substrate* concentration will favor the uncatalyzed reaction and decrease the catalytic acceleration. When the substrate concentration is very low, the latter effect will be negligible. Then the first term on the right-hand side of Eq. (8) dominates and (v_C/v_U) is independent of substrate concentration, because both the standard and enzymic rates are proportional to substrate concentration under these conditions and the effect of substrate concentration cancels from their ratio. At higher substrate concentrations, however, the second term becomes more important. Eventually, the enzyme becomes saturated, and increased substrate concentration will then drive the uncatalyzed reaction but not the enzymic one. Hence, higher substrate concentration—when it has an effect at all—decreases the catalytic acceleration.

The rate-constant ratios, k_E/k_U and k_{ES}/k_U, in Eq. (8) can be related to the free energies of species along the reaction path: G_C^* and G_U^* for the enzymic and standard transition states, respectively, G_E and G_S for the enzyme and the substrate, and G_{ES} for the enzyme–substrate complex:

$$k_E/k_U = \exp\{-[(G_C^* - G_E - G_S)/RT] + [(G_U^* - G_S)/RT]\} \tag{10}$$

$$k_E/k_U = \exp\{-[(G_C^* - G_U^* - G_E)/RT]\} \tag{11}$$

$$k_{ES}/k_U = \exp\{-[(G_C^* - G_{ES})/RT] + [(G_U^* - G_S)/RT]\} \tag{12}$$

$$k_{ES}/k_U = \exp\{-[(G_C^* - G_U^*)/RT] + [(G_{ES} - G_S)/RT]\} \tag{13}$$

In Eq. (13) it is useful to add G_E to the first term in the exponential and subtract it from the second to obtain

$$k_{ES}/k_U = \exp\{-[G_C^* - G_U^* - G_E)/RT] + [(G_{ES} - G_S - G_E)/RT]\} \tag{14}$$

Then it is apparent that in Eq. (11) the free-energy change in the exponential is just that for reaction of the standard transition state with the enzyme to generate the enzymic transition state. In Eq. (14), the first free-energy change is this same one, but the second free-energy change is that for substrate binding to the enzyme. Recalling from above that we used ΔG_b^* for the free energy released on transition-state binding and ΔG_{ES} for the free energy released on substrate binding, we have

$$k_E/k_U = \exp(-\Delta G_b^*/RT) \tag{15}$$

$$k_{ES}/k_U = \exp[(-\Delta G_b^* + G_{ES})/RT] \tag{16}$$

These equations simply express mathematically our graphical conclusion from Fig. 1: Below saturation, the entire transition-state binding energy is available for catalytic acceleration [Eq. (15); cf. Eq. (1)]; above saturation, the free energy wasted in the inhibitory binding of the substrate is lost and must be subtracted from the transition-state binding energy to calculate the catalytic acceleration [Eq. (16); cf. Eq. (2)].

Equations (15) and (16), since they are written for rate-constant ratios, refer to unit-concentration standard states. For a more general result, we return to Eq. (8), restate the concentration terms as $RT \ln E_T$ and $RT \ln S$, thus converting them to free energy or chemical potential, combine Eq. (8) with Eq. (9), and obtain

$$\Delta G_{cat} = RT \ln \{\exp[(-\Delta G_b^* + RT \ln E_T)/RT]$$
$$+ \exp[(-\Delta G_b^* + RT \ln E_T + \Delta G_{ES} - RT \ln S)/RT]\} \tag{17}$$

for the catalytic free energy under any conditions. Equation (17) summarizes all the factors that favor and oppose a large catalytic acceleration (positive ΔG_{cat}). Strongly negative ΔG_b^*, representing strong binding of the transition state, and large values of E_T, which represents the driving of the catalytic reaction through the chemical potential of the enzyme, both favor catalysis. Opposing or inhibitory factors are favorable binding of the substrate, appearing as negative ΔG_{ES}, and high concentration of the substrate (large S) which drives the uncatalyzed reaction after saturation of the enzyme.

These conclusions, it ought to be noted, are free of any assumption of structural relationship between the arbitrarily chosen standard-reaction transition state and the enzymic transition state. Suppose, for example, that a conformation change of the enzyme in one of its complexes were rate determining. Equation (17) would remain a perfectly valid description of the catalytic acceleration, with the choice of the standard reaction remaining arbitrary. It would indeed be natural if some reaction involving covalency changes in the substrate had been chosen as the standard reaction, with these covalency changes in progress at the transition state. If so, the binding energy ΔG_b^* in Eq. (17) would correspond to a sum of free-energy changes. One of these is associated with the conversion of the enzyme from its "resting" form to its transition-state form. A second is associated with alteration of the standard-reaction transition state to the form possessed by the substrate-derived entity in the transition state for the enzymic conformation change. A third, final contributing free-energy change is associated with the combination of these two entities to form the enzymic transition state.

2.4. Example: The Acetylation of Acetylcholinesterase

To illustrate the choice of standard states and the fact that the choice of standard reaction is totally arbitrary, we shall work out the example of the enzymic acceleration of acetylcholine hydrolysis,

$$(CH_3)_3^+ NCH_2CH_2O_2CCH_3 \xrightarrow{\text{H}_2\text{O}} (CH_3)_3\overset{+}{N}CH_2CH_2OH + CH_3CO_2H \tag{18}$$

in some detail. The enzyme acetylcholinesterase[8] catalyzes the hydrolysis of this ester by a mechanism common among hydrolytic enzymes, in the course of which the substrate acetyl group is first transferred to a serine hydroxyl group of the enzyme, with expulsion of the alcohol choline, and then the acetyl group is cleaved hydrolytically from the enzyme in a second step:

$$AcOCh + E-OH \longrightarrow E-OAc + ChOH \longrightarrow E-OH + AcOH \tag{19}$$

At concentrations of substrate low relative to K_m (which is about $9 \times 10^{-5}\ M$ for acetylcholine), the rate-determining step is acetylation of the enzyme and the reaction obeys a simple second-order kinetic law:

$$v_C = k_E E_T S \tag{20}$$

The value of k_E in Eq. (20) is $10^{8.1}\ M^{-1}\ \text{sec}^{-1}$ (as calculated from the data of Froede and Wilson[8]). If a value of the free energy of activation were calculated by direct insertion of the second-order rate constant into the Eyring equation,

$$k = (\mathbf{k}T/h)\exp(-\Delta G^*/RT) \tag{21}$$

it would correspond to a choice of 1-M standard-state concentrations for enzyme, substrate, and transition state (or to a choice of 1 M for either enzyme or substrate and to any arbitrary but equal states for the other two species). Let us maintain the transition-state standard state at 1 M but adjust the standard states for enzyme to a more realistic value of $10^{-7}\ M$ and for substrate to $10^{-5}\ M$. The unrealistic value of 1M for the transition state is unimportant because we need only to use the same value for all transition states. Any other choice would simply introduce a constant factor in all calculations. Under such standard-state conditions, $v_C = 10^{8.1-7-5} = 10^{-3.9}\ M\ \text{sec}^{-1}$, yielding (for 1-$M$ transition state) a free energy of activation of $\Delta G_C^* = 22.6$ kcal/mol. This barrier for the catalyzed reaction may be compared to that for any desired standard reaction of acetylcholine. We shall compare three standard reactions: (1) the neutral hydrolysis, (2) the basic hydrolysis, and (3) the acidic hydrolysis of acetylcholine. The data and calculations are shown in Table I.

Table I

Use of Various Standard Reactions to Calculate the Catalytic Acceleration Produced
by Acetylcholinesterase upon the Hydrolysis of Acetylcholine[a]

Standard reaction	Rate constant	Free energy of activation, ΔG_U^* (kcal/mole)	Free energy of transition-state binding,[b] ΔG_b^* (kcal/mol)	pK_b^*
Neutral hydrolysis of acetylcholine	$(8 \times 10^{-10}\,sec^{-1})^c$	36.5	−13.9	−10.2
Basic hydrolysis of acetylcholine[d]	$2.2\,M^{-1}\,sec^{-1}$	33.2	−10.6	− 7.8
Acidic hydrolysis of acetylcholine[e]	$2.7 \times 10^{-5}\,M^{-1}\,sec^{-1}$	39.9	−17.3	−12.7

[a] Free energies are calculated for standard-state concentrations of $10^{-7}\,M$ for acetylcholinesterase, hydrogen ion, and hydroxide ion and $10^{-5}\,M$ for acetylcholine at 25°.

[b] The free energy of activation for the enzymic reaction is 22.6 kcal/mol.

[c] Estimated from the value for ethyl acetate, which is itself approximate.[11] The acidic hydrolysis of acetylcholine is about 5-fold slower than that of ethyl acetate, while the basic reaction is about 20-fold faster. The neutral reaction was taken as $(20/5)^{1/2}$ or about 2-fold faster.

[d] Reference 9.

[e] Reference 10.

The substrate concentrations are all maintained at the same standard state of $10^{-5}\,M$ in the standard reactions as in the enzymic reaction, the water standard state for the neutral hydrolysis is taken as pure water [thus the first-order rate constant is inserted directly into Eq. (21)], and the hydrogen-ion and hydroxide-ion standard states are taken as $10^{-7}\,M$, which then corresponds to the pH of 7 at which the value of k_E [Eq. (20)] used in the calculations was determined. The free energies of activation of each of these standard reactions, under the chosen standard-state conditions, are listed in Table I. By subtraction of the enzymic barrier of 22.6 kcal/mol from each, the binding energy for each standard transition state is obtained. The transition-state binding equilibrium is given by

$$T_U^N \longrightarrow T_C \tag{22}$$

for the neutral standard reaction, by

$$(T_U^B)^- + E \longrightarrow T_C + HO^- \tag{23}$$

for the basic standard reaction, and by

$$(T_U^A)^+ + E \longrightarrow T_C + H^+ \tag{24}$$

for the acidic standard reaction. In the last column of Table I, the pK for the binding equilibrium is given. With sign reversed, this is equal to the number of orders of magnitude of catalytic acceleration which is effected upon each standard reaction by the action of acetylcholinesterase.

In Fig. 2 the data are presented graphically, and structural hypotheses are shown for the standard-reaction transition states. A structural hypothesis for

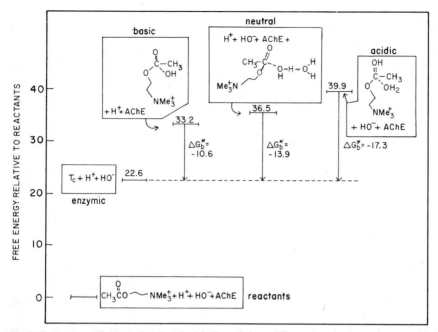

Fig. 2. Free-energy diagram showing the binding of three different standard-reaction transition states to acetylcholinesterase to generate the catalytic transition state for the acetylation of the enzyme. From the left, the standard transition states are those for basic hydrolysis of acetylcholine (binding necessitates the expulsion of a hydroxide ion and liberates 10.6 kcal/mol), neutral hydrolysis of acetylcholine (a water-derived complex is expelled on binding and a net 13.9 kcal/mol is liberated), and acidic hydrolysis of acetylcholine (expulsion of water and proton with release of 17.3 kcal/mol). Other, extensive internal structural changes in the transition states may also occur on binding.

the enzymic transition state will be advanced and discussed in Chapter 10 in this volume. It is not, however, necessary to make or use such hypotheses in order to employ these standard reactions. Notice that none of the binding equilibria represents a simple combination of enzyme with unaltered standard-reaction transition state. Instead, extensive changes, including the expulsion of other materials by the entry of the enzyme, doubtless occur. For example, in the binding of the transition state for basic hydrolysis $[(T_U^B)^-$ of Eq. (23)], the free-energy change is -10.6 kcal/mol. This is the net change for a complex process in which the internal structure of the enzyme and of the substrate-derived portion of the standard transition state must change, the hydroxide ion must be broken away from $(T_U^B)^-$, interactions must form between the substrate-derived and enzyme-derived parts of T_C, and all the hydration changes associated with these processes must occur.

We find, therefore, than *any* standard reaction may be employed for the discussion of enzyme catalysis and that, while the magnitude of the catalytic acceleration depends on the choice, there are no constraints on the choice beyond stoichiometry. Of course, one might for various reasons want to compare the enzymic reaction with a "very closely related" standard reaction (i.e., one with a presumably similar internal structure for the substrate-derived portion of the transition state) in order to allow certain common features to cancel between them. Since the active site of acetylcholinesterase is believed to contain a catalytic imidazole function, one might have chosen imidazole-catalyzed hydrolysis of acetylcholine as a "closer" standard reaction than the three choices presented above (however, see Chapter 10). Indeed, many discussions of the catalytic factors which contribute to enzymic activity constitute, as we shall see below, simply a step-by-step redefinition of the standard reaction until the binding energy of the standard transition state becomes zero. It is then assumed that, in all essential respects, the structures of the enzymic and standard transition states are identical, or—in canonical terms—that all the catalytic factors have been correctly accounted for.

2.5. Poised Structures and Vertical Transition-State Binding

The complexity of the observed free energies of transition-state binding, calculated from experimental data under chosen standard-state conditions, raises the obvious problem of how to proceed with the construction of models for the standard-reaction and enzymic transition states and for the process of their interconversion. Such models will in the end constitute our "understanding" of the phenomenon of enzyme catalysis.

In the most general case, both the enzyme structure and the substrate-derived structure change when the two species bind together. Some of the free energy of binding will therefore be connected with the work of interconverting these structures. Another part of the free energy of binding will be associated with the development of interactions between the enzyme-derived and the substrate-derived parts of the enzymic transition state. With a proper set of definitions, the total free energy of binding can thus be divided into two additive contributions. The first of these may be called the free energy of distortion:

> The *free energy of distortion* is the total work of (1) converting the standard-reaction transition state into exactly the structure it will have in the enzymic transition state, including the breaking away of any extra covalently attached groups, desolvation, etc., and (2) converting the enzyme from its free form into exactly the structure it will possess in the enzymic transition state, including conformation changes, desolvation, etc.

The structures which are generated in this way, for both the enzyme and the substrate-derived entity, may be denoted "poised structures," because they are poised at the form they will have in the actual enzymic transition state.

The second part of the free energy of binding, decomposed according to this model, can be named the vertical binding energy (vertical in the sense that it will not be associated with changes in internal geometrical structure, as is the case with vertical electronic transitions):

The *vertical binding energy* is the net change in free energy that is realized when the poised structures are allowed to come together from their standard-state distributions, combine, and form the enzymic transition state.

The free energy of dilution which is required to bring the enzyme and substrate-derived poised structures from their standard-state distributions into the same region of solution (in a simple case, the "entropy barrier" to their combination) might logically have been included in either the distortion or the vertical binding term. I prefer to place it in the vertical binding energy for the following reason, which is in part philosophical but which also has a practical, procedural component. The model, as described above, need not involve any purely hypothetical entities. Regardless of how improbable the poised structures of enzyme and substrate-derived entities may be when they are separate from one another, it is nonetheless easily conceived that each material (enzyme and substrate) will occasionally (by a "random walk") execute a spontaneous distortion through its poised structure. Thus the poised structures may be regarded as real and extant in the material world, even if their properties may be investigated only by theoretical means. It is then imagined that each of these poised structures is brought, in the course of vertical binding, from its standard-state distribution to the neighborhood of the other poised structure (with no internal structural change) and allowed to combine with it. If, by contrast, one had chosen the alternative model and included the dilution term in the distortion energy, it would have been necessary to postulate purely imaginary structures the properties of which could not be investigated by standard theoretical procedures, because one would have had to imagine the poised structures actually in their combined form but without the vertical binding energy having been released. Since the combination and the energy release are related as cause and effect, the physical system could never perform an excursion through such a configurational circumstance, even in principle.

The algebraic sum of the distortion energy and the vertical binding energy is then just the binding energy of the transition state to the enzyme, which can be calculated from two experimental values, viz., the free energies of activation for the catalyzed and uncatalyzed reactions. Because it can be calculated from experimental data, we shall call the transition-state binding energy the *empirical binding energy* of the transition state to the enzyme. Figure 3 exhibits the dissection of the empirical binding energy into the free energy of distortion and the free energy of vertical binding.

The true utility of this analysis is that it suggests a protocol for the investigation of the origins of enzyme catalytic power. The steps in such a study would be:

1. To determine the structures of the standard transition state and the enzymic transition state (as well as the structure of the enzyme and of any other species involved in the binding equilibrium).
2. To determine the free energies required to distort the enzyme and the standard transition state to their poised structures.

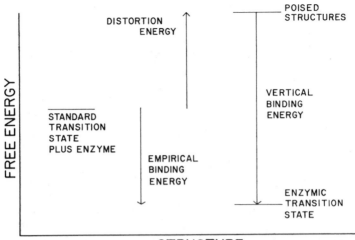

Fig. 3. Dissection of the empirical free energy of binding of the standard-reaction transition state to the enzyme (calculable from the difference of standard-reaction and enzymic-reaction free energies of activation) into a component for reorganization of the standard-reaction transition state and the enzyme into the "poised structures" that they will possess in the enzymic transition state (the free energy of distortion) and a component for the bringing together and interaction of the poised structures to generate the enzymic transition state, not accompanied by any internal structural reorganization (the free energy of vertical binding).

3. To determine the free-energy changes associated with the development of each of the interactions present in the enzymic transition state (favorable and unfavorable) as well as the free energies of dilution referred to above.

These steps will rely on experimental studies, on theoretical information, and on an intimate mixture of the two.[12]

3. TRANSITION-STATE FORMULATIONS OF THE CANONICAL DESCRIPTIONS OF ENZYME CATALYTIC POWER

We are now prepared to establish a general apparatus for the translation of canonical ideas about the origins of enzyme catalytic power into the language of transition-state stabilization, i.e., into Fundamentalist formulations. A major distinction in these two approaches, as we shall see, is that the canonical language requires much greater detail in the postulation of what occurs *between* the reactant and transition states than the Fundamentalist language. Indeed, the Fundamentalist approach completely neglects this aspect of the mechanism, while it is central to the canonical version. Events between reactant and transition states may be called *dynamical* aspects of mechanism, while events restricted to reactant and transition states alone may be called *kinetic* aspects of mechanism (this point is discussed and illustrated by Maggiora and Christoffersen in Chapter 3 in this volume). Thus a high degree of dynamical definition will be found to characterize the canonical discussions of catalysis, while Fundamentalist approaches will be purely kinetic.

3.1. Entropic Contributions to Catalysis

One of the major canonical concepts in current use is the picture of the enzyme as entropy trap. The proximity effect has long been advocated by Bruice as an important source of the catalytic power of enzymes.[1,5,13] In the last few years, Jencks[4] and Page have demonstrated that the magnitude of such contributions may greatly exceed the levels traditionally assumed. Let us first present the simplified model used by Jencks[4] to show how an enzyme could be a powerful catalyst and yet *nothing* but an entropy trap. We shall then show that this is entirely (and necessarily) equivalent to the statement that the enzyme catalyzes the reaction simply by binding the transition state.

The situation is simplest for substrate standard states exceeding K_m, for which the diagram of Fig. 4 (a) has been constructed. Jencks envisions that the entire entropy of activation, in this simplified model, may be expended (with no enthalpy change) in forming a "complex ... in the correct position for reaction" in the course of assembling the standard reaction transition state. To form this complex, the free energy change is therefore $-T\Delta S_U^*$. After complex formation, the enthalpy of activation is supplied, carrying the complex into

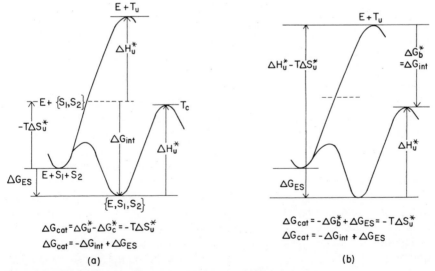

Fig. 4. Free energy diagrams illustrating (a) the canonical version of the enzyme as entropy trap and (b) the equivalent Fundamentalist formulation.

the transition state of the standard reaction, with a free-energy change equal to ΔH_U^*. It ought to be noted that (although this decomposition of the free energy of activation is an absolutely proper procedure) the "complex" is an imaginary species. The reactants might perform a random walk through this structure, as indeed they do in the course of reaction, with expenditure of the entropy of activation, but they could not in reality do so without also requiring the supply of the enthalpy of activation. In any case, having envisioned the hypothetical "complex," one now considers that the enzyme binds this structure with the release of the "intrinsic binding energy of the reactive complex," ΔG_{int}. Part of this binding energy is "utilized" to cancel the unfavorable entropy of formation of the complex, while the remainder is released as net, observed binding energy of substrate, ΔG_{ES}.

"Now let us assume," writes Jencks, "that the enzyme catalyzes the reaction only by bringing the reactants together—that is, there is no 'chemical' catalysis in the usual sense and the enthalpy of activation is the same for the enzymic as for the nonenzymic reaction." Thus the free energy of activation, starting from ES, is given merely by ΔH_U^*. It is now readily seen from Fig. 4 that catalysis will indeed occur and that the catalytic acceleration $\Delta G_{cat} = \Delta G_U^* - \Delta G_C^*$ will be equal to $-T\Delta S_U^*$: The enzyme has acted simply as an entropy trap and has effected a catalysis just equal to the entropy saving. (It ought again to be noticed that the ES complex is here again hypothetical because it contains the species in their "poised structures" but without the expenditure of enthalpy which would automatically accompany their combination.) In addition, since that part of the intrinsic binding energy of the reactive complex which was utilized to cancel the entropy cost of the reaction is necessarily equal to the entropy saving, the catalytic acceleration is also given by the *utilized part* of the intrinsic binding energy. Therefore, one may formulate this description of the catalytic activity of such an enzyme in canonical language as follows: The enzyme utilizes a part of the intrinsic binding energy of the substrate in the form of its reactive complex to cancel the entropy cost of assembling this reactive complex, and the enzymic reaction occurs at a rate corresponding to "cost-free" assembly of the reactive complex.

Let us now reconcile this description with the Fundamentalist version, summarized in Fig. 4 (b). Here, regardless of the structures which may arise between reactants and transition state, the important quantities are just ΔG_b^*, the binding energy of the transition state, and ΔG_{ES}, the binding energy of the reactant state. The former is the origin of the catalytic power, and the latter is an inhibitory contribution. The catalytic acceleration is thus given by their difference [Eq. (2)]: $\Delta G_{cat} = -\Delta G_b^* + \Delta G_{ES}$. It is easy to see, however, that ΔG_b^* is just equal to the intrinsic binding energy of the reactive complex, because the assumption in the canonical model that "the enthalpy of activation is the same for the enzymic as for the nonenzymic reaction" *implicitly requires* that all the stabilizing interactions present in the ES complex also remain fully in force in the enzymic transition state. Since these interactions liberate the intrinsic binding energy ΔG_{int} in the formation of the ES complex, so must they in forming the transition state. Furthermore, when the inhibitory effect of the enzyme on the reactant is subtracted from the transition-state binding energy, we once again find that the catalytic acceleration is given by $\Delta G_{cat} = -\Delta G_{int} + \Delta G_{ES}$, that is, by the utilized part of the intrinsic binding energy (canonical) or by the net catalytic binding energy (Fundamentalist). The "entropy-trap" aspect of the canonical model can be incorporated into the Fundamentalist model, if desired, by noting that although the interactions which generate the intrinsic binding energy occur in both substrate binding and transition-state binding, they are fully expressed only in the transition-state binding. The reason for this is that part of the intrinsic binding energy is cancelled in binding the reactants by the entropic price of assembling them from dilute solution. The enzyme is able to inhibit the reaction only by the quantity ΔG_{ES}, rather than by ΔG_{int}, because an amount $-T\Delta S_U^*$ is cancelled. In the transition state, this cancellation does not occur because the two substrates have already been assembled and no entropy price has to be paid; thus the full intrinsic binding energy is realized. Put even more simply, the transition state, which is one molecule, binds more favorably than the reactants, which are two molecules, and the difference in these free energies is just the entropy cost of assembling two molecules *and* the catalytic free energy.

No conflict is thus in evidence between the two formulations. In the Fundamentalist picture, one notes that both reactants and transition state are stabilized (at high substrate concentrations) by the enzyme, the transition state by a greater amount. Experiments might show that the excess binding energy of the transition state (over the binding energy of the substrate) matches the entropy of activation for the standard reaction and that the entropy of activation from the ES complex is zero. Transition-state structural studies might show that the substrate-derived entities are identical in standard and enzymic transition states and that the enzyme has the same structure in the transition state as in the free form. This, it could be recognized, corresponds to the canonical entropy-trap model, although the true Fundamentalist would note only that the catalysis arises merely from the interactions of enzyme and substrate-derived entity in the catalytic transition state. The invention of the hypothetical "reactive complex" and the definition of its "intrinsic binding energy" go beyond the range of observation.

The entropy-trap model of catalysis illustrates two disadvantages of the canonical language. This language gives no clear picture of what "the enzyme is doing" when the reaction occurs at low substrate concentrations, and it fails to give a simple description of how catalysis of the reverse reaction is occurring. When substrate concentrations are low compared to the K_m values for all substrates, both the enzymic reaction and the nonenzymic reaction proceed from free, unassembled reactants to their respective transition state. The fact that the enzyme can bind the assembled species as a low-concentration intermediate before the transition state is mechanistically irrelevant and cannot produce catalysis. Of course, catalysis does occur: The binding energy released for the "reactive complex" is still in force in the transition state, and the enzymic transition-state free energy is lower than that of the standard-reaction transition state by this amount. In fact, under low-concentration conditions, this *entire* intrinsic binding energy is realized as catalytic free energy [Eq. (1)]. We therefore have the unusual result that (in the canonical model) the entropy-trap enzyme is a better catalyst when it is not saving entropy than when it is!

The reverse reaction presents a similar difficulty, which has been discussed by Jencks.[4] If the enzyme functioned solely as an entropy trap, it could accelerate a bimolecular forward reaction in which there is entropy to be saved, but the corresponding unimolecular reverse reaction *produces* entropy and deprives the enzyme of an obvious role in catalysis (at least in canonical language). Jencks suggests that canonical catalytic factors which have this characteristic may be used as accelerating forces by enzymes with a biological reason for "one-way catalysis" in which (under substrate and product saturation) they favor the forward reaction more than the reverse. In the limit, the transition-state binding energy might also be liberated by binding the product (as with a simple entropy-trap enzyme), which would make the catalytic free energy zero for the reverse reaction ($\Delta G_{cat} = -\Delta G_{int} + \Delta G_{EP} = 0$). However, at product and substrate concentrations well below the saturation level, the enzyme would catalyze both forward and reverse reactions and would catalyze them equally well.

These expository disadvantages of the canonical language relative to the more austere Fundamentalist language constitute a severe disability. It is true that a careful analysis (such as that of Jencks) can always clarify the situation, but the greatest advantage of the canonical approach is presumably its simplicity and lucidity. Much of this advantage is lost if it becomes necessary to change the description of the origins of catalysis as the substrate or product concentrations are altered or as the direction of the reaction is considered.

3.2. Destabilization in the Reactant State

The entropy-trap description of enzyme catalytic power is one among several canonical formulations which hold that the function of the enzyme is to "freeze" the substrate into a reactive form which would otherwise be of negligible concentration. The liberation of intrinsic binding energy is envisioned as the force which brings about this "freezing." Since the bound substrate is held to be in a reactive form, it is by definition "unstable", and such mechanisms are frequently thought of as "reactant-state destabilization mechanisms." The term has a confusing quality, as Jencks has noted, "because the enzyme–substrate complex is generally considered to be *more* stable than the separate enzyme and substrate," under conditions where the complex is kinetically significant. The point is that if the substrate had bound in a less reactive form, the complex would be still more stable. The reactant state is thus destabilized relative to this alternative state.

We shall show that all of these formulations implicitly require a particular choice and a dynamical definition of the standard reaction, often one requiring hypothetical or impossible structures along the reaction path. It is the binding of the special intervening structures which is considered to liberate the intrinsic binding energy. In the simplest models for these cases, "no further change" occurs when the bound substrate is converted to transition state so that, just as in the entropy-trap model, the intrinsic binding energy of the intervening structure also emerges as the binding energy of the transition state. It is noteworthy that it is the intervening structures (the "reactive complexes") which are *required* in the standard reactions that are implicit in the canonical formulations, not a particular transition state—this choice remains totally arbitrary. We notice two points: first, the dynamical character of the canonical approach, to which we have alluded before, and second, that many of the canonical proposals really describe more fully the features of the standard reaction than the features of the enzymic reaction.

We can effectively illustrate the translation of the canonical concepts into Fundamentalist language, and exemplify the requirements for particular choices of standard reaction, by examining three types of reactant destabilization: geometric destabilization (including "rack" and rotamer-freezing mechanisms), electrostatic destabilization, and desolvation. We shall rely heavily for the canonical picture on the detailed and complete descriptions of Jencks.[4] All of these mechanisms (as well as all other "reactant destabilization" hypotheses) share the characteristics shown in Fig. 5 (a). The substrate (or substrates) S are imagined along the standard reaction pathway to pass through a particular unstable form S' (it is the choice of this structure, and thus the corresponding assignment of the dynamical features of the standard reaction, which in effect constitutes the canonical hypothesis about the mechanism of catalysis). The free energy expended by the conversion of S to S' is taken to be ΔG_D, the free energy of destabilization. In the simplest models, in which the enzyme does "nothing except" the destabilization, the free energy which is then required to convert S' to T_U, the standard transition state, $\Delta G'_U$, is also taken to be the free energy of activation from ES for the enzymic reaction, and the substrate-derived part of the enzyme–substrate complex is imagined to have the structure S'. The intrinsic binding energy is that released when the unstable form S' binds to the enzyme. Part of this energy (between the dashed lines) is utilized to repay the destabilization energy, while the remainder appears as net stabilization energy of the ES complex, ΔG_{ES}.

From the Fundamentalist point of view [Fig. 5 (b)] the structure S' is irrelevant so that no dynamical hypothesis about the route between S and T_U for the standard reaction is necessary. The canonical intrinsic binding energy for the simplest models is therefore again equal to the transition-state binding energy, since it is envisioned that all the interactions at work in the ES complex remain fully effective in the catalytic transition state T_C. This is required by the canonical postulation that the free energy of activation from the ES complex is equal to that from S' in the standard reaction. The catalytic acceleration is again just the net or excess transition-state binding energy, i.e., $-\Delta G_b^* + \Delta G_{ES}$, the transition-state binding energy (which is catalytic) minus the substrate binding

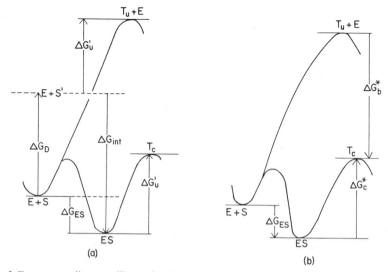

Fig. 5. Free-energy diagrams illustrating (a) the canonical and (b) the Fundamentalist versions of enzyme action involving "destabilization of the reactant state." The canonical formulation requires postulating a particular dynamical path to the standard transition state, through an unstable species S'. This unstable species, with correspondingly enhanced reactivity, is that which is bound to the enzyme in the enzyme-substrate complex, with its instability being cancelled by its intrinsic binding energy. In the Fundamentalist version, no hypothesis is necessary about intervening species along the dynamical pathway since only the transition states are relevant.

energy (which is inhibitory). This translates in the canonical language to the utilized part of the intrinsic binding energy, as it did on the entropy-trap model.

In the postulation that catalysis occurs through *geometric destabilization*, the structure S' is the geometrically destabilized form of the reactants. In the rack mechanism, for example, S' will be a reactant molecule in which a bond, to be broken in the overall reaction, has been stretched to a certain degree. The enzyme is imagined, in the act of binding S, to form interactions with its component parts such that the critical bond is pulled longer, or stretched to its length in S'. The energy liberated in the course of this combination will thus be the sum of the total interaction energy and the distortion energy required to stretch the bond. The total interaction energy is the intrinsic binding energy of S', and the algebraic sum of this and the distortion energy ΔG_D is ΔG_{ES}. The utilized (unliberated part) of the intrinsic binding energy thus is equal to and cancels the destabilization energy. The postulation of the structure S' is central, and, in fact, its postulation constitutes a definition of the intrinsic binding energy.

The Fundamentalist translation of this model, once again, merely takes note of the postulation that the enzyme "does nothing" except stabilize S'. This automatically requires that the transition-state binding energy be equal to the intrinsic binding energy, that correction of this for the inhibitory interaction with the substrate yield the utilized fraction of the intrinsic binding energy and thus the catalytic free energy. Here, too, this is because the full effect of the binding interactions which stretch the bond in the ES complex must still be felt in the transition state; but the bond itself is now broken (or breaking), and thus the energy ΔG_D is returned to the system and appears, essentially, as transition-state stabilization.

Electrostatic destabilization may be considered in terms of a specific model involving the binding of a cationic methyl donor, such as S-adenosylmethionine, near a positive charge of the enzyme structure, with the utilization of intrinsic binding energy to hold it in place:

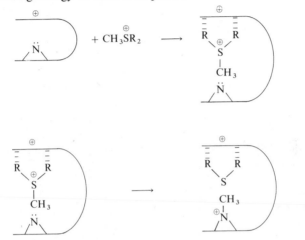

$$(25)$$

As the methyl group is donated to another substrate, the positive charge on the sulfur of the S-adenosylmethionine will move away from the charge on the enzyme and the electrostatic destabilization will be reduced. Such a sequence of events should therefore bring about catalysis of the methyl transfer. Referring again to Fig. 5 (a), we note that S′ in the standard reaction must be a species which is unstable by virtue of having a positive charge near the positive charge of the methyl donor (and which is a more reactive methyl donor for that reason). This requires the selection of the following kind of standard reaction, again complete with detailed dynamical hypotheses. S is taken to represent the methyl donor, the methyl acceptor, and a "dummy cation," all in dilute slution. S′ represents a state in which the methyl donor and the acceptor have come together and assumed the correct orientation for methyl transfer to occur *and* in which the cation has (say, by a random walk) taken a position near the sulfonium positive charge. Such a dynamic picture now allows us to imagine that the enzyme binds this array, ejecting the foreign cation and placing a cationic center of its own structure in just the same location. The usual intrinsic binding energy of the reactive complex is in this case augmented by a term for release of the dummy cation into dilute solution. A term of exactly the opposite magnitude is present in the destabilization energy (corresponding to collecting the cation out of solution in assembling S′) and will cancel the contribution to the binding energy. The enzyme functions to "freeze" a state S′ which is destabilized by two factors in this example: (1) the electrostatic repulsion between the methyl donor and the dummy cation and (2) the unfavorable entropic price of bringing the reactant species together from solution. Both factors contribute to ΔG_D, and both will be cancelled by the utilized fraction of ΔG_{int}. The enzyme thus functions simultaneously "to destabilize the reactant state electrostatically" and as an "entropy trap."

The Fundamentalist model has no requirement for an extensive dynamical hypothesis about the standard reaction. To define the catalytic potential of the enzyme, the energy released by the formation of the bonds holding the methyl donor and the methyl acceptor in their positions in the enzymic transition state for methyl transfer is acknowledged. A part of this may be cancelled by a weak electrostatic repulsion from the two positive charges. The net energy liberated from binding the standard-reaction transition state (the bond-formation energy minus that for the weak repulsion) is the empirical binding energy of the transition state. It will be completely recovered as catalytic free energy when the enzyme is not saturated. When enzyme saturation occurs, the inhibitory effect of binding the substrates must be taken into account. In the simplest model, the bonds formed between enzyme, donor, and acceptor will be equally strong in the enzyme–substrate complex, the enzyme–product complex, and the transition state. The electrostatic repulsion will be greatest in the enzyme–substrate complex, making the enzyme only a weak inhibitor of the forward reaction because the cancellation of the binding energy by electrostatic repulsion will make ΔG_{ES} small. The lesser cancellation in the transition state will permit the enzyme to be a catalyst. The still smaller degree of cancellation at the product stage will make the product the most strongly bound species along the reaction path.

Such an enzyme will be an equally good catalyst for methyl transfer in forward and reverse directions when the enzyme is not saturated with either reactant or product. When saturated with both, it will strongly catalyze the forward reaction but not the reverse reaction. Fundamentally, the enzyme is an inhibitor of the reverse reaction, but it will effectively inhibit the reaction only if its concentration approaches that of the products, which is not the usual situation.

A final example is provided by "catalysis through *desolvation of reactants*." Here we can imagine that S in Fig. 5 consists of two reactants which will undergo bond formation in the standard reaction, one acting as nucleophile and the other as electrophile. To formulate the canonical version of desolvation catalysis, we imagine that S′ [Fig. 5 (a)] is a structure in which both reactants have entered the same region of solution and by a random walk have passed into a configuration such that all solvent is absent between the nucleophilic and electrophilic centers. This species S′ is unstable for two reasons: the unfavorable entropy of assembling its components from free solution and the unfavorable enthalpic price of removing the solvating molecules from the reactive centers. The sum of these destabilization energies gives ΔG_D. It is now envisioned that the enzyme binds this reactive complex (of course, the desolvation of its nucleophilic and electrophilic components has increased their reactivity so that the free energy of activation for their combination is only $\Delta G'_U$) with interactions which produce the intrinsic binding energy ΔG_{int}. Part of this energy may be utilized to cancel the destabilization energy, with the remainder appearing as ΔG_{ES}. If the enzyme now "does nothing else," the free energy of activation starting from the enzyme–substrate complex will be only $\Delta G'_U$ and catalysis will have been achieved.

It is a familiar task at this point to formulate the Fundamentalist translation. Since the stabilizing interactions of the intrinsic binding energy are automatically assumed to carry on into the transition state and since the destabilizing and countervailing effects of the unstable, desolvated electrophilic and nucleophilic centers are removed by partial bond formation between them, it is once again apparent that the catalytic free energy is just given by the binding energy of the transition state and that this is equal to the utilized part of the intrinsic binding energy of the reactive complex employed in the canonical model. One can further see that, in the Fundamentalist model, it is unnecessary to make any postulation about the dynamical character of the standard reaction. The catalysis arises straightforwardly from the free energy released by binding a (largely unsolvated) transition state into a nonsolvating region of the enzyme structure. Whether the standard-reaction reactants passed through a desolvated state in attaining the transition state is irrelevant and need not be considered.

A feature which is common to all these canonical mechanisms is thus the postulation of some unstable species along a particular dynamical pathway of the standard reaction; this species is postulated to be bound by the enzyme in the enzyme–substrate complex, with the intrinsic binding energy being partly (at substrate concentrations above K_m) or wholly (at substrate concentrations below K_m) utilized to cancel the destabilization energy. This allows the species to exist in higher concentration and yet preserve the high reactivity which then appears in the enzymic reaction. Jencks[4] has outlined a variety of types of experimental evidence which can be used to infer, with greater or lesser degrees of reliability, that such mechanisms are actually at work. The problem of making this decision usually resolves into the question of whether an observed, weak binding ought to be regarded as the resultant of two strong and opposing forces or should be considered as "intrinsically weak." Whether such a decision is, even in principle, possible is sometimes questionable.

One approach is to study the enzyme–substrate complex structurally for evidence of the postulated form of destabilization: For example, are two positive charges actually near each other in the complex? An affirmative answer to such a question might be regarded as reasonably meaningful. However, it might be very hard in the end to show that a particular force is to be associated with a particular distance between the charges in the complex environment of the active site. A negative answer, as Fersht[14] has argued, is in any case less meaningful, because even if structural distortion or destabilization is unobservable, it still may be true that forces *tending* to distort (but not succeeding) are at work. Thus the enzyme may function to induce *stress*, which does not result in *strain*—actual movement of atoms. This point, for which quantitative support is cited, emphasizes still further that the character of canonical hypotheses involving distortion and destabilization is commonly idealistic: They depend on intellectual constructs such as "reactive complexes" and countervailing forces which are not directly observable in nature.

Another approach is the "method of bits and pieces," in which one attempts to test for destabilization by omitting some feature of the reactant thought to be involved in destabilizing interactions in the enzyme–substrate complex. If the destabilization hypothesis is true, the omission of the destabilizing feature should permit the intrinsic binding energy to be expressed rather than utilized, so that the altered substrate should be more tightly bound by an amount which measures the destabilization energy. The problem is to show that this interpretation is the correct one when the altered substrate is in fact more tightly bound. Suppose that a positive center is omitted from a potential methyl donor, for which "electrostatic destabilization" is proposed, and that this altered species is more tightly bound by the enzyme by, say, 5 kcal/mol. Can we say with confidence that this observation shows electrostatic destabilization of the unaltered substrate? In view of the structural complexity involved in the substrate, the enzyme, and the enzyme–substrate complex for most systems, the qualitative conclusion (ascription of 5 kcal/mol to the role of "electrostatic destabilization") would constitute a considerable intuitive leap. Nevertheless, this approach and the theoretical and experimental study directly of the interactions present in enzyme–substrate complexes are clearly of the greatest importance for a final understanding of the true nature of enzyme catalytic and inhibitory power.

The real point of linguistic departure between canonical and Fundamentalist workers in the interpretation of these studies will be that while canonical students refer to interactions in the enzyme–substrate complex as "catalytic" through "reactant-state destabilization", these interactions will be considered by Fundamentalists as expressions of enzymic *inhibition*. The "destabilizing" processes simply represent features destructive of the inhibitory power of the enzyme; at saturation, of course, such features permit fuller expression of the catalytic power of the enzyme (arising wholly from transition-state stabilization).

3.3. Other Sources of Enzyme Catalytic Power

The examples discussed in the last two sections cover cases in which the greatest divergence appears to arise between the canonical and Fundamentalist formulations. In many other cases, there is little divergence. For example, so-called "chemical" catalysis, in which active-site functional groups accelerate the reaction by mechanisms familiar from the chemistry of small molecules (such as general acid–base catalysis), tends to be considered a form of transition-state stabilization in canonical as well as Fundamentalist approaches.

One item of canonical terminology which has considerable potential for confusion is the term "transition-state destabilization," as used, for example, by Jencks.[4] Here again, the appropriate question for the puzzled Fundamentalist is, destabilization compared to what? The enzyme cannot, needless to say, be destabilizing the transition state relative to the standard-reaction transition state. Instead, the destabilization referred to canonically is relative to a purely hypothetical *enzymic* transition state in which the "destabilizing" interactions of interest are absent. Both the real enzymic transition state and the hypothetical enzymic transition state of reference are very much *stabilized* relative to the standard-reaction transition state. The destabilization is a small component countervailing against an overwhelmingly larger stabilization, for most common choices of standard reaction. An illustration of this canonical concept arises in the discussion of catalysis by electrostatic destabilization, as presented in the last section. There it was noted that the repulsive interaction between the positive methyl donor and the enzyme positive charge is greatest in the enzyme–substrate complex, less in the transition state, and least in the enzyme–product complex. In canonical terms, one can say that "the reactant state is most destabilized, the transition state is less destabilized, and the product state is least destabilized by the electrostatic effect." The points of reference in each case are hypothetical species in which one or the other positive charge is absent, and thus the energy is lower by the amount of electrostatic interaction in that particular state.

3.4. The Anatomy of Chymotrypsin Catalysis

Finally, it should form a useful exercise to consider a typical complete canonical account of enzyme catalysis and to relate it to its Fundamentalist equivalent. Canonical accounts are commonly lists of contributing factors to the catalytic power of the enzyme. The simplest Fundamentalist translation of such a list usually shows it to amount to a continual redefinition of the standard reaction, until a (nearly always hypothetical) standard reaction is reached for which the empirical binding energy of the transition state is estimated to be zero. It is then concluded, in effect, that this transition state (possessing the same free energy as the enzymic transition state) also possesses the same structure.

The first and most famous canonical account of this type is the "anatomy" of chymotrypsin catalysis of Bender *et al.*[15] They constructed a set of factors, recounted in Table II, which related the hydroxide-promoted hydrolysis of *N*-acetyltryptophanamide to the hydrolysis of this substrate, catalyzed by chymotrypsin. Although this work was done some time ago and various alterations might now be made in the list, it remains remarkably in accord with many of the current concepts of catalysis by this enzyme and will serve nicely for this illustration.

In canonical language, the interpretation of this set of catalytic contributions proceeds as follows. The enzymic reaction is $10^{2.2}$ times faster than the

Table II

Canonical List of Factors Contributing to the Catalytic Power of α-Chymotrypsin
in the Hydrolysis of N-Acetyltryptophanamide[15]

Factor no.	Origin	Magnitude	Rate constant
—	(Estimated rate constant for alkaline hydrolysis)		$3 \times 10^{-4}\,M^{-1}\,\mathrm{sec}^{-1}$
1	Conversion to imidazole catalysis	1.6×10^{-6}	$4.8 \times 10^{-10}\,M^{-1}\,\mathrm{sec}^{-1}$
2	Intramolecularity	$10\,M$	$4.8 \times 10^{-9}\,\mathrm{sec}^{-1}$
3	Alcoholysis, rather than hydrolysis	10^2	$4.8 \times 10^{-7}\,\mathrm{sec}^{-1}$
4	"Freezing" from substrate specificity	10^3	$4.8 \times 10^{-4}\,\mathrm{sec}^{-1}$
5	General acid catalysis	10^2	$4.8 \times 10^{-2}\,\mathrm{sec}^{-1}$
—	(Observed rate constant k_{ES})		$4.4 \times 10^{-2}\,\mathrm{sec}^{-1}$

hydroxide-promoted hydrolysis. It would, however, be better to compare the enzymic reaction to imidazole catalysis (since the active-site catalyst is an imidazole side chain of histidine), a reaction estimated to be slower by $10^{5.8}$-fold. The catalytic acceleration relative to this standard is thus $10^{5.8+2.2} = 10^{8.0}$. The first contribution to enzymic acceleration is taken as entropy saving from the intramolecular character of the enzymic reaction, estimated at 10-fold in rate. Second, the active-site nucleophile is a serine side-chain alcohol function, estimated to react 10^2 times faster than the water nucleophile of the standard reaction. Next, it is estimated that "freezing" of the substrate into a proper transition-state orientation through enzyme interaction with the tryptophan ring produces another entropy saving (and perhaps some enthalpic improvement) that corresponds to a rate factor of 10^3. Last, general acid catalysis by an active-site function is estimated to provide another rate factor of 10^2. Thus the total acceleration of 10^8 is estimated to arise from (1) entropy saving, about $10^{1+3} = 10^4$; (2) enhanced nucleophilic reactivity, 10^2; and (3) electrophilic (general acid) catalysis, 10^2.

The Fundamentalist version of this analysis begins by noting that if hydroxide-promoted hydrolysis is adopted as the standard reaction, the empirical free energy of transition-state binding is $-RT\ln(10^{2.2}) = -3.1$ kcal/mol $(= \Delta G_b^*)$. The next step taken in the canonical analysis was an explicit redefinition of the standard reaction to the imidazole-catalyzed hydrolysis. The barrier for this reaction is higher than that for hydroxide promotion by 8.1 kcal/mol, corresponding to the rate factor $10^{5.8}$. The value of ΔG_b^* for this new standard reaction is thus $-3.1 - 8.1 = -11.2$ kcal/mol. Now, the consideration of the entropy saving tenfold in rate from intramolecularity represents an *implicit* redefinition of the standard reaction to a hypothetical intramolecular reaction, with a barrier lower by 1.4 kcal/mol. Thus ΔG_b^* now becomes $-11.2 + 1.4 = -9.8$ kcal/mol. Further, the effect of active-site alcohol nucleophilicity requires redefinition of the standard reaction to an alcoholysis, with a barrier lower by 2.8 kcal/mol, reducing ΔG_b^* to -7.0 kcal/mol. The "freezing" of the substrate by interaction with enzyme is accounted for by changing to a new standard reaction of "frozen" substrate, further reducing ΔG_b^* to -2.8 kcal/mol. The last step is a change to a standard reaction which incorporates acid catalysis, with a barrier lower by 2.8 kcal/mol. This results in $\Delta G_b^* = 0$, showing that the reaction of such a transition state with enzyme to incorporate it into the active site, with appropriate "exchange" of such functional groups as the enzyme already contains in the active site, would not require or produce any free energy. The conclusion is thus that the structure of the standard-reaction transition state closely approximates that of the enzymic transition state and that the energetic characters of the two are similar. The total binding energy of -11.2 kcal/mol for the original imidazole-catalyzed standard-reaction transition state is postulated to arise from the following sources: (1) Liberation of an imidazole molecule from the standard transition state and replacement by an active-site imidazole as the transition state binds to the enzyme produce -1.4 kcal/mol; (2) liberation of a water molecule and its replacement by an active-site alcohol function generate a stronger transition-state partial bond to this nucleophilic atom, yielding -2.8 kcal/mol; (3) interaction of the tryptophan ring with enzymic components (logically envisioned as a hydrophobic interaction) produces -4.2 kcal/mol; and (4) participation of an acidic group of the active site in C—N bond breaking forms a transition-state partial bond with the liberation of -2.8 kcal/mol.

4. TRANSITION-STATE STABILIZATION RELATED TO ENZYME KINETICS

4.1. Single Rate-Determining Step

When a single step in an enzyme catalytic sequence has a transition state for which the free energy greatly exceeds that of all others along the reaction path, the relationship of kinetic parameters to reactant-state and transition-state free energies is straightforward. The relationships will be illustrated for a few examples, from which the procedure should be sufficiently apparent to permit its application to any other case. Every circumstance in which a single step determines the rate is characterized by a rate (and a rate constant) which relates two free energies: that for a transition state and that for an initial reference state. All experimental probes of such a system then reflect perturbations of these two free energies and thus give information on the particular transition state and the particular initial reference state important under the experimental conditions.

Consider the mechanism

$$E + S \underset{k_{-1}}{\overset{k_1}{\rightleftharpoons}} ES \underset{k_{-2}}{\overset{k_2}{\rightleftharpoons}} EP \overset{k_3}{\longrightarrow} E + P \qquad (26)$$

for which it is easily shown that the kinetic parameters are given by

$$k_{ES} = k_2 k_3 / (k_{-2} + k_2 + k_3) \qquad (27)$$

$$k_E = k_1 k_2 k_3 / [k_3(k_{-1} + k_2) + k_{-1}k_{-2}] \tag{28}$$

We use the following notation: k_{ES} for the first-order rate constant for reaction of the enzyme–substrate complex; $k_E = k_{ES}/K_m$ is the substrate concentration at half-maximal velocity so that k_E is the second-order (or higher-order) rate constant for reaction of enzyme with substrate(s) at low substrate concentration; G_i^*, the free energy of the transition state of the step with rate constants k_i, k_{-i}. Confronted by the complexities of Eqs. (27) and (28), one is tempted to despair of relating the experimental parameters k_E and k_{ES} to the free energies of particular transition states and thus of relating experimental perturbation of these rate constants to the properties of particular transition states. Indeed the situation in the general case is quite complex. However, let us first consider cases in which a single step determines the rate, which will produce some simplification. We refer to the k_1 step of Eq. (26) as "association," to the k_2 step for the interconversion of central complexes as the "chemical step," and to the final, k_3 step as the "product-release" step.

First, we consider *rapid, reversible association with the rate-determining chemical step*. The kinetic relationships which describe such a model are $k_{-1} \gg k_2$, $k_3 \gg k_{-2}$, and $k_3 \gg k_2$. Application of these conditions to Eqs. (27) and (28) yield

$$k_{ES} = k_2 \tag{29}$$

$$k_E = k_1 k_2 / k_{-1} \tag{30}$$

which can be re-expressed in Eyring form as

$$k_{ES} = (\mathbf{k}T/h) \exp[-(G_2^* - G_{ES})/RT] \tag{31}$$

$$k_E = (\mathbf{k}T/h) \exp[-(G_2^* - G_S - G_E)/RT] \tag{32}$$

Note that in arriving at the latter, the free energies of the enzyme–substrate complex and of the transition state of the association step cancel from k_E. It is thus clear that experiments on both k_E and k_{ES} will give information on the transition state of the chemical step. The initial reference state for experiments on k_E is free enzyme and free substrate, while that for experiments on k_{ES} is the enzyme–substrate complex. Experiments are thus to be interpreted as differential perturbations of G_2^* relative to $G_E + G_S$ for k_E and relative to G_{ES} for k_{ES}.

Now we take up the example of a *rapid, reversible association process*, followed by an *irreversible chemical step*, followed by a *rate-determining product-release* step. This is an appropriate model for those practical cases in which the first central complex is rapidly converted to the second with the expulsion of a first product. This will be lost in solution during the initial-rate period, making the chemical step effectively irreversible. The kinetic conditions are $k_{-1} \gg k_2$, $k_2 \gg k_3$, and $k_3 \gg k_{-2}$. Insertion into Eqs. (27) and (28), followed by conversion to Eyring form, produces

$$k_{ES} = k_3 = (\mathbf{k}T/h) \exp[-(G_3^* - G_{EP})/RT] \tag{33}$$

$$k_E = k_1 k_2/k_{-1} = (\mathbf{k}Th)\exp[-(G_2^* - G_E - G_S)/RT] \qquad (34)$$

For this situation, experiments on k_{ES} will yield information on the transition state for product release, referred to the enzyme–product complex as the initial reference state. In contrast, k_E (which as a general matter always corresponds to the overall rate constant for passage through the first irreversible step) provides a probe of the transition state of the chemical step, referred (as is quite generally true of k_E) to the free enzyme and free substrate as initial reference state.

Next, let us consider two cases of *rate-determining product release with reversible interconversion of central complexes.* The two situations differ in the initial reference state at saturation, which depends on the site of the equilibrium connecting the central complexes; the equilibrium may favor ES (endergonic chemical step) or EP (exergonic chemical step).

First we examine *rate-determining product release with reversible association and reversible, exergonic chemical step.* Such a model $(k_{-1} \gg k_2, k_2 \gg k_{-2}, k_{-2} \gg k_3)$ makes the first irreversible step also the rate-determining step so that both kinetic parameters now refer to the same transition state:

$$k_{ES} = k_3 = (\mathbf{k}T/h)\exp[-(G_3^* - G_{EP})/RT] \qquad (35)$$

$$k_E = (k_1 k_2/k_{-1}k_{-2})k_3 = (\mathbf{k}T/h)\exp[-(G_3^* - G_E - G_S)/RT] \qquad (36)$$

Only the initial reference states differ: EP (because the enzyme accumulates in this state at saturation) for k_{ES} and (as always) free enzyme and free substrate for k_E.

Second, we go over to *rate-determining product release with reversible association and reversible, endergonic chemical step.* This model $(k_{-1} \gg k_2, k_{-2} \gg k_2, k_{-2} \gg k_3)$, compared with the last one, illustrates how a shift in the equilibrium of a fast step can change the initial reference state at saturation:

$$k_{ES} = k_2 k_3/k_{-2} = (\mathbf{k}T/h)\exp[-(G_3^* - G_{ES})/RT] \qquad (37)$$

$$k_E = (k_1 k_2/k_{-1}k_{-2})k_3 = (\mathbf{k}T/h)\exp[-(G_3^* - G_E - G_S)/RT] \qquad (38)$$

Here k_E refers to the same two states as before, but because the enzyme accumulates at saturation as ES, this becomes the initial reference state for k_{ES}.

These examples serve to emphasize two points:

1. For all simple kinetic situations, the two parameters of enzyme kinetics, k_E and k_{ES}, relate to the free-energy difference between two states, a transition state and an initial reference state. For k_E, the initial reference state is always the free enzyme and the free substrate(s); for k_{ES}, it is the major form in which the enzyme accumulates at saturation (which might be ES, another complex, EP, etc.). For k_E, the effective transition state is that for the first irreversible step in the reaction sequence. For k_{ES}, the effective transition state is that of highest free energy following the state in which saturated enzyme accumulates (i.e., the transition state of the "overall rate-determining step").

2. For just this reason, the interpretation of these parameters can always

(for such simple situations) be carried out in accord with *kinetic* principles (merely treating the two states in question), and it is never necessary to make a *dynamic* analysis (treating step 1, then step 2, etc.) in which one must carefully consider the characteristics of many species for which the free energy cancels out of the final result.

We now proceed to the consideration of more complex kinetic situations, in the context of the experimental approaches to transition-state structure. As we shall see, these also involve two effective states for each kinetic parameter, but each state may be a weighted average.

4.2. Kinetic Perturbations in the Determination of Transition-State Structures

Essentially all kinetically based probes of reaction mechanism, regardless of the language which may be used, are in effect interpreted by considering how some perturbing variable (concentration, temperature, pressure, substrate structure, isotopic composition) affects the free energy of activation. Because a fair idea of the likely nature of the reactant-state free-energy response is commonly in hand, the interpretation centers on the derived response of the transition-state free energy and the implications of such a response for the structure of the transition state. If the solvent is made more polar and the transition-state free energy radically drops, it is assumed that the transition state must itself be polar, for example. From such experiments, a model of the transition state emerges. The Fundamentalist interpretation of enzyme catalysis derives from exactly these models for standard and enzymic reactions. As we have seen in the previous section, one may easily expect to derive a model from kinetically based experiments which is close in reality to the structure of a single real transition state when a single real transition state along the reaction path determines the rate. But now let us consider the more complex situation when more than a single transition state determines the rate.

We first take as a practical example the circumstance in which k_E is studied for a multistep enzymic reaction. A reasonable mechanistic scheme is

$$E + S \underset{k_{-1}}{\overset{k_1}{\rightleftharpoons}} ES \underset{k_{-2}}{\overset{k_2}{\rightleftharpoons}} EX \underset{k_{-3}}{\overset{k_3}{\rightleftharpoons}} EP \overset{k_4}{\longrightarrow} E + P \qquad (39)$$

Let us consider the last (k_4) step to be effectively irreversible and the initial association to be reversible ($k_{-1} \gg k_2$), but imagine the transition states of of the k_2, k_3, and k_4 steps all to have similar free energies. Thus the limitation of the rate is divided among these steps. Let us ask what, if anything, is learned by experiments on k_E which attempt to delineate the structure of the transition state of the enzymic reaction?

First we obtain, by the usual methods, the steady-state rate constant for k_E:

$$k_E = (k_1 k_2 k_3 k_4 / k_{-1}) / [k_{-2}(k_4 + k_{-3}) + k_3 k_4] \qquad (40)$$

It is informative to invert k_E and to examine its reciprocal,

$$(k_E)^{-1} = (k_{-1}/k_1 k_2) + (k_{-1}k_{-2}/k_1 k_2 k_3) + (k_{-1}k_{-2}k_{-3}/k_1 k_2 k_3 k_4) \quad (41)$$

This shows that k_E (and indeed any rate constant for passage over a series of barriers separating steady-state intermediates) is the harmonic mean of the rate constants for net passage over the contributing barriers. In each of these terms, corresponding to an "overall barrier" along the reaction path (i.e., a product of equilibrium constants for all steps leading up to the barrier, multiplied by a rate constant for traversing the barrier), the free energies of all species between the initial state and the transition state for that barrier will cancel. We can thus cast each of the terms in a simple Eyring form and relate k_E to an apparent free energy of activation ΔG^*_{app} to obtain

$$(h/kT)\exp[+(G^*_{app} - G_E - G_S)/RT] = (h/kT)\exp[+(G^*_2 - G_E - G_S)/RT]$$
$$+ (h/kT)\exp[+(G^*_3 - G_E - G_S)/RT] + (h/kT)\exp[+(G^*_4 - G_E - G_S)/RT] \quad (42)$$

Since $\Delta G^*_{app} = G^*_{app} - G_E - G_S$, the preexponential factors and the terms in G_E and G_S will cancel to provide

$$\exp(G^*_{app}/RT) = \exp(G^*_2/RT) + \exp(G^*_3/RT) + \exp(G^*_4/RT) \quad (43)$$

and thus

$$G^*_{app} = RT \ln(e^{G^*_2/RT} + e^{G^*_3/RT} + eG^*_4/RT) \quad (44)$$

for the apparent free energy of transition state. This apparent free energy is thus a weighted average of the free energies of the various transition states which contribute to determining the rate, with the highest free-energy transition state, as expected, contributing most heavily to the average.

Equation (44) shows us how the apparent free energy of the transition state for a complex process with more than one rate-determining step is related to the free energies of the actual transition states in series along the reaction path. Clearly it is this average free energy which will represent the effect of the enzyme in binding the standard-reaction transition state. The standard-reaction transition state, upon enzyme binding in such a complex process, is in effect converted to an average structure of all the contributing enzymic transition states. We shall return to this point below, but here let us continue from Eq. (44) to a development of the nature of this average transition state. Our conception of the enzymic transition state will be based on the results of experiments designed to find its structure. Thus we must relate the free-energy perturbations observed in such experiments (which will lead to an apparent structure for the enzymic transition state) to the perturbations for each of the contributing transition states in the reaction-path series.

Most perturbations of interest for application in enzymology are small perturbations, because large perturbations always contain the danger of changing the nature of the transition states as well as of reflecting their properties, a danger which is especially notable with the notoriously sensitive enzymic systems. It should therefore be a good initial approximation to treat such perturbations as derivatives. That allows us to discuss the result of a typical experiment of k_E in terms of dG^*_{app}/dx, the apparent perturbation in the free energy of the transition state produced by a change in the experimental variable x, which might be temperature, isotopic composition of the solvent or substrate, etc. Differentiation of Eq. (44) produces

$$dG^*_{app}/dx = (e^{G^*_2/RT}/e^{G^*_{app}/RT})(dG^*_2/dx) + (e^{G^*_3/RT}/e^{G^*_{app}/RT})(dG^*_3/dx)$$
$$+ (e^{G^*_4/RT}/e^{G^*_{app}/RT})(dG^*_4/dx) \qquad (45)$$

This shows that just as the apparent free energy of the transition state is a weighted average so is the perturbation. In fact, we can replace each weighting factor $(e^{G^*_i/RT}/e^{G^*_{app}/RT})$ in Eq. (45) by the symbol w_i to generate

$$dG^*_{app}/dx = w_2(dG^*_2/dx) + w_3(dG^*_3/dx) + w_4/(dG^*_4/dx) \qquad (46)$$

Each of the weighting factors w_i has a simple physical significance. Notice that the sum of the w_i is unity and that each is a positive number less than 1, as befits proper weighting factors in an average. For a given transition state, as G^*_i approaches G^*_{app}, w_i approaches 1 and all other w's approach 0. This corresponds to the limit of a single rate-determining step. If all the G^*_i are equal (all transition states participate equally in determining the rate and the results of mechanistic experiments), then all the w_i will be equal (in this example, all will be $\frac{1}{3}$). Equation (46) thus teaches us that the result of a mechanistic experiment in a complex situation in which more than one transition state is important is such that a weighted average of the expected responses for the various contributing transition states (the dG^*_i/dx) will be obtained and that the weighting factors are simple exponential functions of the free-energy difference separating each individual transition-state free energy from the apparent transition-state free energy.

Thus if one determines a secondary deuterium isotope effect for an enzymic reaction of this complex type, a weighted-average effect will be obtained in accord with Eq. (46). If one continues with further mechanistic experiments, weighted-average results will also be obtained for them. If one is unaware that more than one step determines the rate (and this unawareness is of course the common situation), then one will derive from the collection of a group of mechanistic experiments an *apparent structure* for the transition state. This construct has interesting properties, as we shall see below.

The treatment just completed shows how the weighted-average result of a mechanistic experiment relates to the responses of contributing transition states when these lie along a serial reaction pathway (that is, one after the other). Another common situation is one in which transition states along *parallel* or branched paths combine to determine the observed reaction rate. Thus the reaction occurs simultaneously by two mechanisms (or more, in the general case). It is easy to show, by writing down the appropriate kinetic law and following the same algebraic protocol just used, that the corresponding expression for the weighted-average result in this case is given by

$$dG^*_{app}/dx = (e^{-G^*_1/RT}/e^{-G^*_{app}/RT})(dG^*_1/dx) + (e^{-G^*_2/RT}/e^{-G^*_{app}/RT})(dG^*_2/dx) + \cdots$$

$$(47)$$

This is exactly like Eq. (45) except that the signs of the free energies in the exponentials of the weighting factors have been reversed, because G^*_{app} for several transition states in series is larger than any of the contributing G^*_i (because the net passage over the collection of serial barriers is slower than it would be over any one of them if the other barriers were absent), while G^*_{app} for several transition states in parallel is smaller than any of the contributing G^*_i (because net passage over the collection of parallel barriers is faster than it would be over any one if all the others were shut off). Thus here the *lowest* free-energy transition state is the greatest contributor to the weighted-average result, while with serial transition states it is the highest free-energy transition state which most nearly determines the rate and is the largest contributor. This sort of difference for structures which lie in series along a reaction path as compared to those which lie along a coordinate perpendicular to the reaction path is a general feature of reaction kinetics and is explored by Thornton and Thornton in Chapter 1 in this volume.

4.3. Virtual Transition-State Structures

The burden of our consideration in the last two sections is that if a single transition state lies high in free energy above the others relevant to a given enzyme-kinetic parameter (k_E or k_{ES}), then experiments on the appropriate parameter should, in a straightforward way, produce information leading to a structure for that transition state. The fact that other states are encountered between initial and transition states, and that the observed kinetic parameter may be described by a complex collection of individual rate constants, does not constitute a complication. On the other hand, we saw that if this situation does not hold but if instead the rate is determined by a number of contributing transition states in series or in parallel, then one arrives not at the structure of any one of these contributing transition states but rather at a weighted-average structure. We shall now explore the concept of this weighted-average structure, which may be denoted a *virtual transition-state structure*, and its properties.

It is an amusing characteristic of these virtual transition-state structures that they have a certain antic relationship to resonance hybrids. Resonance hybrid structures, it will be recalled, are descriptions of real molecules for which it is regarded as inconvenient to attempt to draw a single structural formula. Instead one draws structural formulas for a series of imaginary molecules, and the structure of the real ("hybrid") molecule is represented as a weighted average of the structures of the "contributors" to the hybrid. If the energy of the contributors is estimated by some means, then one can specify the weighting factors, for the lowest-energy contributor is taken to contribute most to the hybrid structure and the highest-energy contributor the least. Now a virtual transition-state structure describes an *imaginary* species, for which the contributing structures are those of the *real* transition states. The weighting factors depend on whether the contributors lie along a single, serial pathway or along parallel, competitive pathways. In the latter (parallel) case, the lowest-energy real contributor contributes most to the imaginary hybrid. In the serial case, the highest-energy real transition state contributes most heavily to the average structure of the hybrid, virtual transition state.

The virtual transition-state structure has a certain utility, however. It is a summary of the properties which actually determine the rate of the reaction for which it is derived. That is, it is the structure which if it *were* actually the transition state of some reaction, would have generated the quantities (dG^*/dx) in response to the mechanistic experiments upon which its structure was based. More importantly, if it is a correctly derived structure, it should have correct predictive power for other, as yet unmeasured mechanistic responses. Thus if a series of isotope effects were used to establish the transition-state structure for a particular reaction, one ought to be able to employ that structure in predicting how the reaction rate would respond to other variables such as temperature, pressure, changes in reactant or solvent structure, etc. If several transition states combine to determine the rate, it is then the virtual or weighted-average structure which is in fact required for such a prediction, since the response to any of these variables will be a weighted-average response from all of the contributing transition states.

It may very well be, then, that in the future when many transition-state structures are understood and their use in predicting kinetic responses is well advanced the common technique will be to construct deliberately from known, real individual transition-state structures an appropriate virtual structure for use in rate prediction for a given reaction under certain conditions. In the meantime, however, the problem is likely to be seen from the opposite perspective, and the difficulty will be considered to lie in recognizing when one is dealing with a virtual transition state and when one has a "pure" transition state in hand. One will also desire techniques for dissecting the contributing structures out of the virtual hybrid and establishing the real, individual transition-state structures.

The problem is far more difficult than is at first apparent. One of the few approaches likely to be at all successful is that developed by Northrop, which depends on a comparison of the isotope effects generated by three different isotopes of the same element (in practice, so far, protium, deuterium, and tritium, although there is no in-principle reason that an equivalent approach with ^{12}C, ^{13}C, and ^{14}C could not be used for carbon). Because the potential surfaces for all three isotopic molecules should be the same, a simple numerical relation has been shown both theoretically and experimentally to relate such effects. Violations of this relationship signal a virtual transition state, and adherence to the relationship indicates a "pure" transition state. This method and its applications are discussed in the chapters on transition-state structural determinations by isotope-effect methods elsewhere in this volume.

Techniques which do not depend on isotopic substitution have severe limitations for detecting virtual transition states. For example, consider this approach. One alters a substituent in the leaving group of the substrate for the acylation of an enzyme such as a serine hydrolase and finds a small but definite substituent effect. Is this a true reflection of a slightly broken leaving-group bond in a single, real transition state? Or does it arise from a weighted average of a conformation-change transition state (no substituent effect) and another, "chemical" transition state with a thoroughly broken leaving-group bond (large substituent effect), the weighting factors such as to yield a small average substituent effect? A tempting experiment is to alter another part of the substrate—say, the acyl portion—and again to determine the substituent effect. The hope is, by this stratagem, to alter the weighting factors (i.e., the relative free energies of the two contributing transition states) and thus to change the observed substituent effect. Imagine that this is realized: One now finds a reasonably large substituent effect. The conclusion that the previously observed effect arose from a virtual transition state, while reasonable enough, is not warranted: Alteration of the acyl portion of the substrate might have changed the structure of a single, "pure" transition state from one having only a slightly broken leaving-group bond to one having a much more completely broken leaving-group bond. The result is thus consistent with both hypotheses, virtual or "pure" transition state.

As our experience of transition-state chemistry grows, and particularly as the capacity of chemical theory for investigating those aspects of transition-state structure and energy which are refractory to experiment increases, we may be able to develop new, reliable techniques for distinguishing virtual and real transition-state structures. At the moment, our main source of comfort must be that it is the virtual structures which are both directly derived from experiment and which are in the end desired for a true understanding of the system.

5. SUMMARY

The traditional and customary language of discussions of enzyme catalytic power, here called "canonical," is equivalent to a formulation in which the sole origin of enzyme catalytic power is taken to be stabilization of the transition state, with reactant-state interactions considered expressions of enzyme inhibitory power, a view denoted "Fundamentalist." The choice of an "uncatalyzed" reaction, or standard reaction, with which the enzymic reaction is compared is quite arbitrary, in analogy with the choice of thermodynamic standard states. It is the transition state of the standard reaction which is stabilized by reaction with the enzyme to form the enzymic transition state, and the binding constant for this process is equal to the catalytic acceleration factor. The latter thus differs for different choices of standard reaction. The free energy liberated by the binding of the standard reaction transition state to the enzyme is calculable from experimental data, being the difference between the free energies of activation for the catalyzed and uncatalyzed reactions, and is thus called the *empirical free energy of binding of the transition state*. This empirical binding energy can usefully be separated conceptually into two additive terms: the *free energy of distortion* for distorting the free enzyme and the standard-reaction transition state into the forms they will have in the enzymic transition state ("poised structures") and the *vertical free energy of binding* for bringing the poised structures together from their standard-state distributions and forming the enzymic transition state by their interaction. This suggests a program of investigation of enzymic catalytic power involving (1) determination of standard-reaction and enzymic transition-state structures, (2) investigation of the energy requirements for distortion to the poised structures, and (3) investigation of the free-energy changes associated with combination of the poised structures.

The canonical formulations of catalytic power in which the enzyme is said "to act as an entropy trap" or "to destabilize the reactant state" can readily be re-expressed in Fundamentalist language, whence it emerges that for the simplest models the "intrinsic binding energy" of the canonical pictures refers to the binding energy of the transition state. The canonical models require a more detailed postulation of the standard reaction than the Fundamentalist versions, because the canonical models feature a "reactive complex" along the dynamical pathway connecting the reactant and transition states. The Fundamentalist translation drops the dynamical formulation and concentrates on kinetic characteristics, those features of reactant state and transition state alone which are susceptible to experimental study by kinetic methods. Canonical accounts of enzyme catalysis which consist of a list of factors contributing to catalysis, such as entropy saving, acid–base catalysis, etc., are equivalent to step-by-step reformulations of the standard reaction until the empirical binding energy of the standard-reaction transition state becomes zero. It is then assumed that the structures of the standard-reaction transition state and the corresponding elements of the enzymic transition state are identical.

Whenever a single transition state determines the value of a particular enzyme-kinetic parameter, k_E or k_{ES}, then experiments which relate to these parameters will yield straightforward information about the relevant transition state, regardless of the apparent algebraic complexity of the parameter, because the free energies of states intervening between the initial reference state and the transition state cancel out from the final value of the parameter. In more complex situations where more than one transition state, along with either a serial pathway or various pathways in parallel, determines the rate, the result for perturbations of the parameters is also complex. Perturbation of a kinetic parameter with a view toward determination of transition-state structures generates responses which will lead to a *virtual transition-state structure*, one which is a weighted average of the real, contributing transition-state structures. If the contributing structures lie in series, the structure of highest free energy will contribute most to the average. If the contributors are in parallel, the lowest-energy structure contributes most.

The distinction of real from virtual transition states is difficult, but in the end it is the virtual structure which is of the greatest inherent interest. It is this construct which summarizes the characteristics of the rate process in question and which contains the predictive elements for foretelling its behavior in new situations.

ACKNOWLEDGMENTS

The rather idiosyncratic views advanced in this chapter are ones I have to admit are my own, but a number of people have helped me to arrive at positions less eccentric than would otherwise have been the case. Professor W. P. Jencks, over the years, has shown uncommon forbearance in teaching me the canonical theory of catalysis. That trenchant chemical critic, Professor Joseph L. Kurz, saved me from publishing an incorrect version of the present work. Dr. Janos Südi enlightened me on various problems of the idea of virtual transition states. During a visit at Indiana University, while enjoying the excellent hospitality of Professors E. H. Cordes and V. J. Shiner and their research groups, I found a thoughtful and critical audience provided by the students of Chemistry 648. It is a particular pleasure to thank Professors Attila Szabo and Joseph Gajewski and Dr. Warren Buddenbaum for attending my lectures and discussing them with me in detail. My collaborators and I have been privileged to receive support for our studies of catalysis from the National Science Foundation and the National Institutes of Health of the U.S.A.

REFERENCES

1. T. C. Bruice and S. J. Benkovic, *Bioorganic Mechanisms*, Benjamin, Reading, Mass. (1966), 2 vols.
2. W. P. Jencks, *Catalysis in Chemistry and Enzymology*, McGraw-Hill, New York (1969).
3. M. L. Bender, *Mechanism of Homogeneous Catalysis from Protons to Proteins*, Wiley-Interscience, New York (1971).
4. W. P. Jencks, Binding energy, specificity and catalysis—the Circe effect, *Adv. Enzymol. Relat. Areas Mol. Biol.* **43**, 219–410 (1975).
5. T. C. Bruice, Some pertinent aspects of mechanism as determined with small molecules, *Annu. Rev. Biochem.* **45**, 331–373 (1976).
6. G. E. Lienhard, Enzymatic catalysis and transition-state theory, *Science* **180**, 149–154 (1973).
7. R. V. Wolfenden, Chapter 15 in this volume and references cited there.
8. H. C. Froede and I. B. Wilson, in: *The Enzymes* (P. D. Boyer, ed.), Vol. 5, pp. 87–114, Academic Press, New York (1971).
9. M. Robson Wright, Arrhenius parameters for the alkaline hydrolysis of esters in aqueous solution, *J. Chem. Soc. B* **1968**, 545–547.

10. G. Asknes and J. E. Prue, Kinetic salt effects in the hydrolysis of positively charged esters, *J. Chem. Soc.* **1959**, 103–107.
11. A. J. Kirby, in: *Comprehensive Chemical Kinetics* (C. H. Bamford and C. F. H. Tipper, eds.), Vol. 10, pp. 57–207, Elsevier, Amsterdam (1972).
12. G. M. Maggiora and R. L. Schowen, in: *Bioorganic Chemistry* (E. E. van Tamelen, ed.), Vol. 1, Academic Press, New York (1977).
13. T. C. Bruice, in: *The Enzymes* (P. D. Boyer, ed.), Vol. 2, pp. 217–279, Academic Press, New York (1970).
14. A. R. Fersht, Catalysis, binding and enzyme-substrate complementarity, *Proc. Roy. Soc. London Ser. B* **187**, 397–407 (1974).
15. M. L. Bender, F. J. Kezdy, and C. R. Gunter, The anatomy of an enzymatic catalysis: α-Chymotrypsin, *J. Am. Chem. Soc.* **86**, 3714–3721 (1964).

PART II

Approaches to the Determination of Biochemical Transition-State Structures and Properties

In this section, the longest of the volume, we shall introduce the major investigative techniques of transition-state research. The goals of each author have been to show the powers and the limitations of his technique and to illustrate it with a few applications to biochemically important problems. Further applications can be found in Part III, where individual reaction systems are considered.

More than half of the chapters are devoted to kinetic isotope effects, the chief source of experimental information about transition-state structures. Isotope effects hold this dominant position because they are "noninteracting probes" that illuminate the nature of transition states without at the same time altering them, because the theoretical apparatus that links experimental isotope effects to transition states is the best established theory of any kinetic method and because molecules that differ only isotopically are essentially identical in steric and electronic character.

Three of the four chapters on isotope effects deal with hydrogen (that is, protium–deuterium–tritium) isotope effects. Klinman treats cases in which the isotopic center moves significantly in the reaction coordinate (primary isotope effects). Such effects probe directly into the site of major chemical change in the rate-limiting transition state. Structure at more remote positions of the transition state is reflected in secondary isotope effects, the subject of Hogg's chapter. K. B. J. Schowen describes the effects of whole or partial replacement of the light-water solvent by heavy water (solvent isotope effects), a technique that labels all exchangeable hydrogenic sites of substrates and enzyme. It can thus probe the hydrogen bridges which may be components of the transition-state catalytic interaction.

O'Leary describes heavy-atom isotope effects, in which nonhydrogen centers, chiefly carbon, nitrogen, and oxygen, are labeled. These kinds of atoms are reaction-coordinate participants in most enzymic transition states and always form the structural framework. The method has already been very effective and promises even more for the future.

Readers desiring further discussion of enzymic isotope effects will want to consult the recent volume on this subject, edited by Cleland *et al.**

More unusual approaches, ones we consider worthy of the closest attention, are the substance of the remaining three chapters of this section. These are quantum chemistry, magnetic-resonance studies of enzyme complexes in solution, and crystallographic studies of molecular distortion in the solid state.

Maggiora and Christoffersen discuss what can be learned from quantum chemistry, a field that has always offered the ideal window on the transition state, capable in principle of yielding not only geometric detail but also charge distributions and the strengths of various interactions and much more information. In practice, technical difficulties have for many years blocked its application to any but the smallest and simplest systems. At last, however, advances in computer technology and quantum-chemical methodology conspire to make the investigation of large, complex systems a realistic possibility.

Mildvan describes magnetic-resonance studies that produce three-dimensional pictures of complexes with enzymes in solution. These are analogous to the pictures X-ray crystallography gives of solid-state structures, but they show the complex in its major form in solution, a situation directly relevant to catalysis. Some complexes can simulate features of the transition state, and invaluable information about the geometrical disposition of both substrate-derived and enzyme-derived moieties is obtained.

Shefter treats crystallographic studies of distorted molecules. These also simulate structural features of transition states, which are, after all, distorted molecules. Crystal forces countervail in the solid state against the destabilizing forces that render transition states in solution metastable and short-lived and allow the transition state to be seen in a "frozen" (or fossilized) form.

There are two major omissions from this collection of approaches to biochemical transition-state structure: (1) the X-ray crystallographic study of enzymes and their complexes in the solid state and (2) structure-reactivity relationships. The former has been much reviewed and discussed, and we felt that not much new could be added here. Any reader unfamiliar with this approach will find a fine example of the power of protein crystallography for the structural elucidation of enzymic transition states, and references to the literature of the field, in Blow's† recent paper on chymotrypsin action.† Structure-reactivity studies relevant to biochemical transition states come in two forms: systematic variation of substituents, as in the Hammett, Taft, and Hansch methods, and the "bits and pieces" approach in which the binding or the reactions of various component parts of the substrates with the enzyme are

---

* W. W. Cleland, M. H. O'Leary, and D. B. Northop, eds., *Isotope Effects on Enzyme-Catalyzed Reactions,* University Park Press, Baltimore (1977).
† D. M. Blow, Structure and mechanism of chymotrypsin, *Acc. Chem. Res.* **9**, 145–152 (1976).

examined in order to deduce how much each contributes to the catalytic effect. Enzyme specificities, of course, form a barrier to generalizations about systematic structural-variation studies. Kirsch* has reviewed the area, and here again we think that little new can be said. In Chapters 10 and 11 in Part III of this volume the authors dwell heavily on the particular applications of substituent-effect methods to individual systems. The "bits and pieces" approach is also hard to discuss with generality. Several excellent examples of this method have been treated in very complete detail by Jencks in his article on the Circe effect.†

The order of the chapters in this section progresses from the theoretical through isotope effects (hydrogen, then heavy atom) to the studies of stable structures in solution and in the solid state. We consider this to be a pleasing and effective order of study, but each chapter is in fact accessible by itself. Thus the reader can easily alter the sequence in which he takes up the individual chapters or can omit any chapter he wishes.

* J. F. Kirsch, in: *Advances in Linear Free Energy Relationships* (N. B. Chapman and J. Shorter, eds.), pp. 369–400, Plenum Press, New York (1972).
† W. P. Jencks, Binding energy, specificity and catalysis—the Circe effect, *Adv. Enzymol. Relat. Areas Mol. Biol.* **43**, 219–410 (1975).

3

Quantum-Mechanical Approaches to the Study of Enzymic Transition States and Reaction Paths

Gerald M. Maggiora and Ralph E. Christoffersen

1. INTRODUCTION

Over the last 30 years, transition-state theory has provided a suitable theoretical framework for interpreting a wide range of chemical[1] and biological[2] processes. Nevertheless, it is only within the last few years, due in large measure to the approximately parallel and related development of high-speed digital computers and practical quantum-mechanical procedures for studying the geometric and electronic structure of molecules, that the full potential of transition-state theory is beginning to be realized.

Even with these powerful tools, the interpretation of complex rate processes must rely on the judicious use of approximations. This is especially true in applications of transition-state theory to biochemical processes. For example, in an enzyme-catalyzed reaction, the substrate(s) become bound to the enzyme by covalent and/or noncovalent bonds. At least two questions then arise: (1) What is the importance of the various chemical groups found in the active site to the catalytic mechanism, and (2) can these groups effectively be uncoupled from the remainder of the enzyme? The latter point is an extremely important one with regard to developing models of enzymic transition states which are of manageable proportions for practical theoretical calculations, and it will be examined in detail in subsequent sections.

In the present chapter the approach will be didactic, as we are not primarily interested in describing "state-of-the-art" techniques. Rather, we want to

Gerald M. Maggiora and Ralph E. Christoffersen • Departments of Biochemistry and Chemistry, University of Kansas, Lawrence, Kansas. Correspondence should be directed to GMM.

elucidate important, general principles and, utilizing these principles, to in- indicate how molecular quantum mechanics can be applied to the study of enzymic transition states and reaction paths. Thus, references will be chosen which illustrate the various points, and no attempt at exhaustive coverage will be made.

The discussion to follow is divided roughly into four parts. First, we describe the general dynamical principles of chemical reactions and their relationship to transition states and reaction paths. Second, we analyze the general computational problems encountered in the explicit determination of transition states and reaction paths. Third, we review the general character- istics of a number of quantum-mechanical computational procedures which have been found to be useful for determining transition states and reaction paths. Fourth, we examine the additional difficulties posed by enzyme systems in the determination of transition states and reaction paths, and as an example we analyze a number of the salient features of a typical quantum-mechanical study of an enzyme-catalyzed reaction.

2. MOLECULAR REACTION DYNAMICS

2.1. Potential Energy Surfaces and Pseudominimum Energy Paths

To describe completely the time-dependent structural changes (i.e., the dynamics) that take place during a chemical reaction, it is necessary to solve the quantum-mechanical time-dependent Schrödinger equation.[3] However, as has been discussed by a number of authors,[4-6] quantum effects are generally small, especially in the limit of large particle masses and high energies as is found in most chemical reactions. Therefore, by solving Lagrange's (or Hamilton's) classical equations of motion,[7] an easier but by no means trivial task compared with the quantum-mechanical case, the time-dependent struc- tural behavior of the molecular system can be determined, at least in principle, to a reasonable accuracy.

In either case, solution of the appropriate equations requires a knowledge of the potential energy, V, of the molecular system expressed as a function of its atomic configuration. For an N-atom system, $3N$ coordinates are required to specify its atomic positions with respect to an arbitrary coordinate system. In the absence of external fields, the motion of the center of mass of the system can be uncoupled from the purely internal motions of the atoms with respect to the center of mass. Furthermore, if the three coordinates needed to describe rotation of the system about its center of mass also are uncoupled, $3N - 6$ coordinates remain. These remaining $3N - 6$ coordinates then represent the minimum number of independent coordinates required to specify uniquely an arbitrary atomic configuration of the system in the molecular *configuration space*.[7] The system is said to possess $f = 3N - 6$ degrees of freedom.*

---

* In a purely linear system such as found in a diatomic molecular system $f = 3N - 5$. Also the inclusion of structural constraints which impose restrictions on the dynamical behavior of the system serve to further reduce its degrees of freedom.

Any set of numbers $\{q_1, q_2, \ldots, q_f\}$ which serve to specify uniquely an atomic configuration of the system are examples of *generalized coordinates*.[7] Specific examples include bond lengths, bond angles, torsional angles, etc. Hence, it is always possible* to express the potential energy as a function of the $3N - 6$ generalized coordinates,

$$V = V(q_1, q_2, \ldots, q_f) \tag{1}$$

To illustrate a number of important points regarding the relationship of the potential energy function to the mechanism of a chemical reaction, consider the case of a system possessing only two degrees of freedom (i.e., $f = 2$). In this case, the potential energy function is given by

$$V = V(q_1, q_2) \tag{2}$$

which can be represented graphically as a three-dimensional potential energy surface, as shown in Fig. 1(a). The height of the surface above the q_1-q_2 base plane represents the potential energy of the system for each set of values (q_1, q_2) of the two generalized coordinates. The projection of the surface onto the q_1-q_2 base plane generates the two-dimensional potential energy contour map depicted in Fig. 1(b), where each contour represents a region of constant potential energy (i.e., isopotential energy curves) in the q_1-q_2 configuration space.

The projection onto the q_1-q_2 base plane of the heavy line which connects the various stationary points (i.e., reactants, transition states, intermediates, and products) and moves along the bottom of the minimum energy valley of the potential energy surface represents a *pseudominimum energy path*.

As pointed out by Fukui,[8] this path describes an *orthogonal trajectory*; i.e., the pseudominimum energy path intersects the isopotential energy curves at "right angles." The shape of the path, as well as the shape of the potential energy surface itself, depends on the particular choice of generalized coordinates employed and bears no necessary relationship to any *physical reaction path* (see Section 2.2), which must be coordinate independent, of the system.

The dependence of the shape of the potential energy surface on coordinate choice is due to the nature of the transformations which relate different sets of generalized coordinates. Since the coordinate sets are in general nonorthogonal, transformations between them are in general nonorthogonal. As such nonorthogonal transformations do not preserve angles and distances, the shape of the surface need not remain unchanged. An important consequence of this behavior is that the pseudominimum energy path (an orthogonal trajectory) in one coordinate system will not be transformed into the corresponding pseudominimum energy path (an orthogonal trajectory) in another coordinate system.† Hence, the *two pseudominimum energy paths are not*

---

* While it is always possible to express the potential energy as a function of f generalized coordinates, it may not always be convenient. For a discussion of various functional representations of the potential energy, see, for example, Reference 4.
† See Section 3.4 and Reference 27 for further discussion.

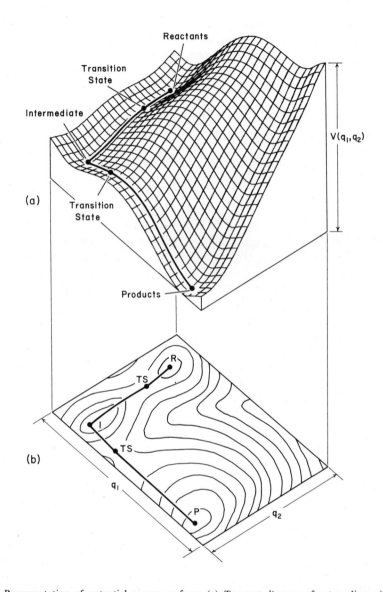

Fig. 1. Representation of potential energy surfaces. (a) *Transect diagram* of a two-dimensional potential energy surface. The heavy dark line along the bottom of the energy valley runs between the reactant, intermediate, and product points and passes over two transition states or col points as indicated in the figure. (b) *Contour map* of a two-dimensional potential energy surface. The contours are generated by projecting the surface depicted in the above transect diagram onto the q_1-q_2 base plane and represent curves of constant potential energy (i.e., isopotential energy curves). The *pseudominimum energy path* connecting the reactant and product points ($R \rightarrow TS \rightarrow I \rightarrow TS \rightarrow P$) is generated by projecting the heavy dark line indicated on the transect diagram onto the q_1-q_2 base plane.

physically equivalent. Only the stationary points on the surface, since they are coordinate independent, possess true physical significance. Nevertheless, pseudominimum energy paths are in widespread use as "guides" to the approximate structural changes that occur during chemical reactions.[5,6,8,9]

The physical or true minimum energy path is related to the dynamics (see Section 2.2) of the reacting system and can be thought of as the lowest-energy physical reaction path which passes through a given transition state. Thus, physical minimum energy paths are invariant to the particular set of generalized coordinates used.

The pseudominimum energy path can be described parametrically by the curve*

$$C_2^{\text{mep}} : \{ q_1(\lambda), q_2(\lambda) \} \tag{3}$$

where the parameter λ represents the *pseudoreaction coordinate*[4-6,8-12] which describes the displacement† of the system along the pseudominimum energy path. As was true for the pseudominimum energy path, the pseudoreaction coordinate depends on the particular set of generalized coordinates used to represent the potential energy surface.

If a plot of potential energy as a function of displacement along the pseudoreaction coordinate is made, the usual "reaction-coordinate diagram"[1] is obtained, as illustrated in Fig. 2. The above approach can be extended easily to dimensions higher than two. Aside from the fact that we can no longer simply visualize the hypersurfaces and hyperplanes in the higher-dimensional space, no serious conceptual problems arise. Note, however, that even in such higher-dimensional spaces, reaction coordinate diagrams analogous to the one depicted in Fig. 2 can still be constructed in principle, although in practice it is quite difficult for large molecular systems. An extensive critical discussion of such diagrams has been presented by Johnston,[12] and additional discussion of pseudominimum energy paths and reaction coordinates is presented in Section 3.3.

To facilitate further discussion of minimum energy paths, reaction paths, reaction coordinates, etc., the adjective "pseudo" will no longer be used, and it will be assumed unless otherwise stated that use of these terms will necessarily imply the presence of the qualifying adjective.

---

* Note that any curve in a space of given dimension can be represented parametrically. For example, in two dimensions consider the simple parabolic function $y = 4px$. If we choose as the parameter the slope of the curve, $\lambda = dy/dx$, then it can easily be shown that $x = x(\lambda) = p/\lambda^2$ and $y = y(\lambda) = (2p)/\lambda$. Other possible parameterizations exist, but the above simple example serves to illustrate the general principle.

† Since the two generalized coordinates in the example may not necessarily represent similar structural features, e.g., a bond length and a bond angle, their units of measure could differ. For example, bond lengths are measured in angstroms and bond angles are measured in degrees or radians. Hence, in such cases the concept of displacement along the reaction coordinate must be interpreted with care since length is not strictly defined. However, under such circumstances a consistent definition of displacement can be made by adapting a unitless system based on the fractional change undergone by both generalized coordinates.

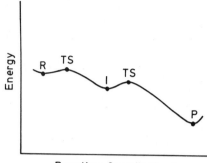

Fig. 2. Schematic representation of a typical *reaction-coordinate diagram* obtained from the data of Fig. 1. The reaction coordinate represents the displacement of the system along the pseudominimum energy path [Fig. 1(b)] measured from the reactant point (R).

2.2. Dynamical Considerations

For a given potential energy function such as that depicted in Fig. 3(a) as a two-dimensional isopotential energy contour map, solution of the classical equations of motion[7] corresponding to a particular set of initial conditions[4-6,13] yields a *classical trajectory* or *physical reaction path* of the system. The two-dimensional classical trajectory can be represented by the curve

$$C_2^{cl} : \{q_1(t), q_2(t)\} \tag{4}$$

where t represents the parametric time dependence of each generalized coordinate. As t increases the system "moves" along the curve C_2^{cl} in the two-dimensional configuration space of the reacting system.

Since we are dealing with a *conservative system*, each trajectory can be viewed as a curve embedded in a two-dimensional plane of constant energy located an appropriate amount in energy above the q_1-q_2 base plane. If the dynamics are handled purely classically (i.e., tunneling is not allowed) none of the trajectories that lie in constant energy planes of lower energy than the *col point* or transition state will lead from reactant(s) to product(s).* This is illustrated in Figs. 3(b) and 3(c) for two trajectories, one embedded in a constant energy plane of greater energy than the col point [Fig. 3(b)] and one embedded in a constant energy plane of lesser energy than the col point [Fig. 3(c)]. The shaded area in the figures represents the classically forbidden regions on the isopotential energy maps.

Examination of several trajectories does not, however, allow one to determine the rate constant for a particular process. To obtain a rate constant, it is necessary typically to examine a large number of trajectories ($\gg 1000$) corresponding to a wide variety of initial conditions. The appropriate initial conditions are usually selected by the use of statistical and/or scanning procedures[4,5,13] such that calculation of a sufficient number of trajectories to determine the reaction cross section and thus its rate constant can be

---

* For a slightly different view of this point, see the discussion on pp. 210–212 of Reference 9.

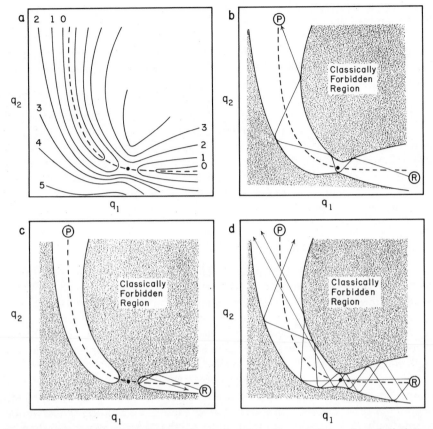

Fig. 3. Representations of classical reaction trajectories with respect to a hypothetical two-dimensional isopotential energy contour map. The transition state is indicated by the closed circle which lies on the pseudominimum energy path (dashed line) connecting reactant(s) R and product(s) P. (a) Two-dimensional isopotential energy contour map. The isopotential energy curves are given in arbitrary energy units. (b) Possible classical reaction trajectory corresponding to a *total* system energy $E = 2$. The trajectory thus lies in a constant energy plane which is co-planar with the q_1-q_2 base plane and intersects the potential energy surface at $E = 2$ [cf. Fig. 1(a)]. The "classically forbidden region" corresponds to that region on the potential energy surface with $E > 2$. Note that the transition state does not in this case fall into a classically forbidden region and that the particular trajectory illustrated does not pass through the transition state. (c) Possible classical reaction trajectory corresponding to a *total* system energy of $E = 1$. As in (b) the trajectory lies in a constant energy plane, but in this case the "classically forbidden region," $E > 1$, contains the transition state. Hence, as indicated, the trajectory does not lead from reactant(s) to product(s), i.e., it is a "nonproductive" trajectory. (d) Series of different classical reaction trajectories corresponding to a *total* system energy of $E = 2$ but to different initial conditions.

accomplished. While this procedure yields the best classical picture of the chemical reaction process, a number of problems arise which make such a detailed dynamical characterization difficult to achieve in practice. Not the least of these problems is the difficulty of obtaining potential energy surfaces of sufficient reliability for systems possessing a large number of degrees of freedom ($f > 20$).

In this regard, it is important to realize that small errors in the shape of a potential energy surface can have significant effects on the molecular dynamics of a chemical reaction.* Such small errors are much less important in, for example, an examination of the gross structural changes which take place along the minimum energy or other suitable reaction paths. Furthermore, while small errors in the shape of the potential energy surface in the region of the transition state can lead to nontrivial errors in the calculated properties of the transition state (e.g., vibrational frequencies), we are dealing only with a very limited region of the total configuration space of the molecular system so that, at least in principle, it is possible to correct these deficiencies by elaborate and highly accurate computational procedures, as discussed in later sections.

2.3. Quasi-Dynamical Aspects of Chemical Reactions

Although each trajectory is dependent on the chosen initial conditions, the shape of the potential energy surface in many instances is such that it is possible to speak of an *average trajectory*. That is, the trajectories corresponding to a large number of different initial conditions are found to lie in approximately the same region in the molecular configuration space. Therefore, this set of trajectories should make a major contribution to the reaction cross section and, hence, to the overall reaction rate. This type of behavior is depicted in Fig. 3(d). Thus, as illustrated in the figure, the average trajectory represents a reaction path which is a reasonably close approximation to the minimum energy path for the system. Under these circumstances the minimum energy path may provide a useful guide to the average structural changes which occur during the reaction process even though (see Section 2.1) direct physical significance cannot be ascribed to the path itself.

Information of this type is *quasi-dynamical* in nature and used with proper care can provide an important bridge between the classical dynamical approach described above and the transition-state theory approach, which describes the reaction process at only two points along the minimum energy path.

Furthermore, it is of interest to note that the quantum-mechanical calculation of potential energy surfaces provides an additional bonus to the energetic and geometric structural information described above. This bonus is in the form of electronic structural information such as molecular charge distributions and bond orders. While such information is basically of a nondynamical nature, knowledge of it can provide important clues into the dynamical and quasi-dynamical behavior of reacting systems.

Finally, before proceeding to the next section it is important to further

---

* Note that the gradient of the potential energy function is proportional to the force on the atoms of the molecular system.

clarify the relationship between potential energy surfaces and dynamical reaction paths. In a number of instances, possible dynamical reaction paths have been represented by curves embedded in a potential energy surface [cf. Fig. 1(a)]. Such paths do not constitute dynamical reaction paths for two reasons in addition to those discussed in Section 2.1. First, as discussed earlier, dynamical reaction paths (i.e., trajectories) represent time-dependent processes which take place at constant total energy and thus lie within constant energy planes parallel to the coordinate base plane. Second, the potential energy surface does not account for mass effects.* The construction of "mass-corrected" potential energy surfaces which approximately represent "dynamical surfaces" have been discussed by a number of authors,[1,14-16] and Hirschfelder[16] has treated the mathematical details. Once such mass-corrected potential energy surfaces have been constructed, the dynamics of the system can be treated approximately by examining the motion of a ball of unit mass as it rolls on the surface. Only, however, in the case of a suitably mass-corrected potential energy surface is this even approximately correct.† Note also that trajectories produced by such surfaces do not directly correspond to minimum energy paths, although they may approximate these paths under suitable conditions.

3. CALCULATION OF TRANSITION STATES AND REACTION PATHS

3.1. Adiabatic Born–Oppenheimer Approximation

Nearly all quantum-mechanical molecular electronic and geometric structure calculations are predicated on the validity of the *Born–Oppenheimer* or *clamped nuclei approximation*,[17] which allows uncoupling of nuclear and electronic motions due to the large mass, and hence velocity, differentials between these two types of particles. Under the Born–Oppenheimer approximation, one then solves a simpler modified *electronic Schrödinger equation*,

$$\hat{H}(\mathbf{r}, \mathbf{q}^0)\,\Phi(\mathbf{r}, \mathbf{q}^0) = E(\mathbf{q}^0)\,\Phi(\mathbf{r}, \mathbf{q}^0) \tag{5}$$

where $\mathbf{r}$ and $\mathbf{q}$ represent the set of electron and nuclear coordinates, respectively. The superscript zero on the nuclear coordinates denotes their parametric role. The *electronic Hamiltonian operator* is given by

$$\hat{H}(\mathbf{r}, \mathbf{q}^0) = \hat{T}(\mathbf{r}) + \hat{V}(\mathbf{r}, \mathbf{q}^0) \tag{6}$$

where $\hat{T}(\mathbf{r})$ and $\hat{V}(\mathbf{r}, \mathbf{q}^0)$ are the corresponding electronic kinetic and potential energy operators. As is clear from Eq. (5), the wave function $\Phi(\mathbf{r}, \mathbf{q}^0)$ depends

---

* This of course is why isotope effects in chemical reactions can be studied using identical potential energy surfaces.

† See Chap. III of Reference 1 for a number of examples and an extensive qualitative discussion of mass-corrected (sometimes referred to a "diagonalized") potential energy surfaces.

on both the electron and nuclear coordinates, while the electronic energy $E(\mathbf{q}^0)$ depends only on the nuclear configuration. $E(\mathbf{q}^0)$ is related to the total potential energy described in Eq. (1) by

$$V(\mathbf{q}^0) = E(\mathbf{q}^0) + V_{nn}(\mathbf{q}^0) \qquad (7)$$

where $V_{nn}(\mathbf{q}^0)$ is the potential energy due to nuclear repulsions.

Equation (5) also assumes that the system behaves *adiabatically*; i.e., the electronic quantum state of the system remains unchanged over the entire potential energy surface.[5] Except in a region of crossing of two potential energy surfaces, breakdown of the adiabatic Born–Oppenheimer approximation is rare, and since very few enzyme-catalyzed reactions have been shown to involve excited electronic states, the possibility of potential surface crossings is very small. Therefore, breakdown of the adiabatic Born–Oppenheimer approximation is not expected to pose a problem in the study of most enzyme-catalyzed reactions.

3.2. Brute-Force Computational Approach

Although the transition-state approach to chemical reaction mechanisms in principle provides much less information than the dynamical or quasi-dynamical approaches outlined above, it has been very successful in explaining the underlying mechanisms of a wide range of rate processes. With regard to computational accessibility and convenience, transition-state theory possesses the extremely convenient attribute of requiring calculations at only two points (i.e., reactant and transition-state configurations) of the entire multi-dimensional $(3N - 6)$ configuration space of the molecular system. However, this simplicity in terms of computational requirements also presents a problem which may not always be easily solved; viz., what is the structure of the transition-state species? We emphasize this point since the long heritage of theoretical and experimental structural chemistry, which provides us with a suitable basis for determining reactant (or product) structure(s), may not be able to provide a similar assistance for transition-state structure.

Of course, the easiest way conceptually to determine transition-state structure is to compute a very "coarse-grained" potential energy surface and then essentially by inspection to select the appropriate structure. Computationally, however, this approach quickly becomes untenable even for systems of modest dimension.

Consider, for example, a molecular system with only ten degrees of freedom (i.e., $f = 10$). To obtain a "coarse-grained" potential energy surface, suppose we compute ten points for each degree of freedom. Thus, to determine the entire grid of points necessary to produce the coarse-grained surface we must compute a total of 10^{10} points. If we assume that we can compute each point in 1 msec, the total time required to generate the surface would be approximately 280 hr. Even for small chemical systems (e.g., the ammonia–formaldehyde adduct with $f = 18$) let alone realistic biochemical systems, such a brute-force approach is currently not possible using quantum-mechanically based computational procedures.

In certain cases, it may be possible to obtain a reasonable "guess" of transition-state structure. For such cases, application of the coarse grid method may be possible. However, in complex enzymic transition states, the dimension of the problem as well as the lack of assistance from a well-developed structural chemistry may preclude this approach.

3.3. Minimum Energy Path Approach

As illustrated in Figs. 1 and 2 and discussed in Section 2.1, knowledge of the minimum energy path is tantamount to a knowledge of any and all transition states along the path.* To take advantage of this important property of the minimum energy path for complex molecular systems, a computationally feasible approach is needed which does not require the calculation of a large number of points on the potential energy surface.

The following approach has been used by most workers. First, define as the *independent variable* a generalized coordinate, q_i, which is expected to be particularly well behaved over the entire course of the reaction.† In particular, q_i should be a monotonically increasing or decreasing function whose range over the entire reaction path is sufficient to guarantee an adequate description of the extent of the reaction. For example, a C—H bond length which changes only 0.01 Å during the reaction or a bond angle which has an oscillatory (i.e., nonmonotonic) behavior would not be desirable independent variables, while the distance of separation of two reacting species which varies from infinity to the bond length of the stable adduct would be acceptable. Unfortunately, it may not always be possible to choose the proper variable *a priori*, although in many cases "chemical intuition" can usually provide a fairly reliable guide.

The independent variable is usually called the reaction coordinate, but this nomenclature should be avoided as it is misleading. Note that the reaction coordinate, λ, as described in Section 2.1 is a function of all $3N - 6$ generalized coordinates, i.e.,

$$\lambda = \lambda(q_1, ..., q_i, ..., q_f) \tag{8}$$

and, hence, only in the most unusual of circumstances would it be equivalent to the independent variable, q_i.

The remaining $3N - 7$ generalized coordinates can then be written as

---

* Note, as discussed in Section 3.4, the conditions which must be satisfied for an extremum point along the minimum energy path to correspond to a transition state or col point. See in particular Eqs. (11), (14), (16), (18), and (19). Satisfaction of these conditions guarantees that the point in question is a transition state and not a maximum or minimum point. However, in most cases where the minimum energy path is known it is not necessary in practice to carry out such detailed mathematical evaluations to confirm the character of an extremum point along the path.

† In some cases it may be necessary to employ more than one independent variable in the determination of the minimum energy path.

functions of the independent variable:

$$q_1 \;\;\; = \xi_1(q_i)$$

$$\vdots \quad \vdots$$

$$q_{i-1} = \xi_{i-1}(q_i)$$

$$\vdots \quad \vdots \tag{9}$$

$$q_{i+1} = \xi_{i+1}(q_i)$$

$$\vdots \quad \vdots$$

$$q_f \;\;\; = \xi_f(q_i)$$

Determination of the minimum energy path then becomes a problem in *constrained minimization.* That is, for a fixed, specified (usually uniform) set of values of the independent variable, the total energy of the system is minimized with respect to the remaining $3N - 7$ dependent variables. The minimization process is constrained since all $3N - 6$ generalized coordinates are not allowed to vary freely during the minimization process.

In actual applications of this technique to real chemical systems further constraints are placed on some of the dependent variables. For example, bond lengths which show little change between reactant(s) and product(s) are usually fixed. Other types of "symmetry constraints" which fix the direction of approach of two reactants or which require that certain groups move in a specific, concerted fashion have also been applied. Mathematically, this can be stated as

$$q_1 = \xi_1^0(q_i)$$

$$q_2 = \xi_2^0(q_i)$$

$$q_3 = \xi_3(q_i) \tag{10}$$

$$\vdots$$

$$q_f = \xi_f(q_i)$$

where ξ_1^0 and ξ_2^0 are *predetermined* and thus are constrained functions of the independent variable, q_i.

In cases where a constrained minimization procedure is used, the possibility of missing part or all of the minimum energy path, and hence the transition states, is very real. Such a situation is illustrated in Fig. 4 for the simple two-dimensional ($f = 2$) case. In the figure, q_2 is taken as the independent variable which is varied in equal increments while minimizing the energy of the system with respect to q_1 for each q_2. The dots (closed circles) in the figure indicate the calculated minimum energy points obtained using this constrained minimization procedure. The solid line connecting the dots represents the constrained minimum energy path. Note, as discussed above, that the path does not pass through the transition state (closed delta), which is located at the approximate center of the diagram in Fig. 4. The actual

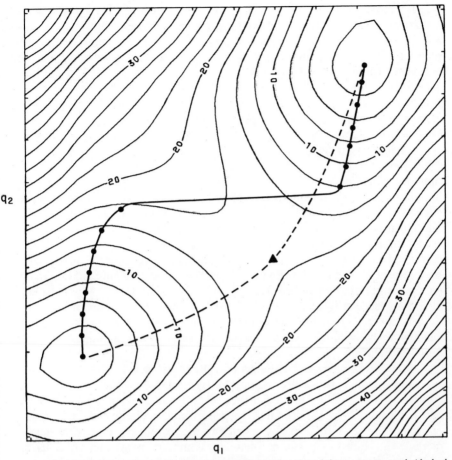

Fig. 4. Example of the difference between two reaction paths: the minimum energy path (dashed line) and a path determined by a constrained minimization procedure (line with dots). The closed circles (dots) represent the coordinate values obtained by the constrained minimization procedure. As indicated in the figure, equal increments of the independent variable, q_2, were taken along the entire path. The value of the dependent variable, q_1, was then determined by standard minimization techniques. The transition state (closed delta) lies on the minimum energy path but not on the path determined by constrained minimization.

minimum energy path (indicated by the dashed line) passes through the transition state and intersects the isopotential energy contours at right angles as required for a true minimum energy path. Note that no such condition is required for the constrained minimum energy path, and, as shown in Fig. 4, it does not intersect the isopotential energy contours at right angles.

The above example illustrates the dangers inherent in the use of constrained minimization procedures for the determination of minimum energy paths and transition states. However, it should not be inferred from this example that constrained minimization will always lead to such disastrous consequences. Furthermore, guidelines do exist which can be of assistance in determining whether the results obtained by a particular constrained minimization procedure realistically represent the structural changes which take place along the minimum energy path. For example, in carrying out the constrained minimization, the reactants should pass "smoothly" to products. This implies two important points which must be carefully assessed in any constrained minimization procedure: (1) Does the procedure lead to the correct product structure with respect to all freely varied dependent variables? (2) Do the values of these dependent variables change smoothly along the computed reaction path?

Close examination of Fig. 4 reveals that, although the chosen minimization procedure does lead to the correct product structure, the constrained minimum energy path exhibits a large, effectively discontinuous change in the dependent variable, q_1. This is an important indication that the constrained minimization procedure being used may be producing a spurious minimum energy path. When this situation occurs, reducing the step size (i.e., the incremental change) of the independent variable in the region of the discontinuity may resolve the difficulty, although, as can be seen in the figure, taking smaller increments of the dependent variable in the region of the "discontinuity" does not necessarily guarantee that the difficulty will be removed. If, after using smaller and smaller increments, the discontinuity persists, a new dependent variable should be selected and the constrained minimization procedure repeated.

Other examples of the dangers inherent in constrained minimization procedures include those discussed by McIver and Komornicki[18] and Dewar and Kirschner.[19-21] Of particular interest is the example provided by Dewar and Kirschner[19] on the orbital symmetry forbidden disrotatory ring opening of cyclobutene to butadiene. By restricting the disrotatory motion of the two methylene groups to occur in phase (a symmetry constraint as described earlier) a potential energy surface consisting of two approximately "blind alley" reaction paths lying on either side of the transition state is obtained. In the usual constrained minimization procedure, in which only one independent variable was used, reactant–product interconversion was not obtained. Not until *two* independent variables were examined simultaneously in the constrained minimization procedure did the apparent character of the reaction path and transition state emerge. Later work by Dewar and Kirschner[21] showed, however, that removal of the symmetry constraints (i.e., not requiring the disrotatory motion of both methylene groups to occur in phase) produced a perfectly "normal" reaction pathway using only one independent variable in the minimization. The above example points up the need to assess carefully the results obtained from constrained minimization procedures, especially procedures which build in symmetry or geometry constraints.

Finally, as alluded to earlier, it may be possible to obtain a reasonable guess of the structure of the transition state based on extensive experimental or theoretical work on analogous systems. An example of this can be found in carbonyl addition reactions, where the presence of a "tetrahedral-like" transition state is well supported by numerous experimental investigations.[22,23] In such circumstances, however, caution always should be exercised lest the outcome of the study by prejudiced by the initial assumptions. At any rate, if sufficient care is exercised, application of constrained minimization procedures can be useful for understanding not only transition-state structure but also the electronic and mechanical changes which take place along the entire minimum energy path.

3.4. Direct Determination of Transition States

Due in large measure to the difficulties described in previous sections, attempts are now being made to determine transition-state structure directly and, with this information, to calculate the properties of the transition state. Such information can then in principle lead to direct calculation of kinetic isotope effects and rate constants which can be compared with the corresponding experimental values. In this way such theoretical studies can provide an additional tool for the study of reaction mechanisms.

The basic approach used involves a *nonlinear search* of the potential energy surface for extremum points which satisfy the appropriate mathematical conditions for transition states. From elementary multivariable calculus[24] it follows that the *extremum points* of the potential energy function $V(q_1, q_2, \ldots, q_f)$ satisfy the conditions

$$\left(\frac{\partial V}{\partial q_i} \right)_{\mathbf{q}^0} = 0, \qquad i = 1, 2, \ldots, f \tag{11}$$

at $\mathbf{q}^0$. However, knowledge of the location of an extremum point on a potential energy surface in itself is not sufficient to characterize the point completely, since such a point may correspond to a minimum, a maximum, or an inflection point on the surface. To assess the character of the surface at an extremum point, it is also necessary to know the *curvature* of the surface at that point, which is a second derivative property.

To explore this further, let us expand the potential energy function about the extreme point $\mathbf{q}^\circ$,

$$V(\mathbf{q}) = V(\mathbf{q}^0) + \sum_i^f \left(\frac{\partial V}{\partial q_i} \right)_{\mathbf{q}^0} \cdot (q_i - q_i^0)$$

$$+ \frac{1}{2} \sum_{ij}^{ff} \left(\frac{\partial^2 V}{\partial q_i \partial q_j} \right)_{\mathbf{q}^0} \cdot (q_i - q_i^0)(q_j - q_j^0) + \text{higher terms} \tag{12}$$

where as indicated all derivatives are evaluated at the extremum point $\mathbf{q}^0$. Since we are at an extremum point, Eq. (11) holds, and if we take (without loss of generality) the location of the extremum point to be the origin of the coordinate system, Eq. (12) simplifies to

$$V(\mathbf{q}) = V(\mathbf{0}) + \frac{1}{2} \sum_{i,j}^{f} F_{ij} q_i q_j + \text{higher terms} \tag{13}$$

where the F_{ij}, which are defined as

$$F_{ij} = \left(\frac{\partial^2 V}{\partial q_i \, \partial q_j} \right)_{\mathbf{q}^0} = F_{ji} \tag{14}$$

are identical to the force constants traditionally used in the theory of molecular vibrations.[25]

Since $V(\mathbf{0})$ represents a scalar constant term, its presence can have no effect on the shape of the potential energy surface, and nonvanishing terms* will determine the shape of the surface in the neighborhood of the extremum point. Thus we need only examine the characteristics of the second-order term

$$V^{(2)}(\mathbf{q}) = \tfrac{1}{2} \mathbf{q}^T \mathbf{F} \mathbf{q} \tag{15}$$

where F is the symmetric f-dimensional matrix

$$\mathbf{F} = \begin{pmatrix} F_{11} & F_{12} & \cdots & F_{1f} \\ F_{21} & F_{22} & \cdots & F_{2f} \\ \vdots & \vdots & & \vdots \\ F_{f1} & F_{f2} & \cdots & F_{ff} \end{pmatrix} \tag{16}$$

and $\mathbf{q}$ can be interpreted as the column vector

$$\mathbf{q} = \begin{pmatrix} q_1 \\ q_2 \\ \vdots \\ q_f \end{pmatrix} \tag{17}$$

$\mathbf{q}^T$ being its transpose.

The shape of the surface around $\mathbf{q}^0$ is then determined by the eigenvalues and eigenvectors of the quadratic force constant matrix[26]

$$\mathbf{F}\mathbf{u}_i = \kappa_i \mathbf{u}_i \tag{18}$$

The eigenvectors $\mathbf{u}_1, \mathbf{u}_2, \ldots, \mathbf{u}_f$ represent the *principal axes of curvature*, and the eigenvalues $\kappa_1, \kappa_2, \ldots, \kappa_f$ represent the *principal curvatures* of the surface along the principal axes of curvature. By standard matrix manipulations, it can be shown that the signs of the eigenvalues correspond to the following

---

* In cases where all quadratic terms vanish, nonvanishing cubic or higher terms determine the shape of the surface in the region of the extremum point.

extremal properties of $V(\mathbf{q}^0)$:

$$\kappa_i > 0, \qquad \text{minimum along the } \mathbf{u}_i \text{ axis at } \mathbf{q}^0 \tag{19a}$$

$$\kappa_i < 0, \qquad \text{maximum along the } \mathbf{u}_i \text{ axis at } \mathbf{q}^0 \tag{19b}$$

$$\kappa_i = 0, \qquad \text{inflection along the } \mathbf{u}_i \text{ axis at } \mathbf{q}^0 \tag{19c}$$

To examine how the above conditions indicate the presence of a transition state rather than a maximum or minimum at $\mathbf{q}^0$, consider the two-dimensional example illustrated in Fig. 5. This example clearly shows that the presence of a transition state or col point requires that only one of the eigenvalues be negative. The case of zero eigenvalues has been described by Stanton and McIver[27] and will not be discussed further in the present treatment.

Murrell and Laidler[28] have proved an important theorem which allows one to generalize the above two-dimensional system to a realistic $(3N - 6)$-dimensional system. Their theorem states that the curvature of $V(\mathbf{q}^0)$ at a col point is negative along one and only one of the principal axes of curvature if none of the remaining principal curvatures vanish. As pointed out by Stanton and McIver,[27] the latter situation is not expected to pose a problem in cases of chemical interest. An examination of Fig. 5(a) clearly shows why the presence of more than one negative principal curvature does not represent a col point even in a $(3N - 6)$-dimensional space. This follows since such an extremum point would correspond to the top of a "local hill," and as a transition state is supposed to be the highest point on the minimum energy reaction path between two stable molecular species, the fact that a lower energy path could be taken "around" the local hill eliminates such an extremum point as a possible transition state.

The eigenvector which corresponds to the one negative eigenvalue has been called the *transition vector* and can be shown by variational arguments to point along the direction of most negative curvature.[27] Since the generalized coordinates used to describe the dynamics of the reacting system are not necessarily orthogonal, the transition vector does not correspond to an actual physical reaction vector. Furthermore, the eigenvalues and eigenvectors of $\mathbf{F}$ do not have a precise physical significance since they are not independent of the coordinate system. To obtain a coordinate-system-independent reaction

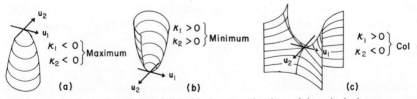

Fig. 5. Schematic illustration of the relationship between the signs of the principal curvatures, κ, and the shape of the potential energy surface in the region of a (a) maximum, (b) minimum, and (c) col point. The vectors $\mathbf{u}_1$ and $\mathbf{u}_2$ are unit vectors along the principal axes of curvature.

vector it is necessary, in a fashion analogous to that used in standard molecular vibration theory,[25] to diagonalize the Wilson **FG** matrix, where **F** is the quadratic force constant matrix defined above and **G** is the mass-dependent transformation matrix which relates the $3N - 6$ generalized (internal) coordinates to the $3N$ Cartesian coordinates needed to describe the system in a laboratory reference frame. The principal axes of curvature and their corresponding principal curvatures derived from the **FG** matrix now have a precise physical significance; namely, they represent the normal modes of vibration and their corresponding frequencies, respectively. The one negative eigenvalue of the transition-state system corresponds to an imaginary vibrational frequency, and the direction of the eigenvector corresponds to the direction of classical mechanical motion of the system at the transition state, and thus it is this vector which could appropriately be designated a *reaction* or *decomposition vector*.

Application of nonlinear search techniques based on the principles described above have been employed by several workers[18,29,30] for the direct determination of transition states. To locate the transition state(s), the norm of the gradient

$$\gamma(\mathbf{q}) = \sum_i^f \left(\frac{\partial V(\mathbf{q})}{\partial q_i} \right)^2 \geq 0 \tag{20}$$

is minimized, and since $\gamma(\mathbf{q})$ is never negative, it is possible to use rather powerful nonlinear search techniques. As is clear from Eq. (11), any extremum point will lead to the vanishing of $\gamma(\mathbf{q})$ at that point. Therefore the curvature criteria described in Eqs. (19) need to be applied at each extremum point obtained. Although determination of the curvature at an extremum point requires knowledge of all quadratic force constants and thus requires added computation,* the F_{ij} so obtained can also be used in the prediction of kinetic isotope effects and rate constants if the point is shown to correspond to a transition state. Hence, the added computational effort required is not entirely lost.

As is the case with all nonlinear search procedures, a good initial guess is important in reducing the amount of computational labor involved in determining a transition state. Furthermore, while certain initial guesses may converge rapidly to an extremum point, they may converge to a relative or global maximum or minimum point, and use of chemical intuition, such as it is for transition-state structure, may provide a helpful guide in determining an appropriate initial guess. The problems associated with and tentative solutions

---

* See Appendix B of Reference 18 for a discussion of the calculation of quadratic force constants. Note that in the approach proposed by Poppinger[30] an *approximate quadratic force constant matrix* is constructed during the nonlinear search procedure. However, this matrix is not in general of sufficient accuracy for use in further calculations requiring accurate force constant data.

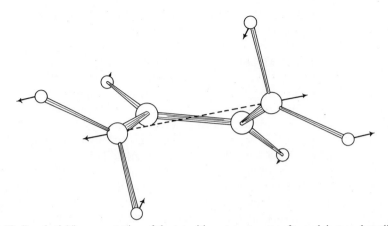

Fig. 6. "Ball-and-stick" representation of the transition-state structure for cyclobutene–butadiene isomerization. The arrows represent the relative motions of the atoms along the reaction (or decomposition) vector at the transition state. Note that the motions of the two carbon atoms of the carbon–carbon bond being broken in the isomerization (dashed line) are colinear with the bond and oppositely directed, as required for a bond-breaking process. (Adapted from McIver and Komornicki[18].)

for making "good" initial guesses have been thoroughly discussed by McIver and Komornicki.[18]

Figure 6 depicts a transition-state structure calculated using the above procedure by McIver and Komornicki[18] for cyclobutene–butadiene isomerization. The arrows represent the "concerted" motion of the particular atoms as the system moves along the reaction vector at the transition state. The dashed line connecting the two carbon atoms represents the cyclobutene carbon–carbon single bond which is broken during the isomerization. Note, in particular, that the direction and magnitude of the motion of these two carbons are such as to favor bond breaking.

4. QUANTUM-MECHANICAL METHODS

4.1. Classification of Methods

As suggested in previous sections, the choice of which quantum-mechanical technique to use in a theoretical study of chemical reactions is a complex one.[31,32] In fact, examination of the various available techniques and the computational procedures necessary for their realization indicates that not only is the choice of a technique a complex one but no one technique can meet all of the accuracy and flexibility criteria that might reasonably be expected if the chosen technique is to be applicable in general to the broad range of potential problems of interest in enzyme chemistry. And although a number of quantum-mechanical theories are available for the calculation of molecular properties and structure, we shall base our discussion primarily on molecular orbital theory techniques,[33] which at the present time are by far the most widely used techniques for the study of large molecular systems.

To consider the available alternatives, it is useful first to classify the various techniques that have been developed, so that their attributes can be assessed and characterized. Broadly speaking, the available techniques can be divided into two classes, namely "*ab initio*" and "semiempirical." While many of the characteristics of these two classes of techniques are similar, their basic difference lies in the manner in which the various terms that arise are treated. In particular, *ab initio* techniques evaluate all terms that arise, using either analytical and/or numerical techniques, with an accuracy that is limited only by the particular computer and/or program that is used. Semiempirical techniques, on the other hand, use experimental data or other criteria to assign or neglect certain terms, thereby usually achieving substantial decreases in the length of time required to carry out a given calculation. It should be noted, however, that the adequacy of a given result is not necessarily improved by using *ab initio* techniques, as might be implied from the above comments. As the discussion below will indicate, the problem is not nearly so simple, and other factors need to be considered as well.

In addition to the above classification, a second characteristic may be of particular significance when chemical reactions are involved. In particular, when chemical bonds are formed and/or broken, it will usually be important to use techniques that can describe the equilibrium as well as the nonequilibrium properties of the molecular system at a number of points (e.g., reactant and transition state) along the reaction path. In theoretical language, this usually implies a need to employ techniques capable of going beyond molecular orbital theory.[34]

To illustrate this concept, as well as to facilitate later comparisons, several comments concerning Hartree–Fock molecular orbital theory and its characteristics and limitations are appropriate. *Molecular orbitals* (MO's) can be defined by

$$\phi_i = \sum_{r=1}^{m} \chi_r c_{ri}, \qquad i = 1, 2, \ldots, m \qquad (21)$$

where the χ_r are known as *basis functions* and the c_{ri} are the MO coefficients that are determined in the calculations and reflect the idiosyncracies of the particular molecular system under consideration. For molecules having $2N$ electrons and a "closed shell" electronic structure (i.e., all electrons are paired), each MO that is occupied will contain two electrons, one with α and one with β spin, i.e.,

$$\psi_i = \phi_i \begin{cases} \alpha \\ \beta \end{cases} \qquad (22)$$

and the *electronic wave function* for the system is given by the Slater determinantal function[33]

$$\Phi_0 = N \det\{\phi_1\alpha(1)\,\phi_2\beta(2)\cdots\phi_N(2N)\,\beta\} \tag{23}$$

where N is a normalization constant.

To determine the MO's, equations of the following form are solved:

$$\hat{H}^{\text{eff}}(1)\,\phi_i(1) = \varepsilon_i\phi_i(1) \tag{24}$$

where ε_i is the energy of the ith MO, ϕ_i, and $\hat{H}^{\text{eff}}(1)$ is an effective one-electron Hamiltonian that contains the interactions between an electron in ϕ_i under the influence of each of the various nuclei and the *average field* produced by all other electrons. Specifically, $\hat{H}^{\text{eff}}(1)$ can be written as

$$\hat{H}^{\text{eff}}(1) = \hat{T}(1) + \hat{V}_{ne}(1) + \hat{V}_{ee}^{\text{eff}}(1) + V_{nn} \tag{25}$$

where $\hat{T}(1)$ represents the kinetic energy operator for electron 1, $\hat{V}_{ne}(1)$ represents the potential energy operator corresponding to attraction of electron 1 to each nucleus, $\hat{V}_{ee}^{\text{eff}}(1)$ is the potential energy operator that describes the interaction of electron 1 with the average field of all the other electrons, and V_{nn} represents the potential energy due to repulsions between the various nuclei [cf. also Eqs. (5)–(7)]. Using Eqs. (21)–(25) and additional analysis, we obtain the *Hartree–Fock equations*,[33,34] shown below in a form appropriate for computational purposes:

$$\sum_{s=i}^{m} (H_{rs}^{\text{eff}} - E_i S_{rs})\,c_{si} = 0, \qquad i = 1, 2, \ldots, m \tag{26}$$

S_{rs} is the overlap integral,

$$S_{rs} = \int \chi_r^*\chi_s\,d\tau \tag{27}$$

and H_{rs}^{eff} is given by

$$H_{rs}^{\text{eff}} = \int \chi_r^*(1)\,\hat{H}^{\text{eff}}(1)\,\chi_s(1)\,d\tau_1 \tag{28}$$

$$= H_{rs}^{\text{core}} + \sum_{t,u}^{m} P_{tu}\{[\chi_r\chi_s|\chi_t\chi_u] - \tfrac{1}{2}[\chi_r\chi_u|\chi_t\chi_s]\} \tag{29}$$

where P_{tu} is an element of the *charge and bond order matrix* defined by

$$P_{tu} = 2\sum_{i}^{\text{occ}} c_{ti}\,c_{ui}^* \tag{30}$$

The "core" matrix element H_{rs}^{core} is defined by

$$H_{rs}^{\text{core}} = \int \chi_r^*(1)\,[\hat{T}(1) + \hat{V}_{ne}(1)]\,\chi_s(1)\,d\tau_1 \tag{31}$$

and the *electron repulsion integrals* are given by

$$[\chi_r\chi_s|\chi_t\chi_u] = \int\int \chi_r^*(1)\,\chi_s(1)\,\frac{1}{r_{12}}\,\chi_t(2)\,\chi_u(2)\,d\tau_1\,d\tau_2 \tag{32}$$

With these definitions and notation, the classification of methods previously mentioned can be easily seen. In particular, the integrals in H_{rs}^{eff} are evaluated directly when *ab initio* techniques are used, but some or all of these integrals are either approximated or neglected when semiempirical procedures are used.

Although Hartree–Fock MO theory is applicable to many types of chemical reactions, in certain cases it *may* be inadequate regardless of whether *ab initio* or semiempirical techniques are employed! For example, closed shell Hartree–Fock theory cannot be used to study chemical reactions in which bond dissociation results in the formation of molecular species with "open shells," i.e., molecular species which contain unpaired electrons. Similarly, reactions which involve the formation of a covalent bond between two open shell molecular species cannot be treated. Furthermore, the inability of Hartree–Fock theory to properly describe the instantaneous interactions between electrons of like spin, generally known as *correlation effects*, can lead to errors of the same order of magnitude as bond energies and thus, if not accounted for, can lead to totally incorrect predictions.[34,35]

To alleviate this difficulty additional methods are needed. At the present time the so-called *configuration interaction* method appears to be the method of choice for large molecular systems.* In the configuration interaction approach the wave function is represented by not one but a series of determinantal functions, Φ_i,

$$\Psi = \sum_{i=0}^{m} C_i\,\Phi_i \tag{33}$$

In most cases Φ_0 is taken as the Hartree–Fock determinantal function defined by Eq. (23), and the additional Φ_i ($i > 0$), which usually have the same form as Φ_0, are obtained by replacing one or more of the ϕ_i in Φ_0 in order to account for the instantaneous correlation of electrons. A more extensive discussion of methods for constructing the Φ_i is given by Schaefer.[34]

Using the above form of Ψ given in Eq. (33), the electronic energy of the system

$$E = \int \Psi^*\hat{H}\Psi\,d\tau \tag{34}$$

where $\hat{H}$ is the Hamiltonian operator described in Eq. (6), is minimized with

---

* Another approach which appears to hold promise for small molecular systems involves the choice of empirical parameters which in themselves contain an implicit dependence on correlation effects even though the Hartree–Fock "framework" may still be employed.

respect to the expansion coefficients, C_j, resulting in the following set of equations [cf. Eq. (26)]:

$$\sum_{j}^{M} (H_{ij} - ES_{ij}) C_j = 0, \qquad i = 1, 2, \ldots, M \tag{35}$$

In the present case [cf. Eqs. (27) and (28)]

$$S_{ij} = \int \Phi_i^* \Phi_j \, d\tau \tag{36}$$

and

$$H_{ij} = \int \Phi_i^* \hat{H} \Phi_i \, d\tau \tag{37}$$

Generally the Φ_i are constructed such that they constitute an orthonormal set of determinantal functions, i.e.,

$$S_{ij} = \begin{cases} 0, & i \neq j \\ 1, & i = j \end{cases} \tag{38}$$

and Eq. (35) then takes on an even simpler form.

4.2. Ab Initio Procedures[36]

Considering first the use of *ab initio* techniques at the Hartree–Fock level, it can be seen from Eq. (21) that different approaches can be obtained by variation in the number (m) of basis functions or in the nature of the basis function themselves (χ_r). In this area, there are several possible approaches that have been taken, all of which use basis functions which are composed of a radial part multiplied by an angularly dependent function.

One approach, used by Huzinaga,[37] Clementi and Davis,[38] Pople et al.,[39] and Whitten,[40] chooses the basis functions by studying isolated atoms. In these studies, the functions that are used are atomic Gaussian orbitals, whose radial dependence is $e^{-\zeta r^2}$. To achieve the angular dependence that is needed, two different approaches have been used. One approach[37-39] uses atomic spherical harmonics or related functions. This approach has the advantage of being related relatively easily to standard calculations on atoms but gives rise to substantial complexities in the evaluation of integrals when orbitals with higher orbital angular momentum are involved. The other approach,[40] using the notion of atomic functions, uses only spherical Gaussians but creates the desired angular dependence by taking appropriate linear combinations of spherical Gaussians that are located off (but nearby) the nucleus of interest. In this manner, "lobe functions" are created. This has the substantial advantage that only integrals over spherical Gaussians are needed. However, this particular manner of achieving angular dependence in the basis is not entirely *rotationally invariant* about the nucleus of interest, and care must be exercised to assure that the results will not be affected substantially by the particular orientation of spherical Gaussians that is chosen.[41]

For large molecular systems such as will typically be of interest in enzyme studies, another kind of basis set has also been devised.[42] This basis set, containing only spherical Gaussian orbitals, eschews the notion that only atomic functions are useful in a molecular context. Instead, the spherical Gaussians are allowed to "float" off the nucleus, and optimized basis functions for use in molecular calculations are determined by studies of molecular fragments instead of atomic systems. This approach has the advantage of allowing attention to be focused on valence electrons, by having spherical Gaussians in bonding and lone-pair regions. Of course, it suffers from the lack of complete local rotational invariance, as does the lobe-function approach, and care must be taken to minimize the nonrotationally invariant components.[41]

In the various techniques, both the size and flexibility of the basis set may have a substantial effect on the results. For example, the most common basis set is a *minimum* (or "single-zeta") *basis set*, where each first-row atom is represented by the Gaussian equivalent of a $1s$, $2s$, $2p_x$, $2p_y$, and $2p_z$ atomic orbital and every hydrogen is represented by a single $1s$ orbital. While this kind of basis set is quite extensively used, it is frequently inadequate and gives rise to artifacts such as a lack of appearance of a transition state. For example, in the study of the S_N2 exchange reaction $F^- + CH_3F \rightarrow FCH_3 + F^-$ carried out by Dedieu and Veillard,[43] the authors pointed out that with a minimum basis set no transition state was obtained. Only when a very extensive basis set was employed did the presumed transition state emerge. Similar conclusions were reached by Baybutt[44] in a more extensive study. Unfortunately, these difficulties are observed in other types of reactions also, most notably carbonyl addition reactions.[45-47] The type of *minimum basis set error* just described arises from an inability of the minimum basis set to describe a localized buildup of negative charge. However, other kinds of minimum basis set errors can also occur (e.g., lack of hybridization flexibility), and care is needed to avoid incorrect conclusions.

A much more satisfactory basis set, which typically contains most of the needed flexibility, is the "double-zeta" basis, in which two functions of each type are used.[48] In addition, "polarization functions" are also sometimes needed to describe the local anisotropy of the charge distribution around nuclei.[42] Unfortunately, such basis sets are usually outside the range of computational feasibility for enzyme systems.* And thus if *ab initio* techniques are to be used, it must be recognized that the size and flexibility of the basis set will almost certainly introduce undesirable effects into the calculation,

---

* The large number of electron repulsion integrals [see Eqs. (29) and (32)] constitutes one of the major computational difficulties associated with *ab initio* calculations on large molecular systems. If N is the number of basis functions, then the number of electron repulsion integrals is of the order $N^4/8$.

and caution must be used in interpreting the results. Even if configuration interaction techniques are used to account for correlation effects, similar difficulties may be expected, since the lack of adequate size and flexibility of the basis set may produce artifacts in configuration interaction studies as well.

4.3. Semiempirical Procedures[49]

As in the case of *ab initio* approaches, a variety of different semiempirical techniques have been developed and used on large systems. For purposes of conciseness, we shall limit our discussion to those techniques capable of dealing with all valence electrons and arbitrary nuclear geometries. Broadly speaking, these methods can be placed into three classes.

The first of these classes involves methods that take into account all overlap effects, but approximate the other effects arising in H_{rs}^{eff}, and are known generally as extended Hückel techniques. The particular form used varies somewhat in different formulations,[50-52] but the general form of the matrix elements can be written as

$$H_{rr}^{\text{eff}} = \alpha_r + (\Delta\alpha_r)Q_A \qquad (r \in A) \tag{39}$$

$$H_{rs}^{\text{eff}} = KS_{rs}(H_{rr}^{\text{eff}} + H_{ss}^{\text{eff}})/2 \tag{40}$$

where α_r is the energy of an electron in the rth atomic orbital of atom A when A is a neutral atom, Q_A is the net charge on atom A, $\Delta\alpha_r$ is the change in α_r per unit change in Q_A, and K is either a constant or chosen to be proportional[51] to S_{rs}. Generally speaking, charge distributions are exaggerated if a self-consistent charge procedure as described in Eqs. (39) and (40) is not used, and barriers and other geometric information are frequently unreliably predicted.[53] Hence, considerable care and testing on prototype systems are needed if this type of technique is to be used.

The second class of methods treats the evaluation of the H_{rs}^{eff} matrix elements directly but ignores some overlaps. These methods, which involve neglect of various differential overlaps,[54,55] are characterized by a variety of acronyms, e.g., CNDO/2,[56] INDO,[57] PRDDO,[58] MINDO/3.[59] As an example of the kind of approach followed when these techniques are used, the following equations represent the expressions used in the INDO formulation:

$$H_{rr}^{\text{eff}} = h_{rr}^{\text{core}} + \sum_{t,u}^{A} P_{tu}[rr/tu] + \sum_{B} \left[\sum_{t}^{B} P_{tt} - Z_B \right] \gamma_{AB} \qquad (r, s, t, u \in A) \tag{41}$$

$$H_{rs}^{\text{eff}} = h_{rs}^{\text{core}} + \sum_{t,u} P_{tu} \left\{ [rs/tu] - \frac{1}{2}[ru/ts] \right\} \qquad (r, s, t, u \in A) \tag{42}$$

and

$$H_{rs}^{\text{eff}} = h_{rs}^{\text{core}} - P_{rs}\gamma_{AB} \qquad (r \in A, s \in B) \tag{43}$$

h_{rr}^{core} is the energy of an electron in the rth atomic orbital, arising from the kinetic energy and potential energy due to the interactions of the nuclei and core electrons, and γ_{AB} is an "average" repulsion between an electron on atom A with one on atom B. It is the nature of the expressions used for H_{rs}^{eff} and the manner of assigning the adjustable parameters (e.g., γ_{AB}) that differentiate the various approaches within the class. Methods in this class tend to be reasonably fast and frequently work well if used on systems and for properties that closely resemble those from which the parameters were determined. If the chemical nature of the system differs substantially from those systems used to determine the parameters or if properties similar to those calculated were not included in the parameterization scheme, unreliable results may be expected. It should be noted that semiempirical methods which seek to emulate minimum basis set *ab initio* calculations will also be subject to minimum basis set errors of the type described earlier.

At least one approach has determined the adjustable parameters so that the results are not restricted to those obtainable using Hartree–Fock theory, even though the formalism used is that of Hartree–Fock theory. In particular, the MINDO approach[59] is one in which the parameters are determined by calibration with experimental data, and not the results of minimum basis set Hartree–Fock calculations. Since the experimental data include all effects, i.e., correlation effects as well as others, the resulting parameters reflect this and allow the procedure to be used even in situations where Hartree–Fock theory will fail. A lack of theoretical consistency results, however, if parameters that contain correlation effects are obtained from equations derived from Hartree–Fock theory. In addition, as with other semiempirical procedures, one must take care to use it only for systems and properties that resemble closely those on which the parameters were determined.

The methods in the last class of semiempirical methods that are available differ from those described above. The PCILO ("perturbative configuration interaction using localized orbitals") procedure[60] begins with a wave function in configuration interaction form [see Eq. (33)], thus allowing, at least in principle, the description of correlation effects as well as those from Hartree–Fock theory. In the individual configurations, valence bond theory[61] is used to construct the functions needed to describe the various bonds, lone pairs, etc. Using these localized orbitals in a configuration interaction wave function, perturbation theory[61] is then employed to evaluate the ground-state energy of the system, utilizing the CNDO/2 integral approximations[56] as an aid in making the procedure efficient. The currently available procedures are quite fast and have been found generally to perform more reliably than CNDO/2 or other semiempirical procedures on conformational problems.[62] Barriers tend to be too small, however, and since the hybridization represented in the localized (e.g., lone pair) functions is fixed, it is expected that difficulties may arise if the procedure is applied to problems in which the hybridization will change (e.g., barriers to inversion involving nonprotonated nitrogen nuclei or bond angle changes).[63]

4.4. Comments and Caveats

As is seen in the previous discussion, there is not a single procedure that can provide the reliability, speed, and flexibility that is needed on problems of interest to enzyme mechanism chemistry, so there is not a "method of choice" at the current time. More positively, however, several of the methods possess quite desirable characteristics in a particular aspect that may make them the method of choice for a particular problem of interest. Thus, when assessing which method(s) might be chosen for use on a given problem, it seems that it is appropriate to ask questions such as the following:

1. Have the parameters been chosen by examining systems similar to those of enzymic interest? This question applies to *ab initio* as well as semiempirical procedures, for the choice of orbital exponents, $e^{-\zeta r^2}$, and/or Gaussian positions will determine the nature of the systems that will be well described by the basis set.
2. What is the computational feasibility of using the procedure on the problem of interest? While the cost of the various methods will vary, depending on programmer and computer idiosyncracies, some methods are also very fast on certain classes of problems (e.g., the PCILO method as applied to conformational studies). If large numbers of nuclear variations are envisioned, this question may be of substantial importance.
3. Can the procedure describe correlation effects; i.e., can it go beyond Hartree–Fock? If the procedure of interest cannot describe correlation effects, then its application must be restricted to studies of systems where correlation effects are relatively constant during the process of interest.
4. Can the sources of error be identified readily, and can the method be improved if the accuracy is inadequate? While this question is certainly of theoretical interest when methods are designed, it should also be of interest to a potential user. In this regard, the choice of a quantum-mechanical method to be used can be likened to the choice of a particular experimental method for a problem of interest, and the choice of which method to be used will certainly depend on the ease of error assessment and the likely size of the errors.

Hence, as mentioned earlier, the choice of method(s) to be used is a complex problem and needs to be considered carefully before embarking on a chemical study.

5. CALCULATION OF ENZYMIC TRANSITION STATES AND REACTION PATHS

5.1. Definition of the Problem

In light of the foregoing discussion of procedures and techniques for the quantum-mechanical calculation of transition states and reaction paths of isolated molecules, let us now consider the problems which arise when these techniques and procedures are applied to entire enzyme–substrate (ES) systems. Figures 7 and 8 provide examples of two well-studied enzymes which possess a typical variety of the types of interactions that occur between a substrate molecule and the specific functional groups of an enzyme.[64,65] The effect of these groups, as well as the remainder of the protein "superstructure," on the catalytic process must be critically assessed in any theoretical model of catalysis.

Due to the relatively large size of most enzyme molecules, straightforward application of the previously discussed techniques and procedures to an entire ES complex is not computationally feasible. However, as is being demonstrated by a growing number of examples, it is possible to emulate many of the properties of complex ES systems (e.g., rate enhancements, pH dependence, substrate specificity, etc.) with much simpler chemical models.[66,67] Thus, it is not unreasonable to assume that properly designed theoretical model systems may also successfully be able to emulate many of the catalytic properties of enzymes. The problem, of course, lies in the proper design of the ES model system.

To develop model systems appropriate for theoretical investigation, it is necessary to consider the type of information sought from the calculation of transition states and reaction paths. First, as detailed in earlier sections, one generally seeks the structure of the reacting species and transition state and, if possible, the structural changes which take place along the minimum energy path. Second, one usually seeks the corresponding electronic (e.g., atomic charges and bond orders) and mechanical (e.g., bond lengths and force constants) properties of the system.

Since most groups in the active-site region of enzymes generally play some, albeit in many cases minor, role in the catalytic process, it is essential that an estimate of the relative magnitude of their effects be made if the development of computationally feasible and biochemically reasonable model ES systems is to be realized. To facilitate such development, it is advantageous to provide as a guide an approximate resolution of the levels of ES interaction, and such a resolution is depicted in Fig. 9. As the guide is only approximate, its suitability must be assessed continually against experimental data derived from studies on model chemical as well as enzyme systems.

It is evident from Fig. 9 that the various types of ES interactions are divided into three categories which are related to structural characteristics of

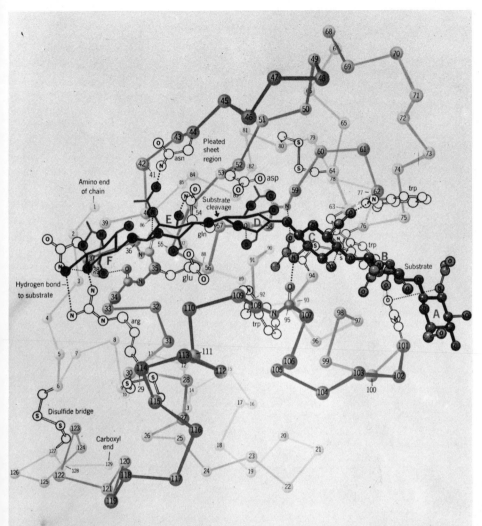

THE FOLDING OF THE MAIN CHAIN IN CHICKEN LYSOZYME. *The crevice that forms the active site runs horizontally across the molecule. The hexasaccharide substrate is shown in a darker color. Rings A, B, and C to the right come from the observed trimer binding. Rings D, E, and F are inferred from model building. The side chains that are believed to interact with the substrate are shown in line. Ilu 98 is so bulky that it helps to prevent NAM, with its large side group, from binding at ring position C, and thus establishes the arrangement on the molecule of the alternating -NAG-NAM- copolymer and points out the locus of cleavage in the active site. (Coordinates by courtesy of Dr. D. C. Phillips, Oxford.)*

Fig. 7. "Ball-and-stick" representation of the ES complex in chicken lysozyme. The site of hydrolytic cleavage of the glycosidic bond in the hexasaccharide substrate is indicated by the arrow, and key hydrogen bonding interactions between the substrate and secondary catalytic groups of the enzyme are indicated by dotted lines. Asp-52 and Glu-35 represent primary catalytic groups which are directly involved in the bond-breaking process. The protein backbone is also indicated in the figure to illustrate the relationship of the bulk enzyme to the substrate and catalytic groups. See Reference 64 for a more complete discussion of the enzyme mechanism. (Reprinted by permission from Dickerson and Geis.[64])

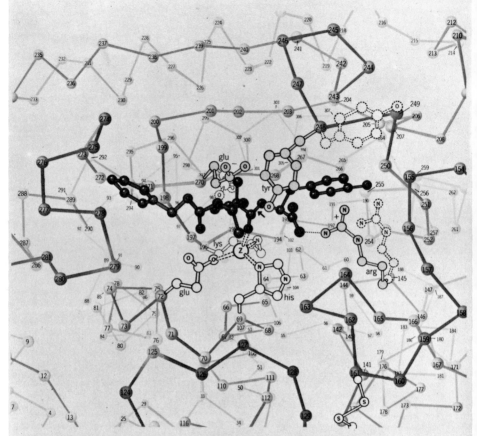

THE ACTIVE SITE WITH SUBSTRATE IN PLACE

*Arg 145, Tyr 248, and Glu 270 are shown before substrate binding
(dotted) and after (solid). Note the tetrahedral coordination of the zinc
atom (Z) to His 69, Glu 72, Lys 196, and the carbonyl oxygen of the
bond to be cleaved in the substrate.*

  Inhibitor

Zn-liganding side chains

Moving side chains without *inhibitor*

Same, with *bound inhibitor*

Fig. 8. "Ball-and-stick" representation of the ES complex in carboxypeptidase A. The site of hydrolytic cleavage of the peptide bond of the peptide substrate is indicated by the arrow. Key catalytic groups are indicated in detail along with a substantial portion of the backbone structure of the bulk enzyme. The tetracoordinate zinc atom ⓩ binds to and polarizes the carbonyl group of the peptide bond cleaved. Note in particular the substantial conformational changes undergone by the various catalytic groups in the absence (indicated by dotted lines and unshaded bonds and atoms) and presence (indicated by solid lines and lightly shaded bonds and atoms) of the substrate. See Reference 64 for a more complete discussion of the enzyme mechanism. (Reprinted by permission from Dickerson and Geis.[64])

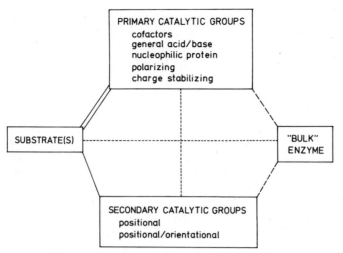

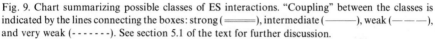

Fig. 9. Chart summarizing possible classes of ES interactions. "Coupling" between the classes is indicated by the lines connecting the boxes: strong ($=\!=\!=\!=$), intermediate (———), weak (— — —), and very weak (- - - - - -). See section 5.1 of the text for further discussion.

the enzyme:

    1. Primary catalytic groups
    2. Secondary catalytic groups
    3. "Bulk" enzyme

Primary catalytic groups are those enzymic groups which exhibit strong to moderate interaction with substrate(s) through the formation of covalent bonds, through the presence of highly charged or polar groups which produce large polarization effects or act to stabilize charged intermediates or transition states, and/or through the donation or acceptance of protons as general acid–base catalysts. The catalytic effects produced by such interactions have been studied successfully in a number of cases using relatively simple model chemical systems.[66]

Secondary catalytic groups are those enzymic groups whose main function is the positioning and orientation of substrates for reaction. Groups which are essentially positional usually involve hydrophobic and/or van der Waals-type interactions. Such interactions generally involve the interaction of a nonpolar or aromatic moiety on the substrate with a "hydrophobic pocket" on the enzyme surface. Orientational effects, on the other hand, are expected to arise primarily from the more directional and specific electrostatic and hydrogen-bonding interactions. These interactions also include "indirect" effects in which water molecules may mediate the interactions between polar groups on the enzyme and substrate.

The remainder of the protein or *"bulk" enzyme* can then be viewed as a "plastic matrix" into which the primary and secondary catalytic groups are embedded. This protein superstructure carries out two principal functions in the catalytic process: (1) It provides a means of arranging the catalytic groups into an appropriate three-dimensional structure for catalysis, and (2) its plasticity provides a mechanism by which, if necessary, rearrangement of these groups can occur during substrate binding or during the entire course of the catalytic process. These two functions of the bulk enzyme raise the following important question: In theoretically modeling ES systems, can the effect of the bulk enzyme on the properties of the substrate and primary and secondary catalytic groups be safely neglected? As described earlier, studies on simpler chemical models suggest that in many cases the answer is yes. However, to provide a more definitive answer it is advantageous to divide the above question further and ask to what extent are the *electronic, vibrational,* and *conformational properties* of the substrate and primary and secondary catalytic groups independent of the bulk enzyme?

First, let us consider *electronic effects.* As is evident from an immense number of theoretical and experimental investigations, electronic effects are usually not propagated through many saturated bonds. Only in conjugated systems are electronic effects expected to be delocalized over a substantial number of chemical bonds. Hence, it should be possible to emulate the electronic effects of primary and secondary catalytic groups through the use of simple model chemical groups, neglecting the remainder of the protein entirely. For example, the effect of the imidazole group of a histidine residue is expected to be effectively modeled using either imidazole or methylimidazole and the effect of the hydroxyl group of a serine residue by methanol.

Second, let us consider *vibrational effects.* In this case the answer is not so obvious since we do not have the same long heritage of "chemical experience" which is available for electronic effects, and thus we need to consider in more detail what molecular properties related to the vibrational properties of the ES complex are of interest. In this regard, the calculation of rate constants and kinetic isotope effects,[31] which in the transition-state theoretical approach requires a knowledge of reactant and transition-state vibrational partition functions,[1,5,14,15,68] is of major importance since it provides a means for uniting proposed enzyme mechanisms with experimental kinetic data.

Specifically, evaluation of vibrational partition functions (in the harmonic approximation) requires a knowledge of the fundamental vibrational frequencies of all the normal modes of vibration. This is not possible for an entire ES system, so that one must decide how much of the structure of the catalytic and substrate groups needs to be considered such that suitable partition functions can be obtained.

Since in the calculation of rate constants and kinetic isotope effects reactant and transition-state vibrational partition functions appear as ratios, normal modes which primarily involve the motions of atoms far from the site of reaction should be little changed between reactants and transition states and thus should effectively cancel. The problem then becomes the determination of suitable "cutoff" procedures. Such procedures have been developed for kinetic isotope effects,[68,69] but so far equivalent developments have not been forthcoming for the calculation of rate constants. Until such procedures are developed for rate constant calculations, the treatment of large molecular systems must be considered somewhat premature.

Nevertheless, in the case of kinetic isotope effects at least, the availability of well-tested "cutoff" procedures indicates that, as was the case for electronic effects, vibrational effects can also be considered to be localized to the region of the ES complex where the reaction is actually taking place. In fact, in many cases, given the correct transition-state structure of the substrate, it may be possible, as a first approximation, to study kinetic isotope effects of enzyme-catalyzed reactions theoretically without explicitly considering the effect of the primary and secondary catalytic groups on the overall vibrational structure of the substrate. Such a simplification will not be possible in cases where the formation or breakdown of a covalent ES complex is the rate-limiting step, but in these cases the functional group(s) on the enzyme (or cofactor) can be modeled in a fashion similar to that described in the discussion of localized electronic effects. However, it may not be possible in some cases to understand the overall enzyme catalytic process in terms of localized vibrational effects alone. As described by two authors,[70,71] the protein superstructure may play an essential role in the "concentration" or "removal" of energy produced by bond making and required by bond breaking processes, although this hypothesis has not as yet been experimentally demonstrated.

Finally, let us consider *conformational effects*. By this we mean the following: To what extent are changes in the structure, position, and orientation of the substrate and catalytic groups independent of the structure of the bulk enzyme? Alternatively, can we safely neglect the effect of the bulk enzyme on the conformational energetics and properties of the catalytic groups?

As an initial approximation, complete neglect of bulk enzyme effects may appear to be too severe, since the catalytic groups are covalently linked to the protein superstructure of the bulk enzyme. However, recent empirical potential energy function calculations by Gelin and Karplus[72] on pancreatic trypsin inhibitor (PTI) provide at least some theoretical evidence that such an approximation is not totally unjustified. In their study they examined the effect of bulk protein relaxations on the calculated rotation barriers of several aromatic amino acid residues located on the surface of the protein. Specifically, they found that barriers to rotation calculated in the presence of a fixed, unrelated bulk protein were about an order of magnitude larger than those calculated for a relaxed bulk protein. Furthermore, the "optimized barriers" were in the range, although slightly larger, of barriers observed for free unsolvated or unbound molecules.[73]

While the above investigation is certainly too limited to draw any general conclusions, it does indicate that, as has in many cases been presumed, the great complexity of the secondary and tertiary structure of proteins allows for considerable "backbone flexibility." Thus "small" motions of particular functional groups can be considered to occur as if they were totally independent of the surrounding bulk protein. However, many more studies similar to that of Gelin and Karplus must be carried out before the limitations of this approximation are thoroughly characterized.

The above discussion clearly indicates that many fundamental problems remain before a chemically reasonable and computationally feasible model of enzyme-catalyzed reactions will be forthcoming. In fact, the discussion of previous sections represents only an initial attempt at the formulation of sufficiently broad principles for use in the development of theoretically sound models of the enzyme catalytic process. The limitations and the refinement of these principles awaits further theoretical and experimental studies, the success of which will depend among other things on a strong interplay between theory and experiment (cf. Maggiora and Schowen[47]).

5.2. An Example: Serine-Protease-Catalyzed Hydrolysis

At the current time, application of quantum-mechanical methods to the study of enzyme catalysis is just beginning, and numerous examples of the principles and procedures are unfortunately unavailable. However, recent studies of serine-protease-catalyzed hydrolysis by Scheiner et al.[74,75] provide an excellent example of the quantum-mechanical approach and clearly indicate the difficulties and limitations as well as the important electronic and mechanical information which can be obtained from such a study.

Serine proteases represent an important and well-studied class of enzymes which catalyze the hydrolysis of peptide bonds. While their primary structures and amino acid compositions are not identical, all enzymes of this class possess almost identical active-site structures, containing a serine residue which acts as a nucleophile in the formation of an *acyl-enzyme intermediate* and a *charge-relay system* which acts as a "proton shuttle."[64-67,76,77]

Figure 10 indicates a proposed catalytic mechanism for α-chymotrypsin (the most well-studied of the serine proteases) as described by Blow et al.[78] in 1969, with the hydrogen-bond chain formed by Asp-102, His-57, and Ser-195 constituting the charge-relay system. As indicated in the figure, the proteolytic reaction takes place in essentially two steps: (1) the *acylation* step [Fig. 10(a)], which results in the formation of an acyl-enzyme intermediate, and (2) the rate-limiting *deacylation* step [Fig. 10(b)] in which the acyl-enzyme intermediate ester bond is hydrolyzed by water.

Although Scheiner et al.[74,75] investigated a reaction path for both the above steps, we shall discuss only the part of their study which deals with the acylation step, as it provides a sufficiently complete example of the approach taken. The structural model used by these workers is based on the X-ray structure of the complex of bovine trypsin with bovine PTI determined by Huber and his associates[79] and is depicted in Figs. 11(a) and (b). The groups indicated in the figure can be considered to be *primary catalytic groups*. Additional optimization of a number of structural degrees of freedom of the

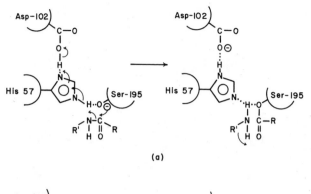

(a)

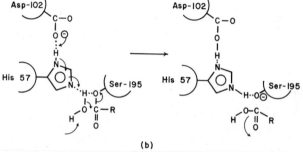

(b)

Fig. 10. Scheme proposed by Blow *et al.*[78] for "charge-relay catalysis" in α-chymotrypsin. This scheme illustrates the basic mechanism of catalysis assumed to operate in all serine proteases: (a) the acylation step and (b) the rate-limiting deacylation step.

model groups did not give rise to major deviations from the X-ray structure and leads to the model (structure A) shown in Fig. 11(b).

During the formation of the acyl-enzyme intermediate [see Fig. 10(a)], the reaction is thought to pass through a *tetrahedral intermediate* as shown in Fig. 11(c). Such intermediates are well established for analogous nonenzymic reactions,[22,23] and a number of kinetic studies indicate[66,80] that such is also true in the enzyme-catalyzed case. Furthermore, the X-ray structure of the trypsin–PTI complex[79] provides more direct structural evidence for the existence of an approximate tetrahedral intermediate and is the structural basis of the present model [Fig. 11(c)].

Figure 12 shows a quantum-mechanically computed two-dimensional potential energy surface for the formation of a tetrahedral ES intermediate based on the above serine protease model. In the figure the α coordinate represents the transfer of H_o from O_s to $N^{\varepsilon 2}$ and the simultaneous, synchronous transfer of $H^{\varepsilon 1}$ from $N^{\varepsilon 1}$ to O_{a2} in terms of the "extent of proton transfer" $(0 \leq \alpha \leq 1)$. The β coordinate represents the even more complex set of molecular motions needed to allow the nucleophilic attack of O_s on the carbon

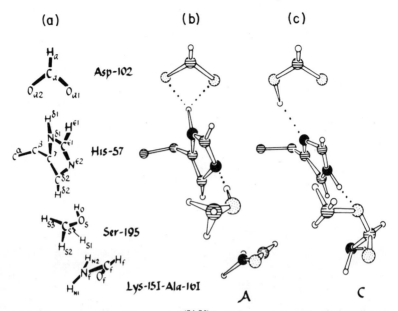

Fig. 11. Model system used by Scheiner *et al.*[74,75] in their quantum-mechanical studies of serine protease catalysis: (a) numbering system used to describe the various residues, (b) structure of the Michaelis ES complex prior to acylation (note the bifurcated hydrogen bond between Asp-102 and His-57), and (c) structure of the tetrahedral intermediate. Note that C^α and C^β of His-57 are indicated merely as reference points for changes in orientation of the imidazole moiety and are not explicitly included in the calculations. (Reprinted by permission from Scheiner and Lipscomb[75].)

(C_f) of the formamide peptide bond.* Due to the large number of degrees of freedom described by the β coordinate, all motions, as was the case for the α coordinate, are required to occur simultaneously and synchronously, so that β measures the extent of completion of the overall geometric process $(0 \le \beta \le 1)$.† The dashed line indicates the minimum energy path determined by the restricted optimization process described above.

A number of important points emerge from the figure and the corresponding calculations. Of primary importance is the *nonconcertedness* predicted for the process. The proton transfer mode is essentially complete ($\alpha \approx 0.8$) before the geometry changes required for formation of the acyl-enzyme intermediate begin ($\beta \approx 0$). This result correlates well with the computed charge on the methoxide group as the reaction proceeds. Initially the charge is only

---

* See References 74 and 75 for details of the motions described by the β coordinate. These motions include (1) the rotation of the methoxide around the C_s—H_{s1} bond axis by 30° to attack C_f, (2) a motion of C_f up out of the carbonyl plane toward the approaching O_s to form the tetrahedral adduct, and (3) the rotation of the imidazole down toward O_s, maintaining the O_s—$N^{\varepsilon 2}$ hydrogen bond. Note also that appropriate bond length changes were included in β.

† Cf. the discussion in Reference 47 on the coupling of "reaction variables."

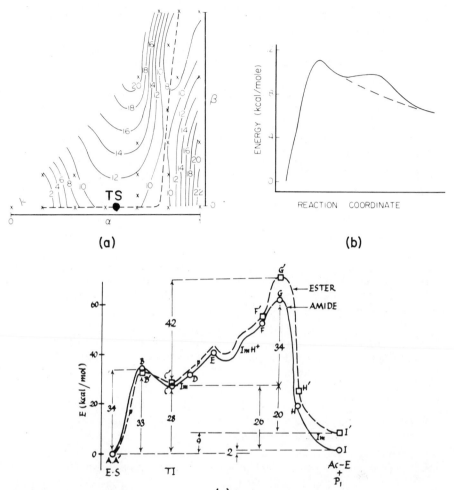

Fig. 12. Energetics of tetrahedral intermediate formation in serine proteases as determined by the quantum-mechanical studies of Scheiner *et al.*[74,75]: (a) Two-dimensional potential energy surface for tetrahedral intermediate formation. α represents the charge-relay proton transfers, and β the complex, concerted motion of the imidazole (His-57), methoxide (Ser-195), and formamide (Lys-16I and Ala-16I) residues needed to form the tetrahedral intermediate. The transition state (TS), indicated by the closed circle, lies on the "minimum energy path" (dashed line). Note the nonconcertedness of the minimum energy path, which is brought about by the high barrier for the concerted process. (b) Reaction coordinate diagram for the formation of the tetrahedral intermediate. The solid line represents the calculated energy as the system moves along the minimum energy path indicated in (a). The dashed line represents the *assumed* energy for the same process (i.e., tetrahedral intermediate formation) allowing independent variation of each of the parameters contained in β. (c) Reaction-coordinate diagram for the complete acylation step. Point C represents the tetrahedral intermediate. The curves in (c) represent more extensive optimizations and more complete correction for minimum basis set effects (see the text and original references for details). (Reprinted from Scheiner *et al.*[74,75].)

-0.34 but changes to ca. -0.65 for $\alpha \approx 0.8$, $\beta \approx 0$. At this point the nucleophilicity of the methoxide group is sufficiently strong to induce the initial structural changes needed ($\beta > 0$) ultimately to form the tetrahedral intermediate. Also of interest is the conclusion, which is consistent with experimental isotope-effect measurements,[80] that proton transfer is the rate-limiting process in the formation of a stable tetrahedral intermediate, and hence the transition-state structure of the formamide moiety shows very little deviation from planarity. This is considerably different from a number of base-catalyzed nonenzymic acyl-transfer reactions in which the transition-state structure approaches the "tetrahedral limit" of the acyl intermediate.[47]

Figure 12(b) depicts the energetics of the above process in a typical "reaction-coordinate diagram" (cf. Figs. 1 and 2), where the reaction coordinate represents the displacement along the minimum energy path shown in Fig. 12(a). As discussed by Scheiner *et al.*,[74] the relative maximum which occurs after the transition state is probably a spurious maximum which arises due to the restrictions placed on the motions of atoms and groups described by the β coordinate. The dashed line represents the *assumed* energy of the process obtained by the independent variation of the structural parameters. The two curves which predict *qualitatively* different behavior, (i.e., the solid line predicts the formation of a metastable intermediate prior to the formation of the tetrahedral intermediate) point out the importance of using as unconstrained an optimization procedure as possible in the determination of a minimum energy path.* This point also re-emphasized the difficulty inherent in studying systems with as many degrees of freedom as encountered in ES systems.

Figure 12(c) presents a recently calculated reaction-coordinate diagram (note that the diagram contains the complete acyl-enzyme formation process for both amides and esters) based on a more extensive optimization with corrections for minimum basis set errors.† In both cases [i.e., Figs. 12(b) and (c)] barriers to tetrahedral intermediate formation exist. However, the energetics differ considerably in the two cases, both with respect to the height of the energy barrier and the energy of the tetrahedral intermediate relative to that of the ES complex. Furthermore, as discussed by these authors,[74,75] the presence of hydrogen-bonding interactions between two residues (*secondary catalytic groups*) on the protein and the carbonyl oxygen (O_f) of the susceptible peptide bond tend to stabilize the developing negative charge on the oxygen brought on by the attack of the highly nucleophilic methoxide (Ser-195). This stabilization due to the presence of secondary catalytic groups tends to

---

* As indicated in the discussion on coupling "reaction variables" in Reference 47, simultaneous and synchronous movement of molecular degrees of freedom is not in general expected.

† Minimum basis set errors arise in the present situation due to the inability of minimum basis set quantum-mechanical calculations to adequately describe highly negatively charged species such as methoxide even in systems which are electronically neutral overall (see also Section 4.2).

make the process more thermoneutral in both cases, although its explicit inclusion is not expected to change materially the nature of the qualitative predictions obtained from the model.

Finally, it is of interest to note that the quantum-mechanical calculations also provide insight into the function of imidazole as a bridging ligand between two acid–base groups.[74] The calculations indicate that imidazole plays a major role in the proton-transfer process of the charge-relay system by providing a means for delocalizing negative charge and by lowering the relative energies of the proton-transfer process through stabilizing hydrogen-bonding interactions. Without the presence of such an electron-delocalizing ligand the removal of a proton from the Ser–OH group would take place effectively in the gas phase and consequently would be extremely difficult. Thus, generation of a strong methoxide-like nucleophile would be severely impeded, and the overall reaction rate would decrease.

In summary, the present example provides an example that illustrates a number of points alluded to in earlier sections:

1. A *qualitatively* reasonable reaction path can be obtained using a simple model containing only primary catalytic groups. Note, however, as discussed in Sections 2.1, 2.2, and 2.3, that such reaction paths may not correspond to actual *physical* reaction paths.
2. Calculated electronic effects correlate well with the structural changes observed during the course of the reaction and thus provide an additional useful tool in theoretical mechanism studies.
3. Proper account must be taken of basis set deficiencies if realistic information is to be obtained.
4. Highly unconstrained optimization and the inclusion of secondary catalytic groups are required for more *quantitative* energetic reaction-path studies.
5. Extensive use is made of experimental data. They are primarily X-ray structural data but also include, for example, any data which bear on the state of protonation of active-site groups through studies of the pH vs. enzyme activity studies.

A number of important points were not examined in the study, for example, treatment of localization of vibrational effects, which is essential for rate-constant and kinetic isotope-effect calculations, as well as a more definitive characterization of bulk enzyme effects. Nevertheless, the study does represent an important step in the quantum-mechanical investigation of enzyme-catalyzed reactions.

For several other examples of quantum-mechanical approaches to the study of enzyme reactions, see, for example, the study by Umeyama *et al.*[81] on α-chymotrypsin and the study by Loew and Thomas[82] on lysozyme.

6. OVERVIEW AND PERSPECTIVES

The mechanism(s) by which enzymes achieve such large rate accelerations ($\sim 10^{10}$) has interested chemists and biochemists for many years. Recently, increasing effort has been directed toward elucidating and assessing the contribution of *specific catalytic effects* (e.g., proximity, "orbital steering," strain, and entropy) to the overall enzymic rate enhancement.[83-89] Unfortunately, the number of proposed effects is mounting,[88] and it appears that soon we shall be inundated by a sea of terms which may drown any hopes we might have of ultimately understanding the molecular basis of enzyme catalysis.

However, the most serious problem with such a proliferation of effects is that many of the effects are not unique (i.e., there is no mutual exclusivity of effects) and that many contain elements common to a number of effects. This is not surprising since rigorous separation of effects would require the corresponding rigorous separation of the classical or quantum-mechanical degrees of freedom of the ES system, a condition which is not likely to be met. Therefore, it appears that to develop a comprehensive molecular theory of enzyme catalysis one should proceed from a general molecular dynamical theory and then, within the framework of the general theory, "dissect" out the specific catalytic effects. In this way, the theory which ultimately emerges will be consistent with known physical and chemical principles, and the degree of independence of the individual catalytic effects can be critically assessed.

Unfortunately, a complete molecular dynamical characterization of enzyme catalysis for even a very simple enzyme system is not possible at the present time. In fact, as discussed by Hopfinger,[73] dynamical characterization of the conformational fluctuations due only to torsional motions about single bonds is beyond current computational capabilities for any biologically realistic protein system. Although refinements of data obtained in X-ray structure studies have been carried out using energy minimization procedures based on empirical potential energy functions for a number of reasonably large proteins,[90,91] it is not expected that such functions will adequately represent the energetics associated with the very anharmonic motions (e.g., bond breaking) of substrate molecules undergoing reaction. This is due primarily to the fact that the parameterization of essentially all empirical potential energy functions designed for large molecular systems is based on molecular properties at or near equilibrium (e.g., vibration, structural, and thermodynamic data).[92] As much less such data are available for transition states and nonequilibrium points along, for example, the minimum energy path, proper reparameterization of the potential energy functions is a nontrivial problem. Application of quantum-mechanical methods effectively trades one problem for another. That is, while quantum-mechanical methods can handle the energetic consequences of the large-amplitude nuclear motions encountered in chemical reactions, they are not currently suitable for the study of large aperiodic macromolecular systems the size of catalytic proteins.

In this regard, transition-state theory offers a theoretically well-defined alternative to the much more complex molecular dynamical theories for studying the mechanism of enzymic catalysis and thus for elucidating and assessing the specific catalytic effects responsible for their large catalytic rate enhancements. As indicated in the earlier discussions, quantum-mechanical procedures can and have been used effectively in the study of transition states and reaction paths of simple chemical reactions, and early evidence indicates that these procedures should also be useful in the study of ES systems. However, many important questions remain unanswered; for example, what are the consequences of neglecting bulk enzyme effects on the vibrational frequencies and conformational energetics of the ES system? Since the answers to these questions have an important bearing on the computational feasibility of the transition-state approach to enzyme-catalyzed reactions, it is important to begin to develop appropriate theoretical models to answer these and related questions. Perhaps it will be necessary to adopt a type of hybrid approach in which quantum-mechanical and empirical potential energy function procedures are merged into a massive computational cycle which iterates between both types of procedures in a coupled fashion until "self-consistency" is obtained, and perhaps in this way some assessment of bulk enzyme effects will be possible.

Even if the transition-state theoretical approach cannot be fully implemented, quasi-dynamical approaches as described earlier for both chemical and enzyme-catalyzed reactions should provide significant electronic and mechanical information over the entire course of the reaction. Such an approach may in many cases be easier to implement computationally. Furthermore, it may be that a fundamental understanding of the catalytic process will not be gained until a detailed understanding of the electronic *and* mechanical properties of the appropriate transition states and/or the changes in these properties as the reacting system moves along the minimum energy reaction path is obtained. Thus, quantum-mechanical procedures may not only be advantageous but may also be essential for reaching any definitive conclusions on the mechanism(s) of enzymic catalysis.

ACKNOWLEDGMENTS

The authors would like to express their appreciation to Professors R. L. Schowen, R. H. Himes, and B. K. Lee for helpful discussions and to the Upjohn Company, Kalamazoo, Michigan, for partial support of this work.

REFERENCES

1. S. Glasstone, K. J. Laidler, and H. Eyring, *The Theory of Rate Processes*, McGraw-Hill, New York (1941).

2. F. H. Johnson, H. Eyring, and J. Polissar, *The Kinetic Basis of Molecular Biology*, Wiley, New York (1954).
3. R. D. Levine, *Quantum Mechanics of Molecular Rate Processes*, Clarendon Press, Oxford (1969).
4. D. L. Bunker, Classical trajectory methods, *Methods Comput. Phys.* **10**, 287–325 (1971).
5. K. J. Laidler, *Theories of Chemical Reaction Rates*, McGraw-Hill, New York (1969).
6. R. D. Levine and R. B. Bernstein, *Molecular Reaction Dynamics*, Oxford University Press, New York (1974).
7. H. Goldstein, *Classical Mechanics*, Addison-Wesley, Reading, Mass. (1950).
8. K. Fukui, A formulation of the reaction coordinate, *J. Phys. Chem.* **74**, 4161–4163 (1970).
9. M. J. S. Dewar and R. C. Dougherty, *The PMO Theory of Organic Chemistry*, Plenum Press, New York (1975).
10. H. Metiu, J. Ross, R. Silbey, and T. F. George, On symmetry properties of reaction coordinates, *J. Chem. Phys.* **61**, 3200–3209 (1974).
11. R. Fuchs and E. S. Lewis, in: *Techniques of Chemistry* (E. S. Lewis, ed.), Vol. 6, Part 1, Chap. 4, Wiley, New York (1974).
12. H. S. Johnston, *Gas Phase Reaction Rate Theory*, Ronald, New York (1966).
13. D. L. Bunker, Simple kinetic models from Arrhenius to the computer, *Acc. Chem. Res.* **7**, 195–201 (1974).
14. H. Eyring and E. M. Eyring, *Modern Chemical Kinetics*, Van Nostrand Reinhold, New York (1963).
15. K. J. Laidler, *Chemical Kinetics*, McGraw-Hill, New York (1965).
16. J. O. Hirschfelder, Coordinates which diagonalize the kinetic energy of relative motion, *Int. J. Quantum Chem.* **3S**, 17–31 (1969).
17. M. Born and K. Huang, *Dynamical Theory of Crystal Lattices*, Oxford University Press, New York (1954).
18. J. W. McIver and A. Komornicki, Structure of transition states of organic reactions. General theory and an application to the cyclobutene–butadiene isomerization using a semiempirical molecular orbital method, *J. Am. Chem. Soc.* **94**, 2625–2633 (1972).
19. M. J. S. Dewar and S. Kirschner, MINDO/2 study of antiaromatic ("forbidden") electrocyclic processes, *J. Am. Chem. Soc.* **93**, 4291–4292 (1971).
20. M. J. S. Dewar and S. Kirschner, Classical and non-classical potential surfaces. The significance of antiaromaticity in transition states, *J. Am. Chem. Soc.* **93**, 4292–4294 (1971).
21. M. J. S. Dewar and S. Kirschner, Nature of transition states in "forbidden" electrocyclic reactions, *J. Am. Chem. Soc.* **96**, 5244–5246 (1974).
22. W. P. Jencks, *Catalysis in Chemistry and Enzymology*, McGraw-Hill, New York (1969).
23. M. L. Bender, *Mechanisms of Homogeneous Catalysis from Protons to Proteins*, Wiley, New York (1971).
24. I. S. Sokolnikoff and R. M. Redheffer, *Mathematics of Physics and Modern Engineering*, McGraw-Hill, New York (1958).
25. E. B. Wilson, J. C. Decius, and P. C. Cross, *Molecular Vibrations*, McGraw-Hill, New York (1955).
26. F. W. Byron and R. W. Fuller, *Mathematics of Classical and Quantum Physics*, Addison-Wesley, Reading, Mass. (1969).
27. R. E. Stanton and J. W. McIver, Group theoretical selection rules for the transition states of chemical reactions, *J. Am. Chem. Soc.* **97**, 3632–3646 (1975).
28. J. N. Murrell and K. J. Laidler, Symmetries of activated complexes, *Trans. Faraday Soc.* **64**, 371–377 (1968).
29. M. J. S. Dewar and S. Kirschner, MINDO/3 study of the thermal conversion of cyclobutene to 1,3-butadiene, *J. Am. Chem. Soc.* **96**, 6809–6810 (1974).
30. D. Poppinger, On the calculation of transition states, *J. Am. Chem. Soc.* **35**, 550–554 (1975).
31. H. Eyring and S. H. Lin, in: *Physical Chemistry: An Advanced Treatise* (H. Eyring, ed.), Vol. 6, Part A, Chap. 3, Academic Press, New York (1975).
32. G. G. Balint-Kurti, in: *Molecular Scattering* (K. P. Lawley, ed.), pp. 137–183, Wiley, London (1975).

33. P. O'D. Offenhartz, *Atomic and Molecular Orbital Theory*, McGraw-Hill, New York (1969).
34. H. F. Schaefer III, *The Electronic Structure of Atoms and Molecules*, Addison-Wesley, Reading, Mass. (1972).
35. A. C. Wahl, Recent progress beyond the Hartree–Fock method for diatomic molecules: The method of optimized valence configurations, *Int. J. Quantum Chem.* **1S**, 123–152 (1967).
36. D. B. Cook, *Ab Initio Valence Calculations in Chemistry*, Wiley, New York (1974).
37. S. Huzinaga, Gaussian-type functions for polyatomic systems. I, *J. Chem. Phys.* **42**, 1293–1302 (1965).
38. E. Clementi and D. R. Davis, Electronic structure of large molecular systems, *J. Comput. Phys.* **1**, 223–244 (1966).
39. W. J. Hehre, R. F. Stewart, and J. A. Pople, Self-consistent molecular-orbital methods. I. Use of Gaussian expansions of Slater-type atomic orbitals, *J. Chem. Phys.* **51**, 2657–2664 (1969).
40. J. L. Whitten, Gaussian lobe function expansions of Hartree–Fock solutions for first-row atoms and ethylene, *J. Chem. Phys.* **44**, 359–364 (1966).
41. G. M. Maggiora and R. E. Christoffersen, *Ab Initio* calculations on large molecules using molecular fragments. Generalization and characteristics of floating spherical Gaussian basis sets, *J. Am. Chem. Soc.* **98**, 8325–8332 (1976).
42. R. E. Christoffersen, D. Spangler, G. G. Hall, and G. M. Maggiora, *Ab initio* calculations on large molecules using molecular fragments. Evaluation and extension of initial procedures, *J. Am. Chem. Soc.* **95**, 8526–8536 (1973).
43. A. Dedieu and A. Veillard, *Ab initio* calculation of the activation energy for an S_N2 reaction, *Chem. Phys. Lett.* **5**, 328–330 (1970).
44. P. Baybutt, The molecular orbital description of S_N2 reactions at silicon centers, *Mol. Phys.* **29**, 389–403 (1975).
45. G. Alagona, E. Scrocco, and J. Tomasi, An *ab initio* study of the amidic bond cleavage by OH^- in formamide, *J. Am. Chem. Soc.* **97**, 6976–6983 (1975).
46. H. Bürgi, J. D. Dunitz, J. M. Lehn, and G. Wipff, Stereochemistry of reaction paths at carbonyl centers, *Tetrahedron* **30**, 1563–1572 (1974).
47. G. M. Maggiora and R. L. Schowen, in: *A Survey of Bioorganic Chemistry* (E. E. van Tamelen, ed.), Vol. I, pp. 173–229, Wiley, New York (1977).
48. L. C. Snyder and H. Basch, *Molecular Wavefunctions and Properties*, Wiley, New York (1972).
49. K. Jug, On the development of semi-empirical methods in the MO formalism, *Theor. Chim. Acta* **14**, 91–135 (1969).
50. R. Hoffman, An extended Hückel theory. I. Hydrocarbons, *J. Chem. Phys.* **39**, 1397–1412 (1963).
51. L. C. Cusachs and J. W. Reynolds, Selection of molecular matrix elements from atomic data, *J. Chem. Phys.* **43**, S160–S164 (1965).
52. R. Rein, N. Fukuda, H. Win, G. A. Clarke, and F. E. Harris, Iterated extended Hückel theory, *J. Chem. Phys.* **45**, 4743–4744 (1966).
53. L. C. Allen, in: *Sigma Molecular Orbital Theory* (O. Sinanoğlu and K. B. Wiberg, eds.), pp. 227–248, Yale University Press, New Haven, Conn. (1970).
54. J. A. Pople and D. L. Beveridge, *Approximate Molecular Orbital Theory*, McGraw-Hill, New York (1970).
55. J. N. Murrell and J. A. Harget, *Semi-Empirical Self-Consistent-Field Molecular Orbital Theory of Molecules*, Wiley, New York (1972).
56. J. A. Pople, D. P. Santry, and G. A. Segal, Approximate self-consistent molecular orbital theory. I. Invariant procedures, *J. Chem. Phys.* **43**, S129–S135 (1965).
57. J. A. Pople, D. L. Beveridge, and P. A. Dobosh, Approximate self-consistent molecular orbital theory. V. Intermediate neglect of differential overlap, *J. Chem. Phys.* **47**, 2026–2033 (1967).
58. T. A. Halgren and W. N. Lipscomb, Self-consistent-field wavefunctions for complex molecules. The approximation of partial retention of diatomic differential overlap, *J. Chem. Phys.* **58**, 1569–1591 (1973).
59. R. C. Bingham, M. J. S. Dewar, and D. H. Lo, Ground states of molecules. XXVI. MINDO/3

calculations for hydrocarbons, for CHON species, and for compounds containing third row elements, *J. Am. Chem. Soc.* **97**, 1294–1318 (1975).

60. S. Diner, J. P. Malrieu, and P. Claverie, Localized bond orbitals and the correlation problem. I. The perturbation calculation of the ground state energy, *Theor. Chim. Acta* **13**, 1–17 (1969).

61. F. L. Pilar, *Elementary Quantum Chemistry*, McGraw-Hill, New York (1968).

62. D. Perahia and A. Pullman, Success of the PCILO method and failure of the CNDO/2 method for predicting conformations in some conjugated systems, *Chem. Phys. Lett.* **19**, 73–75 (1973).

63. J. Langlet and H. van der Meer, Calculation of molecular geometry with the PCILO method, *Theor. Chim. Acta* **21**, 410–412 (1971).

64. R. E. Dickerson and I. Geis, *The Structure and Action of Proteins*, Harper & Row, New York (1969).

65. R. Henderson and J. H. Wang, Catalytic Configurations, *Ann. Rev. Biophys. Bioeng.* **1**, 1–26 (1972).

66. T. H. Fife, Physical organic model systems and the problem of enzymatic catalysis, *Adv. Phys. Org. Chem.* **11**, 1–122 (1975).

67. Y. Iwakura, K. Uno, F. Toda, S. Onozuka, K. Hattori, and M. L. Bender, The stereochemically correct catalytic site on cyclodextrin resulting in a better enzyme model, *J. Am. Chem. Soc.* **97**, 4432–4434 (1975).

68. W. A. Van Hook, in: *Isotope Effects in Chemical Reactions* (C. J. Collins and N. S. Bowman, eds.), pp. 1–89, Van Nostrand Reinhold, New York (1970).

69. E. K. Thornton and E. R. Thornton, in: *Isotope Effects in Chemical Reactions* (C. J. Collins and N. S. Bowman, eds.), pp. 213–285, Van Nostrand Reinhold, New York (1970).

70. P. E. Phillipson, On the possible importance of relaxation processes in enzyme catalysis, *J. Mol. Biol.* **31**, 319–321 (1968).

71. M. V. Vol'kenshtein, *Enzyme Physics*, Plenum Press, New York (1969).

72. B. R. Gelin and M. Karplus, Side chain torsional potentials and motion of amino acids in proteins: Bovine pancreatic trypsin inhibitor, *Proc. Nat. Acad. Sci. USA* **72**, 2002–2006 (1975).

73. A. J. Hopfinger, *Conformational Properties of Macromolecules*, Academic Press, New York (1973).

74. S. Scheiner, D. A. Kleier, and W. N. Lipscomb, Molecular orbital studies of enzyme activity: I: Charge relay system and tetrahedral intermediate in acylation of serine proteinases, *Proc. Nat. Acad. Sci. USA* **72**, 2606–2610 (1975).

75. S. Scheiner and W. N. Lipscomb, Molecular orbital studies of enzyme activity: Catalytic mechanism of serine proteinases, *Proc. Nat. Acad. Sci. USA* **73**, 432–436 (1976).

76. L. Polgár and B. Asbóth, On the stereochemistry of catalysis by serine proteases, *J. Theor. Biol.* **46**, 543–558 (1974).

77. Papers on the structure and function of proteins at the three-dimensional level, *Cold Spring Harbor Symp. Quant. Biol.* **36**, 63–150 (1972).

78. D. M. Blow, J. J. Birktoft, and B. S. Hartley, Role of a buried acid group in the mechanism of action of chymotrypsin, *Nature (London)* **221**, 337–340 (1969).

79. A. Rühlmann, D. Kukla, P. Schwager, K. Bartels, and R. Huber, Structure of the complex formed by bovine trypsin and bovine pancreatic trypsin inhibitor, *J. Mol. Biol.* **77**, 417–436 (1973).

80. M. L. Bender and J. V. Kilheffer, Chymotrypsins, *CRC Crit. Rev. Biochem.* **1**, 149–199 (1973).

81. H. Umeyama, A. Imamura, C. Nagata, and M. Hanano, A molecular orbital study on the enzymic reaction mechanism of α-chymotrypsin, *J. Theor. Biol.* **41**, 485–502 (1973).

82. G. H. Loew and D. D. Thomas, Molecular orbital calculations of the catalytic effect of lysozyme, *J. Theor. Biol.* **36**, 89–104 (1972).

83. T. C. Bruice, in: *The Enzymes*, 3rd ed. (P. D. Boyer, ed.), Vol. II, pp. 217–279, Academic Press, New York (1970).

84. D. E. Koshland, K. W. Carraway, G. A. Dafforn, J. D. Gass, and D. R. Storm, The importance of orientation factors in enzymatic reactions, *Cold Spring Harbor Symp. Quant. Biol.* **36**, 13–20 (1971).

85. T. C. Bruice, Views on approximation, orbital steering, and enzymatic model reactions, *Cold Spring Harbor Symp. Quant. Biol.* **36**, 21–27 (1971).
86. D. M. Chipman and N. Sharon, Mechanism of lysozyme action, *Science* **165**, 464–465 (1969).
87. C. Delisi and D. M. Crothers, The contribution of proximity and orientation to catalytic reaction rates, *Biopolymers* **12**, 1689–1704 (1973).
88. W. P. Jencks and M. I. Page, "Orbital steering," entropy, and rate accelerations, *Biochem. Biophys. Res. Commun.* **57**, 887–892 (1974).
89. A. R. Fersht, Catalysis, binding, and enzyme-substrate complementarity, *Proc. Roy. Soc. London Ser. B* **187**, 397–407 (1974).
90. P. K. Warme and H. A. Scheraga, Refinement of the X-ray structure of lysozyme by complete energy minimization, *Biochemistry* **13**, 757–767 (1974).
91. M. Levitt, Energy refinement of hen egg-white lysozyme, *J. Mol. Biol.* **82**, 393–420 (1974).
92. B. S. Hudson, in: *Neutron, X-Ray, and Laser Spectroscopy in Biology and Chemistry* (S. Chen and S. Yip, eds.), pp. 119–144, Academic Press, New York (1974).

4

Primary Hydrogen Isotope Effects

Judith P. Klinman

1. INTRODUCTION

The important relationship between enzyme catalysis and the enhanced affinity of an enzyme for its substrate in the transition state was first noted by Pauling in 1948.[1] In recent years, attention has been focused on the design of stable analogs of transition states of enzyme reactions, since the structure of high-affinity analogs can provide considerable insight into the structure and energy of enzyme transition states. The factors which give rise to enhanced transition-state analog binding can be complex, however, and many of the analogs which have been tested mimic intermediates rather than true transition states along the enzyme reaction path.[2-5]

Although a number of kinetic probes of transition-state structure are available for reactions in solution, application of these methods to enzyme systems is frequently limited by the conformational and specificity requirements of enzymes. The substitution of hydrogen by a deuterium or tritium comes as close as possible to a noninteracting probe, since vibrational force constants are independent of isotopic substitution. This fact, together with the large number of enzyme systems which catalyze hydrogen abstraction reactions, makes the application of primary hydrogen isotope effects to the study of enzyme transition-state structure particularly attractive. In the past a major application of isotope effects has been to determine whether hydrogen transfer is occurring in the rate-determining step(s) of an enzyme reaction; in addition, hydrogen isotope effects have been put to elegant use in determining the stereochemical pathway of enzyme reactions and in demonstrating the presence of intermediates in enzyme-catalyzed C—H cleavage reactions.[6-11]

Judith P. Klinman • The Institute for Cancer Research, Fox Chase Cancer Center, Philadelphia, Pennsylvania.

An objective of this chapter is to point out the application and limitation of kinetic primary hydrogen isotope effects in providing information on transition-state structure in enzyme-catalyzed reactions.

2. ORIGIN AND INTERPRETATION

The abstraction of hydrogen is a fundamental process in both chemical and enzyme-catalyzed reactions. Substitution of hydrogen by deuterium will normally slow down the rate of such reactions by a large factor, e.g., $k_H/k_D = 2$–10, where k_H is the rate constant for hydrogen abstraction and k_D is the rate constant for deuterium abstraction. In addition to kinetic effects, isotopic substitution can give rise to primary hydrogen equilibrium isotope effects, which may be normal (>1) or inverse (<1). In this chapter we shall be concerned primarily with the transfer of hydrogen from carbon to either another carbon center or a hetero atom such as oxygen, nitrogen, or sulfur rather than the transfer of hydrogen between two hetero atoms, which is normally considered within the context of solvent isotope effects (cf. Chapter 6 in this volume). Although the factors which can contribute to kinetic isotope effects are complex, models have been proposed relating the magnitude of the observed isotope effect to the structure of the transition state.

2.1. Formulation of the Kinetic Isotope Effect

The theory of kinetic isotope effects is generally developed within the framework of transition-state theory.[12-17] The ground state of a reaction is considered to be in equilibrium with its transition state, and $K^{\ddagger}$ is the equilibrium constant defining this process. In the case of the transfer of hydrogen between two atomic centers, $A\!-\!H + B \rightleftharpoons [A\text{-}\text{-}H\text{-}\text{-}B]^{\ddagger}$, the equilibrium constant for this process can be expressed in terms of partition functions representing the contribution of vibrational (Q_V), rotational (Q_R), and translational (Q_T) motions to the total energy of the transition state ($[A\text{-}\text{-}H\text{-}\text{-}B]^{\ddagger}$) and ground state ($A\!-\!H + B$):

$$K^{\ddagger} = \frac{Q_V^{\ddagger}Q_R^{\ddagger}Q_T^{\ddagger}e^{-\Delta E/kT}}{Q_{V(AH)}Q_{R(AH)}Q_{T(AH)}Q_{V(B)}Q_{R(B)}Q_{T(B)}} \tag{1}$$

The exponential in Eq. (1) contains ΔE, the difference in electronic potential energy between the transition state and ground state [Fig. 1(a)]. As stated by the Born–Oppenheimer approximation, *this potential energy difference is independent of nuclear masses and ΔE is assumed to be unchanged upon isotopic substitution of reacting molecules.*

The transition state differs from ground-state molecules in that one of its vibrational modes is the reaction coordinate. The reaction coordinate, which corresponds to a *translation* of H from A to B, is described by a vibrational

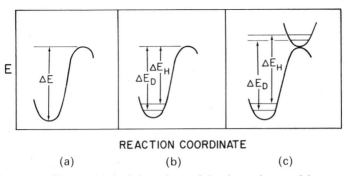

REACTION COORDINATE

(a)   (b)   (c)

Fig. 1. Diagrams to illustrate (a) the independence of the electronic potential energy surface of isotopic substitution, (b) the origin of maximal isotope effects due to the loss of a ground-state stretching frequency, and (c) the cancellation of ground-state by transition-state zero-point energy leading to small kinetic isotope effects.

frequency $\gamma_L^{\ddagger}$. The magnitude of $\gamma_L^{\ddagger}$ is a measure of the curvature of the energy barrier and can be thought of as the frequency of vibration of a particle in a stable potential energy well, except that this vibration has a negative rather than a positive restoring force. From Hooke's law the frequency of a vibrating particle is $\gamma = (1/2\pi)\sqrt{F/M_R}$, where F is the force constant and M_R the reduced mass; a negative force constant leads to an expression containing $\sqrt{-1}$, and $\gamma_L^{\ddagger}$ is referred to as the imaginary frequency. Factoring the partition function for the reaction coordinate out of Eq. (1) gives $K^{\ddagger} = K(h\gamma_L^{\ddagger}/\mathbf{k}T)$, where $\mathbf{k}$, T, and h are Boltzmann's constant, temperature in $°K$, and Planck's constant, respectively.

The kinetic isotope effect k_H/k_D is equal to the ratio $K_H^{\ddagger}/K_D^{\ddagger}$, which is considerably simplified relative to $K^{\ddagger}$, since partition functions for non-isotopically labeled reactants and the exponential, $e^{-\Delta E/\mathbf{k}T}$, cancel. By use of the Redlich–Teller product rule, which permits one to describe rotational and translational partition functions by vibrational frequencies, Bigeleisen's formulation[12] for the isotope effect is obtained for nonlinear molecules of N atoms, where $\mu = h\gamma/\mathbf{k}T$:

$$\frac{k_H}{k_D} = \left(\frac{\gamma_{HL}^{\ddagger}}{\gamma_{DL}^{\ddagger}}\right)\underbrace{\left(\frac{\overset{3N-6}{\underset{}{\prod}}\frac{\mu_{Di}}{\mu_{Hi}}}{\overset{3N-7}{\underset{}{\prod}}\frac{\mu_{Di}^{\ddagger}}{\mu_{Hi}^{\ddagger}}}\right)\left(\frac{\overset{3N-6}{\underset{}{\prod}}\frac{(1-e^{-\mu_{Hi}})}{(1-e^{-\mu_{Di}})}}{\overset{3N-7}{\underset{}{\prod}}\frac{(1-e^{-\mu_{Hi}^{\ddagger}})}{(1-e^{-\mu_{Di}^{\ddagger}})}}\right)}_{\text{PEF}}\underbrace{\left(\frac{\exp\overset{3N-6}{\underset{}{\sum}}(\mu_{Hi}-\mu_{Di})/2}{\exp\overset{3N-7}{\underset{}{\sum}}(\mu_{Hi}^{\ddagger}-\mu_{Di}^{\ddagger})/2}\right)}_{\text{EF}} \quad (2)$$

This expression for the isotope effect is conveniently divided into a pre-exponential factor (PEF) and an exponential factor (EF). Included in the PEF term is the contribution of molecular rotations, translations, and excited-state vibrations to the isotope effect. At or near room temperature the contribution of excited-state molecular vibrations is negligible; furthermore, isotopic substitution of hydrogen is not expected to affect appreciably rotational and translational motions, and PEF $\simeq 1$. The temperature dependence of this term is small, and the PEF contribution to k_H/k_D is estimated to fall within the range of $0.5–1.4$.[17] The exponential factor in Eq. (2), which contains the zero-point vibrational frequencies of the ground state and transition state, is generally concluded to be the single most important factor in determining the magnitude of primary hydrogen isotope effects. Vibrational frequencies are conveniently expressed as wave numbers (ω) rather than cycles per second (γ), $\omega = \gamma/c$, where c is the speed of light:

$$\frac{k_H}{k_D} = \exp\frac{hc}{2kT}\left[\sum_{(reactant)}^{3N-6}(\omega_{Hi} - \omega_{Di}) - \sum_{\substack{(transition \\ state)}}^{3N-7}(\omega_{Hi}^{\ddagger} - \omega_{Di}^{\ddagger})\right] \qquad (3)$$

A consideration of the unimolecular decomposition of a diatomic molecule provides a simple illustration of the contribution of zero-point energy to the isotope effect. The conversion of A—H to $[\text{A}----\text{H}]^{\ddagger}$ involves the loss of a single ground-state stretching vibration; the vibration of the transition state is the reaction coordinate and does not contribute to the isotope effect:

$$\frac{k_H}{k_D} = \exp\frac{hc}{2kT}(\omega_H - \omega_D) \qquad (4)$$

The frequency of the ground-state vibration, $\omega = (1/2\pi c)\sqrt{F/M_R}$, is dependent on a force constant, F, which is independent of isotopic substitution, and the reduced mass of A—H vs. A—D, $M_R = M_A M_{H(D)}/[M_A + M_{H(D)}]$. When the mass of A is large relative to H or D, the reduced mass of A—H(D) is approximated by the mass of H(D), and $\omega_D \simeq \omega_H/\sqrt{2}$. Figure 1(b) illustrates the decrease in zero-point energy and increase in activation energy which result from the larger reduced mass and smaller vibrational frequency of A—D relative to A—H. Predicted maximum kinetic isotope effects due to the loss of a single ground-state stretching frequency are summarized in Table I. The

Table I
Maximum Kinetic Isotope Effects Due to
the Loss of a Ground-State Stretching
Frequency, 25°C

Bond	$\omega(\text{cm}^{-1})$	k_H/k_D	k_H/k_T
C—H	3000	6.8	15.8
N—H	3100	7.5	18.2
O—H	3300	8.5	21.8
S—H	2500	5.5	11.7

interrelationship between deuterium and tritium isotope effects reflects primarily the difference in mass between deuterium and tritium, as stated by the Swain equation[18]:

$$\left(\frac{k_H}{k_D}\right)^{1.44} = \frac{k_H}{k_T} \tag{5}$$

2.2. Force Constant Model Relating the Magnitude of the Kinetic Isotope Effect to the Extent of Hydrogen Transfer in the Transition State

Primary hydrogen isotope effects are frequently observed to be less than the maximal effects summarized in Table I. Although small isotope effects will arise in complex reactions characterized by several rate-limiting steps, many simple hydrogen abstraction reactions are characterized by isotope effects less than 6–9. Westheimer[19] and others[13,20] have analyzed the contribution of transition-state vibrational frequencies to the magnitude of observed isotope effects in the linear, three-atom system: $A—H + B \rightleftharpoons [A{-}{-}H{-}{-}B]^{\ddagger}$. The ground state, $A—H + B$, is characterized by a single vibrational degree of freedom. In the transition state, $[A{-}{-}H{-}{-}B]^{\ddagger}$, the translational degrees of freedom of one of the reactants have been frozen out; consequently, of $3N$ degrees of freedom five correspond to rotations and translations and there are four vibrational degrees of freedom. The transition-state vibrational modes correspond to an asymmetric and symmetric stretch and two degenerate bending modes:

$$A{-}{-}{-}H{-}{-}{-}B \qquad A{-}{-}{-}H{-}{-}{-}B$$
$$\leftarrow O \qquad O\rightarrow \quad \leftarrow O \qquad\qquad O \qquad O \qquad O$$

asymmetric stretch   bends (degenerate)

$$A \overset{F_1}{-\!-} H \overset{F_2}{-\!-} B$$
$$O\rightarrow \qquad \leftarrow O \qquad \leftarrow O$$

symmetric stretch

A simplified mathematical treatment, which neglects the two bending modes, leads to an expression for the frequency of the asymmetric and symmetric stretching modes in terms of force constants, F_1, F_2, and F_{12}, and the masses of A, B, and H.[19] The force constants F_1 and F_2 refer to the $A{-}{-}{-}H$ and $H{-}{-}{-}B$ stretch, respectively. Unlike ground-state molecules, the stretching of $A{-}{-}{-}H$ and $H{-}{-}{-}B$ is strongly coupled in the transition state, and F_{12} is the coupling constant. The asymmetric stretch corresponds to the reaction coordinate and as such is expected to be characterized by an imaginary frequency. As a limiting approximation Westheimer let $F_1F_2 = (F_{12})^2$, which is equivalent to a frequency of zero for the reaction coordinate. The expression for the symmetric stretching frequency which results is

$$\omega_S^{\ddagger} = \frac{1}{2\pi c} \sqrt{\frac{F_1}{M_A} + \frac{F_2}{M_B} + \frac{F_1 + F_2 - 2F_{12}}{M_H}} \tag{6}$$

When the forces acting on the hydrogen are equal, e.g., $F_1 = F_2 = F$, Eq. (6) simplifies:

$$\omega_S^{\ddagger} = \frac{1}{2\pi c} \sqrt{F\left(\frac{1}{M_A} + \frac{1}{M_B}\right)} \tag{7}$$

This situation corresponds to a transition state in which the hydrogen is approximately symmetrically placed between A and B. The symmetrical stretch is independent of the mass of hydrogen and consequently isotopic substitution. The sole contribution to k_H/k_D is the ground-state zero-point energy [Fig. 1(b)], and the isotope effect is predicted to be maximal. Alternatively, when the force constants F_1 and F_2 are unequal, e.g., $F_1 \gg F_2$,

$$\omega_S^{\ddagger} = \frac{1}{2\pi c} \sqrt{F_1\left(\frac{1}{M_A} + \frac{1}{M_H}\right)} \simeq \frac{1}{2\pi c} \sqrt{\frac{F_1}{M_H}} \tag{8}$$

Since $M_A > M_H$, the dependence of $\omega_S^{\ddagger}$ on the mass of hydrogen is large in this instance; the magnitude of $\omega_S^{\ddagger}$ is a function of the force constant F_1, which reflects the extent to which the transition state resembles the reactant, A—H. For reactant-like $(F_1 \gg F_2)$ or product-like $(F_2 \gg F_1)$ transition states, $\omega_{HS}^{\ddagger} - \omega_{DS}^{\ddagger}$ will be large, resulting in considerable cancellation of ground-state by transition-state zero-point energy [Fig. 1(c)] and small kinetic isotope effects.

A graphical illustration of the Westheimer model is given by the solid line in Fig. 2, which plots the magnitude of the isotope effect, k_H/k_D, as a function of the extent of hydrogen transfer in the transition state, X. For negligible differences in vibrational frequencies between A—H and B—H, i.e., neglecting equilibrium isotope effects, the kinetic isotope effect approaches 1 for transition states which resemble either the reactant, $X = 0$, or product, $X = 1$. Intermediate transition-state structures are characterized by increasingly large isotope effects. The isotope effect reaches a maximal value for a symmetrical transition state, $X = 0.5$.

The assumptions in the Westheimer model, which include a linear transition state, the neglect of bending vibrations, and the assignment of a zero frequency to the reaction coordinate, have been examined by numerous investigators. Nonlinear transition states are expected to give rise to small isotope effects, which are relatively insensitive to the position of hydrogen in the transition state.[22] In a nonlinear transition state, the reaction coordinate

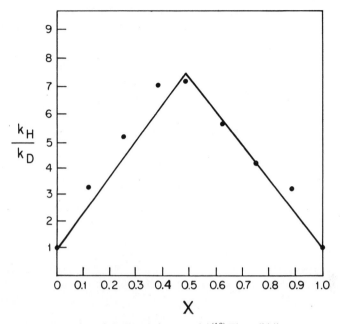

Fig. 2. Graphical illustration of the Westheimer model.[19] The solid line represents the predicted relationship between the magnitude of the isotope effect, k_H/k_D, and the extent of hydrogen transfer in the transition state, X. Isotope effects calculated for the transfer of hydrogen between a secondary carbon and methoxide ion[21] are represented by the closed circles.

would correspond to the loss of one of the bending modes rather than the asymmetric stretching mode. The contribution of both stretching frequencies to the transition state will largely offset the zero-point energy of the ground state, resulting in small kinetic isotope effects.

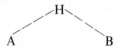

Although bent transition states are likely to be important for intramolecular cyclic reactions, deviations from linearity for most intermolecular hydrogen transfers are probably not very great.[23-27]

As indicated above, a linear triatomic model of hydrogen transfer involves the formation of a transition state which is characterized by two stretching and two bending modes. Since the presence of bending vibrations in the transition state which are absent in the ground state will reduce the magnitude of the observed isotope effect, a number of investigators have attempted to evaluate the role of these bending vibrations in determining the magnitude of the isotope effect. More O'Ferrall and Kouba[21] computed isotope effects for the transfer of hydrogen between a secondary carbon and methoxide ion, C—C--H--O—C. For such a model, both the reactant and product are characterized by a pair of ground-state degenerate bending vibrations. When the bending frequencies for the transition state were computed they were found to be similar to ground-state bending frequencies, leading to the conclusion that transition-state and ground-state bending frequencies will cancel. In Fig. 2, calculated isotope effects are compared to the predictions of Westheimer. The isotope effect is found to pass through a maximum at $X \simeq 0.45$, rather than $X = 0.50$, a result of the greater stretching frequency of an O—H vs. a C—H bond.

In the case of hydrogen transfer from a secondary carbon to a halide ion, the product, HX, is devoid of bending frequencies. Computed isotope effects for these reactions indicate equilibrium isotope effects significantly greater than 1 and broad, more product-like maxima for the kinetic isotope effect. In the direction of hydrogen transfer from HX to a carbanion, bending vibrations will increase in the transition state and decrease the magnitude of the observed isotope effect. The acid-catalyzed hydrolysis of ethyl vinyl ether provided an experimental system for examining the contribution of bending vibrations to the isotope effect. Kinetic studies of this reaction had indicated a nearly symmetrical transition state; when Kresge and Chiang measured the isotope effect for HF catalysis it was found to be small, $k_H/k_D = 3.35$, in contrast to a deuterium isotope effect of 6.8 for formic acid catalysis.[28] The low observed isotope effect for HF catalysis is consistent with the presence of bending vibrations in the transition state which are absent in the ground state, since HF is diatomic; a frequency of 1100 cm^{-1} was calculated for the transition-state bending mode, which is similar to observed ground-state bending vibrations of bonds to hydrogen. While the assumption can be made that bending vibration will cancel for most hydrogen-transfer reactions between polyatomic centers, this assumption is not valid when reactant and product are characterized by markedly different bending frequencies.

The dependence of the magnitude of the isotope effect on the symmetry of the transition state for nonzero values of the frequency of the reaction coordinate is most difficult to assess without a knowledge of the curvature of the energy barrier for a range of transition-state structures. Willi and Wolfsberg have shown that the isotope effect will be maximal and invariant except for very reactant- or product-like transition states when the frequency of the reaction coordinate is large, e.g., $\omega_L^{\ddagger} = 1756i - 3515i$ cm^{-1}.[29] Computed frequencies by Caldin[30] for reactions believed to be characterized by proton tunneling, and hence especially curved energy barriers [cf. Eqs. (13) and (14)], indicate $\omega_L^{\ddagger} \simeq 1000i$ cm^{-1}. More O'Ferrall and Kouba let $\omega_L^{\ddagger}$ vary systematically from $800i - 1100i$ cm^{-1} for a symmetrical transition state to $\omega_L^{\ddagger} = 0$ for reactant- or product-like transition states.[19] As illustrated in Fig. 2 and pointed out by Albery,[31] these values for the imaginary frequency of the reaction coordinate lead to a dependence of the isotope effect on transition-state structure which is similar to the function obtained by Westheimer.

2.3. Relationship between the Observed Kinetic Isotope Effect and the Thermodynamics of Hydrogen Abstraction Reactions

The Hammond postulate provides a conceptual basis for an experimental investigation of the relationship of the magnitude of the isotope effect to the position of hydrogen in the transition state. According to this postulate, the structure of the transition state is related to the thermodynamic driving force of a reaction: For highly exoenergetic or endoenergic reactions the structure of the transition state will resemble that of the reactant or product; in the case of isoenergetic reactions the transition state is predicted to be symmetrical.[32]

Bell and his co-workers undertook a systematic investigation of the relationship of the observed isotope effect to the difference in pK between carbon acids, $pK(CH)$, and a variety of oxygen and nitrogen bases of varying strength, $pK(BH)$:

$$\Delta pK = pK(C—H) - pK(B—H) \tag{9}$$

A monotonic increase in the observed isotope effect for abstraction of a proton from ethyl α-methylacetoacetate ($pK = 12.7$) by bases varying in strength from H_2O ($\Delta pK = 14$) to HPO_4^{2-} ($\Delta pK = 5.6$) was reported.[33] The lower pK's and slow rates of exchange of nitroalkanes provided an experimental system for extending the range of ΔpK from $+14$ to -8.3.[34] Cockerill observed a maximum isotope effect for the E2 reaction of 2-phenylethyldimethyl-sulfonium bromide by hydroxide ion in aqueous dimethyl sulfoxide[35]; the use of mixed solvent systems to perturb the pK of hydroxide ion permits the measurement of isotope effects for a single carbon acid and base over a wide range of pK's, and studies of the ionization of nitroethane were reported in

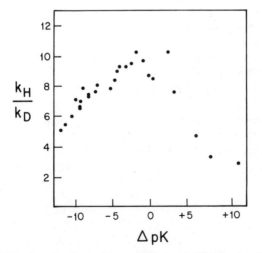

Fig. 3. Relationship between the observed isotope effect and difference in pK for proton abstraction from nitroalkanes by oxygen and nitrogen bases. The data points have been corrected for small secondary isotope effects, as described by Bordwell.[34,36,38]

dimethyl sulfoxide.[36] A compilation of the observed isotope effects for the ionization of carbonyl and nitro compounds has been published.[37] The relationship between k_H/k_D and ΔpK for nitroalkane ionization, corrected for small 2° isotope effects, is shown in Fig. 3. Pryor and Kneipp studied the dependence of the isotope effect on the heat of reaction for the abstraction of hydrogen from mercaptans by radicals, observing a relationship between k_H/k_D and ΔH which is qualitatively analogous to the results for nitroalkane ionization.[39]

Consistent with the predictions of the Westheimer model, Fig. 3 illustrates a maximum isotope effect near $\Delta pK = 0$; large changes in ΔpK are required to give variations in the isotope effect, however. Bordwell has interpreted the low sensitivity of the isotope effect to changes in nitroalkane structure in terms of a two-step mechanism involving slow proton removal to form a pyramidal nitrocarbanion, followed by a rapid rehybridization to a planar nitronate ion[38]:

$$
B^- + H\!\!-\!\!\overset{|}{\underset{|}{C}}\!\!-\!\!NO_2 \xrightarrow{\text{slow}}
\left[
\begin{array}{c}
\overset{\delta-}{} \\
B\text{-}\text{-}H\text{-}\text{-}\text{-} \\
\overset{\delta}{\underset{\underset{H}{|}}{\overset{O}{\underset{H}{C}}}}\!\overset{+}{N}\!\!\!=\!\!\overset{O^-}{\underset{O}{}}
\end{array}
\right]^{\ddagger}
\xrightarrow{\text{fast}} BH + \;\;\overset{}{\underset{}{}}C\!\!=\!\!\overset{+}{N}\!\!\!\overset{O^-}{\underset{O^-}{}} \quad (10)
$$

$$\qquad\qquad 1 \qquad\qquad\qquad\qquad 2 \qquad\qquad\qquad\qquad\qquad 3$$

According to the above scheme, substituent effects on the equilibrium between 3 and 1 are reflected in ΔpK, whereas kinetic isotope effects refer to the formation of the transition state 2 from 1.

Marcus' theory factors a hydrogen-transfer reaction into work terms for the encounter of reactants, W_R, diffusion away of products, $-W_P$, and free-energy changes for the chemical reaction within the encounter process, ΔG_R[40]:

$$
AH + B \xrightleftharpoons{W_R} A\text{---}H\text{------}B \xrightleftharpoons{\Delta G_R} A\text{------}H\text{---}B \xrightleftharpoons{-W_P} A + BH \quad (11)
$$

In the formulation of the isotope effect according to this theory, the magnitude of the isotope effect is related to ΔG_R within the encounter complex.[41] The sensitivity of the isotope effect to ΔG_R is a function of the intrinsic energy barrier to hydrogen transfer, $\Delta G^{\ddagger}$:

$$
\ln(k_H/k_D) = \ln(k_H/k_D)_{max}\left[1 - (\Delta G_R/4\Delta G^{\ddagger})^2\right] \quad (12)
$$

$\Delta G^{\ddagger}$ is defined as the kinetic barrier independent of a thermodynamic drive or hindrance and, as such, is equal to the free energy of activation when $\Delta G_R = 0$. From the observed relationship between k_H/k_D and ΔpK for the ionization of nitroalkanes, Kresge has estimated that $\Delta G^{\ddagger} = 7.5 \pm 0.8$ kcal/mol. Although maximum isotope effects may not be expected to occur precisely at $\Delta pK = 0$, a consequence of the intrinsic differences in bond strength and reactivity of

C—H vs. B—H bonds,[42] the observation of a maximum at $\Delta pK \simeq -2$ rather than $pK = 0$ for nitroalkane ionization has been attributed to a larger work term for the dissolvation of A and BH than for AH and B.[41]*

2.4. Tunneling

Kinetic hydrogen isotope effects are quantal in origin, and a basic tenet of quantum mechanics is the uncertainty principle. Extremely small particles such as electrons are characterized by large wavelengths and large uncertainties in their positions. In the case of proton-transfer reactions the wavelength of the proton is comparable to the expected width of the energy barrier (1–2 Å), leading to the prediction that there is a finite probability that molecules which possess less energy than the energy barrier will react. The mathematical expression for the tunnel correction, Γ, depends on the shape of energy barrier; treatment of the reaction barrier as a parabola leads to[43]

$$\Gamma = \frac{v}{v_{\text{class}}} = \left[\frac{\frac{1}{2}\mu}{\sin \frac{1}{2}\mu} \right]$$
$$- \left[\mu \exp\left(\frac{E}{kT} \right) \left(\frac{y}{2\pi - \mu} - \frac{y^2}{4\pi - \mu} + \frac{y^3}{6\pi - \mu} - \cdots \right) \right] \quad (13)$$

where v and v_{class} are the velocity of a reaction with or without tunneling, respectively; E is the height of the energy barrier; and $\mu = hv/kT$ and $y = \exp(-2\pi E/hv)$. At room temperature the first term in Eq. (13) is normally concluded to be dominant, and Γ is a function of the frequency of the energy barrier. The frequency of the energy barrier can be expressed in terms of

$$v = \frac{s^{1/2}}{2\pi M^{1/2}} \quad (14)$$

which illustrates that the frequency will be large when the curvature at the top of the energy barrier, s, is large and when the mass of the particle, M, is small.

The anticipated phenomenologic manifestations of tunneling have been reviewed[17,30,37,44-46] and include (1) values of the isotope effect, k_H/k_D, and differences in activation energy, $E_D - E_H$, which exceed the maximum values predicted from a loss of ground-state zero-point energy (Table I) and (2) isotope effects on the Arrhenius parameter [PEF factor in Eq. (2)] which fall outside the no-tunneling range, $0.5 < A_H/A_D < 1.4$. Stern and Weston concluded as the result of model calculations that the Swain relationship, Eq. (5), is essentially correct even when tunneling is important.[46] Caldin[30] and Bell[37] have compiled data for a number of reactions which are concluded to exhibit one or more of the above criteria for tunneling. One class of reactions characterized by exceptionally large isotope effects involves hydrogen transfer from

---

* Since $2.3R\,\Gamma\Delta pK = \Delta G = \Delta G_R + W_R - W_P$.

nitroalkanes to nitrogen bases; for example, $k_H/k_D = 45$ for proton abstraction from p-nitrophenylnitromethane by tetramethylguanidine in toluene.[47] Caldin et al. have studied the effect of variations in the nitrogen base and solvent on the magnitude of the observed isotope effect for p-nitrophenylnitromethane ionization, concluding that the major factors contributing to a large tunnel effect are an sp^2 hybridization state at nitrogen, steric crowding in the nitrogen base, an apparent lack of configuration change coupled to proton transfer, and low solvent polarity.[48]

A model in which *tunneling determines the magnitude of the observed isotope effect* has been introduced by Bell et al.[49] Using a charge cloud triatomic model, the contribution of the symmetrical stretching frequency to the isotope effect was calculated to be small and invariant for large changes in free energy, $\Delta G = -27.5$ to $+25.1$ kcal/mol. The major cause of the isotope effect was concluded to be a tunnel correction; the dependence of the isotope effect on ΔG was attributed to the second term in Eq. (13), which contains E, the height of the energy barrier. The rationale for relating the magnitude of the isotope effect to E is illustrated in Fig. 4. The area under the curve which is available for tunneling is considered from the thermodynamically favored direction, since by the principle of microscopic reversibility the same amount of tunneling must be available for a reaction in both directions. When this is done, both endoenergetic and exoenergetic reactions are concluded to have less region available for tunneling than the isoenergetic situation.

Although Bell et al. conclude that their calculated isotope effects are consistent with the range of isotope effects observed for proton abstraction reactions, e.g., Fig. 3, their model retains transition-state bending vibrations. If ground-state and transition-state bending vibrations are assumed to cancel, their calculated isotope effects are too large, suggesting a significant overestimate of the tunnel correction. Calculations of a tunnel correction in the five center model calculations of More O'Ferrall and Kouba indicate a maximum value of $\Gamma_H/\Gamma_D = 1.3$–1.5 for a symmetrical transition state, which decreases for asymmetric transition states. These authors conclude that the contribution of tunneling is small and amplifies the dependence of the isotope effect on transition-state structure due to force constant changes.[21]

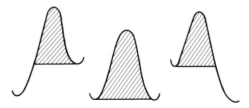

Fig. 4. Region available for tunneling as a function of the overall free-energy change of reaction.[17] (Reprinted from R. P. Bell, *The Proton in Chemistry*. Second edition © 1973 by R. P. Bell. Used by permission of Cornell University Press.)

Saunders and his co-workers have investigated carbon and hydrogen isotope effects on E2 reactions of 2-phenylethyldimethylsulfonium and -trimethylammonium ions in aqueous dimethyl sulfoxide. Both carbon and hydrogen isotope effects are found to vary with solvent in distinctly different ways, consistent with calculations based on stretching force constant changes. Although the observed carbon isotope effect is normal at the region where k_H/k_D is maximal instead of slightly inverse as predicted from models, indicating a tunnel contribution to the isotope effect of approximately 1.015–1.025 for carbon and 1.5–2.0 for hydrogen, the data are consistent with force constant changes being the primary determinant of the observed variations in the isotope effects.[50]

2.5. Equilibrium Isotope Effects

Discussions of primary hydrogen isotope effects normally focus on kinetic rather than equilibrium effects, due to the much greater magnitude of the former. In analogy with kinetic isotope effects, equilibrium effects arise from changes in vibrational frequencies in proceeding between two states. An important distinction is that equilibrium effects do not involve the conversion of a real vibrational mode to a translation [cf. Eqs. (2) and (3)]. To a large extent, reactant and product vibrational frequencies are expected to cancel, leading to relatively small values for K_H/K_D.

Hartshorn and Shiner[51] have computed equilibrium isotope effects for the transfer of hydrogen between two carbon centers using force fields developed from small-molecule spectra data. These workers express their isotope effects in terms of the following isotope exchange reaction:

$$R\text{—}H + H\text{—}C\equiv C\text{—}D \rightleftharpoons H\text{—}C\equiv C\text{—}H + R\text{—}D \qquad (15)$$

This isotope exchange is conveniently defined in terms of a fractionation factor, ϕ:

$$\phi = \frac{R\text{—}D}{R\text{—}H} \times \frac{H\text{—}C\equiv C\text{—}H}{H\text{—}C\equiv C\text{—}D} \qquad (16)$$

By defining

$$H\text{—}C\equiv C\text{—}H/H\text{—}C\equiv C\text{—}D = 1 \qquad (17)$$

ϕ simplifies to give $R\text{—}D/R\text{—}H$. For $\phi > 1$, deuterium will be enriched in R—H relative to acetylene, indicating that R—H is a tighter bond. Alternatively, when $\phi < 1$, the deuterium concentration in R—H will be depleted relative to acetylene. Computed fractionation factors (summarized in References 51 and 52) indicate ϕ values that vary from 0.987 for $FC\equiv CD$ to 1.993 for CF_3D.

These differences in fractionation factors result primarily from changes in the frequencies of bending vibrations at a C—H bond and appear to be largely dependent on the nature of the functional groups directly attached to the carbon bearing the C—H bond of interest.

Experimentally determined equilibrium isotope effects for the transfer of a hydrogen from C-4 of the dihydronicotinamide ring of NAD(P)H to C-2 of oxaloacetate and acetone or C-1 of acetaldehyde indicate $K_H/K_D = 0.85$[53] and 0.89[54], respectively, consistent with the computations of Hartshorn and Shiner. Meloche *et al.* have compiled ϕ values for the exchange of isotopic hydrogen from water into sp^3 hybridized C—H bonds of biochemical interest[55]; in conjunction with the fractionation factors of Hartshorn and Shiner, their values can be extended to include sp^2 and sp hybridized C—H bonds.[56] In addition, the available fractionation factors for oxygen, nitrogen, and sulfur functional groups relative to water[56,57] permit one to estimate the magnitude of primary equilibrium isotope effects for a variety of base-catalyzed C—H bond cleavage reactions.

3. APPLICATION TO ENZYME SYSTEMS

In recent years it has become increasingly clear that many enzymes are "fine-tuned" to the point where the chemical interconversion step is no longer rate limiting. Kinetic studies of such enzyme systems under steady-state conditions may involve the measurement of isomerization or product-release steps, rather than the chemical interconversion step(s). Early investigations of isotope effects in enzyme-catalyzed C—H cleavages indicated small primary hydrogen isotope effects,[11] raising the possibility that enzyme reactions were characterized by asymmetric transition states; subsequent work has indicated that in most instances the observation of small hydrogen isotope effects results from the multistep nature of these reactions (cf. Reference 58).

3.1. Steady-State Kinetic Parameters

3.1.1. V_{max} vs. V_{max}/K_m

Steady-state kinetic investigations of enzyme reactions provide us with two fundamental parameters: V_{max} is the rate of an enzyme reaction under conditions of substrate saturation; V_{max}/K_m, where K_m is the Michaelis constant for substrate, measures the rate at substrate concentrations far below the K_m. Cleland has introduced the concept of net rate constants as a simple method for deriving V_{max} and V_{max}/K_m in terms of rate constants for individual steps in the overall reaction.[59] In the case of a two-substrate enzyme reaction, assuming ordered addition of substrates to enzyme and neglecting enzyme isomerizations, one obtains the following expressions for V_{max} and V_{max}/K_m at early percent conversion when the release of products from enzyme can be considered to be irreversible:

$$E + A \underset{k_2}{\overset{k_1}{\rightleftharpoons}} EA + B \underset{k_4}{\overset{k_3}{\rightleftharpoons}} EAB \underset{k_6}{\overset{k_5}{\rightleftharpoons}} ECD \xrightarrow{k_7} ED + C \xrightarrow{k_9} E + D \tag{18}$$

$$V_{\max} = \frac{(k_5 k_7 k_9) E_T}{k_5 k_7 + k_5 k_9 + k_6 k_9 + k_7 k_9} \tag{19}$$

$$\frac{V_{\max}}{K_B} = \frac{(k_3 k_5 k_7) E_T}{k_4 k_7 + k_4 k_6 + k_5 k_7} \tag{20}$$

For an ordered kinetic mechanism, $V_{\max}/K_A$ is equal to the rate of addition of A to E and does not contain rate constants for the chemical interconversion steps, k_5 and k_6, whereas rate constants for these steps appear in both the numerator and denominator of Eqs. (19) and (20). The relationship of the observed isotope effect on $V_{\max}$ and $V_{\max}/K_m$ to an intrinsic isotope effect on k_5 is summarized in Table II for three limiting cases: $k_{5(6)} \ll k_7, k_9$; $k_7 \ll k_{5(6)}, k_9$; and $k_9 \ll k_{5(6)}, k_7$. The assumption has been made that the contribution of isotope effects to binding steps is negligible. Clearly, only when the release of both products is fast will the isotope effect on $V_{\max}$ reflect the full intrinsic isotope effect of the chemical conversion step. For rate-determining product release, the presence of a primary equilibrium isotope effect for the interconversion of ternary complex, $K_{eq} = k_5/k_6$, may give rise to relatively small normal or inverse isotope effects on $V_{\max}$ and $V_{\max}/K_m$.

The useful relationship between isotope effects observed for $V_{\max}$ vs. $V_{\max}/K_m$ is illustrated in Table II. When $V_{\max}$ is limited by a single rate-determining hydrogen transfer step (case I), the isotope effect on $V_{\max}/K_m$ is expected to vary from one to the intrinsic isotope effect and provides information on the *partitioning* of EAB between EA (k_4) and products (k_5). This approach has been used to calculate substrate dissociation constants from Michaelis constants in the yeast alcohol dehydrogenase-catalyzed reduction of aldehydes[60]; in the direction of alcohol oxidation, Michaelis constants were shown to vary from steady-state constants for the oxidation of protonated alcohol ($k_4 \simeq k_5$) to dissociation constants for the oxidation of deuterated alcohols ($k_4 > k_5$).[61] A comparison of isotope effects on $V_{\max}$ and $V_{\max}/K_m$ in a study of glyoxylase led Vander Jagt and Han to conclude that $V_{\max}/K_m$ was equal to the rate of addition of substrate to enzyme.[62]

A striking difference between isotope effects on $V_{\max}$ and $V_{\max}/K_m$ may occur when the release of second product limits $V_{\max}$ (case III). Although $V_{\max}$ is generally expected to be independent of isotope in this kinetic situation, isotope effects on $V_{\max}/K_m$ may be as large as the intrinsic value for a single step. In drawing mechanistic conclusions from observed isotope effects, it is essential that a measurement distinguish between $V_{\max}$ and $V_{\max}/K_m$. In a study of isotope effects in the horse liver alcohol dehydrogenase reaction,

Table II

Relationship of Observed Isotope Effects on V_{max} and V_{max}/K_B to the Intrinsic Isotope Effect on a Single Step[a]

Limiting case	V_{max}		V_{max}/K_B	
	Term	Observed isotope effects	Term	Observed isotope effects
I: $k_{5(6)} \ll k_7, k_9$	k_5	Intrinsic	$k_3k_5/(k_4 + k_5)$	$k_4 \gg k_5$: intrinsic $k_4 \ll k_5$: one
II: $k_7 \ll k_{5(6)}, k_9$	$k_7[1 + (k_6/k_5)]$	Reflects an equilibrium effect	$k_3k_5k_7/(k_4k_6 + k_5k_7)$	$k_4 \gg k_7$: equilibrium $k_4 \ll k_7$: one
III: $k_9 \ll k_{5(6)}, k_7$	k_9	One	$k_3k_5k_7/(k_4k_7 + k_4k_6 + k_5k_7)$	$k_4, k_7 \gg k_{5(6)}$: intrinsic $k_4, k_7 \ll k_{5(6)}$: reflects an equilibrium effect

[a] Equations (18)–(20) in the text.

which is known to be characterized by a rate-limiting release of NADH* in the direction of alcohol oxidation, a small normal isotope effect was attributed to an isotope effect on the dissociation of NADH(D) from enzyme.[63] It has been pointed out that at the concentration of substrate used for this measurement (about $14 \times K_m$) the observed result could be the consequence of a large isotope effect on V_{max}/K_m.[64] Deuterium isotope effects determined at a single substrate concentration for the dopamine-β-hydroxylase-catalyzed hydroxylation of $[2R - {}^2H]$ phenylethylamine indicated an isotope effect of 5.0. Following the addition of fumarate, an activator which has been reported to reduce the K_m of substrates, the observed isotope effect was 2.0.[65] Although the authors suggest that fumarate reduces the activation energy for the C—H bond cleavage, the results are also consistent with a larger isotope effect on V_{max}/K_m ($-$fumarate) than on V_{max} ($+$fumarate).

The above considerations illustrate several important generalities concerning the magnitude of isotope effects on V_{max} vs. V_{max}/K_m (cf. Reference 66). Isotope effects on V_{max} will be unaffected by the rate of substrate binding and release but reflect all kinetically important steps between the ES complex and regeneration of free enzyme. In the case of V_{max}/K_m, (1) the rate of substrate release from the ES complex will influence the magnitude of an observed effect—tightly bound substrates give small isotope effects, and (2) the magnitude of an observed isotope effect will be independent of slow steps subsequent to the first irreversible step.

3.1.2. Deuterium vs. Tritium Isotope Effects

Most measurements of deuterium isotope effects involve the use of highly enriched isotopically labeled substrate; an accurate comparison of the rates of conversion of protonated vs. deuterated substrates requires that the substrates be free of contaminants which may inhibit the enzyme reaction. A tritium isotope effect is a necessarily competitive measurement, since tritium is present in trace amounts, and this circumvents the problem of contaminating inhibitors in isotopically labeled substrates. However, the two isotopic measurements are not equivalent. In contrast to the independent measurement of deuterium isotope effects on V_{max} and V_{max}/K_m, tritium isotope effects can only provide information concerning V_{max}/K_m.

The relationship between the tritium isotope effect and V_{max}/K_m can be demonstrated by considering the protonated and tritiated substrates, S_H and S_T, as competitive inhibitors of one another[11,67]:

$$v_H = \frac{V_H S_H}{S_H + K_H[1 + (S_T/K_T)]} \tag{21}$$

---

* Abbreviations used: NADH, reduced nicotinamide-adenine dinucleotide; NAD, nicotinamide-adenine dinucleotide; and ATP, adenosine triphosphate.

$$v_T = \frac{V_T S_T}{S_T + K_T[1 + (S_H/K_H)]} \tag{22}$$

In Eqs. (21) and (22) $v_{H(T)}$ is the observed velocity, $V_{H(T)}$ is the maximal velocity, and $K_{H(T)}$ is the Michaelis constant. Since S_T is very much smaller than K_T and S_H, these equations simplify:

$$\frac{v_H}{v_T} = \frac{(V/K)_H\, S_H}{(V/K)_T\, S_T} \tag{23}$$

A comparison between V_{max} and V_{max}/K_m for an ordered two-substrate enzyme reaction (Table II) indicates that the tritium isotope effect may be less than or considerably greater than an observed deuterium isotope effect on V_{max}, cases I and III, respectively.

As part of a study of the interaction of chloroalanine with D-amino acid oxidase, Walsh et al. measured both deuterium and tritium isotope effects for the oxidation of serine, isotopically labeled at C-2, $V_H/V_D = 1.43$ vs. $(V/K)_H/(V/K)_T = 5.0$.[71] Although the C—H cleavage step is concluded to contribute to V_{max}, the large tritium isotope effect indicates a partially rate-determining step subsequent to the first irreversible step. Subsequent stopped flow kinetic studies of valine oxidation indicated a deuterium isotope effect of 3 for a fast half-reaction, in contrast to $V_H/V_D = 1.1\text{--}1.5$ under steady-state conditions.[72] The kinetic mechanism of D-amino acid oxidase has been discussed in detail in a recent review by Bright and Porter.[73]

The ethanolamine ammonia lyase catalyzed conversion of ethanolamines, deuterated or tritiated at C-1, to acetaldehyde and ammonia provides an extremely interesting comparison of tritium and deuterium isotope effects. Weisblat and Babior reported that $V_H/V_D = 7.4$ and $(V/K)_H/(V/K)_T = 4.7$.[74] At first glance the smaller tritium isotope effect suggests that substrate does not dissociate rapidly from the ES complex. However, this reaction (which is coenzyme B_{12} dependent) has been demonstrated to proceed by two sequential hydrogen-transfer steps in which the coenzyme B_{12} functions as an intermediate hydrogen carrier:

$$CH_2OHCH_2NH_2 + \boxed{\overset{\overset{\displaystyle CH_2R}{\displaystyle |}}{Co}} \xrightarrow{1} CH_3R + \boxed{\overset{\overset{\displaystyle CHOHCH_2NH_2}{\displaystyle |}}{Co}} \xrightarrow{2} CH_3CHO + \boxed{\overset{\overset{\displaystyle CH_2R}{\displaystyle |}}{Co}} + NH_3 \tag{24}$$

An analysis of isotope effects in the half-reaction involving transfer of hydrogen from labeled coenzyme to product (step 2) indicated that this step was rate determining in the overall reaction, $V_H/V_D = 7.4$. The smaller tritium isotope effect was concluded to represent an irreversible transfer of tritium from substrate to coenzyme in the first half-reaction (step 1). The irreversibility of tritium transfer in step 1 results both from a dilution of tritium by two hydrogens in the formation of tritiated coenzyme and from an excessively large, as yet unexplained, tritium isotope effect of 160 on step 2.

3.1.3. Calculation of Intrinsic Isotope Effects

Northrop has shown that intrinsic isotope effects can be calculated for single steps in enzyme-catalyzed reactions by comparing the observed deuterium and tritium isotope effects on V_{max}/K_m.[66] The expression for the isotope effect on V_{max}/K_m can be represented in terms of two parameters, a, the intrinsic isotope effect, and b, a collection of rate constants for steps preceding and subsequent to the bond cleavage step:

$$\frac{(V/K)_H}{(V/K)_D} = \frac{a + b}{1 + b} \tag{25}$$

$$\frac{(V/K)_H}{(V/K)_T} = \frac{a^{1.44} + b}{1 + b} \tag{26}$$

For reactions in which there is a negligible equilibrium isotope effect, the relationship between the observed deuterium and tritium isotope effects is independent of b:

$$\frac{[(V/K)_H/(V/K)_D] - 1}{[(V/K)_H/(V/K)_T] - 1} = \frac{a - 1}{a^{1.44} - 1} \tag{27}$$

A graphical representation of the function given in Eq. (27) is illustrated in Fig. 5 and points out the need for precise measurements of isotope effects, since a range of deuterium isotope effects from 2 to 10 corresponds to a variation from 0.59 to 0.34 in the left-hand side of Eq. (27).

As indicated earlier, equilibrium isotope effects cannot, in general, be assumed to be unity. Since the b term in Eqs. (25) and (26) contains rate constants for steps which occur after the bond cleavage step, equilibrium isotope effects preclude the factoring of b from these equations. Schimerlik *et al.* have modified Northrop's treatment to include equilibrium isotope effects; although it is no longer possible to solve for a unique value of a, it is possible to determine a range for the intrinsic isotope effect.[75] Application of their approach to the malic enzyme-catalyzed oxidation of malate, deuterated or tritiated at C-2, indicated small isotope effects on V_{max}/K_m, e.g., $(V/K)_H/(V/K)_D = 1.47 \pm 0.08$ and $(V/K)_H/(V/K)_T = 2.02 \pm 0.06$ and a range for the intrinsic isotope effect of $4.9 \pm 2.4 - 7.9 \pm 1.6$. Somewhat disappointingly, the intrinsic isotope effect for malate oxidation could only be concluded to fall between 3 and 9. However, the authors point out that for reactions characterized by small equilibrium isotope effects and large observed isotope effects on V_{max}/K_m it should be possible to establish intrinsic isotope effects within a much more narrow range of values.

The basis for Northrop's analysis is the anticipated apparent breakdown of the Swain relationship [Eq. (5)] in complex reactions characterized by several rate-determining steps. Albery and Knowles[76] have considered the

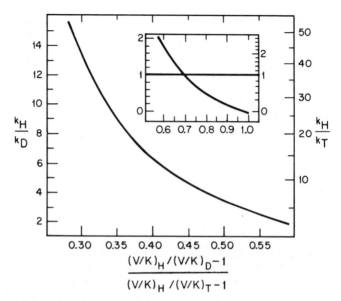

Fig. 5. Values of the intrinsic isotope effect as a function of the observed isotope effects on V_{max}/K_m for enzyme-catalyzed hydrogen abstraction reactions.[66] (Reprinted with permission from D. B. Northrop, *Biochemistry* **14**, 2644 (1975), copyright by the American Chemical Society.)

percentage breakdown of this relationship as a function of the extent to which hydrogen transfer is rate limiting and the intrinsic isotope effect for this step. For the simple case

$$E + S \underset{k_2}{\overset{k_1}{\rightleftharpoons}} ES \xrightarrow{k_3} EP \tag{28}$$

the deviation of an observed tritium isotope effect from one calculated from an observed deuterium isotope effect has been shown to be a function of k_2/k_3 and the intrinsic isotope effect for k_3. The breakdown is normally small, a maximum deviation of 31% being observed when $k_{3)H}/k_{3)D} = 10$ and $k_2/k_3 = 0.13$. As Albery and Knowles point out, it is necessary to determine whether deuterium and tritium isotope effects have been measured with the requisite accuracy before attempting to calculate intrinsic isotope effects. The authors note that when the magnitude of an observed deuterium isotope effect is 2–3, a range commonly observed for enzymes, breakdown of the Swain relationship is maximal. For example, in the case of intrinsic isotope effects of 10 and 5, the percentage breakdown is maximal at $k_2/k_3 = 0.13$ and 0.11; these ratios correspond to observed deuterium isotope effects of 2.2 and 2.0, respectively.

3.2. Experimental Studies of Isolated C—H Cleavage Steps in Enzyme Systems

A number of investigators have explored methods for studying isolated steps in enzyme reactions. Several of these approaches described below include (1) the study of enzyme partial reactions by such methods as stopped flow and

Table III
"Intrinsic" Hydrogen Isotope Effects for Enzyme-Catalyzed Proton Abstractions

Enzyme reaction	k_H/k_D	Ref.
1. Pyruvate kinase, enolization of pyruvate (pD 7.7) activated by[a]		68
Fluorophosphate	9.0	
Methylphosphonate	7.0	
Phosphate	7.0	
2. CP-aldolase, condensation of dihydroxyacetone phosphate and glyceraldehyde[b]	7.0	69
3. Enolase, exchange of the proton at C-2 of phosphoglyceric acid (pH 6.5)	7.6	70

[a] Calculated from tritium isotope effects, using the Swain equation.
[b] Dihydroxyacetone phosphate, deuterated at C-1 in the transferable position.

equilibrium exchange kinetics, (2) chemical or enzymic modification of enzymes in an effort to alter the relative contribution of the chemical conversion step, and (3) the use of slowly reacting substrates.

3.2.1. Proton-Activating Enzymes

Pyruvate Kinase. Robinson and Rose observed that the pyruvate-kinase-catalyzed conversion of tritiated phosphoenolpyruvate and ADP to pyruvate and ATP was accompanied by a release of tritium into water, under conditions of irreversible trapping of free pyruvate. This loss of tritium requires that the addition of a proton to enolpyruvate be followed by both a rotation of the methyl group and reabstraction of tritium to form enolpyruvate, prior to the release of pyruvate from the enzyme[68]:

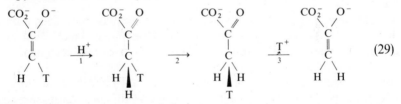 (29)

Depending on the intrinsic isotope effect for step 3, the rate of tritium loss could underestimate the rate of proton exchange, and a study of isotope effects in the enzyme-catalyzed enolization of pyruvate was carried out in the presence of either ATP or ATP analogs.

In the presence of ATP the tritium isotope effect for pyruvate enolization was found to be 4.0, whereas activation by fluorophosphate ($pK = 4.8$), methylphosphonate ($pK = 7.7$), and phosphate ($pK = 6.7$) gave tritium isotope effects of 23.7, 16.8, and 16.8, respectively, at pD = 7.7 (Table III). The small isotope effect for ATP was concluded to be consistent with the observed detritiation of PEP, i.e., indicative of a slow product-release step in the direction of pyruvate formation. In contrast, the isotope effects for ATP analogs are slightly greater than the maximal value predicted from the loss of a ground-state stretching vibration (Table I). The large size of these isotope effects suggests a symmetrical transition state and, by analogy with the observed relationship between the isotope effect and ΔpK for proton abstraction reactions (Fig. 3), a difference in pK between an active-site base ($pK \simeq 3{-}10$) and the bound substrate which is close to zero. Since the pK of the methyl protons of free pyruvate is >20, these considerations imply a reduction in pK of ≥ 10 pH units upon formation of a productive enzyme pyruvate complex.

Recent studies on pyruvate kinase indicate a dual divalent metal requirement for activity. The effect of divalent metal ion on the rate of pyruvate detritiation has been attributed to the enhancement by a nucleotide-bound metal of an electrophilic effect of the terminal phosphorus of ATP or ATP analogs in polarizing the carbonyl of pyruvate.[77] A possible electrophilic role of ATP and its analogs in reducing the pK of the protons of the methyl group of pyruvate is

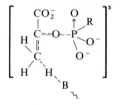

Electrophilic catalysis by ATP analogs of varying pK could lead to different intrinsic isotope effects. However, the smaller isotope effects observed with methylphosphonate and phosphate as compared to fluorophosphate most likely reflect the contribution of more than a single step to the observed rate of detritiation, since the tritium isotope effect for methylphosphonate increased from 17 at pD 7.7 to 26 at pD 9.

Aldolase. The kinetic properties of muscle aldolase have been found to be dramatically altered by carboxypeptidase treatment, which removes three carboxy-terminal tyrosine residues.[69] Studies of isotope exchange at equilibrium indicated that both modified (CP-aldolase) and native aldolases catalyze the exchange of glyceraldehyde-3-phosphate (G3P) into fructose diphosphate (FDP) at the same rate. Since the exchange of dihydroxyacetone phosphate (DHAP) into FDP is reduced 33-fold, carboxypeptidase treatment is concluded to alter steps which involve the protonation of the carbanion and/or release of DHAP from enzyme, steps 4 and 5 in

$$E + FDP \underset{1}{\rightleftharpoons} EFDP \underset{2}{\rightleftharpoons} E^{DHAP^-}_{G3P} \underset{3}{\overset{G3P}{\rightleftharpoons}} EDHAP^- \underset{4}{\overset{H^+}{\rightleftharpoons}} EDHAP \underset{5}{\rightleftharpoons} E + DHAP$$

$$(30)$$

As the result of measurements of deuterium isotope effects on V_{max} for the condensation of glyceraldehyde with DHAP containing deuterium in the transferable position at C-1 ($V_H/V_D = 7.0$ with CP-aldolase compared to

$V_H/V_D = 1.0$ for native aldolase), step 4 was concluded to be rate determining for modified enzyme.

Biellmann et al.[78] extended the studies on modified enzyme to include measurements of secondary isotope effects. The condensation of DHAP, containing tritium in the nontransferable position at C-1, with G3P to form FDP indicated a secondary tritium isotope effect of 1.15. In the direction of FDP cleavage, large secondary tritium isotope effects were observed for the cleavage of 3- and 4-tritiated FDP. The tritium isotope effect of 1.33, observed for the cleavage of 3-tritiated FDP, could reflect either a kinetic or equilibrium isotope effect for the conversion of FDP to the enol of DHAP, depending on the rate of G3P release relative to C—C cleavage, and can be considered to be a lower limit for the equilibrium isotope effect for the conversion of FDP to the enol of DHAP. The equilibrium isotope effect for the conversion of DHAP to its enol can be estimated to be ≥ 1.24 using the fractionation factors of Hartshorn and Shiner[51] to estimate an equilibrium isotope effect of 1.07 for the conversion of FDP to DHAP:

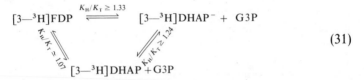

$$(31)$$

A comparison of kinetic (1.15) to equilibrium (≥ 1.24) secondary isotope effects suggests that the hybridization change which occurs in the conversion of DHAP to the enol of DHAP is $\leq 60\%$ complete in the transition state.* Two independent measurements, a secondary isotope effect and the large, near maximal primary hydrogen isotope effect, are consistent with a nearly symmetrical transition state for the proton abstraction reaction catalyzed by aldolase. Muscle aldolase has been demonstrated to involve Schiff base formation[79]; a role for a protonated Schiff base in reducing the pK of the enzyme-bound DHAP close to that of an active-site base is

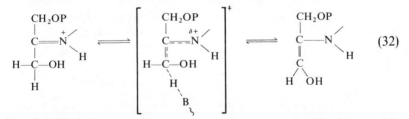

$$(32)$$

Enolase. The enolase-catalyzed dehydration of phosphoglyceric acid (PGA) involves the cleavage of both a C—H and a C—OH bond to form phosphoenolpyruvate (PEP):

---

* A discussion of the relationship of secondary hydrogen isotope effects to transition-state structure can be found in Chapter 5 in this volume.

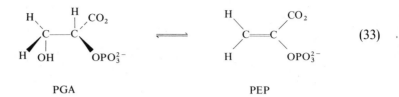

$$PGA \qquad\qquad\qquad PEP$$

Under conditions where substitution of hydrogen by deuterium at C-2 has a small effect on V_{max},[80] Dinovo and Boyer observed that hydrogen exchanges from C-2 sixfold faster than the exchange of oxygen at C-3. These data could result from a sequestration of oxygen, which reduces its rate of exchange; however, secondary isotope effects in the direction of PEP hydration led to the proposal that PGA dehydration involves a rapid abstraction of a proton from C-2, followed by a predominantly rate-determining loss of oxygen from C-3.[70] A measurement of the deuterium isotope effect for the enolase-catalyzed exchange of the proton at C-2 of PGA indicated a value of 7.6 at pH 6.5. Although a mechanism for a large decrease in the pK of the C-2 proton of bound PGA is difficult to imagine, especially in the context of a carbanion mechanism, the near-maximal isotope effect suggests a symmetrical transition state.

The magnitude of the observed "intrinsic" isotope effects for the pyruvate kinase, aldolase, and enolase reactions (Table III) raises the possibility (1) that large near-maximal isotope effects may be observed for numerous enzyme-catalyzed proton abstractions, assuming a single step can be measured for these reactions, and (2) that the magnitude of these effects reflects a difference in pK between the bound substrate and an active-site base which is close to zero.

3.2.2. Dehydrogenases

Yeast Alcohol Dehydrogenase (YADH). Early studies on the reaction mechanism of this enzyme showed a small primary hydrogen isotope effect on V_{max} for acetaldehyde reduction[81] consistent with a partially rate-determining product-release step. The observation that YADH will interconvert aromatic substrates at a reduced rate relative to acetaldehyde led to the investigation of both structure-reactivity correlations and deuterium isotope effects in this system:

$$X-\langle O \rangle-C\overset{H}{\underset{O}{\diagdown}} + NADH(D) \rightleftharpoons X-\langle O \rangle-\overset{H}{\underset{OH}{C}}-H(D) + NAD \qquad (34)$$

Large primary isotope effects were observed for aldehyde reduction $k_{R)H}/k_{R)D} = 3$–5.4 (Table IV) and alcohol oxidation, $k_{O)H}/k_{O)D} = 3.2$–4.8.[61] Despite the relative insensitivity of the isotope effect to change in substrate structure, rate constants in the direction of aldehyde reduction were found to vary more than 100-fold. The linear relationship between $\log k_R$ and electronic substituent

Table IV
"Intrinsic" Hydrogen Isotope Effects for Dehydrogenases

Enzyme	k_H/k_D	Ref.
1. Yeast alcohol dehydrogenase reduction of para-substituted benzaldehydes by NADH[a]:		60
p-Br	3.5	
-Cl	3.3	
-H	3.0	
-CH$_3$	5.4	
-CH$_3$O	3.4	
2. Horse liver alcohol dehydrogenase, oxidation of alcohols by NAD:		
Ethanol	5.2[b]	82
Benzyl alcohol	3.6,[c] 4.3[d]	83, 84
3. Glutamate dehydrogenase reduction of para-substituted dinitrobenzene sulfonates by NADH[e]:		
p-NO$_2$	4.3–6.3	85, 86
-CN	5.4	86
-CONH$_2$	5.7	86
-CF$_3$	4.9	86
-CO$_2$	4.8	86

[a] NADH, deuterated in the A position at C-4 of the dihydronicotinamide ring.
[b] This isotope effect contains secondary as well as primary effects since ethanol-d_5 was studied.
[c] Benzyl alcohol-1,1-d_2, modified enzyme, steady-state kinetics.
[d] Benzyl alcohol-1,1-d_2, native enzyme, pre-steady-state kinetics.
[e] NADH, dideuterated at C-4 of the dihydronicotinamide ring.

constants, together with the large observed isotope effects, led to the conclusion that C—H cleavage is fully rate determining in the direction of aromatic aldehyde reduction.[60] If the same argument is to pertain to alcohol oxidation, isotope effects in this direction must be related to isotope effects for aldehyde reduction by an equilibrium effect:

$$\frac{k_{R)H}/k_{R)D}}{k_{O)H}/k_{O)D}} = \frac{K_H}{K_D} \tag{35}$$

The equilibrium isotope effect in the direction of NADH oxidation is $K_H/K_D = 0.89 \pm 0.03$.[54] Since the oxidation studied was of dideuterated rather than monodeuterated alcohols, Eq. (35) was expanded to include a secondary kinetic deuterium isotope effect:

$$\frac{k_{R)H}/k_{R)D}}{k_{O)H}/k_{O)D}} = \frac{K_H}{K_D} \frac{k_{\alpha D}}{k_{\alpha H}} \tag{36}$$

Assuming a range of 0.77–1 for $k_{\alpha)D}/k_{\alpha)H}$,[87] the right-hand side of Eq. (36) is 0.69–0.89. With the exception of the p-CH$_3$ substrate (which appears to be characterized by a partially rate-determining product-release step in the direction of alcohol oxidation), the observed ratios of 0.78–1.06 support the conclusion that a single C—H cleavage step is rate determining and indicate that the measured isotope effects reflect intrinsic values.[61]

Multiple linear regression analyses of the effect of para-substituents on the measured kinetic parameters have permitted a determination of the contribution of hydrophobic, steric, and electronic factors to substrate binding and turnover: (1) Electronic factors contribute to aldehyde binding consistent with a polarization of the carbonyl of bound aldehyde, (2) a hydrophobic pocket is important in alcohol binding and possibly catalysis in the direction of aldehyde reduction, and (3) the transition state is characterized by a structure that is uncharged relative to bound alcohol. Importantly, the same transition-state structure was deduced in the direction of either aldehyde reduction or alcohol oxidation, corroborating the conclusion based on the magnitude of the observed isotope effects that a single hydrogen-transfer step is rate determining.[61]

It has been proposed that the observed lack of charge development at C-1 of substrate, together with a role for an active-site residue of $pK = 8.25$, is consistent with a mechanism involving concerted acid–base catalysis of hydride transfer[88]:

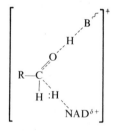

Although concerted acid catalysis by a protonic base of pK 8.25 would meet the requirements for concerted catalysis described by Jencks[89] in that the conversion of an aldehyde ($pK \simeq -3$ to -7)[90] to an alcohol ($pK \simeq 15$)[91] involves a large change in pK and the pK of the catalyst ($pK = 8.25$) is intermediate between that of substrate and product, recently determined deuterium solvent isotope effects in the direction of alcohol oxidation argue against this mode of catalysis, since $k_{H_2O}/k_{D_2O} = 1.20 \pm 0.09$.[92] In contrast to the absence of a solvent isotope effect for alcohol oxidation, aldehyde reduction is characterized by an inverse isotope effect, $k_{H_2O}/k_{D_2O} = 0.50$ and 0.58 for reduction by NADH and NADD, respectively.[92] The inverse isotope effects on k_{cat} for aldehyde reduction implicate an intermediate, formed before a rate-determining C—H cleavage, in the direction of aldehyde reduction. The nature of this intermediate has an important bearing on the mode of hydrogen transfer between carbon centers for the yeast alcohol dehydrogenase reaction, and two highly dissimilar mechanisms have been proposed to be consistent with the available kinetic data.[92]* These mechanisms involve either a preequilibrium displacement of a zinc-bound water by substrate to form an inner sphere coordination complex followed by a rate-determining hydride transfer (structure A) or a preequilibrium transfer of a proton plus one electron to form a protonated radical intermediate followed by a rate-determining hydrogen atom transfer (structure B):

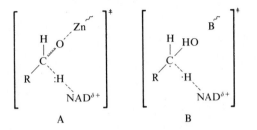

A carbonium-ion mechanism involving a preequilibrium transfer of a proton to the aldehyde carbonyl appears unlikely because the transition state is uncharged and because of the large difference in pK between a protonated carbonyl and an active-site residue of pK = 8.25.[92] It has been proposed that model reactions involving 1,4-dihydronicotinamides proceed through the formation of kinetically significant, possibly radical, intermediates.[96-98]

Liver Alcohol Dehydrogenase (LADH). LADH has been reported to be characterized by a rate-limiting release of coenzyme under steady-state conditions.[99] Studies of ethanol oxidation under pre-steady-state conditions reveal that the reaction proceeds in two phases; the first phase corresponds to the production of enzyme-bound NADH and is characterized by a deuterium isotope effect of 5.2.[82] The ability to isolate the hydrogen-transfer step under pre-steady-state conditions has led to several investigations of structure-reactivity correlations in the LADH-catalyzed interconversion of aromatic substrates.[84,100] Hardman et al.[84] studied the oxidation of a series of benzyl alcohols and reported a deuterium isotope effect of 4.3 for the oxidation of unsubstituted benzyl alcohol; the effect of variations in electronic substituent was small, $\rho = -0.7$ ($\rho^+ \simeq -0.3$).

An alternative approach to studying the hydrogen-transfer step in LADH has been explored by Plapp and his co-workers, who found that chemical modification of lysine side chains increases the rate of coenzyme dissociation, leading to large deuterium isotope effects on V_{max} under steady-state conditions.[101] Recent studies of structure-reactivity correlations and isotope effects in the interconversion of aromatic substrates by hydroxybutyrimidylated LADH indicate $V_H/V_D = 3.6$ for benzyl alcohol oxidation and $\rho^+ = -0.2$.[83] Both pre-steady-state studies on native enzyme and steady-state studies on modified enzyme reveal a relatively uncharged transition-state structure.

---

* Yeast alcohol dehydrogenase is a zinc–metalloenzyme,[93-95] and an active-site metal or protonic base (possibly Zn-OH$_2$) could function in electrophilic catalysis.

Furthermore, the absence of a normal solvent isotope effect in the direction of benzyl alcohol oxidation under pre-steady-state conditions indicates that O—H and C—H bond cleavages are uncoupled.[102] Although the molecular weight and turnover numbers of yeast and horse liver alcohol dehydrogenases are markedly different, these enzymes appear to be characterized by similar transition-state structures. A long-standing question concerning the mechanism of horse liver alcohol dehydrogenase is the role of zinc vs. zinc–water in the chemical reaction. A high-resolution X-ray structure of this enzyme supports the view that substrate is directly coordinated to an acitve-site zinc.[103,104] In contrast, nuclear magnetic-resonance studies of cobalt-substituted enzyme indicate second-sphere complexes between substrate and metal.[105]

Glutamate Dehydrogenase (GDH). The initial finding that GDH catalyzes the reduction of trinitrobenzenesulfonate by NADH,[106] a reaction which occurs readily in the absence of enzyme, has facilitated a comparison of the enzymic and nonenzymic reactions:

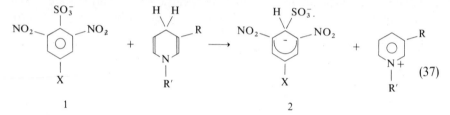

In a study by Brown and Fisher,[85] rate constants for hydrogen transfer from a series of 1,4-dihydropyridines to trinitrobenzenesulfonate were found to correlate with dissociation constants (K_d) for the cyano complexes formed from the oxidized pyridinium compounds and cyanide ion, slope = 0.57; the isotope effect for the oxidation of NADH, dideuterated at C-4 of the dihydronicotinamide ring, was 3.4. In the GDH-catalyzed reaction, both the isotope effect for NADH oxidation, $V_H/V_D = 4.9$, and the slope of the correlation between V_{max} and K_d, 0.72, were found to be increased relative to the model reactions, and the authors suggest that the transition state for the enzyme-catalyzed reaction is characterized by a greater amount of C—H cleavage than the reaction which proceeds in solution.

Extensive studies of structure-reactivity correlations and isotope effects in the nonenzymic and GDH-catalyzed reduction of para-substituted dinitrobenzene sulfonates by NADH have been carried out by Kurz and Frieden.[86,107] The isotope effects for the GDH reaction are given in Table IV; similar values have been observed for the nonenzymic reaction ($k_H/k_D = 4.7$–4.9), with the exception of the p-CF$_3$ substituent ($k_H/k_D = 3.2$). Structure-reactivity correlations indicate a strong dependence on electron-withdrawing substituents, $\rho = 4.97$ both in the presence and absence of GDH. The magnitude of the kinetically determined ρ, in relationship to an estimated $\rho \geq 11$ for the equilibrium conversion of 1 to 2 in Eq. (37), has been concluded to be consistent with a direct transfer of a hydride ion from NADH to benzenesulfonates.[86] In contrast to the work of Brown and Fisher, the studies of Kurz and Frieden indicate similar if not identical transition-state structures for the nonenzymic and GDH-catalyzed reactions. The relatively small enzyme rate acceleration of ~ 10 M at pH 8.0[86] may be due solely to a reduction in the entropy of activation which results from the juxtaposition of coenzyme and benzenesulfonate at the enzyme active site.

The isotope effects which have been observed for YADH, LADH, and GDH are summarized in Table IV. To the extent that these effects represent intrinsic values, their relatively small magnitude and insensitivity to changes in substrate reactivity may indicate that hydrogen activation occurs via a hydride ion. Swain *et al.*[108] have pointed out some fundamental differences between hydride- and proton-transfer reactions: Whereas hydride transfer occurs between two electron-deficient centers, A—:H—B, proton transfer occurs between two electron-rich centers, A:—H—:B; the presence of a single pair of electrons in the hydride-transfer reaction serves to "cement" A, H, and B together, leading to relatively strong, short, and nonpolarizable bonding in the transition state. As the result of such considerations, together with measured isotope effect for the oxidation of secondary alcohols by bromine, it was proposed that the magnitude of the isotope effect will be characterized by an especially low sensitivity to changes in substrate reactivity for hydride-relative to proton-transfer reactions.

3.3. Construction of a Free-Energy Profile: Triose Phosphate Isomerase

Experiments by Rose on the interconversion of specifically tritiated substrates by sugar isomerases have implicated the presence of a common enediol intermediate in these reactions, illustrated below for the triose phosphate isomerase (TPI) reaction[109]:

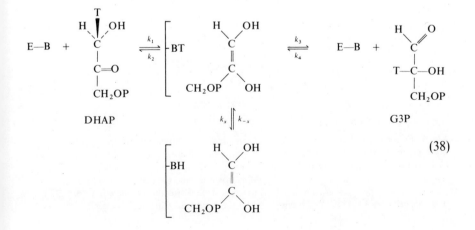

(38)

In the TPI-catalyzed conversion of tritiated dihydroxyacetone phosphate (DHAP) to glyceraldehyde-3-phosphate (G3P) 94–97% of the tritium initially present in DHAP is lost to the water, 3–6% being transferred to G3P. These results are consistent with the presence of an intermediate, EBT, which exchanges tritium with water fairly rapidly relative to free G3P formation.[110]

Both deuterium and tritium isotope effects have been explored at great length by Knowles and his co-workers for the TPI reaction. By converting substrate to product in tritiated water, product tritium isotope effects of 1.3 and 9 were determined for the formation of G3P and DHAP, respectively. These results are consistent with the tritium-transfer experiments indicating a slow release of G3P from enzyme. In the direction of DHAP formation the large observed isotope effect suggests a fairly fast rate for the dissociation of DHAP.[110] The incorporation of tritium from solvent into unreacted substrate was also investigated, under conditions of irreversible trapping of product. For both GAP → DHAP and DHAP → GAP, exchange was found to occur at one-third the rate of product formation. The apparently contradictory observation of the same exchange-to-conversion ratios for GAP → DHAP and DHAP → GAP results from the differential isotope effects for the formation of GAP and DHAP from the enediol; in addition, exchange requires C—T bond formation, whereas conversion measures C—H production.[111,112] In conjunction with tritium measurements, the effect of deuterium substitution on V_{max} for the reaction of [2-^{2}H]GAP and [1R-^{2}H]DHAP was determined. In the direction of GAP → DHAP the isotope effect was 1, indicating a rapid equilibration of the enediol intermediate and bound GAP.

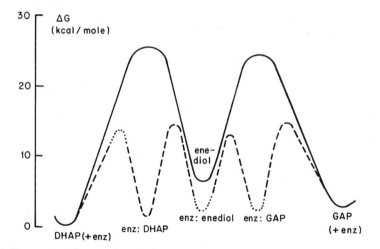

Fig. 6. Free-energy diagrams for the interconversion of dihydroxyacetone phosphate and D-glyceraldehyde-3-phosphate in an uncatalyzed reaction (solid line) and triosephosphate-isomerase-catalyzed reaction (dashed line). The dotted lines refer to parts of the free-energy profile that are less well defined. A standard state of 40 μM is taken for the enzyme-catalyzed reaction.[114,115] (Reprinted with permission from W. J. Albery and J. R. Knowles, *Biochemistry* **15**, 5627 (1976), and *Biochemistry* **14**, 4348 (1975); copyright by the American Chemical Society.)

In contrast, an isotope effect of 2.9 was observed for DHAP → GAP. The observation of this effect reflects a relatively slow hydrogen abstraction from DHAP and a rapid exchange of substrate-derived proton or deuterium from the enediol intermediate. Although the steady-state formation of GAP from bound DHAP is largely limited by GAP release, the deuterium isotope effect is independent of all steps subsequent to enediol formation.[113]

Utilizing the available deuterium and tritium isotopic measurements, a free-energy profile for the triose phosphate isomerase reaction has been constructed[114] and compared to a profile obtained from a detailed study of the DHAP and GAP interconversions in solution[115] (Fig. 6). Although the highest barrier on the enzyme profile corresponds to GAP binding, Fig. 6 illustrates that no single step is fully rate determining. A comparison of the free energy of the enediol intermediate and the transition states leading to this intermediate for the enzymic and nonenzymic reactions indicates that triose phosphate isomerase stabilizes its transition state by approximately 7 kcal/mol more than the enediol intermediate.[115] This result is consistent with the anticipated enhanced affinity of an enzyme for its transition state, and it also provides a reasonable explanation for the relatively low affinity of triose phosphate isomerase for phosphoglycollate, a transition-state analog resembling the enediol intermediate[116]:

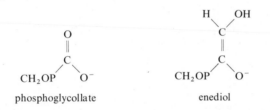

phosphoglycollate   enediol

As discussed by Reynolds *et al.*,[117] the K_i for phosphoglycollate is almost identical to the K_m for the active (keto) form of GAP. Cleland has suggested, on the basis of the differential pH dependencies of substrate K_m and inhibitor binding, that the inhibitory form of phosphoglycollate contains a protonated carboxyl group and a doubly ionized phosphate group.[118] Although these considerations lead to a 1600-fold decrease in the estimated K_i for phosphoglycollate, the enhanced binding of phosphoglycollate relative to GAP remains small compared to an enzyme rate acceleration of greater than 10^9.[115]

REFERENCES

1. L. Pauling, Chemical achievement and hope for the future, *Am. Sci.* **36**, 51–58 (1948).
2. R. Wolfenden, Analog approaches to the structure of the transition state in enzyme reactions, *Acc. Chem. Res.* **5**, 10–18 (1972).
3. G. E. Lienhard, Enzymatic catalysis and transition-state theory, *Science* **180**, 149–154 (1973).
4. W. P. Jencks, Binding energy, specificity, and enzymic catalysis: The Circe effect, *Adv. Enzymol. Relat. Areas Mol. Biol.* **43**, 219–410 (1976).

5. K. Schray and J. P. Klinman, The magnitude of enzyme transition state analog binding constants, *Biochem. Biophys. Res. Commun.* **57**, 641–648 (1974).

6. J. H. Richards, in: *The Enzymes*, 3rd ed. (P. D. Boyer, ed.), Vol. II pp. 321–333, Academic Press, New York (1970).

7. I. A. Rose, in: *The Enzymes*, 3rd ed. (P. D. Boyer, ed.), Vol. II pp. 281–320, Academic Press, New York (1970).

8. G. Popjak, in: *The Enzymes*, 3rd ed. (P. D. Boyer, ed.), Vol. II pp. 115–215, Academic Press, New York (1970).

9. J. F. Kirsch, Mechanism of enzyme action, *Ann. Rev. Biochem.* **42**, 205–234 (1973).

10. J. W. Cornforth, The logic of working with enzymes, *Chem. Soc. Rev.* **2**, 1–20 (1973).

11. H. Simon and D. Palm, Isotope effects in organic chemistry and biochemistry, *Angew Chem. Int. Ed. Engl.* **5**, 920–933 (1966).

12. J. Bigeleisen and M. Wolfsberg, Theoretical and experimental aspects of isotope effects in chemical kinetics. *Adv. Chem. Phys.* **1**, 15–76 (1958).

13. L. Melander, *Isotope Effects on Reaction Rates*, Ronald, New York (1960).

14. H. S. Johnston, *Gas Phase Reaction Rate Theory*, Ronald, New York (1966).

15. M. Wolfsberg, Isotope effects, *Ann. Rev. Phys. Chem.* **20**, 449–473 (1969).

16. C. J. Collins and N. S. Bowman, eds., *Isotope Effects in Chemical Reactions*, Van Nostrand Reinhold, New York (1971).

17. R. P. Bell, *The Proton in Chemistry*, 2nd ed. Cornell University Press, Ithaca, N.Y. (1973).

18. C. G. Swain, E. C. Stivers, J. F. Reuwer, Jr., and L. J. Schaad, Use of hydrogen isotope effects to identify the attacking nucleophile in the enolization of ketones catalyzed by acetic acid, *J. Am. Chem. Soc.* **80**, 5885–5893 (1958).

19. F. H. Westheimer, The magnitude of the primary kinetic isotope effect for compounds of hydrogen and deuterium, *Chem. Rev.* **61**, 265–273 (1961).

20. J. Bigeleisen, Correlation of kinetic isotope effects with chemical bonding in three-centre reactions, *Pure Appl. Chem.* **8**, 217–223 (1964).

21. R. A. More O'Ferrall and J. Kouba, Model calculations of primary hydrogen isotope effects, *J. Chem. Soc. B* **1967**, 985–990.

22. R. A. More O'Ferrall, Model calculations of hydrogen isotope effects for non-linear transition studies, *J. Chem. Soc. B* **1970**, 785–790.

23. J. Donahue, in: *Structural Chemistry and Molecular Biology* (A. Rich and N. Davidson, eds.), pp. 443–465, W. H. Freeman, San Francisco (1968).

24. W. C. Hamilton, in: *Structural Chemistry and Molecular Biology* (A. Rich and N. Davidson, eds.), pp. 466–483, W. H. Freeman, San Francisco (1968).

25. M. S. Lehmann, T. F. Koetzle, and W. C. Hamilton, Precise neutron diffraction structure determination of protein and nucleic acid components, *J. Am. Chem. Soc.* **94**, 2657–2660 (1972).

26. P. A. Kollman, and L. C. Allen, The theory of the hydrogen bond, *Chem. Rev.* **72**, 283–303 (1972).

27. R. Yamdagni and P. Kebarle, Gas phase basicities of amines. Hydrogen bonding in proton-bound amine dimers and proton-induced cyclization of α, ω diamines, *J. Am. Chem. Soc.* **95**, 3504–3510 (1973).

28. A. J. Kresge and Y. Chiang, The effect of bending vibrations on the magnitude of hydrogen isotope effects, *J. Am. Chem. Soc.* **91**, 1025–1026 (1969).

29. A. V. Willi and M. Wolfsberg, The influence of "bond making and bond breaking" in the transition state on hydrogen isotope effects in linear three center reactions, *Chem. Ind. London* **1964**, 2097–2098.

30. E. F. Caldin, Tunneling in proton-transfer reactions in solution, *Chem. Rev.* **68**, 135–156 (1968).

31. W. J. Albery, Isotope effects in proton transfer reactions, *Trans. Faraday Soc.* **63**, 200–206 (1967).

32. G. S. Hammond, A correlation of reaction rates, *J. Am. Chem. Soc.* **77**, 334–338 (1955).

33. R. P. Bell and J. E. Crooks, Kinetic hydrogen isotope effects in the ionization of some ketonic substances, *Proc. Roy. Soc. London Ser. A* **286**, 285–299 (1965).

34. R. P. Bell, F. R. S. Goodall, and D. M. Goodall, Kinetic hydrogen isotope effects in the ionization of some nitroparaffins, *Proc. Roy. Soc. London Ser. A* **294**, 273–297 (1966).
35. A. F. Cockerill, Mechanisms of elimination reactions, *J. Chem. Soc. B* **1967**, 964–969.
36. R. P. Bell and B. G. Cox, Primary hydrogen isotope effects on the rate of ionization of nitroethane in mixtures of water and dimethyl sulfoxide, *J. Chem. Soc. B* **1971**, 783–785.
37. R. P. Bell, Liversidge lecture, recent advances in the study of kinetic hydrogen isotope effects, *Chem. Soc. Rev.* **3**, 513–544 (1974).
38. F. G. Bordwell and W. J. Boyle, Jr., Kinetic isotope effects for nitro-alkanes and their relationship to transition state structure in proton-transfer reactions, *J. Am. Chem. Soc.* **97**, 3447–3452 (1975).
39. W. A. Pryor and K. G. Kneipp, Primary kinetic isotope effects and the nature of hydrogen-transfer transition states, *J. Am. Chem. Soc.* **93**, 5584–5586 (1971).
40. R. A. Marcus, Theoretical relations among rate constants, barriers, and brønsted slopes of chemical reactions, *J. Phys. Chem.* **72**, 891–899 (1968).
41. A. J. Kresge, *Sixth Steenbock Symposium on Isotope Effects on Enzyme Catalyzed Reactions* (W. W. Cleland, M. H. O'Leary, and D. B. Northrop, eds.), pp. 37–63, University Park Press, Baltimore (1977).
42. W. P. Jencks, *Catalysis in Chemistry and Enzymology*, McGraw-Hill, New York (1969), p. 243.
43. R. P. Bell, The tunnel effect correction for parabolic potential barriers, *Trans. Faraday Soc.* **55**, 1–4 (1959).
44. M. J. Stern and R. E. Weston, Jr., Phenomenologic manifestations of quantum mechanical tunnelling. I. Curvature in Arrhenius plots, *J. Chem. Phys.* **60**, 2803–2807 (1974).
45. M. J. Stern and R. E. Weston, Jr., Phenomenologic manifestations of quantum mechanical tunnelling. II. Effect on Arrhenius pre-exponential factors for primary hydrogen kinetic isotope effects, *J. Chem. Phys.* **60**, 2808–2814 (1974).
46. M. J. Stern and R. E. Weston, Jr., Phenomenologic manifestations of quantum mechanical tunnelling. III. Effects on relative tritium–deuterium kinetic isotope effects, *J. Chem. Phys.* **60**, 2815–2821 (1974).
47. E. F. Caldin and S. Mateo, Kinetic isotope effects in various solvents for the proton-transfer reactions of 4-nitrophenylnitromethane with basis, *J. Chem. Soc. Chem. Commun.* **1973**, 854–855.
48. E. F. Caldin and C. J. Wilson, Structure and solvent influences on tunnelling in reactions of 4-nitrophenylnitromethane with nitrogen bases in aprotic solvents, *Faraday Symp. Chem. Soc.* **10**, 121–131 (1975).
49. R. P. Bell, W. H. Sachs, and R. L. Tranter, Model calculations of isotope effects in proton transfer reactions, *Trans. Faraday Soc.* **67**, 1995–2003 (1970).
50. J. Banger, A. Jaffe, An-Chung Lin, and W. H. Saunders, Jr., Carbon isotope effects on proton transfers from carbon, and the question of hydrogen tunneling, *J. Am. Chem. Soc.* **97**, 7177–7178 (1975).
51. S. R. Hartshorn and V. J. Shiner, Calculation of H/D, $^{12}C/^{13}C$, and $^{12}C/^{14}C$. Factors from valence force fields derived from a series of simple organic molecules, *J. Am. Chem. Soc.* **94**, 9002–9012 (1972).
52. W. E. Buddenbaum and V. J. Shiner, Jr., in: *Sixth Steenbock Symposium on Isotope Effects on Enzyme Catalyzed Reactions* (W. W. Cleland, M. H. O'Leary, and D. B. Northrop, eds.), pp. 1–36, University Park Press, Baltimore (1977).
53. P. F. Cook and W. W. Cleland, Deuterium and tritium isotope effects for liver alcohol dehydrogenase using cyclohexanol, *Fed. Proc. Fed. Am. Soc. Exp. Biol.* **36**, 2078 (1977).
54. J. P. Klinman, unpublished results.
55. H. P. Meloshe, C. T. Monti, and W. W. Cleland, Magnitude of the equilibrium isotope effects on carbon–tritium bond synthesis, *Biochem. Biophys. Acta* **480**, 517–519 (1977).
56. R. L. Schowen, in: *Sixth Steenbock Symposium on Isotope Effects on Enzyme Catalyzed Reactions* (W. W. Cleland, M. H. O'Leary, and D. B. Northrop, eds.), pp. 64–99, University Park Press, Baltimore (1977).

57. R. L. Schowen, Mechanistic deductions from solvent isotope effects, *Prog. Phys. Org. Chem.* **9**, 275–332 (1972).

58. W. W. Cleland, M. H. O'Leary, and D. B. Northrop, eds., *Sixth Steenbock Symposium on Isotope Effects on Enzyme Catalyzed Reactions*, University Park Press, Baltimore (1977).

59. W. W. Cleland, Partition analysis and the concept of net rate constants as tools in enzyme kinetics, *Biochemistry* **14**, 3220–3224 (1975).

60. J. P. Klinman, The mechanism of enzyme-catalyzed reduced nicotinamide adenine dinucleotide-dependent reductions: Substituent and isotope effects in the yeast alcohol dehydrogenase reaction, *J. Biol. Chem.* **247**, 7977–7987 (1972).

61. J. P. Klinman, Isotope effects and structure-reactivity correlations in the yeast alcohol dehydrogenase reaction. A study of the enzyme catalyzed oxidation of aromatic alcohols, *Biochemistry* **15**, 2018–2026 (1976).

62. D. L. Vander Jagt and L. P. B. Han, Deuterium isotope effects and chemically modified coenzymes as mechanistic probes of yeast alyoxylase-I, *Biochemistry* **12**, 5161–5166 (1973).

63. K. Bush, V. J. Shiner, Jr., and H. R. Mahler, Deuterium isotope effects on initial rates of the liver alcohol dehydrogenase reaction, *Biochemistry* **12**, 4802–4805 (1972).

64. W. W. Cleland, What limits the rate of an enzyme-catalyzed reaction, *Acc. Chem. Res.* **8**, 145–151 (1975).

65. L. Bachan, C. B. Storm, J. W. Wheeler, and S. Kaufman, Isotope effects in the hydroxylation of phenylethylamine by dopamine-β-hydroxylase, *J. Am. Chem. Soc.* **96**, 6799–6800 (1974).

66. D. B. Northrop, Steady state analysis of kinetic isotope effects in enzymatic reactions, *Biochemistry* **14**, 2644–2651 (1975).

67. R. H. Abeles, W. R. Frisell, and C. G. Mackenzie, A dual isotope effect in the enzymatic oxidation of deuteromethyl sarcosine, *J. Biol. Chem.* **235**, 853–856 (1960); corrected **235**, 1544 (1960).

68. J. L. Robinson and I. A. Rose, The proton transfer reactions of muscle pyruvate kinase, *J. Biol. Chem.* **247**, 1096–1105 (1972).

69. I. A. Rose, E. L. O'Connell, and A. H. Mehler, Mechanism of the aldolase reaction, *J. Biol. Chem.* **240**, 1758–1765 (1965).

70. E. C. Dinovo and P. D. Boyer, Isotopic probes of the enolase reaction mechanism, *J. Biol. Chem.* **246**, 4586–4593 (1971).

71. C. T. Walsh, A. Schonbrunn, and R. H. Abeles, Studies on the mechanism of action of D-amino oxidase, *J. Biol. Chem.* **246**, 6855–6866 (1971).

72. K. Yagi, M. Nishikimi, A. Takai, and N. Ohishi, Mechanism of enzyme action. VI. Kinetic isotope effect on D-amino acid oxidase reaction, *Biochim. Biophys. Acta* **321**, 64–71 (1973).

73. H. J. Bright and D. J. T. Porter, in: *The Enzymes*, 3rd ed. (P. D. Boyer, ed.), Vol. XII, pp. 421–505, Academic Press, New York (1976).

74. D. A. Weisblat and B. M. Babior, The mechanism of action of ethanolamine ammonialyase, a B_{12}-dependent enzyme, *J. Biol. Chem.* **246**, 6064–6071 (1971).

75. M. I. Schimerlik, C. E. Grimshaw, and W. W. Cleland, The use of isotope effects to determine the rate limiting steps for malic enzyme, *Biochemistry* **16**, 571 (1977).

76. W. J. Albery and J. R. Knowles, The determination of the rate-limiting step in a proton transfer reaction from the breakdown of the Swain–Schaad relation, *J. Am. Chem. Soc.* **99**, 637–638 (1977).

77. R. K. Gupta, R. M. Oesterling, and A. S. Mildvan, Dual divalent cation requirement for activation of pyruvate kinase: Essential roles for both enzyme-bound and nucleotide-bound metal ions, *Biochemistry* **15**, 2881–2887 (1976).

78. J. F. Biellmann, E. L. O'Connell, and I. A. Rose, Secondary isotope effects in reactions catalyzed by yeast and muscle aldolase, *J. Am. Chem. Soc.* **91**, 6484–6488 (1969).

79. E. Grazi, T. Cheng, and B. L. Horecker, The formation of a stable aldolase–dihydroxyacetone phosphate complex, *Biochem. Biophys. Res. Commun.* **7**, 250–253 (1962).

80. T. Y. S. Shen and E. W. Westhead, Divalent cation and pH-dependent primary isotope effects in the enolase reaction, *Biochemistry* **12**, 3333–3337 (1973).

81. H. R. Mahler and J. Douglas, Mechanisms of enzyme-catalyzed oxidation–reduction reactions I., *J. Am. Chem. Soc.* **79**, 1159–1166 (1957).

82. J. D. Shore and H. Gutfreund, Transients in the reactions of liver alcohol dehydrogenase, *Biochemistry* **9**, 4655–4659 (1970).

83. R. T. Dworschack and B. V. Plapp, pH, isotope and substituent effects on the interconversion of aromatic substrates, catalyzed by hydroxybutyrimidylated liver alcohol dehydrogenase, *Biochemistry* **16**, 2716–2725 (1977).

84. G. J. Hardman, L. F. Blackwell, C. R. Boswell, and P. D. Buckley, Substituent effects on the pre-steady state kinetics of oxidation of benzyl alcohols by liver alcohol dehydrogenase, *Eur. J. Biochem.* **50**, 113–118 (1974).

85. A. Brown and H. F. Fisher, A comparison of the glutamate dehydrogenase catalyzed oxidation of NADPH by trinitrobenzenesulfonate with the uncatalyzed reaction, *J. Am. Chem. Soc.* **98**, 5682–5688 (1976).

86. L. C. Kurz and C. Frieden, Comparison of the structure of enzymatic and non-enzymatic transition states. The reductive desulfonation of 4-X-2,6-dinitrobenzenesulfonates by NADH, *Biochemistry* **16**, 5207–5216 (1977).

87. L. do Amaral, M. P. Bastos, H. G. Bull, and E. H. Cordes, Secondary deuterium isotope effects for addition of nitrogen nucleophiles to substituted benzaldehydes, *J. Am. Chem. Soc.* **95**, 7369–7374 (1973).

88. J. P. Klinman, Acid–base catalysis in the yeast alcohol dehydrogenase reaction, *J. Biol. Chem.* **250**, 2569–2573 (1974).

89. W. P. Jencks, General acid–base catalysis of complex reactions in water, *Chem. Rev.* **72**, 705–718 (1972).

90. R. Stewart, A. L. Gatzke, M. Macke, and K. Yates, Deuterium isotope effects in organic cations, *Chem. Ind.* **1959**, 331–332.

91. P. Ballinger and F. A. Long, Acid ionization constants of alcohols. I. Trifluoroethanol in the solvents H_2O and D_2O, *J. Am. Chem. Soc.* **81**, 1050–1053 (1959).

92. J. P. Klinman, K. Welsh, and D. J. Creighton, in: *Solvent Isotope Effects in the Yeast Alcohol Dehydrogenase Reaction in Alcohol and Aldehyde Metabolizing Systems* (R. G. Thurman, T. Yonetani, J. R. Williamson, and B. Chance, eds.), Vol. II, Academic Press, New York, in press.

93. B. L. Vallee and F. L. Hock, Zinc, a component of yeast alcohol dehydrogenase. *Proc. Nat. Acad. Sci. USA* **41**, 327–338 (1955).

94. C. Veillon and A. J. Sytkowski, The intrinsic zinc atoms of yeast alcohol dehydrogenase, *Biochem. Biophys. Res. Commun.* **67**, 1494–1500 (1976).

95. J. P. Klinman and K. Welsh, The zinc content of yeast alcohol dehydrogenase, *Biochem. Biophys. Res. Commun.* **70**, 878–884 (1976).

96. J. J. Steffens and D. M. Chipman, Reactions of dihydronicotinamides. I. Reduction of trifluoroacetophenone by 1-substituted dihydronicotinamides, *J. Am. Chem. Soc.* **93**, 6694–6696 (1971).

97. D. J. Creighton, J. Hajdu, G. Mooser, and D. S. Sigman, Model dehydrogenase reactions. Reduction of *N*-methylacridinium ion by reduced nicotinamide adenine dinucleotide and its derivatives, *J. Am. Chem. Soc.* **95**, 6855–6857 (1973).

98. R. F. Williams, S. Shinkai, and T. C. Bruice, Radical mechanism for 1.5-dihydroflavin reduction of carbonyl compounds, *Proc. Nat. Acad. Sci. USA* **72**, 1763–1767 (1975).

99. H. Sund and H. Theorell, in: *The Enzymes* (P. D. Boyer, H. Lardy, and K. Myrback, eds.), Vol. VII, pp. 25–83, Academic Press, New York (1963).

100. J. W. Jacobs, J. T. McFarland, I. Wainer, D. Jeanmaier, C. Ham, K. Hamm, M. Wnuk, and M. Lam, Electronic substituent effects during liver alcohol dehydrogenase catalyzed reduction of aromatic alcohols, *Biochemistry* **13**, 60–64 (1974).

101. B. V. Plapp, R. L. Brooks, and J. D. Shore, Horse liver alcohol dehydrogenase, amino groups and rate-limiting steps in catalysis, *J. Biol. Chem.* **248**, 3470–3475 (1973).

102. J. McFarland, personal communication.

103. C. I. Branden, H. Jornvall, H. Eklund, and B. Furugren, in: *The Enzymes*, 3rd ed. (P. D. Boyer, ed.), Vol. XI, pp. 104–190, Academic Press, New York (1975).

104. H. Eklund, B. Nordstrom, E. Zeppezauer, G. Soderland, I. Ohlsson, T. Boiwe, B. O. Soderberg, O. Tapia, C. J. Branden, and A. Akeson, Three-dimensional structure of horse liver alcohol dehydrogenase at 2.4 Å resolution, *J. Mol. Biol.* **102**, 27–59 (1976).

105. D. L. Sloan, M. M. Young, and A. S. Mildvan, NMR studies of substrate interaction with cobalt substituted alcohol dehydrogenase from liver, *Biochemistry* **14**, 1998–2008 (1975).

106. D. J. Bates, B. R. Golden, and C. Frieden, A new reaction of glutamate dehydrogenase: The enzyme-catalyzed formation of trinitrobenzene from TNBS in the presence of reduced coenzyme, *Biochem. Biophys. Res. Commun.* **39**, 502–507 (1970).

107. L. C. Kurz and C. Frieden, A model dehydrogenase reaction. Charge distribution in the transition state, *J. Am. Chem. Soc.* **97**, 677–679 (1975).

108. C. G. Swain, R. A. Wiles, and R. F. W. Bader, Use of substituent effects on isotope effects to distinguish between proton and hydride transfers. Part I. Mechanism of oxidation of alcohols by bromine in water, *J. Am. Chem. Soc.* **83**, 1945–1950 (1961).

109. I. A. Rose, Mechanism of the aldose–ketose isomerase reactions, *Adv. Enzymol. Relat. Areas Mol. Biol.* **43**, 491–517 (1976).

110. J. M. Herlihy, S. G. Maister, W. J. Albery, and J. R. Knowles, Energetics of triophosphate isomerase: The fate of the 1(R)-^{3}H label of tritiated dehydroxyacetone phosphate in the isomerase reaction, *Biochemistry* **15**, 5601–5607 (1976).

111. S. G. Maister, C. P. Pett, J. W. Albery, and J. R. Knowles, Energetics of triosephosphate isomerase: The appearance of solvent tritium in substrate dihydroxyacetone phosphate and in product, *Biochemistry* **15**, 5607–5612 (1976).

112. S. J. Fletcher, J. M. Herlihy, W. J. Albery, and J. R. Knowles, Energetics of triosephosphate isomerase: The appearance of solvent tritium in substrate glyceraldehyde 3-phosphate and in product, *Biochemistry* **15**, 5612–5617 (1976).

113. P. F. Leadlay, W. J. Albery, and J. R. Knowles, Energetics of triosphosphate isomerase: Deuterium isotope effects in the enzyme-catalyzed reaction, *Biochemistry* **15**, 5617–5620 (1976).

114. W. J. Albery and J. R. Knowles, Free-energy profile for the reaction catalyzed by triose-phosphate isomerase, *Biochemistry* **15**, 5627–5631 (1976).

115. A. Hall and J. R. Knowles, The uncatalyzed rates of enolization of dihydroxyacetone phosphate and of glyceraldehyde 3-phosphate in neutral aqueous solution. The quantitative assessment of the effectiveness of an enzyme catalyst, *Biochemistry* **14**, 4348–4352 (1975).

116. R. Wolfenden, Transition state analogues for enzyme catalysis, *Nature* (London) **223**, 704–705 (1969).

117. S. J. Reynolds, D. W. Yates, and C. I. Pogson, Dihydroxyacetone phosphate: Its structure and reactivity with α-glycerophosphate dehydrogenase, aldolase and triose phosphate isomerase and some possible metabolic implications, *Biochem. J.* **122**, 285–297 (1971).

118. W. W. Cleland, Determining the chemical mechanisms of enzyme catalyzed reactions by kinetic studies, *Adv. Enzymol. Relat. Areas Mol. Biol.* **45**, 273–387 (1977).

5

Secondary Hydrogen Isotope Effects

John L. Hogg

1. INTRODUCTION

Encouraged by the vast amount of knowledge gathered in years of study of all types of chemical catalysis and armed with the plethora of information available concerning chemical reactions of every conceivable nature, scientists have recently begun to explore the most elegant of all catalytic systems—enzymes. Only in the relatively recent past have scientists gained the skills and confidence necessary to tackle such a formidable task as the elucidation of the mechanism of enzymic catalysis. And a formidable task it is! The elucidation of the mechanism of many "simple" bimolecular reactions has not been an easy task. And what a different goal it is to study reactions between two (or more) relatively simple, low-molecular-weight compounds containing relatively few atoms or to study an enzyme mechanism. Even the simplest enzymes are proteins containing hundreds or thousands of atoms. It is little wonder then that 50 years after Sumner's[1] demonstration that urease was a protein still relatively little is known about the mechanism of enzymic catalysis.

As Wolfenden[2] has recently pointed out, great advances have been made in the development of model reactions suggesting possible mechanisms of catalysis, in the determination of the amino acid sequence and crystal structure of enzymes, in the isolation and characterization of relatively stable intermediates, and in the detection of short-lived events, but all of these approaches have their limitations.

One of the most sophisticated approaches currently in use is that of transition-state structure determination. The transition state has the potential for providing the most significant information about the dynamic and structural requirements of the enzymic catalytic mechanism. Wolfenden[2] and Lienhard[3]

John L. Hogg • Department of Chemistry, Texas A & M University, College Station, Texas.

have recently discussed the problem of analog approaches in the determination of transition-state structure in some detail.

Although there are numerous dynamic approaches one might take to study the importance of transition-state structure and stabilization, many of them have a serious drawback in that they require a nonnatural substrate or chemically modified enzyme. Very little is known about the changes brought about in the catalytic mechanism when such probes are employed.

A technique that most nearly avoids such problems involves the use of isotopes as probes of reaction mechanism. Sensitive probes of transition-state structure may be made using isotopically substituted compounds. Isotopic experiments do not perturb the sensitive enzyme systems because the uncharged neutrons of isotopically different compounds do not affect the electronic nature of the transition state. Many probes announce their presence to the enzyme and hence may observe an energetically modified mechanism. Isotopes, being less obvious in their presence, are able to observe the true mechanism as they inconspicuously ride along the reaction path. Hence, isotopes may be thought of as voyeurs of reaction mechanism. In this chapter we shall deal with the utility of secondary hydrogen isotope effects as probes of transition-state structure.

2. ORIGIN AND MAGNITUDE—KINETIC AND EQUILIBRIUM EFFECTS

There are several excellent discussions of the origin and magnitude of secondary isotope effects in the literature[4] so the basic concepts will be presented in limited detail here.

A secondary isotope effect is a difference in reaction rate or position of equilibrium brought about by isotopic substitution of atoms to which no bonds are broken or formed during the course of reaction or attainment of equilibrium. Such effects are further classified as α, β, or remote (γ, δ) depending on the position of the isotopic label relative to the position at which bonding changes are occurring. This terminology is not to be confused with the common use of α, β, γ to designate carbon atoms.

The finite zero-point energy difference between isotopically labeled reactants can be altered in one of two ways in reaching the transition state (or equilibrium state) depending on the manner in which the significant force constants describing the isotopic vibrational modes change. This is illustrated in Fig. 1 for protium (^{1}H) vs. deuterium (^{2}H) substitution, but the same approach will work for any isotopes.* If the force constant decreases (case 1), the curvature

---

* Throughout this chapter the symbols ^{1}H, ^{2}H, and ^{3}H will be used to refer to protium, deuterium, and tritium, respectively. However, isotope effects will be written as k_H/k_D or k_H/k_T for simplicity. The use of the symbol D has been avoided elsewhere in order to avoid confusion with configuration assignments of biological compounds.

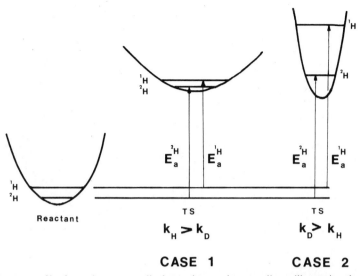

Fig. 1. Energy profiles for a plane perpendicular to the reaction coordinate illustrating the manner in which the zero point energy difference between the ^{1}H- and ^{2}H-labeled reactant changes in going to the transition state (TS). The energy of activation is represented by E_a. Case 1 would give rise to a *normal* secondary deuterium isotope effect, while case 2 would give an *inverse* secondary deuterium isotope effect.

of the potential energy well is decreased, forcing the ^{1}H and ^{2}H energy levels closer together in the transition state (or equilibrium product). Thus, the energy of activation will be less for the ^{1}H compound (i.e., $E_a^{1H} < E_a^{2H}$), and the rate constant (or equilibrium constant) will be larger for the ^{1}H compound than for the ^{2}H compound. This is referred to as a normal isotope effect. If the force constant increases (case 2), the opposite effect is observed, and an inverse isotope effect results. The trivial case of no force constant changes in vibrational modes sensitive to isotopic substitution gives, of course, no isotope effect.

Secondary α-deuterium effects are generally thought to arise from changes in bending frequencies of the isotopic bonds in reaching the transition state (or equilibrium state). The origin of secondary β-deuterium effects is most likely hyperconjugative, with the elegant experiment of Shiner and Humphrey[5] being probably the most convincing evidence for this. Remote effects may arise from a combination of many effects depending on the exact system studied. Steric size differences between protium, deuterium, and tritium have also been used to explain many remote isotope effects.[4c] Regardless of the explanation used to explain a given effect the basic concept of differences in vibrational frequencies is responsible for all effects.

Figure 2 illustrates how α- and β-deuterium probes may be used to detect changes in geometry and charge distribution in nucleophilic reactions at carbonyl carbon, a biochemically significant reaction type. In the reactants there is a partial positive charge at the carbonyl carbon which can be stabilized

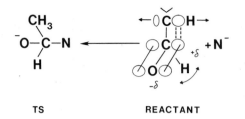

Fig. 2. Illustration to show how α- and β-deuterium substitution can be used to probe the transition state (TS) geometry and charge distribution in the addition of a nucleophile (N⁻) to a carbonyl group. The carbonyl group is represented here as being aldehydic.

to some extent by the $C—^1H(^2H)$ bonds of the adjacent methyl group through hyperconjugation. However, in the tetrahedral transition state the $C—^1H(^2H)$ bonds are tightened relative to the reactants since hyperconjugation is no longer possible. The net result for the beta $C—^1H(^2H)$ bonds has been to go to a "tighter" transition state so an inverse isotope effect is observed.

For an alpha $C—^1H(^2H)$ bond one must consider bending frequency changes, since hyperconjugation is not possible for this bond. In going from reactant to transition state the net result is to make the bending of the $C—^1H(^2H)$ bond more difficult (i.e., a higher bending frequency) so an inverse isotope effect is observed here also. These arguments may be used in reverse to predict the isotope effects for going from a tetrahedral reactant to a trigonal transition state.

Secondary α-deuterium isotope effects may approach a maximum of about 40%, while secondary β-deuterium isotope effects of about 30% per deuterium may be possible. However, these maximally predicted effects are almost never observed for numerous reasons. β effects can be especially dependent on conformation and are generally not quite cumulative for this reason.[6] Remote effects[7] and steric effects[8] are generally significantly smaller in magnitude and require rather more specific explanations for their origin in any given system.

It is frequently reported that $α-^2H$ substitution in an S_N2 reaction has essentially no effect on reaction rate because the bending frequencies of the isotopic bonds are unchanged in going from reactants to transition state.[9] The reaction of isopropyl bromide with sodium ethoxide, which gives $k_H/k_D = 1.00$, is usually cited in support of this statement.[10] However, a very wide range of isotope effects in such reactions is seen. Seltzer and Zavitsas have tabulated the effects observed up to 1967.[11] Inverse effects as large as $k_H/k_D = 0.87$ were observed in nucleophilic displacement of iodine from deuterated methyl iodide by water at 70°,[12] while a normal effect as large as $k_H/k_D = 1.21$ has been observed for the displacement of dimethyl-$^2H_6$ sulfide from trimethyl-$^2H_9$ sulfonium halides by phenoxide at 76°.[13] Clearly there are more complicating factors in the latter effect; however, subsequent experimental studies and theoretical calculations on such isotope effects and their temperature dependence have indicated that the relative degrees of bond breaking and bond formation as well as temperature play significant roles in determining the magnitude of the observed isotope effect.[14]

The observation of secondary hydrogen isotope effects in carbanion-forming or radical-forming reactions is not so commonly reported, although Halevi does not consider these possibilities.[4a] The effects are predicted to be similar to those seen in carbonium-ion formation. Fry, in a recent review,[15] has considered the possibility of secondary hydrogen isotope effects in elimination reactions.

Recent papers dealing with theoretical considerations and temperature dependences of secondary hydrogen effects (as well as others) for many mechanistic types are especially relevant to the above discussion.[16] Space does not permit a discussion of these important papers, but the serious reader should certainly consult them.

The relationship between tritium and deuterium isotope effects is given as

$$k_H/k_T = (k_H/k_D)^{1.44} \quad \text{or} \quad \log(k_H/k_T) = 1.44 \log(k_H/k_D) \tag{1}$$

and has been experimentally verified in several instances.[17]

3. REPRESENTATIVE EFFECTS IN MODEL SYSTEMS

The first studies done in this area were those of Bender and his co-workers.[18] The rates of hydrolysis of ethyl acetate-$^2H_3$, acetyl chloride-$^2H_3$, and acetic anhydride-$^2H_6$ and their protium analogs were compared. The inverse secondary β-deuterium effect of $k_H/k_D = 0.90$ for ethyl acetate hydrolysis at 25° was consistent with loss of hyperconjugation in going from reactants to transition state. The solvolysis of acetyl chloride in ethanol–cyclohexane or 5% water–tetrahydrofuran showed almost no effect of deuterium substitution. Acetic anhydride solvolysis in water showed essentially no isotope effect either. These results indicate little or no development of an acylium ion or tetrahedral addition intermediate. However, at $-22°$ the hydrolysis of acetyl chloride in 10% and 20% water–acetone gave secondary β-deuterium isotope effects of $k_H/k_D = 1.51$ and 1.62, respectively. It was postulated that development of an acylium ion gave rise to increased hyperconjugative stabilization representing a transition from a tight to weaker potential in the transition state. The positive charge should be more fully developed in the medium of higher dielectric constant giving rise to larger isotope effects.

Subsequent to this study a significant dependence of the ethyl acetate isotope effects on temperature was discovered.[19] Over an easily accessible range of temperatures the effect was seen to vary in the following manner: k_H/k_D (T); 1.00 (0°), 0.90 (25°), 0.93 (35°), and 1.15 (65°). It was suggested that this unusual temperature dependence offered a significant challenge to the application of general isotope-effect theory to secondary effects.

A reinvestigation of this phenomenon using methyl acetate-$^2H_3$ was undertaken in the hope of eliminating this problem.[20] The measured isotope effects were k_H/k_D (T), 1.009 ± 0.024 (0°), 0.953 ± 0.014 (15°), 0.898 ± 0.022 (25.0°), 0.922 ± 0.028 (35°), and 1.027 ± 0.099 (50°). These results are indistinguishable from the previous results on a plot of k_H/k_D vs. temperature.

However, Eyring plots of log (k_2/T) vs. $1/T$ for the individual rate constants obtained in the present study show significant downward curvature at higher temperatures. Such curvature is indicative of a change in rate-limiting step.[21] The Eyring plots for the $^1H_3$ and $^2H_3$ compounds intersect and cross over at about 45°, the temperature at which the isotope effect changes from inverse to normal.

This result implies that the unusual isotope-effect behavior is mechanistic in origin. If some effect such as desolvation is rate limiting at higher temperatures, one might observe a normal isotope effect since the isotopic bonds should be in a weaker potential once the restriction of the solvent cage is removed. Desolvation would be followed by rapid attack of hydroxide at these high temperatures, which should give no isotope effect for this fast step. As the temperature is lowered below 45°, the rate-limiting step probably now becomes attack of hydroxide on the carbonyl carbon, which shows the expected inverse isotope effect. Since the isotope effect becomes less inverse at temperatures below 25°, it is conceivable that another mechanistic change is beginning to take place. This is hinted at by the shape of the Eyring plots, but additional data are required to be sure of this. A similar analysis of the data of Halevi and Margolin was not possible since the individual rate constants were not reported.

The mechanistic origin of this anomalous temperature dependence is also supported by recent Wolfsberg–Stern-type calculations of various isotope effects for addition of hydroxide ion to acetaldehyde.[20] No unusual temperature dependencies or crossovers were discovered for the β-deuterium isotope effect calculated at 0°, 25°, and 50° for this theoretical model as a function of the bond order, n, of the hydroxide oxygen–carbonyl carbon bond. The magnitude of the isotope effect was found to vary linearly with bond order also. Similar results were found in the theoretical calculations on the secondary α-deuterium isotope effect.

Experimental values for the secondary β-deuterium effect observed in the acid-catalyzed hydrolysis of methyl acetate were essentially $k_H/k_D = 0.93 \pm 0.01$ over the temperature range 0°–42°.[20] These effects are consistent with the observed solvent deuterium isotope effect,[22a] which was interpreted in terms of rate-limiting attack of water on the protonated substrate to give a tetrahedral transition state.[22b] The lack of any significant temperature dependence is also consistent with the other postulates.

The equilibrium secondary isotope effects of $K_H/K_D = 0.958$ and 0.913 determined for methanol hemiketal formation with acetone-$^2H_6$ and cyclopentanone-$^2H_4$, respectively, are in the direction expected for loss of hyperconjugation in forming the tetrahedral hemiketal.[18b] It should be noted that the dissociation constant ratios are reported in the original paper. The observed effect is larger for cyclopentanone even though fewer deuteria are present presumably because hyperconjugative stabilization of the carbonyl group is more significant due to the restricted conformational mobility. The hydration of 1,3-dichloracetone-$^2H_4$ has been found to give $K_H/K_D = 0.83 \pm 0.02$

over the temperature range $15°-46°$.[20] This effect is consistent with the same explanation, and the lack of any significant temperature dependence is to be expected for an equilibrium isotope effect. Several other kinetic and equilibrium effects for β-deuteration have been reported for similar systems.[23] Recently, related studies have been reported for carboxylic acid ionizations[24] and for deuterium substitution in the leaving group of trifluoroacetates.[25]

The transition-state structure for both the acid-catalyzed and alkaline hydrolysis of several formyl-transfer reactions has recently been probed with the secondary α-deuterium isotope-effect technique.[26] The value of k_D/k_H for alkaline hydrolysis of methyl formate was 1.05 ± 0.01. The value for the previously determined methoxyl-O-18 kinetic isotope effect,[27] this value, and a simple theoretical model for the anionic tetrahedral intermediate were used to estimate the bond orders of the transition state. These parameters indicated a transition state which had progressed about 36% of the way toward the tetrahedral intermediate. Another study has indicated that such α-deuterium effects are approximately linear with the bond order of the attacking nucleophile.[20] A similar effect of $k_D/k_H = 1.10 \pm 0.01$ for ethyl formate hydrolysis is consistent with a more nearly tetrahedral transition state for this ester. Values of $k_D/k_H = 1.23 \pm 0.02$ for acid hydrolysis of methyl formate, 1.23 ± 0.01 for rate-limiting formation of an anionic tetrahedral intermediate in the hydrazinolysis of methyl formate, and 1.02 ± 0.01 for rate-limiting breakdown of this intermediate were determined. The former three values are consistent with a tetrahedral transition state, with the latter being indicative of a trigonal transition state.

Cordes and co-workers have determined a number of secondary deuterium kinetic and equilibrium isotope effects for addition of nitrogen nucleophiles to carbonyl compounds[28] and for the hydrolysis of Schiff bases.[29] Similar studies have been conducted on acetal and orthoester hydrolysis.[30] Although these studies represent some of the most informative, they will not be discussed here since Cordes and Bull have considered them in Chapter 11 in this volume.

Kinetic deuterium isotope effects have been measured for the peracid epoxidation of p-phenylstyrene.[31] The kinetic secondary effects (k_H/k_D) were 0.99, 0.82, and 0.82 at $0°$ for the α-^2H, β, β-$^2H_2$, and α, β, β-$^2H_3$ compounds, respectively. The α-^2H effect for p-nitrostyrene was 0.98 at $25°$. These and other data suggest an epoxidation transition state in which there is substantial C_β—O bond formation but negligible changes in bonding or hybridization at C_α. A new transition-state structure for epoxidation reactions is proposed.

Partition coefficients for the equilibrium phase transfer of p-nitrophenyl acetate-$^2H_3$ and $^1H_3$ from water to cyclohexane have been determined at $25°$.[32] The ratio, $K =$ (ester in organic phase)/(ester in aqueous phase), was $K_H/K_D \simeq 1.15 \pm 0.04$. This effect is consistent with the isotopically labeled bonds being in a weaker potential in the organic phase than in the aqueous phase. Such an effect could arise from interactions of the carbonyl dipole in water which are relieved on transfer to cyclohexane and may have implications for enzymic reactions where transfer of labeled substrates from aqueous environments to relatively nonaqueous enzymic active sites could give rise to similar effects.

4. SECONDARY HYDROGEN ISOTOPE EFFECTS IN ENZYMIC SYSTEMS

Excellent discussions are available dealing with isotope effects of all kinds in biological systems[33] and on studies with deuterated drugs.[34] The emphasis here will be on studies dealing with secondary hydrogen isotope effects encountered with labeled substrates in enzymic systems.

In general there are two potential ways in which a secondary hydrogen isotope effect may manifest itself in enzymic systems. Rose[35] and Richards[36] have briefly discussed them with special emphasis on secondary effects, while Jencks[33b] has given a similar discussion with emphasis on a variety of isotope effects (solvent, primary, and secondary). The two significant parameters of an enzymic reaction are V_{max}, the maximal velocity of an enzymic reaction, and K_m, the Michaelis constant. In an extreme oversimplification one may assume that a secondary isotope effect on K_m reflects a difference in the strength of enzyme–substrate binding. Effects on V_{max} are more difficult to interpret since V_{max} and K_m are related. Thus an effect on V_{max} may be a true kinetic effect, a combination of a binding effect and a kinetic effect, or entirely a binding effect.

Northrop[37] has recently developed a method for determining the relative contributions of different reaction steps to V_{max} and V_{max}/K_m. While the emphasis is on primary isotope effects, the discussion is pertinent here since in many complex systems the primary isotope effects measured approach the magnitude of secondary effects due to the fact that no single enzymic step is entirely rate limiting. Cleland and co-workers[38] have developed a new technique for measuring isotope effects in enzymic systems. The technique eliminates many of the problems normally encountered in such studies and allows the determination of extremely small isotope effects. The technique, equilibrium perturbation by isotopic substitution, will work for both primary and secondary effects for hydrogen and other isotopes. The technique should aid in the distinction between kinetic and equilibrium effects in enzymic systems. Obviously before a detailed explanation of any given effect is put forth all of these factors should be considered. Jencks points out that steric effects could possibly be due to the slightly different effective radii of protium vs. deuterium. However, these effects would probably be negligible due to a slight flexibility of enzymes except in the case of multiple isotopic labeling where a cumulative effect might be noticeable. Strain and polarizability might give observable isotope effects, especially in an enzymic system where a very restricted active site and tight binding cause detectable frequency changes of isotopic bonds. Forcing the substrate to resemble the transition state could induce significant frequency changes also. To reiterate, *any* phenomenon which causes significant frequency changes of isotopic bonds could give rise to secondary hydrogen isotope effects.

4.1. Acetylcholinesterase[39]

Acetylcholinesterase (AChE) plays a central role in the nervous system by hydrolyzing acetylcholine (ACh), which functions in the transmission of impulses across a synapse as shown in Eq. (2):

$$CH_3CO_2CH_2CH_2N^+(CH_3)_3 + H_2O \xrightleftharpoons{AChE} CH_3CO_2H$$
$$+ HOCH_2CH_2N^+(CH_3)_3 \quad (2)$$

The active site of AChE consists of two subsites, an anionic site which serves to bind the quaternary ammonium group of the substrate and an esteratic site which contains the amino acid residues responsible for the ester hydrolysis. The mechanism of hydrolysis is thought to involve formation of an acetyl enzyme (acetylation of serine hydroxyl) with release of choline followed by deacetylation. By analogy with model reactions (see the previous section) the acetylation and deacetylation steps of the mechanistic sequence should pass through a tetrahedral intermediate. The transition states for these processes should be characterizable using secondary hydrogen isotope effects. Since enzymes catalyze reactions by lowering the free energy of transition states, a detailed knowledge of transition-state structure should lend a great deal of insight into the mechanism of enzymic catalysis. Using acetyl-$^2H_3$-choline, one should be able to detect changes occurring at the carbonyl carbon of the substrate via a secondary β-deuterium isotope effect study. N-$(C^2H_3)_3$-acetylcholine, acetylcholine-1, 1-$^2H_2$, and acetylcholine-2, 2-$^2H_2$ should be useful in detecting remote isotope effects incurred through interactions with the enzyme. Isotope effects could arise from steric restrictions, hydrophobic interactions, solvation changes, or rotational restrictions provided significant frequency changes of the isotopic bonds were incurred.

The variously deuterated acetylcholines discussed above were prepared and subjected to enzymic hydrolysis by AChE at pH 7.5 at 25°.[20] The studies were done under conditions such that acetylation was the rate-limiting step. The results of Table I show that k_H/k_D was essentially 1.00, within experimental error, for all of the deuterated substrates. This interesting result is coupled with the result of Belleau,[40] who found the K_m values to be the same for the various deuterated acetylcholines. These isotope effects of 1.00 mean that no significant changes in the vibrational states of the labeled sites have occurred in reaching the transition state for the rate-determining step. In other words, there has been no significant change in the acetyl C—^{1}H(^{2}H) bonds due to loss of hyperconjugation, as might have been expected. Nor has any restriction of the stretching, bending, or torsional frequencies of any of the isotopic bonds been severe enough to give a detectable isotope effect. Acetylation of AChE by p-nitrophenylacetate-$^2H_3$ gave a value of $k_H/k_D = 0.992 \pm 0.020$.[41]

Table I

Secondary Deuterium Isotope Effects for the
Acetylcholinesterase-Catalyzed Hydrolysis of
Deuterated Acetylcholine Perchlorates[a]

Substrate	k_H/k_D
$CD_3CO_2CH_2CH_2\overset{+}{N}(CH_3)_3ClO_4^-$	0.99 ± 0.025
$CH_3CO_2CD_2CH_2\overset{+}{N}(CH_3)_3ClO_4^-$	1.00 ± 0.067
$CH_3CO_2CH_2CD_2\overset{+}{N}(CH_3)_3ClO_4^-$	1.01 ± 0.029
$CH_3CO_2CH_2CH_2\overset{+}{N}(CD_3)_3ClO_4^-$	1.05 ± 0.065

[a] Reactions were carried out at $25.00 \pm 0.03°$ at pH 7.5. Error limits are standard deviations.

There appear to be three possible explanations for the difference observed in the isotope effect with model systems and the AChE system. Using literature values for various kinetic parameters and making reasonable estimates of other kinetic parameters, one can calculate a value for k_1, the rate constant for combination of the enzyme and substrate. Reasonable estimates give values of about 10^8–10^9 M^{-1} s^{-1} for k_1. This value is just on the lower limit for a diffusion-controlled reaction. A diffusion-controlled rate-limiting step would be consistent with $k_H/k_D = 1.00$ since no significant frequency changes should occur during diffusion. However, the rate constant for *p*-nitrophenylacetate hydrolysis by AChE is 10^3 times slower, clearly well below the limit for a diffusion-controlled reaction, but the same isotope effect is seen.

The rate-limiting step could be a conformation change of the enzyme. To be consistent with the isotope effects this conformation change would have to involve no significant frequency changes of isotopic bonds.

A third possible explanation is that the enzymic mechanism involves a reactant-like transition state as opposed to the tetrahedral-like transition state of the model reactions. A reactant-like transition state would involve only minor frequency changes and give k_H/k_D values near 1.00. However, one is left with the question of explaining how the enzyme stabilizes a reactant-like transition state, which it must do to catalyze the reaction, without also stabilizing the reactants. Enzymes may have this capability, but at the present time this explanation seems unlikely. We are then left with the rate-determining conformation change as the most likely explanation of these results.

A crude but useful calculation showed that the magnitude of the isotope effect for restriction of rotation about the ethano bridge of ACh should be around $k_H/k_D = 0.9$. An isotope effect of this magnitude would surely have been detected.

Secondary kinetic isotope effects were also determined for the hydroxide-promoted hydrolysis of the deuterium-labeled acetylcholines at pH 11 at

$25°.$[20] As would be expected, only the acetate-labeled compound gave any secondary effect. The measured value was about $k_H/k_D = 0.96$.

4.2. α-Chymotrypsin,[42] Elastase,[43] Thrombin,[44] Trypsin[45]

Zero-order rate constants for the deacetylation of various $^1H_3$- and $^2H_3$-acetyl hydrolases have been measured at $25°.$[41] Secondary β-deuterium isotope effects of $k_H/k_D = 0.940$, 0.981, 0.982, and 0.976 were found for α-chymotrypsin, elastase, thrombin,* and trypsin, respectively. Only α-chymotrypsin shows an isotope effect consistent with significant loss of hyperconjugation in reaching the transition state even though all of these enzymes are serine hydrolases.

4.3. L-Asparaginase[46]

Secondary β-deuterium isotope effects determined for the hydrolysis of succinamic acid and L-asparagine catalyzed by various L-asparaginases are given in Table II.[47] These effects were determined at 37°. The normal isotope effects indicate that the β-protium or β-deuterium atoms are in a weaker potential in the transition state than in the reactant state. Rate-limiting tetrahedralization of the acyl carbon would have been expected to give inverse isotope effects due to loss of hyperconjugation. Effects of this type have been observed in hydroxide-promoted amide hydrolysis.[48] It is postulated that rate-limiting desolvation and/or conformation changes may be responsible for the normal effects seen in the enzymic system. In the hydroxide-promoted hydrolysis of acetamide-$^2H_3$ and acetamide-$^1H_3$ values of $k_H/k_D = 0.952 \pm 0.030$ and 0.982 ± 0.018 at 25° and 37°, respectively, were determined.[48]

---

* It is not known that deacetylation is rate limiting for thrombin.

Table II

Secondary Deuterium Isotope Effects for the L-Asparaginase-Catalyzed Hydrolysis of Succinamic Acid or L-Asparagine

Enzyme	Substrate	$V_0, {}^1H/V_0, {}^2H$
Erwinia L-asparaginase	Succinamic acid	1.017 ± 0.019
	Succinamic acid	1.019 ± 0.030
	L-Asparagine	1.086 ± 0.019
	L-Asparagine	1.084 ± 0.036
	L-Asparagine	1.011 ± 0.057
E. coli L-asparaginase	L-Asparagine	1.045 ± 0.013
	L-Asparagine	1.040 ± 0.020
Proteus vulgaris L-asparaginase	L-Asparagine	1.030 ± 0.020

4.4. Cathechol-O-methyltransferase

Deuterium substitution in the methyl group transferred from S-adenosyl-methionine (AdoMet) to 3,4-dihydroxyacetophenone (DAH) causes an inverse secondary α-deuterium isotope effect on the maximum velocity of $k_H/k_D =$ 0.832 ± 0.045 at 37° for the rat liver catechol-O-methyltransferase-catalyzed reaction.[49] The K_m values are equal within experimental error for the labeled and unlabeled substrates. These results are consistent with a trigonal–bipyramidal transition-state structure which could arise from rate-determining methyl transfer directly from AdoMet to DAH or methyl transfer from AdoMet to enzyme followed by enzyme-to-DAH transfer with either or both steps rate limiting. Many S_N2 reactions involve isotope effects of the magnitude seen here.

4.5. Lysozyme[50] and Related Systems

Dahlquist et al.[51] have successfully used α-deuterium isotope effects in the study of lysozyme catalysis of the hydrolysis of aryl-4-O-(2-acetamido-2-deoxy-β-D-glucopyranosyl)-β-D-glucopyranosides-1-^{1}H(^{2}H) (Fig. 3, **I**) and in the acid- and base-catalyzed hydrolysis of phenyl-β-D-glucopyranose-1-^{1}H(^{2}H) (Fig. 3, **II**).

The acid-catalyzed hydrolysis of **II** gave a secondary α-deuterium isotope effect of $k_H/k_D = 1.13$, which is consistent with a value of $k_H/k_D = 1.14$ for typical S_N1 reactions.[4a] The methoxide-catalyzed hydrolysis of **II**, thought to involve an S_N2 mechanism, gave $k_H/k_D = 1.03$, which is consistent with the very small isotope effects seen in many S_N2 reactions.

The lysozyme-catalyzed hydrolysis of **I** gave $k_H/k_D = 1.11$, suggesting an

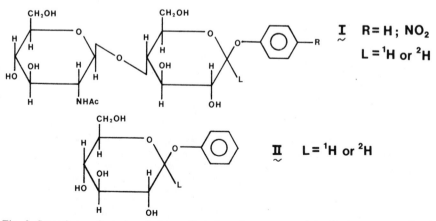

Fig. 3. Secondary α-deuterium isotope effects have been measured in the lysozyme-catalyzed hydrolysis of **I** and in the acid- and base-catalyzed hydrolysis of **II**.

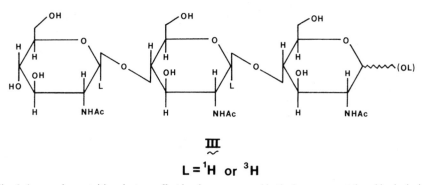

Fig. 4. A secondary α-tritium isotope effect has been measured in the lysozyme-catalyzed hydrolysis of **III**.

intermediate with considerable carbonium-ion character in the enzymic reaction. A subsequent study showed this effect to be pH independent from pH 3.1 to 8.3.[52]

In a related system it was shown that the β-glucosidase-catalyzed hydrolysis of phenyl-β-D-glucopyranose-1-$^1H(^2H)$ (**II**) gave $k_H/k_D = 1.01$, which was in agreement with recent chemical evidence indicating a displacement mechanism for this enzyme.[52]

There has been some disagreement over the interpretation of these results, however. Wang and Touster[53] have proposed that ion-pair and covalent intermediates could give α-deuterium isotope effects similar to those observed in the proposed carbonium-ion processes above. This proposal is based on results of studies with β-glucuronidase.

More recent studies of the mechanism of catalysis by lysozyme have concentrated on the use of a substrate more natural than **I**.[54] The secondary α-tritium kinetic isotope effect was found to be $k_H/k_T = 1.19$ for the lysozyme-catalyzed hydrolysis of 1, 1′, 1″-$[^3H]$chitotriose (Fig. 4, **III**). This effect (equivalent to an ~14% deuterium effect) again indicates a carbonium-ion-like transition state.

Another related study,[55] although done on model systems, is discussed here because of its close relation to the above. The secondary α-tritium isotope effect was determined in the aqueous, acid-catalyzed hydrolysis of a series of α- and β-glycopyranosides of N-acetyl-D-glucosamine of the general structure shown in Fig. 5. The isotope effect varied from $k_H/k_T = 1.038 \pm 0.014$ for $R = OC_6H_5NO_2$ (β anomer) to 1.136 ± 0.013 for $R = -CH_2CH_2CH_3$ (α anomer). The magnitude and change in the isotope effect are consistent with a mechanism involving rate-determining carbonium-ion formation. As the leaving-group ability increases the carbonium-ion character of the transition state decreases. The difference in the tritium isotope effects for α and β anomers is discussed in terms of acetamido group participation in the β series, which decreases the carbonium-ion character of the transition state ($k_H/k_T = 1.097 \pm 0.010$). An isotope effect of $k_H/k_T = 1.103 \pm 0.016$ for the pH-inde-

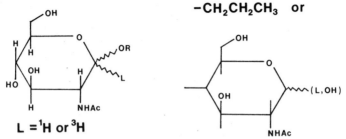

$$R = -C_6H_5NO_2; -C_6H_5; -CH_3;$$

$$-CH_2CH_2CH_3 \quad \text{or}$$

L = ¹H or ³H

Fig. 5. Secondary α-tritium isotope effects have been determined in the aqueous, acid-catalyzed hydrolysis of a series of α- and β-glycopyranosides of N-acetyl-D-glucosamine having the R groups illustrated.

pendent hydrolysis of the β-p-nitrophenyl compound is inconsistent with nucleophilic displacement of the leaving group by the acetamido group, and a stereospecific ion pair appears to be implicated.

4.6. Galactosidase[56]

Sinnott and Souchard[57] have determined secondary α-deuterium isotope effects for the β-galactosidase-catalyzed hydrolysis of a series of aryl 2', 3', 4', 6'-tetra-O-acetyl-β-D-galactopyranosides (Fig. 6) labeled at C-1'. The effects were determined under steady-state conditions, and a correlation of k_H/k_D with k_{cat} and K_m showed no simple dependence on leaving-group acidity, while k_H/k_D varied from about 1.25 for rate-limiting degalactosylation with dinitrophenyl compounds to about 1.00 for slow substrates such as the 4-bromo- and 4-cyanophenyl compounds. The k_H/k_D values for "slow" aryl galactosides were interpreted in terms of a rate-limiting enzyme conformation change rather than an S_N2 reaction for reasons discussed in detail in the paper. The k_H/k_D values of about 1.25 for the dinitrophenyl compounds were interpreted in terms of a transition state having more carbonium-ion character than the galactosyl enzyme itself. The net mechanism involves substrate binding followed by a protein conformation change (which could give $k_H/k_D \sim$ 1.00) and generation of a galactopyranosyl cation in the vicinity of an anionic

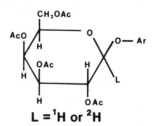

L = ¹H or ²H

Fig. 6. Secondary α-deuterium isotope effects have been measured in the β-galactosidase catalyzed hydrolysis of a series of aryl tetra-O-acetyl-β-D-galactopyranosides of the general structure shown here.

group. This ion pair then rapidly and reversibly collapses to a majority of a covalent α-D-galactopyranosyl enzyme. Reaction with the acceptor is, however, proposed to occur through the ion-paired species rather than the covalent tetrahedral species. Detailed supporting arguments are given in the paper, with three other possible mechanisms being considered and rejected.

Evidence obtained by Sinnott and Withers[58] in a subsequent study of the β-galactosidase-catalyzed hydrolysis of some β-D-galactopyranosyl pyridinium salts was used as further support for the above mechanism. It was postulated, based on k_{cat} values, that bond breaking and not a conformation change should be rate limiting for these substrates. If this is true, then k_{cat} should show a dependence on the pK_a of the parent pyridine, and the secondary α-deuterium isotope effects should be indicative of a carbonium-ion mechanism. In marked contrast to the absence of any correlation between pK_a and k_{cat} seen in the previous study the pyridinium compounds gave a very good correlation. It was proposed that k_{cat} depends more on the shape of the leaving group than its pK_a for the aryl galactosides. The k_H/k_D value for the 4-bromo-isoquinolinium bromide compound was 1.187 ± 0.046, while that for the pyridinium bromide compound was 1.136 ± 0.040. These effects are consistent with the formation of a glycopyranosyl cation. The secondary α-deuterium kinetic isotope effect of 1.098 ± 0.033 for the hydrolysis of β-D-galactopyranosyl azide and a comparison of k_{cat} values allowed an estimate of the acceleration of departure of the azide ion due to general acid catalysis. The authors also propose that the value of $k_H/k_D = 1.01 ± 0.01$ for the hydrolysis of phenyl-β-D-glucopyranoside by β-glucosidase may be indicative of a rate-limiting conformation change and not an S_N2 reaction, as originally proposed by Dahlquist et al.[52]

4.7. Polysaccharide Phosphorylase[59]

The polysaccharide phosphorylase reaction occurs with cleavage of the C—O bond of α-D-glucose 1-phosphate with retention of configuration.[60] Secondary α-deuterium isotope effects were determined in the acid-catalyzed hydrolysis of the C-1-labeled compound, which is thought to proceed through formation of a carbonium ion, and in several phosphorylase reactions.[61] The value of $k_H/k_D = 1.13$ observed at 24° in the acid-catalyzed hydrolysis is consistent with a carbonium-ion mechanism. E. coli maltodextrin phosphorylase and rabbit muscle phosphorylase b gave $k_H/k_D = 1.09$ and 1.10, respectively, at 10°. These effects were also interpreted as being consistent with a carbonium-ion mechanism. The bacterial alkaline phophatase reaction which is known to involve P—O bond cleavage[60] was carried out with the same substrates and gave the expected value of $k_H/k_D = 1.00$.

Firsov et al.,[62] concerned that the experimental procedure used by Tu et al.[61] might have introduced a possible selective inhibition factor by one of the isotopically substituted substrates, used a double isotopic labeling technique to determine similar isotope effects with phosphorylase b. The secondary α-deuterium and α-tritium isotope effects were both 1.00 within experimental error for both glycogen phosphorolysis and glycogen synthesis. This result is in direct disagreement with the value of $k_H/k_D = 1.10$ determined by Tu et al.[61] As a control the secondary α-tritium effect was determined to be $k_H/k_T = 1.21 \pm 0.03$ for the acid hydrolysis of α-D-glucose-1-phosphate at 20° and 1.01 ± 0.04 for the alkaline phosphatase-catalyzed hydrolysis. These two controls are in good agreement with the previous study, but the significant difference in the phosphorylase b system is puzzling. Firsov et al. argue that the absence of any significant isotope effect (i.e., $k_H/k_D = k_H/k_T = 1.00$) is consistent not with a carbonium-ion mechanism but with a double displacement mechanism proposed by Koshland.[63] To this author's knowledge this conflict has not been resolved.

4.8. Fumarase[64]

Secondary α-isotope effects have been determined in the dehydration of L-malate catalyzed by fumarase.[65] Substitution at C-2 gave a secondary α-deuterium isotope effect, k_H/k_D, of 1.09 (calculated from $k_H/k_T = 1.12$), while isotopic substitution at C-3 gave a secondary β-deuterium isotope effect of 1.09 (calculated from $k_H/k_D = 1.12$) for the hydrogen not lost in the elimination. The α effect was interpreted in terms of carbonium-ion formation at C-2 as the hydroxyl is lost. However, the magnitude is not consistent with development of a full positive charge, which may indicate some concomitant nucleophilic participation.

The β effect is consistent with hyperconjugative stabilization of the developing positive charge by the β-hydrogen not lost in elimination. The equilibrium isotope effect of $K_H/K_T = 1.24 \pm 0.03$ was also consistent with that expected for conversion of an sp^3 center to an sp^2 center. No primary isotope effect was observed in the dehydration of 3-(R)-3-^{2}H-2-(S)-malate.[66]

Subsequent studies on various exchange rates in the above system[67] indicated that slow dissociation of an enzyme-bound proton was also a principal rate-limiting step in fumarate formation. The observation of a secondary β-deuterium effect of $k_H/k_D = 1.15$ during equilibrium ^{18}O exchange between ^{18}O-malate and water also supports a carbonium-ion mechanism. These combined studies seem then to support a carbonium-ion mechanism as opposed to an alternative suggestion offered by Rose.[68]

4.9. Aldolase[69]

Secondary tritium isotope effects have been used to study the yeast and muscle aldolase-catalyzed aldol cleavage of fructose-1,6-diphosphate (FDP).[70] The enzyme-catalyzed condensation of (1R)-[1-^{3}H]-dihydroxyacetone phosphate (DHAP) with D-glyceraldehyde-3-phosphate (G3P) to give [3-^{3}H]-FDP was found to give a normal secondary tritium isotope effect under conditions which had previously indicated proton abstraction to be rate determining. Substitution of tritium at C-3 or C-4 of FDP gave normal secondary tritium isotope effects for both yeast and muscle aldolases. These results were interpreted to mean that either C—C bond cleavage to give a kinetic isotope effect or G3P release to give an equilibrium isotope effect was the rate-limiting step for formation of G3P.

4.10. Enolase[71]

Enolase, which reversibly catalyzes the hydration–dehydration of the 2-phosphoglycerate-phosphoenolpyruvate system in the presence of divalent metal ions, has been the subject of another isotope-effect study.[72] Initial and equilibrium measurements of ^3H, ^2H, ^{18}O, and ^{14}C exchange and of ^3H secondary isotope effects have been made. One might have predicted an inverse isotope effect for the addition of water to phosphoenolpyruvate (PEP) for tritium substitution at C-3 since carbon hybridization changes from sp^2 to sp^3, and indeed such an effect was observed. A kinetic secondary α-tritium isotope effect, k_H/k_T, of 0.79 was observed, while the equilibrium effect was $K_H/K_T = 1.40$ for the equilibrium formation of phosphoenolpyruvate from 2-phosphoglycerate (PGA).* This effect is equivalent to $K_H/K_T = 0.71$ for the hydration equilibrium and is also consistent with the expected effect for sp^2 to sp^3 conversion. These secondary isotope effects have been interpreted with other data to indicate rate-determining addition of OH^- to C-3 to give a carbanion at C-2 which may involve metal-ion stabilization. This effect is to be contrasted with the fumarase effects.

4.11. Aspartase[73]

The secondary α-deuterium isotope effect of $k_H/k_D = 1.11 \pm 0.02$ was obtained in the aspartase-catalyzed conversion of 2S-[2-^2H]-aspartate to [2-^2H]-fumarate.[74] This effect, coupled with the absence of a primary kinetic isotope effect using 2S,3R-[3-^2H]-aspartate, was suggestive of rate-determining C—N bond cleavage to give an intermediate with considerable carbonium-ion character provided that product release is not rate limiting. The somewhat low magnitude of the effect might indicate that additional steps are partly rate limiting or participation of some other group in the transition state.

4.12. Propionyl Coenzyme A Carboxylase[75]

Isotope effects of $k_H/k_D = 1.061 \pm 0.025$ and 1.131 ± 0.020 were observed in the enzymic carboxylation of (S)- and (R)-2-deuteriopropionyl coenzyme A

---

* This equilibrium is expressed as the ratio of equilibrium constants, K, for the conversion of PGA to PEP [i.e., $K = (PEP)/(PGA)$]. This point caused some confusion to this author due to a lack of familiarity with the usual manner in which doubly labeled isotopic data are expressed. For an excellent discussion regarding this point, see Dahlquist et al.[52]

A, respectively, by propionyl coenzyme A carboxylase.[76] These isotope effects could not be enhanced under a variety of conditions, indicating that they were indeed secondary isotope effects on V_{max}. Stereochemically pure (R)-2-^{3}H substrate gave no isotope effect whatever. Even though the α-carbon–hydrogen bond is being broken, no primary isotope effect is incurred. Both (R)- and (S)-deuterated substrates gave secondary isotope effects since a partly racemic mixture was used. These results were quite similar to those of Arigoni et al.,[77] who obtained a 16% isotope effect using 2,2-dideuteriopropionyl-CoA. The results were interpreted in terms of proton removal concerted with carboxylation with retention of configuration.

4.13. β-Hydroxydecanoyl Thioester Dehydrase[78]

The reversible enzymic dehydration of $D(-)$-β-hydroxydecanoyl-N-acetylcysteamine to trans-2-decenoyl- and cis-3-decenoyl-N-acetyl-cysteamine has been studied by Rando and Bloch.[79] The following isotope effects were determined:

$$\underset{\text{RCH}_2-\overset{\overset{\displaystyle\text{OH}}{|}}{\text{CH}}-\text{CD}_2-\overset{\overset{\displaystyle\text{O}}{\|}}{\text{C}}-\text{SX}}{} \longrightarrow \underset{\text{R}-\text{CH}_2\text{CH}=\text{CD}-\overset{\overset{\displaystyle\text{O}}{\|}}{\text{C}}-\text{SX},}{} \qquad k_H/k_D = 2.25$$

$$\underset{\text{RCD}_2\overset{\overset{\displaystyle\text{OH}}{|}}{\text{CH}}-\text{CH}_2-\overset{\overset{\displaystyle\text{O}}{\|}}{\text{C}}-\text{SX}}{} \longrightarrow \underset{\text{RCD}_2\text{CH}=\text{CH}-\overset{\overset{\displaystyle\text{O}}{\|}}{\text{C}}-\text{SX},}{} \qquad k_H/k_D = 1.00$$

The distinct primary kinetic isotope effect observed in the dehydration of the α-^{2}H$_2$ compound indicates that stretching of the α-carbon–hydrogen bond is rate limiting. The unchanged rate for the γ-deuterated compound excludes any reaction at C_γ in the rate-limiting step for α,β-decenoate formation. These results are consistent with direct α,β-decenoate formation either by a carbanion or an E2 elimination mechanism. It is interesting that no β secondary isotope effect is observed with the γ-deuterated substrate since a carbonium-ion mechanism might have given a 15–20% normal isotope effect and an 11% normal isotope effect has been reported for carbanion formation on a carbon adjacent to a methyl group labeled with three deuteria.[80] Cis-3-decenoyl-N-acetylcysteamine was shown to be formed only by isomerization of the conjugated enoate.

4.14. Cytochrome-P-450-Dependent Enzymes

Cytochrome-P-450-dependent monoxygenase enzymes are responsible for fixing molecular oxygen into an organically bound form which is usually an epoxide or aliphatic hydroxyl group. Recent comparisons have been made with organic epoxidation by peracids.[81] Recently both substituent and isotope-effects data have been obtained for the enzymic epoxidation of some styrene derivatives.[82] Inverse secondary hydrogen isotope effects of $k_H/k_D = 0.935 \pm 0.03$ were obtained for the epoxidation of the p-CH_3 and p-phenyl-styrenes labeled with deuteria at the α and both β positions of the vinyl group. However, k_H/k_D values of 1.00 ± 0.03 were obtained for the compounds labeled with deuteria at the β positions only. These effects indicate that the oxygen is not symmetrically transferred to the olefin π bond, suggesting that C_α—O bond formation must be rate limiting. A mechanism has been proposed which represents the rate-limiting step as being closure of the oxirane ring. This is supported by the isotope effect and substituent effects obtained in the same study. These effects are in direct contrast to those obtained in a model system.[31]

4.15. Transaminases

In studies of the stereochemistry of enzymic transamination using deuterium labels in the substrates, small secondary deuterium effects have been observed with apoglutamate-oxaloacetate transaminase and pyridoxamine-pyruvate transaminase systems.[83] These effects have been interpreted in terms of differences in K_m for the labeled and unlabeled substrates.

Michaelis constants and maximum velocities for the transamination of variously deuterated L-glutamic acids have been determined by Fang *et al.*[84] In addition to a significant primary effect of $k_H/k_D = 1.85$ on V_{max} seen for L-$[2$-$^2H]$ glutamic acid, the following secondary effects were observed: $k_H/k_D = 1.26$ for L-$[3,3,4,4$-$^2H]$- and 1.85 for L-$[2,3,3,4,4$-$^2H]$ glutamic acid. All of the labeled substrates showed inverse isotope effects of about 0.76–0.80 on K_m. These results are rather unusual in that a 26% secondary effect is seen for the 3,4-tetradeuterated compound, but the compound with deuterium at C-2 showed the same isotope effect of 1.85 regardless of the presence (or absence) of deuteria at C-3 and C-4. No comment is made with respect to this point in the paper. The effects were interpreted to indicate that only elimination of the α-hydrogen occurred, with a secondary isotope effect of 26% being seen for C-3 and C-4 deuteration.

4.16. Isocitrate Dehydrogenase

Secondary deuterium isotope effects on V_{max} ranging from $k_H/k_D = 1.15$–1.30 were found at various pH values for NADP-specific isocitrate-dehydrogenase-catalyzed reactions of isocitrate labeled at C-2.[85] The analogous reaction with NAD-specific enzyme gave $k_H/k_D = 0.95$. These effects are inconsistent with C—H bond cleavage at C-2 being the rate-limiting step but may indicate rate-limiting proton abstraction from C-2 hydroxyl. Significant isotope effects (in opposing directions for the two enzymes) on K_m were also seen.

4.17. Methanogenic Bacterium M.O.H.

Kinetic studies of methyl transfer promoted by the methanogenic bacterium M.O.H. using either protium- or deuterium-labeled methylcobalamin gave $k_H/k_D = 1.38 \pm 0.04$ for three deuteria.[86] Photolysis of the same substrate under anaerobic conditions gave $k_H/k_D = 1.26 \pm 0.01$. These and other data appear to support methyl radical formation during the cobalamin-dependent methyl transfer in methane synthesis.

5. CONCLUSION

Secondary hydrogen isotope effects have been determined for a significant number of reactions of biochemical interest, and these studies have significantly added to the knowledge about the transition states for these reactions. The remaining potential sources of application are unlimited, and the technique promises to be a valuable calibration tool in studies of enzymic reactions.

REFERENCES

1. (a) J. B. Sumner, The isolation and crystallization of the enzyme urease, *J. Biol. Chem.* **69**, 435–441 (1926). (b) J. B. Sumner, Note: The recrystallization of urease, *J. Biol. Chem.* **70**, 97–98 (1926).
2. R. Wolfenden, Analog approaches to the structure of the transition state in enzyme reactions, *Acc. Chem. Res.* **5**, 10–18 (1972).
3. (a) G. E. Lienhard, Transition state analogs as enzyme inhibitors, *Annu. Rep. Med. Chem.* **7**, 249–258 (1972). (b) G. E. Lienhard, Enzymic catalysis and transition-state theory, *Science* **180**, 149–154 (1973).
4. (a) E. A. Halevi, Secondary isotope effects, *Prog. Phys. Org. Chem.* **1**, 109–221 (1963). (b) C. J. Collins and N. S. Bowman, eds., *Isotope Effects in Chemical Reactions,* Van Nostrand Reinhold, New York (1970). (c) W. P. Jencks, *Catalysis in Chemistry and Enzymology,* McGraw-Hill, New York (1969), pp. 253ff.
5. V. J. Shiner, Jr., and J. S. Humphrey Jr., The effect of deuterium substitution on the rates of organic reactions. IX. Bridgehead β-deuterium in a carbonium ion solvolysis *J. Am. Chem. Soc.* **85**, 2416–2419 (1963).
6. E. K. Thornton and E. R. Thornton, in: *Isotope Effects in Chemical Reactions* (C. J. Collins and N. S. Bowman, eds.), Chap. 4, pp. 274ff., Van Nostrand Reinhold, New York (1970).
7. (a) V. J. Shiner, Jr., in: *Isotope Effects in Chemical Reactions* (C. J. Collins and N. S. Bowman, eds.), Chap. 2, pp. 151ff., Van Nostrand Reinhold, New York (1970). (b) D. E. Sunko and S. Borcic, in: *Isotope Effects in Chemical Reactions* (C. J. Collins and N. S. Bowman, eds.), Chap. 3, p. 172. Van Nostrand Reinhold, New York (1970).
8. R. E. Carter and L. Melander, Experiments on the nature of steric isotope effects, *Adv. Phys. Org. Chem.* **10**, 1–27 (1973).
9. J. H. Richards, in: *The Enzymes* (P. D. Boyer, ed.), Vol. II, p. 330, Academic Press, New York (1970).
10. V. J. Shiner, Jr., Substitution and elimination rate studies on some deuterio-isopropyl bromides, *J. Am. Chem. Soc.* **74**, 5285–5288 (1952).
11. S. Seltzer and A. A. Zavitsas, Correlation of isotope effects in substitution reactions with

nucleophilicities. Secondary α-deuterium isotope effect in the Iodide-131 exchange of methyl-d_3 iodide, *Can. J. Chem.* **45**, 2024–2031 (1967).

12. J. A. Llewellyn, R. E. Robertson, and J. M. W. Scott, Some deuterium isotope effects. I. Water solvolysis of methyl-d_3 esters, *Can. J. Chem.* **38**, 222–232 (1960).

13. C.-Y. Wu and R. E. Robertson, Normal and inverse secondary α-deuterium isotope effects with trimethylsulfonium ions, *Chem. Ind. London* **1964**, 1803–1804.

14. (a) A. V. Willi and C. M. Won, The kinetic deuterium isotope effect in the reaction of methyl iodide with cyanide ion in aqueous solution from 0 to 40°, *J. Am. Chem. Soc.* **90**, 5999–6001 (1968). (b) C. M. Won and A. V. Willi, Kinetic deuterium isotope effects in the reactions of methyl iodide with azide and acetate ions in aqueous solution, *J. Phys. Chem.* **76**, 427–432 (1972). (c) J. Bron, Model calculations of kinetic isotope effects in nucleophilic substitution reactions, *Can. J. Chem.* **52**, 903–909 (1974).

15. A. Fry, Isotope effect studies of elimination reactions, *Chem. Soc. Rev.* **1**, 163–210 (1972).

16. (a) P. C. Vogel and M. J. Stern, Temperature dependences of kinetic isotope effects, *J. Chem. Phys.* **54**, 779–796 (1971). (b) S. R. Hartshorn and V. J. Shiner, Jr., Calculation of H/D, $^{12}C/^{13}C$, and $^{12}C/^{14}C$ fractionation factors from valence force fields derived for a series of simple organic molecules, *J. Am. Chem. Soc.* **94**, 9002–9012 (1972). (c) M. E. Schneider and M. J. Stern, Additivity of contributions to secondary deuterium kinetic isotope effects and their Arrhenius preexponential factors. Toward a method of fitting model calculations to experimental data, *J. Am. Chem. Soc.* **95**, 1355–1365 (1973).

17. C. G. Swain, E. C. Stivers, J. F. Reuwer, Jr., and L. J. Schaad, Use of hydrogen isotope effects to identify the attacking nucleophile in the enolization of ketones catalyzed by acetic acid, *J. Am. Chem. Soc.* **80**, 5885–5893 (1958).

18. (a) M. L. Bender and M. S. Feng, Secondary deuterium isotope effects in the reactions of carboxylic acid derivatives, *J. Am. Chem. Soc.* **82**, 6318–6321 (1960). (b) J. M. Jones and M. L. Bender, Secondary deuterium isotope effects in the addition equilibria of ketones, *J. Am. Chem. Soc.* **82**, 6322–6326 (1960).

19. E. A. Halevi and Z. Margolin, Temperature dependence of the secondary isotope effect on aqueous alkaline ester hydrolysis, *Proc. Chem. Soc. London* **1974**, 174.

20. J. L. Hogg, Transition state structures for catalysis by serine hydrolases and for related organic reactions, Ph.D. thesis, University of Kansas, Lawrence (1974).

21. W. P. Jencks, *Catalysis in Chemistry and Enzymology*, McGraw-Hill, New York (1969), p. 605.

22. (a) W. E. Nelson and J. A. V. Butler, Experiments with heavy water on the acid hydrolysis of esters and the alkaline decomposition of diacetone alcohol, *J. Chem. Soc.* 957–962 (1938). (b) R. L. Schowen, Mechanistic deductions from solvent isotope effects, *Prog. Phys. Org. Chem.* **9**, 275–332 (1972).

23. (a) V. F. Raen, T. K. Dunham, D. D. Thompson, and C. J. Collins, Steric origin of some secondary isotope effects of deuterium, *J. Am. Chem. Soc.* **85**, 3497–3499 (1963). (b) V. F. Raen and C. J. Collins, Steric origin of some secondary isotope effects, *Pure Appl. Chem.* **8**, 347–355 (1964). (c) M. A. Winnik, V. Stoute, and P. Fitzgerald, Secondary deuterium isotope effects in the Baeyer–Villiger reaction, *J. Am. Chem. Soc.* **96**, 1977–1979 (1974). (d) W. I. Congdon and J. T. Edward, Thiohydantoins. XII. Secondary deuterium isotope effects in the acid- and base-catalyzed hydrolysis of 1-acetyl-5,5-dimethyl-2-thiohydantoin, *Can. J. Chem.* **50**, 3921–3923 (1972). (e) P. Geneste, C. Lamaty, and J. P. Roque, Réaction d'addition nucléophile sur les cétones. Addition de l'ion sulfite: Mise en évidence de l'hyperconjuigaison par la mesure de l'effet isotopique secondaire du deuterium, *Tetrahedron* **27**, 5539–5559 (1971). (f) P. Geneste, G. Lamaty, and J. P. Roque, Réactions d'addition nucléophile sur les cétones. Addition de l'hydroxylamine: Mise en évidence des facteurs stériques par la mesure de l'effet isotopique secondaire du deuterium, *Tetrahedron* **27**, 5561–5578 (1971).

24. D. J. Barnes, P. D. Golding, and J. M. W. Scott, Isotope effects on chemical equilibria. 2. Some further secondary isotope effects of a second kind, *Can. J. Chem.* **52**, 1966–1972 (1974).

25. D. J. Barnes, M. Cole, S. Lobo, J. G. Winter, and J. M. W. Scott, Studies in solvolysis. Part V. Further investigations concerning the solvolysis of primary, secondary, and tertiary trifluoroacetates, *Can. J. Chem.* **50**, 2175–2181 (1972).

26. Z. Bilkadi, R. deLorimier, and J. F. Kirsch, Secondary α-deuterium kinetic isotope effects and transition-state structures for the hydrolysis and hydrazinolysis of formate esters, *J. Am. Chem. Soc.* **97**, 4317–4322 (1975).

27. C. B. Sawyer and J. F. Kirsch, Kinetic isotope effects for reactions of methyl formate–methoxyl-^{18}O, *J. Am. Chem. Soc.* **95**, 7375–7381 (1973).

28. (a) L. doAmaral, H. G. Bull, and E. H. Cordes, Secondary deuterium isotope effects for carbonyl addition reactions, *J. Am. Chem. Soc.* **94**, 7579–7580 (1972). (b) L. doAmaral, M. P. Bastos, H. G. Bull, and E. H. Cordes, Secondary deuterium isotope effects for addition of nitrogen nucleophiles to substituted benzaldehydes, *J. Am. Chem. Soc.* **95**, 7369–7374 (1973).

29. J. Archila, H. Bull, C. Lagenaur, and E. H. Cordes, Substituent and secondary deuterium isotope effects for hydrolysis of Schiff bases, *J. Org. Chem.* **36**, 1345–1347 (1971).

30. (a) H. Bull, T. C. Pletcher, and E. H. Cordes, Secondary deuterium isotope effects for hydrolysis of acetals and orthoformates, *Chem. Commun.* **1970**, 527–528. (b) H. G. Bull, K. Koehler, T. C. Pletcher, J. J. Ortiz, and E. H. Cordes, Effects of α-deuterium substitution, polar substituents, temperature, and salts on the kinetics of hydrolysis of acetals and ortho esters, *J. Am. Chem. Soc.* **93**, 3002–3011 (1971).

31. R. P. Hanzlik and G. O. Shearer, Transition state structure for peracid epoxidation. Secondary deuterium isotope effects, *J. Am. Chem. Soc.* **97**, 5231–5233 (1975).

32. D. Quinn, J. F. Mata-Segreda, C. Olomon, and R. L. Schowen, personal communication.

33. (a) J. J. Katz and H. L. Crespi, in: *Isotope Effects in Chemical Reactions* (C. J. Collins and N. S. Bowman, eds.), pp. 286–363, Van Nostrand Reinhold, New York (1970). (b) W. P. Jencks, *Catalysis in Chemistry and Enzymology*, McGraw-Hill, New York (1969), pp. 274–281.

34. M. I. Blake, H. L. Crespi, and J. J. Katz, Studies with deuterated drugs, *J. Pharm. Sci.* **64**, 367–391 (1975).

35. I. A. Rose, in: *The Enzymes* (P. D. Boyer, ed.), Vol. II, pp. 285–287, Academic Press, New York (1970).

36. J. H. Richards, in: *The Enzymes* (P. D. Boyer, ed.), Vol. II, pp. 329–333, Academic Press, New York (1970).

37. D. B. Northrop, Steady-state analysis of kinetic isotope effects in enzymic reactions, *Biochemistry* **14**, 2644–2651 (1975).

38. M. I. Schimerlik, J. E. Rife, and W. W. Cleland, Equilibrium perturbation by isotope substitution, *Biochemistry* **14**, 5347–5354 (1975).

39. H. C. Froede and I. B. Wilson, in: *The Enzymes* (P. D. Boyer, ed.), Vol. V, pp. 87–114, Academic Press, New York (1971).

40. B. Belleau, in: *Isotopes in Experimental Pharmacology* (L. J. Roth, ed.), Chap. 36, University of Chicago Press, Chicago (1965).

41. J. P. Elrod, Comparative mechanistic study of a set of serine hydrolases using a proton inventory technique and beta-deuterium probe, Ph.D. thesis, University of Kansas, Lawrence (1975).

42. G. P. Hess, in: *The Enzymes* (P. D. Boyer, ed.), Vol. III, pp. 213–248, Academic Press, New York (1971).

43. B. S. Hartley and D. M. Shotton, in: *The Enzymes* (P. D. Boyer, ed.), Vol. III, pp. 323–373, Academic Press, New York (1971).

44. S. Magnusson, in: *The Enzymes* (P. D. Boyer, ed.), Vol. III, pp. 278–321, Academic Press, New York (1971).

45. B. Keil, in: *The Enzymes* (P. D. Boyer, ed.), Vol. III, pp. 250–275, Academic Press, New York (1971).

46. J. C. Wriston, Jr., in: *The Enzymes* (P. D. Boyer, ed.), Vol. IV, pp. 101–121, Academic Press, New York (1971).

47. J. Bayliss, J. Fried, D. M. Quinn, R. D. Gandour, and R. L. Schowen, personal communication.

48. D. M. Quinn and R. L. Schowen, personal communication.

49. M. F. Hegazi, R. T. Borchardt, and R. L. Schowen, S_N2-like transition state for methyl

transfer catalyzed by catechol-O-methyltransferase, *J. Am. Chem. Soc.* **98**, 3048–3049 (1976).

50. T. Imoto, L. N. Johnson, A. C. T. North, D. C. Phillips, and J. A. Rupley, in: *The Enzymes* (P. D. Boyer, ed.), Vol. VII, pp. 666–868, Academic Press, New York (1972).

51. F. W. Dahlquist, T. Rand-Meir, and M. A. Raftery, Demonstration of carbonium ion intermediate during lysozyme catalysis, *Proc. Nat. Acad. Sci. USA* **61**, 1194–1198 (1968).

52. F. W. Dahlquist, T. Rand-Meir, and M. A. Raftery, Application of secondary α-deuterium kinetic isotope effects to studies of enzyme catalysis. Glycoside hydrolysis by lysozyme and β-glucosidase, *Biochemistry* **8**, 4214–4221 (1969).

53. C. C. Wang and O. Touster, Studies of catalysis by β-glucuronidase. The effect of structure on the rate of hydrolysis of substituted phenyl-β-D-glucopyranosiduronic acids, *J. Biol. Chem.* **247**, 2650–2656 (1972).

54. L. E. H. Smith, L. H. Mohr, and M. A. Raftery, Mechanism of lysozyme-catalyzed hydrolysis, *J. Am. Chem. Soc.* **95**, 7497–7500 (1973).

55. L. E. H. Smith, L. H. Mohr, and M. A. Raftery, α-Secondary tritium isotope effects in the aqueous hydrolysis of glycopyranosides of *N*-acetyl-β-D-glucosamine, *Arch. Biochem. Biophys.* **159**, 505–511 (1973).

56. K. Wallenfels and R. Weil, in: *The Enzymes* (P. D. Boyer, ed.), Vol. VII, pp. 618–663, Academic Press, New York (1972).

57. M. L. Sinnott and I. J. L. Souchard, The mechanism of action of β-galactosidase. Effect of aglycone nature and α-deuterium substitution on the hydrolysis of aryl galactosides, *Biochem. J.* **133**, 89–98 (1973).

58. M. L. Sinnott and S. G. Withers, The β-galactosidase-catalyzed hydrolyses of β-D-galactopyranosyl pyridinium salts. Rate-limiting generation of an enzyme-bound galactopyranosyl cation in a process dependent only on aglycone formation, *Biochem. J.* **143**, 751–762 (1974).

59. (a) D. J. Graves and J. H. Wang, in: *The Enzymes* (P. D. Boyer, ed.), Vol. VII, pp. 435–482, Academic Press, New York (1972). (b) J. J. Mieyal and R. H. Abeles, in: *The Enzymes* (P. D. Boyer, ed.), Vol. VII, pp. 515–532, Academic Press, New York (1972).

60. M. Cohn, Mechanisms of cleavage of glucose-1-phosphate, *J. Biol. Chem.* **180**, 771–781 (1949).

61. J.-I. Tu, G. Jacobson, and D. Graves, Isotopic effects and inhibition of polysaccharide phosphorylase by 1,5-gluconolactone. Relationship to the catalytic mechanism, *Biochemistry* **10**, 1229–1236 (1971).

62. L. M. Firsov, T. I. Bogacheva, and S. E. Bresler, Secondary isotope effect in the phosphorylase reaction, *Eur. J. Biochem.* **42**, 605–609 (1974).

63. D. Koshland, Stereochemistry and the mechanism of enzymatic reactions, *Biol. Rev. Cambridge Philos. Soc.* **28**, 416–436 (1953).

64. R. L. Hill and J. W. Teipel, in: *The Enzymes* (P. D. Boyer, ed.), Vol. V, pp. 539–571, Academic Press, New York (1971).

65. D. E. Schmidt, Jr., W. G. Nigh, C. Tanzer, and J. H. Richards, Secondary isotope effects in the dehydration of malic acid by fumarate, *J. Am. Chem. Soc.* **91**, 5849–5854 (1969).

66. (a) H. Fisher, C. Frieden, J. S. M. Mckee, and R. A. Alberty, Concerning the stereospecificity of the fumarase reaction and the demonstration of a new intermediate, *J. Am. Chem. Soc.* **77**, 4436 (1955). (b) R. A. Alberty, W. G. Miller, and H. F. Fisher, Studies of the enzyme fumarase. VI. Study of the incorporation of deuterium into L-malate during the reaction in deuterium oxide, *J. Am. Chem. Soc.* **79**, 3973–3977 (1957).

67. J. N. Hansen, E. C. Dinovo, and P. D. Boyer, Relationships of pH to exchange rates and deuterium isotope effects in the fumarase reaction, *Bioorg. Chem.* **1**, 234–242 (1971).

68. I. A. Rose, in: *The Enzymes* (P. D. Boyer, ed.), Vol. II, p. 304, Academic Press, New York (1970).

69. B. L. Horecker, O. Tsolas, and C. Y. Lai, in: *The Enzymes* (P. D. Boyer, ed.), Vol. VII, pp. 213–258, Academic Press, New York (1972).

70. J. F. Biellmann, E. L. O'Connell, and I. A. Rose, Secondary isotope effects in reactions catalyzed by yeast and muscle aldolase, *J. Am. Chem. Soc.* **91**, 6484–6488 (1969).

71. F. Wold, in: *The Enzymes* (P. D. Boyer, ed.), Vol. V, pp. 499–538, Academic Press, New York (1971).

72. E. C. Dinovo and P. D. Boyer, Isotopic probes of the enolase reaction mechanism—initial and equilibrium isotope exchange rates: Primary and secondary isotope effects, *J. Biol. Chem.* **246**, 4586–4593 (1971).

73. K. R. Hanson and E. A. Havir, in: *The Enzymes* (P. D. Boyer, ed.), Vol. VII, pp. 75–166, Academic Press, New York (1972).

74. T. B. Dougherty, V. R. Williams, and E. S. Younathan, Mechanism of action of aspartase. A kinetic study and isotope rate effects with ^{2}H, *Biochemistry* **11**, 2493–2498 (1972).

75. A. W. Alberts and P. R. Vagelos, in: *The Enzymes* (P. D. Boyer, ed.), Vol. VI, pp. 37–82, Academic Press, New York (1972).

76. D. J. Prescott and J. L. Rabinowitz, The enzymatic carboxylation of propionyl coenzyme A. Studies involving deuterated and tritiated substrates, *J. Biol. Chem.* **243**, 1551–1557 (1968).

77. D. Arigoni, F. Lynen, and J. Retey, Stereochemie der enzymatischen carboxylierung von (2R)-2-^{3}H-propionyl coenzyme A, *Helv. Chim. Acta* **49**, 311–316 (1966).

78. K. Bloch, in: *The Enzymes* (P. D. Boyer, ed.), Vol. V, pp. 441–464, Academic Press, New York (1971).

79. R. R. Rando and K. Bloch, Mechanism of action of β-hydroxydecanoyl thioester dehydrase, *J. Biol. Chem.* **243**, 5627–5634 (1968).

80. A. Streitweiser, Jr., and D. E. Van Sickle, Acidity of hydrocarbons. IV. Secondary deuterium isotope effects in exchange reactions of toluene and ethylbenzene with lithium cyclohexylamide, *J. Am. Chem. Soc.* **84**, 254–258 (1962).

81. (a) V. Ullrich, Enzymatic hydroxylations with molecular oxygen, *Angew. Chem. Int. Ed. Engl.* **11**, 701–712 (1972). (b) G. A. Hamilton, in: *Molecular Mechanisms of Oxygen Activation* (O. Hayaishi, ed.), p. 405, Academic Press, New York (1974).

82. R. P. Hanzlik and G. O. Shearer, personal communication.

83. C. Dunathan, L. Davis, P. G. Kury, and M. Kaplan, The stereochemistry of enzymatic transamination, *Biochemistry* **7**, 4532–4537 (1968).

84. S.-M. Fang, H. J. Rhodes, and M. I. Blake, Deuterium isotope effects in enzymatic transamination, *Biochim. Biophys. Acta* **212**, 281–287 (1970).

85. N. Ramachandran, M. Durbano, and R. F. Colman, Kinetic isotope effects in the NAD- and NADP-specific isocitrate dehydrogenases of pig heart, *FEBS Lett.* **49**, 129–133 (1974).

86. M. W. Penley and J. M. Wood, Mass spectrometry studies of substituted methanes formed from deutero- and fluoromethylcobalamins, *Biochim. Biophys. Acta* **273**, 265–274 (1972).

6

Solvent Hydrogen Isotope Effects

Katharine B. J. Schowen

1. SOLVENT ISOTOPE EFFECTS AND BIOLOGICAL SYSTEMS

Biological and biochemical processes are affected by the partial or complete replacement of solvent protium oxide (H_2O) by deuterium oxide (D_2O)[1] Although some very simple organisms (certain algae, bacteria, etc.) have survived complete replacement of protium by deuterium, differences in morphology and metabolism are very apparent. Not surprisingly, deuterium substitution affects virtually all aspects of metabolism and physiological function. Higher plants, after experiencing impaired chlorophyll synthesis and inhibition of cell division, will cease to grow after $\sim 70\%$ D exchange. Mammals (unable to survive much more than 25% D incorporation) show evidence of severe difficulty with protein synthesis, decreased enzyme levels, decreased erythrocyte production, impaired carbohydrate metabolism, hormone imbalance, central nervous system disturbance, difficulty with mitosis, and increasingly impaired reproductive capability. Most of these consequences are directly or indirectly attributable to changes in the *rates* of biochemical reactions.

A number of such individual biochemical processes, isolated from the organism for *in vitro* laboratory study, have shown pronounced changes in rate with a change in solvent from H_2O to D_2O. These rate differences are conventionally expressed and reported as the ratio k_{H_2O}/k_{D_2O} of the respective rate constants for the reaction in light and heavy water; this ratio defines the term *kinetic solvent (hydrogen) isotope effect*.* In Table I are some recently

---

* Hydroxylic solvents other than water also give rise to kinetic solvent isotope effects. Studies with these other solvents are relatively rare and will be mentioned infrequently in this chapter; the general principles discussed, however, will apply.

Katharine B. J. Schowen • Department of Chemistry, Baker University, Baldwin City, Kansas.

Table I
Solvent Isotope Effects for Some Biochemical Reactions[a]

	Isotope effect[b]	Reference
Monovalent-cation-induced tetramerization of formyltetrahydrofolate synthetase subunits	∼3	2
Polymerization of beef brain tubulin	5	3
Conformational isomerization of ribonuclease	4.3; 4.8	4, 5
Deacetylation of acetyl-α-chymotrypsin	2.4; 2.7	6, 7
Acylation of α-chymotrypsin by N-acetyl-L-tryptophanamide	1.9; 2.0	8, 7
Deacetylation of acetyltrypsin	1.4	8
Deacylation of α-N-benzoyl-L-arginyltrypsin	2.6	8, 9, 10
Acylation of trypsin by α-N-benzoyl-L-phenylalanyl-L-valyl-L-arginyl p-nitroanilide	3.9	10, 11
Deacylation of α-N-benzoyl-L-arginylthrombin	2.9	8b, 12
Deacetylation of acetylelastase	2.2	8
Acylation of elastase by α-N-carbobenzyloxy-L-alanyl p-nitrophenyl ester	1.8	8b, 12
Acylation of elastase by N-acetyl-L-alanyl-L-prolyl-L-alanyl p-nitroanilide	2.1	8b
Acylation of α-lytic protease by N-acetyl-L-alanyl-L-prolyl-L-alanyl p-nitroanilide	2.8	8b
Glutaminase-catalyzed hydrolysis of glutamine	1.8[c]	8a, 10, 13
Asparaginase-catalyzed hydrolysis of asparagine	2.9[c]; 2.6[d]	8a, 10, 13
Inorganic pyrophosphatase-catalyzed hydrolysis of inorganic pyrophosphate	1.9	14

[a] Reactions for which complete proton inventories have been carried out.
[b] Ratio of rate constants (k_{H_2O}/k_{D_2O}) or of velocities (v_{H_2O}/v_{D_2O}).
[c] Enzyme obtained from E. coli.
[d] Enzyme obtained from Erwinia carotovora.

available values for the kinetic solvent isotope effects (KSIE's) observed for a variety of biochemical reactions (enzyme-catalyzed hydrolyses, enzyme–substrate reactions, protein conformation changes, and protein subunit associations), some of which will be discussed in detail in Section 8. Systems chosen for inclusion in Table I are limited to those for which a proton inventory (Section 5) has been carried out. The list is meant to be illustrative and not an exhaustive collection of solvent isotope effects reported for biological reactions.

2. ISOTOPE EFFECTS AND TRANSITION-STATE STRUCTURE

The magnitude of an experimentally observed kinetic solvent isotope effect should, in principle, provide information about the transition state for the particular reaction under investigation. Since the rate constant of a chemical reaction is determined by the free energy of activation [$\Delta G^{\ddagger}$, the difference in free energy between the transition-state-activated complex and the reactant ground state, Eq. (3)], the isotope effect k_{H_2O}/k_{D_2O} is an expression involving the difference between the free energy of activation for the reaction in light solvent and heavy solvent:

$$k = \frac{kT}{h} e^{-\Delta G^{\ddagger}/RT} \tag{1}$$

$$k_{H_2O}/k_{D_2O} = e^{(\Delta G_{D}^{\ddagger} - \Delta G_{H}^{\ddagger})/RT} = e^{\delta \Delta G^{\ddagger}/RT} \tag{2}$$

where

$$\Delta G^{\ddagger} = G^{\ddagger} - G_r \tag{3}$$

and

$$\delta\Delta G^{\ddagger} = \Delta G^{\ddagger}_D - \Delta G^{\ddagger}_H = (G^{\ddagger} - G_r)_D - (G^{\ddagger} - G_r)_H \tag{4}$$

Illustrated by means of a very simple energy diagram (Fig. 1) is a possible situation where the overall reactant-state *zero-point energy difference* in the two solvents *alters* (decreases in this case) on going to the transition state, a situation which arises when the bonding to an isotopic hydrogen atom becomes weaker or "looser." If the bonding of more than one hydrogen atom is changed on activation, a common situation in solvent isotope-effect studies, the overall isotope effect will be a net reflection of all the bonding changes, some perhaps "tighter," some "looser"; it will be the product of the isotope effects for all the bonds to hydrogen that undergo change on activation. For the situation illustrated in Fig. 1, $\Delta G^{\ddagger}_D > \Delta G^{\ddagger}_H$ and $k_{H_2O}/k_{D_2O} > 1$ (reaction faster in light solvent). The conclusion is that at least one bond to a hydrogen atom (in water or in the reactant as a result of hydrogen exchange) is looser or undergoing transfer in the transition state of the rate-determining step of the reaction. It is important to realize that it is only when the isotopic free-energy difference is changed on going from ground state to transition state that an isotope rate effect is observed. Since this change will only occur when a bond to an isotopic atom undergoes some change between ground state and transition state, isotope effects (provided we have independent knowledge about reactant-state structure) can provide information about transition-state structure in the neighborhood of the isotopic atom(s).

Certainly it is true that isotopic substitution is the least disturbing structural change that can be made in a system in order to gain mechanistic information. Nuclear charge, electron distribution, and potential energy remain un-

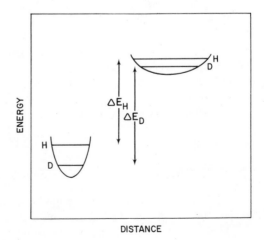

Fig. 1. Diagram showing that the zero-point energy difference for bonds to H and to D decreases on going to the transition state for a reaction in which bonding to hydrogen becomes looser, resulting in $E_H < E_D$ and a normal isotope effect $k_H/k_D > 1$.

changed; reactions of isotopically substituted substances take place on the same potential surface. Since the only change introduced upon isotopic substitution is one of mass, the distinguishing properties of isotopic species are those consequently affected: translational energy of the molecule as a whole and the rotational and vibrational energies of the bonds to the isotopic atoms. As the translational and rotational effects can be expressed in terms of the vibrational effects according to the Redlich–Teller rule, the entire kinetic isotope effect can be expressed by the Bigeleisen[15] equation,

$$\frac{k}{k'} = \frac{v_L}{v'_L}\left\{ \overset{\text{real transition-state frequencies}}{\prod_i} \left(\frac{v_i}{v'_i}\right) e^{-\Delta u_i/2} \frac{1 - e^{-u'_i}}{1 - e^{-u_i}}\right\}$$

$$\left\{ \overset{\text{reactant frequencies}}{\prod_j} \left(\frac{v_j}{v'_j}\right) e^{-\Delta u_j/2} \frac{1 - e^{-u'_j}}{1 - e^{-u_j}}\right\} \quad (5)$$

in terms of changes in the vibrational energies alone. In Eq. (5), $\mu = hv/kT$, $\Delta u = u - u'$ (where the prime refers to the heavy isotope), and v_L/v'_L is the imaginary frequency ratio for the reaction coordinate of the activated complex. The last term is factored out because it has no corresponding zero-point energy $(e^{-\Delta u/2})$ or excitation factor $[(1 - e^{-\mu'})/(1 - e^{-\mu})]$. If the appropriate reactant- and transition-state frequencies are available (from infrared or Raman spectra for the former, from estimation by vibrational analysis of possible transition-state structures, a computationally complex undertaking, for the latter), an isotope effect can be calculated. Equation (5) can be expressed in terms of the reactant-state contribution (RSC) and the transition-state contribution (TSC) to the isotope effect,

$$k/k' = \text{TSC}/\text{RSC} \quad (6)$$

Rearrangement to

$$\text{TSC} = (k/k')(\text{GSC}) \quad (7)$$

emphasizes the point of this and earlier chapters, namely, that information about the transition state is available from experimental isotope-effect measurements. Because of the dependence of TSC on vibrational frequencies, the information should be expressible in terms of transition-state force constants, bond lengths, and bond angles. It is our purpose in this chapter to show how transition-state structures and reaction mechanisms can be deduced from those isotope effects that arise when reactions of biochemical interest are carried out in isotopic solvents.

3. THE ORIGIN OF KINETIC SOLVENT ISOTOPE EFFECTS

Solvent isotope-effect theory has been presented in a number of excellent reviews by Laughton and Robertson,[16] Arnett and McKelvey,[17] Gold,[18,19]

Schowen,[20] Thornton and Thornton,[21] Kresge,[22] and Albery[23] and earlier by Wiberg,[24] Gold and Satchell,[25] and Bader.[26] Consequently only the briefest outline will be attempted here. It must of course first be recognized that a number of different phenomena contribute to produce a net observed solvent isotope effect. Changing the hydroxylic hydrogen atoms of water from protium (^1_1H) to deuterium (^2_1H) will bring about rate differences in reactions occurring in those solvents if any or all of the following undergo *change* on going from reactant to transition state: (1) differences in bulk solvent properties, (2) differences in solute–solvent interactions, (3) differences in zero-point energy of O—L bonds (where L is an unspecified isotope of hydrogen) of reacting solvent water molecules, or (4) differences in zero-point energy of solute bonds to (exchangeable) H which have become labeled by rapid exchange with solvent. The first two arise from changes in the medium and may be termed "transfer" effects and "solvation" effects, respectively. The latter two arise from actual bond changes in reacting molecules. These may be further classified as *primary effects* if the reaction involves rate-determining proton (deuteron) transfer from either reactant water or solute and as *secondary effects* if the exchangeable H in either water or solute is involved in the reaction but not undergoing rate-determining transfer. Explanations of these various contributions to the overall solvent isotope effect will generally involve either (1) the effect of isotopic substitution on internal (that is, pertaining to the reacting molecules) vibration (stretching, bending) frequencies of bonds to H (D) and the changes in these frequencies on going from reactant state to transition state or (2) the effect of isotopic substitution on the librational (that is, the hindered rotational) frequencies of external solvent water molecules and the changes in these frequencies on activation.[27]

The overall observed KSIE is the *product* of each of these various isotope-effect contributions, that is,

$$\left(\frac{k_{H_2O}}{k_{D_2O}}\right)_{total} = \left(\frac{k_H}{k_D}\right)_{pri} \left(\frac{k_H}{k_D}\right)_{sec} \left(\frac{k_{H_2O}}{k_{D_2O}}\right)_{medium} \tag{8}$$

recognizing that $(k_H/k_D)_{sec}$ may itself be the product of more than one secondary isotope effect. One can then, as shown below, examine the transition state for a proposed reaction mechanism, estimate the primary and secondary SIE contributions, and compare the product with the experimentally determined value.

3.1. Transfer and Solvation Isotope-Effect Contributions

Heavy water is not the same solvent as its light counterpart; there are differences in bulk structure (partly reflecting differences in intermolecular hydrogen bonding) and, of course, in molecular weight and, consequently, in physical properties. At a given temperature, for example, the isotopic solvents differ in viscosity, density, and dielectric constant and in their ability to solvate ionic and nonpolar solutes. In Table II are given some important physical properties of light and heavy water.

It would be interesting to know what effect a change of solvent from light to heavy water would have on the rate of a reaction *per se* (because of a change in bulk dielectric constant, for example) and *in the absence of any specific solute–solvent interaction*. Such information is difficult to obtain; the effects most certainly are small, and it is generally assumed that such "transfer" effects (resulting from the intact transfer of a reaction from one solvent to another) on reaction rates are safely neglected. These effects and their possible contribution to the net solvent isotope effect have been discussed in some detail by Gold.[18]

The contribution from the medium to the overall isotope effect that *is* observed results from solute–solvent interactions, that is, from solvation of the reacting solutes by the solvent water molecules. Such "solvation" isotope effects are estimated to be modest, ranging from 1.2 to 1.3.[21] The solute–solvent interactions giving rise to these purely "solvation" isotope effects are a consequence of the structure of water. Although the precise structure of water is not exactly known, a degree of understanding sufficient to provide workable models for the calculation and prediction of solvation isotope effects exists. Water is known[1,17] to have a great deal of internal three-dimensional structure arising from intermolecular hydrogen bonding; its properties depend to a large extent on the degree and nature of this structure. At least up to 35°C heavy water appears to have more structure, i.e., more intermolecular hydrogen bonding than light water. Evidence for the preceding is provided by the facts of poorer solubility of ions in D_2O,[17] the SIE's on ionic conductivity,[28] the greater temperature of maximum density, and the much greater viscosity of D_2O. The structure of water decreases with increasing temperature; the rate of this decrease is greater with D_2O. Solution of nonpolar nonelectrolytes

Table II
Some Physical Properties of Light and Heavy Water[a]

Property (unit)	H_2O	D_2O
Molecular weight (^{12}C scale)	18.015	20.028
Melting point (°C)	0.00	3.81
Boiling point (°C)	100.00	101.42
Density at 25° (g/cm³)	0.9970	1.1044
Temperature of maximum density (°C)	3.98	11.23
Dielectric constant, 25°	78.39[b]	78.06[b]
Viscosity at 25° (cP)	0.8903	1.107
Viscosity at 20° (cP)	1.005[c]	1.25[c]
Ionization constant at 25°	1×10^{-14}	1.95×10^{-15}

[a] Data taken from more extensive tables in Reference 17, pp. 392–394.
[b] Reference 57.
[c] Reference 17, p. 347.

in water is accompanied by a large loss in entropy, evidence for an increase in water structure in the neighborhood of the solute. This structure-making effect is somewhat greater in D_2O than H_2O (presumably because of the greater strength of the H bond in the former solvent). Hydrophobic bonding in biological polymers should, as a consequence, be favored in D_2O. Ionic substances when dissolved in water appear to effect a slight net decrease in the overall solvent structure. This structure-breaking effect is greater in D_2O (presumably because there is more structure to begin with).

Based on this understanding of the structure of water and its observed effect on the solvation of various kinds of solutes, two fairly detailed physical models to explain solvation isotope effects have been proposed. That of Bunton and Shiner[29] emphasizes changes in internal stretching frequencies of bonds to hydrogen in water molecules or labeled solutes as the basicity or acidity of solutes change on activation, while Swain et al.[30] have favored the librational (hindered external rotational) degrees of freedom of the water molecules as the major origin of solvent isotope effects of this type. According to the latter theory, water molecules are treated as hindered rotors 4-coordinated by hydrogen bonding to four adjacent water molecules in the ground state. Most of the solvation isotope effect arises from changes in the librational frequencies (as opposed to internal vibrations and translations) of the water molecules solvating each reactant molecule upon activation. Derivation from this theory of a relationship that can be used to calculate the isotope effect for simple ionic solubilities is provided by Thornton and Thornton.[21] Isotope effects for solubilities of salts range from 1.01 to 1.33.[17] From such treatments it is expected that ion-forming reactions will have a solvation isotope-effect contribution to the overall solvent isotope effect greater than 1, consistent with a net decrease in librational frequencies due to a net increase in structure breaking on going from reactants to transition state. k_{H_2O}/k_{D_2O} contributions of less than 1 are expected for ion-destroying reactions. Unfortunately, in spite of the importance of understanding the role of the solvent in contributing to solvent isotope effects, there is a regrettable lack of data allowing for the separation of solvation effects from specific effects where water is a reactant or exchangeable proton transfer is occurring. Experimental solvent isotope effects for the solvolysis of alkyl halides vary from around 1.2 to 1.3 at temperatures between 80° and 90°C.[21] Both the secondary isotope effect associated with nucleophilic attack and the solvation isotope effect (since the reaction is ion forming and ions are structure breaking) should be greater than 1, implying that the solvation isotope effect is a small but not negligible contributor to the overall observed solvent isotope effect. Thus it is clear that the contribution of the structural differences of solvent light and heavy water to the stability of reactants and activated complexes by virtue of their different solvating ability should be considered, specifically the thermodynamics of transfer $(\Delta H_t^\circ, \Delta S_t^\circ)$ from light to heavy water. It is at present not easy to make safe predictions about the relative magnitude of these effects.

3.2. Primary Isotope-Effect Contributions

A primary kinetic hydrogen isotope effect is observed for a particular reaction when an isotopically substituted hydrogen involved in the reaction (that is, one whose bond is being made or broken) is *in motion* along the reaction coordinate at the transition state. The theory of primary isotope effects and their relation to transition-state structure have been discussed in detail by Westheimer,[31] Bigeleisen,[32] Bell,[33] Willi and Wolfsberg,[34] More O'Ferrall and Kouba,[35] Albery,[23] Thornton and Thornton,[21] Schowen,[20] and More O'Ferrall.[37] Treatment particularly relevant to the elucidation of biochemical transition states appears in Chapter 4 in this volume. A simple treatment using Westheimer's three-atom transition-state model[31] has been applied by Schowen[20] to make quick approximate primary isotope-effect calculations and will be summarized in Section 4.3.

3.3. Secondary Isotope-Effect Contributions

Secondary hydrogen isotope effects are observed when an isotopically substituted hydrogen atom undergoes some change in bonding on going from reactant to transition state but is not actually moving or in the process of transfer. Excellent theoretical treatments of the origin of such isotope effects have been made available by Thornton and Thornton[21] and in Chapter 5 in this volume. Secondary isotope effects originate mainly from the difference in zero-point energy as a result of changes in force constants upon going from reactants to the transition state. Smaller force constants for bonds to isotopic atoms in the transition state than in the reactant state (bond "loosening"), decreasing the zero-point energy difference, tend to make the secondary contribution, $k_H/k_D > 1$ and vice versa. Secondary isotope effects can be calculated using a computer to arrive at moments of inertia and normal vibrations for assumed geometries and various force constants.[38-42] They can also be estimated less rigorously but more rapidly using fractionation factors, as described in the following sections.

4. CALCULATION OF SOLVENT ISOTOPE EFFECTS

The ultimate purpose of determining a KSIE is usually to gain information about the transition state for the rate-determining step of the reaction in question. The way to relate an experimentally determined KSIE to a transition-state structure is to propose a number of reasonable structures, assign to each an expected isotope effect, and compare the actual result to the model. Any method of isotope-effect estimation will usually suffice: rigorous calculation from theory, approximations using simplifying assumptions, or simply by analogy with known reaction mechanisms.

To rigorously calculate the solvent isotope effect expected for a given transition state a complete vibrational analysis is needed, a difficult and time-consuming process. A more approximate but simpler method has been described by Schowen[20] using isotopic fractionation factors to obtain the secondary isotope-effect contribution and the simple triatomic proton-transfer transition state of Westheimer[31] to obtain the primary contribution.

4.1. Ground-State Fractionation Factors[18,22,43,44]

Placement of a molecule SH with one exchangeable hydrogen atom in a solvent mixture of ROH and ROD results in an equilibrium described by

$$SH + ROD \rightleftharpoons SD + ROH \qquad (9)$$

with an equilibrium constant given by

$$K = \frac{[SD][ROH]}{[SH][ROD]} = \frac{[SD]/[SH]}{[ROD]/[ROH]} = \phi \qquad (10)$$

This equilibrium constant is identical with and defines the term ϕ, the isotopic fractionation factor for the hydrogenic position of SL. ϕ is an expression of the preference of the hydrogenic position in SL for deuterium over protium relative to the preference of the hydrogenic position of ROL for deuterium over protium. In more general terms, an isotopic fractionation factor for any particular site in a molecule is defined as the ratio of its preference for deuterium over protium relative to the similar preference of a single site in a solvent molecule. For water, it is assumed that the deuterium preference of one of the sites is independent of whether the other site is occupied by H or D, i.e., that the "rule of the geometric mean" holds. Although the rule does not hold entirely, any errors so introduced have generally been considered to be slight.[45,46] Albery[23] discusses this point more fully, implying that corrections may on occasion be required. It is likewise assumed that for substrates with more than one exchangeable site the fractionation factor for a given position is independent of the H or D content at any other sites. It is generally true that the heavier isotope will accumulate in those sites of an equilibrium such as Eq. (9) in which bonds with the larger force constants (i.e., stronger, tighter bonds) are found. For a fractionation factor greater than 1, Eq. (10) implies that D prefers the particular solute site relative to a solvent site, that is, that D will preferentially accumulate in the ith solute site, indicating that hydrogens are more tightly bound in SL than in the solvent. Likewise, for ϕ values less than 1, the implication is that H prefers the particular solute site relative to solvent and that the particular SL bond is weaker than the solvent bonds to hydrogen. Not only do we assume that fractionation factors are independent of isotopic substitution at other positions in the same molecule but also that they remain the same for the same bond types, i.e., that they are unaffected by the particular molecular environment.[47] A number of fractionation factors have been experimentally determined and are collected in Table III. It will be noted that

Table III
Isotopic Fractionation Factors
Relative to Water[a]

Functional group	Fractionation factor, ϕ
$\diagdown$O—L	1.0
—CO_2—L	1.0
$\diagdown$C$\diagup$O—L	1.23–1.28
$\diagdown\overset{+}{O}$—L	0.69
$\bar{O}$—L	0.47–0.56
$\diagdown$N—L	0.92
$\overset{+}{N}$—L	0.97
$\diagdown$S—L	0.40–0.46
$\diagdown$C—L	0.62–0.64 (sp) 0.78–0.85 (sp^2) 0.84–1.18 (sp^3)

[a] Data are taken from similar tables in Schowen[20] and Schowen.[10] Original sources and discussion of the values reported are contained in those references.

the isotopic fractionation factor for an —O—L position is 1.0; i.e., deuterium or hydrogen is randomly distributed among such bonds.

To see how fractionation factors may be used in isotope-effect calculations, consider a situation in which protiated and deuterated reactants RH and RD undergo chemical reaction to give protiated and deuterated products PH and PD,

$$RH \rightleftharpoons PH, \quad K_H \tag{11}$$

$$RD \rightleftharpoons PD, \quad K_D \tag{12}$$

It is likely that the reactant fractionation factor ϕ^R and the product fractionation factor ϕ^P are different. It is readily seen that the equilibrium isotope effect, K_H/K_D, for this process is given by the fractionation factor ratio ϕ^R/ϕ^P:

$$\frac{K_H}{K_D} = \frac{[PH]/[RH]}{[PD]/[RD]} = \frac{[RD]/[RH]}{[PD]/[PH]}$$

$$= \frac{[RD]/[RH]}{[ROD]/[ROH]} \bigg/ \frac{[PD]/[PH]}{[ROD]/[ROH]} = \frac{\phi^R}{\phi^P} \tag{13}$$

For substrates with multiple exchangeable hydrogenic positions, the equilibrium isotope effect is given by

$$\frac{K_H}{K_D} = \prod_i^{\substack{\text{reactant} \\ \text{sites}}} \phi_i^R \bigg/ \prod_j^{\substack{\text{product} \\ \text{sites}}} \phi_j^P \tag{14}$$

where a ϕ value is included for every hydrogenic site, even if identical with another. Thus, using the ground-state fractionation factors as given in Table III, one can easily calculate equilibrium isotope effects. As an example, consider the equilibrium

$$2H_2O \rightleftharpoons H_3O^+ + OH^- \tag{15}$$

The isotope effect K_H/K_D roughly calculated from fractionation factors should be

$$K_H/K_D = (\phi)^4_{O-L} / (\phi)^{\frac{3}{3}}_{O-L} (\phi)_{\bar{O}-L}$$

$$= 1.0/(0.69)^3 (0.5) \simeq 6.1 \tag{16}$$

Experimental values range from 5.2 to 7.4.[48-52] The necessary ϕ values can be obtained experimentally most conveniently by NMR techniques.[53-55]

4.2. Transition-State Fractionation Factors and Calculation of Approximate Secondary Solvent Isotope Effects

By the reasoning outlined in the foregoing section, one should be able to define a transition-state fractionation factor which could then be used to calculate kinetic secondary hydrogen isotope effects. If PH in Eq. (11) were a transition state, TH, the kinetic isotope effect would be given by the ratio of reactant- and transition-state fractionation factors,

$$k_H/k_D = \phi^R/\phi^T \tag{17}$$

For substrates with multiple exchangeable hydrogenic positions the kinetic isotope effect is given by

$$\frac{k_H}{k_D} = \prod_i^{\substack{\text{reactant} \\ \text{sites}}} \phi_i^R \bigg/ \prod_j^{\substack{\text{transition-} \\ \text{state sites}}} \phi_i^T \tag{18}$$

in terms of the reactant- and transition-state isotopic fractionation factors *for every separate exchangeable hydrogenic site.*

The problem of obtaining ϕ^T values now arises. In particular, considering secondary effects, how do we obtain fractionation factors for those bonds that are intermediate in character to those given in Table III (H—$O^{\delta+}$, for example). (Partially formed or broken bonds where the hydrogen is being transferred give primary isotope effects and can be so treated.)

Assume for the process

$$SL \rightleftharpoons TL \rightleftharpoons PL \tag{19}$$

that the isotope effect on the transition-state free energy is a weighted average of the isotope effects on reactant and product free energies.[56] Then, where χ is a weighting factor describing the transition-state structure ($\chi = 0$ for an exactly reactant-like transition state and $\chi = 1$ for an exactly product-like transition state) and $\delta\Delta G^0$ is the double difference in zero-point energy between reactant SL and product PL and given by [see Eq. (2)]

$$K_H/K_D = \phi^R/\phi^P = e^{\delta\Delta G^0/RT} \tag{20}$$

then

$$\delta\Delta G^{\ddagger} = \chi\,\delta\Delta G^0 \tag{21}$$

and from Eqs. (2) and (20)

$$k_H/k_D = \phi^R/\phi^T = e^{\delta\,\Delta G^{\ddagger}/RT} = e^{\chi\,\delta\Delta G^0/RT} = (\phi^R/\phi^P)^{\chi} \tag{22}$$

The transition-state fractionation factor is then given by

$$\phi^T = (\phi^R)^{1-\chi}(\phi^P)^{\chi} \tag{23}$$

If fractionation factors ϕ^R and ϕ^P are known for the substances on either side of a transition state, ϕ^T can then be calculated provided an estimate of χ is available, from Brønsted α or β values, for example.[20] Also, knowing an experimental value for k_H/k_D in addition to the ground-state fractionation factors, estimates for χ can be obtained—and thus information about the transition-state structure.

4.3. Calculation of Approximate Primary Solvent Isotope Effects

Approximate values of primary solvent isotope effects for hydrogen transfer may be obtained by the method described by Schowen[20] and briefly recapitulated here. Consider a transition state for the transfer of a proton from O_1 to O_2

$$O_1 \overset{1-\chi}{\text{-----}} H \overset{\chi}{\text{-----}} O_2$$

where χ defines the bond order for the forming H—O bond. For $\chi = 0$, the transition state is like reactants, and there is no isotope effect; i.e., $k_H/k_D = 1$. For $\chi = 1$, the transition state is like products, and the isotope effect is equal to the equilibrium isotope effect,

$$k_H/k_D = K_H/K_D = \phi^R/\phi^P \tag{24}$$

where R and P refer to the immediate reactants and products of the rate-determining step. A third value of χ is χ_{max}, defined as the bond order which gives a maximum isotope effect. This occurs when the force constants on the two partial transition-state $O \cdots H$ bonds are equal. An expression for χ_{max} is

$$\chi_{max} = 1/[1 + (v_P^H/v_R^H)^2] \tag{25}$$

where v_P^H and v_R^H are the stretching frequencies in reciprocal centimeters for the reactant and product bonds in question (OH in this case).

The idea is to construct a plot of isotope effect, k_H/k_D, vs. χ for the three cases $\chi = 0$, $\chi = 1$, and $\chi = \chi_{max}$. The maximum isotope effect is roughly given by

$$(k_H/k_D)_{max} = e^{-\Delta u_R/2} \tag{26}$$

resulting in, assuming $v_R^H/v_R^D \sim \sqrt{2}$,

$$\log(k_H/k_D) = (3.2 \times 10^{-4}) v_R^H \tag{27}$$

From the resulting plot one can read off values for the primary isotope effect, $(k_H/k_D)_{pri}$. Alternatively, one can use the relationship

$$\left(\frac{k_H}{k_D}\right)_{pri} = 1 + \left[\left(\frac{k_H}{k_D}\right)_{max} - 1\right]\left(\frac{\chi}{\chi_{max}}\right) \tag{28}$$

for values of $\chi \leq \chi_{max}$, and

$$\left(\frac{k_H}{k_D}\right)_{pri} = \left(\frac{\phi_R}{\phi_P}\right) + \left[\left(\frac{k_H}{k_D}\right)_{max} - \left(\frac{\phi_R}{\phi_P}\right)\right]\left(\frac{1-\chi}{1-\chi_{max}}\right) \tag{29}$$

for $\chi \geq \chi_{max}$ to arrive at an estimate of the primary isotope effect.

Occasionally it is true that proton transfer occurs late in a reaction sequence. This means that the actual proton-donating reactant is different from the initial reactant with consequent differences in fractionation factors. The isotope effect calculated by means of Eqs. (28) and (29) must be corrected by a term ϕ_R/ϕ_R', where ϕ_R' is the fractionation factor for the site in the species immediately preceding the rate-determining proton-transfer step:

$$\left(\frac{k_H}{k_D}\right)_{pri} = \left(\frac{\phi_R}{\phi_R'}\right)\left(\frac{k_H}{k_D}\right)_{pri \, [from \, Eqs. \, (28) \, and \, (29)]} \tag{30}$$

4.4. Estimation of the Overall KSIE

We have seen that the major contributions to a solvent isotope effect are either primary (from transferring hydrogen) or secondary (from hydrogens not undergoing transfer but experiencing some change in bonding). The overall solvent isotope effect can be expressed as

$$\frac{k_{H_2O}}{k_{D_2O}} = \left(\frac{k_H}{k_D}\right)_{pri} \left(\frac{k_H}{k_D}\right)_{sec} \tag{31}$$

Its magnitude can be approximated using fractionation factors in Table III and Eq. (22) and Eq. (28) or (29) as described in the preceding sections to obtain estimates for the secondary and primary contributions, respectively. A procedure to follow is:

1. Postulate a transition-state structure, indicating bond orders.
2. Estimate or look up fractionation factors for all exchangeable hydrogens not being transferred.
3. Calculate $(k_H/k_D)_{sec}$ from Eq. (22).
4. Estimate ϕ'_R for the transferring hydrogen in the species just prior to the rate-determining step, and calculate ϕ_R/ϕ'_R.
5. Calculate $(k_H/k_D)_{pri}$ from Eq. (28) or (29) and then Eq. (30).
6. Calculate k_{H_2O}/k_{D_2O} from Eq. (31).

The calculated isotope effect can then be compared to the experimental value, and necessary modifications in the structure of the transition state proposed in step 1 can be made. Numerous examples illustrating the ease, accuracy, and utility of this method are given by Schowen.[20]

It should be noted that the method just outlined is but one way to estimate KSIE's. As stated in the beginning of Section 4, any method used to predict an isotope effect for a proposed mechanism, including simple analogy with a known reaction, is usually acceptable. The estimate may then be compared with the experimental KSIE and the proposed transition-state structure confirmed or modified as necessary.

In summary, then, we have seen that kinetic solvent isotope effects are determined by the double difference in zero-point energy as a result of bonding changes to H (D) on going from reactant to transition state, that the magnitude of these effects is determined by and provides information about the structure of the transition state (insofar as it involves bonds to hydrogen) and to some extent about its solvational environment, and that values for these effects can be estimated from theory, fractionation factors, and hypothetical transition-state models. These calculated values can then be compared to those obtained after experimentally obtaining the rate constants for the reaction in each of two solvents, pure H_2O and pure D_2O, and the transition-state model accepted or rejected. The technique is generally rapid, simple, and reliable. While a degree of ambiguity will exist for more complex systems, i.e., those with numerous exchangeable hydrogens (see below), the isotope effects obtained from the pure isotopic solvents alone can always be used to gain rough initial transition-state information.

5. SOLVENT ISOTOPE EFFECTS IN MIXTURES OF LIGHT AND HEAVY SOLVENTS: THE PROTON INVENTORY TECHNIQUE

The foregoing method for deducing transition-state structures from solvent isotope-effect calculations and experimental data is most useful, accurate, and unambiguous when the system is relatively simple and there are few exchangeable hydrogens. As the number of exchangeable hydrogens in the reactants increases so does the number of possible transition-state structures. The origin of the isotope effect now becomes less certain. To what extent, for example, is the net effect now the result of rate-determining proton transfer (primary effect) or of many many smaller secondary changes in bonding to H multiplying together to give the observed effect?

This is, of course, precisely the situation we face in studying biochemical reactions. Biological molecules have abundant exchangeable hydrogens, those bonded to oxygen, nitrogen, and sulfur atoms, to carbon atoms located in the α position relative to a carbonyl group, etc. The number of such exchangeable sites on proteins and other biopolymers is enormous. Not only may a very large number of protons be undergoing bonding changes on activation but conformation changes (thus altering reactant shape, energy, etc.) may be induced on dissolution in D_2O as a result of exchange of H for D at sites throughout the molecule. Solvent isotope-effect studies of enzyme-catalyzed reactions or reactions otherwise involving biological macromolecules are then a particular problem, although the determination of their reaction mechanisms and transition-state structures is crucially important for fundamental and practical (see Part IV) reasons.

What is clearly needed for these biochemical reactions is (1) a "list" of all hydrogens undergoing a change of state on going to the transition state, i.e., of those generating the net isotope effect, and (2) the isotope effect for each. Such a list is known as a *proton inventory* and is obtained from reaction-rate studies in mixtures of light and heavy water. A complete proton inventory should give information about the number of hydrogens undergoing transfer or secondary bond changes in the transition state. One can then apply methods similar to those outlined in the previous sections to arrive at reasonable transition-state structures.

5.1. The Gross–Butler Equation

Forty years ago LaMer and Chittum[58] and then Gross et al.[59] and Butler[60] reported that the solvent isotope effect, k_0/k_n, for a variety of proton-transfer reactions in mixtures of light and heavy water did not vary linearly with n, the atom fraction of D in the mixture. The reason for this behavior rests in the fact that not all exchangeable protons responsible for the isotope effect have the same isotopic composition as the solvent; i.e., their fractionation factors are not necessarily equal to unity. Theoretical treatments of this phenomenon led to a mathematical formulation known as the Gross–Butler equation. The form of this equation currently is that given by

$$K_n = K_0 \prod_i^v (1 - n + n\phi_i^P) \bigg/ \prod_j^v (1 - n + n\phi_j^R) \tag{32}$$

$$k_n = k_0 \prod_i^v (1 - n + n\phi_i^T) \bigg/ \prod_j^v (1 - n + n\phi_j^R) \tag{33}$$

for equilibrium and rate processes, respectively. The ϕ's are the fractionation factors defined in Sections 4.1 and 4.2 and n is the atom fraction D in the solvent and varies from 0 (pure protiated solvent) to 1 (pure deuterated solvent). K_0 and k_0 are the respective equilibrium and rate constants in pure protiated solvent ($n = 0$); K_n and k_n are the corresponding constants obtained for a solvent in which the atom fraction D is equal to n. A simple derivation of the Gross–Butler relationship for rate processes follows.[61]

5.2. Simple Derivation of the Gross–Butler Relation

The rate constant k_n for a reaction carried out in a mixture of D_2O (mole fraction n) and H_2O (mole fraction $1 - n$) will equal that for a reaction in pure water, k_0, multiplied by a series of correction factors, J_i^n,

$$k_n = k_0 \prod_i^v J_i^n \tag{34}$$

one for each of the v exchangeable hydrogenic positions which contribute to the solvent isotope effect. The correction factor for a particular hydrogen will depend on the value of n, i.e., the amount of D_2O relative to H_2O in the solvent mixture, and on the differences in bonding of the particular hydrogen atom in the reactant and transition state. This correction factor will equal the equilibrium fraction (χ) of protium present in the particular hydrogenic site of the reactant divided by the fraction for that particular site in the transition state.

$$J_i^n = \left[\frac{\chi_H^R/(\chi_H^R + \chi_D^R)}{\chi_H^T/(\chi_H^T + \chi_D^T)} \right]_i^n = \frac{1 + (\chi_D^T/\chi_H^T)_i^n}{1 + (\chi_D^R/\chi_H^R)_i^n} \tag{35}$$

It can be seen that if the particular (ith) site being considered prefers deuterium more in the reactant than in the transition state, $J_i^n < 1$ and $k_n < k_0$. This is the situation encountered when bonding to hydrogen is weakened on going to the transition state, resulting in "normal" isotope effects, $k_0/k_n > 1$. Deuterium, it will be recalled, prefers that side of an equilibrium in which bonding to hydrogen is tighter.

For a given hydrogenic site i and a particular solvent composition n, the fraction χ_D^R/χ_H^R, for example, in Eq. (35), is equivalent to the concentration ratio [SD]/[SH] for the equilibrium of Eq. (9) and is thus related to the isotopic fractionation factor, as shown

$$\phi^R = \frac{[SD]/[SH]}{[D_2O]/[H_2O]} = \frac{\chi_D^R/\chi_H^R}{n/(1-n)} \tag{36}$$

$$(\chi_D^R/\chi_H^R)_i = \phi_i^R n/(1-n) \tag{37}$$

The expression for J_i^n in Eq. (35) can thus be written in terms of these fractionation factors,

$$J_i^n = \frac{1 + [\phi_i^T n/(1-n)]}{1 + [\phi_i^R n/(1-n)]} = \frac{1 - n + \phi_i^T n}{1 - n + \phi_i^R n} \tag{38}$$

and Eq. (34) becomes

$$k_n = k_0 \prod_i^v (1 - n + n\phi_i^T) \Big/ \prod_i^v (1 - n + n\phi_i^R) \tag{39}$$

This, then, is the Gross–Butler equation [cf. Eq. (33)] relating rate constants to mole fraction D_2O for reactions in mixtures of H_2O and D_2O. Equation (33) is the more general form of the equation, showing that the number of hydrogenic sites in the reactant and transition states need not be counted as equal.

Equation (33) or (39) can be expressed in terms of a reactant-state contribution (RSC) and a transition-state contribution (TSC),

$$k_n = k_0 [(TSC)_n/(RSC)_n] \tag{40}$$

and, as has been argued earlier, since k_n and k_0 are accessible from experiment and RSC can be evaluated from known or estimated reactant-state fractionation factors, the TSC can be determined

$$(TSC)_n = (k_n/k_0)(RSC)_n \tag{41}$$

and information about the number of ϕ's and their values obtained. A list of all the ϕ_i^T obtained by fitting $k_n(n)$ with known or assumed ϕ_j^R gives an inventory of transition-state protons and the isotope effect associated with each one.

In Section 6 we shall discuss some of the experimental considerations necessary for proton inventory work, and in Section 7 we shall discuss how the rate data so generated may be analyzed so as to provide mechanistic information.

6. EXPERIMENTAL PROCEDURES[62]

6.1. Preparation and Analysis of Mixtures of Light and Heavy Water

Heavy water is readily available from commercial sources. The usual 99.7–99.8% purity is entirely adequate for proton inventory work since the deuterium content of the experimental solutions will generally be independently determined. Distillation over barium nitrate is recommended if the pD is greater than 7; otherwise the D_2O may be used directly. Ordinary distilled water may be passed through an ion-exchange column and/or boiled to remove CO_2. Manipulations involving D_2O should normally be carried out in a dry box under dry nitrogen. Solutions can be stored in flasks equipped with serum caps. Transfer from one container to another can then be effected by means of a syringe. Contamination of D_2O solutions upon *brief* (3–4 min) exposure to atmospheric moisture has been demonstrated to be negligible.[63]

Mixtures of H_2O and D_2O can be prepared gravimetrically, or somewhat less accurately, volumetrically, using the molecular weights and densities given in Table II to obtain the desired values of *n*, mole fraction D_2O. The various H_2O–D_2O mixtures thus obtained may then be used to prepare solutions of the required ionic strength (by dilution of KCl, for example) or of the required pH (by dilution of various buffer pairs). (In the latter case account must be taken of any exchangeable protons which will dilute the mixture with protium and decrease the atom fraction of D.) Such preparation of salt or buffer solutions directly from the mixed solvent tends, however, to be wasteful of D_2O, still relatively expensive. One can therefore make each solution directly each time from weighed amounts of buffers and/or salt, diluting to volume by the addition of successive, weighed amounts of D_2O and H_2O. Again account must be taken of exchangeable hydrogen in the buffer. As this procedure tends to be tedious, alternative methods using stock H_2O solutions containing buffer and/or salt and dilution of these with the proper amount of pure H_2O and pure D_2O are therefore preferable. In calculating the final atom fraction of D and of H, the actual weight of these substances must be computed and account taken of any H added via buffer. Section 6.2 contains more information about buffers.

Experimental verification of these calculations is essential. It is desirable to have this verification for each solvent mixture used and preferably both at the start and at the end of an experimental run. Selected solutions may most conveniently be sent off for deuterium analysis* or analyzed by means of NMR techniques, one of which is described in the following paragraph. Selected references to other analytical methods are given in Reference 62.

Small amounts of H_2O and D_2O may be determined[62] by adding small weighed amounts of pure H_2O and pure *p*-dioxane (or acetonitrile, etc.) to a weighed quantity of H_2O–D_2O mixture, the composition of which is to be determined, and determining the NMR spectrum of the resulting solution. The amount of dioxane and total H_2O should be adjusted so that the NMR peak heights will be similar. The ratio of the integrated peak heights for H_2O and dioxane is related to the total number of moles of H_2O present:

---

* Joseph Nemeth, 303 W. Washington, Urbana, Illinois 61801 performs deuterium analyses using the falling drop method.

$$\frac{8 \,(\text{integrated } H_2O \text{ peak height})}{2 \,(\text{integrated dioxane peak height})} = \frac{\text{moles } H_2O \text{ total}}{\text{moles dioxane}} \qquad (42)$$

Simple subtraction from the moles of H_2O added will give the moles of H_2O in the original H_2O-D_2O mixture.

6.2. pL Measurements and Preparation of Buffer Solutions in Light and Heavy Water

In the course of a kinetic investigation, particularly of acid- or base-catalyzed or enzyme-catalyzed reactions, it will become necessary to know the exact pH dependence of the particular reaction under consideration. If a reaction rate is pH controlled and solvent isotope or proton inventory studies are to be made, it is, of course, important not only to be able to measure the pL accurately in each solvent mixture but also to be able to stay at the same relative position on the pH rate curve from solvent mixture to solvent mixture. The former is accomplished by means of a pH meter and the latter usually by using appropriate and correctly prepared buffer solutions as the reaction medium.

Accurate measurement of pL using a pH meter is described in Sections 6.2.1 and 6.2.2. Establishment and maintenance of a desired pL during a reaction is discussed in Sections 6.2.3–6.2.5.

6.2.1. Interpretation of pH-Meter Readings

The apparent pH of an aqueous solution of given $[L^+]$ or $[OL^-]$, as determined with a glass electrode and read on a pH meter, is dependent on the mole fraction of D_2O. Thus, for example, a solution known to be 0.001 M in L^+ may give a reading on a pH meter of 3.00 in pure H_2O, 2.85 when the mole fraction of D_2O is 0.5, and 2.61 in pure D_2O.

Glasoe and Long,[64] in their empirical approach to the standardization of a pK scale for solution in D_2O, found that when a pH meter with a glass and a calomel electrode is calibrated with standard pH solutions in H_2O and is then used for the measurement of the pD of solutions of known DCl (molar) concentration in pure D_2O a constant difference of 0.4 pH unit is observed at 25°C:

$$pD = (\text{meter reading}) + 0.4 \qquad (43)$$

or

$$(\Delta pH)_{n=1} = pD - (\text{meter reading}) = 0.4 \qquad (44)$$

Numerous independent investigations have demonstrated the essential correctness of the value 0.4 (± 0.02 pH unit) and that it holds for different electrolytes at various pH's and temperatures as well as for different commercial glass electrodes from different manufacturers.[48,63-66]

In a similar manner ΔpH_n values have been obtained for different values of n. From their data, obtained at 25° from pH meter readings on solutions 0.01 M and 0.1 M in LCl and 0.01 M in $LClO_4$ using mixed $H_2O–D_2O$ solvents, Salomaa et al.[48] derived by the method of least squares the empirical equation

$$\Delta pH_n = 0.076n^2 + 0.3314n \qquad (45)$$

which can then be used to obtain the true pL value for any $H_2O–D_2O$ solvent mixture of known mole fraction of D_2O, n. A similar relationship which holds at both 25° and 37° was derived by Schowen et al.[66] from data using 0.001- to 0.0007-M LCl solutions in mixtures of light and heavy water:

$$\Delta pH_n = (0.17324)n^2 + (0.22076)n + 0.00009 \qquad (46)$$
$$\pm 0.01298 \qquad \pm 0.01216$$

The difference in ΔpH as calculated from either Eq. (45) or (46) is at most 0.03 pH unit.

It may be deemed desirable to prepare a calibration curve of meter reading vs. n for the particular system, pH, instrument, and electrodes to be used in a given investigation. The measurements can easily be made following a procedure similar to that described in the following section.

6.2.2. Experimental Determination of Relationship between pL and n_{D_2O}

The pH meter is calibrated with standard aqueous buffers and adjusted to the desired temperature controlled to $\pm 0.1°C$. The electrodes are immersed in a known volume of solvent (pure H_2O, pure D_2O, and at least five mixtures of known mole fraction of D_2O) contained in a thermostatted vessel. The solution and electrodes are allowed to reach thermal equilibrium (about 5 min) at which point a known volume (the same for each solvent mixture) of standard aqueous HCl is added. It is unnecessary to soak or equilibrate the electrode in the particular solvent being used for a longer period of time.[64,66] After stirring, the pH meter is read. In this fashion a calibration curve of pH meter reading vs. n_{D_2O} (corrected for the added titrant) is obtained as shown in Fig. 2, and ΔpH_n for any mole fraction D_2O can thus be read with a good degree of confidence from such a curve. If desired, ΔpH_n vs. n values obtained from this curve can be subjected to polynomial regression analysis[62,67] to obtain values for the coefficients for the quadratic equation

$$\Delta pH_n = an^2 + bn + c \qquad (47)$$

6.2.3. Ionization of Weak Acids in Light and Heavy Water

The ionization of weak acids decreases by a factor of 3–6 in D_2O.[16] The difference $pK_D - pK_H$ or ΔpK_a for most acids is about 0.5 to 0.7 unit and has been shown by Rule and LaMer[68] and Bell[69] to increase with pK_a

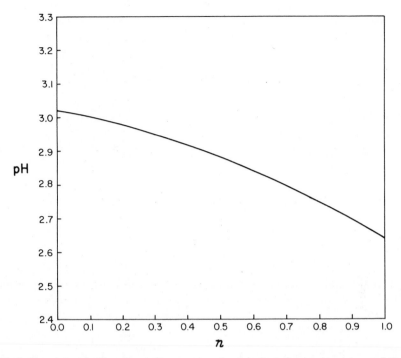

Fig 2. Experimental pH meter reading vs. atom fraction deuterium, n, in mixtures of H_2O and D_2O at 25°.

according to the relationship

$$\Delta pK_a = 0.41 + 0.020 pK_H \tag{48}$$

Note that this predicts that acids of pK_a 3–10 should all have ΔpK_a in the range 0.54 ± 0.07. An excellent collection of pK_H and ΔpK_a values for a large number of weak acids has been compiled by Laughton and Robertson,[16] and in fact at least 70% of the acids listed with pK_a's between 3 and 10 have ΔpK_a values between 0.46 and 0.61.

In mixtures of H_2O and D_2O, the change in ionization constant of a weak acid follows a nonlinear relationship with the mole fraction of D_2O.[16,18,22] The particular form of this relationship can be derived by considering the equilibrium for the ionization of a weak acid,

$$H\!-\!A + H\!-\!O\!-\!H \; \overset{K_a}{\rightleftharpoons} \; H\!-\!\overset{+}{\underset{\underset{H}{|}}{O}}\!-\!H + A^- \tag{49}$$

and using the Gross–Butler equation given in Section 5.1 [Eq. (32)] for equilibrium processes:

$$K_n = K_0 \left[\overbrace{\prod_i^{v} (1 - n + n\phi_i^P)}^{\substack{\text{product} \\ \text{sites}}} \bigg/ \overbrace{\prod_j^{v} (1 - n + n\phi_j^R)}^{\substack{\text{reactant} \\ \text{sites}}} \right] \tag{50}$$

There are on each side of Eq. (49) three hydrogens undergoing a change of state: on the reactant side, the acidic hydrogen of the weak acid and the two (identical) hydrogens of water, and on the product side, the three identical $\overset{+}{O}$—H hydrogens of the hydronium ion. Equation (50) then becomes

$$K_n = K_0 \frac{(1 - n + nl)^3}{(1 - n + n\phi_{H-A})(1 - n + n\phi_{H_2O})^2} \tag{51}$$

where K_n is the acid dissociation constant in an H_2O–D_2O mixture in which the mole fraction D_2O is equal to n; K_0 is the corresponding constant in pure H_2O, i.e., K_H; and l is the fractionation factor for the $\overset{+}{O}$—H hydrogens of the hydronium ion, 0.69. Since the fractionation factor for O—H bonds is equal to 1.0 (Table III), the squared term in the denominator of Eq. (51) disappears. Setting $n = 1$ (pure D_2O), we see that

$$K_n/K_0 = K_D/K_H = l^3/\phi_{H-A} \tag{52}$$

and

$$\phi_{H-A} = l^3/(K_D/K_H) \tag{53}$$

where K_H and K_D are the acid dissociation constants for the weak acid in pure H_2O and pure D_2O, respectively. Equation (51) then becomes, for all values of n,

$$K_n = K_0 \frac{(1 - n + nl)^3}{1 - n + nl^3 K_H/K_D} \tag{54}$$

Thus if one has experimentally determined pK_H and pK_D in the pure solvents, the value K_H/K_n and therefore ΔpK_a can be calculated from Eq. (54) for any value of n. Equation (54) assumes that K_H and K_D are independent of the medium, i.e., independent of changes in the standard states for the ionization process. It also assumes that the proton is monosolvated—hence the term l^3. If nonspecific solvation of the proton is assumed, an equation linear in n follows. Gold and Lowe[50] and Salomaa et al.[48] concluded that the cubic equation gave better agreement with experimental K_H/K_n values than either assuming the proton was nonspecifically solvated or attributing the change in K_H/K_n to medium effects, although Halevi et al.[70] found little to distinguish between the two. For further discussion of this equation, its assumption, limitations, derivation, and alternate forms, see References 16 and 18.

6.2.4. Selection of a Suitable Buffer

To study most conveniently a pH-dependent reaction under the same conditions from solvent to solvent, a buffering system can be selected such that the ΔpK_a of the acidic buffer component is the same as that (or those) for the reagent, catalyst, enzyme, etc., responsible for the pH dependence. In the following paragraphs we shall attempt to rationalize this general statement.

If not already available, a pH-rate profile for the desired reaction should be determined experimentally in both light and heavy water; profiles in H_2O–D_2O mixtures are not required. The results when plotted may appear as sketched in Fig. 3. pD values must be calculated before plotting by adding 0.4 to the experimental "pH" readings, as discussed in Section 6.2.1. As shown in Fig. 4, the pH-rate curves are simply segments of "titration" curves, i.e., plots of concentration (measured here by the rate) of reactive form (protonated, nonprotonated, etc.) of the pH-determining reaction component vs. pH.

Since the existence of a rate variation with pH for a reaction may imply that different concentrations of the reactive form of a reagent are present at different pH's, curvature such as in Fig. 3 may be observed because of the presence of an ionizable group at, for example, an enzyme active site, one form of which is more catalytically active than another. If the following equilibria are possible for an enzyme, with form EH being catalytically active,

$$EH_2 \rightleftharpoons EH \rightleftharpoons E \qquad (55)$$
$$+ \qquad +$$
$$H^+ \qquad H^+$$

curves such as those in Figs. 3 and 4 are expected.

The displacement of the curve in D_2O is observed because the ionization constants are different in light and heavy water (see Section 6.2.3). The values for ΔpK_1 and ΔpK_2 differ very slightly (refer to the Bell equation in Section 6.2.3 and to the fact that ΔpK_a values of almost all acids with pK's between

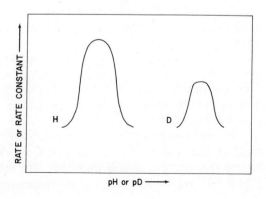

Fig. 3. Hypothetical pH-rate profile for a reaction in H_2O (curve H) or D_2O (curve D).

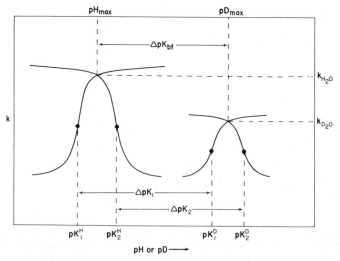

Fig. 4. pH vs. rate constant curves for a hypothetical reaction in H_2O and in D_2O showing differences in pK_a's.

3 and 10 are in the range 0.54 ± 0.07); the difference between the pH and the pD values at the maximum of each curve, $pD_{max} - pH_{max}$, will be approximately equal to ΔpK_1, and ΔpK_2, or, more exactly, a weighted average of ΔpK_1 and ΔpK_2. We are particularly interested in the pH (or pD) values at the maxima of these pH-rate curves since it will be at these pH's that the reaction should be studied. At those pH's the reaction rate is least sensitive to small deviations in pH, and errors of measurement will be minimized. Consider that a buffer is chosen for which the acid component has, referring to Fig. 4, a pK_a equal to pK_1, pK_2, or, preferably, pH_{max} for the reaction in question and that the reaction will be studied at a pH equal to pH_{max}. Then the change in buffer pK_a, ΔpK_{bf}, is such that (assuming always the same buffer ratio) a shift to pure D_2O or *any* H_2O–D_2O mixture will ensure that one will always be at the same relative point, in this case the maximum, of the pH-rate curve for the solvent mixture.

In practice, then, one selects a buffer (from Laughton and Robertson's table, for example[16]) whose acid component has a ΔpK_a close to the difference $pD_{max} - pH_{max}$ (ΔpK_{bf} of Fig. 4) obtained from experimental pH-rate data for the reaction in light and heavy water. Use of the same concentration and ratio (i.e., the same amounts) of such a buffer acid and its conjugate base for each solvent mixture will assure that one is always working at the same region of the pH-rate curve. The value of pL will be different from solution to solution, but the *relative proportion of reactive species will be the same.*

It should be noted that it is not necessary to determine the pH-rate profile in D_2O. One can simply choose a buffer with a pK_{bf}^H close to pK_1^H or pK_2^H for the reaction, or, lacking even that information, with a pK_{bf}^H in the usual 0.54 ± 0.07 region. This will include most common buffering systems, including the frequently used Tris buffers. A table in Bates,[71] giving pD values for three buffers (citrate, phosphate, and carbonate, prepared by dissolving the protiated salts in pure D_2O) at various temperatures, may be useful in providing an independent check of the expected pD in D_2O when these buffers are used.

6.2.5. pH Stat Method for Maintaining Constant pH

In some cases constant pH is maintained during the course of a reaction in the absence of buffers by the pH Stat method. The desired pH is set on the instrument and titrant is automatically dispensed according to the appearance–disappearance of acid or base as the reaction proceeds. The rate of production of acetic acid in the acetylcholinesterase-catalyzed hydrolysis of acetylcholine, for example, may be followed by recording the rate of addition of standard NaOH needed to keep the pH at the constant preset value.[72,73] To do a proton inventory study using the pH Stat technique one would need to determine experimentally at least a section of the pH rate profile of the reaction in *each* $H_2O–D_2O$ mixture in order to find the pH reading for the same region on the curve.

6.3. Determination of Ground-State Fractionation Factors[55,62]

An experimental technique using nuclear magnetic resonance spectrometry for the determination of ground-state fractionation factors has been used and described by Gold and Kessick,[53,74] Kresge and Allred,[54] and Mata-Segreda.[55] The technique is based on the fact that hydrogen exchange between a hydroxylic solvent and many solutes is rapid relative to the NMR time scale. Consequently the only observed "exchangeable-H" signal has a chemical shift equal to the weighted (according to mole fraction) average chemical shift of *each* solute and solvent exchangeable-proton signal[75]:

$$\delta_{obs} = \frac{\delta_1\chi_1 + \delta_2\chi_2 \cdots}{\chi_1 + \chi_2 \cdots} = \frac{\sum \delta_i\chi_i}{\sum \chi_i} \tag{56}$$

Here χ_i is the mole fraction of the ith kind of exchangeable hydrogen and δ_i is the NMR chemical shift for the ith proton when not exchanging with other kinds of protons. The term $\sum \chi_i$ will equal 1 in pure protiated solvent. Separation of solvent from solute contributions in Eq. (56) gives

$$\delta_{obs} = \delta_{ROH}\chi_{ROH}/\sum \chi_j + \sum_i^s \delta_i\chi_i \Big/ \sum \chi_j \tag{57}$$

where the first term gives the contribution from the solvent exchangeable protons, s means the number of exchangeable substrate hydrogenic sites, and $\sum_i^s$ means a sum over all s exchangeable substrate protons. For dilute solutions, in which nonideal behavior of the solution such as specific solvent–solute interactions (e.g., protonation) or solute–solute interactions (e.g., dimerization) is at a minimum, the mole fraction of substrate exchangeable protons is negligible compared to solvent; thus $\chi_{ROH} \simeq \sum \chi_j$, and the relationship becomes

$$\delta_{obs} = \delta_{ROH} + \sum_i^s \delta_i\chi_i \Big/ \sum \chi_j \qquad (58)$$

For mixed H_2O–D_2O solvents, and again for very dilute solutions, the total atom fraction of exchangeable protium is essentially that for the solvent H_2O, that is $1 - n$, where n is the mole fraction of D_2O. Equation (58) thus becomes

$$\delta_{obs} = \delta_{H_2O} + \sum_i^s \delta_i\chi_i^H \Big/ (1 - n) \qquad (59)$$

Recalling that the isotopic fractionation factor for the ith hydrogenic site in the solute is given by [see Eq. (36)],

$$\phi_i = (\chi_i^D/\chi_i^H)/[n/(1 - n)] \qquad (60)$$

and defining the total mole fraction of hydrogens of the ith kind as χ_i^t, where

$$\chi_i^t = \chi_i^D + \chi_i^H \qquad (61)$$

then algebraic manipulation of Eq. (60) will give an expression for the mole fraction of the ith exchangeable protium:

$$\chi_i^H = (1 - n)\chi_i^t/(1 - n + \phi_i n) \qquad (62)$$

Substituting the right-hand side of Eq. (62) into Eq. (57) gives an expression for the observed chemical shift in terms of the total mole fraction of exchangeable substrate hydrogen of the ith type:

$$\delta_{obs} = \delta_{H_2O} + \sum_i^s \delta_i\chi_i^t \Big/ (1 - n + \phi_i n) \qquad (63)$$

Since the total mole fraction of the ith kind of exchangeable substrate hydrogen, χ_i^t, is equal to the number of positions of the ith type, μ_i, times the mole fraction of solute, χ_s, we can write a final expression for the chemical shift expected for dilute solutions of a solute with i types of hydrogens in mixtures of light and heavy water:

$$\delta_{obs} = \delta_{H_2O} + \sum_i^s \frac{\delta_i\mu_i}{1 - n + \phi_i n}\chi_s \qquad (64)$$

According to this equation, for low concentrations of solute in a particular H_2O–D_2O mixture, the slope of a plot of the observed chemical shift vs. mole fraction of solute should be linear. For two experiments using two different solvents (but otherwise identical) the ratio of slopes is given by

$$\frac{slope\ 1}{slope\ 2} = \frac{(1 - n + \phi_i n)_2}{(1 - n + \phi_i n)_1} \qquad (65)$$

If experiment 1 is carried out in pure H_2O, $n = 0$ and Eq. (65) becomes

$$\frac{\text{slope (H}_2\text{O)}}{\text{slope (L}_2\text{O)}} = 1 - n + \phi_i n = 1 - n + \Phi_s n \tag{66}$$

The term Φ_s is defined as the overall fractionation factor for a particular solute and is thus readily available from NMR experiments. If the solute has one or more *identical* exchangeable hydrogenic sites, Φ is the fractionation factor, ϕ, for each site. If a given solute has two or more different exchangeable sites, Φ is an overall or "average" fractionation factor for that solute.

In practice, then, one measures the observed chemical shift for at least three, but preferably four or more, solutions of pure H_2O and for which the solute mole fraction is in the range 0.01–0.1. Lower χ_s may result in difficulty in correctly determining the position of the NMR signal; higher χ_s may introduce undesired complications resulting from solute–solute or solute–solvent interaction. Chemical shift readings should be reproducible to ± 0.1 Hz. The temperature is generally maintained constant, although it appears that ϕ values remain relatively unaffected over a small range in temperature. This procedure may then be repeated for mixed H_2O–D_2O solutions, prepared gravimetrically, for which n, the mole fraction of D_2O, is quite high, about 0.95. The experimental δ_{obs} vs. χ_s data may then be plotted or submitted for linear least-squares analysis and Φ calculated from Eq. (66). The percent error in Φ is given by

$$\% \text{ error in } \Phi = [(\% \text{ error in slope } H_2O)^2 + (\% \text{ error in slope } L_2O)^2$$
$$+ (\% \text{ error in } n)^2]^{1/2} \tag{67}$$

It should be noted that the technique just described for determining reactant-state (average) fractionation factors has not been used for proteins or other macromolecules. It is essential for this NMR method to be reliable that the solute be concentrated enough to effect an observable change in the chemical shift from that of pure solvent yet not so concentrated that chemical shift changes due to solute–solvent or solute–solute interactions will occur. It is not certain that these conditions can be met in the case of macromolecules, but perhaps more experimentation is justified. (See Section 6.4.2 for further discussion.)

6.4. Macromolecular Systems: Special Considerations [62]

Enzymes, proteins, and biological macromolecules in general present some special problems when involved in proton inventory studies. Transfer of such substances from protium oxide to deuterium oxide may induce a number of changes in behavior. Some of these changes may arise from the difference in the physical properties of the two solvents (see Table II) but more noticeably from exchange of labile protium for deuterium. An average-sized protein, for example, will have hundreds of exchangeable hydrogens including those of the backbone —NH— and the side-chain functional groups containing bonds to oxygen, nitrogen, and sulfur. These hydrogens play an important role in maintaining the secondary and tertiary structure and the state of ionization necessary for catalytic action (in the case of enzymes), self-association, binding to small molecules, etc. Replacement of these hydrogens by deuterium will alter the lengths and strengths of hydrogen bonds, pK_a values, and the extent of ionization, with possible resultant changes in conformation, extent of association, binding equilibria, catalytic power, etc. It is important to recognize that all of the above are examples of isotope effects and as such can be analyzed and understood. The following considerations are important in connection with the proton inventory technique.

6.4.1. Rate of Hydrogen Exchange

In carrying out a proton inventory study it is necessary to know the atom fraction of deuterium in the system, n, and how it is distributed, i.e. the reactant-state fractionation factors, ϕ_i^R, at the time the reaction is taking place. It is desirable, therefore, that all exchange of protein labile protons take place instantly or within a few minutes. Hvidt and Nielsen[76] review in detail methods for determining these exchange rates; Richards and Wyckoff[77] review results for ribonuclease. In most cases it appears that while exchange of backbone amide N—H may take place over several hours or more,[78] exchange of side-chain hydrogens is complete within a few minutes.

In the absence of precise exchange-rate information for the various kinds of exchangeable sites for a given protein a series of simple experiments such as those described by Bender and Hamilton[79] for α-chymotrypsin can be performed. In their case it was found that the same kinetic results were obtained in D_2O if either a stock solution of enzyme in D_2O or a stock solution of enzyme in H_2O was used to initiate the reaction under investigation. Similarly in H_2O, the kinetics was unaffected by use of either stock enzyme solution. Such results would allow one to conclude that the exchange of (at least) those labile hydrogens *involved in the reaction* was very rapid. It was also found in their case that the reaction rate was the same when using freshly prepared D_2O enzyme solutions or similar solutions which had been prepared several days previously. Such results would suggest that there are no very slowly exchanging hydrogens involved in the reaction. Hydrogen exchange occurring at intermediate rates, i.e., during the course of a reaction being studied, will show up as departure from the expected kinetic order, again, if those hydrogens are involved in the reaction under investigation.

6.4.2. Macromolecular Ground-State Fractionation Factors

Average fractionation factors for the exchangeable protons of macromolecules have been obtained from hydrogen-exchange studies in the form of the equilibrium isotope effect used to correct experimental deuterium or tritium exchange data to the numbers of exchangeable hydrogens. The general subject of hydrogen exchange in proteins has been reviewed by Hvidt and Nielsen[76] and the equilibrium isotope effect for exchange determined for ribonuclease, human serum albumin and many other proteins, insulin and synthetic polypeptides, small peptides, and nucleic acids (see Reference 5, footnote 3, for many specific references). It was found in these cases that the average fractionation factor (that is, the inverse equilibrium isotope effect for converting macromolecular exchangeable protium to deuterium from the solvent) was either unity (no discrimination) or small, ~ 1.14 (deuterium favored). Of course, since what is required in proton inventory work are the reactant-state fractionation factors, ϕ_i^R, for the specific sites *which change on conversion of reactants to the transition state*, and since proteins and other biomacromolecules have so many exchangeable sites, such average values are of much less use than for substrates with very few and often identical hydrogenic sites. There is at present no straightforward technique that may be used to determine fractionation factors for individual specific sites in macromolecules. Protons in the active site of an enzyme are in principle more accessible than others, and NMR techniques such as those described by Robillard and Shulman[80] for studying the active-site exchanging protons of α-chymotrypsin may find use for this purpose.

What is generally done at present is to assume that all protein ϕ_i^R's are equal to unity or do not change on activation. Since most N—H and O—H sites are known to have fractionation factors close to 1.0 (see Table III) and since it is further known that fractionation factors for protein side-chain exchangeable sites are 1.0 except for sulfhydryl (0.40–0.46), the assumption should be justified. *Small* departures from this assumption will affect the results quantitatively but not qualitatively (i.e., conclusions regarding the number of protons undergoing change in the transition state should remain unchanged).

6.5. Precision Required in Rate Measurements

In a proton inventory study rates are measured in pure protiated solvent, in pure deuteriated solvent, and in solvents of mixed isotopic composition. The rate constants are then plotted against atom fraction deuterium and the resulting curve examined for linearity or curvature (see Fig. 5 and the discussion in Section 7). The minimum number of points (solvent compositions) needed to define the curve will be three, most conveniently those for the pure solvents ($n = 0$, $n = 1$) and for an intermediate case, $n = 0.5$, for example. For confidence, depending on the difficulty involved and the precision of the data, more points for different values of n are desirable. Most studies in the literature[2-14] report the use of at least three and more often six to nine different solvent compositions, although Albery[23] has argued for the use of only the one intermediate point where $n = 0.5$. It is, of course, essential that the precision or reproducibility of the experimental rate constant at a given n be accurately known (by performing replicate experiments).

Table IV
Approximate Precision in Rate Constants Required for Proton Inventory Work[a]

Solvent isotope effect, k_0/k_1	Number of protons	Gross–Butler equation, k_n/k_0	$k_{0.5}/k_0$	Precision required[b]
1.5	1	$1 - n + n/1.5$	0.833	
	2	$[1 - n + n/(1.5)^{1/2}]^2$	0.825	1%
	3	$[1 - n + n/(1.5)^{1/3}]^3$	0.822	0.4%
2.0	1	$1 - n + n/2.0$	0.750	
	2	$[1 - n + n/(2.0)^{1/2}]^2$	0.729	2.8%
	3	$[1 - n + n/(2.0)^{1/3}]^3$	0.721	1.1%
4.0	1	$1 - n + n/4.0$	0.625	
	2	$[1 - n + n/(4.0)^{1/2}]^2$	0.563	10.4%
	3	$[1 - n + n/(4.0)^{1/3}]^3$	0.541	4.0%
10	1	$1 - n + n/10$	0.55	
	2	$[1 - n + n/(10)^{1/2}]^2$	0.433	23.7%
	3	$[1 - n + n/(10)^{1/3}]^3$	0.393	10.1%
25	1	$1 - n + n/25$	0.52	
	2	$(1 - n + n/5)^2$	0.36	36%
	3	$[1 - n + n/(25)^{1/3}]^3$	0.302	18%
60	1	$1 - n + n/60$	0.508	
	2	$[1 - n + n/(60)^{1/2}]^2$	0.319	46%
	3	$[1 - n + n/(60)^{1/3}]^3$	0.248	25%

[a] The numerical percentages calculated assume no reactant-state terms in the Gross–Butler equation and no uncertainty in k_0 and k_1 and are, therefore, only approximate.
[b] The first number in this column for a given isotope effect is the precision required to distinguish a two-proton from a one-proton mechanism; the second number is that precision required to distinguish a three-proton from a two-proton mechanism.

To distinguish a linear (one proton in the transition state) plot from a curved (two-proton) plot from an even more curved (three-proton) plot, etc., the required precision in experimental data must be possible. In Table IV is listed the precision necessary in the rate constant at $n = 0.5$ to distinguish one-proton, two-proton, and three-proton mechanisms for different overall solvent isotope effects. Two conclusions are clear from examining the table. (1) The precision required goes down as the isotope effect, k_0/k_1, goes up. (2) The precision required goes up as one tries to distinguish one- and two-, two- and three-, three- and four-proton, etc., mechanisms. Greater precision will be required for values of $0 < n < 0.5$ and $0.5 < n < 1.0$. The exact numerical percentage values in Table IV were calculated assuming no reactant-state terms in the Gross–Butler equation and no uncertainty in k_0 and k_1 and will therefore not apply to all cases with the same isotope effects. It is obvious that extremely precise data are necessary to draw mechanistic conclusions from a proton inventory for reactions with small solvent isotope effects ($< \sim 2$). Fortunately most enzymic solvent isotope effects encountered lie between 2 and 5, and some very interesting ones are very, very large.

7. ANALYSIS AND INTERPRETATION OF RESULTS

After carrying out a proton inventory study, the data that are obtained and which must be interpreted are a series of rate constants as a function of n, $k_n(n)$. From these data it will be possible to determine the form of the Gross–Butler equation

$$k = k_0 \frac{\text{TSC}}{\text{RSC}} = k_0 \frac{(1 - n + n\phi_1^T)(1 - n + n\phi_2^T) \cdots (1 - n + n\phi_v^T)}{(1 - n + n\phi_1^R)(1 - n + n\phi_2^R) \cdots (1 - n + n\phi_v^R)} \tag{68}$$

for the particular reaction under investigation. It will be recalled that there is a $1 - n + n\phi$ term in the numerator and in the denominator of Eq. (68) for every exchangeable proton in the system. Since most of these protons are not changing state on activation, their reactant- and transition-state terms will cancel. Terms for which $\phi = 1$ will likewise disappear. What remains, then, is an equation from which the number of $1 - n + n\phi$ terms and the corresponding ϕ values can be discovered. For the frequently encountered cases for which RSC = 1 the number of terms in the numerator gives the number of protons involved in the transition state, while the precise ϕ^T values provide information about the nature of these protons. For example,

$$k_n = k_0(1 - n + 0.3n)(1 - n + 0.7n) \tag{69}$$

is for a two-proton reaction for which $\phi_1^T = 0.3$, $\phi_2^T = 0.7$, and $\phi_1^R = \phi_2^R = 1$. In this section we shall attempt to describe just how one can obtain other such equations from $k_n(n)$ data and then how they may be interpreted in terms of transition-state structures.

7.1. Preliminary Examination of Data: Shape of k_n vs. n Plots

The data to be examined consist of a series of rate constants for different mixtures of isotopic solvents. The overall solvent isotope effect k_H/k_D (or k_0/k_1) can be calculated immediately. It can be noted whether it is normal or inverse, and preliminary mechanistic deductions can be made as suggested in Section 4. Following this, all the rate constants should be plotted (ordinate) against their corresponding values of n (abscissa) and the shape of the resulting function examined. The appearance of this plot can immediately provide a clue as to the general form of the Gross–Butler equation for the reaction. It can suggest, for example, whether there is one or more than one TSC term in the numerator and whether RSC terms are present or absent in the denominator.

Let us consider some of the possibilities sketched in Fig. 5 for a hypothetical series of reactions (1–6), each of which has a solvent isotope effect, k_0/k_1, of 2.5. The isotope effect is normal ($k_0 > k_1$), TSC < RSC, so all plots go down toward the right. One, however, appears linear, while others are bent upward or downward.

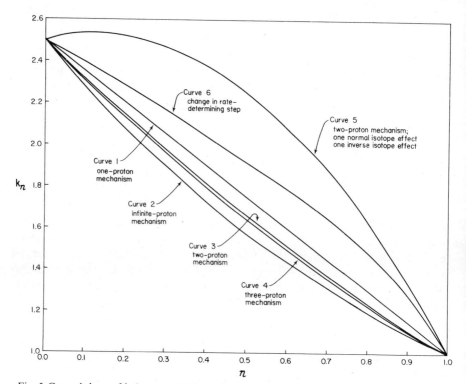

Fig. 5. General shape of k_n (rate constant) vs. n (atom fraction deuterium) plots for some hypothetical reactions possessing the same kinetic solvent isotope effect but proceeding by different mechanisms. Curve 1: a one-proton transition state, normal isotope effect; curve 2: an infinite number of transition-state protons, each contributing a very small normal isotope effect; curve 3: a two-proton transition state, two identical normal isotope effects; curve 4: a three-proton transition state, three identical normal isotope effects; curve 5: a two-proton transition state, one normal and one inverse isotope effect; curve 6: two one-proton transition states in sequence, normal isotope effects, rate-determining step changes with n.

7.1.1. Linear Plot, Curve 1

This is by far the easiest situation for interpretation. There is only one simple or practical explanation for such a curve, namely, that the solvent isotope effect arises from a single hydrogenic site in the transition state. In other words, it means that one proton only is undergoing change in the rate-determining step of the reaction. It is reasonable that such a one-proton mechanism should give a linear plot; the rate constant k_n will be the mole-fraction-weighted average of k_0 (for pure H_2O) and k_1 (for pure D_2O) as in

$$k_n = k_0(1 - n) + k_1(n) = k_0\left(1 - n + n\frac{k_1}{k_0}\right) = k_0\left(1 - n + n\frac{k_D}{k_H}\right) \quad (70)$$

This is equivalent to the linear form of the Gross–Butler equation:

$$k_n = k_0(1 - n + n\phi_1^T) \tag{71}$$

The reactant-state fractionation factor ϕ_1^R for the hydrogenic site in question is equal to 1.0, and ϕ_1^T is given by the inverse isotope effect for the reaction. A rapid check for linearity or the absence thereof can be made early in a proton inventory study. Since the degree of curvature of a $k_n(n)$ plot will be greatest where $n = 0.5$, the general shape of the curve becomes most apparent after only three rate constants have been determined, those for $n = 0$, 0.5, and 1. Thus a fair amount of mechanistic information is available once rates have been measured in the pure solvents and a 50:50 mixture of the two. The observed value of $k_{0.5}$ should coincide with the arithmetic mean of k_0 and k_1. This is easily shown by evaluating Eq. (70) at $n = 0.5$:

$$k_n = k_0\left(1 - n + n\frac{k_1}{k_0}\right) \tag{72}$$

$$k_{0.5} = k_0\left(0.5 + 0.5\frac{k_1}{k_0}\right) = \frac{1}{2}(k_0 + k_1) \tag{73}$$

If the experimental $k_{0.5} = \frac{1}{2}(k_0 + k_1)$ (or if all the experimental points fall on a line), a linear dependence of k_n on n is indicated.

It is worthwhile to be aware that at least two additional, more complicated kinds of interpretations of linear $k_n(n)$ dependences are possible.

1. If there are two or more parallel reaction routes, at least one of which proceeds via a one-proton transition state and none of which proceeds via a $(n > 1)$-proton transition state, a linear $k_n(n)$ relationship will apply. Consider, for example, a system with two parallel pathways going through two one-proton transition states:

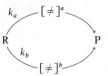

Consider, too, the following equations which describe this system,

$$k_n = k_n^a + k_n^b \tag{74}$$

$$k_n = k_0^a(1 - n + n\phi_a) + k_0^b(1 - n + n\phi_b) \tag{75}$$

$$\frac{k_n}{k_0} = \frac{k_0^a}{k_0}(1 - n + n\phi_a) + \frac{k_0^b}{k_0}(1 - n + n\phi_b) \tag{76}$$

$$= f_0^a(1 - n + n\phi_a) + f_0^b(1 - n + n\phi_b) \tag{77}$$

$$= (f_0^a + f_0^b)(1 - n) + n(f_0^a \phi_a + f_0^b \phi_b) \tag{78}$$

$$= 1 - n + n\Phi \tag{79}$$

where k_0^a and k_0^b are the rate constants for reactions a and b in pure H_2O, respectively; where f_0^a and f_0^b are the fractions of the total reaction that proceed via $[\neq]^a$ and $[\neq]^b$, respectively; where other terms have their usual meanings; and where Φ, the inverse observed solvent isotope effect, is the average of the two "actual" ϕ's for the two transition-state protons, weighted according to the fraction of reaction proceeding through those transition states:

$$\Phi = (f_0^a \phi_a + f_0^b \phi_b) \tag{80}$$

Equation (79) is obviously linear. Φ is the fractionation factor for the proton in the "average" or *virtual transition state* for the reaction.[81] Since a virtual transition state for a multipathway process is indistinguishable from a single transition state for a single pathway process, it is best to interpret the data in the simplest way possible.

2. The possibility of linear dependences resulting from a fortuitous cancellation of terms in the Gross–Butler equation was recognized by Kresge.[82] If there are $1 - n + n\phi$ terms in the numerator and the denominator of the actual Gross–Butler equation describing a system which by coincidence cancel each other and therefore disappear, a linear dependence may result. One can imagine, for example, a two-proton reaction correctly described by

$$k_n = k_0 \frac{(1 - n + n_1^T)(1 - n + n\phi_2^T)}{(1 - n + n_1^R)(1 - n + n\phi_2^R)} \tag{81}$$

If $\phi_2^R = 1.0$ and if $\phi_1^R = \phi_2^T \neq 1.0$, the expression will reduce to Eq. (71). One can likewise imagine a large number of very small effects in the reactant state combining to offset and cancel one transition-state term (five ϕ^R's of 0.95 will multiply to cancel one ϕ^T of 0.77, for example). Such coincidental exact cancellation of terms is conceivable but not highly probable. Since by these arguments there is an infinite number of alternative interpretations of this plot, or indeed one of any other shape, it again seems best in the absence of contrary information to invoke the simplest interpretation.[6,10]

There exists a further possibility for a one-proton reaction for which TSC < RSC. If a reactant-state proton of $\phi^R > 1$ undergoes change on activation to give a transition-state proton of $\phi^T = 1$ and all other ϕ^R's and ϕ^T's are unity, then

$$k_n^{-1} = k_0^{-1}(1 - n + n\phi_1^R) \tag{82}$$

and a plot of k_n^{-1} vs. n will be linear, going down toward the right (normal isotope effect).

7.1.2. "Bulging Down" Plots: Curves 2–4

Plots of this shape almost always indicate the existence of a multi-proton transition state. The Gross–Butler equation is usually of the form of

$$k_n = k_0 \prod_i^v (1 - n + n\phi_i^T) \tag{83}$$

with all $\phi_i^T < 1$ and all RSC terms equal to 1. The $k_n(n)$ plots are polynomials in n of order v:

$$k = k_0 + c_1 n + c_2 n^2 + \cdots + c_v n^v \tag{84}$$

It is also possible for multiproton transition states to have TSC $= 1$ (all ϕ_i^T's equal to unity) and RSC $\neq 1$. For such cases

$$k_n = k_0 \left/ \prod_j^v (1 - n + n\phi_j^R) \right. \tag{85}$$

and it is $1/k_n$ which is a polynomial in n:

$$k_0^{-1} = k_0^{-1} \prod_j^v (1 - n + n\phi_j^R) \tag{86}$$

Thus k_n^{-1} vs. n plots should routinely be constructed and analyzed as described below, especially if it is suspected that TSC $= 1$ or known that RSC $\neq 1$.

The order v of the polynomials of Eqs. (84) and (86) gives the number of hydrogenic sites involved on activation and may be any integral number between 2 and ∞. Before engaging in polynomial regression analysis (Section 7.2) to determine v, a little preliminary data plotting can provide clues as to its magnitude.

A downward-bowing curve could indicate the other extreme of the linear curve, namely, a reaction involving an infinite number of protons all contributing very small, approximately equal normal isotope effects. The ϕ_i^T values will be less than but very nearly equal to 1. If this is in fact the case, (1) a plot of $\ln k_n$ vs. n will be linear and (2) the $k_{0.5}$ value will be given by $\sqrt{k_0 k_1}$, that is, the geometric mean of k_0 and k_1. The following arguments justify these last two statements. If there is a very large number of protons, m, all contributing about equally to an observed solvent isotope effect, k_0/k_1, we have

$$k_1/k_0 = \phi^m \tag{87}$$

and

$$k_n = k_0(1 - n + n\phi)^m \tag{88}$$

Since the value of ϕ will be less than but very nearly equal to 1, $\phi = 1 - \chi$, where χ is very small. Equation (88) now becomes

$$k_n = k_0[1 - n + n(1 - \chi)]^m = k_0(1 - n\chi)^m \tag{89}$$

and

$$\ln(k_n/k_0) = m \ln(1 - n\chi) \tag{90}$$

Since $n \leq 1$ and χ is small, $n\chi$ is a small number and $\ln(1 - n\chi) \simeq -n\chi$.

Equation (90) can therefore be written

$$\ln(k_n/k_0) \simeq -mn\chi \tag{91}$$

or

$$\ln k_n \simeq (-m\chi)n + \ln k_0 \tag{92}$$

Thus, as m becomes very large, $\ln k_n(n)$ becomes linear. The rate constant $k_{0.5}$ is evaluated as follows. From Eq. (87) and the relationship $\phi = 1 - \chi$ we can write

$$\ln(k_1/k_0) = m \ln \phi = m \ln(1 - \chi) = m(-\chi) \tag{93}$$

Combining Eq. (93) with Eq. (91), we have

$$\ln(k_n/k_0) \simeq (-m\chi)n = \left[\ln(k_1/k_0)\right]n \tag{94}$$

Evaluating Eq. (94) for $n = 0.5$ gives

$$\ln(k_{0.5}/k_0) = \tfrac{1}{2}\ln(k_1/k_0) \tag{95}$$

$$k_{0.5}/k_0 = (k_1/k_0)^{1/2} \tag{96}$$

$$k_{0.5} = k_0\sqrt{k_1}/\sqrt{k_0} = \sqrt{k_1 k_0} \tag{97}$$

demonstrating that the value of $k_{0.5}$ is the geometric mean of the k_1 and k_0 values if the solvent isotope effect arises from the small equivalent contributions of a very large number of protons in the transition state. Curve 2 shows the appearance of a $k_n(n)$ plot for just such a situation.

In summary, then, one can easily differentiate between the two mechanistic extremes (one active proton vs. a very large number of active protons) by means of two simple tests:

$$1\ H: \qquad k_n \text{ vs. } n \text{ linear}, \quad k_{0.5} = \frac{k_1 + k_0}{2}$$

$$\infty\ H: \qquad \ln k_n \text{ vs. } n \text{ linear}, \quad k_{0.5} = \sqrt{k_1 k_0}$$

The degree of departure from these two extremes can be helpful in qualitatively estimating the number of TSC terms.

Data generating plots which bulge downward but lie between the line (curve 1, 1 H) and the extreme (curve 2, ∞ H) are consistent with a mechanism in which two or three or some reasonably small number of protons have undergone change on activation. [At this point one can carry out a rapid check for the possibility of a *two*-proton mechanism in which each proton contributes the same isotope effect, i.e., the case where

$$k_n = k_0(1 - n + n\phi_1^T)(1 - n + n\phi_2^T) \tag{98}$$

and $\phi_1^T = \phi_2^T$, and therefore

$$k_n = k_0(1 - n + n\phi_1^T)^2 \tag{99}$$

A plot of $(k_n/k_0)^{1/2}$ vs. n will be linear for such a reaction, and ϕ^T will be given by the square root of the inverse solvent isotope effect, $(k_1/k_0)^{1/2}$. If this procedure does not give a linear plot, the data must be subjected to polynomial regression analysis as described in Section 7.2 to obtain an accurate proton count and individual ϕ_i^T values.] Curve 3 in Fig. 5 has been calculated for a two-proton mechanism assuming the overall solvent isotope effect to be 2.5 and both ϕ^T's equal to $(1/2.5)^{1/2} = 0.63$, and curve 4 for a three-proton reaction, ϕ^T's $= (1/2.5)^{1/3} = 0.74$. As may be seen in Fig. 5, the curvature will be more pronounced as the number of protons increases until the limit is reached. It is also clear from Fig. 5 that it is very difficult to distinguish a two-proton mechanism from a three-proton mechanism, pointing out the need for precision in kinetic measurements (see also Table IV). The downward curvature implies that the multiple-site effects are all in the same direction (all ϕ^T's < 1, a normal isotope effect contribution by each "active" H). The curve bows down since even at small n more than one site can be deuterated and the rate will drop off faster than for a one-proton case. The RSC for reactions generating such plots may (initially at least) be assumed to be unity; that is, hydrogen equally stable in H or D solvent, ϕ^R's $= 1.0$.

There are other less common but conceivable types of mechanisms which could account for bulging down $k_n(n)$ plots. One of these will be mentioned here. A mechanism involving two or more parallel competing pathways will generate $k_n(n)$ data resulting in "bulged down" plots *if* at least one of the transition states involves at least two protons. The polynomial [Eq. (84)] will be quadratic if none of the competing transition states has more than two protons, cubic if three is the highest number of protons in any one transition state, etc. In such cases, the "apparent" ϕ values obtained by fitting the data to the appropriate (quadratic, cubic, etc.) form of the Gross–Butler equation, as in

$$k_n = k_0(1 - n + n\text{``}\phi\text{''}_a^T)(1 - n + n\text{``}\phi\text{''}_b^T) \qquad (100)$$

may turn out to be complex numbers (real and imaginary). It is also possible to get complex ϕ's from poor data or experimental error. If the "ϕ" values are not complex, the situation cannot be distinguished from a single-pathway mechanism.

7.1.3. "Bulging Up" Plots, Curves 5–6

Bowing upward when the net SIE is normal is caused when (1) a large normal effect is partially offset by a smaller inverse effect or (2) a change in rate-determining step from one sequential step to another occurs as n is altered. These two different interpretations are not visually distinguishable.

A situation described by the first category is one in which one proton has become "looser" and a second "tighter" on activation. The one will give rise to a normal isotope effect ($\phi_1^T = 0.2$, say) with linear falloff of rate with increasing n, and the other to an inverse isotope effect ($\phi_2^T = 2$, for example):

$$k_n = k_0(1 - n + 0.2n)(1 - n + 2n) \qquad (101)$$

The second effect will partially offset the first, thus decreasing the falloff in rate with increasing n and generating an upward bulge in the $k_n(n)$ plot. Since the second effect is smaller than the first, the overall SIE will remain normal and the plot will still go down to the right. Curve 5 in Fig. 5 has been calculated for this example. A second situation that may be considered in this same category is one arising from opposing contributions from transition-state sites and reactant-state sites. An example of such a reaction is

$$\text{RSH} + :\text{B} \longrightarrow [\text{RS} \cdots \text{H} \cdots \text{B}] \rightarrow \text{RS}^- + \text{BH}^+ \tag{102}$$

This is a one-proton reaction with $\phi^R_{\text{SH}} \simeq 0.5$ and, say, $\phi^T = 0.2$. One then has

$$k_n = k_0 \frac{1 - n + 0.2n}{1 - n + 0.5n} \tag{103}$$

and the net solvent isotope effect is 2.5. If ϕ^T had not been partially offset by $\phi^R = 0.5$, the net solvent isotope effect would have been larger than observed, $k_0/k_1 = 1/0.2 = 5$; if ϕ^R had not been cancelled by ϕ^T, i.e., if $\phi^T = 1$, the net solvent isotope effect would have been inverse, $k_0/k_1 = 0.5/1 = 0.5$. Thus, because of the TSC, increasing D in the solvent reduces the rate, but the opposing effect of the RSC term causes it to be reduced more slowly than otherwise — hence upward bowing. (The example just cited serves to illustrate another point: ϕ^R_j values of less than unity will cause an SIE to appear smaller than otherwise.)

Let us consider a hypothetical example from the second category of "bulging up" plots: a change in rate-determining step from one sequential reaction step to another as n is changed. Consider a reaction with an overall SIE of 2.5 and two steps in sequence: step (a) and step (b). Step (a) is 10% rate determining in H_2O with an isotope effect $k_H/k_D = 10$; step (b) is 90% rate determining in H_2O, and the isotope effect is 1.67. The system is represented by

$$\text{R} \xrightarrow[k_a]{(a)} \text{I} \xrightarrow[k_b]{(b)} \text{P} \tag{104}$$

$$k_n = k_n^a k_n^b / (k_n^a + k_n^b) \tag{105}$$

$$1/k_n = 1/k_n^a + 1/k_n^b \tag{106}$$

$$\frac{1}{k_n} = \frac{1}{k_0^a(1 - n + n/10)} + \frac{1}{k_0^b(1 - n + n/1.67)} \tag{107}$$

$$\frac{k_0}{k_n} = \frac{k_0}{k_0^a(1 - n + 0.1n)} + \frac{k_0}{k_0^b(1 - n + 0.6n)} \tag{108}$$

Since both k_0^a and k_0^b are larger than the overall rate in H_2O (k_0), k_0/k_0^a and k_0/k_0^b represent the fraction of the total that that step is rate determining. Thus

$$\frac{k_0}{k_n} = \frac{f_0^a}{1 - n + 0.1n} + \frac{f_0^b}{1 - n + 0.6n} \tag{109}$$

Evaluation of k_n at various values of n will result in a bulged up plot, as shown by curve 6 in Fig. 5.

Analogous information about transition-state structure is conveyed by the shapes of $k_n(n)$ plots for reactions with net *inverse* solvent isotope effects (where TSC > RSC). Again k_n^{-1} vs. n should be plotted for occasions when TSC might be unity and RSC < 1.

1. A *linear* plot indicates a *"one-proton"* transition state.
2. Plots *bulging upward* indicate transition states with *two or more protons* each contributing an inverse effect. A linear $\ln k_n$ vs. n plot means small inverse contributions from a very large number of protons; a linear $\sqrt{k_n/k_0}$ vs. n plot is evidence for a two-proton mechanism, $\phi_1^T = \phi_2^T$. Plots of this shape can also point to a change in rate-determining step.
3. Plots *bulging downward* indicate a transition state where larger inverse isotope effects are being partially offset by smaller normal isotope effects. As before, the partial cancellation can come from RSC terms, TSC terms, or both. Cautions and the alternative, less common explanations mentioned for normal isotope effects will also apply here.

7.2. Determination of the Number of Active Protons: Polynomial Regression[62]

For any nonlinear $k_n(n)$ plot the polynomial regression procedure is recommended. For this procedure to be useful either RSC or TSC must be known. In a large number of cases one has good reason to suppose that the reactant state does not contribute to the isotope effect, that is, that all $\phi_j^R = 1$. One then has a situation described by Eq. (83). The common examples of such cases are (1) hydrolysis reactions involving substrates with no exchangeable hydrogens or hydrogens with fractionation factors of 1 and (2) reactions involving enzymes whose exchangeable protons (except those of sulfhydryl groups) are expected to have reactant fractionation factors and hence RSC values of unity (see the discussion of this point in Section 6.4.2). Where RSC is known or suspected to be other than unity, it must be evaluated from knowledge of the individual ϕ_j^R values. This is generally not difficult: The reactant-state structure is known and the appropriate fractionation factors are available, e.g., from Table III. If not, they can be independently determined (see Section 6.3). The appropriate equation then is

$$k_n(\text{RSC}) = k_0 \prod_i^v (1 - n + n\phi_i^T) \tag{110}$$

Care should be taken to consider various perhaps unexpected $\phi^R \neq 1$ contributions to RSC particularly if the data are not easily fit assuming unit RSC. For example, in the methoxide-catalyzed methanolysis reaction

$$CH_3OH + CF_3C\overset{\overset{\textstyle CH_3}{|}}{\underset{\overset{\textstyle \|}{O}}{N}}C_6H_4X \xrightarrow{CH_3O^-} CF_3COCH_3 + CH_3NHC_6H_4X \tag{111}$$

the reactant state consists of the triply solvated methoxide ion $CH_3O^-(CH_3OH)_3$ with $\phi_{CH_3OH} = 0.74$, giving an RSC term of $(1 - n + 0.74n)^3$.[61]

Occasionally, as mentioned in Section 7.1, another situation is encountered: The isotope effect arises from reactant protons of $\phi_j^R \neq 1$ becoming transition-state protons of $\phi_j^T = 1$. In this case TSC is unity, and one has a system described by Eq. (85) or (86). An example of such a reaction is

$$\text{RSH} + \text{\Large $>$}\text{C}{=}\text{O} \longrightarrow \left[\text{RS} \cdots \text{\Large $>$}\text{C}{=\!=\!=}\text{O}{-}\text{H}\right]^{\neq} \longrightarrow \text{RS}{-}\overset{|}{\underset{|}{\text{C}}}{-}\text{OH} \qquad (112)$$

where $\phi^R \simeq 0.45$ and $\phi^T = 1.0$.

The three equations [Eqs. (83), (110), and (86)] become, respectively,

$$k_n = k_0 + c_1 n + c_2 n^2 + c_3 n^3 + \cdots + c_v n^v \qquad (113)$$

$$k_n(\text{RSC}) = k_0 + c_1 n + c_2 n^2 + c_3 n^3 + \cdots + c_v n^v \qquad (114)$$

$$k_n^{-1} = k_0^{-1} + c_1 n + c_2 n^2 + c_3 n^3 + \cdots + c_v n^v \qquad (115)$$

in which k_n, $k_n(\text{RSC})$, or k_n^{-1} is a polynomial in n, the order of which (v) specifies the number of protons whose state has been altered on going from ground state to transition state. The coefficients of this polynomial in principle allow for the calculation of ϕ values, i.e., the isotope effect for each such proton. In this way the experimental $k_n(n)$ data, along with some measure or hypothesis concerning $\text{RSC}(n)$, can give the proton inventory.

A polynomial regression computer program (available at most computation centers; one such is BMDO5R of the UCLA Health Sciences Computing Facility[67]) is now used to fit the appropriate left-hand side of Eq. (113), (114), or (115) to polynomials in n of ever-increasing order. Each succeeding term in the polynomial is tested for significance, for example, by the F test.[83] The exponent of n in the last statistically significant term gives the number of active protons consistent with the experimental data. The results, of course, depend not only on the system studied but also on the number and precision of the measurements. One should recognize, therefore, that hidden in the "noise" of the data may be additional active hydrogens. As always, as previously mentioned and as discussed by Kresge,[82] the (slight but real) possibility exists that small RSC contributions are exactly cancelling and thus concealing an additional TSC term and proton. Such a possibility can best be discarded if similar results are obtained for the same reaction with somewhat different substrates or conditions.[8a]

7.3. Extraction of Fractionation Factor Values

The isotope-effect contribution, or ϕ values, for each active proton could be found from the coefficients c_i of the best-fit polynomial. However, since the polynomial-fitting program pivots around the c_i and not the ϕ values, the procedure does not give desirable results. The preferred procedure is to construct a reasonable model of the transition state based on the results of the polynomial regression analysis. Reasonable ϕ^T values can be postulated for each active proton and the data fitted by any least-squares method directly to the appropriate equation [Eq. (113), (114), or (115)].

8. EXAMPLES FROM THE RECENT LITERATURE

In this section we shall give some examples, chosen for their variety and illustrative character, showing how proton inventory and solvent isotope effect data may be used to gain information about transition-state structures. The format will be to give (1) an equation or other description of the particular reaction; (2) the experimental rate data and/or the net KSIE; (3) a description or actual plot of the $k_n(n)$ function, together with conclusions drawn from its general appearance; (4) a structure for the proposed transition state; (5) the final form of the Gross–Butler equation with the fractionation factor values resulting from data fitting; and (6) comments and remarks about the chosen system as necessary. Further details may be obtained from the original references cited.

8.1. Example I[84]

The reaction is the intramolecular carboxylate-catalyzed hydrolysis of o-dichloroacetylsalicylic acid anion:

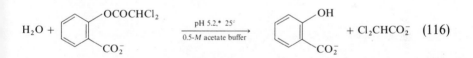

$$H_2O + \quad \xrightarrow[\text{0.5-}M \text{ acetate buffer}]{\text{pH 5.2.* } 25°} \quad + \; Cl_2CHCO_2^- \quad (116)$$

From the rate data presented in Table V the KSIE is seen to be normal: $k_{H_2O}/k_{D_2O} = 2.17$; thus TSC < RSC. Small KSIE's of around 1.5–3 are generally considered to be inconsistent with a primary isotope-effect contribution; the proton is thus not "in flight" but more likely engaged in "solvation catalysis" and therefore not on the reaction coordinate.[47] Figure 6 is a plot of k_n vs. n and shows clearly that the experimental points fall on a line sloping down to the right, with the conclusion that a single proton is undergoing change in the transition state of the reaction. The dashed line calculated for a two-proton mechanism for the same KSIE of 2.17 is seen to be outside of experi-

---

* This and other pH values given in this section are for $n = 0$, pure H_2O; it is understood that the equivalent pH was used for solvent mixtures (see Section 6.2).

Table V

First-Order Rate Constants for Hydrolysis
of 2×10^{-4} M o-Dichloroacetylsalicylate Anion
in Sodium Acetate–Acetic Acid Buffers in Mixtures
of H_2O and D_2O at $25°$[84]

n	$10^5 k_n \, (\text{sec}^{-1})$
0.0000	7510 ± 129[a]
0.0955	6994 ± 135
0.1984	6622 ± 61
0.2991	6173 ± 24
0.4012	5882 ± 38
0.5020	5438 ± 26
0.5981	4949 ± 29
0.6945	4699 ± 16
0.8009	4281 ± 54
0.8890	3886 ± 60
0.9964	3355 ± 26

[a] Error limits are standard deviations. Rate constants were obtained by a least-squares fit of absorbance-time data.

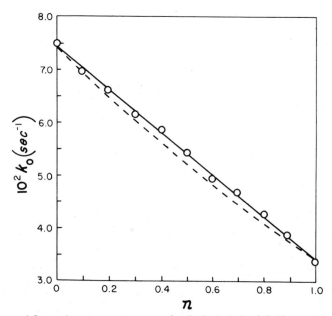

Fig. 6. Observed first-order rate constants vs. n for the hydrolysis of dichloroacetylsalicylate ion. The solid line is a plot of Eq. (117); the dashed line is a plot assuming two protons each contributing the same isotope effect. (Reproduced with permission from S. S. Minor and R. L. Schowen.[84])

mental error. When subjected to linear least-squares fitting the rate data give

$$10^5 k_n = (7432 \pm 33) [1 - n + (0.460 \pm 0.008) n] \qquad (117)$$

which when plotted on Fig. 6 gives the solid line through the experimental points.

The proposed transition-state structure **1** has the bridging proton H_1 responsible for the isotope effect. H_2, giving no isotope effect, is assumed to be undergoing a smooth transition from a water proton ($\phi_w^R = 1.0$) to a hydroxyl proton (also with $\phi = 1.0$).

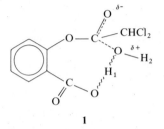

1

The final form of the Gross–Butler equation which results from the experimental data [Eq. (117)] must therefore have been reduced from

$$k_n = k_0 \frac{(1 - n + n\phi_1^T)(1 - n + n\phi_2^T)}{(1 - n + n\phi_w^R)^2} \qquad (118)$$

where $\phi_w = 1.0$, $\phi_2^T = 1.0$, and $\phi_1^T = 0.46$, thus giving the linear relationship

$$k_n = k_0(1 - n + 0.46n) \qquad (119)$$

8.2. Example II[6]

The reaction is the deacetylation of acetyl-α-chymotrypsin that occurs during the α-chymotrypsin-catalyzed hydrolysis of p-nitrophenyl acetate:

$$\text{enzyme—O—} \overset{\overset{\displaystyle O}{\|}}{\text{C}} \text{—CH}_3 + \text{H}_2\text{O} \xrightarrow[\text{Tris buffer}]{\overset{25°}{\text{pH 7.5}}} \text{enzyme—OH} + \text{CH}_3\text{CO}_2^- \qquad (120)$$

The experimental rate data (Table VI) give a net normal KSIE, $k_{\text{H}_2\text{O}}/k_{\text{D}_2\text{O}} = 2.4$, TSC < RSC. A plot of zero-order rate constants (velocities) vs. n is linear (Fig. 7), suggesting a single-proton mechanism. Since the reactant fractionation factors are all expected to be unity and the $v_n(n)$ plot is linear, the system must be described by

$$k_n = k_0(1 - n + n\phi_1^T) \qquad (121)$$

which, when fitted to the data, gives a value for ϕ_1^T of 0.42.

Table VI

Zero-Order Rate Constants for Deacetylation of
Acetyl-α-chymotrypsin at pH 7.5 and Equivalent
in Mixtures of H_2O and D_2O at $25°$[6]

n	$10^{11}v_n\,(M\ sec^{-1})$
0.000	2649 ± 3
0.175	2425 ± 11
0.261	2239 ± 4
0.398	2126 ± 11
0.485	1878 ± 7
0.497	2064 ± 32
0.583	1791 ± 14
0.745	1479 ± 13
0.765	1459 ± 10
0.995	1102 ± 24

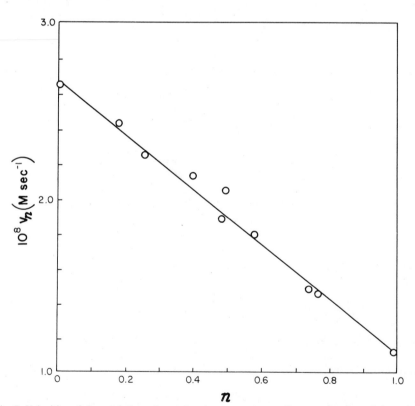

Fig. 7. Velocities of deacetylation of acetyl-α-chymotrypsin vs. the atom fraction of deuterium in the solvent. The data are from Table VI. (Reproduced with permission from E. Pollock *et al.*[6])

By analogy with Example I and other base-catalyzed ester hydrolyses[85] where $k_{H_2O}/k_{D_2O} \simeq 2-3$ and $\phi^T \simeq 0.45$, and taking into account what is known about the active site of this[86] and similar serine proteases, transition state **2**, representing the formation of the tetrahedral intermediate, is reasonable. Again solvation catalysis rather than proton-transfer catalysis by H_1 is as-

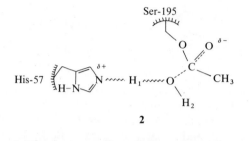

2

sumed to give rise to the isotope effect. Other transition-state structures, for example, **3** or **4**, representing decomposition of the tetrahedral intermediate, are not ruled out by the data, nor is a rate-determining one-proton conformation change on the part of the acyl enzyme.

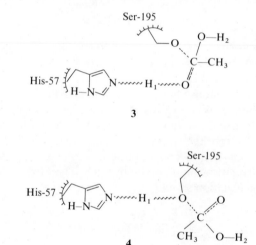

For acetyl enzyme and water going to transition state **2** (or **3** or **4**) the Gross–Butler equation would have the form

$$k_n = k_0 \frac{(1 - n + n\phi_1^T)(1 - n + n\phi_2^T)}{(1 - n + n\phi_w)^2} \tag{122}$$

assuming the fractionation factors for the exchangeable protons in the enzyme all to be unity. With $\phi_w = 1$ and $\phi_2^T = 1$, the final form is that of Eq. (121).

8.3. Example III

Three closely related systems will be considered here. The first reaction is the acylation step,

$$\text{enzyme—OH} + \text{R—}\overset{\overset{\displaystyle O}{\|}}{\text{C}}\text{NH—}\underset{}{\bigcirc}\text{—NO}_2 \xrightarrow[\text{Tris buffer}]{\substack{25° \\ \text{pH} \sim 7.4}} \text{enzyme—O}\overset{\overset{\displaystyle O}{\|}}{\text{C}}\text{R} + \text{H}_3\overset{+}{\text{N}}\text{—}\underset{}{\bigcirc}\text{—NO}_2$$

$$(123)$$

in the trypsin-catalyzed hydrolysis of N-benzoyl-L-phenylalanyl-L-valyl-L-arginyl p-nitroanilide, an analog of a tetrapeptide studied by Quinn *et al.*[10,11] Trypsin, like α-chymotrypsin, is a serine hydrolase with an active site consisting of serine, histidine, and aspartate residues. Both are believed to use a charge-relay mechanism[86] involving coupled transfer of two protons when acting on their physiological polypeptide substrates.

$$R = \overset{\overset{\displaystyle R_1}{|}}{\text{—CH—NH—}}\overset{\overset{\displaystyle O}{\|}}{\text{C}}\text{—}\overset{\overset{\displaystyle R_2}{|}}{\text{CH—NH—}}\overset{\overset{\displaystyle O}{\|}}{\text{C}}\text{—}\overset{\overset{\displaystyle R_3}{|}}{\text{CH—NH—}}\overset{\overset{\displaystyle O}{\|}}{\text{C}}\text{—C}_6\text{H}_5$$

(R_1, R_2, and R_3 are the side-chain groups of the amino acids arginine, valine, and phenylalanine, respectively.)

Under saturation conditions $V_{H_2O}/V_{D_2O} = 4$, and a nonlinear bowed-down k_n vs. n plot is obtained. When $k_n^{1/2}$ is plotted against n a straight line results, implying that the transition state involves two protons, each with an identical fractionation factor. From the slope of the line a value of $\phi_1^T = \phi_2^T = 0.5$ was obtained. Each proton is thus contributing an isotope effect, k_H/k_D, of $1/0.5$ or 2, consistent with solvation catalysis by each. The Gross–Butler equation for this two-proton acylation should be

$$k_n = k_0 \frac{(1 - n + n\phi_1^T)(1 - n + n\phi_2^T)}{(1 - n + n\phi_w)^2} \tag{124}$$

which is reduced to

$$k_n = k_0(1 - n + 0.5n)^2 \tag{125}$$

The results of this experiment are consistent with a charge-relay mechanism as in transition-state structure **5**, involving coupled solvation catalysis by protons 1 and 2. It is of great interest to note that when R is smaller, i.e., less "physiological," for both acylation and deacylation steps of serine hydrolase reactions, the two-proton process uncouples and the mechanism becomes that

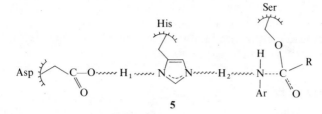

5

of one-proton catalysis [see Example II and entries 5–7 and 9–11 in Table I, all of which give linear $k_n(n)$ plots and KSIE's of ~ 1.5–3].

A recently published report by Hunkapiller *et al.*[73] for a similar substrate, acetyl-L-alanyl-L-prolyl-L-alanyl *p*-nitroanilide, and the serine proteases elastase and α-lytic protease at 25° and pH 8.75 gave entirely analogous results. They report normal net solvent isotope effects of $k_{H_2O}^{cat}/k_{D_2O}^{cat}$ of 2.8 (α-lytic protease) and 2.1 (elastase), with bulging down $k_n(n)$ plots, and linear $(k_n^{cat})^{1/2}$ vs. n plots giving values of $\phi_1^T = \phi_2^T = 0.60$ (α-lytic protease) and 0.68 (elastase). See Fig. 8 for the α-lytic protease data.

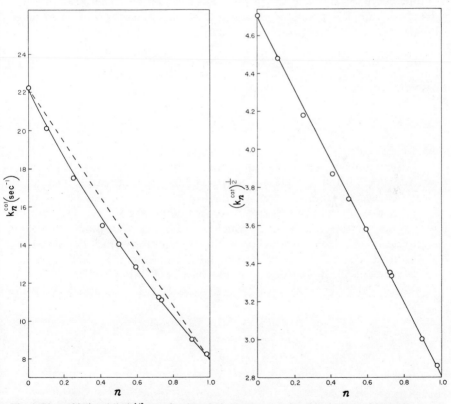

Fig. 8. Plots of k_n^{cat} and $(k_n^{cat})^{1/2}$ vs. n for the α-lytic protease-catalyzed hydrolysis of Ac-L-ala-L-pro-L-ala *p*-nitroanilide (acylation step). (Reproduced with permission from M. W. Hunkapillar.[73])

Their stopped flow kinetic data provide evidence that, for the substrate chosen, the tetrahedral intermediate for the acylation step accumulates, that its decomposition is the rate-limiting step, and that the transition-state structure is analogous to structure **5**.

8.4. Example IV[14]

The hydrolysis of inorganic pyrophosphate catalyzed by magnesium ion and inorganic pyrophosphatase, represented most simply by

$$P_2O_7^{-4} + H_2O \xrightarrow[\substack{pH\ 7 \\ PPase}]{\substack{Mg^{2+} \\ 30}} 2HPO_4^{2-} \tag{126}$$

has been shown by Konsowitz and Cooperman to generate a most interesting example of a nonlinear bowed-up $k_n(n)$ plot (Fig. 9) and a net k_{H_2O}/k_{D_2O} of 1.90. The appearance of Fig. 9 implies that the transition state contains more than one proton and (see Section 7.1) either that (1) a large normal isotope effect generated by one of the protons is being offset by a smaller inverse isotope effect or (2) one of the protons has a ϕ^R of <1. (The possibility of a change in rate-determining step with increasing n was not considered here.) In fact the data fit two two-proton equations about equally well:

$$k_n = k_0(1 - n + 0.37n)(1 - n + 1.37n) \tag{127}$$

describes situation 1 and

$$k_n = k_0 \frac{1 - n + 0.4n}{1 - n + 0.9n} \tag{128}$$

describes situation 2.

Three transition-state structures were proposed consistent with either Eq. (127) or (128) and with present understanding about the enzyme- and phosphoryl-transfer reactions. It was assumed that the isotope effect arises solely from the two water protons. Structure **6** corresponds to Eq. (127). The reactant-state water protons each have the usual $\phi_w = 1.0$. One transition-state proton is transferring and one is attached to an oxygen atom which, because of its ability to delocalize electrons toward the phosphorus atom and to restrict rotation about the P—O bond, should have a ϕ value similar to that for gem-diols and hemiacetals (~ 1.25; see Table III). The value $\phi_2^T = 1.37$ for

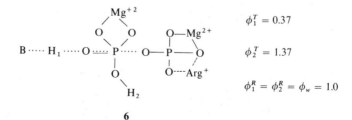

$$\phi_1^T = 0.37$$

$$\phi_2^T = 1.37$$

$$\phi_1^R = \phi_2^R = \phi_w = 1.0$$

6

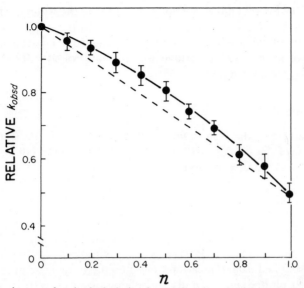

Fig. 9. Relative k_n vs. n for the hydrolysis of pyrophosphate by inorganic pyrophosphatase. (Reproduced with permission from L. M. Konsowitz and B. S. Cooperman.[14])

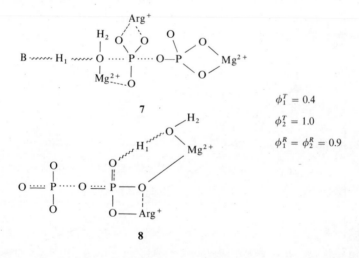

$$\phi_1^T = 0.4$$
$$\phi_2^T = 1.0$$
$$\phi_1^R = \phi_2^R = 0.9$$

a "tighter-than-bulk-solvent" H_2 is quite plausible. Structures **7** and **8** are each consistent with Eq. (128). The reactant water molecule is assumed to be co-ordinated to Mg^{2+}. The resulting partial positive charge on oxygen should give ϕ^R values for the water protons somewhere between

$$1.0\,(H\!-\!O\!-\!H) \quad \text{and} \quad 0.69\,(H\!-\!\overset{+}{O}\!-\!)$$
$$\underset{H}{\big|}$$

The value of 0.9 for ϕ_1^R and ϕ_2^R is reasonable. One of the water protons is seen in both transition states to remain fully attached to the oxygen with little or no charge, thus allowing a value of unity to be assigned to ϕ_2^T. The other proton is seen as a solvation–catalytic bridge, and the value of $\phi_1^T = 0.4$ ($k_H/k_D = 2.5$) that emerges is most satisfying.

8.5. Example V [5]

This example is the isomerization of ribonuclease A, the conversion of the E_1H form of the protein (with $pK_a > 8$ in H_2O) to the conformationally isomeric E_2H form ($pK_a = 6.1$ in H_2O):

$$E_1H \underset{\substack{25° \\ pH\ 7.6}}{\overset{k}{\rightleftharpoons}} E_2H \tag{129}$$

The solvent isotope effect is 4.7 ± 0.4 as measured by Wang et al.[5] The magnitude of this normal effect suggests a primary isotope-effect contribution. The values of k (obtained from relaxation times measured by the temperature-jump method) when plotted against n give a bulging down curve (Fig. 10). Such a

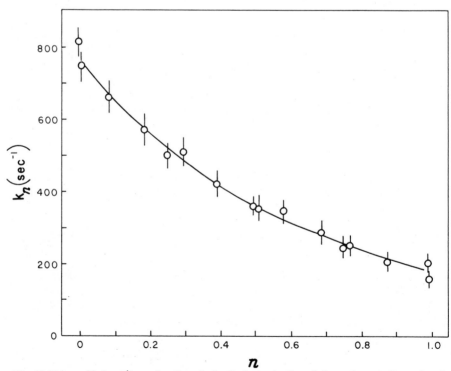

Fig. 10. Values of k_n (sec^{-1}) as a function of n for the isomerization of ribonuclease A. (Reproduced with permission from M.-S. Wang et al.[5]

curve is clearly indicative of a multiproton transition state. In treating the data it was first assumed that the ϕ_j^R's are equal to unity or do not change on activation. The 16 data points of Fig. 10 were then subjected to polynomial regression and nonlinear least-squares fits to polynomials [Eq. (113)] of linear, quadratic, cubic, and higher order. The best fit of the data is to the cubic polynomial:

$$k_n(\text{sec}^{-1}) = 772 + (-1248 \pm 144)\,n + (1082 \pm 354)\,n^2 + (-440 \pm 234)\,n^3 \tag{130}$$

The linear and quadratic terms were shown to be statistically significant at the 99.9% confidence limit by the F test, and the cubic terms at the 90% confidence limit. Arguments based on the magnitude of the KSIE, data from the earlier work on this system by French and Hammes,[4] and other considerations led the authors to conclude that the most reasonable rate-limiting step is the ionization of a histidinium residue,

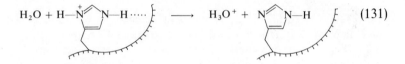

$$\tag{131}$$

prior to a conformational change. The transition state for this ionization would be expected to resemble structure **9**. The Gross–Butler equation for

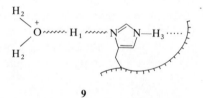

9

such a process should take the form

$$k_n = k_0 \frac{(1 - n + n\phi_1^T)(1 - n + n\phi_2^T)^2(1 - n + n\phi_3^T)}{(1 - n + n\phi_1^R)(1 - n + n\phi_w)^2(1 - n + n\phi_3^R)} \tag{132}$$

For the transfer of a proton between bases of very different basicity, the transition state is expected to be asymmetric, with the transferring proton closer to the weaker base.[87] For transition state **9**, then, the hydronium ion is expected to be almost fully formed, with $\phi_2^T \simeq 0.69$ (for H_3O^+). Still assuming all $\phi_j^R = 1.0$ and setting $\phi_3^T = 1.0$ [since $\phi(\overset{+}{N}—H) = \phi(N—H) = 1.0$, Table III], we obtain a simplified equation:

$$k_n = k_0(1 - n + n\phi_1^T)(1 - n + 0.69n)^2 \tag{133}$$

A test for the correctness of this equation was performed by plotting

$k_n/(1 - n + 0.69n)^2$ vs. n. The resulting plot was linear, giving a value of k_0 consistent with experiment and a ϕ_1^T value of 0.46 for the transferring proton H_1 in **9**. This corresponds to a k_H/k_D value of 2.2 for that proton, precisely what is expected for such an asymmetric transition state. The data thus are consistent with transition-state structure **9**, and the most likely fractionation factor values are $\phi_1^T = 0.46$, $\phi_2^T = 0.69$, $\phi_3^T = 1.0$, and $\phi_1^R = \phi_w = \phi_3^R = 1.0$.

8.6. Example VI [2]

This example deals with rate and equilibrium effects in subunit association of formyltetrahydrofolate synthetase. Contain monovalent cations (K^+, Cs^+, NH_4^+) bind to the inactive monomeric form of formyltetrahydrofolate synthetase (from *Clostridium cylindrosporum*), resulting in the formation of an active tetramer containing two of the cations:

$$4M + 2C^+ \underset{pH \sim 8}{\overset{K}{\underset{20°}{\rightleftharpoons}}} M_4C_2^+ \tag{134}$$

Both rate and equilibrium constants of association were found in the studies of Harmony and Himes to be increased in D_2O with $k_{D_2O}/k_{H_2O} = 3.3$ and $K_{D_2O}/K_{H_2O} \simeq 50$. It was shown that the rate effect came solely from a step in which cation and monomer rapidly and reversibly combine:

$$M + C^+ \underset{pH \sim 8}{\overset{k}{\underset{20°}{\rightleftharpoons}}} MC^+ \tag{135}$$

[The rates were measured at the optimum pL by using a buffer (Tris) with a ΔpK value close to the ΔpH of ~ 0.4 found from the experimental pH-rate profiles shown in Fig. 11. See Section 6.2.] When plotted vs. n the rate data (Table VII) give a curve (Fig. 12) going up to the right (inverse isotope effect,

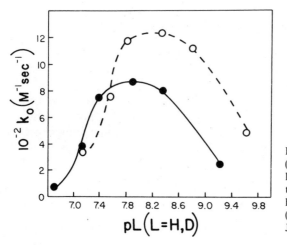

Fig. 11. Dependence on pL in H_2O (solid line) and in 90% D_2O (dashed line) of the rate of K^+-induced tetramerization of formyltetrahydrofolate synthetase monomers. (Reproduced with permission from J. A. K. Harmony *et al.* [2])

Table VII
Rate and Equilibrium Constants for the
Potassium-Ion-Induced Tetramerization of
Formyltetrahydrofolate Synthetase
Subunits at 20° and pH 7.9 (or Equivalent)
in Mixtures of H_2O and $D_2O^{(2)}$

n	$k_n\,(M^{-1}\sec^{-1}) \times 10^{-3}$	$K\,(M^{-3}) \times 10^{-19}$
0.00	2.45	0.04
0.10		0.05
0.20	2.90	0.07
0.25		0.08
0.35	3.35	0.10
0.40		0.12
0.50	3.88	0.16
0.60		0.27
0.70	4.51	0.32
0.80	5.60	0.66
0.90	7.11	1.25
0.99	8.60	1.60

TSC > RSC) and bulging down, thus suggesting a multiproton mechanism. Any of the following multiproton possibilities (it will be recalled from Section 7.1) can give rise to curves of the shape of Fig. 12: (1) large inverse isotope effect(s) partially offset by smaller normal effect(s), (2) opposing reactant- and transition-state contributions (for this case of an inverse isotope effect both ϕ^T and ϕ^R would be >1 with $\phi^T > \phi^R$), and (3) a change in rate-determining step with n. A fourth possibility, always to be considered, is that the entire contribution may come from one or more reactant-state protons such that all ϕ_j^T's are unity and that (for this case of a net inverse isotope effect where RSC < TSC) one or more $\phi_i^R < 1$. It was found that a plot of $1/k_n$ vs. n was linear (see Fig. 12) with $\phi^R = 0.33$,

$$k_n^{-1} = k_0^{-1}(1 - n + n\phi_1^R) \tag{136}$$

clear evidence for a one-proton mechanism corresponding to the fourth possibility just cited. Thus the simplest (but in the absence of more information about the enzyme not necessarily the only or the correct) mechanistic model is one involving a single loosely bound reactant proton of $\phi_1^R = 0.33$ being expelled from a cationic binding site and changing in the transition state to a more normal proton of $\phi_1^T = 1.0$ as the cation binds, for example,

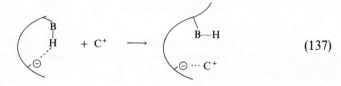

$$\tag{137}$$

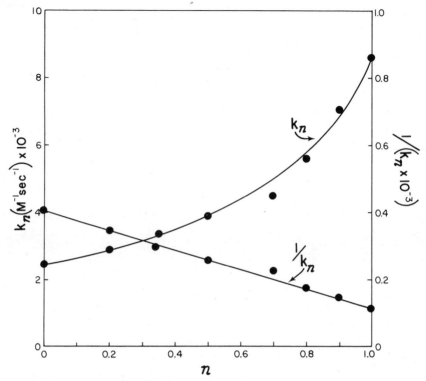

Fig. 12. Plot of k_n and $1/k_n$ vs. n for formyltetrahydrofolate synthetase subunit association. Data from Table VII. (Reproduced with permission from J. A. K. Harmony *et al.*[2])

When the equilibrium constant K for Eq. (134) is plotted vs. n a curve going up to the right and bulging down results. One can imagine a process analogous to that just proposed to account for the rate data: Each of the two cations which bind the tetramer do so by occupying simultaneously two cationic binding sites (one on each monomer). The binding is accomplished by the expulsion from each site of one of the protons of $\phi^R \simeq 0.3$ proposed above. The equilibrium solvent isotope effect would thus arise from the conversion of four reactant protons of $\phi^R \simeq 0.3$ to four product protons of $\phi^P \simeq 1.0$:

$$K_n = K_0 \frac{(1 - n + 1.0n)^4}{(1 - n + n\phi^R)^4} \tag{138}$$

$$K_n^{-1} = K_0^{-1}(1 - n + n\phi^R)^4 \tag{139}$$

A plot of $(K_n^{-1})^{1/4}$ vs. n is linear, generating a value for ϕ^R of 0.36, which agrees well with the 0.33 obtained from the kinetic data, and giving a value of $K_{D_2O}/K_{H_2O} \simeq 60$.

At this point there is no proof that the subunit-association SIE's arise as suggested above. Nevertheless, this example shows how such results can be analyzed and interpreted and that very large SIE's are easily accounted for by reasonable chemical mechanisms.

It is to be hoped that the preceding six examples and the chapter as a whole will prove both informative and useful in the determination of biochemical transition-state structures. To date relatively few instances of the application of the proton inventory–solvent isotope technique exist, yet their variety and the results obtained indicate the power of the method. It will indeed be interesting to see further biologically interesting mechanisms emerge as a result of its use.

REFERENCES

1. J. J. Katz and H. L. Crespi, in: *Isotope Effects in Chemical Reactions* (C. J. Collins and N. S. Bowman, eds.), pp. 286–363, Van Nostrand Reinhold, New York (1970).
2. J. A. K. Harmony, R. H. Himes, and R. L. Schowen, The monovalent cation-induced association of formyltetrahydrofolate synthetase subunits: A solvent isotope effect, *Biochemistry* **14**, 5379–5386 (1975).
3. L. L. Houston, J. Odell, Y. C. Lee, and R. H. Himes, Solvent isotope effects on microtubule polymerization and depolymerization, *J. Mol. Biol.* **87**, 141–146 (1974).
4. T. C. French and G. G. Hammes, Relaxation spectra of ribonuclease. II. Isomerization of ribonuclease at neutral pH values, *J. Am. Chem. Soc.* **87**, 4669–4673 (1965).
5. M.-S. Wang, R. D. Gandour, J. Rodgers, J. L. Haslam, and R. L. Schowen, Transition-state structure for a conformation change of ribonuclease, *Bioorg. Chem.* **4**, 392–406 (1975).
6. E. Pollock, J. L. Hogg, and R. L. Schowen, One-proton catalysis in the deacetylation of acetyl-α-chymotrypsin, *J. Am. Chem. Soc.* **95**, 968 (1973).
7. M. L. Bender, G. E. Clement, F. J. Kézdy, and H. D'A. Heck, The correlation of the pH (pD) dependence and the stepwise mechanism of α-chymotrypsin-catalyzed reactions, *J. Am. Chem. Soc.* **86**, 3680–3690 (1964).
8. (a) J. P. Elrod, R. D. Gandour, J. L. Hogg, M. Kise, G. M. Maggiora, R. L. Schowen, and K. S. Venkatasubban, Proton bridges in enzyme catalysis, *Faraday Symp. Chem. Soc.* **10**, 145–153 (1975). (b) J. P. Elrod, Comparative mechanistic study of a set of serine hydrolases using the proton inventory technique and beta-deuterium probe, Ph.D. thesis, University of Kansas, Lawrence (1975).
9. R. Mason and C. A. Ghiron, An isotope effect on the esterase and protease activity of trypsin, *Biochim. Biophys. Acta* **51**, 377–378 (1961).
10. R. L. Schowen, in: *Isotope Effects on Enzyme Catalyzed Reactions* (W. W. Cleland, M. H. O'Leary, and D. B. Northrup, eds.), pp. 64–99, University Park Press, Baltimore (1977).
11. D. M. Quinn, M. Patterson, R. Jarvis, G. Ranney, and R. L. Schowen, Changes in catalytic coupling in the action of amidohydrolases and serine hydrolases with changes in substrate structure (paper in preparation).
12. G. M. Maggiora and R. L. Schowen, in: *Bioorganic Chemistry* (E. E. van Tamelen, ed.), Vol. 1, pp. 173–229, Academic Press, New York (1977).
13. K. S. Venkatasubban, Proton bridges in enzymic and nonenzymic amide solvolysis, Ph.D. thesis, University of Kansas, Lawrence (1975).
14. L. M. Konsowitz and B. S. Cooperman, Solvent isotope effect in inorganic pyrophosphatase-catalyzed hydrolysis of inorganic pyrophosphate, *J. Am. Chem. Soc.* **98**, 1993–1995 (1976).

15. J. Bigeleisen and M. Wolfsberg, Theoretical and experimental aspects of isotope effects in chemical kinetics, *Adv. Chem. Phys.* **1**, 15–76 (1958).

16. P. M. Laughton and R. E. Robertson, in: *Solute–Solvent Interactions* (J. F. Coetzee and C. D. Ritchie, eds.), pp. 400–538, Dekker, New York (1969).

17. E. M. Arnett and D. R. McKelvey, in: *Solute–Solvent Interactions* (J. F. Coetzee and C. D. Ritchie, eds.), pp. 344–399, Dekker, New York (1969).

18. V. Gold, in: *Advances in Physical Organic Chemistry* (V. Gold, ed.), Vol. 7, pp. 259–331, Academic Press, New York (1969).

19. V. Gold, in: *Hydrogen-Bonded Solvent Systems* (A. K. Covington and P. Jones, eds.), pp. 295–300, Taylor and Francis, Ltd., London (1968).

20. R. L. Schowen, Mechanistic deductions from solvent isotope effects, *Prog. Phys. Org. Chem.* **9**, 275–332 (1972).

21. E. K. Thornton and E. R. Thornton, in: *Isotope Effects in Chemical Reactions* (C. J. Collins and N. S. Bowman, eds.), pp. 213–286, Van Nostrand Reinhold, New York (1970).

22. A. J. Kresge, Solvent isotope effect in H_2O–D_2O mixtures, *Pure Appl. Chem.* **8**, 243–258 (1964).

23. J. Albery, in: *Proton-Transfer Reactions* (E. Caldin and V. Gold, eds.), pp. 263–315, Chapman & Hall, London (1975).

24. K. B. Wiberg, The deuterium isotope effect, *Chem. Rev.* **55**, 713–743 (1955).

25. V. Gold and D. P. N. Satchell, The principles of hydrogen isotope exchange reactions in solution, *Q. Rev. Chem. Soc.* **9**, 51–72 (1955).

26. R. F. W. Bader, Ph.D. thesis in Organic Chemistry, Massachusetts Institute of Technology, Cambridge (1958).

27. R. A. More O'Ferrall, G. W. Koeppl, and A. J. Kresge, Solvent isotope effects upon proton transfer from the hydronium ion, *J. Am. Chem. Soc.* **93**, 9–20 (1971).

28. R. L. Kay and D. F. Evans, The conductance of the tetraalkylammonium halides in deuterium oxide solutions at 25°, *J. Phys. Chem.* **69**, 4216–4221 (1965).

29. (a) C. A. Bunton and V. J. Shiner, Acid–base equilibria in deuterium oxide solutions, *J. Am. Chem. Soc.* **83**, 42–47 (1961). (b) C. A. Bunton and V. J. Shiner, Isotope effects in deuterium oxide solution. Part II. Reaction rates in acid, alkaline and neutral solution, involving only secondary solvent effects, *J. Am. Chem. Soc.* **83**, 3207–3214 (1961). (c) C. A. Bunton and V. J. Shiner, Isotope effects in deuterium oxide solution. Part III. Reactions involving primary effects, *J. Am. Chem. Soc.* **83**, 3214–3220 (1961).

30. (a) C. G. Swain and R. F. W. Bader, The nature of the structure difference between light and heavy water and the origin of the solvent isotope effect, *Tetrahedron* **10**, 182–199 (1960). (b) C. G. Swain, R. F. W. Bader, and E. R. Thornton, A theoretical interpretation of isotope effects in mixtures of light and heavy water, *Tetrahedron* **10**, 200–211 (1960). (c) C. G. Swain and E. R. Thornton, Calculated isotope effects for reactions of lyonium ion in mixtures of light and heavy water, *J. Am. Chem. Soc.* **83**, 3884–3889 (1961). (d) C. G. Swain and E. R. Thornton, Calculated isotope effects for reactions of lyoxide ion or water in mixtures of light and heavy water, *J. Am. Chem. Soc.* **83**, 3890–3896 (1961).

31. F. Westheimer, The magnitude of the primary kinetic isotope effect for compounds of hydrogen and deuterium, *Chem. Rev.* **61**, 265–273 (1961).

32. J. Bigeleisen, Correlation of kinetic isotope effects with chemical bonding in three-centre reactions, *Pure Appl. Chem.* **8**, 217–242 (1964).

33. R. P. Bell, Isotope effects and the nature of proton-transfer transition states, *Discuss. Faraday Soc.* **39**, 16–24 (1965).

34. A. V. Willi and M. Wolfsberg, The influence of "bond making and bond breaking" in the transition state on hydrogen isotope effects in linear three-centre reactions, *Chem. Ind. (London)* **1964**, 2097–2098.

35. R. A. More O'Ferrall and J. Kouba, Model calculations of primary hydrogen isotope effects, *J. Chem. Soc. B* **1967**, 985–990.

36. W. J. Albery, Isotope effects in proton transfer reactions, *Trans. Faraday Soc.* **63**, 200–206 (1967).

37. R. A. More O'Ferrall, in: *Proton-Transfer Reactions* (E. Caldin and V. Gold, eds.), pp. 201–261, Chapman & Hall, London (1975).
38. R. E. Weston, Jr., Transition-state models and hydrogen isotope effects, *Science* **158**, 332–342 (1967).
39. M. J. Goldstein, Kinetic isotope effects and organic reaction mechanisms, *Science* **154**, 1616–1621 (1966).
40. A. V. Willi, Predictions of isotope effects in S_N2 reactions, *Z. Phys. Chem. (Frankfurt am Main)* **66**, 317–328 (1969).
41. (a) M. Wolfsberg and M. J. Stern, Validity of some approximation procedures used in the theoretical calculation of isotope effects, *Pure Appl. Chem.* **8**, 225–242 (1964). (b) M. Wolfsberg and M. J. Stern, Secondary isotope effects as probes for force constant changes, *Pure Appl. Chem.* **8**, 325–338 (1964).
42. M. J. Stern and M. Wolfsberg, Simplified procedure for the theoretical calculation of isotope effects involving large molecules, *J. Chem. Phys.* **45**, 4105–4124 (1966).
43. P. Salomaa, Solvent deuterium isotope effects on acid–base reactions. Part I. Thermodynamic theory and its simplifications, *Acta Chem. Scand.* **23**, 2095–2106 (1969).
44. P. Salomaa, A. Vesala, and S. Vesala, Solvent deuterium isotope effects on acid–base reactions. Part II, Variation of the first and second acidity constants of carbonic and sulfurous acids in mixtures of light and heavy water, *Acta Chem. Scand.* **23**, 2107–2115 (1969).
45. V. Gold, Rule of the geometric mean: Its role in the treatment of thermodynamic and kinetic deuterium solvent isotope effects, *Trans. Faraday Soc.* **64**, 2770–2779 (1968).
46. W. J. Albery and M. H. Davies, Effect of the breakdown of the rule of the geometric mean on fractionation factor theory, *Trans. Faraday Soc.* **65**, 1059–1065 (1969).
47. C. G. Swain, D. A. Kuhn, and R. L. Schowen, Effect of structural changes in reactants on the position of hydrogen—bonding hydrogens and solvating molecules in transition states. The mechanism of tetrahydrofuran formation from 4-chlorobutanol, *J. Am. Chem. Soc.* **87**, 1553–1561 (1965).
48. P. Salomaa, L. L. Schaleger, and F. A. Long, Solvent deuterium isotope effects on acid–base equilibria, *J. Am. Chem. Soc.* **86**, 1–7 (1964).
49. A. K. Covington, R. A. Robinson, and R. G. Bates, The ionization constant of deuterium oxide from 5 to 50°, *J. Phys. Chem.* **70**, 3820–3824 (1966).
50. V. Gold and B. M. Lowe, Measurement of solvent isotope effects with the glass electrode. Part I. The ionic product of D_2O and D_2O–H_2O mixtures, *J. Chem. Soc. A* **1967**, 936–943.
51. M. Goldblatt and W. M. Jones, Ionization constants of T_2O and D_2O at 25° from cell emf's. Interpretation of the hydrogen isotope effects in emf's, *J. Chem. Phys.* **51**, 1881–1894 (1969).
52. L. Pentz and E. R. Thornton, Isotope effects on the basicity of 2-nitrophenoxide, 2,4-dinitrophenoxide, hydroxide, and imidazole in protium oxide–deuterium oxide mixtures, *J. Am. Chem. Soc.* **89**, 6931–6938 (1967), and references cited.
53. V. Gold, The fractionation of hydrogen isotopes between hydrogen ions and water, *Proc. Chem. Soc. London* **1963**, 141–143.
54. A. J. Kresge and A. L. Allred, Hydrogen isotope fractionation in acidified solutions of protium and deuterium oxide, *J. Am. Chem. Soc.* **85**, 1541 (1963).
55. J. F. Mata-Segreda, Chemical models for aqueous biodynamical processes, Ph.D. thesis, University of Kansas, Lawrence (1975).
56. J. E. Leffler, Parameters for the description of transition states, *Science* **117**, 340–341 (1953).
57. G. A. Vidulich, D. F. Evans, and R. L. Kay, The dielectric constant of water and heavy water between 0 and 40°, *J. Phys. Chem.* **71**, 656–662 (1967).
58. V. K. LaMer and J. P. Chittum, The conductance of salts (potassium acetate) and the dissociation constant of acetic acid in deuterium oxide, *J. Am. Chem. Soc.* **58**, 1642–1644 (1936).
59. (a) P. Gross, H. Steiner, and F. Krauss, On the decomposition of diazoacetic ester catalysed by protons and deuterons, *Trans. Faraday Soc.* **32**, 877–879 (1936). (b) P. Gross and A. Wischin, On the distribution of picric acid between benzene and mixtures of light and heavy water, *Trans. Faraday Soc.* **32**, 879–883 (1936). (c) P. Gross, H. Steiner, and H. Suess, The inversion of cane sugar in mixtures of light and heavy water, *Trans. Faraday Soc.* **32**, 883–889 (1936).

60. (a) J. C. Hornel and J. A. V. Butler, The rates of some acid- and base-catalysed reactions, and the dissociation constants of weak acids in "heavy" water, *J. Chem. Soc.* **1936**, 1361–1366. (b) W. J. C. Orr and J. A. V. Butler, The kinetic and thermodynamic activity of protons and deuterons in water–deuterium oxide solutions, *J. Chem. Soc.* **1937**, 330–335. (c) W. E. Nelson and J. A. V. Butler, Experiments with heavy water on the acid hydrolysis of esters and the alkaline decomposition of diacetone alcohol, *J. Chem. Soc.* **1938**, 957–962.

61. C. R. Hopper, R. L. Schowen, K. S. Venkatasubban, and H. Jayaraman, Proton inventories of the transition states for solvation catalysis and proton-transfer catalysis. Decomposition of the tetrahedral intermediate in amide methanolysis, *J. Am. Chem. Soc.* **95**, 3280–3283 (1973).

62. J. Elrod, R. D. Gandour, M. Hegazi, J. L. Hogg, J. Mata, D. Quinn, K. B. Schowen, R. L. Schowen, and K. S. Venkatasubban, Notes on the proton inventory technique, a handbook prepared by the Bio-organic Chemical Dynamics Group, Department of Chemistry, University of Kansas, Lawrence (1974).

63. K. Mikkelsen and S. O. Nielsen, Acidity measurements with the glass electrode in H_2O and D_2O mixtures, *J. Phys. Chem.* **64**, 632–637 (1960).

64. P. K. Glasoe and F. A. Long, Use of glass electrodes to measure acidities in deuterium oxide, *J. Phys. Chem.* **64**, 188–191 (1960).

65. A. K. Covington, M. Paabo, R. A. Robinson, and R. G. Bates, Use of the glass electrode in deuterium oxide and the relation between the standardized pD (pa_D) scale and the operational pH in heavy water, *Anal. Chem.* **40**, 700–706 (1968).

66. K. B. J. Schowen, J. K. Lee, and R. L. Schowen, Dynamical mechanism of action of the glass electrode, unpublished paper (No. 121) presented at the 10th Midwest Regional Meeting of the American Chemical Society, University of Iowa, Iowa City, November 7–8, 1974.

67. W. J. Dixon, ed., *Biomedical Computer Programs*, University of California Press, Berkeley and Los Angeles (1968), pp. 289–296, Program BMDO5R Polynomial Regression.

68. C. K. Rule and V. K. LaMer, Dissociation constants of deutero acids by e.m.f. measurements, *J. Am. Chem. Soc.* **60**, 1974–1981 (1938).

69. R. P. Bell, *The Proton in Chemistry*, Cornell University Press, Ithaca, N.Y. (1959), Chap. XI.

70. E. Halevi, F. A. Long, and M. A. Paul, Acid–base equilibria in solvent mixtures of deuterium oxide and water, *J. Am. Chem. Soc.* **83**, 305–311 (1961).

71. (a) R. G. Bates, *Determination of pH: Theory and Practice*, 2nd ed., Wiley-Interscience, New York (1973), Chap. 8, pp. 211–253. (b) M. Paabo and R. G. Bates, Standards for a practical scale of pD in heavy water, *Anal. Chem.* **41**, 283–285 (1969).

72. K. B. Schowen, E. E. Smissman, and W. F. Stephen, Jr., Base–catalyzed and cholinesterase-catalyzed hydrolysis of acetylcholine and optically active analogs, *J. Med. Chem.* **18**, 292–300 (1975).

73. M. W. Hunkapiller, M. D. Forgac, and J. H. Richards, Mechanism of action of serine proteases: Tetrahedral intermediate and concerted proton transfer, *Biochemistry* **15**, 5581–5588 (1976).

74. V. Gold and M. A. Kessick, Proton transfer to olefins, *Discuss. Faraday Soc.* **39**, 84–93 (1965).

75. M. D. Zeidler, in: *Water: A Comprehensive Treatise* (F. Franks, ed.), Vol. 2, pp. 529–584, Plenum Press, New York (1973); see pp. 540–541.

76. A. Hvidt and S. O. Nielsen, in: *Advances in Protein Chemistry* (C. B. Anfinsen, Jr., M. L. Anson, J. T. Edsall, and F. M. Richards, eds.), Vol. 21, pp. 287–386, Academic Press, New York (1966).

77. F. M. Richards and H. W. Wyckoff, in: *The Enzymes*, 3rd ed. (P. D. Boyer, ed.), Vol. 4, pp. 647–806, Academic Press, New York (1971).

78. J. S. Scarpa, D. D. Mueller, and I. M. Klotz, Slow hydrogen–deuterium exchange in a non-α-helical polyamide, *J. Am. Chem. Soc.* **89**, 6024–6030 (1967).

79. M. L. Bender and G. A. Hamilton, Kinetic isotope effects of deuterium oxide on several α-chymotrypsin-catalyzed reactions, *J. Am. Chem. Soc.* **84**, 2570–2576 (1962).

80. G. Robillard and R. G. Shulman, High resolution nuclear magnetic resonance study of the histidine–aspartate hydrogen bond in chymotrypsin and chymotrypsinogen, *J. Mol. Biol.* **71**, 507–511 (1972).

81. R. L. Schowen, Chapter 2 in this volume.
82. A. J. Kresge, Solvent isotope effects and the mechanism of chymotrypsin action, *J. Am. Chem. Soc.* **95**, 3065–3067 (1973).
83. R. A. Fisher and F. Yates, *Statistical Tables for Biological, Agricultural and Medical Research,* 5th ed., Oliver & Boyd, Edinburgh (1957), pp. 47–55.
84. S. S. Minor and R. L. Schowen, One-proton solvation bridge in intramolecular carboxylate catalysis of ester hydrolysis, *J. Am. Chem. Soc.* **95**, 2279–2281 (1973).
85. S. L. Johnson, General base and nucleophilic catalysis of ester hydrolysis and related reactions, *Adv. Phys. Org. Chem.* **5**, 237–330 (1967).
86. D. M. Blow, J. J. Birktoft, and B. S. Hartley, Role of a buried acid group in the mechanism of action of chymotrypsin, *Nature (London)* **221**, 337–340 (1969).
87. G. S. Hammond, A correlation of reaction rates, *J. Am. Chem. Soc.* **77**, 334–338 (1955).

Heavy-Atom Isotope Effects in Enzyme-Catalyzed Reactions

Marion H. O'Leary

1. INTRODUCTION

Various methods are used for studies of enzyme mechanisms. Heavy-atom isotope effects are unique among kinetic methods in the extent to which they provide specific information about the rate-determining step* and other relatively slow steps in the mechanism. Because of this specificity, they are complementary to the usual steady-state and rapid-kinetics techniques.

Both the measurement and the interpretation of heavy-atom isotope effects must be approached with caution. Substitution of, for example, carbon-13 for carbon-12 will cause a change in rate of no more than a few percent. Consequently, highly sophisticated experimental methods are necessary for determination of heavy-atom isotope effects. The usual precision of a few percent encountered in enzyme kinetic measurements is enough to obscure a heavy-atom isotope effect altogether. Interpretation of these effects frequently rests on small differences and requires a clear understanding of the various factors which determine the magnitudes of isotope effects.

Since the 1940's, heavy-atom isotope effects have been studied by physical chemists interested in theories of chemical reactions and by organic chemists interested in reaction mechanisms.[1-5] Interest of biochemists has lagged because of experimental and theoretical difficulties, but activity in this field

---

* It is important to define the term "rate-determining step" in the context of isotope effects. For the present purpose, the rate-determining step is the reaction step whose transition state is of the highest free energy. Thus, it is the rate of passage through this step which will ultimately control the rate of the reaction.

Marion H. O'Leary • Department of Chemistry, University of Wisconsin, Madison, Wisconsin.

is increasing. Several previous reviews have been devoted to isotope effects in enzymic reactions.[6-8]

2. THE MAGNITUDES OF HEAVY-ATOM ISOTOPE EFFECTS IN ORGANIC REACTIONS*

Heavy-atom isotope effects, like other isotope effects, are interpreted in terms of the structure difference between the ground state and the transition state for the reaction in question.[1-4,9] To a good approximation, one need consider only the atoms near the isotopic atom; long-range effects are small. An isotope effect is likely to be observed only to the extent that the isotopic atom undergoes a change in bonding on going from ground state to transition state.

The theoretical prediction of heavy-atom isotope effects from first principles is a subject which has occupied numerous investigators. The procedure used is to compare an experimental isotope effect with isotope effects predicted for various possible transition-state structures. However, for enzyme-catalyzed reactions such a procedure is usually of little value, and we shall not consider it further. Instead, we shall compare isotope effects obtained in enzyme-catalyzed reactions with those obtained in similar organic reactions. Assuming that the mechanisms of the latter reactions are known, this comparison can provide information about mechanisms of enzymic reactions.

2.1. Carbon Isotope Effects

Isotope effects for ^{13}C are usually in the range $k^{12}/k^{13} = 1.00-1.07$ near room temperature. Like all heavy-atom isotope effects, they become smaller at higher temperatures, although the precise nature of this temperature dependence is still a matter of discussion.[10,11] The comparison of ^{13}C isotope effects with those for ^{14}C is given by[12]

$$k^{12}/k^{14} - 1 = 1.9(k^{12}/k^{13} - 1) \tag{1}$$

Decarboxylation reactions have been studied extensively. The decarboxylation of malonic acid proceeds through a cyclic transition state

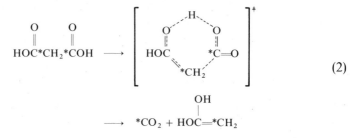

$$(2)$$

---

* Throughout this section, an asterisk on an atom indicates a position of isotopic substitution.

In a subsequent step the enol produced above is converted to acetic acid. Decarboxylation of carboxyl-^{14}C-malonic acid has a carbon isotope effect $k^{12}/k^{14} = 1.064$ at 154°.[13] The methylene carbon shows a carbon isotope effect of 1.076 under the same conditions.[13] Both these isotope effects are typical of those observed when extensive bond change at the isotopic atom is occurring at the transition state. The fact that a large isotope effect is observed for the methylene carbon even though the bond breaking at that site is to some extent being compensated by bond formation indicates that this carbon is much less strongly bound at the transition state than in the ground state.

Phenylmalonic acid shows a carboxyl carbon isotope effect $k^{12}/k^{14} = 1.088$ at 163°.[14] The larger isotope effect in this case is observed because the phenyl group makes the transition state more enol-like.

The decarboxylation of azulene-1-carboxylic acid occurs in two steps: The first is protonation of the aromatic ring, and the second is decarboxylation:

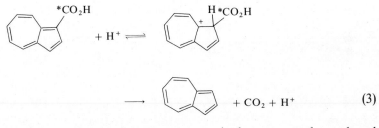

$$\longrightarrow \quad + CO_2 + H^+ \tag{3}$$

Protonation of the aromatic ring may occur whether or not the carboxyl group is protonated. However, decarboxylation can occur only from the ring-protonated intermediate in which the carboxyl group is ionized. Under conditions of high acidity, decarboxylation is rate determining, and $k^{12}/k^{13} = 1.043$ at 25°. At low acidity, protonation is rate determining, and the carbon isotope effect is 1.008.[15]

Carbon isotope effects have also proved useful in distinguishing between nucleophilic substitution reactions occurring by the S_N1 and by the S_N2 mechanism. The methanolysis of 1-bromo-1-phenylethane proceeds by way of a carbonium-ion intermediate:

$$\underset{\text{Br}}{\text{Ph–*CH–CH}_3} \longrightarrow [\text{Ph–*CH–CH}_3]^+ + \text{Br}^-$$

$$\xrightarrow{\text{CH}_3\text{OH}} \underset{\text{OCH}_3}{\text{Ph–*CH–CH}_3} + \text{HBr} \tag{4}$$

The rate-determining step is formation of the carbonium-ion intermediate, and the transition state is presumably very similar in structure to the intermediate. The reaction shows a carbon isotope effect $k^{12}/k^{13} = 1.0065$ at 25°.[16] Substituent effects and temperature effects in this reaction have also been reported.[16-18] The small size of the carbon isotope effect indicates that although a carbon–bromine bond is broken in the transition state, carbon–carbon bonds are actually strengthened, thus offsetting the bond-breaking effect.

Substitution reactions which proceed by S_N2 mechanisms show larger carbon isotope effects than those in S_N1 reactions. The reaction of 1-bromo-1-phenylethane with methoxide,

$$Ph-*CH-CH_3 + {}^-OCH_3 \longrightarrow \left[\begin{array}{c} Ph \\ | \\ CH_3O{-}{-}{-}{-}*CH{-}{-}{-}{-}Br \\ | \\ CH_3 \end{array} \right]^{\ddagger}$$

(5)

$$\longrightarrow Ph-*\overset{\displaystyle OCH_3}{\underset{}{C}}H-CH_3 + Br^-$$

shows an isotope effect $k^{12}/k^{13} = 1.032$ at 25° in methanol.[19]

2.2. Chlorine Isotope Effects

Chlorine isotope effects can also be used to distinguish between S_N1 reactions and S_N2 reactions. However, unlike carbon isotope effects, chlorine isotope effects are larger for S_N1 than for S_N2 reactions. The methanolysis of *t*-butyl chloride, known to be an S_N1 reaction, shows a chlorine isotope effect $k^{35}/k^{37} = 1.0106$ at 20°.[20] The S_N2 reaction of *n*-butyl chloride with phenylmercaptide ion in methanol shows an isotope effect $k^{35}/k^{37} = 1.0089$ at 20°.[20] The difference between the isotope effects observed in the two mechanisms occurs because the carbon–chlorine bond is more completely broken at the transition state in the S_N1 than in the S_N2 reaction. For bond-breaking reactions, the greater the decrease in bonding to the leaving group at the transition state, the greater the observed isotope effect. A number of other chlorine isotope effects have been reported.[21,22]

2.3. Oxygen Isotope Effects

Heavy-atom isotope effects have been used to probe a variety of reactions of carboxylic acid derivatives. A key question in such studies is invariably the role of one or more tetrahedral intermediates such as, for reaction of an ester with a nucleophile (N),

$$\begin{array}{c} OH \\ | \\ RC-OR' \\ | \\ N \end{array}$$

The isotope effects observed in such reactions depend on whether the formation of the intermediate or its decomposition has the transition state of higher free energy. If the first step is entirely rate determining, no leaving-group isotope effect is expected.* If the second step is entirely rate determining, a leaving-group isotope effect of the normal magnitude may be expected, and carbonyl oxygen exchange may be observed simultaneously if water or hydroxide is the nucleophile. If neither formation nor decomposition of the tetrahedral intermediate is entirely rate limiting, a leaving-group isotope effect may be observed, but it will be smaller than the actual isotope effect on the second step because of the slowness of the first step.

Oxygen isotope effects on reactions of methyl formate with a number of nucleophiles have been reported by Sawyer and Kirsch.[23] Reaction of methyl formate with hydrazine in aqueous solution at 25°, pH 7.85,

$$
\underset{\displaystyle \text{HC---*OCH}_3 + \text{H}_2\text{NNH}_2}{\overset{\displaystyle \text{O} \atop \displaystyle \|}{}} \longrightarrow \underset{\displaystyle \text{HC---NHNH}_2 + \text{H*OCH}_3}{\overset{\displaystyle \text{O} \atop \displaystyle \|}{}} \qquad (6)
$$

shows a large isotope effect: $k^{16}/k^{18} = 1.062$. Independent kinetic data have shown that under these conditions a tetrahedral intermediate is formed and that loss of methoxide from this tetrahedral intermediate is the rate-determining step.[24] At pH 10 the isotope effect is $k^{16}/k^{18} = 1.017$ because under these conditions the reaction proceeds primarily by a pathway in which proton transfer is rate determining.

In the same study, Sawyer and Kirsch[23] measured oxygen isotope effects on the acid-catalyzed hydrolysis ($k^{16}/k^{18} = 1.0009$), the alkaline hydrolysis (1.0091), and the general-base-catalyzed hydrolysis (1.0115) of methyl formate. The fact that the isotope effects in all these cases are small but significantly different from unity probably indicates that the transition states for formation and decomposition of the tetrahedral intermediate are of very similar energy. That is, carbon–oxygen bond breaking is partially, but not entirely, rate determining. The presence of a small amount of carbonyl oxygen exchange during the hydrolysis[23] is consistent with this possibility.

The alkaline hydrolysis of N-acetyl-L-tryptophan methyl ester shows an oxygen isotope effect $k^{16}/k^{18} = 1.007$ at 25° in water containing 3% acetonitrile.[25] The alkaline hydrolysis of methyl benzoate shows an oxygen isotope effect of 1.007 at 25° in water.[25]

Oxygen isotope effects on the hydrolysis of benzoyl chloride have been used to demonstrate the transition from a displacement mechanism to an ionization mechanism.[26] In 95% dioxane–5% water, the solvent oxygen isotope effect is $k^{16}/k^{18} = 0.999$ and the chlorine isotope effect is $k^{35}/k^{37} = 1.008$. The latter value is similar to chlorine isotope effects which have been observed in S_N2 reactions (Section 2.2). No oxygen exchange between the

---

* There is one possible exception to this statement. The carbon–nitrogen bond in amides is significantly stronger than an ordinary carbon–nitrogen bond. This added strength would disappear on formation of a tetrahedral intermediate. Thus, there might be a significant nitrogen isotope effect on the first step of an amide hydrolysis, and a nitrogen isotope effect would be observed even if the first step were wholly rate limiting.

carbonyl oxygen and the solvent occurs during the hydrolysis.[26-28] In this case the hydrolysis appears to occur by a direct displacement mechanism not involving a tetrahedral intermediate of appreciable stability.

In 50% dioxane–water, the solvent oxygen isotope effect on the hydrolysis of benzoyl chloride is $k^{16}/k^{18} = 1.009$ at 25°. This is similar to the value of $k^{16}/k^{18} = 1.008$ for the hydrolysis of 2,4,6-trimethylbenzoyl chloride, a reaction known to proceed by way of a cationic intermediate[29]:

$$Ar—\overset{+}{C}{=}O$$

The oxygen isotope effect is larger in this mechanism than in the case of the displacement mechanism because of the difference in carbon–oxygen bond order at the transition state between the two cases. In the ionic mechanism, the transition state resembles the cationic intermediate, and there is only a small degree of bond formation at the transition state. In the displacement mechanism, there is a much greater degree of bond formation. With all other things held constant, heavy-atom isotope effects for entering groups become smaller (and eventually become inverse) as the degree of bonding at the transition state increases.

The carbonyl oxygen atom also shows appreciable oxygen isotope effects in acyl-transfer reactions. The reaction of p-bromophenyl benzoate with sodium methoxide in methanol at 29.5° shows a carbonyl oxygen effect $k^{16}/k^{18} = 1.018$.[30] Similar experiments with phenyl benzoate gave an isotope effect of 1.024.[30] These results indicate that these methanolysis reactions proceed via a transition state involving extensive carbonyl oxygen participation. The carbonyl carbon–oxygen bond must be appreciably weaker than a double bond at the transition state.

2.4. Nitrogen Isotope Effects

Isotope effects have been reported for a variety of reactions in which a nitrogen-containing substituent acts as a leaving group. Among the easiest of these effects to understand are those in the decomposition of benzene-diazonium ion, a reaction believed to proceed by way of a highly unstable phenyl cation:

$$Ph—N{=}N^+ \longrightarrow [Ph^+] + N_2 \xrightarrow{H_2O} Ph—OH + H^+ \qquad (7)$$

Because of the instability of the phenyl cation intermediate, it is logical to assume that the transition state for the reaction resembles the phenyl cation and has an almost broken carbon–nitrogen bond. Brown and Drury[31] reported the nitrogen isotope effect on the decomposition of benzenediazonium fluoborate in aqueous solution at 40.5° to be $k^{14}/k^{15} = 1.022$. However, this value represents the mean of the isotope effects for the two nitrogens of the diazonium salt. In subsequent work, Swain et al.[32] measured, by a specific labeling procedure, the nitrogen isotope effects for the two nitrogens individually. The nitrogen adjacent to the ring shows an effect $k^{14}/k^{15} = 1.038$, whereas the nitrogen remote from the ring shows an effect of 1.011. Thus, there are extensive changes in bonding to both nitrogens in going to the transition state.

Elimination reactions of β-phenylethyltrimethylammonium salts and β-phenylethyldimethylsulfonium salts have been extensively studied by means of hydrogen, nitrogen, carbon, and sulfur isotope effects.[33] The transition state for this reaction involves breaking of a carbon–hydrogen bond and a carbon–nitrogen bond,

$$\text{PhCH}_2\text{CH}_2\text{N}^+(\text{CH}_3)_3 + \text{OH}^- \longrightarrow \left[\begin{array}{c} ^-\text{OH} \\ / \\ \text{H} \\ | \\ \text{PhCH}\text{---}\text{CH}_2\text{-----}^+\text{N}(\text{CH}_3)_3 \end{array} \right]^{\ddagger}$$

$$\longrightarrow \text{H}_2\text{O} + \text{PhCH}{=}\text{CH}_2 + \text{N}(\text{CH}_3)_3$$

(8)

and a variety of transition states may be expected according to the degree of breaking of these bonds.

Smith and Bourns[34] reported that the nitrogen isotope effect on the reaction of β-phenylethyltrimethylammonium ion with hydroxide ion in water at 97° is $k^{14}/k^{15} = 1.0078$. This indicates that the carbon–nitrogen bond is being broken in the rate determining step. At 60°, the hydrogen being removed by base shows an isotope effect $k^H/k^D = 2$, indicating that carbon–hydrogen bond breaking is also occurring.[35] The carbon isotope effect for the carbon adjacent to the aromatic ring is $k^{12}/k^{13} = 1.021$ at 60°.[36] Thus, all three isotope effects indicate that the bond changes are highly concerted.

Similar isotope-effect results have been obtained for other elimination reactions in aqueous solution and in alcoholic solution.[33] Substituent effects in the aromatic ring indicate that in elimination reactions of β-phenylethyltrimethylammonium salts both the carbon–nitrogen bond and the carbon–hydrogen bond become less broken at the transition state as the electron-withdrawing power of the substituent increases.[37]

2.5. Equilibrium Isotope Effects

Heavy-atom isotope effects on chemical equilibria have been extensively examined both experimentally and theoretically.[38] The experimental interest results largely from the use of such isotope effects in isotopic separation processes. The theoretical interest results from the ability of theoreticians to make accurate and confirmable statistical-mechanical calculations for all species present in the system—an advantage not shared by kinetic isotope effects.

By comparison with kinetic isotope effects, most equilibrium isotope effects are expected to be small. Both kinetic and equilibrium effects reflect the difference in bonding between the two states involved. In the case of carbon, the tendency to remain tetravalent is so strong that only very small equilibrium carbon isotope effects are observed. The carbon isotope effect on the equilibrium between carbon dioxide vapor and bicarbonate ion in aqueous solution,

$$^{13}CO_2(g) + H^{12}CO_3^- (aq) \rightleftharpoons ^{12}CO_2(g) + H^{13}CO_3^- (aq) \tag{9}$$

is $K = 1.0077$ at $25°$.[39a] A small portion of that difference is simply due to the difference in solubility between $^{12}CO_2$ and $^{13}CO_2$, the latter being more soluble in water by a factor of 1.0009 at $25°$.[39a]

The carboxyl group of carboxylic acids may show a small carbon isotope effect in a number of processes. The carbon isotope effect on the acid dissociation constant of benzoic acid, K^{12}/K^{13}, is 1.003 in water at $10°$ and 1.001 at $25°$.[39b] The equilibrium carboxyl carbon isotope effect on the decarboxylation of isocitric acid was measured by equilibration of isocitrate, α-ketoglutarate, and CO_2 in the presence of NADPH, $NADP^+$, and isocitrate dehydrogenase. At pH 7.5, $25°$, the equilibrium isotope effect is 1.0027, with ^{13}C concentrating in the carboxyl group.[39c] This value represents the fractionation between CO_2 and the ionized form of the carboxylic acid substrate. If all carboxylic acids show approximately equal equilibrium isotope fractionations on ionization and decarboxylation, we can estimate that the isotopic equilibrium between CO_2 and the neutral form of a carboxylic acid would show an isotope effect of approximately 1.004 at $25°$.

Oxygen and nitrogen are more likely to show large equilibrium isotope effects than is carbon because of the ease with which they form species with different numbers of bonds. The nitrogen isotope effect, for example, on the basicity of ammonia,

$$^{15}NH_3 + {}^{14}NH_4^+ \rightleftharpoons {}^{14}NH_3 + {}^{15}NH_4^+ \tag{10}$$

has recently been observed to be $K = 1.0393 \pm 0.0004$,[40] in qualitative agreement with earlier results of Thode *et al.*[41]

There is also a large oxygen fractionation in the equilibration between carbon dioxide and water. For the exchange

$$H_2^{18}O + C^{16}O_2 \rightleftharpoons H_2^{16}O + C^{16}O^{18}O \tag{11}$$

the equilibrium constant* $K = 2.076$ at $25°$.[42,43]

2.6. Isotope Effects on Physical Processes

Physical processes, such as melting, vaporization, solubility, and other processes, can also show small heavy-atom isotope effects. A variety of such effects have been summarized by Bigeleisen *et al.*[38] These isotope effects are generally of little use in enzymology, but they provide a seemingly endless source of pitfalls for scientists interested in measuring isotope effects. Because of the existence of these effects, it is imperative that all physical and chemical

---

* That the equilibrium constant is approximately 2 for this reaction, rather than 1, is a result of the fact that CO_2 has two possible positions for the isotopic oxygen, whereas H_2O has only one.

steps which intervene between the preparation of a sample and its measurement be thoroughly understood.

Even the simplest systems may be subject to isotope effects. For example, the solubility of ^{3}He in water is less than that of ^{4}He by 1.2% at 0°.[44] The vapor pressure of methanol is increased slightly by ^{13}C substitution and decreased slightly by ^{18}O substitution.[45]

3. THE THEORY OF ISOTOPE EFFECTS IN ENZYME-CATALYZED REACTIONS

Enzymic reactions are like organic reactions in that they proceed through a series of reactive intermediates of varying stabilities and lifetimes. These intermediates are separated by high-energy configurations of reactants, commonly called transition states. In the case of enzymes, the intermediates and transition states are bound to a specific locus on the surface of the enzyme (the "active site"), but with this exception, it seems probable that the properties of these intermediates and transition states are rather the same in enzymic systems as they are in organic systems. The kinetics of enzymic reactions are more complex than those of organic reactions because of the greater number of reaction steps and because of the occurrence of steps in which substrates bind to and dissociate from the enzyme. These binding and dissociation steps are usually assumed to occur at rates which are rapid compared to the rates of the chemical steps, but this may not invariably be so.[46]

In spite of the large number of separate steps which may be involved in an enzyme-catalyzed reaction, the number of kinetic parameters which can be obtained in a particular case is severly limited. For a single substrate reaction, steady-state kinetics can provide a value of V_{max}, the velocity of the reaction at high substrate concentration, and K_m, the Michaelis constant, numerically equal to the concentration of substrate at which the velocity is half its maximum value. From these two parameters can be obtained the ratio V_{max}/K_m, which represents the bimolecular rate constant for combination of enzyme with substrate extrapolated to zero substrate concentration. Reactions which involve more than a single substrate can provide a larger set of bimolecular rate constants, but steady-state kinetics cannot provide rate constants for individual steps in the reaction mechanism except in special cases.

Stopped flow and temperature-jump methods can often provide additional information about the kinetics. These methods yield rate constants for steps prior to the rate-determining step which proceed at rates at least five- or tenfold faster than the rate-determining step. However, these methods are limited to comparatively rapid steps prior to the rate-determining step, and they provide no information about steps in the mechanism whose rates are similar to that of the rate-determining step.

Heavy-atom isotope effects provide information which is complementary to steady-state and pre-steady-state kinetic data. In most enzymic reactions, the occurrence of a heavy-atom isotope effect reflects the occurrence of an isotope effect on one particular step in the reaction mechanism. This step may be the sole rate-determining step, but more often this step is only partially rate determining, and it is possible from the measured isotope effect to estimate the rate of this step relative to other steps in the mechanism.

3.1. Transition-State Structure

We have seen that in the case of organic reactions heavy-atom isotope effects are useful indicators of transition-state structure. Enzyme-catalyzed reactions, however, are more complicated than their organic counterparts, and most enzymic reactions do not have a single rate-determining step. Rather, several steps have transition states of similar energies and are partially rate determining. For this reason, the ability of heavy-atom isotope effects to provide information about transition-state structure in enzymic reactions is severely limited. The more useful approach, and the one taken here, is to use models from organic chemistry for predicting approximate isotope effects on individual steps in enzymic reactions. The observed isotope effect in the enzymic reaction can then be used to provide information about the relative rates of various steps in the reaction sequence.

However, this procedure is not without its hazards, and it is wise not to rely strictly on the size of the isotope effect. It is far better to consider also the variation of an isotope effect with pH, substrate structure, temperature, or some other factor.

3.2. A Caveat on Rate-Determining Steps

It is often assumed that the presence of a heavy-atom isotope effect in a chemical reaction is evidence that a change in bonding to the isotopic atom is occurring in (or before) the rate-determining step. Such an interpretation is correct and adequate in the case of organic reactions, but it is not entirely correct in the case of enzymic reactions.

Heavy-atom isotope effects are almost always measured by competitive methods—a mixture of two isotopic species of substrate is reacted, and the change in isotopic composition of substrate or product is measured over the course of the reaction. In the Michaelis–Menten terminology, such an isotope effect represents the effect on V_{max}/K_m, which is the rate of the bimolecular reaction between enzyme and substrate at zero substrate concentration. Such measurements do not provide the isotope effect on V_{max}—that isotope effect can be obtained only by noncompetitive methods. The concept of "rate-determining step" in the usual parlance refers to V_{max}, and thus to the rate of passage through the transition state of highest energy, rather than to any other expression of the enzymic rate. Thus, it is possible that a heavy-atom isotope effect might be observed even in a case where the change in bonding to the isotopic atom does not occur in the rate-determining step. The chymotrypsin-catalyzed hydrolysis of N-acetyl-L-tryptophan ethyl ester, for example, shows an oxygen isotope effect $k^{16}/k^{18} = 1.018$, in spite of the fact that the rate-determining step is the deacylation of the acyl enzyme.[47] The isotope effect reflects only the acylation steps in the reaction mechanism.

Therefore, the following statement may be made concerning the relationship between isotope effects and rate-determining steps: If, based on adequate model studies, an isotope effect is expected in a particular step and no isotope effect is observed, then that step is not rate determining. If an isotope effect is observed, the step in question may or may not be rate determining.

3.3. General Mathematical Form of Isotope Effect*

Enzyme-catalyzed reactions generally involve a sequence of steps which may be represented by

$$E + S \underset{k_2}{\overset{k_1}{\rightleftharpoons}} ES_1 \underset{k_4}{\overset{k_3}{\rightleftharpoons}} \cdots \rightleftharpoons ES_n \underset{k_{i+1}}{\overset{k_i}{\rightleftharpoons}} EP \rightleftharpoons E + P \qquad (12)$$

The first step is the binding of the substrate to the enzyme, and the last step is the dissociation of the product from the enzyme. Reactions which have more than a single substrate or more than a single product differ in detail but not in principle. The number of steps in the mechanism may or may not be known.

Heavy-atom isotope effects are localized effects and are usually not influenced by chemical changes more than about two bonds removed from the isotopic atom. For most cases it is adequate to assume that one step in the reaction mechanism will be subject to a heavy-atom isotope effect. This step, which we shall call the "key step," will be assumed to be irreversible, and the isotope effect on this reaction step will be denoted by k_i/k_i^*, where the asterisk denotes the heavier isotope. The observed isotope effect for the enzymic reaction, which we shall denote by k/k^*(observed), has been derived previously[48,49] and is given by

$$k/k^*(\text{observed}) = \frac{k_i/k_i^* + R}{1 + R} \qquad (13)$$

The factor R, which will be called the "partitioning factor," is a measure of the rate of the key step relative to the rates of all preceding steps in the reaction sequence. The partitioning factor depends on the rate constants $k_2, k_3, \ldots, k_{i-1}$, which are independent of isotopic substitution, and upon rate constant, k_i, the rate constant for the key step for the lighter isotope.

---

* Throughout this section, the * is used to indicate the heavier isotope. A normal isotope effect is one for which $k/k^* > 1$; conversely, an inverse isotope effect is one for which $k/k^* < 1$.

3.4. A Compromise Approach to Eq. (13)

Equation (13) tells us that a measured heavy-atom isotope effect depends on two factors, the isotope effect on the key step and the partitioning factor. When the partitioning factor is small, the intermediate immediately preceding the key step is in equilibrium with the starting state, and the observed isotope effect is equal to the isotope effect on the key step. When the partitioning factor is not small, this equilibrium condition does not hold, and the observed isotope effect is smaller than the isotope effect on the key step.

It is obviously not possible to obtain both the partitioning factor and the isotope effect on the key step from a single isotope-effect experiment. The best compromise under these circumstances is usually to assume a value (or range of values) for the isotope effect on the key step and then calculate a value (or range of values) for the partitioning factor. Such an assumption can often be made easily by use of data from analogous systems in organic chemistry.

Because of the necessity of making this assumption, heavy-atom isotope effects are relatively useless as probes of transition-state structure unless other information is available which permits estimation of the partitioning factor. Thus, the question of whether transition states in enzyme-catalyzed reactions are really like those in organic reactions remains to some extent unanswered.

3.5. The Partitioning Factor

To derive Eq. (13), and thus the partitioning factor, for any mechanism of interest, it is necessary to write the expression for V_{max}/K_m for that mechanism. Alternatively, the partitioning factor can be derived by the method of net rate constants.[50] In this section we shall give equations for R for a number of commonly encountered mechanisms.

For a simple two-step mechanism

$$E + S \underset{k_2}{\overset{k_1}{\rightleftharpoons}} ES \xrightarrow{k_3} E + P \tag{14}$$

in which k_3 is the key step, the partitioning factor is given by

$$R = k_3/k_2 \tag{15}$$

This same form applies even if a number of rapid steps precede the slow steps. In the case of

$$E + S \rightleftharpoons ES_1 \rightleftharpoons \cdots \underset{k_2}{\overset{k_1}{\rightleftharpoons}} ES_n \xrightarrow{k_3} E + P \tag{16}$$

if the rate constants for all unnumbered steps are large compared to k_1, then Eq. (15) still applies.

If two slow steps precede the key step, as in

$$E + S \underset{k_2}{\overset{k_1}{\rightleftharpoons}} ES_1 \underset{k_4}{\overset{k_3}{\rightleftharpoons}} ES_2 \xrightarrow{k_5} E + P \tag{17}$$

in which k_5 is the key step, the partitioning factor is given by

$$R = \frac{k_5}{k_4}\left[1 + \frac{k_3}{k_2}\right] \tag{18}$$

Again, prior rapid steps need not be considered. The rate constant for the key step (k_5 in this case) which is included in the equation for R is always the rate constant for the lighter of the two isotopic species of reactants.

The equations above apply only when the key step is irreversible under the conditions of the isotope-effect experiment. This is a kinetic, rather than thermodynamic, criterion. That is, the irreversibility may be due either to a negligible rate constant for the reverse of the key step or it may be due to rapid product removal following the key step.

3.6. General Form of the Partitioning Equation

The equations in the preceding sections can be used to relate the observed isotope effects to the isotope effects on key steps in most reactions commonly studied by means of heavy-atom isotope effects. In this section we shall give a more complete expression which allows for the possibility that (1) the key step is reversible under the reaction conditions and (2) there is an isotope effect on the equilibrium constant for the overall reaction.

For a reversible reaction in which the ith step is the key step, the observed isotope effect in the forward reaction is given by

$$k/k^*(\text{observed}) = \frac{k_i/k_i^* + R + R'K/K^*}{1 + R + R'} \tag{19}$$

in which K is the overall equilibrium constant (products/reactants) for the light substrates and K^* is the corresponding equilibrium constant for the heavy substrates.

The partitioning factor R reflects steps up through the key step and corresponds to the partitioning factors used in the preceding section. In the general case, R is given by

$$R = \frac{k_i}{k_{i-1}}\left[1 + \frac{k_{i-2}}{k_{i-3}}\left(1 + \frac{k_{i-4}}{k_{i-5}}\{\cdots\}\right)\right] \tag{20}$$

in which the continued product is taken for all rate constants from k_i back to k_2. When the key step is irreversible, $R' = 0$, and the equation reduces to its previous form.

Partitioning factor R' includes all rate constants from k_{i+1} to the first irreversible step (which may be a product-release step or some other irreversible step) and is given by

$$R' = \frac{k_{i+1}}{k_{i+2}}\left[1 + \frac{k_{i+3}}{k_{i+4}}\left(1 + \frac{k_{i+5}}{k_{i+6}}\{\cdots\}\right)\right] \tag{21}$$

in which the continued product is taken from k_{i+1} to the first irreversible step.

It may on occasion be possible to measure the isotope effect on the reverse reaction in such cases. That isotope effect is given by

$$k/k^*(\text{observed}) = \frac{(K^*/K)(k_i/k_i^*) + (K^*/K)R + R'}{1 + R + R'}$$

3.7. Isotope Effects and Changes in Reaction Conditions

Our approach to isotope effects suffers from the fact that in many cases neither the isotope effect on the key step in the reaction nor the value of the partitioning factor is known precisely. The usefulness of isotope effects as mechanistic tools can be increased by comparing isotope effects obtained under a variety of conditions. Within that framework, it is frequently possible to predict the change in the isotope effect on the key step which occurs as a result of a change in reaction conditions. If that prediction is possible, then useful information can be obtained concerning the variation of the partitioning factor with reaction conditions. A number of such approaches are described in the following sections.

3.8. Effects of Temperature

The theoretical basis for the effect of temperature on heavy-atom isotope effects is well known,[10,11] and, although a number of aspects of this theory are in dispute, the limited temperature range available in studies of enzymes makes such disputes of no consequence in studies of enzyme mechanisms. In fact, it is probably adequate for most purposes to assume that the isotope effect on the key step is invariant with temperature. Any observed variations in isotope effect can then be attributed to variations in the partitioning factor; such variations might be large even within the narrow temperature range which is generally accessible in such studies.

3.9. Effects of Substrate, Coenzyme, and Metal Ion

Many enzymes are active with a variety of closely related substrates or coenzyme analogs, and a number of metal-ion-dependent enzymes function with varying efficiencies with different metals. The expected variation in the isotope effect on the key step as a result of such changes is often predictable from principles of physical organic chemistry. If this prediction is possible, then conclusions concerning the variation of the partitioning factor with different variables can be made. This approach has been used in the case of isocitrate dehydrogenase (Section 5.4) and arginine decarboxylase (Section 5.3).

3.10. Effect of Enzyme Source

Different enzymes catalyzing the same reaction should in general have very similar structures for the transition state for the key step. Thus, comparison of such isotope effects would provide useful information about the variation in partitioning factor from enzyme to enzyme. No such studies appear to have been reported for heavy-atom isotope effects.

3.11. Effect of pH

Under many circumstances, it is probably safe to assume that the isotope effect on the key step of an enzymic reaction is independent of pH, even though the rate of the reaction itself may be quite sensitive to pH. This insensitivity arises in the following way: The transition state for the key step (which is often at least partially rate determining) has a structure which has been optimized over evolutionary time to provide the highest possible rate for the enzymic reaction. This transition state contains a certain number of protons—obviously, the number of protons which allows the step to proceed at an optimum rate. If the transition state is changed by adding protons to it or by removing protons from it, the energy of that transition state will probably be raised considerably, and the rate of passage of enzyme–substrate complexes through that transition state will be effectively zero, compared to the rate of passage through the normally constituted transition state. Thus, the key step can occur only when the preceding intermediate has the proper number of protons, only one structure is possible for the transition state for that step, and the isotope effect on that step will be pH independent.

Thus, if a heavy-atom isotope effect on an enzyme-catalyzed reaction is observed to be pH dependent, this probably reflects a pH dependence in the partitioning factor R. A number of such isotope effects have been observed, and their pH dependence is discussed in the following paragraphs.

The usual explanation for pH dependence in steady-state kinetics is that enzyme, substrate, or one or more complexes contain ionizable groups which must be in a proper protonation state in order for reaction to occur. Such a scheme might be illustrated as

$$
\begin{array}{ccc}
E + S & ES & E + P \\
\updownarrow & \updownarrow & \updownarrow \\
HE + S \rightleftharpoons & HES \longrightarrow & HE + P \\
\updownarrow & \updownarrow & \updownarrow \\
H_2E + S & H_2ES & H_2E + P
\end{array}
\tag{22}
$$

Obviously, other examples containing additional intermediates can be given. However, the isotope effect predicted for the above mechanism is not pH dependent because neither the isotope effect on the key step nor the partitioning factor can vary with pH. That is, all reaction steps occur via transition states containing the same number of protons, and as long as that constancy holds, the isotope effect cannot be pH dependent.

To obtain a pH-dependent partitioning factor, a pathway or partial pathway must exist in which the transition states contain a different number of protons. A simple mechanism by which this might occur is

$$
\begin{array}{ccc}
E + S \rightleftharpoons ES & E + P \\
\updownarrow \quad\quad \updownarrow & \updownarrow \\
HE + S \rightleftharpoons HES \longrightarrow HE + P & (23) \\
\updownarrow \quad\quad \updownarrow & \updownarrow \\
H_2E + S \rightleftharpoons H_2ES & H_2E + P
\end{array}
$$

For this mechanism, the partitioning factor R is quite small at high pH and at low pH. At intermediate pH, the concentration of intermediate HES is larger than at the pH extremes and the partitioning factor is larger than at the pH extremes. Under these circumstances the observed isotope effect will be small at the pH optimum of the enzyme and will increase away from the optimum pH.

It should be emphasized that the R factor in a complex mechanism such as Eq. (23) is no longer simply a function of the rate constants for the principal reaction pathway but involves as well the rate constants for the auxiliary pathways and the pK_a values for the various ionizations in the mechanism.

4. EXPERIMENTAL METHODS

The rate difference caused by substituting one heavy isotope for another in an organic molecule is always small—never more than a few percent, and often a percent or less. Consequently, considerable care is required in making reliable measurements of heavy-atom isotope effects. Particularly in the case of enzymes, which provide a number of new complications not present in organic reactions, the time required to set up and "debug" a system is often greater than the time required to make the actual measurements.

A variety of sources can give rise to spurious isotope effects, including impure enzyme preparations, improperly purified reagents, inadequately controlled measurements of isotopic abundances, isotopic fractionations occurring during product isolation or analysis, inadequate control of extent of reaction, and others. The literature is replete with reports of poorly conducted experiments, and the reader must always be prepared to read critically all experimental procedures and make his or her own judgment about whether the reported effects are valid.

In this section, we shall discuss the methods which have been commonly used for the determination of heavy-atom isotope effects, and we shall describe two new methods which are likely to find increasing use in the future.

4.1. The Competitive Method

Most heavy-atom isotope effects have been measured by observing the isotopic composition of a particular atom in the starting material or product as a function of the extent of reaction. For example, if a decarboxylation reaction is subject to a significant carbon isotope effect ($k^{12}/k^{13} > 1$), the carbon dioxide isolated early in the reaction will be depleted in ^{13}C compared to the carboxyl carbon of the starting compound. Although direct analysis of the carboxyl carbon is often inconvenient, the same end can be accomplished by complete decarboxylation of a sample of substrate and analysis of the carbon dioxide thus produced.

Similar procedures are used when it is desirable to perform the isotopic analysis on the starting material, rather than on the product. Measurement of the isotopic composition of the appropriate site in the starting material as a function of extent of reaction provides the isotope effect directly.[2,4]

When nonradioactive isotopes are used in the competitive method, it is frequently possible to use the small natural abundance of carbon-13 (1.1%), nitrogen-15 (0.4%), or oxygen-18 (0.2%). This procedure is possible only if very high-precision measurements of isotopic abundances are possible— for carbon, for example, a precision for $^{13}C/^{12}C$ of ± 0.000002. Although such precision is readily available on isotope-ratio mass spectrometers specifically designed for making such measurements, ordinary multipurpose mass spectrometers are unable to produce data of the required accuracy. The problem is further complicated by the fact that isotope-ratio machines are designed to alternate with a frequency of 1–5 min between a sample of unknown isotopic composition and a chemically identical sample of known isotopic composition. This can be accomplished only if the material being analyzed is very volatile; only CO, CO_2, N_2, and CH_3Cl have been widely used.

An alternative procedure is to use isotopically enriched substrates. This approach somewhat alleviates the necessity for high-precision isotopic analyses, but this alleviation is to some extent illusory. In addition to the obvious problems of synthesis, this method is much more seriously affected by errors due to contamination than is the natural abundance method.

4.2. Radiochemical Methods

Radioactive isotopes can also be used in the determination of isotope effects. In principle, the method is the same as for stable isotopes, except that the isotopic ratio measurements are made by measurement of specific activities in a liquid scintillation counter, rather than by mass spectrometry. However, the determination of specific activities at the necessary precision level is difficult because of the problems of quenching in liquid scintillation counting. In addition, convenient radioactive isotopes of nitrogen and oxygen are not available.

4.3. Direct Methods

The methods discussed above involve the measurement of ratios of isotopic species, and they are only capable of providing measurements of the isotope effect on V_{max}/K_m. In theory, it should be possible to measure directly the kinetic parameters for the two isotopic species, provided that highly enriched substrates are available. Such a procedure would make it possible to obtain isotope effects on V_{max} and would make a variety of studies possible which are not possible by present methodology. The precision of commercially available spectrophotometers is such that direct measurements should be possible, but the problem of obtaining enzyme kinetic data accurate to $\pm 0.1\%$ are severe, particularly since under most conditions enzymes do not conform to simple rate laws. Problems of concentration, volume measurement, temperature, impurities, enzyme stability, and other things are severe, and it is probable that the ultimate limit of precision of this method is an order of magnitude worse than the best competitive method. Nonetheless, this technique has been applied to organic reactions,[30,51] and at least one enzymic reaction has been studied by this technique.[52] An interesting variant of this technique in which the two isotopic species are compared directly in the two beams of a double-beam spectrophotometer has recently been described.[53] Direct methods are useful especially in cases where direct product analysis by mass spectrometry is not possible, such as when using labeled coenzymes.

4.4. Equilibrium Perturbation

The mathematical and experimental treatments of isotope effects given above have generally assumed that the key step in the reaction is irreversible. A very interesting method has recently been described by Schimerlik et al.[54] for treating reversible reactions. Called the "equilibrium perturbation" method, it can be used for any enzymic reaction in which appreciable amounts of starting materials and products are present at equilibrium.

One substrate or product is highly labeled with an isotope. Prior to addition of the enzyme, all substrates and products are adjusted to their equilibrium concentrations. When enzyme is added, the system temporarily goes away from equilibrium because the isotopic compound scrambles through the system at a different rate than its unlabeled counterpart. Eventually isotopic equilibrium is reached, and the system returns to its initial equilibrium state. The magnitude of the temporary deviation from equilibrium can be used to calculate the isotope effect.

5. ISOTOPE EFFECTS IN ENZYMIC REACTIONS

In this section we shall describe some results which have been obtained in studies of heavy-atom isotope effects on enzyme-catalyzed reactions. A number of the isotope effects reported in the literature, both enzymic and nonenzymic, are of questionable validity. The author has endeavored to avoid those and present only cases for which reasonable supporting data are available.

5.1. Oxalacetate Decarboxylase

Seltzer et al.[48] reported that the metal-ion-catalyzed decarboxylation of oxalacetic acid,

$$^{-}O_2{}^*CCH_2CCO_2^- \xrightarrow[\text{or enzyme}]{M^{2+}} {}^*CO_2 + CH_3CCO_2^- \tag{24}$$

shows a carboxyl carbon isotope effect $k^{12}/k^{13} = 1.06$, whereas the enzymic reaction shows an isotope effect of 1.002. The value for the model reaction is typical of cases where decarboxylation is the rate-determining step. Clearly, in the enzymic reaction decarboxylation is not the rate-determining step. The simplest explanation of these results might be that the enzyme operates by a two-step mechanism [Eq. (14)] and that the partition factor for this mechanism is quite large. That is, substrate binding is virtually irreversible.

A second possible mechanism takes into account the fact that the decarboxylation step is reversible[55] and that the reaction is inhibited to a slight extent by added carbon dioxide.[48] The lack of an isotope effect might be attributed to the fact that decarboxylation is reversible and is not rate determining. Under those conditions, the observed carbon isotope effect actually represents an equilibrium effect. Unfortunately, a distinction between these two mechanisms is not possible at the present time.

5.2. Acetoacetate Decarboxylase

The decarboxylation of acetoacetic acid is catalyzed by primary amines by way of a mechanism involving an intermediate Schiff base,

$$CH_3CCH_2{}^*CO_2^- + RNH_2 \rightleftharpoons CH_3CCH_2{}^*CO_2^- + H_2O$$

$$CH_3C{=}CH_2 + {}^*CO_2 \longleftarrow CH_3CCH_2{}^*CO_2^- \tag{25}$$

$$CH_3CCH_3 + RNH_2$$

The primary amine-catalyzed decarboxylation shows a pH-dependent carbon isotope effect because the rate-determining step changes with pH.[49] At low pH, nucleophilic attack on the carbonyl carbon of acetoacetate is rate determining, whereas at high pH, decarboxylation is rate determining. Conclusions based on isotope effects are consistent with those based on other kinetic studies.[56]

Acetoacetate decarboxylase functions by a mechanism analogous to that shown above. The primary amino group is provided by a lysine residue of the enzyme.[57] The carbon isotope effect on the enzymic reaction is $k^{12}/k^{13} = 1.018$ and is pH independent. Assuming that the isotope effect on the key step is 1.04–1.06, this means that the partitioning factor for the reaction is near unity, a conclusion consistent with oxygen exchange results of Hamilton.[58] Thus, both Schiff base formation and decarboxylation are partially rate determining.

5.3. Decarboxylases Requiring Pyridoxal 5'-Phosphate

Glutamate decarboxylase contains covalently bound pyridoxal 5'-phosphate as a cofactor. The general outline of the mechanism (Fig. 1) has been deduced by analogy with other pyridoxal phosphate-dependent enzymes. Extensive kinetic studies have been reported by O'Leary et al.[59] and by Fonda.[60] The maximum activity of the enzyme occurs around pH 4.5, but there is little change in activity with pH down to pH 3.5. The activity diminishes rapidly above pH 5.0.

The carboxyl carbon isotope effect on the enzymic decarboxylation of glutamic acid was shown by Hoering[61a] to be $k^{12}/k^{13} = 1.026$ at 20° at an unstated pH. In a more extensive study, O'Leary et al.[59] found that in the pH range 3.5–4.5 the isotope effect is about $k^{12}/k^{13} = 1.015$. Above pH 4.5 the isotope effect increases, reaching a value of 1.022 at pH 5.5.

The carbon isotope effects are significantly smaller than those which

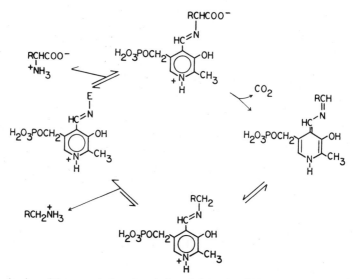

Fig. 1. Mechanism of the enzymic decarboxylation of glutamic acid by the pyridoxal 5'-phosphate-dependent glutamate decarboxylase.

would be expected if decarboxylation of the Schiff base intermediate were entirely the rate-determining step. Further, if that were true, the isotope effect would not be expected to be pH dependent. The more attractive alternative is that the partitioning factor is approximately unity and varies with pH. The probable explanation of this pH variation lies in the step in which the Schiff base interchange takes place. This interchange can probably take place independent of whether or not the pyridine nitrogen is protonated. Decarboxylation, on the other hand, requires that the nitrogen be protonated. Thus, at low pH the observed isotope effect reflects partitioning of the Schiff base intermediate in which the pyridine nitrogen is protonated. At higher pH Schiff base formation occurs without protonation of the pyridine nitrogen, whereas decarboxylation requires protonation, and the partitioning factor decreases. Unfortunately, it is not possible to determine the pK_a of the pyridine nitrogen from these isotope effect data.

The arginine decarboxylase from *E. coli* contains pyridoxal 5'-phosphate and presumably operates by a mechanism like that described for glutamate decarboxylase. The carbon isotope effect on the decarboxylation of arginine at pH 5.25 decreases smoothly with increasing temperature from a maximum of 1.027 at 5° to a minimum of 1.012 at 50°.[61b] This temperature dependence is much larger than is expected for a single-step decarboxylation and must reflect large changes in the partitioning factor with temperature. Such changes will only influence the isotope effect strongly if the partitioning factor is not too different from unity and thus the isotope effect on the decarboxylation step is in the predicted 1.04–1.06 range.

Further evidence for such an isotope effect on the decarboxylation step comes from studies of substrate analogs. Norarginine and homoarginine are both decarboxylated by arginine decarboxylase about a 100-fold more slowly than is arginine, although all three substrates have similar K_m values. Carbon isotope effects for homoarginine and norarginine at 25°, pH 5.25, are 1.053 and 1.043, respectively.[61b] Thus it appears that the low efficiency of decarboxylation of these substrates is primarily due to the reduction in rate of the decarboxylation step.

5.4. Isocitrate Dehydrogenase

This enzyme catalyzes the oxidative decarboxylation of isocitric acid in a mechanism which involves at least two chemical steps;

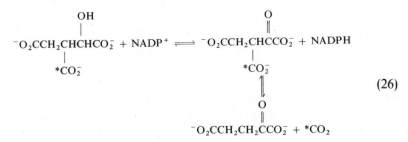

Enzyme-bound oxalosuccinate is assumed to be an intermediate in the reaction. Substrate binding is very tight; the Michaelis constants for isocitrate and for NADP$^+$ are below 1 μM.[62] Under optimum conditions, the carbon isotope effect on the decarboxylation is $k^{12}/k^{13} = 0.999$,[63,64] the hydrogen isotope effect on V_{max} for the hydrogen being transferred to the nucleotide is $k^H/k^D = 1.09$,[64,65] and the hydrogen isotope effect on V_{max}/K_M is near unity.[64,65] The lack of substantial carbon and hydrogen isotope effects indicates that neither decarboxylation nor hydride transfer is rate determining in the overall sequence; the partitioning factors controlling both isotope effects must be quite large. That is, once the substrate is bound to the enzyme, reaction virtually always occurs. Unlike the case of oxalacetate decarboxylase cited above, this conclusion is reasonable, considering the very tight binding of substrates to the enzyme. Rate-determining product release may be a frequently encountered phenomenon with enzymes which bind their substrates and products very tightly.[46]

Isocitrate dehydrogenase requires a metal ion for activity. Use of nickel rather than magnesium reduces the catalytic activity of the enzyme by approximately a factor of 10.[66] The carbon isotope effect in the presence of nickel is $k^{12}/k^{13} = 1.005$, the hydrogen isotope effect on V_{max} is $k^H/k^D = 0.98$, and that on V_{max}/K_m is 1.11.[64] Thus, even when the velocity of the reaction has been substantially reduced, the chemical steps in the mechanism have not become rate determining. The kinetic differential between rates of chemical steps and rates of substrate and product release must be quite large.

Malic enzyme catalyzes the oxidative decarboxylation of malic acid by a mechanism which is in many ways similar to that of isocitrate dehydrogenase. In this case, a hydrogen isotope effect $k^H/k^D = 1.45$ and a carbon isotope effect $k^{12}/k^{13} = 1.031$ have been determined by the equilibrium perturbation technique.[54]

5.5. Pyruvate Decarboxylase

A single measurement of the carbon isotope effect on the decarboxylation of pyruvic acid catalyzed by crude yeast pyruvate decarboxylase at an unstated pH and temperature yielded a value of $k^{12}/k^{13} = 1.002$.[67] Recent measurements in the author's laboratory have yielded a value of $k^{12}/k^{13} = 1.008$ at pH 7.5, 25°.[68] Thus, the decarboxylation step is not rate determining in the overall reaction. Decarboxylation is known to be irreversible in this case.[69a] The Michaelis constant for pyruvate is approximately 0.025 M, so it is unlikely that the release of pyruvate from the enzyme–substrate complex is slow. The more likely possibility that formation of the thiamine pyrophosphate–pyruvate adduct is rate determining is supported by the observation of DeNiro and Epstein that the enzymic decarboxylation shows a carbon isotope effect of 1.014 for C-2 of pyruvate.[69b]

In a recent investigation, Schellenberger et al. have measured the nitrogen isotope effect of the amino group of the coenzyme thiamine pyrophosphate in the same reaction by a direct kinetic technique.[53] The nitrogen isotope effect is $k^{14}/k^{15} = 0.994$ at 22°, pH 6.4. Qualitatively, this indicates that the amino nitrogen is more tightly bound in the transition state than it is in the ground state. This is consistent with the suggestion above that the rate-determining step is the attack of the carbon of thiamine pyrophosphate on the carbonyl carbon of pyruvate.

5.6. Chymotrypsin

The natural substrates of chymotrypsin are peptides, and the preferred sites of hydrolysis are adjacent to hydrophobic amino acids.[70] However, to a lesser extent the enzyme also catalyzes the hydrolysis of a variety of amides and esters.

Nitrogen isotope effects on the chymotrypsin-catalyzed hydrolysis of *N*-acetyl-L-tryptophanamide have been studied by O'Leary and Kluetz.[71,72] At the pH optimum of the enzyme (pH 8) the isotope effect is $k^{14}/k^{15} = 1.010$. This isotope effect is significantly different from unity, indicating that carbon–nitrogen bond cleavage plays some part in the overall rate. The isotope effect is similar to those observed in organic reactions where carbon–nitrogen bond cleavage is rate determining (Section 2.4), indicating that carbon–nitrogen bond cleavage might be solely the rate-limiting step.

However, it is not clear what nitrogen isotope effect should be expected in a case like this. The only model reactions of amides are unpublished studies of Harbison,[72] and these results suggest that the effects may often be in the range $k^{14}/k^{15} = 1.00–1.01$, even when carbon–nitrogen bond cleavage is entirely rate limiting. This result is not unexpected if amide hydrolysis involves a tetrahedral intermediate of the form

$$
\begin{array}{c}
O^- \\
| \\
R\!-\!C\!-\!NH_2 \\
| \\
OE
\end{array}
$$

If this intermediate exists, then by the Hammond postulate, we expect that the transition state for decomposition of the intermediate will resemble the intermediate; that is, the degree of carbon–nitrogen bond breaking at the transition state will be small, and the nitrogen isotope effect will be correspondingly small. On the other hand, the large nitrogen isotope effects observed in the papain-catalyzed hydrolysis of an amide[73] indicate that at least under certain circumstances amide hydrolysis may give rise to large nitrogen isotope effects.

One additional factor complicates the interpretation of nitrogen isotope effects in amide hydrolysis. The basicity of ammonia (and therefore, presumably, that of other amines) shows a large nitrogen isotope effect (Section 2.5). In most cases, the leaving group in amide hydrolyses is neutral ammonia or an amine, rather than a negatively charged nitrogen species. Thus, the transition state for decomposition of the tetrahedral intermediate should be represented as

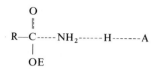

in which a proton is being transferred from the acid HA to the nitrogen. In the limit, this proton might have been completely transferred prior to the beginning of the carbon–nitrogen bond breaking. The important point is that the proton transfer to the nitrogen will make the observed isotope effect smaller; the greater the degree of proton transfer, the smaller the observed isotope effect. For this reason, it is impossible for the present to make accurate predictions of the nitrogen isotope effects expected in amide hydrolysis.

In the case of chymotrypsin, the presence of an appreciable nitrogen isotope effect indicates that carbon–nitrogen bond cleavage is to some extent the rate-limiting step. The nitrogen isotope effect is pH dependent, decreasing from $k^{14}/k^{15} = 1.010$ at pH 8.0 to 1.006 at pH 6.7 and at pH 9.4. By the arguments given previously (Section 3.11), it is most likely that the pH dependence of the isotope effect reflects a change in the partitioning factor with pH. The most likely steps to contribute to this partitioning factor are formation and decomposition of a tetrahedral intermediate, and the pH dependence indicates that both steps are partially rate determining. The same line of reasoning indicates that the actual value for the nitrogen isotope effect on the carbon–nitrogen bond-breaking step is even larger than the maximum value of 1.010 observed at pH 8, indicating that a large degree of carbon–nitrogen bond breaking has occurred and that this effect is incompletely compensated by the proton transfer which is presumably occurring simultaneously.

Ether oxygen isotope effects on the chymotrypsin-catalyzed hydrolysis of N-acetyl-L-tryptophan ethyl ester and N-carbomethoxy-L-tryptophan ethyl ester have been studied by Sawyer and Kirsch.[47] The former compound gives an oxygen isotope effect $k^{16}/k^{18} = 1.018$, and the latter, 1.012, at pH 6.8, 25°. An ether oxygen isotope effect $k^{16}/k^{18} = 1.007$ has recently been obtained by O'Leary and Marlier[25] for the chymotrypsin-catalyzed hydrolysis of N-acetyl-L-tryptophan methyl ester at pH 8, 25°.

The magnitude of the oxygen isotope effects in chymotrypsin-catalyzed ester hydrolysis indicates that carbon–oxygen bond breaking is partially or wholly rate determining in the acylation of the enzyme. The absence of a pH dependence in the chymotrypsin-catalyzed hydrolyses of the two ethyl esters mentioned above between pH 4.5 and pH 9.0[74] is consistent with the suggestion of Sawyer and Kirsch[47] that the carbon–oxygen bond-breaking step is entirely rate limiting in ester hydrolysis.

It is interesting to note that in the case of enzyme-catalyzed ester hydrolysis there remains no convincing evidence that a tetrahedral intermediate occurs. Isotope effect data and substituent effects can be accommodated without recourse to such an intermediate.

5.7. Papain

Nitrogen isotope effects on the papain-catalyzed hydrolysis of amides provide an interesting contrast to chymotrypsin-catalyzed hydrolysis. Papain is a protease which is similar to chymotrypsin, except that the active site contains a functional sulfhydryl group, whereas chymotrypsin contains a hydroxyl group.

The papain-catalyzed hydrolysis of N-benzoyl-L-argininamide shows nitrogen isotope effects $k^{14}/k^{15} = 1.021$ at pH 8.0, 1.024 at pH 6.0, and 1.023 at pH 4.0 at 25°.[73,75] The magnitudes of these isotope effects compared with those in model reactions and compared with those observed for chymotrypsin indicate that the partitioning factor must be very small in this case; that is, the carbon–nitrogen bond-breaking step must be essentially totally rate determining in the acylation of the enzyme. This is consistent with the observed lack of a pH dependence in the isotope effect. Although these results can be rationalized in terms of a tetrahedral intermediate, no compelling evidence for tetrahedral intermediates exists, and it is possible that papain-catalyzed amide hydrolysis does not proceed by way of such an intermediate.

5.8. Aspartate Transcarbamylase

Carbon and oxygen isotope effects on the reaction of carbamyl phosphate with aspartate catalyzed by aspartate transcarbamylase have been studied by Stark.[76] The carbon isotope effect for carbamyl phosphate-^{14}C is $k^{12}/k^{14} = 1.01$ or slightly larger in the range pH 6–9. At pH 10, the isotope effect is about 1.05. The oxygen isotope effect on the same reaction was measured by a double-label technique using ^{14}C-^{18}O-carbamyl phosphate. Isotope ratio measurements could then be made by scintillation counting of ^{14}C in the product. The oxygen isotope effect is $k^{16}/k^{18} = 1.02$ at pH 6 and approaches 1.00 at higher pH. It appears that transfer of the carbamyl group is partially rate determining at the extremes of pH but not at the pH optimum of the enzyme.

5.9. Catechol-O-methyltransferase

Hydrogen and carbon isotope effects on the reaction of labeled S-adenosylmethionine with 3,4-dihydroxyacetophenone catalyzed by catechol-O-methyltransferase have been measured by Hegazi et al.[52] The hydrogen isotope effect for S-adenosylmethionine-methyl-D$_3$ is $k^{H}/k^{D} = 0.83$ at 37°, pH 7.6. The carbon isotope effect for methyl-^{13}C-S-adenosylmethionine is $k^{12}/k^{13} = 1.09$ under the same conditions. Both these isotope effects indicate that the methyl transfer is a classical S_N2 reaction. The displacement step is clearly rate determining.

5.10. Carbon Isotope Effects in Carboxylation Reactions

Two different photosynthetic mechanisms exist in plants. C_3 plants incorporate carbon dioxide by carboxylation of ribulose diphosphate. C_4 plants incorporate carbon dioxide by carboxylation of phosphoenolpyruvate. It has been known for some time that plants which use the C_4 pathway are richer in carbon-13 than plants grown in the same atmosphere which use the C_3 pathway. This difference in carbon isotope content can be used to distinguish between the two types of plants.[77]

A number of measurements of the carbon isotope effect associated with the enzymic reaction of carbon dioxide with phosphoenolpyruvate have been reported. For the most part these fall into the range $k^{12}/k^{13} = 1.002$–1.003.[78,79a] The carbon isotope effect associated with carboxylation of ribulose diphosphate is clearly much larger than this; various measurements give values in the range $k^{12}/k^{13} = 1.02$–1.03.[79b,79c] There is clearly a large difference in carbon isotope effect between carbon fixed by phosphoenolpyruvate carboxylase and carbon fixed by ribulose diphosphate carboxylase, and this difference is probably the principal controlling factor responsible for carbon isotope fractionation in plants.

However, the carbon isotope effects reported to date have been measured with impure enzyme preparations under conditions where the fractionation between CO_2 and HCO_3^- may be a significant source of error. In addition, isotopic analyses have been made on combusted samples of the carboxylation products. Even if the carbon isotopic compositions of the starting materials are accurately known, this procedure can produce large errors. Better measurements of these important isotope effects are needed.

5.11. Nitrogen Fixation

In connection with various geochemical and oceanographic studies, a number of nitrogen isotope effects have been measured for important processes of nitrogen metabolism. In all cases the measurements have been made with crude cell preparations, and it is possible that purified enzymes might give somewhat different results. Nonetheless, these results provide some indication of the fractionation factors which might be observed during nitrogen fixation and denitrification.

Hoering and Ford[80] measured nitrogen isotope effects on the fixation of molecular nitrogen by various species of *Azotobacter*. They found $k^{14}/k^{15} = 1.000$. Thus, some chemical step other than reduction of the nitrogen–nitrogen triple bond is rate determining.

Crude nitrate reductase from denitrifying bacteria reduces nitrate to molecular nitrogen. A nitrogen isotope effect $k^{14}/k^{15} = 1.02$ was observed for this reaction at $25°$.[81] Other nitrogen fractionation factors have also been reported.[81]

5.12. Other Enzymes

Carbon isotope effects on the hydrolysis of urea catalyzed by urease have been studied by at least two independent groups.[82-84] It is clear that the isotope effects are near $k^{12}/k^{13} = 1.01$, but unexplained variations of isotope effect with enzyme preparation and lack of correspondence between carbon-13 effects and carbon-14 effects make the precise values uncertain. A much larger effect, $k^{12}/k^{13} = 1.055$, is observed in the acid-catalyzed hydrolysis of urea at 100°.[85]

An oxygen isotope effect $k^{16}/k^{18} = 1.013$ has been reported for cytochrome oxidase using succinate as substrate.[86] The oxidation of cresol by tyrosinase shows an oxygen isotope effect $k^{16}/k^{18} = 1.010$.[86]

The decarboxylation of formic acid by formate dehydrogenase from *Clostridium* shows a carbon isotope effect $k^{12}/k^{13} = 1.026$ at 20°.[61a]

A large nitrogen isotope effect, $k^{14}/k^{15} = 1.047$, has been observed for the glutamate-dehydrogenase-catalyzed reaction of ammonia with α-keto-glutarate in the presence of NADPH at pH 8.1.[54] The rate-determining step under these conditions is probably the reaction of ammonia with α-keto-glutarate. This isotope effect seems very large, particularly considering that this is a bond-forming reaction, and such reactions often give rise to quite small isotope effects. However, the solution to this dilemma probably lies in the fact that the starting state for the reaction is the ammonium ion, whereas the nitrogen in the transition state is much less charged. By analogy with the large nitrogen isotope effect on the basicity of ammonia (Section 2.5), this change in bonding to nitrogen is expected to produce a large effect.

6. CONCLUSIONS

Enzymes operate by complex, multistep mechanisms, and in most cases more than one of these steps contribute in an important way to the observed rate. As a result, heavy-atom isotope effects observed in enzymic reactions reflect not only the structure of the transition state for the step in which the isotopic atom undergoes transformation but also the rate of this step relative to other steps in the mechanism.

Useful information about the structures of transition states cannot usually be obtained from studies of heavy-atom isotope effects on enzyme-catalyzed reactions. Instead, the most expeditious approach is often to assume an isotope effect on the key step and use that as a starting point for obtaining additional kinetic information. This technique is strengthened by comparison of the isotope effects obtained under a variety of reaction conditions.

The usefulness of heavy-atom isotope effects for understanding the details of enzyme-catalyzed reactions increases in direct proportion to the amount of other information which is available about the enzyme mechanism. In favorable cases subtle changes may usefully be probed by means of heavy-atom isotope effects. Small changes such as allosteric effects should be particularly amenable to study by means of isotope effects, but such studies have not yet been reported.

The most serious limitation on the use of heavy-atom isotope effects at the present time is probably the experimental limitation. The competitive method using an isotope-ratio mass spectrometer will continue to be the best method, but the cost of such machines and the difficulty of preparing samples for analysis in any but the simplest cases will continue to limit the usefulness of the method. Radiochemical methods, though more convenient, are unlikely to come into widespread use because of lack of precision. The equilibrium perturbation method represents a significant advance that should find widespread application. Direct kinetic methods are probably the most difficult of all, but they may find increasing use because of their versatility.

In closing, it is worthwhile to point out once more the many potential sources of error in measurement of these small but very useful effects. It is incumbent upon authors describing isotope-effect experiments to convince their readers that their results are correct. It is incumbent upon readers to make this evaluation in every case.

ACKNOWLEDGMENT

Work in the author's laboratory was supported by the National Science Foundation, the National Institutes of Health, and the University of Wisconsin Graduate School. The author is grateful to Professors J. F. Kirsch and R. L. Schowen for permission to quote unpublished work.

REFERENCES

1. A. Fry, in: *Isotope Effects in Chemical Reactions* (C. J. Collins and N. S. Bowman, eds,), pp. 364–414, Van Nostrand Reinhold, New York (1970).
2. J. Bigeleisen and M. Wolfsberg, Theoretical and experimental aspects of isotope effects in chemical kinetics, *Adv. Chem. Phys.* **1**, 15–76 (1958).
3. L. Melander, *Isotope Effects on Reaction Rates*, Ronald, New York (1960).
4. A. MacColl, Heavy-atom kinetic isotope effects, *Annu. Rep. Chem. Soc. B* **71**, 77–101 (1974).
5. M. J. Stern and M. Wolfsberg, "Heavy-Atom Kinetic Isotope Effects, an Indexed Bibliography," NBS Special Publication No. 349, U.S. Department of Commerce, Washington, D.C. (1972).
6. H. Simon and D. Palm, Isotope effects in organic chemistry and biochemistry, *Angew. Chem. Int. Ed. Engl.* **5**, 920–933 (1966).
7. J. H. Richards, in: *The Enzymes*, 3rd ed. (P. D. Boyer, ed.), Vol. 2, pp. 321–333, Academic Press, New York (1970).
8. (a) J. F. Kirsch, Mechanism of enzyme action, *Annu. Rev. Biochem.* **42**, 205–234 (1973). (b) W. W. Cleland, M. H. O'Leary, and D. B. Northrop, *Isotope Effects on Enzyme-Catalyzed Reactions*, University Park Press, Baltimore (1977). (c) M. H. O'Leary, in: *Bioorganic Chemistry* (E. E. van Tamelen, ed.), pp. 259–275, Academic Press, New York (1977).
9. W. A. Van Hook, in: *Isotope Effects in Chemical Reactions* (C. J. Collins and N. S. Bowman, eds.), pp. 1–89, Van Nostrand Reinhold, New York (1970).

10. P. C. Vogel and M. J. Stern, Temperature dependences of kinetic isotope effects, *J. Chem. Phys.* **54**, 779–796 (1971).
11. T. T.-S. Huang, W. J. Kass, W. E. Buddenbaum, and P. E. Yankwich, Anomalous temperature dependence of kinetic carbon isotope effects and the phenomenon of crossover, *J. Phys. Chem.* **72**, 4431–4446 (1968).
12. M. J. Stern and P. C. Vogel, Relative $^{14}C-^{13}C$ kinetic isotope effects, *J. Chem. Phys.* **55**, 2007–2013 (1971).
13. G. A. Ropp and V. F. Raaen, A comparison of the magnitudes of the isotope intermolecular effects in the decarboxylations of malonic-1-C^{14} acid and malonic-2-C^{14} acid at 154°, *J. Am. Chem. Soc.* **74**, 4992–4994 (1952).
14. A. Fry and M. Calvin, The C^{14} isotope effect in the decarboxylation of α-naphthyl- and phenylmalonic acids, *J. Phys. Chem.* **56**, 901–905 (1952).
15. H. H. Huang and F. A. Long, The decarboxylation of azulene-1-carboxylic acid. II. Carbon-13 isotope effects, *J. Am. Chem. Soc.* **91**, 2872–2875 (1969).
16. J. Bron and J. B. Stothers, Carbon-13 kinetic isotope effects. III. Temperature dependence of k^{12}/k^{13} for 1-bromo-1-phenylethane alcoholysis, *Can. J. Chem.* **46**, 1435–1439 (1968).
17. J. B. Stothers and A. N. Bourns, Carbon-13 kinetic isotope effects in the solvolysis of 1-bromo-1-phenylethane, *Can. J. Chem.* **38**, 923–935 (1960).
18. J. Bron and J. B. Stothers, Carbon-13 kinetic isotope effects. V. Substituent effects on k^{12}/k^{13} for alcoholysis of 1-phenyl-1-bromoethane, *Can. J. Chem.* **47**, 2506–2509 (1969).
19. J. Bron and J. B. Stothers, Carbon-13 kinetic isotope effects. IV. The effect of temperature on k^{12}/k^{13} for benzyl halides in bimolecular reactions, *Can. J. Chem.* **46**, 1825–1829 (1968).
20. C. R. Turnquist, J. W. Taylor, E. P. Grimsrud, and R. C. Williams, Temperature dependence of chlorine kinetic isotope effects for aliphatic chlorides, *J. Am. Chem. Soc.* **95**, 4133–4138 (1973).
21. D. G. Graczyk and J. W. Taylor, Chlorine kinetic isotope effects in nucleophilic substitution reactions. Support for the ion pairs mechanism in the reactions of *p*-methoxybenzyl chloride in 70% aqueous acetone, *J. Am. Chem. Soc.* **96**, 3255–3261 (1974).
22. E. P. Grimsrud and J. W. Taylor, Chlorine kinetic isotope effects in nucleophilic displacements at a saturated carbon, *J. Am. Chem. Soc.* **92**, 739–741 (1970).
23. C. B. Sawyer and J. F. Kirsch, Kinetic isotope effects for reactions of methyl formate-methoxyl-$^{18}O^{1}$, *J. Am. Chem. Soc.* **95**, 7375–7381 (1973).
24. G. M. Blackburn and W. P. Jencks, The mechanism of the aminolysis of methyl formate. *J. Am. Chem. Soc.* **90**, 2638–2645 (1968).
25. M. H. O'Leary and J. F. Marlier, unpublished results.
26. M. H. O'Leary, unpublished results.
27. C. A. Bunton, T. A. Lewis, and D. R. Llewellyn, The mechanism of hydrolysis at carbonyl carbon, *Chem. Ind. (London)* **1954**, 1154–1155.
28. M. L. Bender and R. D. Ginger, Solvent effects in the hydrolysis of an ester, an amide, and an acid chloride involving carbonyl-oxygen exchange, *Suom. Kemistil. B* **33**, 25–30 (1960).
29. M. L. Bender and M. C. Chen, Acylium ion formation in the reactions of carboxylic acid derivatives. III. The hydrolysis of 4-substituted-2,6-dimethylbenzoyl chlorides, *J. Am. Chem. Soc.* **85**, 30–36 (1963).
30. C. G. Mitton and R. L. Schowen, Oxygen isotope effects by a noncompetitive technique: The transition-state carbonyl stretching frequency in ester cleavage, *Tetrahedron Lett.* **1968**, 5803–5806.
31. L. L. Brown and J. S. Drury, Nitrogen isotope effects in the decomposition of diazonium salts, *J. Chem. Phys.* **43**, 1688–1691 (1965).
32. C. G. Swain, J. E. Sheats, and K. G. Harbison, Nitrogen isotope effects in the hydrolysis of benzenediazonium salts, *J. Am. Chem. Soc.* **97**, 796–798 (1975).
33. A. Fry, Isotope effect studies of elimination reactions, *Chem. Soc. Rev.* **1**, 163–210 (1972).
34. P. J. Smith and A. N. Bourns, Isotope effect studies of elimination reactions. VI. The mechanism of the bimolecular elimination reaction of 2-arylethylammonium ions, *Can. J. Chem.* **48**, 125–132 (1970).

35. J. Banger, A. Jaffe, A.-C. Lin, and W. H. Saunders, Jr., Carbon isotope effects on proton transfers from carbon and the question of hydrogen tunneling, *J. Am. Chem. Soc.* **97**, 7177–7178 (1975).
36. J. Banger, A. Jaffe, A.-C. Lin, and W. H. Saunders, Jr., Carbon-13 isotope effects on proton transfers from carbon, *J. Am. Chem. Soc.* **97**, 7177–7178 (1975).
37. P. J. Smith and A. N. Bourns, Isotope effect studies on elimination reactions. IX. The nature of the transition state for the E2 reaction of 2-arylethylammonium ions with ethoxide in ethanol, *Can. J. Chem.* **52**, 749–760 (1974).
38. J. Bigeleisen, M. W. Lee, and F. Mandel, Equilibrium isotope effects, *Annu. Rev. Phys. Chem.* **24**, 407–440 (1973).
39. (a) H. G. Thode, M. Shima, C. E. Rees, and K. V. Krishnamurty, Carbon-13 isotope effects in systems containing carbon dioxide, bicarbonate, carbonate, and metal ions, *Can. J. Chem.* **43**, 582–595 (1965). (b) J. W. Bayles, J. Bron, and S. O. Paul, Secondary carbon-13 isotope effect on the ionization of benzoic acid, *J. Chem. Soc. Faraday Trans. 1* **7**, 1546–1552 (1976). (c) M. H. O'Leary and C. J. Yapp, *Biochem. Biophys. Res. Commun.*, in press.
40. M. H. O'Leary and A. P. Young, unpublished results.
41. H. G. Thode, R. L. Graham, and J. A. Ziegler, A mass spectrometer and the measurement of isotope exchange factors, *Can. J. Res. Sect. B* **23**, 40–47 (1945).
42. M. Cohn and H. C. Urey, Oxygen exchange reactions of organic compounds and water, *J. Am. Chem. Soc.* **60**, 679–687 (1938).
43. I. Dostrovsky and F. S. Klein, Mass spectrometric determination of oxygen in water samples, *Ind. Eng. Chem. Anal. Ed.* **24**, 414–415 (1952).
44. R. F. Weiss, Helium isotope effect in solution in water and seawater, *Science* **168**, 247–248 (1970).
45. J. L. Borowitz and F. S. Klein, Vapor pressure isotope effects in methanol, *J. Phys. Chem.* **75**, 1815–1820 (1971).
46. W. W. Cleland, What limits the rate of an enzyme-catalyzed reaction?, *Acc. Chem. Res.* **8**, 145–151 (1975).
47. C. B. Sawyer and J. F. Kirsch, Kinetic isotope effects for the chymotrypsin catalyzed hydrolysis of ethoxyl-^{18}O labeled specific ester substrates, *J. Am. Chem. Soc.* **97**, 1963–1964 (1975).
48. S. Seltzer, G. A. Hamilton, and F. H. Westheimer, Isotope effects in the enzymatic decarboxylation of oxalacetic acid, *J. Am. Chem. Soc.* **81**, 4018–4024 (1959).
49. M. H. O'Leary and R. L. Baughn, Acetoacetate decarboxylase. Identification of the rate-determining step in the primary amine catalyzed reaction and in the enzymic reaction, *J. Am. Chem. Soc.* **94**, 626–630 (1972).
50. W. W. Cleland, Partition analysis and the concept of net rate constants as tools in enzyme kinetics, *Biochemistry* **14**, 3220–3224 (1975).
51. D. G. Gorenstein, Oxygen-18 isotope effect in the hydrolysis of 2,4-dinitrophenyl phosphate. A monomeric metaphosphate mechanism, *J. Am. Chem. Soc.* **94**, 2523–2525 (1972).
52. M. F. Hegazi, R. T. Borchardt, and R. L. Schowen, unpublished results.
53. G. Hübner, H. Neef, G. Fischer, and A. Schellenberger, ^{15}N isotope effect as direct evidence for the direct participation of the amino group of thiamine pyrophosphate in the enzymatic decarboxylation of α-ketoacids by yeast pyruvate decarboxylase, *Z. Chem.* **15**, 221 (1975).
54. M. I. Schimerlik, J. E. Rife, and W. W. Cleland, Equilibrium perturbation by isotope substitution, *Biochemistry* **14**, 5347–5354 (1975).
55. L. O. Krampitz, H. G. Wood, and C. H. Werkman, Enzymatic fixation of carbon dioxide in oxalacetate, *J. Biol. Chem.* **147**, 243–253 (1943).
56. J. P. Guthrie and F. Jordan, Amine-catalyzed decarboxylation of acetoacetic acid. The rate constant for decarboxylation of a β-imino acid, *J. Am. Chem. Soc.* **94**, 9136–9141 (1972).
57. S. Warren, B. Zerner, and F. H. Westheimer, Acetoacetate decarboxylase. Identification of lysine at the active site, *Biochemistry* **5**, 817–823 (1966).
58. G. A. Hamilton, Studies on the mechanism of enzymatic decarboxylation, Ph.D. dissertation, Harvard University, Cambridge, Mass. (1959).

59. M. H. O'Leary, D. T. Richards, and D. W. Hendrickson, Carbon isotope effects on the enzymatic decarboxylation of glutamic acid, *J. Am. Chem. Soc.* **92**, 4435–4440 (1970).

60. M. L. Fonda, Glutamate decarboxylase. Substrate specificity and inhibition by carboxylic acids, *Biochemistry* **11**, 1304–1309 (1972).

61. (a) T. C. Hoering, The carbon isotope effect on the rate of enzymatic decarboxylation of formic and glutamic acid, *Carnegie Inst. Washington Pap. Geophys. Lab.* **1363**, 200–201 (1960–1961). (b) M. H. O'Leary and G. J. Piazza, *J. Am. Chem. Soc.*, in press, and unpublished work.

62. M. L. Uhr, V. W. Thompson, and W. W. Cleland, The kinetics of pig heart triphosphopyridine nucleotide–isocitrate dehydrogenase, *J. Biol. Chem.* **249**, 2920–2927 (1974).

63. M. H. O'Leary, The rate-determining step in the oxidative decarboxylation of isocitric acid, *Biochim. Biophys. Acta* **235**, 14–18 (1971).

64. M. H. O'Leary and J. A. Limburg, Isotope effect studies of the role of metal ions in isocitrate dehydrogenase, *Biochemistry* **16**, 1129–1135 (1977).

65. N. Ramachandran, M. Durbano, and R. F. Colman, Kinetic isotope effects in the NAD- and NADP-specific isocitrate dehydrogenases of pig heart, *FEBS Lett.* **49**, 129–133 (1974).

66. D. B. Northrop and W. W. Cleland, The kinetics of metal ion activators for TPN-isocitrate dehydrogenase, *Fed. Proc. Fed. Am. Soc. Exp. Biol.* **19**, 408 (1970).

67. P. H. Abelson and T. C. Hoering, Carbon isotope fractionation in formation of amino acids by photosynthetic organisms, *Proc. Nat. Acad. Sci. USA* **47**, 623–632 (1961).

68. M. H. O'Leary, Carbon isotope effect on the decarboxylation of pyruvic acid, *Biochem. Biophys. Res. Commun.* **73**, 614–618 (1976).

69. (a) K. Burton and H. A. Krebs, Free-energy changes with the individual steps of the tricarboxylic acid cycle, glycolysis and alcoholic fermentation, and with the hydrolysis of the pyrophosphate groups of adenosine triphosphate, *Biochem. J.* **54**, 94–107 (1953). (b) M. J. DeNiro and S. Epstein, Mechanism of carbon isotope fractionation associated with lipid synthesis, *Science* **197**, 261–263 (1977).

70. G. P. Hess, in: *The Enzymes*, 3rd ed. (P. D. Boyer, ed.), Vol. 3, pp. 213–248, Academic Press, New York (1971).

71. M. H. O'Leary and M. D. Kluetz, Identification of the rate-limiting step in the chymotrypsin-catalyzed hydrolysis of N-acetyl-L-tryptophanamide, *J. Am. Chem. Soc.* **92**, 6089–6090 (1970).

72. M. H. O'Leary and M. D. Kluetz, Nitrogen isotope effects on the chymotrypsin-catalyzed hydrolysis of N-acetyl-L-tryptophanamide, *J. Am. Chem. Soc.* **94**, 3585–3589 (1972).

73. M. H. O'Leary, M. Urberg, and A. P. Young, Nitrogen isotope effects on the papain-catalyzed hydrolysis of N-benzoyl-L-argininamide, *Biochemistry* **13**, 2077–2081 (1974).

74. C. B. Sawyer and J. F. Kirsch, personal communication.

75. M. H. O'Leary and M. D. Kluetz, The rate-determining step in the acylation of papain by N-benzoyl-L-argininamide, *J. Am. Chem. Soc.* **94**, 665 (1972).

76. G. R. Stark, Aspartate transcarbamylase, *J. Biol. Chem.* **246**, 3064–3068 (1971).

77. B. N. Smith, Natural abundance of the stable isotopes of carbon in biological systems, *BioScience* **22**, 226–231 (1972).

78. T. Whelan, W. M. Sackett, and C. R. Benedict, Enzymatic fractionation of carbon isotopes by phosphoenolpyruvate carboxylase from C_4 plants, *Plant Physiol.* **51**, 1051–1054 (1973).

79. (a) P. H. Reibach and C. R. Benedict, Fractionation of stable carbon isotopes by phosphoenolpyruvate carboxylase from C_4 plants, *Plant Physiol.* **59**, 564–568 (1977). (b) R. Park and S. Epstein, Carbon isotope fractionation during photosynthesis, *Geochim. Cosmochim. Acta* **21**, 110–126 (1960). (c) J. T. Christeller, W. A. Laing, and J. H. Troughton, Isotope discrimination by ribulose 1,5-diphosphate carboxylase, *Plant Physiol.* **57**, 580–582 (1976).

80. T. C. Hoering and H. T. Ford, The isotope effect in the fixation of nitrogen by *Azotobacter*, *J. Am. Chem. Soc.* **82**, 376–378 (1960).

81. Y. Miyake and E. Wada, The isotope effect on the nitrogen in biochemical oxidation–reduction reactions, *Rec. Oceanogr. Works Jpn.* **11**, 1–6 (1971).

82. K. R. Lynn and P. E. Yankwich, ^{13}C kinetic isotope effects in the urease-catalyzed hydrolysis of urea. I. Temperature dependence, *Biochim. Biophys. Acta* **56**, 512–530 (1962).

sis of urea. II. Influence of reaction variables other than temperature, *Biochim. Biophys. Acta* **81**, 533–547 (1964).

84. J. A. Schmitt, A. L. Myerson, and F. Daniels, Relative rates of hydrolysis of urea containing C^{14}, C^{13}, and C^{12}, *J. Phys. Chem.* **56**, 917–920 (1952).

85. J. A. Schmitt and F. Daniels, The carbon isotope effect in the acid hydrolysis of urea, *J. Am. Chem. Soc.* **75**, 3564–3566 (1953).

86. D. E. Feldman, H.T. Yost, Jr., and B. B. Benson, Oxygen isotope fractionation in reactions catalyzed by enzymes, *Science* **129**, 146–147 (1959).

8

Magnetic-Resonance Approaches to Transition-State Structure

Albert S. Mildvan

1. INTRODUCTION

Measurement of nuclear magnetic relaxation rates in the presence of paramagnetic probes, an NMR method which determines distances from the individual atoms of a molecule in solution to a nearby paramagnetic reference point, has emerged in the past decade as a useful approach to the study of the conformation and arrangement of enzyme-bound substrates in solution.[1-8] Such NMR studies in solution, like X-ray diffraction in the crystalline state, detect individual atoms. The obvious advantage of observing complexes in solution, with directly measurable kinetic and thermodynamic properties, is that the relevance of such complexes to catalysis is testable.

By use of the NMR method most thoroughly tested by comparison of some 13 distances determined in solution[5-7] with the corresponding crystallographic distances, namely the measurement of paramagnetic effects on the longitudinal relaxation rates $(1/T_1)$ of substrate nuclei at several magnetic fields, distances $\leq 14\,\text{Å}$ from a paramagnetic center have been calculated with precisions of $\pm 10\%$, and conformations of enzyme-bound substrates with molecular weights as large as 824 (propionyl-coenzyme A)[8] have been determined. Limitations of the method include the limited range of accurately measurable distances and the complexity of the analysis when more than one paramagnetic species is present.[8] The major mechanistic disadvantage of the paramagnetic probe–$1/T_1$ method, which requires fast exchange between free and bound substrates, is that elevated ground states of bound substrates

Albert S. Mildvan • The Institute for Cancer Research, Fox Chase Cancer Center, Philadelphia, Pennsylvania.

and of bound metal–substrate complexes rather than transition states are observed. From the conformations of the bound substrates and the arrangements of bound metals and substrates, only inferences may be made about the structure of transition states. The correctness of these inferences depends on the extent to which the conformations of enzyme-bound substrates approximate the transition states for the catalytic process.[9-12] A more direct approach may be achieved by magnetic-resonance studies of complexes of enzymes with transition-state analogs of substrates.[13-17] However, thus far only limited information has been obtained even by this approach.

A comprehensive review of the advantages and limitations of NMR and of other methods for determining the conformations and arrangements of enzyme-bound substrates has been written.[18] This review will briefly consider the applications of the paramagnetic probe–$1/T_1$ method to the transition-state structures and mechanisms in three classes of enzyme reactions: (1) carbonyl polarization, (2) elimination reactions, and (3) phosphoryl- and nucleotidyl-transferring reactions.

2. METHODOLOGY

Since the development of the theory of paramagnetic effects on nuclear relaxation rates[19-23] and its first application to metal–substrate interactions on enzymes,[1-4] several reviews[4,5,24] and textbooks[25-27] have treated the techniques for measuring relaxation rates and the quantitative interpretation of such data. We shall therefore provide here only a qualitative summary of the experimental approach.

The longitudinal $(1/T_1)$ and transverse $(1/T_2)$ relaxation rates of a population of magnetic nuclei represent the first-order rate constants for the equilibration of the magnetization of the spins along the magnetic field and the dephasing of the spins in a plane perpendicular to the magnetic field, respectively. Pulsed NMR methods have long been used for measuring the relaxation rates of solvent protons,[28] and continuous-wave NMR methods have been used to measure the relaxation rates of fluorine [1,2] protons,[3] and phosphorus atoms[29,30] of enzyme-bound substrates. The introduction of Fourier transform methods has greatly facilitated the latter studies[29-33] and has enlarged the slope of detectable nuclei to include ^{13}C.[32]

From the theory, [4-6,18-27] the paramagnetic effect of an unpaired electron, for example, an enzyme-bound transition metal or spin label, on the longitudinal relaxation rate $(1/T_{1p})$ of a nearby magnetic nucleus, for example, a proton or phosphorus atom of a substrate which is exchanging into the paramagnetic enzyme complex, depends predominantly on four parameters: the lifetime of the complex (τ_m), the relative stoichiometry or coordination number (q) of the substrate and paramagnet in the complex, the correlation time for electron–nuclear dipolar interaction (τ_c), and the distance (r) from the unpaired electron to the nucleus in the complex. The corresponding paramagnetic effect on the transverse relaxation rate $(1/T_{2p})$ of this nucleus is a different function of these same four parameters and in addition contains a contact contribution and chemical shift contribution resulting from contact and dipolar interactions. The paramagnetic contribution to the transverse relaxation rate $(1/T_{2p})$ sets a lower limit to $1/\tau_m$, the pseudo-first-order rate constant for dissociation of the substrate from the paramagnetic complex,[30] and measurements of $1/T_{2p}$ as a function of temperature and frequency may be used, in suitable cases, to evaluate $1/\tau_m$ and provide kinetic information on the observed complex.[4,34] When $1/T_{2p}$ exceeds $1/T_{1p}$ by an order of magnitude or more it may safely be concluded that the rate constant for dissociation of the substrate $(1/\tau_m)$ greatly exceeds $1/T_{1p}$, i.e., that the lifetime of the complex (τ_m) contributes little to T_{1p}, eliminating one of the four unknown parameters in $1/T_{1p}$.[30] The correlation time can be evaluated by measurements of $1/T_{1p}$ at several magnetic fields. The two remaining parameters q and r cannot be separately evaluated by measurements of $1/T_{1p}$ alone. However, an independent evaluation of the stoichiometry (q) of the paramagnet and substrate in the complex by appropriate binding studies permits a direct calculation of the distance (r).

This approach provides a highly precise method for determining distances from paramagnetic centers to substrates on enzymes. Since $1/T_{1p}$ is inversely proportional to the sixth power of r, errors in the measurement of $1/T_{1p}$ and in the evaluation of τ_c are truncated, resulting in errors $\leq 10\%$ in the determination of r.

Conversely, an independent evaluation of the distance r by X-ray analysis of model complexes permits a direct calculation of q,[29,35] the coordination number for rapidly exchanging water ligands on an enzyme-bound metal, and permits the estimation of changes in q as substrates bind to the enzyme–metal complex. In this type of calculation there is no truncation of errors resulting in uncertainties as great as 25–50% in estimates of q. Paramagnetic effects on nuclear relaxation thus provide a precise method for determining distances but only an approximate method for determining coordination numbers for rapidly exchanging water ligands.

When a set of distances from a paramagnetic center to several nuclei of a bound substrate has been determined from $1/T_{1p}$ the conformation of the substrate in the paramagnetic complex is determined by model building, which is usually done in three stages. First a stick model of the substrate is constructed, and its conformation is adjusted to be consistent with the measured distances from the substrate atoms to the paramagnetic reference point.[29,30] In this process one can often gain qualitative insight into the uniqueness of the model. Next, a space-filling model of the substrate is constructed to test for van der Waals overlap and to further explore the uniqueness of the conformation. Finally a systematic computer search among thousands of conformations is carried out that eliminates those which require significant van der Waals overlap or which require distances to the paramagnetic reference point beyond the values and errors of those measured experimentally.[8,36-39]

Since many important biochemical reactions involve the simultaneous interaction of two or more substrates with an enzyme, the separate determinations of the conformations of the individual substrates bound at their respective active sites, while necessary, are often insufficient to clarify an enzyme mechanism. In addition, one or more intersubstrate distances may be needed. For such measurements a paramagnetic analog of one of the substrates is required,[40] and its effects on the longitudinal relaxation rates of nuclei of the other substrate on the enzyme are measured and analyzed as outlined above.[41] The first measurements of intersubstrate distances were made on dehydrogenases using a spin-labeled analog of NAD.[41-43] More recently tempo–ATP[44] (a paramagnetic analog of ATP), Cr^{3+}–ATP (a substitution inert paramagnetic analog of MgATP[45]), and R · CoA (a paramagnetic ester of coenzyme A[46,47]) have been used to determine intersubstrate distances on other enzymes.

3. APPLICATIONS OF MAGNETIC-RESONANCE METHODS

3.1. Carbonyl-Polarizing Enzymes

3.1.1. Biotin Carboxylases

Transcarboxylase, which has been purified and characterized by Wood and co-workers,[48,49] catalyzes the transfer of a carboxyl group from methyl malonyl-CoA to pyruvate via an enzyme-bound biotin. The enzyme has been found to contain Cu^{2+} in addition to Zn^{2+} and Co^{2+}.[50] The total metal content is ~ 12 g ions per mole or ~ 2 metal ions per biotin.[50] From EPR studies at liquid He temperature, the Co^{2+} appears to be in a slightly distorted octahedral environment, while Cu^{2+} is in a symmetric environment. From $1/T_1$ of water protons at various frequencies, two rapidly exchanging water ligands on the enzyme-bound Co^{2+} were detected, while the Cu^{2+} was inaccessible to water. Formation of the enzyme pyruvate complex decreased the number of fast-exchanging water ligands on Co^{2+} by 1. However, the Co^{2+} to pyruvate distances, as calculated from $1/T_1$ of the carbon atoms and protons of pyruvate by ^{13}C and proton NMR (5.0–6.3 Å), indicated a second sphere complex (Fig. 1).[50] Hence pyruvate binding in the second coordination sphere appears to have immobilized an inner-sphere water ligand so that it no longer exchanges rapidly, suggesting that the metal-bound water ligand might polarize the carbonyl group of pyruvate by hydrogen bonding or protonation rather than by direct metal coordination (Fig. 2). Similar results were obtained with the Mn^{2+} metalloenzyme pyruvate carboxylase,[32] which catalyzes the same half-reaction as transcarboxylase (Fig. 1). In the binary Mn–pyruvate complex in solution, monodentate carboxyl coordination is observed (Fig. 1).[32] Hence the active sites of these enzymes alter the metal substrate interaction, preventing direct coordination.

The effects of three preparations of transcarboxylase, containing varying mole ratios of Zn^{2+}, Co^{2+}, and Cu^{2+} at the 12 metal sites on the longitudinal

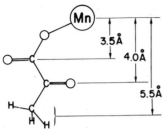
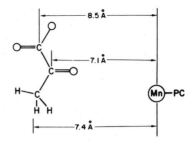
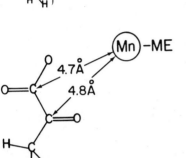
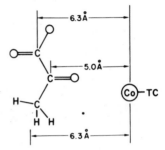
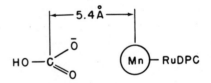
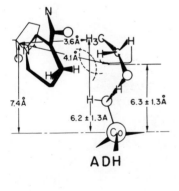
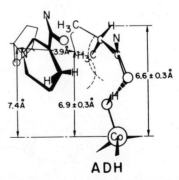

Fig. 1. Comparison of binary Mn^{2+}–pyruvate complex with conformations of five second-sphere complexes in solution as determined by nuclear relaxation. ADH, alcohol dehydrogenase[51]; TC, transcarboxylase[50]; PC, pyruvate carboxylase[32]; RuDPC, ribulose diphosphate carboxylase[52]; ME, malic enzyme.[53]

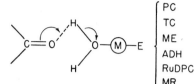

PC
TC
ME
ADH
RuDPC
MR

Fig. 2. General mechanism of carbonyl polarization by enzyme-bound metal ions based on the data of Figs. 1 and 8. MR, mandelate racemase.

$(1/T_1)$ and transverse $(1/T_2)$ relaxation rates of 12 protons and the 3 phosphorus atoms of the bound substrate, propionyl CoA, were measured at 100 and 40.5 MHz.[8] The paramagnetic effects of Co^{2+} bound at the active site, on $1/T_1$ of these nuclei, were obtained by solving the appropriate simultaneous equations. The resulting $1/T_1$ values and the correlation time (2.2 psec) determined from the frequency dependence of $1/T_1$ of water protons yielded absolute distances from the bound Co^{2+} to 7 protons (6.5–8.7 Å) as well as lower-limit distances to 5 additional protons (≥ 7.4 Å) and to the 3 phosphorus atoms (≥ 9.4 Å) of bound propionyl CoA. These 15 distances were used in a computer

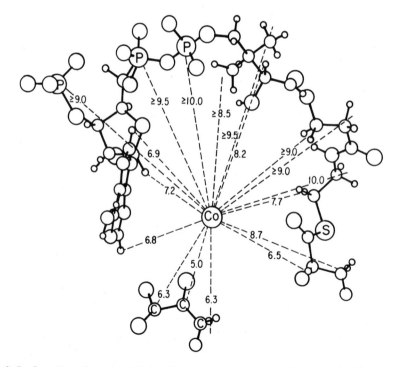

Fig. 3. Conformation of propionyl CoA and pyruvate on transcarboxylase determined by computer using the indicated distances from $1/T_1$ data (angstrom units).[8]

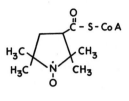

R · CoA

Fig. 4. Structure of R · CoA.[46]

search among 47,000 rotamers for that conformation of propionyl CoA which provided the best fit to the distances and minimized van der Waals overlaps. The best-fit structure shows a U shape about the Co^{2+} with an anti adenine–ribose conformation (Fig. 3).[8] From the distance between the adenine H_2 proton to the propionyl methylene carbon of this structure it is estimated that propionyl CoA is ~ 22% unfolded when bound to transcarboxylase.

To determine whether pyruvate and propionyl CoA interact with the same metal or at different metal sites, the intersubstrate distance between pyruvate and propionyl CoA was determined using R · CoA, a spin-labeled thioester of CoA (Fig. 4).[47] R · CoA binds to the propionyl CoA site of trans-carboxylase with a stoichiometry of 0.7 ± 0.2 R · CoA molecules bound per biotin residue and a $K_D = 0.33 \pm 0.12$ mM. Moreover, the dissociation constant of propionyl CoA (0.25 ± 0.10) determined by displacement of R · CoA agrees with the K_I of propionyl CoA as a competitive inhibitor against methyl-malonyl CoA (0.15 ± 0.05 mM) when redetermined under the conditions of the magnetic resonance experiments, indicating active-site binding of

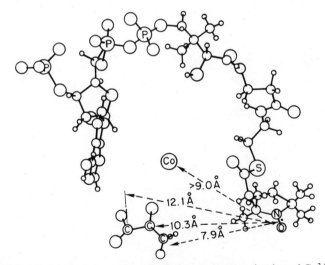

Fig. 5. Intersubstrate distances from R · CoA to pyruvate and to bound Co^{2+} on transcarboxylase.[47]

R · CoA.[47] The paramagnetic effects of bound R · CoA on the $1/T_1$ of the methyl protons and carbon atoms of bound pyruvate[47] establish the presence of a ternary complex in accord with kinetic data.[54] The $1/T_{1p}$ values of the protons of pyruvate at 40.5 MHz and 100 MHz are not exchange limited and yield a correlation time of $6 \pm 2 \times 10^{-9}$ sec and a R · CoA to pyruvate distance of 7.9 ± 0.7 Å to the methyl protons, 10.3 ± 0.8 Å to the carbonyl carbon, and 12.1 ± 0.9 Å to the carboxyl carbon.[47] The absence of an effect of the enzyme-bound Co^{2+} on the EPR spectrum of bound R · CoA indicates a lower-limit distance, ≥ 9.0 Å. These distances indicate that pyruvate and propionyl CoA are bound near the same metal ion on transcarboxylase, as shown in Fig. 5.[47]

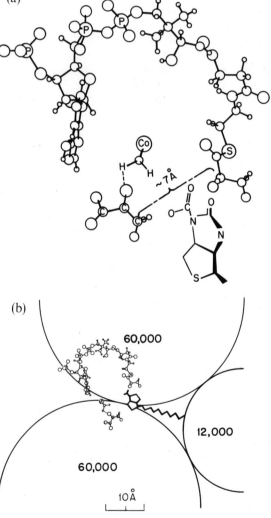

Fig. 6. Active site of transcarboxylase.[8,47] (a) Arrangement and conformations of pyruvate, propionyl CoA, and carboxybiotin derived from Fig. 5. (b) Diagram to scale of transcarboxylase subunits,[49] showing the role of the 14-Å arm of biotin.[47]

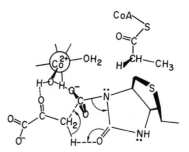

Fig. 7. Mechanism for carboxylation of pyruvate on transcarboxylase consistent with geometry and steric constraints.[47]

Using these distances as well as 3 Co^{2+} to pyruvate distances and the 15 Co^{2+} to propionyl CoA distances, a composite model of the bound substrates on transcarboxylase has been derived (Fig. 3).[8] From this model it is concluded that carboxybiotin moves, at most, only $\sim 7\,\text{Å}$ in transferring CO_2 from methyl malonyl-CoA to pyruvate during catalysis [Fig. 6(a)].[47] This close proximity was unexpected since biotin is on a 14-Å-long arm when extended and the two half-reactions of transcarboxylase occur on different subunits. The role of this long arm thus appears to be to place the transferred carboxyl group at the end of a long probe, permitting it to transverse the gap which occurs at the interface of three subunits and to insinuate itself between the CoA and pyruvate sites [Fig. 6(b)].[47] Independent evidence for the proximity of the two substrates of transcarboxylase has been obtained by Rose et al., who detected a small amount of enzyme-catalyzed tritium transfer between pyruvate and propionyl CoA.[55] These results also provide the first evidence suggesting that biotin transfers protons as well as carboxyl groups between the substrates.

A mechanism showing the role of biotin in carboxyl-transfer reactions consistent with the geometric data, with steric considerations, and with the recent isotopic studies is given in Fig. 7.[47,55,56]

3.1.2. Mandelate Racemase

Another example of the orientational effect of an enzyme on a metal–substrate interaction is provided by the isomerase, mandelate racemase, which catalyzes the interconversion of D- and L-mandelate. The binding of the activator, Mn^{2+}, the substrate D-mandelate, and competitive inhibitors with mandelate racemase have been studied by EPR and by $1/T_1$ of the protons of water and the ^{13}C-enriched C_1 and C_2 carbon atoms of mandelate.[57] In the binary Mn^{2+}–mandelate complex, the Mn^{2+} to carbon distances indicate bidentate chelation, while on the enzyme a linear second-sphere complex is detected [Fig. 8(a)]. The number of fast-exchanging water ligands on the enzyme-bound Mn^{2+} decreases from 3 to 2 when the substrate binds, suggesting the intervening ligand to be a water molecule. A Michael-type mechanism

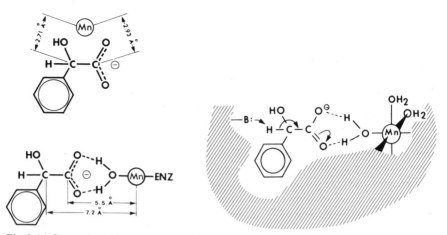

Fig. 8. (a) Comparison of structures in solution of binary Mn^{2+}–mandelate complex and ternary mandelate racemase–Mn^{2+}–mandelate complex.[57] (b) Mechanism of mandelate racemase[57] consistent with the geometry of Fig. 8(a).

[Fig. 8(b)] with carbonyl polarization in the transition state is consistent with the linear complex of Fig. 8(a).[57]

3.1.3. Other Carbonyl-Polarizing Enzymes

Second-sphere complexes are also detected on three other enzymes which use divalent cations to polarize carbonyl groups of their substrates: malic enzyme, which activates pyruvate;[53] ribulose diphosphate carboxylase,[52] which activates CO_2; and alcohol dehydrogenase,[51] which activates aldehydes. From these six examples (Figs. 1 and 8), it appears that a general mode of activating carbonyl groups of substrates by enzyme-bound metals is not by direct coordination but rather by protonation via an intervening water ligand (Fig. 2). The participation of a metal-bound water in the alcohol dehydrogenase reaction is strongly supported by the detection of a small inverse secondary kinetic solvent isotope effect[58] in accord with a model study of a known de-protonation of a metal-bound water ligand.[59] Solvent isotope effects, discussed elsewhere in this volume,[60] should be examined with other metal-activated carbonyl-polarizing enzymes as well. A second-sphere mechanism is probably also applicable to yeast aldolase,[61] from the upper-limit metal–substrate distances, and to carboxypeptidase,[62] since the esterase activity of this enzyme is preserved when Co^{2+} at the active site is oxidized to substitution inert Co^{3+}, although the peptidase activity is lost. This generality is not, however, universal since it does not apply to the D-xylose isomerase–Mn–α-D-xylose complex where we now obtain Mn^{2+} to substrate distances of 9 Å based on properly determined correlation times.[63] Hence, contrary to our previous view,[64] the required divalent cation in D-xylose isomerase does not interact directly with the bound substrate but plays a more distant structural role.[63]

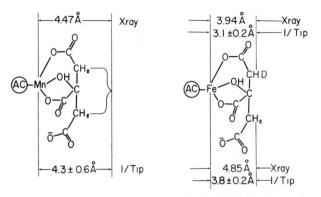

Fig. 9. Comparison of distances determined in solution by NMR in the aconitase–Mn–citrate and aconitase–Fe^{2+}–citrate complexes[65a] with distances in the binary crystalline metal–citrate complexes.[65b]

3.2. Elimination Reactions

In a previous review[64] we suggested that enzymes utilized metal ions to promote elimination reactions by having the metal coordinate the leaving basic group at the α-carbon atom or a basic group on a carbon atom adjacent to or conjugated to the β-carbon atom:

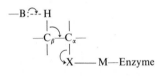

The structure of the inner sphere Fe^{2+}–citrate complex on aconitase (Fig. 9)[65a] and of one of the two alternative Mn–substrate second-sphere complexes on enolase (Fig. 10)[29] are consistent with this view. The interaction of a water

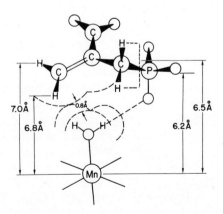

Fig. 10. Structure of the enolase–Mn–substrate complex.[29]

ligand of the enzyme-bound Mn^{2+} with the *re* face of the vinylic substrate of enolase would be consistent with the microscopic reverse of such a mechanism. Indirect evidence based on water relaxation data using the asymmetric product, 3-phosphoglyceric acid, supports this alternative.[29]

3.3. Phosphoryl- and Nucleotidyl-Transfer Enzymes

3.3.1. Muscle Pyruvate Kinase

This enzyme, the first studied by $1/T_1$ measurements of substrates,[1,2] catalyzes reversible phosphoryl-transfer reactions between P-enolpyruvate and ADP as well as irreversible phosphoryl transfer from ATP to F^-, NH_2OH, and P-glycolate. Nuclear relaxation studies of the active enzyme–Mn–pyruvate-P_i complex by ^{13}C, ^{31}P,[32] and ^1H NMR[65c]; of the active enzyme–Co^{2+}–P-enolpyruvate complex[66]; and of the active enzyme–Mn–ATP complex by ^1H and ^{31}P NMR[67] revealed second-sphere complexes of the transferable phosphoryl group (Fig. 11). In fact, none of the three phosphoryl groups of ATP were directly coordinated by the enzyme-bound divalent cation but formed second-sphere complexes. The nucleotide conformation in the pyruvate kinase–Mn–ATP complex differs greatly from that in the binary Mn–ATP complex, emphasizing the danger of using distances in a binary complex to predict those on an enzyme (Fig. 11).[45]

Cr^{3+}–ATP, a substitution inert metal–ATP complex, has been shown to be active in promoting the enolization of pyruvate in the presence of pyruvate kinase and an enzyme-bound divalent cation.[45] In addition to the enzyme-bound divalent cation, the nucleotide-bound cation has been shown by kinetic studies to be essential for catalysis of the enolization of pyruvate.[68] The rate of enolization depends on the electronegativity of the nucleotide-bound metal ion as reflected in the pK_a of its coordinated water ($\rho = -0.2$).[68] The intersubstrate distance on pyruvate kinase has been determined by the paramagnetic effects of Cr–ATP on the relaxation rates of the $^{13}C_1$, the $^{13}C_2$, and the methyl protons of pyruvate in the active complex (Fig. 11).[45] The availability of 13 distances from the enzyme-bound divalent cation to the substrates pyruvate, P-enolpyruvate, and ATP; to the activators P_i; and to the monovalent cations and the 3 intersubstrate distances permitted the construction of a model of the conformation and arrangement of the substrates at the active site.[69] This model (Fig. 12) revealed molecular contact between the phosphorus of the γ-phosphoryl group of ATP and the carbonyl oxygen of pyruvate, consistent with direct phosphoryl transfer, requiring no phosphoenzyme or metaphosphate intermediate, although such intermediates are not strictly excluded.[69] Crystallographic studies of cat muscle pyruvate kinase and its binary substrate complexes at 6-Å resolution[70] indicate proximity of the binding sites for the enzyme-bound divalent cation, P-enolpyruvate, and ATP in accord with the magnetic-resonance data.

Second-sphere distances from the enzyme-bound Mn^{2+} to phosphorus

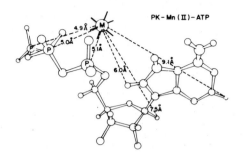
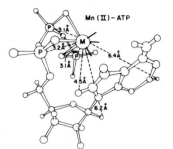
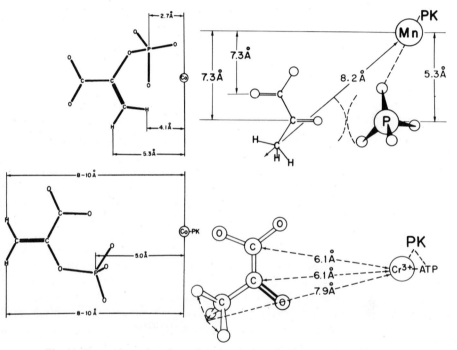

Fig. 11. Comparison of conformations in solution of active complexes of pyruvate kinase[32,45,65c,66,67] with that of the binary Mn–ATP complex.[67]

atoms of interacting ligands are also found by $1/T_1$ methods in the phosphoryl-transferring enzymes phosphoglucomutase[71] and the $(Na^+ + K^+)$ = activated ATPase,[72] although only inactive substrate analogs were used in these cases. Similarly, a predominantly ($\geq 85\%$) second-sphere complex has been detected between the enzyme-bound Mn^{2+} at the active site of DNA polymerase I and the reaction center α-phosphorus atom of dTTP,[39] suggesting that second-sphere complexes may be a general mode of activating phosphoryl groups for nucleophilic attack (Fig. 13).[69]

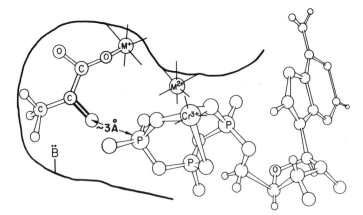

Fig. 12. Arrangement and conformations of substrates at the active site of pyruvate kinase based on the data of Fig. 11.[69]

3.3.2. Creatine Kinase

A more direct NMR approach to the structure of transition states[13-17] has been made possible by the discovery[73] that the presence of both substrates, ADP + creatine, raises the affinity of creatine kinase for planar trigonal mono-anions such as nitrate or formate and conversely that these monoanions raise the affinity of creatine kinase for the substrates ADP and creatine. On the enzyme these planar trigonal monoanions may be functioning either as analogs of metaphosphate or, more likely, as part of a trigonal bipyramidal transition state for an associative (S_N2) displacement.[74] Magnetic-resonance studies of the transition-state complexes of creatine kinase by Cohn and co-workers[13-17] confirm the mutual tightening of binding of substrates by the monoanions nitrate or formate. From EPR and water relaxation studies the enzyme-bound Mn^{2+} in such complexes is in a highly asymmetric environment and is highly shielded from the solvent.[13-16] Distance measurements from Mn^{2+} to creatine

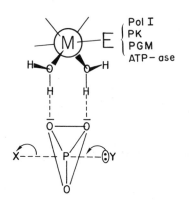

Fig. 13. General mechanism of phosphoryl activation based on distances on the enzymes DNA polymerase, Pol I[39]; pyruvate kinase, PK[66,67]; phosphoglucomutase, PGM[71]; and the $Na^+ + K^+$ ATPase.[72]

indicate that formate decreases the distance from Mn^{2+} to the methyl group of creatine by ≥ 1.2 Å, indicating structural adjustments as the transition state is approached.[17] The important distance from Mn^{2+} or Co^{2+} to the proton of formate has not been precisely determined, but its upper limit ($\leq 5.1 \pm 0.5$ Å)[14,16,17] is consistent either with direct coordination or with a second-sphere complex.

3.3.3. DNA Polymerase

DNA polymerase I from *E. coli* catalyzes the *in vitro* synthesis of DNA (Fig. 14).[75,76] Using four deoxynucleoside triphosphate substrates, this enzyme accurately copies a DNA template, elongating the DNA primer chain according to the Watson–Crick base-pairing scheme, making less than one mistake in 10^5 nucleotides. Each chain elongation step results from the nucleophilic displacement of pyrophosphate on the substrate by the 3'OH group of the preceding sugar of the growing chain. DNA polymerases are Zn–metalloenzymes,[77-79] and the enzyme-bound Zn appears to interact with the DNA template–primer complex.[78] These enzymes also require a divalent cation such as Mg^{2+} or Mn^{2+} for activity which binds tightly to the enzyme.[80] Using Mn^{2+} at this site as a paramagnetic probe, we have mapped the conformation of the substrates dTTP and dATP by ^{31}P and proton relaxation rates.[39] In the binary Mn–dTTP and ternary enzyme–Mn–dTTP complexes the calculated distances were used to construct molecular models (Fig. 15). The uniqueness of these models was tested, as described above, by a computer search among 47,000 conformations, rejecting those structures which produced a total van der Waals overlap greater than 0.4 Å or which required distances which exceeded our error limits of 5.5% and 7.5% for the binary and ternary complexes, respectively. By these tests the structures of Fig. 15 provide a highly unique fit to our data.[39]

Two major differences are noted between the binary and ternary complexes. First, in the binary complex all three of the phosphoryl groups of dTTP are directly coordinated to Mn^{2+}. On DNA polymerase only the γ-phosphoryl group remains coordinated by the enzyme-bound Mn^{2+}. The distance from

Fig. 14. Reaction catalyzed by DNA polymerase.

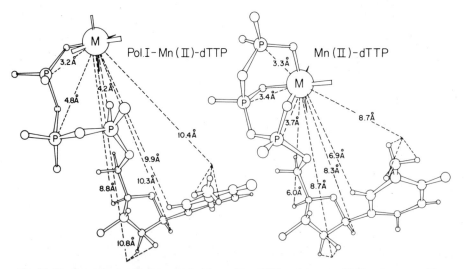

Fig. 15. Conformations and distances in binary Mn–dTTP and ternary DNA polymerase–Mn–dTTP complexes in solution.[39]

Mn^{2+} to the β-phosphorus atom (4.8 Å) indicates no direct coordination. The intermediate distance to the reaction center α-phosphorus atom (4.2 Å) is most simply explained by the rapid averaging of $\leq 15\%$ inner-sphere coordination with $\geq 85\%$ second-sphere coordination, possibly with an intervening water ligand. The resulting polyphosphate conformation is puckered and somewhat strained. Hence an important role of the divalent cation activator in catalysis is to assist the departure of the leaving pyrophosphate group by γ coordination and possibly to facilitate nucleophilic attack on the α-phosphorus atom by strain and by hydrogen bonding through a coordinated water ligand.[39]

A second important difference between the binary and ternary complexes is in the conformation about the thymine–deoxyribose bond of dTTP. Such glycosidic conformations are quantitatively described by the torsion angle χ, in the present case the dihedral angle between N_1—C_6 of thymine and C_1'—O_1' of deoxyribose when viewed along the glycosidic N_1—C_1' bond. The χ value of $40 \pm 5°$ in the binary complex increases to $90 \pm 5°$ in the ternary complex (Fig. 15). Interestingly, the latter torsion angle of 90° is indistinguishable from that found for deoxynucleotidyl units in double-helical DNA. Hence the binding of the substrate Mn–dTTP to the enzyme, DNA polymerase, in the absence of template, has changed the substrate conformation to that of a nucleotidyl unit in the product—double-helical DNA. Similarly, a 90° torsion angle is also found for the purine nucleotide substrate Mn–dATP when bound to DNA polymerase (Fig. 16).[39]

When the structure of enzyme-bound Mn–dTTP (Fig. 15) is superimposed by computer onto the double-helical structure of DNA B (Fig. 17), the resulting location of the α-phosphorus atom and the leaving pyrophosphate group of

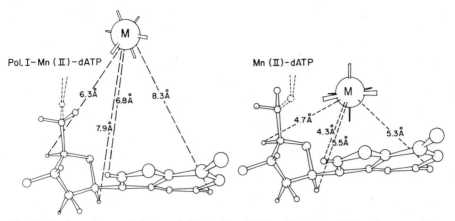

Fig. 16. Conformations and distances in binary Mn–dATP and in ternary DNA polymerase–Mn–dATP complexes in solution.[39]

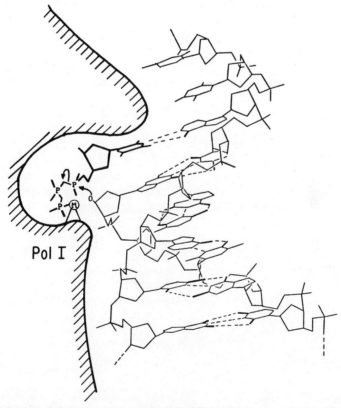

Fig. 17. Mechanism of chain elongation catalyzed by DNA polymerase consistent with the NMR data of Figs. 15 and 16.[39]

the bound substrate relative to the attacking 3'-OH group of the preceding nucleotide unit is consistent only with an in-line nucleophilic displacement on the α-phosphorus.[39] Hence the biosynthesis of nucleic acids, like their hydrolysis,[81] appears to proceed by an in-line mechanism.

The selection by the enzyme of those substrate conformations that fit into the double helix would amplify the Watson–Crick base-pairing scheme and could thereby serve to prevent errors in template copying.[39] Such error-preventing mechanisms are required by DNA polymerases since enzymes from eukaryotic cells lack an error-correcting exonuclease activity yet synthesize DNA with an accuracy at least two orders of magnitude greater than predicted by the thermodynamic[82] and kinetic effects of base-pairing alone.[83]

4. CONCLUSIONS

1. Numerous examples have been provided which establish the point that the average conformation of a flexible substrate, when bound to an enzyme, generally differs from that of the free substrate in solution [Figs. 1, 8(a), 11, 15, and 16].
2. Second-sphere enzyme–metal–(H_2O)–substrate complexes are used by enzymes to polarize carbonyl groups (six examples, Figs. 1, 2, and 8) and to position phosphoryl groups for nucleophilic attack (four examples, Figs. 11, 13, and 15). In the case of pyruvate kinase, an additional metal interacts directly with the ATP (Fig. 12).
3. On two-substrate enzymes, such as dehydrogenases (Fig. 1), kinases (Fig. 12), and even a biotin enzyme (Figs. 5 and 6), close proximity of the two bound substrates is observed, in some cases approaching molecular contact.
4. The structures of the elevated ground states of various enzyme-bound substrates determined by NMR may be used to infer possible structures of the transition states in the catalytic processes.

ACKNOWLEDGMENT

This study was supported by grants from the National Institutes of Health (AM-13351) and by the National Science Foundation (PCM74-03739), by grants to this institute from the National Institutes of Health (CA-06927, RR-05539), and by an appropriation from the Commonwealth of Pennsylvania.

REFERENCES

1. A. S. Mildvan, M. Cohn, and J. S. Leigh, Jr., in: *Magnetic Resonance in Biological Systems* (A. Ehrenberg, B. G. Malmstrom, and T. Vanngard, eds.), pp. 113–117, Pergamon, Elmsford, N.Y. (1967).

2. A. S. Mildvan, J. S. Leigh, Jr., and M. Cohn, Kinetic and magnetic studies of pyruvate kinase. III. The enzyme–metal–phosphoryl bridge complex in the fluorokinase reaction, *Biochemistry* **6**, 1805–1818 (1967).

3. A. S. Mildvan and M. C. Scrutton, Pyruvate carboxylase. X. The demonstration of direct coordination of pyruvate and α-ketobutyrate by the bound manganese and the formation of enzyme–metal–substrate bridge complexes, *Biochemistry* **6**, 2978–2994 (1967).

4. A. S. Mildvan and M. Cohn, Aspects of enzyme mechanisms studied by nuclear spin relaxation induced by paramagnetic probes, *Adv. Enzymol.* **33**, 1–70 (1970).

5. A. S. Mildvan and J. L. Engle, Nuclear relaxation measurements of water protons and other ligands, *Methods Enzymol.* **26C**, 654–682 (1972).

6. A. S. Mildvan, T. Nowak, and C. H. Fung, Nuclear relaxation studies of the role of the divalent cation in the mechanism of pyruvate kinase and enolase: Inner sphere and second sphere complexes, *Ann. N.Y. Acad. Sci.* **222**, 192–210 (1973).

7. D. L. Sloan and A. S. Mildvan, Magnetic resonance studies of the geometry of bound nicotanamide adenine dinucleotide and isobutyramide on spin-labeled alcohol dehydrogenase, *Biochemistry* **13**, 1711–1718 (1974).

8. C. H. Fung, R. J. Feldmann, and A. S. Mildvan, ^1H and ^{31}P Fourier transform magnetic resonance studies of the conformation of enzyme-bound propionyl coenzyme A on transcarboxylase, *Biochemistry* **15**, 75–84 (1976).

9. L. Pauling, Molecular architecture and biological reactions, *Chem. Eng. News* **24**, 1375–1377 (1946).

10. R. Wolfenden, Analog approaches to the structure of the transition state in enzyme reactions, *Acc. Chem. Res.* **5**, 10–18 (1972).

11. G. E. Lienhard, Enzymic catalysis and transition-state theory, *Science* **180**, 149 (1973).

12. K. Schray and J. P. Klinman, The magnitude of enzyme transition state analog binding constants, *Biochem. Biophys. Res. Commun.* **57**, 641–648 (1974).

13. G. H. Reed and M. Cohn, Structural changes induced by substrates and anions at the active site of creatine kinase, *J. Biol. Chem.* **247**, 3073–3081 (1972).

14. G. H. Reed and A. C. McLaughlin, Structural studies of transition state analog complexes of creatine kinase, *Ann. N.Y. Acad. Sci.* **222**, 118–129 (1973).

15. M. Cohn, J. S. Leigh, Jr., and G. H. Reed, Mapping active sites of phosphoryl-transferring enzymes by magnetic resonance methods, *Cold Spring Harbor Symp. Quant. Biol.* **36**, 533–540 (1971).

16. M. Cohn, in: *Enzymes Structure and Function*, (J. Drenth, R. A. Osterbaan, and C. Veeger, eds.), Proceedings, 8th FEBS Meeting, Vol. 29, p. 59–70, American Elsevier/North-Holland, Amsterdam (1972).

17. A. C. McLaughlin, J. S. Leigh, and M. Cohn, Magnetic resonance study of the three-dimensional structure of creatine kinase–substrate complexes, *J. Biol. Chem.* **251**, 2777–2787 (1976).

18. (a) A. S. Mildvan, *Acc. Chem. Res.* **10**, 246–252 (1977). (b) A. S. Mildvan and R. K. Gupta, *Methods in Enzymology*, in press (1978).

19. T. J. Swift and R. E. Connick, NMR-relaxation mechanisms of O^{17} in aqueous solutions of paramagnetic cations and the lifetime of water molecules in the first coordination sphere, *J. Chem. Phys.* **37**, 307–320 (1962).

20. Z. Luz and S. Meiboom, Proton relaxation in dilute solutions of cobalt (II) and nickel (II) ions in methanol and the rate of methanol exchange of the solvation sphere, *J. Chem. Phys.* **40**, 2686–2692 (1964).

21. I. Solomon, Relaxation processes in a system of two spins, *Phys. Rev.* **99**, 559–565 (1955).

22. I. Solomon and N. Bloembergen, Nuclear magnetic interactions in the HF molecule, *J. Chem. Phys.* **25**, 261–266 (1956).

23. N. Bloembergen and L. O. Morgan, Proton relaxation times in paramagnetic solutions. Effects of electron spin relaxation, *J. Chem. Phys.* **34**, 842–850 (1961).

24. M. Cohn and J. Reuben, Paramagnetic probes in magnetic resonance studies of phosphoryl transfer enzymes, *Acc. Chem. Res.* **4**, 214–222 (1971).

25. J. W. Emsley, J. Feeney, and L. H. Sutcliffe, in: *High Resolution NMR Spectroscopy*, Vol. 1, p. 520, Pergamon, Elmsford, N.Y. (1965).

26. T. L. James, in: *NMR in Biochemistry*, p. 177, Academic Press, New York (1975).
27. R. A. Dwek, in: *NMR in Biochemistry*, p. 174, Clarendon Press, Oxford (1973).
28. (a) M. Cohn and J. S. Leigh, Magnetic resonance investigations of ternary complexes of enzyme–metal–substrate, *Nature (London)* **193**, 1037–1040 (1962). (b) M. Cohn, Magnetic resonance studies of metal activation of enzymic reactions of nucleotides and other phosphate substrates, *Biochemistry* **2**, 623–629 (1963).
29. T. Nowak, A. S. Mildvan, and G. L. Kenyon, Nuclear relaxation and kinetic studies of the role of Mn^{2+} in the mechanism of enolase, *Biochemistry* **12**, 1690–1701 (1973).
30. T. Nowak and A. S. Mildvan, Nuclear magnetic resonance studies of the function of potassium in the mechanism of pyruvate kinase, *Biochemistry* **11**, 2819–2828 (1972).
31. (a) R. R. Ernst and W. A. Anderson, Application of Fourier transform spectroscopy to magnetic resonance, *Rev. Sci. Instrum.* **37**, 93–102 (1966). (b) R. L. Vold, J. S. Waugh, M. P. Klein, and D. E. Phelps, Measurement of spin relaxation in complex systems, *J. Chem. Phys* **48**, 3831–3832 (1968).
32. C. H. Fung, A. S. Mildvan, A. Allerhand, R. Komoroski, and M. C. Scrutton, Interaction of pyruvate with pyruvate carboxylase and pyruvate kinase as studied by paramagnetic effects on ^{13}C relaxation rates, *Biochemistry* **12**, 620–629 (1973).
33. G. G. McDonald and J. S. Leigh, Jr., A new method for measuring longitudinal relaxation times, *J. Magn. Reson.* **9**, 358–362 (1973).
34. M. C. Scrutton and A. S. Mildvan, Pyruvate carboxylase: Nuclear magnetic resonance studies of the enzyme–manganese–oxalacetate and enzyme–manganese–pyruvate bridge complexes, *Arch. Biochem. Biophys.* **140**, 131–151 (1970).
35. J. Reuben and M. Cohn, Magnetic resonance studies of manganese(II) binding sites of pyruvate kinase, *J. Biol. Chem.* **245**, 6539–6546 (1970).
36. C. D. Barry, J. Glasel, R. J. P. Williams, and A. V. Xavier, Quantitative determination of conformations of flexible molecules in solution using lanthanide ions as nuclear magnetic resonance probes: Application to adenosine-5′-monophosphate, *J. Mol. Biol.* **84**, 471–502 (1974).
37. C. D. Barry, D. R. Martin, R. J. P. Williams, and A. V. Xavier, Quantitative determination of the conformation of cyclic 3′,5′-adenosine monophosphate in solution using lanthanide ions as nuclear magnetic resonance probes, *J. Mol. Biol.* **84**, 491–502 (1974).
38. B. Furie, J. H. Griffen, R. J. Feldmann, A. Sokoloski, and A. N. Schechter, The active site of staphylococcal nuclease: Paramagnetic relaxation of bound nucleotide inhibitor nuclei by lanthanide ions, *Proc. Nat. Acad. Sci. USA* **71**, 2833–2837 (1974).
39. D. L. Sloan, L. A. Loeb, A. S. Mildvan, and R. L. Feldmann, Conformation of deoxynucleoside triphosphate substrates on DNA polymerase I from *Escherichia coli* as determined by nuclear magnetic relaxation, *J. Biol. Chem.* **250**, 8913–8920 (1975).
40. H. Weiner, Interaction of a spin-labeled analog of nicotinamide–adenine dinucleotide with alcohol dehydrogenase. I. Synthesis, kinetics, and electron paramagnetic resonance studies, *Biochemistry* **8**, 526–533 (1969).
41. A. S. Mildvan and H. Weiner, Interaction of a spin-labeled analogue of nicotinamide adenine dinucleotide with alcohol dehydrogenase, *J. Biol. Chem.* **244**, 2465–2475 (1969).
42. A. S. Mildvan and H. Weiner, Interaction of a spin-labeled analog of nicotinamide–adenine dinucleotide with alcohol dehydrogenase. II. Proton relaxation rate and electron paramagnetic resonance studies of binary and ternary complexes, *Biochemistry* **8**, 552–562 (1969).
43. A. S. Mildvan, L. Waber, J. J. Villafranca, and H. Weiner, in: *Structure and Function of Oxidation Reduction Enzymes* (Å. Åkeson and Å. Ehrenberg, eds.), p. 745–754, Pergamon, Elmsford, N.Y. (1972).
44. T. R. Krugh, Proximity of the nucleotide monophosphate and triphosphate binding sites on deoxyribonucleic acid polymerase, *Biochemistry* **10**, 2594–2599 (1971).
45. R. K. Gupta, C. H. Fung, and A. S. Mildvan, Chromium(III)–adenosine triphosphate as a paramagnetic probe to determine intersubstrate distances on pyruvate kinase, *J. Biol. Chem.* **251**, 2421–2430 (1976).
46. S. W. Weidmann, G. R. Drysdale, and A. S. Mildvan, Interaction of a spin-labeled analog of

acetyl coenzyme A with citrate synthase. Paramagnetic resonance and proton relaxation rate studies of binary and ternary complexes, *Biochemistry* **12**, 1874–1883 (1973).

47. (a) C. H. Fung, R. K. Gupta, and A. S. Mildvan, Magnetic resonance studies of the proximity and spatial arrangement of propionyl coenzyme A and pyruvate on a biotin–metalloenzyme, transcarboxylase, *Biochemistry* **15**, 85–92 (1976). (b) R. M. Oesterling and A. S. Mildvan, unpublished observations.

48. (a) H. G. Wood, H. Lochmüller, C. Riepertinger, and F. Lynen, Transcarboxylase. IV. Function of biotin and the structure and properties of the carboxylated enzyme, *Biochem. Z.* **337**, 247–266 (1963). (b) D. B. Northrop and H. G. Wood, Transcarboxylase. VII. Exchange reactions and kinetics of oxalate inhibition, *J. Biol. Chem.* **244**, 5820–5827 (1969).

49. F. Ahmad, B. Jacobson, B. Chuang, W. Brattin, and H. G. Wood, Isolation of peptides from the carboxyl carrier subunit of transcarboxylase. Role of the non-biotinyl peptide in assembly, *Biochemistry* **14**, 1606–1611 (1975).

50. C. H. Fung, A. S. Mildvan, and J. S. Leigh, Jr., Electron and nuclear magnetic resonance studies of the interaction of pyruvate with transcarboxylase, *Biochemistry* **13**, 1160–1169 (1974).

51. D. L. Sloan, J. M. Young, and A. S. Mildvan, Nuclear magnetic resonance studies of substrate interaction with cobalt substituted alcohol dehydrogenase from liver, *Biochemistry* **14**, 1998–2008 (1975).

52. H. Miziorko and A. S. Mildvan, Electron paramagnetic resonance, ^1H, and ^{13}C nuclear magnetic resonance studies of the interaction of manganese and bicarbonate with ribulose 1,5-diphosphate carboxylase, *J. Biol. Chem.* **249**, 2743–2750 (1974).

53. R. Y. Hsu, A. S. Mildvan, G. G. Chang, and C. H. Fung, Mechanism of malic enzyme from pigeon liver: Magnetic resonance and kinetic studies of the role of Mn^{2+}. *J. Biol. Chem.* **251**, 6574–6583 (1976).

54. D. B. Northrop, Transcarboxylase. VI. Kinetic analysis of the reaction mechanism, *J. Biol. Chem.* **244**, 5808–5819 (1969).

55. I. A. Rose, E. L. O'Connell, and F. Solomon, Intermolecular tritium transfer in the transcarboxylase reaction, *J. Biol. Chem.* **251**, 902–904 (1976).

56. J. Retey and F. Lynen, Zur biochemischen funktion des biotins. IX. Der sterische verlauf der carboxylierung von propionyl-CoA, *Biochem. Z.* **342**, 256–271 (1965).

57. E. T. Maggio, G. L. Kenyon, A. S. Mildvan, and G. D. Hegeman, Mandelate racemase from *Pseudomonas putida*. Magnetic resonance and kinetic studies of the mechanism of catalysis, *Biochemistry* **14**, 1131–1139 (1975).

58. J. P. Klinman, in: *Isotope Effects on Enzyme Catalyzed Reactions* (W. W. Cleland, M. H. O'Leary, and D. B. Northrop, eds.), pp. 176–208, University Park Press, Baltimore (1977).

59. R. K. Gupta and A. S. Mildvan, Nuclear relaxation studies on human methemoglobin. Observation of cooperativity and alkaline Bohr effect with inositol hexaphosphate, *J. Biol. Chem.* **250**, 246–253 (1975).

60. K. B. J. Schowen, Chapter 6 in this volume.

61. A. S. Mildvan, R. D. Kobes, and W. J. Rutter, Magnetic resonance studies of the role of the divalent cation in the mechanism of yeast aldolase, *Biochemistry* **10**, 1191–1204 (1971).

62. E. P. Kang, C. B. Storm, and F. W. Carson, Cobalt(III) carboxypeptidase A: Preparation and esterase activity, *Biochem. Biophys. Res. Commun.* **49**, 621–625 (1972).

63. J. M. Young, K. J. Schray, and A. S. Mildvan, Proton magnetic relaxation studies of the interaction of D-Xylose and xylitol with D-xylose isomerase, *J. Biol. Chem.* **250**, 9021–9027 (1975).

64. A. S. Mildvan, in: *Bioinorganic Chemistry*, A.C.S. Advances in Chemistry, Vol. 100, pp. 390–412 (1971).

65. (a) J. J. Villafranca and A. S. Mildvan, The mechanism of aconitase action. III. Detection and properties of enzyme–metal–substrate and enzyme–metal–inhibitor bridge complexes with manganese(II) and iron(II), *J. Biol. Chem.* **247**, 3454–3463 (1972). (b) H. L. Carrell and J. P. Glusker, Manganous citrate decahydrate, *Acta Crystallogr. Sect. B* **29**, 638–640 (1973). (c) T. James and M. Cohn, Structural aspects of manganese–pyruvate kinase substrate and in-

hibitor complexes deduced from proton magnetic relaxation rates of pyruvate and a phosphoenolpyruvate analog, *J. Biol. Chem.* **249**, 3519–3526 (1974).

66. E. Melamud and A. S. Mildvan, Magnetic resonance studies of the interaction of Co^{2+} and phosphoenolpyruvate with pyruvate kinase, *J. Biol. Chem.* **250**, 8193–8201 (1975).

67. D. L. Sloan and A. S. Mildvan, Nuclear magnetic relaxation studies of the conformation of adenosine 5'-triphosphate on pyruvate kinase from rabbit muscle, *J. Biol. Chem.* **251**, 2412–2420 (1976).

68. R. K. Gupta, R. M. Oesterling, and A. S. Mildvan, *Biochemistry* **15**, 2881–2887 (1976).

69. A. S. Mildvan, D. L. Sloan, C. H. Fung, R. K. Gupta, and E. Melamud, Arrangement and conformations of substrates at the active site of pyruvate kinase from model building studies based on magnetic resonance data, *J. Biol. Chem.* **251**, 2431–2434 (1976).

70. D. K. Stammers and H. Muirhead, Three-dimensional structure of cat muscle pyruvate kinase at 6 Å resolution, *J. Mol. Biol.* **95**, 213–225 (1975).

71. W. J. Ray and A. S. Mildvan, Arrangement of the phosphate- and metal-binding subsites of phosphoglucomutase. Intersubsite distance by means of nuclear magnetic resonance measurements, *Biochemistry* **12**, 3733–3743 (1973).

72. C. M. Grisham and A. S. Mildvan, Magnetic resonance and kinetic studies of $Na^+ + K^+$ ATPase, *J. Supramol. Struct.* **3**, 304–313 (1975).

73. E. J. Milner-White and D. C. Watts, Inhibition of adenosine 5'-triphosphate-creatine phosphotransferase by substrate–anion complexes. Evidence for the transition-state organization of the catalytic site, *Biochem. J.* **122**, 727–740 (1971).

74. A. S. Mildvan and C. M. Grisham, The role of divalent cations in the mechanism of enzyme catalyzed phosphoryl and nucleotidyl transfer reactions, *Struct. Bonding (Berlin)* **20**, 1–21 (1974).

75. (a) T. Kornberg and A. Kornberg, in: *The Enzymes*, 3rd ed. (P. D. Boyer, ed.), Vol. X, p. 119–144, Academic Press, New York (1974). (b) A. Kornberg, *DNA Synthesis*, W. H. Freeman, San Francisco (1974).

76. L. A. Loeb, in: *The Enzymes*, 3rd ed. (P. D. Boyer, ed.), Vol. X, p. 173–209, Academic Press, New York (1974).

77. J. P. Slater, A. S. Mildvan, and L. A. Loeb, Zinc in DNA polymerases, *Biochem. Biophys. Res. Commun.* **44**, 37–43 (1971).

78. C. F. Springgate, A. S. Mildvan, R. Abramson, J. L. Engle, and L. A. Loeb, *Escherichia coli* deoxyribonucleic acid polymerase I, a zinc metalloenzyme, *J. Biol. Chem.* **248**, 5987–5993 (1973).

79. (a) B. J. Poiesz, N. Battula, and L. A. Loeb, Zinc in reverse transcriptase, *Biochem. Biophys. Res. Commun.* **56**, 959–964 (1974). (b) B. J. Poiesz, G. Seal, and L. A. Loeb, Reverse transcriptase: Correlation of zinc content with activity, *Proc. Nat. Acad. Sci. USA* **71**, 4892–4896 (1975).

80. J. P. Slater, I. Tamir, L. A. Loeb, and A. S. Mildvan, The mechanism of *Escherichia coli* deoxyribonucleic acid polymerase I, *J. Biol. Chem.* **247**, 6784–6794 (1972).

81. D. A. Usher, E. S. Erenrich, and F. Eckstein, Geometry of the first step in the action of ribonuclease-A, *Proc. Nat. Acad. Sci USA* **69**, 115 (1972).

82. A. S. Mildvan, Mechanism of enzyme action, *Ann. Rev. Biochem.* **43**, 357–399 (1974).

83. E. C. Travaglini, A. S. Mildvan, and L. Loeb, Kinetic analysis of *Escherichia coli* deoxyribonucleic acid polymerase I, *J. Biol. Chem.* **250**, 8647–8656 (1975).

Mapping Reaction Pathways from Crystallographic Data

Eli Shefter

To appreciate biochemical processes at the molecular level, it is necessary to have a basic understanding of the stereochemical course of reactions. Though much data have been accumulated on a variety of reaction mechanisms from kinetic studies, the detailed course of the structural transformations that take place during a reaction has eluded the experimentalist. At present, experimental techniques are not capable of directly measuring the geometrical changes that occur during a reaction, apart from characterizing the products, reactants, and stable intermediates. Quantum-mechanical calculations, which are of value in deriving such information for very simple reactions, are beset with numerous practical difficulties when complex organic systems are to be treated.

Energy–reaction coordinate diagrams are found adorning many texts (an example is shown in Fig. 1). Though the activation energy can be determined experimentally, the definition of the reaction coordinate is more difficult to assess. It is usually not a single parameter but represents multidimensional structural changes. For example, bonds may lengthen and angles may be altered in the course of a reaction. To clarify the reaction coordinate for certain reaction types, a method using X-ray crystallographic data was devised.[1-5]

One way of obtaining accurate structural information about a molecule is by use of single-crystal X-ray diffraction techniques. These techniques have provided vast amounts of geometrical data on a wide gamut of molecular entities. With such information it has been possible to formulate standard values for various classes of bonds, both bonded and nonbonded.[6,7] Reference

Eli Shefter • Department of Pharmaceutics, School of Pharmacy, State University of New York, Amherst, New York.

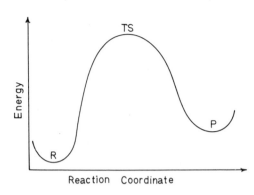

Fig. 1. Energy–reaction path profile.

values have also been presented for bond angles and torsion angles. These standard values are useful to the crystallographer searching for structural explanations for certain experimental observations, for geometric distortions can have a profound effect on the spectral, thermal, and biological properties of a molecule.

If it were possible to freeze (crystallize) the molecules at various points along the reaction path of Fig. 1 and characterize their structures by X-ray analysis, one would possess a method of directly delineating the reaction coordinate. The displacements from the ground-state geometry of the reactants and products could be easily measured. However, since this is not experimentally possible, another approach using structures in molecular and crystal environments which artificially fix a chemical fragment at a point on the reaction path was designed.[1-4]*

The basic idea behind the B–D approach is to search for structural correlations that exist within a certain chemical subunit, e.g., tetrahedral centers, in a wide variety of environments. These environments will each have a perturbing effect on the subunit as a result of the complicated interplay of inter- and intramolecular forces. It is assumed that no matter what the nature of the perturbing forces on the subunit the geometrical alterations of this entity should occur along a potential energy valley that is characteristic of its parameter space. Therefore, by examining the geometry of a particular chemical subsystem exposed to different strains by virtue of its being a part of extremely varied molecular frameworks and different crystal-packing environments, its minimum-energy pathway can, in certain instances, be mapped. Both experimental error and the different perturbations acting on the chemical subunit in its varied environments will result in some scatter from the smooth curve which would be expected to define the path.

---

* Credit for the development of this treatment of X-ray structural data must be given to Drs. H. B. Bürgi and J. D. Dunitz of the Swiss Federal Institute of Technology (E.T.H.), Zurich, for they were the first to make extensive use of "static" X-ray data to extract "dynamic" reaction coordinate information. I shall be referring to the approach used by them by the prefix B–D.

1. NUCLEOPHILIC SUBSTITUTION: S_N2

In the course of a bimolecular nucleophilic substitution reaction (S_N2 type), a bond between the central atom and the leaving group (L) is cleaved with the simultaneous formation of a bond between the nucleophile (N) and the central atom.[8] Such a reaction at carbon is thought to exhibit a single maximum in its energy–reaction coordinate diagram, corresponding to a unique transition-state geometry. Bürgi examined a series of trigonal bipyramidal-type complexes of Cd^{II}–ion to glean information about the structural changes that could occur for this class of reactions.[1]

Compounds with a central carbon that exhibit the geometry resembling the intermediates along the S_N2 reaction have not been crystallized. This may be due to the magnitude of the activation energy for these systems, which must be overcome by imposed intramolecular and/or intermolecular constraints. Therefore, direct application of the B–D approach to the Walden inversion has yet to be demonstrated experimentally.

The complexes examined by Bürgi all contained three equatorial sulfur ligands with iodide, sulfur, or alcoholic oxygens in the axial positions (N and L). It was observed that the Cd atom did not generally reside in the plane

$$S—\overset{\overset{\textstyle L}{|}}{\underset{\underset{\textstyle N}{|}}{Cd}}\overset{\displaystyle S}{\underset{\displaystyle S}{\big\langle}}$$

determined by the three equatorial sulfurs. The parameter which describes the magnitude and direction of this displacement (measured in the direction of the axial ligand L) is denoted by Δz. To compare this displacement with the axial bonding, it was necessary to convert the axial bond lengths measured from the crystallographic studies into comparable-type distances. This was accomplished by subtracting the appropriate sum of the covalent radii* from the Cd–N and Cd–L lengths. These values will be referred to as ΔN and ΔL, respectively.

Intuition about an S_N2 reaction suggests that as the nucleophile (N) approaches the electrophile, distortion of the tetrahedral center containing the leaving group (L) should be directly correlatable with Δz. A plot showing the changes observed for Δz as a function of ΔN and ΔL for a number of Cd complexes is shown in Fig. 2. At small values of ΔL, the corresponding values

---

* All covalent, ionic, and van der Waals radii are taken from Reference 6.

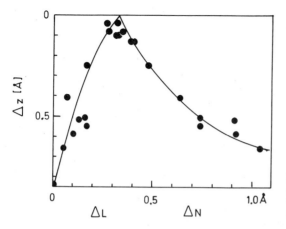

Fig. 2. Plot showing the course of nucleophilic substitution at Cd^{II}. Δz vs. ΔN, ΔL. (Reprinted with permission from Bürgi.[1])

of ΔN and Δz are greatest, as would be expected. With decrease in the covalent nature of the Cd–L bond, i.e., increase in ΔL, there is a corresponding decrease in ΔN and Δz. The intermediate state (transition state) where Δz is zero gives a value of 0.32 Å for both ΔN and ΔL. At this point the geometry of the complex is trigonal bipyramidal, with the axial bond lengths equal to the sum of their ionic radii (which are approximately 0.3 Å greater than their covalent radii).

An analytical function of the type denoted by Pauling[9] to relate changes in interatomic distance ($\Delta d = d - d_0$) to bond number (n) was used to describe the trends in Fig. 2:

$$\Delta d = c \log n, \qquad 0 < n < 1 \tag{1}$$

The bond number (n) is defined by Pauling as the number of electron pairs participating in a bond. The bond lengths d and d_0 refer to the observed and the reference length of a single bond, respectively. The value of the constant c is dependent on the nature of the pair of atoms involved. The equation can be adapted to the measured crystallographic data by expressing n in terms of structural coordinates.

This expression is purely an empirical one, whose theoretical justification has never been clearly delineated. However, it has been pointed out[10] that if the bond number can be identified with the attractive portion of the Morse equation, n will be an exponential function of the bond distance. It has also been suggested[5] that there is a relationship between Hückel bond orders and the Pauling bond number. In any event, the Pauling exponential function is only used to arrive at correlations between the geometric parameters observed.

The equations for the Cd complexes were derived by expressing n in terms of a fractional distortion from tetrahedral geometry. The equations obtained are

$$\Delta L = f(\Delta z) = -1.05 \log[(\Delta z + 0.84)/1.68] \text{ Å} \tag{2}$$

$$\Delta N = f(-\Delta z) = -1.05 \log[(-\Delta z + 0.84)/1.68] \, \text{Å} \tag{3}$$

where

$$n_L = (\Delta z + 0.84)/1.68, \qquad 0 < n_L < 1 \tag{4}$$

$$n_N = (-\Delta z + 0.84)/1.68, \qquad 0 < n_N < 1 \tag{5}$$

A maximum displacement (Δz) was assumed at 0.84 Å for an ideal CdS_4 tetrahedron. These expressions imply an invariance of the sum of the bond numbers, i.e., $n_N + n_L = 1$, for every point along the reaction path.

The angles made by the axial groups with the equatorial ligands also change in a characteristic manner along the reaction path. As the nucleophile approaches the Cd atom via the tetrahedral face perpendicular to the Cd–L bond, the angle it makes with the three sulfurs should increase, for initially, the L–CdS and N–CdS angles should be 109.5° and 70.5°, respectively, with both tending toward 90° at the transition state. Geometrical arguments show that the average values of these angles are related to Δz and therefore follow the trend expected.*

Without experimental data on carbon compounds one can only postulate that a similar sequence of structural changes will occur for these compounds. There is support of this in the results of quantum-mechanical calculations on the reaction of the nucleophiles F^- and H^- with fluoromethane.[11] The axial bonds in the trigonal-bipyramidal transition state were calculated to be longer than the corresponding lengths in tetrahedral compounds. It appears reasonable to assume that similar analytical expressions might be used to describe the general class of S_N2 reactions.

2. NUCLEOPHILIC SUBSTITUTION: S_N1

The unimolecular nucleophilic substitution reactions (S_N1) of tetrahedral molecules are characterized by a kinetic process where the rate-determining step is the dissociation of the leaving group (L). As L departs, the molecule tends toward planarity.[8] One approach to mapping such a process is to examine distorted tetrahedral structures which exhibit C_{3v} symmetry. In this manner the relationship of the axial bond length (r_2) can be correlated with the geometric parameters describing the system.

---

* The lines in Fig. 2 can be expressed in terms of the average N-CdS (α_N) and L-CdS (α_L) angles by replacing n_L and n_N in Eq. (2) with the following expressions: $n_L = \frac{1}{2} + \frac{3}{2} \cos \alpha$ and $n_N = \frac{1}{2} + \frac{3}{2} \cos \alpha_N$.

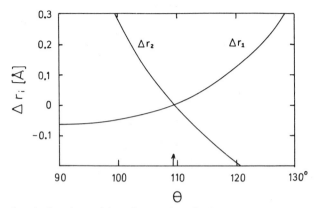

Fig. 3. Curves found when Δr_1 and Δr_2 of approximately C_{3v}-symmetrical tetrahedral molecules are plotted against θ. Data points left out for clarity. (Reproduced with permission from Murray-Rust *et al.*[4])

Murray-Rust *et al.*[4] examined a large number of C_{3v} structures determined by X-ray diffraction to delineate this relationship. Specifically, the following types of species were scrutinized: $LAlCl_3$, LSO_3, LPO_3, OPX_3, $LSnCL_3$, $LGeCl_3$, $LSiCl_3$, $LSnBr_3$, $LSnPh_3$, LPF_3, $LPCl_3$, and $LPPh_3$, where L represents a wide assortment of ligands.

For comparative purposes all the structural data were referred to a common reference point. A parameter Δr_2 was defined as the difference between the axial bond length observed in the C_{3v} structure (r_2) and that found in a T_d structure of that ligand with the same central atom (ML_4). Similarly, Δr_1 is the average bond length for the three basal substituents ($\langle r_1 \rangle$) minus the bond distance in a T_d species (MX_4). When these parameters are plotted against the average angle between the axial and basal bonds (θ), two distinct curves are observed, one for Δr_2 and the other for Δr_1 (see Fig. 3). Approximately 200 data points were used to derive the two curves shown in this figure (data points have been left out of the figure here).*

The curves illustrated in Fig. 3 can be represented extremely well by a Pauling-type relationship [Eq. (1)], with

$$n_2 = 9 \cos^2 \theta \tag{6}$$

$$n_1 = \tfrac{4}{3} - 3 \cos^2 \theta \tag{7}$$

The equations are

$$\Delta r_1 = -0.5 \log(\tfrac{4}{3} - 3 \cos^2 \theta) \tag{8}$$

$$\Delta r_2 = -0.5 \log(9 \cos^2 \theta) \tag{9}$$

Inherent in these equations are the assumptions that the displacement of the

---

* The deviations of data points from the curves are generally less than $3°$ in θ and less than 0.03 Å in the r values.

central atom from the plane defined through the three X atoms (basal ligands)* is a function of n and that the total bond order of the tetrahedral compounds is 4 $(n_2 + 3n_1 = 4)$. The two curves intersect at the T_d symmetry point where $\theta = 109.47$ and $\Delta r = 0$.

The relationships show that as the r_2 bond weakens and is ultimately cleaved, there is concurrent strengthening of the basal bond (decrease in r_1) as they become coplanar. After the cleavage, a trigonal molecule, such as AlCl$_3$ or SO$_3$, is formed, and its geometry is represented by the point at $\theta = 90°$ and the Δr_1 curve.

Though structural information on the deformation of tetrahedral carbon entities is very scanty, the bonding parameters measured for t-butyl chloride[12] are consistent with the curves observed. Since this type of molecule can undergo S$_N$1 substitution, it seems reasonable to assume that the reaction path for such species is approximately given by the equations above.

3. NUCLEOPHILIC ADDITION

One of the important reactions in biochemistry is the addition of nucleophiles to carbonyl centers. Such reactions are responsible for the hydrolysis and formation of a host of biomolecules, including esters, proteins, and glycosides. This class of reactions has been extensively studied, and vast amounts of data exist on their kinetics and mechanisms.

The general course of these reactions shows that a tetrahedral species is formed. There are certain instances where the tetrahedral state is stable. However, in general, the tetrahedral intermediate formed will undergo further reaction by an elimination process to yield a new trigonal compound. The reaction paths in these sequences are known to be dependent on structural as well as stereoelectronic factors.

The B–D approach was applied to a number of structures which contain a tertiary amino group and a carbonyl function which are suitably disposed to interact with each other.[2] The two moieties are connected by carbon chains of varying length (three or four methylene residues) in molecular frameworks that give a wide range of N to C=O lengths. A long N···C=O interaction is found in methadone[13] (2.91 Å), and a short one is observed in N-brosylmitomycin A (1.49 Å)[14] which exhibits tetrahedral stereochemistry. The proper ordering of these and other data should provide a description of the nucleophilic addition path of an amine nitrogen to a carbonyl group.

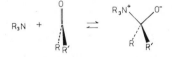

---

* The displacement of the central atom from the plane of the three basal bonds (Δ) is related to n_2 by $n_2 = (\Delta/\Delta_t)^2 = 9 \cos^2 \theta$, where Δ_t is the displacement of the central atom in a T_d molecule.

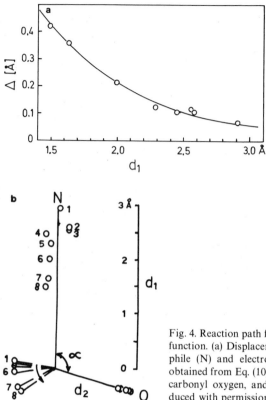

Fig. 4. Reaction path for nucleophilic addition to a carbonyl function. (a) Displacement (Δ) vs. distance between nucleophile (N) and electrophilic carbon atom. Smooth curve obtained from Eq. (10). (b) Reaction path showing nitrogen, carbonyl oxygen, and inclination of RCR′ plane. (Reproduced with permission from Bürgi et al.[2])

The closer the nitrogen atom is to the electrophilic center, the greater is the observed pyramidalization of the carbonyl function. This can be seen [Fig. 4(a)] in the displacement (Δ) of the carbonyl carbon from the plane defined by its three bonded neighbors (R, R′, and O) toward the nucleophile. The magnitude of Δ increases as the N$\cdots$C distance (d_1) decreases. This correlation can be described by a logarithmic equation (1). The relationship between the N$\cdots$C distance and bond number was found to be

$$d_1 = -0.85 \log n_1 + 1.48 \, \text{Å} \tag{10}$$

with

$$n_1 = (\Delta/\Delta_t)^2 \tag{11}$$

The out-of-plane displacement for a tetrahedral structure where d_1 equals the standard C—N bond length (1.48 Å) is (Δ_t) 0.44 Å. A reasonably good fit was also obtained with the Pauling expression for the changes in the carbonyl bond length as a function of bond order:

$$d_2 = 0.71 \log(2 - n_1) + 1.43 \, \text{Å} \tag{12}$$

These expressions indicate that at every point along the reaction path the sum of the C—O and C—N bond numbers has a value of 2.

The pyramidalization of the carbonyl carbon was observed to exhibit similar deformations in the presence of oxygen nucleophiles.[3] However, the degree of distortion found at the same stage of reaction, i.e., same nucleophile to electrophile distance, is much greater in the case of the nitrogen. The mean displacements (Δ) observed for the O$\cdots$C=O interactions are only about one third of those found in the nitrogen examples, which is consistent with the idea that alcoholic and ketonic oxygens are much weaker nucleophiles than an amine nitrogen.

A most interesting observation is that the angle at which the nucleophile approaches the carbonyl center is relatively constant over the range of N to C distances examined. The average value observed for this angle (α), 107°, is very close to a tetrahedral angle [see Fig. 4(b)]. The large number of O$\cdots$C=O interactions examined also produced a peak distribution of α angles between 100° and 110°.[3]

In a series of studies on sterically "rigid" molecules which undergo intra-molecular lactonization, Storm and Koshland[15] associated the optimum reaction rates with molecules exhibiting a O$\cdots$C=O angle of approximately 98°. Since this value was measured from molecular models of the compounds studied, its value may be considered as consistent with those observed in the structure determinations.

A series of quantum-mechanical calculations were carried out on a model system (hydride addition to formaldehyde)[10] to check on the geometric relationships derived from the X-ray data. At some selected H$^-$ to C distances the total energy was minimized by variation of Δ and d (C=O). The energy along the minimum-energy path was found to continuously decrease and to reach a minimum for the tetrahedral methanolate anion (no transition state observed). The nucleophile approached the formaldehyde molecule in a circuitous route (see Fig. 5). At H$^-$ to C distances greater than 3 Å the path is along the H—C—H bisector of formaldehyde in the plane of the molecule, following a course dictated by dipole interactions. As a result of repulsive interactions (between H$^-$ and the hydrogens on formaldehyde) it then changes course until at approximately 2.5 Å the optimal direction for addition is set. From this point to the formation of the tetrahedral anion, the α angle was found to range between 120° and 109°. In this stage of the reaction scheme the carbon atom starts to become significantly pyramidalized. The relation-ship of the displacement (Δ) of the carbon from the plane of the oxygen and two bonded hydrogens with respect to the distance d_1 (H$^-$ to C length) ob-tained from these calculations is remarkably similar to that derived for the nitrogen nucleophiles, the only difference of consequence being the term for the standard bond length, which is 1.13 Å in this reaction. These calculations essentially support the structural features of the nucleophilic addition reaction pathway obtained using the B–D approach.

In a quest to gather information on the decomposition of tetrahedral

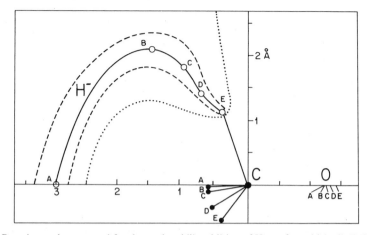

Fig. 5. Reaction path computed for the nucleophilic addition of H^- to formaldehyde. Points A to E correspond to d_1 lengths of 3.0, 2.5, 2.0, 1.5, and 1.12 Å, respectively. The dashed and dotted lines represent paths that are 0.6 and 6 kcal/mol higher than minimum-energy path. (Reproduced with permission from Bürgi et al.[10])

intermediates, structural data on tetrahedral subunits where two oxygens are attached to the same carbon atom were scrutinized for correlations.[3] The data were extracted from X-ray studies on diols, ketals, acetals, and hemi-acetals. Two trends in the data, relevant to the breakdown of the intermediate, were noted.

1. A correlation was observed between Δd (difference in carbon–oxygen bond lengths) and $\Delta\beta$ (a measure of distortion from tetrahedral geometry; see Fig. 6). The relationship noted in this figure corresponds to the antisymmetric stretching and bending distortion of a tetrahedron with C_{2v} symmetry. In

terms of a reaction path, this suggests that the leaving group of the inter-mediate (O_L) does not simply depart in the direction of its bond. Instead, there is a bending to maintain the α angle (O—C—O_L) at about 110°, while one carbon–oxygen bond shortens and the other lengthens. The pattern of angular change observed for the dioxo compounds lends further credence to the path observed for the addition of an amino group to a carbonyl center.

The extension of this analysis to orthoesters and amide acetals was hampered by the availability of accurate structural data. However, the bonding distortions observed in the few orthoesters reported[3] are consistent with the coupling of the antisymmetric bending and stretching formations expected for C_{3v} molecules.

2. The conformational angles about the C—O bonds in dioxo compounds

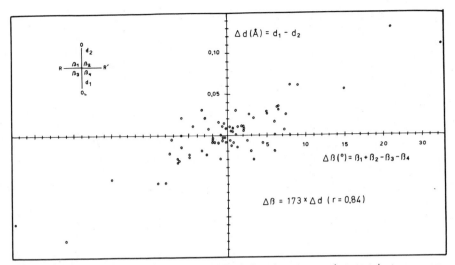

Fig. 6. Scatter plot of Δd against $\Delta \beta$ for tetrahedral carbon centers that contain two oxygen substituents. (Reproduced with permission from Bürgi *et al.*[3])

appear[3,16] to influence the C—O bond lengths. The C—O' bond in the grouping R(H)—O—C—O' exhibited a tendency to be longest when the torsion angle about the C—O bond is synclinal ($\pm 60°$) and shortest for antiperiplanar

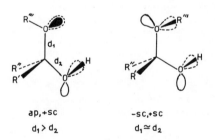

(180°) conformations. A similar trend was also found for the glycosidic bond in nucleosides and nucleotides.[17] When the glycosidic bond is approximately antiperiplanar to one of the furanose ring oxygen's lone-pair orbitals, its bond length is maximized.

Deslongchamps and co-workers[18-23] have developed a stereoelectronic theory for the hydrolysis of tetrahedral intermediates based on experiments with conformationally constrained orthoesters. It was proposed that the C—O bond most easily cleaved in an orthoester has an electron-pair orbital on each of the nonleaving oxygen functions antiperiplanar to it. This is in concordance with the observed conformational dependency of the C—O and C—N bond length in tetrahedral species. Reinforcement of the stereoelectronic effects of bond cleavage in tetrahedral intermediates is found in the nonempirical molecular orbital calculations performed on $CH_2(OH)_2$,[16,24] $CH(OH)_3$,[24] $C(OH)_4$,[24] and $CH(NH_2)(OH)_2$.[25] These computations also provide theoretical justification for the difference in bond lengths observed in the solid-state structures mentioned above.

A number of experimental observations reported in the literature can be cited as supportive evidence of the reaction paths mapped for nucleophilic addition reactions. One such example is found in two studies[26,27] on the reduction of acid anhydrides by sodium borohydride. The products isolated in these reactions are explainable in terms of the optimum trajectory for the H^- anion attack (as suggested above) being hindered by substituents on the anhydrides. The intramolecular thermal rearrangements of phenylazotribenzoylmethane[28,29] in solution and in the solid state is another reaction consistent with the general ideas presented. The products formed in this reaction are easily deduced by examining the $N \cdots C{=}O$ and $O \cdots C{=}O$ interactions present in crystals of the compound.[3,30] Only those intramolecular interactions that are favorably disposed lead to intermediates of consequence.

An area where the stereochemistry of the minimum-energy reaction pathway should prove to be extremely useful is enzyme catalysis, for such information will provide insight into the specificity and efficiency of enzymically controlled reactions. Toward this end it is most encouraging that recent X-ray diffraction studies of serine proteinases and their natural inhibitors[31-35] suggest reaction paths that are in agreement with those proposed above. It was found that the nucleophile (serine-195-hydroxyl) in the trypsin enzymes can be put into an orientation for optimal attack of the electrophile by rotation about the adjacent bond (C^α—C^β).

In the above discussion three classical organic reactions were shown to be amenable to the B–D approach for mapping reactions. Recently, crystal structure data on 1,6-methano-[10] annulenes and some related compounds were used to map the reaction path of a pericyclic ring closure–ring opening system.[5,36] The conclusions drawn from this particular analysis were once again consistent with experimental measurements and molecular orbital calculations. Processes which do not necessarily result in the formation of a new bond but involve an intramolecular transformation have also been examined by use of crystallographic data. For example, pseudorotation in five-membered rings[37] and the Berry mechanism[5,38] were mapped by reference of geometric distortions found in appropriate crystal structures to idealized models. Thus, systematic studies of chemical processes along the lines described are just starting to come through with valuable information. The advent of "automatic" X-ray diffraction units and "direct methods" for solving crystal structures will no doubt provide the impetus for mapping out other reaction schemes. Together with some of the other techniques discussed in this volume, one can look forward to a dramatic increase in our understanding of biochemical processes in the near future.

REFERENCES

1. H. B. Bürgi, Chemical reaction coordinates from crystal structure data. I, *Inorg. Chem.* **12**, 2321–2325 (1973).
2. H. B. Bürgi, J. D. Dunitz, and E. Shefter, Geometrical reaction coordinates. II. Nucleophilic addition to a carbonyl group, *J. Am. Chem. Soc.* **95**, 5065– 5067 (1973).
3. H. B. Bürgi, J. D. Dunitz, and E. Shefter, Chemical reaction paths. IV. Aspects of O C=O interactions in crystals, *Acta Crystallogr. Sect. B* **30**, 1517–1527 (1974).
4. P. Murray-Rust, H. B. Bürgi, and J. D. Dunitz, Chemical reaction paths. V. The S_N1 reaction of tetrahedral molecules, *J. Am. Chem. Soc.* **97**, 921–922 (1975).
5. H. B. Bürgi, Stereochemistry of reaction paths as determined from crystal structure data—a relationship between structure and energy, *Angew. Chem. Int. Ed. Engl.* **14**, 460–473 (1975).
6. L. Pauling, *The Nature of the Chemical Bond*, 3rd ed., Cornell University Press, Ithaca, N.Y. (1963).
7. A. Bondi, Van der Waals volumes and radii, *J. Phys. Chem.* **68**, 441–451 (1964).
8. C. K. Ingold, *Structure and Mechanism in Organic Chemistry*, Cornell University Press, Ithaca, N.Y. (1953).
9. L. Pauling, Atomic radii and interatomic distances in metals, *J. Am. Chem. Soc.* **69**, 542–553 (1947).
10. H. B. Bürgi, J. D. Dunitz, J. M. Lehn, and G. Wipff, Stereochemistry of reaction paths at carbonyl centers, *Tetrahedron* **30**, 1563–1572 (1974).
11. (a) A. Dedieu and A. Veillard, A comparative study of some S_N2 reactions through *ab initio* calculations, *J. Am. Chem. Soc.* **94**, 6730–6738 (1972). (b) J. Duke and R. F. W. Bader, A Hartree–Fock SCF calculation of the activation energies for two S_N2 reactions, *Chem. Phys. Lett.* **10**, 631–635 (1971).
12. R. L. Hildebrandt and J. D. Wieser, Average structures of *t*-butyl chloride and 9D-*t*-butyl chloride determined by gas-phase electron diffraction, *J. Chem. Phys.* **55**, 4648–4654 (1971).
13. H. B. Bürgi, J. D. Dunitz, and E. Shefter, Methadone, *Cryst. Struct. Commun.* **2**, 667 (1973).
14. A. Tulinsky and J. H. van den Hende, The crystal and molecular structure of *N*-brosylmitomycin A, *J. Am. Chem. Soc.* **89**, 2905–2911 (1967).
15. D. R. Storm and D. E. Koshland, Jr., Effect of small changes in orientation on reaction rate, *J. Am. Chem. Soc.* **94**, 5815–5825 (1972).
16. (a) G. A. Jeffrey, J. A. Pople, and L. Radom, The application of *ab initio* molecular orbital theory to the anomeric effect. A comparison of theoretical predictions and experimental data on conformations and bond lengths in some pyranoses and methyl pyranosides, *Carbohydr. Res.* **25**, 117–131 (1972). (b) G. A. Jeffrey, J. A. Pople, and L. Radom, The application of *ab initio* molecular orbital theory to structural moieties of carbohydrates, *Carbohydr. Res.* **38**, 81–95 (1974).
17. A. Lo, E. Shefter, and T. Cochran, Analysis of *N*-glycosyl bond length in crystal structures of nucleosides and nucleotides, *J. Pharm. Sci.* **64**, 1707–1710 (1975).
18. P. Deslongchamps, C. Moreau, D. Frehel, and P. Atlani, The importance of conformation in the ozonolysis of acetals, *Can. J. Chem.* **50**, 3402–3404 (1972).
19. P. Deslongchamps, P. Atlani, D. Frehel, and A. Malaval, The importance of conformation of the tetrahedral intermediate in the hydrolysis of esters. Selective cleavage of the tetrahedral intermediate controlled by orbital orientation, *Can. J. Chem.* **50**, 3405–3408 (1972).
20. P. Deslongchamps, C. Lebreux, and R. J. Taillefer, The importance of conformation of the tetrahedral intermediate in the hydrolysis of amides. Selective cleavage of the tetrahedral intermediate controlled by orbital orientation, *Can. J. Chem.* **51**, 1665–1669 (1973).
21. P. Deslongchamps, R. Chenevert, R. J. Taillefer, C. Moreau, and J. K. Saunders, The hydrolysis of cyclic orthoesters. Stereoelectronic control in the cleavage of hemiorthoester tetrahedral intermediates, *Can. J. Chem.* **53**, 1601–1615 (1975).
22. P. Deslongchamps, S. Dube, C. Lebreux, D. R. Patterson, and R. J. Taillefer, The hydrolysis of imidate salts. Stereoelectronic control in the cleavage of the hemiorthoamide tetrahedral intermediate, *Can. J. Chem.* **53**, 2791–2807 (1975).

23. P. Deslongchamps, Stereoelectric control in the cleavage of tetrahedral intermediates in the hydrolysis of esters and amides, *Tetrahedron* **31**, 2463–2490 (1975).

24. J. M. Lehn, G. Wipff, and H. B. Bürgi, Stereoelectronic properties of tetrahedral species derived from carbonyl groups. *Ab initio* study of the hydroxymethanes, *Helv. Chim. Acta* **57**, 493–496 (1974).

25. J. M. Lehn and G. Wipff, Stereoelectronic properties and reactivity of the tetrahedral intermediate in amide hydrolysis. Nonempirical study of aminodihydroxymethane and relation to enzyme catalysis, *J. Am. Chem. Soc.* **96**, 4048–4050 (1974).

26. J. B. P. A. Wijnberg and W. N. Speckamp, New total synthesis of eserine-type alkaloids via regioselective NaBH₄-reduction of imides, *Tetrahedron Lett.* **1975**, 4035–4038.

27. D. M. Bailey and R. E. Johnson, Reduction of cyclic anhydrides with NaBH₄. Versatile lactone synthesis, *J. Org. Chem.* **35**, 3574–3576 (1970).

28. D. Y. Curtin and L. L. Miller, 1,3-Acyl migrations in unsaturated triad (allyloid) systems. Rearrangements of N-(2,4-dinitrophenyl)benzimidoyl benzoates, *J. Am. Chem. Soc.* **89**, 637–645 (1967).

29. D. Y. Curtin, S. R. Byrn, and D. B. Pendergrass, Thermal rearrangement of arylazotribenzoylmethanes in the solid state. Examination with differential thermal analysis, *J. Org. Chem.* **34**, 3345–3349 (1969).

30. D. B. Pendergrass, D. Y. Curtin, and I. C. Paul, X-ray crystal structure and solid state rearrangement of phenylazotribenzoylmethane and the X-ray crystal structure of α-p-bromophenylazo-β-benzoyloxybenzalacetophenone, *J. Am. Chem. Soc.* **94**, 8722–8730 (1972).

31. R. M. Sweet, H. T. Wright, J. Janin, C. H. Chothia, and D. M. Blow, Chemical structure of the complex of porcine trypsin with soybean trypsin inhibitor (Kunitz) at 2.6-Å resolution, *Biochemistry* **13**, 4212–4228 (1974).

32. S. A. Bizzozero and B. O. Zweifel, The importance of the conformation of the tetrahedral intermediate for the α-chymotrypsin-catalyzed hydrolysis of peptide substrates, *FEBS Lett.* **1975**, 105–108.

33. A. Rühlmann, D. Kukla, P. Schwager, K. Bartels, and R. Huber, Bovine pancreatic trypsin inhibitor. Crystal structure determination and stereochemistry of the contact region, *J. Mol. Biol.* **77**, 417–436 (1973).

34. R. Huber, D. Kukla, W. Steigemann, J. Deisenhofer, and A. Jones, in: *Proteinase Inhibitors* (H. Fritz, H. Tschesche, L. J. Greene, and E. Truscheit, eds.), Bayer Symposium V, p. 497–512, Springer-Verlag New York, Inc., New York (1974).

35. D. Blow, in: *Proteinase Inhibitors* (H. Fritz, H. Tschesche, L. J. Greene, and E. Truscheit, eds), Bayer Symposium V, p. 473–483, Springer-Verlag New York, Inc., New York (1974).

36. H. B. Bürgi, E. Shefter, and J. D. Dunitz, Chemical reaction paths—VI. A pericyclic ring closure, *Tetrahedron* **31**, 3089–3092 (1975).

37. C. Altona, H. J. Geise, and C. Romers, Conformation of non-aromatic ring compounds—XXV. Geometry and conformation of ring D in some steroids from X-ray structure determinations, *Tetrahedron* **24**, 13–32 (1968).

38. E. L. Muetterties and L. J. Gruggenberger, Idealized polytopal forms. Description of real molecules referenced to idealized polygons or polyhedra in geometric reaction path form, *J. Am. Chem. Soc.* **96**, 1748–1756 (1974).

PART III

Studies of Transition-State Properties in Enzymic and Related Reactions

A comprehensive review of transition-state information for biochemically interesting reactions would be very useful but not possible in a book this size. Instead we shall provide here an illustrative selection, particularly chosen to exemplify the methods discussed or alluded to in Part II.

Most of the authors of the chapters in this section make liberal use of information about both enzymic and nonenzymic transition states, with the goal of understanding the enzymic process in detail. The exception is Gandour's Chapter 14, which concentrates on the role of nonenzymic data ("model reactions") in the study of enzymic transition states. He distinguishes "mimetic" models, which attempt to imitate enzymic features, from "nonmimetic" models, which attempt to illuminate general principles which will apply to both enzymic and nonenzymic reactions. The point is illustrated by the information on intramolecular reactions, solely for nonenzymic systems.

In Chapters 10 through 13 we shall be concerned with particular reaction classes, both mimetic and nonmimetic models being used in conjunction with enzymic results in a pragmatic search for a clear picture of the enzymic transition states. Group-transfer processes, including hydrolytic reactions, dominate. Methyl transfer (Chapter 10 by Hegazi, Quinn, and Schowen), hydrolysis at oxyalkyl centers capable of carbonium-ion formation (Chapter 11 by Cordes and Bull), and transfer of acyl centers susceptible to nucleophilic addition (Chapter 10) and of phosphoryl centers with similar susceptibility (Chapter 13 by Benkovic and Schray) are accorded discussion. Pollack, in Chapter 12, considers the extrusion of carbon dioxide in two different model reactions and in one enzymic system.

The choice of reactions for treatment in this section is of course arbitrary, but we believe it to reflect reasonably well the state of the field. Some obvious omissions are to a degree remedied elsewhere in the book, notably in Wolfenden's Chapter 15 and in Klinman's Chapter 4 on primary hydrogen isotope effects where hydrogen-transfer reactions are discussed.

10

Transition-State Properties in Acyl and Methyl Transfer

Mohamed F. Hegazi, Daniel M. Quinn, and Richard L. Schowen

1. INTRODUCTION

1.1. Scope of This Chapter

Our aspirations for this chapter are to give a brief discussion of the mechanistic background for acyl and methyl transfer (two types of group transfer which share some characteristics but make a useful contrast in other ways), to show how some of the methods of Part II of this volume have been applied to acyl-transfer and methyl-transfer enzymes and to indicate what conclusions can be drawn at present about transition-state properties for the particular enzymes discussed. Only transfer by nucleophilic displacement at the intact groups will be treated.

The consideration of acyl-transfer enzymes gives us the opportunity to illustrate structure-reactivity relations in enzymic transition-state characterization. This technique was explicitly omitted from Part II for two reasons. First, an excellent and recent treatment by Kirsch[1] already exists. Second, there are real difficulties in the way of a perfectly general treatment of structure-reactivity approaches to enzymic transition states, because the essence of a structure-reactivity study is to vary structure and measure reactivity, while the essence of enzyme catalysis is a combination of specificity and efficiency which bespeaks a very intimate and energetically sensitive transition-state interaction between enzyme and a *single, particular* structure (namely, that

---

Mohamed F. Hegazi, Daniel M. Quinn, and Richard L. Schowen • Department of Chemistry, University of Kansas, Lawrence, Kansas.

derived from the specific substrate). This seems to present a "double bind" from which the experimenter is unlikely to emerge with useful information. The situation in particular cases may, however, not be so unfavorable as this makes it seem. Such appears to be the case with some structure-reactivity studies of acyl-transfer enzymes.[2]

Acyl transfer and methyl transfer are members of the general class of group-transfer reactions, encompassing the transfer from a donor to an acceptor of entities as simple as the proton (some might include the electron) and as complex as macromolecules. An interesting view of the importance of certain group-transfer transition states in biological energetics has been given by Douglas.[3] We consider acyl transfer and methyl transfer an especially interesting pair of processes to compare and contrast. Both reactions in the form to be considered here (nucleophilic displacement at carbonyl and at methyl) have been subjected, in the nonenzymic cases, to very long and intensive mechanistic study.[4-15] In spite of this, lively controversy abounds with respect to both mechanisms. On the enzymic side, acyl transfer is perhaps the most exhaustively investigated example of an enzyme catalytic mechanism,[16] while the mechanistic examination of enzymic methyl transfer is at a primitive state of development.[17] Our examination here should thus give us a feeling for what is known about enzymic transition states in two group-transfer reactions, both with a strong background of nonenzymic mechanistic information, one at an advanced point of enzymological investigation and the other at the outset.

1.2. General Mechanism Chemistry of Acyl and Methyl Transfer

As in all other group-transfer processes, the primary components of chemical change in acyl transfer,

$$
\begin{array}{ccc}
\overset{\text{O}}{\underset{\|}{\text{R—C—X}}} + \text{Y}: & \longrightarrow & \overset{\text{O}}{\underset{\|}{\text{R—C—Y}}} + \text{X}:
\end{array}
\tag{1}
$$

and methyl transfer,

$$
\text{CH}_3\text{—X} + \text{Y}: \longrightarrow \text{CH}_3\text{—Y} + \text{X}:
\tag{2}
$$

are the *formation* of a new bond (C—Y) and the *fission* of an old bond (C—X). The bond orders of these two bonds may be taken as independent reaction variables.[18] If it is then assumed that all other reaction variables are relaxed to their minimum-energy values, a contour map or three-dimensional surface can be used to represent every structure intervening between reactants and products, with the energy of each structure plotted as the third dimension. These figures have been used in mechanistic contexts by a number of scientists and have been christened "maps of alternate routes" (MAR's) by Bruice.[19] Figure 1 shows schematic MAR's for acyl and methyl transfer.

No energy contours appear in Fig. 1 because we wish to consider *various* possibilities of reaction route and transition-state structure. In a real case,

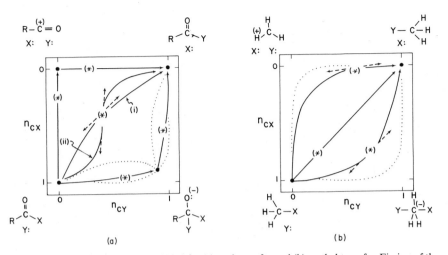

Fig. 1. Maps of alternate routes (MARs) for (a) acyl transfer and (b) methyl transfer. Fission of the "old bond" is plotted along the ordinate and formation of the "new bond" along the abscissa. Stable structures (reactant, product and intermediate states) are marked by heavy dots while activated complexes (transition states) are marked by asterisks. Reaction routes leading across the center of a diagram are dynamically coupled; those leading around the edges are dynamically uncoupled. The dashed arrows at transition-state points indicate the reaction-coordinate motion; if both reaction variables (n_{CX} and n_{CY}) contribute, they are kinetically coupled, and if only one contributes, they are kinetically uncoupled. For acyl transfer, four types of situations are portrayed. Two dynamically uncoupled routes are shown, one through the acylonium intermediate and the other through the tetrahedral adduct. Transition states are indicated for formation and destruction of each of the intermediates. Two dynamically coupled routes are also shown. The transition state for route (i) has the reaction variables kinetically coupled. The transition state for route (ii) is shown as having the same geometrical structure but with a different force field so that the reaction variables are kinetically uncoupled. Note that only one of these two routes can be a true route, not both simultaneously. The dotted lines encompass the range of transition state structures likely to be important for the reactions under consideration here. For methyl transfer, both dynamically coupled and uncoupled routes are shown. One dynamically uncoupled route is shown with a transition state having strong kinetic coupling; the other dynamically uncoupled route has a transition state with essentially no kinetic coupling. The dotted lines enclose transition-state structures which are reasonable to consider; nearly the whole diagram is included.

a line corresponding to the minimum-energy path from reactants (lower left-hand corner) to products (upper right-hand corner) can be traced on the diagram. Along such a path there will be one or more transition states (local energy maxima: marked by asterisks), and there may be one or more intermediates (local energy minima: marked by heavy dots). Reaction routes which pass generally along the diagonal directly linking reactants and products are ones upon which the changes in the two reaction variables n_{CX} and n_{CY} are strongly coupled. This coupling is a property of the entire dynamical path, rather than of some local region along it, and such paths are thus said to show *dynamical coupling*.[18] Paths which are *dynamically uncoupled*, by contrast, pass around the edges of this diagram. They may encounter one of the (usually) high-energy structures at the corners of the diagram: the *cationic intermediate* (generated by fission of the "old bond" before formation of the "new bond"; upper left-hand corner) or the *anionic intermediate* (generated by formation of the "new bond" in advance of fission of the "old bond"; lower right-hand corner). In the case of acyl transfer [Fig. 1(a)], the cationic intermediate is the acyl cation[11] or acylonium ion, and the anionic intermediate is the familiar tetrahedral intermediate.[10]

The dot representing the tetrahedral intermediate is put at $n_{CX} = n_{CY} = 0.9$ in recognition of the fact that these bonds will probably be somewhat weaker in the adduct than in the reactant and product acyl species. In methyl transfer [Fig. 1(b)], the cationic intermediate is the methyl cation, so that a dynamically uncoupled route through this species would represent an S_N1 reaction[4,7] or one of the various "ion-pair" mechanisms.[20,21] The anionic intermediate for methyl transfer would be a species with ten electrons in the carbon valence shell; it is thus likely that no minimum will exist in the energy surface very near this corner but that the energy in this region will rise smoothly, continuously, and probably sharply.

The different reaction paths sketched in Fig. 1 emphasize the distinction between *reaction path* and *transition-state structure*. The reaction path is defined by the entire minimum-energy *line* which links reactants and products along the potential energy surface. Its properties can be considered dynamical ones. As explained above, dynamically coupled paths are short ones, along the diagonal, and dynamically uncoupled ones are longer, around the edges. The structure of the transition state, on the other hand, is defined by a *point* on the MAR. This point must lie on the reaction path, but there is no way to infer the shape of the reaction path from a knowledge of the transition-state structure alone. The dashed arrows associated with some of the transition states in Fig. 1 represent the corresponding reaction-coordinate motion. This is a property of the transition state, being one of its vibrational normal coordinates, and the degree to which the two reaction variables are coupled in the reaction coordinate can be called the *kinetic coupling*. Figure 1(a) shows two transition states for hypothetical carbonyl displacement mechanisms, both having the same geometrical structure; they lie near the center of the diagram and are represented by a single asterisk. Although their geometrical structures are the same, they are hypothesized to have different force fields (note that only one of these transition states will exist in nature; we are considering two mutually exclusive possibilities). One has a reaction coordinate with a high degree of kinetic coupling between CX bond fission and CY bond formation (dashed arrow at $45°$). The other, with a nearly vertical dashed arrow, has a kinetically uncoupled reaction coordinate consisting almost entirely of CX bond fission. In this case, the formation of the CY bond is accomplished in part before the transition state occurs and in part after decomposition of the transition state. If one knew the geometrical structure alone of such a transition state, no conclusion could be drawn about the extent of even the kinetic coupling, much less the dynamic coupling. If one knew the geometrical structure and also the force field (including some information about off-diagonal force constants), then the magnitude of kinetic coupling could be deduced. It is possible that certain detailed isotope-effects experiments will allow such deductions.[22] Even then, however, the dynamical coupling remains a mystery. Any number of paths could be drawn through a given transition state tangent to a given reaction coordinate. This dynamical information will have to come from studies beyond the regime of transition-state theory.

The same considerations, of course, apply to the methyl transfer reaction diagrammed in Fig. 1(b). Shown there are three reaction paths. One (across the center) is dynamically coupled and passes through a kinetically coupled transition state. The second, passing toward the lower right, shows a smaller degree of dynamic coupling but strong kinetic coupling in the transition state. The latter has a "tight" structure with the total bond order at carbon exceeding unity and the carbon thus bearing a partial negative charge. The third path (around the upper left) is dynamically uncoupled in considerable degree and has a transition state of very low kinetic coupling. The transition state has a "loose" structure, total bond order to carbon being only about one half. The carbon thus bears some positive charge.

The MAR's of Fig. 1 refer to "crude" reactions, either omitting refinements from solvent or catalyst influence or assuming some form of coupling of these features to those represented. To a first approximation, they should describe the systems adequately because the covalent bonding changes chosen as the major reaction variables ought to dominate the internal energy changes. Using these MAR's, we shall summarize the results of general mechanistic experience for these reactions, on the "crude" level, and then proceed to consider the influence of catalysts.

The highly uncoupled route through the anionic (tetrahedral) intermediate in Fig. 1(a) is an adequate description of the acyl transfer reactions we shall consider here.[8-12] These involve esters (X = OR) or amides (X = NR_1R_2) as the acyl substrates. The very electronegative leaving groups of these substrates would have high energy as anions, and this energy would have to be added to the already great requirement for acylonium-ion formation to generate structures near the cationic intermediate. The energy surface of Fig. 1(a) thus tends to rise sharply in the region of the cationic intermediate, forcing minimum-energy reaction paths toward the anionic side. Other substrates with better leaving groups, such as acyl halides, have a lower-energy surface on the cationic side and may react through the acylonium intermediate.[11] An acylonium mechanism can be achieved even with esters by simultaneously (1) reducing the energy of the alkoxide leaving group by providing a strongly acidic medium to protonate it before it departs and (2) increasing the energy of the tetrahedral intermediate through a severe steric constriction about the substrate carbonyl group. Thus one observes acylonium reaction in the fission of 2,6-dialkyl benzoate esters in superacid solutions.[11] The acylonium mechanism is, however, unimportant for ordinary esters and amides under mild conditions. There is no evidence for any acyl-transfer reaction which compels or strongly suggests a coupled route, avoiding an intermediate. Acyl-transfer data can generally be accounted for by the uncoupled route through the tetrahedral intermediate, the uncoupled route through the acylonium ion, or a combination of the two.*

The tetrahedral-intermediate route is generally, as shown in Fig. 1(a), a two-step process with a first transition state for CY bond formation and a second transition state for CX bond fission, separated by the intermediate, which is commonly assumed to survive for a considerable period. The survival time is, however, sometimes too short for reaction with other species and may occasionally be very short indeed.[14] Either of the two transition states may be of higher energy, and thus determine the rate alone, or both may be of similar energy. Clearly it is the relative ease of separating X: and Y: from the intermediate which decides this question, but the situation is not simple, because when Y departs, X is left behind with the developing acyl group and vice versa. Generally speaking, however, a greater stability of the leaving group as a free species lowers the energy of the transition state in which it is being freed.

Methyl-transfer reactions appear generally to proceed along reaction paths of at least some degree of coupling and thus to occur by the classical S_N2 route. All evidence points against the true intermediacy of either anionic or cationic species.[21] It is also clear, however, that perfect coupling is not the rule. α-Deuterium isotope effects suggest a considerable variation in "tightness" of transition-state structure.[23] Values range from quite inverse effects ($k_{3H}/k_{3D} \sim 0.85$), indicating that the methyl hydrogens are experiencing a restricted bending potential in the transition state, to quite normal effects ($k_{3H}/k_{3D} \sim 1.25$ or even larger[24]) that show a decrease in constriction of hydrogen motion in the transition state, relative to the tetrahedral environment of the methyl substrate. In general, exergonic reactions will be expected to have transition-state structures lying nearer reactants than products (i.e., in the "southwest") and endergonic reactions vice versa ("northeast"). Greater stability of both X: and Y: as free species or an environment which favors cationic charge development will cause transition-state structure to move "northwest," and a "southeast" drift will be elicited by opposite influences.[25]

Thus it appears reasonable to assume that gross transition-state structures for reactions of interest here will fall within the regions outlined by the dotted lines of Figs. 1(a) and (b). In Fig. 1(a), the acyl-transfer transition states are shown as restricted to within rather narrow limits around the uncoupled anionic

---

* It is extremely difficult to distinguish experimentally between a single transition state with kinetic coupling of bond formation and fission and a virtual transition state (see Chapter 2) with nearly equal contributions from different transition states in which bond formation and fission are occurring in kinetically uncoupled fashion. O'Leary cites data in Chapter 7 in this volume that are certainly strongly consistent with a single kinetically coupled transition state in the hydrolysis of benzoyl chloride. It may require, in this and other cases showing weakened bonds to both entering and leaving groups in carbonyl displacement, detailed isotope-effect experiments[22] to sort out the possibilities just mentioned.

route. In the methyl-transfer case of Fig. 1(b), the dotted lines encompass nearly the entirety of the map.

We can now go on to the consideration of MAR's for more complex reactions of particular relevance to questions of enzyme catalysis.

1.3. Acid–Base Catalysis of Acyl and Methyl Transfer[8-15]

Although enzymes employ many devices for the stabilization of transition states, none comes more readily to mind than acid–base catalysis. A corresponding wealth of effort has been devoted to the study of this phenomenon in nonenzymic systems supposed to be related to enzymic reactions. Clearly acid–base functions of importance to proton-transfer reaction in enzymes must have pK_a's near 7 in order to make protolytic interactions possible at physiological pH's. We shall, therefore, concentrate on features which allow catalysis by acids and bases with pK's ~ 7.

The particular acyl-transfer reactions we shall consider can be summarized by

$$R_1OH + R_2COXR_3 \longrightarrow R_2COOR_1 + R_3XH \tag{3}$$

This shows acyl transfer to either alcohol (R_1 = alkyl, as in the acylation of serine enzymes) or water (R_1 = H, as in the hydrolysis of acyl enzymes) from esters ($X = O$) or amides ($X = NH$). Notice that in addition to transfer of the acyl group from X to O, a proton is transferred from O to X. The particular methyl-transfer processes of interest are those summarized by

$$R_1OH + CH_3Y \longrightarrow R_1OCH_3 + Y^{(-)} + H^+ \tag{4}$$

The leaving groups (Y) of interest (thioethers in the enzymic case) are so weakly basic that transfer of the methyl group from Y to O is accompanied by *release* of a proton. If Y is initially neutral, it becomes an anion; if initially positively charged, it becomes neutral.

As these equations show, both acyl transfer and methyl transfer involve shifts in the location of protons in the course of the group-transfer reactions. Some or all transition states along the reaction path will thus have full or partial bonds to protons, and this suggests that acids and bases might catalyze these reactions by interacting with or at such protonic centers. This proves to be a major feature of acyl-transfer reactions, about which much is currently known, and a puzzling aspect of methyl-transfer reactions.

1.3.1. Protolytic Potentials in Acyl Transfer[8-10,13-15]

Consider the nucleophilic attack step of acyl transfer

$$R_1OH + R_2COXR_3 \longrightarrow \left[\begin{array}{c} R_1 \diagdown \overset{\delta+}{} \overset{O^{\delta-}}{\underset{\|}{C}} \diagdown \\ O \underset{R_2}{|} XR_3 \\ H \end{array} \right] \longrightarrow \begin{array}{c} R_1 \diagdown \overset{O^-}{\underset{|}{C}} \\ O \overset{+}{\underset{R_2}{}} XR_3 \\ H \end{array} \tag{5}$$

1

The reactant molecule ROH is a weak acid (typical $pK_a = 16$), while the product as shown should be about as strongly acidic as a protonated ether[26] (say, $pK = -2$ after appropriate corrections[27]). At pH 7, therefore, proton release from the reactant is endergonic ($\Delta G_R^0 = +12.2$ kcal/mol) and from the adduct is exergonic ($\Delta G_{A1}^0 = -12.2$ kcal/mol). The acidity of the transition state **1**, which should be intermediate between these limits, depends on its structure.[28] If its pK_a increases linearly with n_{CO}, the order of the new CO bond (i.e., if the reaction variables pK_a and n_{CO} are highly coupled), then when $n_{CO} = \frac{1}{2}$, $\Delta G_{T1}^0 = 0$. If $n_{CO} > \frac{1}{2}$, $\Delta G_{T1}^0 < 0$, and proton release from the transition state becomes exergonic at pH 7:

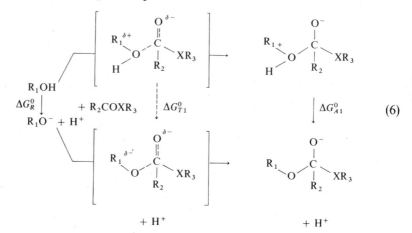

$$(6)$$

Smaller values of n_{CO} correspond to transition states which will not spontaneously release a proton from this site. Spontaneous release of a proton from the transition state at pH 7 ($pK_a < 7$) implies that bases in this pK range could accelerate the reaction by accepting the acidic proton. We thus have a relation which defines which transition-state structures should be susceptible to base catalysis and which should not. This is in essence an extension, or more detailed delineation, of the limits that Jencks has worked out for acid–base catalysis.[13]

However, even if the proton is spontaneously released from the nucleophilic site for $n_{CO} > 0.5$, there is another transition-state site which is basic and may simply "retrieve" the proton: That is the developing oxyanion center being formed from the carbonyl oxygen. Thus one ought to consider, in addition to dissociation of the transition state **1** as in Eq. (6), its *isomerization*, as in

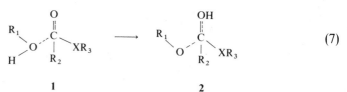

$$(7)$$

to structure **2**. The free-energy change for the isomerization of Eq. (7) will be given by

$$\Delta G_{12} = RT(pK_a^1 - pK_a^2) \tag{8}$$

The pK_a of **2** should be between that of an ester (amides will be considered below), which is[26] about -7, and that of the tetrahedral adduct, estimated by the method of Fox and Jencks[27] to be about 13. Assuming that n_{CO} is coupled to both pK_a's *and* assuming that n_{CO} is the same in both **1** and **2**,* we can write

$$pK_a^1 = 16 - 18n_{CO} \tag{9a}$$

$$pK_a^2 = 20n_{CO} - 7 \tag{9b}$$

$$\Delta G_{12} = RT(23 - 38n_{CO}) \tag{9c}$$

From these relations, we deduce that:

1. Structure **2** will be preferred over structure **1** whenever $n_{CO} > \frac{23}{28} \sim 0.6$.

2. When structure **2** prevails ($n_{CO} > 0.6$), the transition-state pK_a will be between 5 and 13.

3. When structure **1** prevails ($n_{CO} < 0.6$), the transition-state pK_a will be between about 5 and 16.

4. For $n_{CO} = 0.5$–0.6, structure **1** prevails, $pK_a = 5$–7, and thus exergonic proton loss and base catalysis near neutral pH can occur.

5. For $n_{CO} = 0.6$–0.7, structure **2** prevails, $pK_a = 5$–7, and base catalysis near neutral pH can occur.

We find therefore only a very narrow range of transition-state structures ($n_{CO} = 0.5$–0.7) in which catalysis by exergonic proton release near pH 7 is possible. Although we have considered only the nucleophilic attack transition states, the situation will be unchanged for ester substrates if leaving-group expulsion is considered. The two processes are microscopic reverses, and the same relationships will hold. We shall now consider leaving-group expulsion with amide substrates.

With rate-limiting leaving-group expulsion, as expected with amides, the transition state will lie between structural limits of the adduct and the products,

---

* This assumption, that the heavy-atom framework structure of the transition state is independent of its state of protonation, is, first, necessary for the correctness of the predictions of the kind of treatment being carried through here and, second, almost surely not true in detail (see below). It may emerge, however, that the alterations in structure on protolysis are insufficient to affect the rather broad-brush picture being painted here.

as in

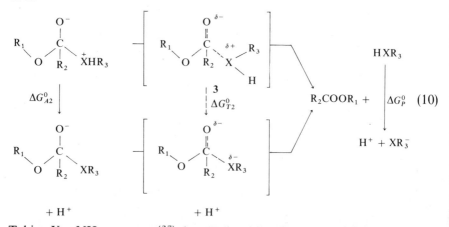

$$+ H^+ \qquad\qquad\qquad + H^+$$

Taking $X = NH$, we expect[27] the pK_a for the zwitterionic adduct to be about 13. The product pK_a should be about 35 (for an amine). Thus *no* transition-state structure for species **3** should exergonically lose a proton near pH 7. The isomerization of **3** to transition-state **4** can also be considered:

$$R_1O\underset{\underset{R_2}{|}}{\overset{\overset{O}{\|}}{C}}XHR_3 \longrightarrow R_1O\underset{\underset{R_2}{|}}{\overset{\overset{OH}{\|}}{C}}XR_3 \qquad (11)$$

$$\qquad\quad \mathbf{3} \qquad\qquad\qquad\qquad \mathbf{4}$$

The pK_a of **4** should lie between 13 (see below) and the pK_a of a protonated ester (about -7). Again assuming coupling of these pK_a's to n_{CX},

$$pK_a^3 = 35 - 22n_{CX} \qquad (12a)$$

$$pK_a^4 = 20n_{CX} - 7 \qquad (12b)$$

$$\Delta G_{34} = RT(42 - 42n_{CX}) \qquad (12c)$$

we can deduce the following features of this system:

1. Structure **3** will essentially always prevail.
2. Structure **3** cannot spontaneously release a proton near pH 7.
3. In the very limited range where structure **4** is permitted ($n_{CX} \simeq 1$), its pK_a is around 13, and it also cannot spontaneously release a proton near pH 7.

Thus acceleration of amide fission through base catalysis in the range of pH 7 is not to be anticipated.

In addition to acceleration through proton removal, acceleration by proton donation to the transition state is possible. Thus a proton could be added to transition states **1** to form **5** or to **3** to form **6**:

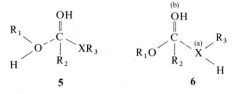

5 **6**

The pK_a's of these two species (the carbonyl-formed protons) can be estimated as before,[27] leading to

$$pK_a^5 = 15n_{CO} - 7 \tag{13}$$

$$pK_a^6 = 15n_{CX} - 7 \tag{14}$$

These pK_a's are equal for similar transition-state structures, and unless n_{CO} or n_{CX} is greater than $14/15$ (~ 0.9), then the pK_a's are less than 7 and no effective acid catalysis can occur near pH 7.

The overall conclusions are thus that bases with pK_a's (for their conjugate acids) in the neighborhood of 7 can accelerate acyl-transfer reactions only when the transition state has a partial bond to an alcohol-like entering or leaving group with a bond order of about 0.5–0.7 and that acids with pK_a's around 7 can catalyze acyl transfer only if the transition state is essentially equivalent to the tetrahedral adduct.

1.3.2. Protolytic Potentials in Methyl Transfer

Application of the same considerations to methyl transfer, for the cases of interest here, is rather straightforward. Transition state **7** should have a pK_a between that of an alcohol (16) and that of a protonated ether (say -4)[26]:

7

Assuming coupling of pK_a to n_{CO}, $pK_a = 16 - 20n_{CO}$, and proton transfer will be favored to bases of pK 7 for $n_{CO} > \frac{9}{20} \sim 0.5$.

1.3.3. General Catalysis

In these cases, we have been considering transition-state stabilization through proton removal or addition. The full stabilization indicated by the transition-state pK can be realized only by complete loss or gain of the proton. Enzymes may make use of catalytic mechanisms of this kind through production of microenvironmental pH's different from those of the medium or through microenvironmental alteration of transition-state pK's, with enzyme functional groups providing local supplies and sinks of protons. However, many enzymic acid–base catalytic interactions almost surely involve direct transition-state bridging by enzyme functional groups to protons which may or may not eventually be gained or lost but which are not wholly transferred in the transition state. This is analogous to *general* acid–base catalysis in nonenzymic reactions.* If the driving forces involved in transition-state stabilization are merely those summarized by transition-state pK's, then the structural limitations are the same as for specific catalysis and the energy gains to be realized as catalytic power are necessarily smaller, corresponding to some fraction (equivalent to a Brønsted coefficient) of the potential gain given by the pK. However, there may be other stabilizing factors in general catalysis. The question merits consideration in terms of the details of transition-state coupling features.

General catalysis can be envisioned with two extremes of kinetic coupling between the proton transfer (PT) to or from the general catalyst and the heavy-atom reorganization (HAR) of the substrate framework. In one extreme, precise kinetic coupling is imagined. The reaction-coordinate motion for, say, general-acid-catalyzed leaving-group expulsion from a tetrahedral intermediate would look like this:

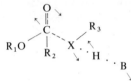

The other extreme hypothesis posits completely *uncoupled* PT and HAR, with the two processes occurring in successive transition states, either PT, HAR,

or HAR, PT,

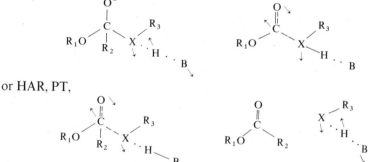

These hypotheses are shown on MAR diagrams in Fig. 2 for both acyl transfer [Fig. 2(a)] and methyl transfer [Fig. 2(b)]. The purely uncoupled routes [(i) and (ii)] run along the edges of the diagrams. These are both dynamically and kinetically uncoupled. The purely (both dynamically and kinetically) coupled routes, labeled (iii), run on the diagonals.

---

* Recent work in the field of general catalysis has shown the importance of many phenomena in the reactions of small molecules in solution that are of far smaller significance in enzyme catalysis.[14] These include the participation of transport processes, preassociation and spectator mechanisms, etc., not treated in detail in this chapter. A particularly nice paper linking up these concepts with chymotrypsin catalysis is that of A. C. Satterthwait and W. P. Jencks, "The Mechanism of the Aminolysis of Acetate Esters," *J. Am. Chem. Soc.* **96**, 7018–7031 (1974).

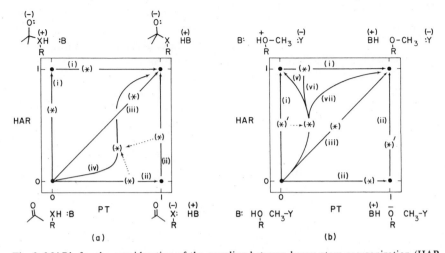

Fig. 2. MAR's for the consideration of the coupling between heavy-atom reorganization (HAR, plotted along the ordinate) and proton transfer (PT, plotted along the abscissa) for base-catalyzed carbonyl addition (a) and base-catalyzed methyl transfer (b). The second stage of acyl transfer, the decomposition of the tetrahedral adduct, is also described by diagram (a), taken in reverse. For both reactions, a variety of dynamically coupled and dynamically uncoupled routes are shown. On both diagrams, routes (i) and (ii) represent sequential, dynamically uncoupled HAR and PT; routes (i) correspond to the order HAR/PT, and routes (ii) to the order PT/HAR. At least with methyl transfer, the primed transition states, for HAR, would be expected to determine the rate, although, in principle, either the HAR or the PT transition state might be of higher free energy. Routes (iii) in both cases exhibit complete dynamic coupling of HAR and PT. Route (iv) for acyl transfer is dynamically well coupled, but its transition state has a reaction coordinate involving only HAR (thus kinetically uncoupled). The dotted arrows show the hypothetical reactions in which a "pure PT" or "pure HAR" transition state could be rearranged into that for route (iv). If both routes (ii) and (iv) are open channels on the true potential surface, this kind of hypothetical reaction will be endergonic if the acyl-transfer reaction follows route (ii) and exergonic if the acyl-transfer reaction follows route (iv). On the diagram for methyl transfer, routes (v), (vi), and (vii) are possible routes through a kinetically uncoupled transition state in which the reaction-coordinate motion consists only of HAR but in which PT is about 25% advanced. Route (v) envisions a post-transition-state retrogression of PT to generate the oxonium intermediate, route (vi) a Choi–Thornton pathway[34] leading into a PT transition state, and route (vii) a post-transition-state relaxation to complete the PT process. The dotted arrow again shows a hypothetical reaction in which the "pure HAR" transition state forms a hydrogen bond to generate a transition state in which PT is partially advanced.

In their pure forms, these two hypotheses make testable predictions. If a kinetically coupled route is followed, primary H isotope effects and primary heavy-atom (O, C, and X) isotope effects should be *simultaneously* observed. If a kinetically uncoupled route is followed, *either* H or heavy-atom primary isotope effects should be observed (depending on whether the PT or HAR transition state is rate determining) but not both. Insufficient data are available for a general conclusion to be drawn, although most isotope-effect information favors the view that PT and HAR are kinetically uncoupled in acyl and methyl transfer.

Thus the rule is that small H isotope effects are observed in general-catalyzed acyl-transfer reactions; acyl-transfer reactions also on occasion show primary heavy-atom effects. However, careful studies of both in the same system have not been done. The simplest conclusion is that the two processes are largely kinetically uncoupled and that HAR dominates in the rate-determining transition state. In some cases, PT may dominate.[29]

In methyl transfer (or alkyl transfer in general), general catalysis is rare. In one established case, the cyclization of 4-chlorobutanol,[30]

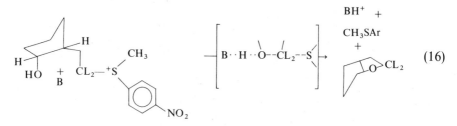

$$ \tag{15} $$

the solvent isotope effects (k_{H_2O}/k_{D_2O}) were extremely small (1.28 for $B = H_2O$, 1.07 for $B = HO^-$). At the same time, the chlorine isotope effects[31] $(k_{35}/k_{37} = 1.0080 \pm 0.0001$ for $B = H_2O$, 1.0076 ± 0.0002 for $B = HO^-$) clearly show that heavy-atom reorganization is in progress. In a second case of special relevance to the enzymic reaction discussed below, the cyclization of Eq. 16

$$ \tag{16} $$

discovered by Coward and his group,[32] the solvent isotope effect is only 1.37 ($B = H_2O$), while the α-D isotope effect[33] of 1.17 ($L = H$ or D) shows considerable progress of heavy-atom reorganization. Apparently, as in acyl transfer, PT and HAR are uncoupled for these two cases at least.

It is important to note that isotope-effect evidence of this kind, indicating kinetic uncoupling, does not imply anything about the degree of dynamic coupling. Indeed, various indications show that it is *not* generally true that *no* PT has occurred in transition states for general catalysis with dominant HAR in the reaction coordinate. This can be understood in terms of reaction paths like (iv), (v), (vi), and (vii) in Fig. 2. In all of the transition states for these routes, HAR and PT are kinetically uncoupled: The reaction coordinate (tangent to the reaction path at the transition state) has no PT component and is thus "pure HAR." Paths (iv) and (vii) show considerable dynamical coupling, however, and in all paths some PT has been accomplished at the transition state: about 60% for (iv) and about 25% for (v)–(vii). This would be consistent with the frequently observed Brønsted β values of, say, 0.2–0.8 for such reactions. Paths (iv) and (vii) show PT being accomplished partly before and partly after the transition state but not as a component of the reaction coordinate and without the intervention of a second transition state on the reaction path. Path (v) refers to the formation of a hydrogen bond which is only extant in the transition state, so that the proton reverts to its original position upon completion of HAR and is then transferred subsequently. Path (vi) leads out of the HAR transition state directly into the PT transition state, so that although the two processes are uncoupled and occur in separate transition states, no intermediate arises between them. The reaction is thus a "one-step" reaction with two transition states (Choi–Thornton process).[34]

An interesting problem of general catalysis in which some degree of PT has been achieved in the transition state, but in which PT is not a component of the reaction coordinate, is the question of what stabilizes such transition states against dissociation. As shown in

$$ \text{B} \cdots \text{H} \cdots \text{O} -- \text{CH}_2 --- \text{Y} + \text{H}_2\text{O} \longrightarrow \text{H}_2\text{O} \cdots \text{H} \cdots \text{O} -- \text{CH}_2 -- \text{Y} \qquad (17) $$
$$ + \text{B} $$

a general catalysis transition state with base B can dissociate to the transition state for catalysis by water and the free catalyst. Consider a case in which the reaction coordinate is pure HAR; then the $\text{B} \cdots \text{H} \cdots \text{O}$ bond is simply a hydrogen bond, and the catalysis effected by B is occasioned by the release of hydrogen-bonding energy or energy of specific solvation by B as opposed to water ("solvation catalysis").[30] The strength of such a hydrogen bond must be effectively much greater than that which is usually seen in ordinary, stable molecules in order for any appreciable part of the reaction to proceed through the catalyzed transition state. For example, if the Brønsted coefficient is 0.5, then an amine catalyst (pK_a of $BH^+ \sim 10$) will dissociate from the transition state with a dissociation constant* of only $10^{-5.9}$ M. This means that at 10^{-3}-M amine, over 90% of the transition states will be complexed by the amine! Bases cannot compete with water for hydrogen bond sites in any stable molecule with anything approaching this success. Put differently, the hydrogen bond in the catalytic transition state, for the case just mentioned, must have a strength greater than that of the hydrogen bond to water by 8 kcal/mol.

The origin of this large interaction strength must be something that is peculiar to or commonly met with in transition states but is rare or unknown for ordinary reactant-state molecules in dilute aqueous solution. Swain and his co-workers, who first called attention to this point, suggested the possibility that an extraordinary electron polarizability might characterize both these hydrogen bonds in transition states, and the nearby reacting orbitals, thus providing a basis for strong interaction through mutual polarization.[30] Although the question remains open, there is much to be said for this view. The reacting-orbital electrons, being less tightly bound than those in ordinary

---

* The dissociation constant $K = k_{\text{water}}/k_{\text{amine}}$; by the Brønsted law this is $(10^{-10}/10^{+1.74})^{0.5} = 10^{-5.9}$. Also $K = (\text{TS}_{\text{water}})(\text{B})/(\text{TS}_{\text{amine}})(\text{H}_2\text{O})$ so that if amine is 10^{-3} M and water $10^{+1.74}$ M, $\text{TS}_{\text{amine}}/\text{TS}_{\text{water}} = 18$, or the transition state is 95% complexed.

bonds, ought to be easily polarized.[35] The hydrogen bond which holds the catalyst may well also be of a highly polarizable type. Zundel and his collaborators have found many examples of unusually polarizable hydrogen bonds.[36] Among these are bonds to rather basic centers, such as methoxide or hydroxide ions, which are known[37] to have deuterium fractionation factors of around 0.7–0.75. Formation of such a bridge from catalyst to a transition-state reaction site of transiently high basicity could give a solvent isotope-effect contribution of 1.3–1.4, just in the range observed for alkyl transfer.* Kreevoy and his collaborators[38] have observed unusual hydrogen bonds also, probably of the single-minimum variety, which possess deuterium fractionation factors around 0.5. Formation of such bridges could generate the solvent isotope-effect contribution of around 2 that is so typical of acyl transfer.[8,39]

A second factor which may greatly stabilize hydrogen bonds in catalytic transition states is *high plasticity of the nuclear framework*. The partially bonded structures of transition states should be more readily reorganized, under weaker influences, than the fully bonded frames of stable molecules. Formation of a catalytic hydrogen bond may thus alter the general structural features of the transition state in a stabilizing way. That this kind of change occurs can be seen from the chlorine isotope effects of Cromartie and Swain,[31] cited above. Further evidence comes from the extensive structure-reactivity studies of Gravitz and Jencks.[40-42] For example, the expulsion of a series of alcohol leaving groups from tetrahedral adducts was observed with different general acid catalysts. With catalysis by acetic acid, more acidic alcohols reacted more rapidly (Brønsted slope -0.23). With hydrogen ion as catalyst, the reverse was true (Brønsted slope $+0.24$). Therefore a change of catalyst from one acid to another was able sufficiently to reorganize the transition-state structure to invert the charge on the leaving-group oxygen. It is important to keep in mind that a structural sensitivity of one part of the transition state to features in another part does *not* imply a coupling of motions between these parts in the reaction coordinate (kinetic coupling), so no inference about this point should be drawn from observations such as those just cited. Whether these factors offer the true explanation of the stability of general catalysis transition states, relative to their "water catalysis" counterparts, is yet to be seen. The subject remains worthy of further investigation, particularly from the theoretical viewpoint.†

---

* This transiently high basicity (and acidity) in transition-state sites may give rise to hydrogen bonds of strengths unknown in relaxed, ground-state molecules, a factor that Jencks[14] has argued dominates in the stability of these transition states.

† It is, of course, true that enzyme-binding functions can maintain general catalytic groups at the appropriate place in the transition states of enzyme-catalyzed reactions without the necessity of any special stability for the interactions. This means that, on the one hand, weaker interactions may appear as components of enzyme catalysis than would be observable in solution externally and that, on the other hand, interactions of specially large strength may be unusually effective in enzyme catalysis since other binding forces can maintain the transition-state integrity, liberating the entire effect of the interaction for catalysis.

1.4. Enzymic Acyl and Methyl Transfer

When an enzyme intervenes in a group-transfer process as catalyst, two mechanisms immediately come to mind for the principal feature of its action. In one mechanism, it may serve to bring together the group donor and the group acceptor in the active site at the same time. It may then promote direct transfer from donor to acceptor. In the customary notation,[43] this may be symbolized as

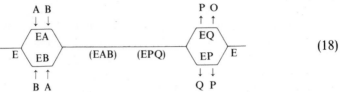

$$(18)$$

This is the "sequential" mechanism of the type "Random Bi Bi." This means that the enzyme collects the donor A and acceptor B (in either order, as indicated by the branched path) to form the *central complex* (one having a completely filled active site) EAB. The group is transferred, converting A to P and B to Q, thus generating the new central complex EPQ, from which P and Q dissociate in either order ("randomly"). Both forward and reverse reactions are "bireactant," involving two nonenzymic species—thus "Bi Bi." Different versions of this would require the reactants to be assembled in a certain order, or the products to be released in a certain order, or both. The second major mechanistic possibility for group transfer is one in which the group is transferred from the donor to the enzyme, forming a modified enzyme, and thence to the acceptor:

$$\begin{array}{ccccccc} A & & P & & B & & Q \\ \downarrow & & \uparrow & & \downarrow & & \uparrow \\ \hline E & (EA) & (FP) & F & (FP) & (EQ) & E \end{array}$$

$$(19)$$

Here F represents the modified enzyme. This kind of mechanism, in which the enzyme oscillates between two forms, E and F, is known as a "Ping-Pong" mechanism. In the form shown, with F generated as a free species by departure of P before arrival of B, kinetic and inhibition experiments are in principle capable of distinguishing Ping-Pong from sequential mechanisms.[43] A more difficult problem is presented by this mechanism involving a modified enzyme:

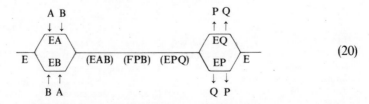

$$(20)$$

Here the enzyme is intermittently transformed to the group-modified form F but without dissociation of the donor–acceptor-derived moieties. This is kinetically equivalent to the Random Bi Bi mechanism above [Eq. (18)] and in fact would be so classified.

Another modified Ping-Pong process of importance is that for hydrolytic reactions, in which water is not customarily classified as substrate*:

$$
\begin{array}{cccccccc}
& A & & P & & Q & & \\
& \downarrow & & \uparrow & & \uparrow & & \\
\hline
E & (EA) & (FP) & & F & (EQ) & & E
\end{array}
\qquad (21)
$$

The best evidence currently at hand indicates that the hydrolytic acyl-transfer reactions to be discussed in this chapter proceed by the mechanism of Eq. (21). Frequently the acyl enzyme F can be isolated and studied and trapping vs. rate experiments show that the acyl enzyme is a true reaction-path intermediate in enzymic acyl transfer.

In methyl transfer, the best indications are for the mechanism of Eq. (18). The kinetic picture is ambiguous, but recent studies by Floss and his collaborators[45] show that, at least in methyl transfer to C, methyl transferases lead to stereochemical inversion in the chiral CHDT group. Retention would be expected for either Eq. (19) or (20).

2. SERINE PROTEASES[44,46]: ACYLATION OF α-CHYMOTRYPSIN

2.1. General Mechanistic Features

The serine proteases are enzymes which catalyze the hydrolysis of the peptide bond of proteins and other peptides:

$$
R_1CONHCHR_2CONHR_3 + H_2O \longrightarrow R_1CONHCHR_2CO_2^- + H_3NR_3^+ \qquad (22)
$$

They occur at all evolutionary levels and are important in various physiological events.[44] Perhaps the most famous examples are the mammalian digestive enzymes chymotrypsin[50] and trypsin. The serine proteases are single-subunit globular proteins of molecular weight about 25,000. They share a common mechanistic scheme, and their differing specificities arise from detailed differences in the binding regions of their active sites.

The catalytic process involves an initial manifold of steps leading to an

---

* In this scheme, A can represent a typical acyl substrate such as an ester or an amide. It combines with the enzyme to form the enzyme–substrate complex EA. This undergoes the acylation reaction in which the acyl group is transferred to the enzyme to generate the acyl enzyme F, initially complexed with the first product (alcohol or amine) P in the form of FP. The latter dissociates (first-product release) to F and P. Then the acyl enzyme reacts with a water molecule (taken as already present) to deacylate and form the second product Q (carboxylic acid) and the original enzyme as the complex EQ. This then dissociates (second-product release) to E and Q.

acyl enzyme:

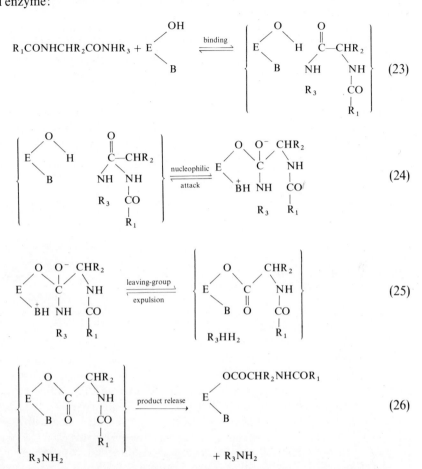

$$R_1CONHCHR_2CONHR_3 + E \overset{OH}{\underset{B}{\diagdown}} \overset{binding}{\rightleftharpoons} \left\{ \cdots \right\} \quad (23)$$

$$\quad (24)$$

$$\quad (25)$$

$$\quad (26)$$

The first of these steps [Eq. (23)] is the binding of substrate to the enzyme to form a complex, which at high substrate concentrations (typically $> 10^{-5} - 10^{-3}$ M) constitutes a large fraction of the total enzyme present. The next event [Eq. (24)] is nucleophilic attack by an active-site serine hydroxyl upon the substrate carbonyl group. It is this serine that lends its name to the enzyme class. This attack is assisted by an active-site basic unit (B in the structures), and the result of nucleophilic attack is the tetrahedral adduct of the carbonyl group. Next the leaving group is expelled [Eq. (25)], probably with general acid catalysis, and finally [Eq. (26)] product release provides the acyl enzyme as a free species. In principle, protonation of the amine product to form the ammonium ion that predominates under physiological conditions could occur either before or after product release. In this scheme it is assumed to occur in solution after release.

The second stage of catalysis consists in regeneration of the enzyme from the acyl enzyme with production of the carboxylate product. The reaction is a three-step process analogous to that above:

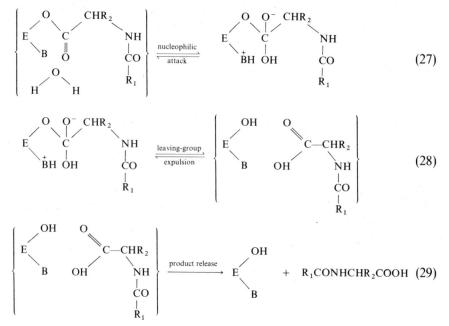

$$(27)$$

$$(28)$$

$$\text{product release} \longrightarrow \text{E} \overset{OH}{\underset{B}{<}} \quad + \quad R_1CONHCHR_2COOH \quad (29)$$

This manifold is truncated by one step relative to that for acylation because no explicit step for binding of water is necessary (although it is possible that some conformation change would be required to bring the water molecule into a reactive position). Thus the first step [Eq. (27)] is base-assisted nucleophilic attack, and the second [Eq. (28)] is leaving-group expulsion with general acid catalysis. Equation (29) shows release of the un-ionized carboxylic acid product, although here, as before, protolytic equilibrium could in principle be established before rather than after desorption from the enzyme surface.

The enzymic environment in which these processes are promoted is shown, a bit schematically, in Fig. 3. Although the details shown are taken from the structure of α-chymotrypsin,[50] the picture is quite representative of the situation in other serine proteases and very likely in a number of other enzymes not even related by biological evolution to the serine proteases.

The heart of the enzyme's mechanistic strategy is the nucleophilic hydroxyl group of Ser-195, shown as an open sphere near the top center of the structure. Four other regions of the active site enhance catalysis of reactions of specific substrates by transition-state interaction with features of such substrates or catalysis in general by increasing the transition-state stability at the nucleophilic hydroxyl or substrate carbonyl. The first such region, just clockwise from the nucleophilic hydroxyl in Fig. 3, the "oxanion hole," is a binding pocket which provides two peptide NH bonds (Asp-194 and Gly-193) so fixed as to form highly directed hydrogen bonds to the partially negative oxygen of

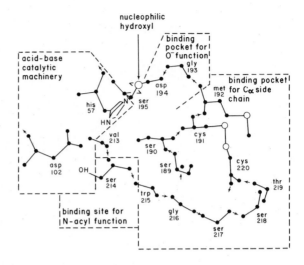

Fig. 3. The active site of α-chymotrypsin, a representation based on the cystallographic work of Blow and his collaborators.[50] The heavy dots show the positions of the main-chain atoms, with arrows indicating the sense of progression along the peptide chain from N terminus toward C terminus. The large open circles to the right are sulfur atoms; that at top center is the hydroxyl group of Ser 195. The positions of the side-chain atoms of some residues also appear.

the carbonyl group in the transition states for both nucleophilic attack [Eqs. (23) and (27)] and leaving-group expulsion [Eqs. (28) and (25)].

Further clockwise is found the large hydrophobic binding pocket, which leads to specially strong stabilization of transition states for substrates with hydrophobic α substituents such as derivatives of tryptophan, tyrosine, phenylalanine, valine, etc. This pocket is bounded at the "bottom" by Ser-189 and Ser-190 and has a "lid" formed by Met-192. Its "walls" consist of Cys-191, the disulfide link to Cys-220 and the chain back to Val-213. The hydrophobic side chains enter the pocket such that, if aromatic, their rings are roughly perpendicular to the page and are sandwiched between Cys-191 and Trp-215. This pocket lends α-chymotrypsin its own peculiar specificity, and it is in this region that structural variation among the serine proteases produces their individuality.

Beyond this region of the active site, still moving clockwise, one encounters the peptide bond linking Ser-214 to Trp-215. When an *N*-acyl α-amino acid derivative of S configuration at C_α binds such that its carbonyl function is presented to Ser-195 and its α-substituent enters the hydrophobic pocket, then its *N*-acyl group forms a natural dual hydrogen bond to the Ser-214 site. The stereochemistry of these three elements, the nucleophilic hydroxyl, the hydrophobic pocket, and the *N*-acyl-binding site, therefore confers upon α-chymotrypsin its configurational specificity for the S configuration at C_α. The (S) transition state can be fully stabilized by all three interactions while the (R) transition state can enjoy optimally only two of the three.

The final region of the active site to be considered is the acid–base catalytic machinery shown at the upper left of the figure. When the serine hydroxyl forms a partial bond to the carbonyl carbon in the nucleophilic attack step of enzyme acylation, this transition state can be stabilized by whole or partial donation of the serine proton to a base. The imidazole ring of His-57 is per-

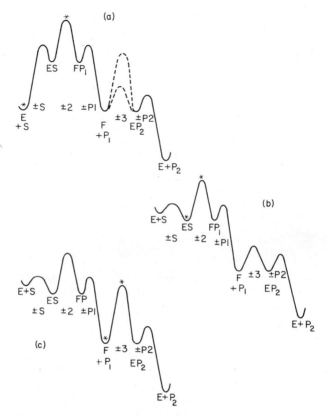

Fig. 4. Diagrams of free energy vs. reaction progress to illustrate the variation of initial reference state and rate-determining transition state with substrate concentration in such enzyme systems as the serine hydrolases. The enzyme E and substrate S are shown as combining in a step ±S to form the enzyme–substrate complex ES. This is transformed in a "chemical step" ±2 to the complex of acyl enzyme with first product, FP_1. The complex dissociates to acyl enzyme F and free first product P_1 (a "physical step" ±P1). Now the acyl enzyme is deacylated in a "chemical step" ±3 to give the complex of enzyme and second product, EP_2. In a final "physical step," ±P2, this complex dissociates to E and P_2, and the system is prepared for a new catalytic cycle. An important point is that *initial rates* are to be considered, so that the free energy of P_1 and P_2 are to be taken as extremely low. In (a), very low substrate concentrations are considered; the free energy of S and thus of E + S is very low, and the major form of the enzyme is free enzyme E. The rate constant measured will be k_E. The initial reference state is E + S and the rate-determining transition state will be that of highest free energy *preceding the state F + P_1*. No subsequent transition state can be of higher free energy because the initial-rate condition requires P_1 to be nearly zero in concentration and thus of such low free energy that regardless of the height of any subsequent *barrier* its transition state free energy would still be lower than that of any transition state preceding the state F + P_1. Here it is assumed that the transition state for step ±2 is highest in free energy and thus rate determining; it and the initial reference state are marked by asterisks. Two circumstances are considered for the rapid, subsequent ±3 step, corresponding to the dashed barriers. One of these barriers is lower than the *barrier* for the rate-determining ±2 step (the barrier of this *step* being the free-energy difference between the transition state and ES), the other is higher. Which of these circumstances prevails has no relevance under low-substrate concentrations, where the rate-determining transition state must be located in the acylation manifold.

fectly positioned to serve this purpose. Furthermore, the carboxylate of Asp-102 is precisely located so as to form a hydrogen bond to the NH of the same His-57 imidazole. This chain of two hydrogen bonds from the nucleophilic hydroxyl constitutes the famous "charge-relay chain"[51] and offers an effective mechanism for transition-state stabilization:

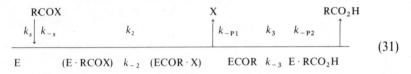

$$\text{(30)}$$

The stabilization portrayed in Eq. (30) should operate whether or not the protons are "in flight," i.e., participating in the reaction coordinate in the transition state. It is possible that this hydrogen-bond chain may be even more extensive, with the hydroxyl group of Ser-214 hydrogen-bonding to the Asp-102 carboxylate.

The kinetic picture for serine protease action, from which information about transition states must derive, can be represented by addition of rate constants to the scheme of Eq. (21), as in

$$
\begin{array}{ccccc}
\text{RCOX} & & \text{X} & & \text{RCO}_2\text{H} \\
k_s \downarrow k_{-s} & k_2 & \Big\uparrow k_{-\text{P1}} \quad k_3 \quad k_{-\text{P2}} & & \Big\uparrow \\
\hline
\text{E} & (\text{E} \cdot \text{RCOX}) \quad k_{-2} \quad (\text{ECOR} \cdot \text{X}) & \text{ECOR} \quad k_{-3} \quad \text{E} \cdot \text{RCO}_2\text{H}
\end{array}
\tag{31}
$$

At a later point, an even further resolution of the rate constants into those for individual steps will be wanted, but at present this level is sufficient. The Michaelis–Menten expression obtained from this scheme by the steady-state treatment has parameters k_{cat} and K_m, which can be expressed as k_{ES} ($\equiv k_{cat}$) and k_E ($\equiv k_{cat}/K_m$) (as discussed in Chapters 2) as in Eqs. (32a) and (32b):

Parts (b) and (c) refer to conditions of high substrate concentration so that the free energy of S and thus of E + S are very high. Part (b) is for the circumstance in which the ±3 barrier is low and part (c) is for the circumstance in which this barrier is high, both relative to the ±2 barrier. The free energy of S is so high in both cases that the enzyme is saturated, and some bound form of the enzyme will be the major enzyme species. This form will be the lowest free-energy form which precedes the highest barrier; material will accumulate before the highest barrier (the "bottleneck") and in the lowest-free-energy state available there. In part (b), the highest barrier is that for the ±2 step; the lowest free-energy form preceding it is ES. ES is therefore the initial reference state. The rate-determining transition state will be that of highest free energy after the initial reference state. In (b), this is the transition state of the ±2 step. For a system of this kind, the major enzyme form at saturation will be ES, this complex will slowly undergo acylation, and then all subsequent processes, leading back to ES again, will be rapid. The rate constant measured is k_{ES}, and it is said that "acylation is rate limiting." This is true, for example, with amide substrates of the serine proteases.

Part (c) portrays the situation when the deacylation barrier (±3) is higher than that for acylation (±2). Now the enzyme accumulates in the form of the acyl enzyme F, which is then the initial reference state. The highest-free-energy subsequent transition state is that for the ±3 step and it therefore limits the rate. This is the case in which "deacylation limits the rate for k_{ES}," as is seen with ester substrates of the serine proteases.

$$k_{ES} = \frac{[k_2 k_{-P1}/(k_{-2} + k_{-P1})][k_3 k_{-P2}/(k_{-3} + k_{-P2})]}{[(k_2 k_{-P1})/(k_{-2} + k_{-P1})] + [(k_3 k_{-P2})/(k_{-3} + k_{-P2})]} \qquad (32a)$$

$$+ [(k_2 k_3 [k_{-P1} + k_{-P2}])/(k_{-2} + k_{-P1})(k_{-3} + k_{-P2})]$$

$$k_E = \frac{(k_s k_2 k_{-P1})/(k_{-2} + k_{-P1})}{[k_2 k_{-P1}/(k_{-2} + k_{-P1})] + k_{-s}} \qquad (32b)$$

Note that the constants $k_{\pm s}$ are for the process of enzyme binding, the $k_{\pm 2}$ constants are for (actually multistep) acylation starting at the enzyme–substrate complex, the k_{-P1} step is for release of the first product (leaving group) from the enzyme, the $k_{\pm 3}$ step (again multistep in reality) is for conversion of acyl enzyme to enzyme–carboxylic acid complex, and the k_{-P2} step is for release of carboxylic acid (second product). From Eq. (32a), it is apparent that any one of, or any combination of, the last two "physical steps" k_{-P1} and k_{-P2} or the "chemical steps" $(k_{\pm 2}, k_{\pm 3})$ can determine the value of k_{ES}. Only the $k_{\pm s}$ step cannot contribute because binding has occurred before the initial reference state for k_{ES}. Similarly, Eq. (32b) shows that any one of, or any combination of, the earlier two "physical steps" $k_{\pm s}$ and k_{-P1} or the earlier "chemical step" $k_{\pm 2}$ can contribute to the value of k_E.

An important distinction is this: k_E can only be determined by processes leading *to the acyl enzyme* and thus *is the rate constant for acylation of the enzyme*, while k_{ES} may refer to either the acylation or the deacylation process, whichever is slower. This point is illustrated by free-energy diagrams in Fig. 4. Its operational significance is this:

1. Experimental quantities such as isotope effects, substituent effects, etc., determined for k_E necessarily relate the free substrate and enzyme as initial reference state to the transition state (or states) for *acylation*.
2. Such quantities determined for k_{ES} may relate either ES or the acyl enzyme, as initial reference state, to the transition state(s) for either acylation or deacylation; slow acylation leads to effects arising from conversion of ES to the acylation transition state(s); slow deacylation leads to effects from conversion of acyl enzyme to deacylation transition state(s).

In Section 2.3, we shall want to make use of structure-reactivity effects on k_E to derive information about the transition state(s) for *acylation* of α-chymotrypsin. These should therefore be one or more of the transition states for Eqs. (23)–(26). In the next section, we shall consider what expectations are generated for the transition states of such reactions from the study of related nonenzymic ("model") reactions.

2.2. Chemical Expectations for the Acylation Process

In this and the next section, we want to make use of structure-reactivity data for the acylation of chymotrypsin to learn about the transition state(s) for this process. This immediately raises the fundamental question, alluded to previously, whether meaningful information may be gained from the rate effects of varying substrate structure in enzymic reactions. In the case of chymotrypsin, the detailed picture of the active site, generated by crystallographic studies, makes this question easier to answer.

Typical natural and near-natural substrates of chymotrypsin may be represented $R_1CONHCHR_2COX$ [cf. Eq. (22)], showing that structural variation in the substrate, for purposes of a structure-reactivity study, can easily be envisioned in (1) the N-acyl region (R_1), (2) the C_α region (R_2), and (3) the leaving-group region (X). As Fig. 3 emphasizes, the N-acyl region and the C_α substituent both interact with specific binding sites in the enzyme, the C_α substituent being intimately enclosed in the hydrophobic pocket. This suggests that structural variation in these regions, particularly in the C_α substituent, may lead to rate changes which reflect both interactions within the substrate-derived "core" of the transition state *and* less informative interactions with the enzymic framework.

The leaving group, however, presents a different picture. Although some catalytic interaction is expected here, both the crystallographic evidence and the results of varying the leaving-group structure of peptide substrates suggest that strong, direct, and intimate binding does not occur between leaving-group structure and enzyme.[50,52] One may therefore hope, by variation of leaving-group electronic structure, to learn about electronic features of the transition state(s) for acylation through effects on the value of k_E.

First, some simple measure of leaving-group electronic character is needed. We shall use the *free energy of dissociation of the doubly protonated leaving group* XH_2^+, as represented by the pK_a of this species. Thus, if we are dealing with an amide (leaving group X = NH_2^-), we shall use the pK_a of NH_4^+, etc. Table I gives a list of these constants, measured or estimated, for groups of interest. The table also includes rate data to be used below.

To interpret such measurements, one needs a calibration of the effects in characterized systems. The requirement is to know quantitatively how the rate of reaction *should* change as leaving-group structure is systematically varied for (1) binding of the substrate to the enzyme [Eq. (23)], (2) nucleophilic attack by the serine hydroxyl on the substrate carbonyl group with general catalysis by the charge-relay function [Eq. (24)], (3) expulsion of the leaving group with general catalysis by the charge-relay function [Eq. (25)], and (4) release of first product from acyl enzyme [Eq. (26)]. The data to be considered in the enzymic case will consist of a plot of log k_E vs. pK_a. Now a large pK_a indicates a leaving group readily capable of bearing positive charge, while a small pK_a indicates readiness to bear negative charge. Therefore, a positive slope (increasing rate with increasing pK_a) for such a plot shows an increased

Table I
Rate Constants for Acylation of Chymotrypsin by Various Substrates at pH $\sim$ 8, 25°

	Substrate structure[a]		$k_E = k_{cal}/K_m$ ($M^{-1} \ sec^{-1}$)	Reference	Protonated leaving group	pK_a	Reference
	$P_3{}^b$	P_2	P_1					
1.	Cbz	Trp	$OC_6H_4NO_2(p)$	9.0×10^7	53	$p\text{-}NO_2C_6H_4OH_2^+$	(−10.3)	c
2.	Cbz	Trp	$OC_6H_4COCH_3(p)$	5.3×10^7	53	$p\text{-}CH_3COC_6H_4OH_2^+$	(−9.1)	c
3.	Cbz	Trp	$OC_6H_4Cl(p)$	3.5×10^7	53	$p\text{-}ClC_6H_4OH_2^+$	(−7.5)	c
4.	Cbz	Trp	$OC_6H_4OCH_3(p)$	1.6×10^7	53	$p\text{-}CH_3OC_6H_4OH_2^+$	(−6.0)	c
5.	Ac	Tyr	OCH_3	5.8×10^5	54	$CH_3OH_2^+$	(−0.7)	c
6.	Ac	Trp	OCH_3	4.2×10^5	55	$CH_3OH_2^+$	(−0.7)	c
7.	Ac	Trp	OCH_2CH_3	2.8×10^5	56	$CH_3CH_2OH_2^+$	−0.5	63
8.	Ac	Phe	OCH_3	1.6×10^5	54	$CH_3OH_2^+$	(−0.7)	c
9.	Ac	Phe	OCH_3	4.2×10^4	55	$CH_3OH_2^+$	(−0.7)	c
10.	Bz	Tyr	$NHC_6H_4NO_2(p)$	2.3×10^3	57	$p\text{-}NO_2C_6H_4NH_3^+$	1.0	64
11.	Bz	Tyr	$NHC_6H_4NO_2(m)$	3.1×10^2	57	$m\text{-}NO_2C_6H_4NH_3^+$	2.45	64
12.	Bz	Tyr	$NHC_6H_4NO_2(o)$	2.5×10^2	57	$o\text{-}NO_2C_6H_4NH_3^+$	−0.28	64
13.	Ac	Tyr	$NHC_6H_4NO_2(p)$	7.4×10^1	58	$p\text{-}NO_2C_6H_4NH_3^+$	1.0	64
14.	Ac	Tyr	$NHC_6H_4NO_2(m)$	2.1×10^1	58	$m\text{-}NO_2C_6H_4NH_3^+$	2.45	64
15.	Ac	Tyr	$NHC_6H_4Cl(m)$	1.3×10^1	59	$m\text{-}ClC_6H_4NH_3^+$	3.32	64
16.	Ac	Tyr	$NHC_6H_4Cl(p)$	2.1×10^1	59	$p\text{-}ClC_6H_4NH_3^+$	3.81	64
17.	Ac	Tyr	$NHC_6H_4OCH_3(m)$	7.8	59	$m\text{-}CH_3OC_6H_4NH_3^+$	4.20	64
18.	Ac	Tyr	$NHC_6H_4CH_3(p)$	6.7	59	$p\text{-}CH_3C_6H_4NH_3^+$	5.07	64
19.	Ac	Tyr	$NHC_6H_4OCH_3(p)$	1.8×10^1	59	$p\text{-}CH_3OC_6H_4NH_3^+$	5.29	64
20.	Ac	Tyr	$NHCH_2CONH_2$	2.8×10^1	54	$NH_2COCH_2NH_3^+$	(7.73)	d
21.	Ac	Phe	$NHCH_2CONH_2$	9.6	54	$NH_2COCH_2NH_3^+$	(7.73)	d
22.	Ac	Phe	NH_2	1.0×10^1	60	NH_4^+	9.25	64
23.	Ac	Tyr	NH_2	5.0	61	NH_4^+	9.25	64
24.	Ac	Tyr	$NHCH_3$	1.0×10^{-2}	62	$CH_3NH_3^+$	10.62	64

[a] P_1 = leaving group. P_2 = acyl compound. P_3 = N substituent.   [b] Cbz = carbobenzyloxy. Ac = acetyl. Bz = benzoyl.   [c] Estimated relative to pK_a for $C_6H_5OH_2^+$ (−6.7)[26] assuming the substituent effect on ionization to be the same as for RNH_3^+.   [d] Assumed equal to pK_a for methyl ester.[64]

positive charge on the leaving group in the transition state, and a negative slope an increased negative charge. A zero slope indicates no change from the initial-state charge distribution. The initial reference state will be the free reactant molecule since the rate constant employed is k_E.

How, then, can we calibrate a plot of log k_E vs. pK_a for the four transition-state possibilities: binding, nucleophilic attack, leaving-group expulsion with general catalysis, and product release from acyl enzyme? Clearly we require either direct information from the enzymic system or information from suitable nonenzymic models. In the case of binding and product release, enzymic information is used. For nucleophilic attack and leaving-group departure, one relies on model reactions. The strategy and results of the calibration are outlined in Table II, and some of the requisite data are collected in Table III.

Binding. A theoretical estimate[65] of the diffusional rate of approach of a small molecule to a chymotrypsin-like macromolecule yields a rate constant of 10^9 M^{-1} sec^{-1}, essentially independent of small-molecule electronic structure. Since the rate-determining step in binding may well not be diffusion itself but molecular reorganization of the enzyme structure in the presence of the substrate, a reduction to about 10^8 M^{-1} sec^{-1} is prudent. Indeed, NMR methods yield an "on" rate constant for the inhibitor N-trifluoroacetyl-D-tryptophan with α-chymotrypsin of 1.5×10^7 M^{-1} sec^{-1}. If the specific substrate structure to some degree induces the enzyme reorganization, then it may occur (say) tenfold faster, again yielding about 10^8 M^{-1} sec^{-1}. Finally, if the leaving-group electronic structure is roughly identical in the initial state and the transition state for this reorganization, the rate of binding should be independent of leaving-group pK_a. In other words, the slope m of a plot of log k_E vs. pK_a should be zero. Thus for rate-limiting binding of substrate to enzyme, the chemical expectation is

$$k_E \sim 10^8 \ M^{-1} \sec^{-1}, \qquad m \sim 0$$

Nucleophilic Attack. Here the problem is to find model acyl-transfer reactions in which it has been reliably established that nucleophilic attack,

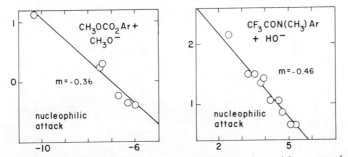

Fig. 5. Structure-reactivity relations for leaving-group variation in model systems where nucleophilic attack is known to be rate limiting. The ordinate is log (second-order rate constant in M^{-1} sec^{-1}), and the abscissa is the pK_a of the doubly protonated leaving group, $ArOH_2^+$ or $ArNCH_3H_2^+$. The rate constants are from Table III (k_a was taken for the anilides) and the pK_a's from Table I.

Table II

Chemical Expectations for the Slope m of a Plot of log k_E vs. pK_a for α-Chymotrypsin with Various Rate-Determining Processes

Rate-determining process	Expectation	Data or model system	Remarks
Binding	$k_E \sim 10^8 \ M^{-1} \ sec^{-1}$	Theory $(10^9 \ M^{-1} \ sec^{-1})^a$	Reduced by tenfold for molecular reorganization of enzyme after collision
		k_{on} for N-trifluoroacetyl-D-tryptophan[66] $(1.5 \times 10^7 \ M^{-1} \ sec^{-1})$	Increased by ∼ tenfold for substrate induction of enzyme reorganization[b]
	$m \sim 0$	—	Electronic structure of substrate leaving group assumed unaltered in diffusion and reorganization transition states
Nucleophilic attack	$m \sim -0.3$ to -0.5	Methoxide attack on aryl methyl carbonates ($m = -0.36$, Fig. 5)	Nucleophilic attack shown to be rate determining by lack of isotope exchange[68]
		Hydroxide attack on N-methyl trifluoro-acetanilides ($m = -0.46$, Fig. 5)	Change in rate-determining step allows isolation of nucleophilic attack[69]
Leaving-group expulsion: (a) Good leaving groups	$m \sim -0.6$ to -1.0	Water-catalyzed expulsion from $$CF_3\overset{\displaystyle O^-}{\underset{\displaystyle OH}{C}}-NCH_3Ar$$ ($m = -0.6$, Fig. 6)	Change in rate-determining step allows isolation of leaving-group expulsion[69]
		Hydroxide-catalyzed expulsion from $$CF_3\overset{\displaystyle O^-}{\underset{\displaystyle OH}{C}}-NCH_3Ar$$ ($m = -0.8$, Fig. 6)	Same; change in pathway exposes both water and hydroxide catalysis[69]
		Methoxide-catalyzed methanolysis of CF_3CONCH_3Ar[70] ($m = -1.0$, Fig. 6)	Similarity of results to aqueous case suggests expulsion is rate determining[70]

Rate-determining process	Expectation	Data or model system	Remarks
(b) Poor leaving groups	$m \sim -0.2$ to 0	Water and hydroxide catalysis both have $m = 0$; methanolysis has $m \sim -0.2$ (Fig. 6)	See text: different transition states for good and poor leaving groups[69,70]
Product release from acyl enzyme	$m \sim -0.7$	Combination of k_E for acylation[59] and k_{-2} for for leaving-group attack on acyl enzyme[52] gives k_E/k_{-2}, the equilibrium constant for enzyme acylation ($m \sim -0.7$, Fig. 7)	Electronic structure of product departing from enzyme assumed unaltered

[a] Estimated from Debye's expression for "slowly diffusing reactants and reactions where charge interactions are not favorable... approximates the expected upper limit for most enzyme–substrate reactions."[65]
[b] The evolutionarily favored substrate may, in effect, catalyze the enzymic reorganization. Renard and Fersht[67] estimate k_{on} to be 6×10^7 M^{-1} sec^{-1} for N-acetyl-L-tryptophan p-nitrophenyl ester with δ-chymotrypsin.

rather than leaving-group expulsion, is actually the rate-determining step. There are several examples available. We choose two from our laboratory: the methoxide-catalyzed methanolysis of aryl methyl carbonates (Table III, first column) and the attack of hydroxide ion on N-methyltrifluoroacetanilides (Table III, third column).

In the methanolysis of aryl methyl carbonates

$$CH_3O^- + CH_3^*OCO_2Ar \rightleftharpoons CH_3O \underset{OCH_3^*}{\overset{\overset{\displaystyle O^-}{\underset{\displaystyle |}{C}}}{|}} OAr \longrightarrow (CH_3O)_2CO + ArO^- \quad (33)$$

the rate-determining step is known to be attack of the methoxide because there is no exchange of tritium-labeled methoxy from substrate into solvent.[68] If leaving-group expulsion were rate limiting with the first step reversible, such an exchange would have occurred (note that no ambiguity arises from proton-transfer questions, as in O^{18} exchange in the hydrolysis of esters). A plot of these data vs. pK_a yields $m = -0.36$ (Fig. 5).

The rate-determining step for basic hydrolysis of N-methyltrifluoro-acetanilides,

$$HO^- + CF_3CONCH_3Ar \rightleftharpoons HO \underset{CF_3}{\overset{\overset{\displaystyle O^-}{\underset{\displaystyle |}{C}}}{|}} NCH_3Ar \quad \left| \begin{array}{l} \xrightarrow{HO^-} \\ \xrightarrow{H_2O} \end{array} \right. \quad \begin{array}{l} CF_3CO_2^- \\ + \\ CH_3NHAr \end{array} \quad (34)$$

Table III

Data for Establishing Chemical Expectations in Chymotrypsin Acylation[a]

Substituent, X	Basic methanolysis of $CH_3OCO_2C_6H_4X$[68]	Basic methanolysis of $CF_3CONCH_3C_6H_4X$[70]	Hydrolysis of $CF_3CONCH_3C_6H_4X$[69] k_a	k_1	k_2^b	Aminolysis of N-acetyl-tyrosyl chymotrypsin by $NH_2C_6H_4X$[52]	Equilibrium constants for acylation of chymotrypsin by N-acetyltyrosylanilides[c]
p-NO$_2$	13.5	—	—	—	—	—	—
m-NO$_2$	—	5.75	140	25.7	78,100	—	—
m-Br	—	0.52	31	6.7	21,700	—	—
m-Cl	—	0.51	30	6.4	19,400	1.1	11.8
p-Br	1.95	0.35	26	4.1	9,090	—	—
p-Cl	1.66	0.27	23	4.0	8,790	5.0	4.2
m-OCH$_3$	—	0.11	11	3.3	4,050	2.7	2.9
H	0.58	0.10	11	2.2	2,500	—	—
m-CH$_3$	—	0.083	6.9	2.2	2,460	—	—
p-CH$_3$	0.43	0.073	5.7	2.2	2,460	7.8	0.86
p-OCH$_3$	0.40	0.056	5.7	1.9	2,190	36	0.50

[a] Unless otherwise noted, entries are second-order rate constants, M^{-1} sec^{-1}, at 25°.

[b] Third-order rate constants, M^{-2} sec^{-1}, for the term $k_2[HO^-]^2[S]$.

[c] Dimensionless, pH 7.8–8.0, calculated as k_E/k_{-2}, with k_E from Table I, entries 15–19, and k_{-2} from adjacent column in this table.

depends on the base concentration, because leaving-group expulsion is base catalyzed.[69] At high hydroxide concentrations, this expulsion is rapid and the rate constant k_a for nucleophilic attack can be measured (Table III, third column). A plot of $\log k_a$ vs. pK_a shows $m = -0.46$ (Fig. 5).

The conclusion should therefore be that nucleophilic attack will produce $m \sim -0.3$ to -0.5. The possible enzymic involvement of the charge-relay acid–base catalytic machinery in serine nucleophilic attack may alter this prediction, although a large effect is not expected and its direction cannot reliably be given.

Leaving-Group Expulsion. To form a model for enzymic leaving-group expulsion with the presumed protolytic assistance from the acid–base machinery, we need a system where, first, leaving-group expulsion has reliably been established as rate limiting and, second, protolytic assistance is occurring. Such a system is provided by the hydrolysis of N-methyltrifluoroacetanilides [Eq. (34)] at *low* hydroxide concentrations. The second stage of Eq. (34) is rate determining under these circumstances,[69] so that both water- and hydroxide-catalyzed leaving-group expulsion can be seen. Indications from substituent-effect and solvent isotope-effect similarities are that the basic methanolysis reaction[70] of the same substrates also has this stage rate determining. Protolytic assistance surely plays a role because both reactions have substantial normal contributions to the solvent isotope effect[69,71] in spite of consuming either a strongly solvated methoxide ion or hydroxide ion, a factor which if it dominated would cause inverse solvent isotope effects.[72] The detailed significance of the solvent isotope effects is discussed below.

The data (Table III, second, fourth, and fifth columns) are plotted in Fig. 6.

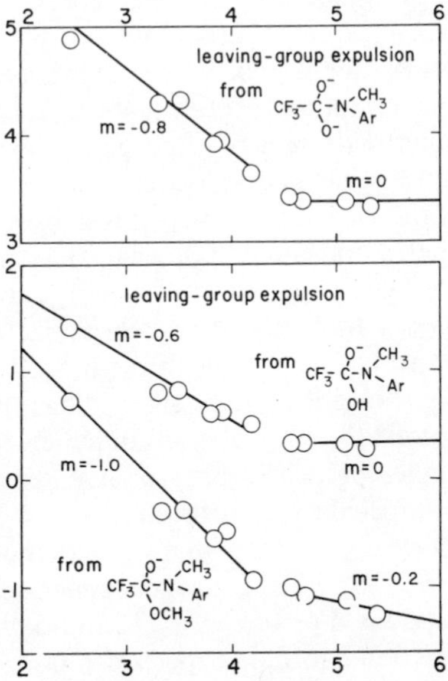

Fig. 6. Structure-reactivity relations for leaving-group variation in model systems in which leaving-group expulsion is known to be rate limiting. The ordinate is $\log$ (second-order rate constant in M^{-1} $\sec^{-1}$), and the abscissa is the pK_a of $ArNCH_3H_2^+$. The rate constants are from Table III and the pK_a's from Table I. The rate constants for the topmost plot are the k_2 and those of the middle plot are the k_1, both for anilide hydrolysis, and the rate constants of the bottom plot are those for anilide methanolysis, all in Table III.

A major feature is that the dependence is biphasic, a large slope m of -0.8 to -1.0 dominating in the range $pK_a \gtrsim 4$ and a very small slope $m \sim -0.2$ to 0 in the range of higher pK_a. This has previously been explained[69] by postulation of two parallel pathways for protolytically assisted leaving-group expulsion with valence–isomeric transition states **8** and **9**:

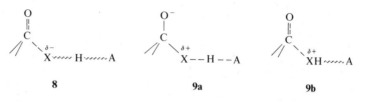

Transition state **8** arises for "good" leaving groups of $pK_a \gtrsim 4$ (for the structures involved here), i.e., ones capable of supporting negative charge. Structure **8** has negative charge on the leaving group and is favored for such leaving groups (typically oxygen leaving groups as in esters and anilines containing electron-withdrawing substituents). The reaction-coordinate motion consists of heavy-atom reorganization, i.e., fission of the C—X bond and formation of the carbonyl π bond. The HA moiety catalyzes through formation of a strong hydrogen bond which participates minimally in reaction-coordinate motion ("solvation catalysis").[30]

Transition states **9a** and **9b** are expected for "poor" leaving groups of higher pK_a, which are better able to stabilize the partial positive charge in these structures than the partial negative charge in structure **8**. The conversion from the pathway of transition state **8** to that of **9** seems to come at around pK_a 4.5 for the reactions considered in Fig. 6, as shown by the "concave-upward" structure-reactivity break, diagnostic of changes in reaction pathway. However, with other acyl-group structures and nucleophilic reagents, the break might well come elsewhere. Transition state **9a** is one in which the transfer of the proton from the catalyst to the leaving group is the reaction-coordinate motion. Reorganization of the heavy-atom framework is fast and must occur in a subsequent, lower-energy transition state. Transition state **9b** is the variant in which this subsequent process is assumed to have a higher-energy transition state than that for proton transfer; then the proton motion has been *previously* and rapidly accomplished. For **9b**, the reaction-coordinate motion is heavy-atom reorganization with the catalyst as a spectator[69] or possibly functioning through "solvation catalysis."

Solvent isotope-effect data also favor the view that transition state **8** dominates at $pK_a < 4$, and they further select **9a** as the likely structure at higher pK_a. The proton inventory technique was used to separate out the isotope-effect contribution of the hydrogen being donated from solvent to the leaving group in the methanolysis reaction.[71] For the m-NO$_2$ anilide (pK_a 2.5), this hydrogen generates $k_H/k_D \sim 2.6$, as expected for solvation catalysis. For the p-OCH$_3$ anilide (pK_a 5.3), k_H/k_D is 7.4, fully consistent with reaction-coordinate proton motion. The intermediate case of the p-Cl compound (pK_a 3.8), which probably reacts through both transition states at once, gives k_H/k_D 3.5.

An alternative formulation can also be made according to which only one pathway exists but possesses a transition state for which the structure is gradually transformed from **8** to **9** as the leaving-group pK_a rises. The curved structure-reactivity plot reflects this transformation. Both heavy-atom reorganization and proton transfer are then assumed to be coupled together in the reaction-coordinate motion, although not necessarily to equal degrees. Thus in one limit, the reaction-coordinate motion might be dominated by heavy-atom motion and in another limit by proton motion, but both would be considered to contribute throughout. Various arguments have been given on the two (or more) sides of this question, but no rigorous solution to the problem is yet in hand.[14,29,30]

Even so, the chemical expectation for this type of transition state in chymotrypsin action is clear: Leaving groups which prefer to stabilize negative charge should do so, generating $m \sim -0.8$ to -1.0, while those which prefer to stabilize positive charge (those of higher pK_a) should shift over to $m \sim -0.2$ to 0.

Product Release. Now we need to establish the expected shape of log k_E vs. pK_a if the leaving group were rapidly produced on the enzyme surface by formation of acyl enzyme and if desorption of the leaving group into solution were rate limiting. We shall assume the leaving-group electronic structure in this product-release transition state to be the same as in free solution. In that case, the substituent effect for product release,

$$\text{EOH} + \text{RCOX} \rightleftharpoons \text{EOCOR} \cdot \text{XH} \quad -[\text{EOCOR} \cdots \text{XH}] \rightarrow \text{EOCOR} + \text{XH} \quad (35)$$

should be the same as that for equilibrium formation of acyl enzyme,

$$\text{EOH} + \text{RCOX} \rightleftharpoons \text{EOCOR} + \text{XH} \quad (36)$$

Information is available in the literature for calculation of the equilibrium constants for Eq. (36). The forward rate constant is k_E ($= k_{cat}/K_m$), which has been measured for a series of substituted N-acetyltyrosylanilides (entries 15–19 in Table I). The corresponding reverse rate constant is that for reaction of the anilines with N-acetyltyrosylchymotrypsin (k_{-2}), which has been determined under similar conditions by Fersht *et al.*[52] The ratio of these is the equilibrium constant, k_E/k_{-2} or $k_{cat}/K_m k_{-2}$. Values are given in the final column of Table III and are plotted vs. pK_a in Fig. 7. The expected slope is thus $m = -0.7$.

Table II summarizes these chemical expectations in terms of leaving-group structure-reactivity relations for the various transition states that may arise in chymotrypsin acylation.

2.3. Structure-Reactivity Studies

It now remains only to construct the plot of log k_E vs. pK_a for the acylation of chymotrypsin and to deduce from comparisons of the slope with the chemi-

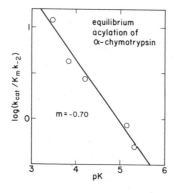

Fig. 7. Structure-reactivity relation for leaving-group variation in equilibrium acylation of α-chymotrypsin. The data were calculated as shown in Table III and explained in the text. The pK's are from Table I.

cal expectations whatever one can about the important transition states. The data ($k_E = k_{cat}/K_m$) for 24 N-acyl derivatives of tryptophan, tyrosine, and phenylalanine are given in Table I. They are plotted in Fig. 8, where the points are numbered as in the table.

Obviously, the general shape of the plot exhibits a negative dependence of log k_E on pK_a ("worse" leaving groups—less able to bear negative charge—react more slowly), but it is a nonlinear and, indeed, an irregular dependence. The irregularity might be thought to arise from unsystematic interactions of leaving-group structure and enzyme structure, but we believe the irregularity is systematic, that it can be described by the solid line drawn through or near the points, and that it is revelatory of the structures of various transition states along the reaction path for acylation of chymotrypsin.

In what follows, it may be didactically helpful to bear in mind the following characteristics of structure-reactivity plots. In any such plot, curvature

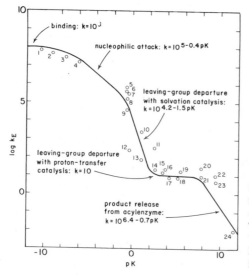

Fig. 8. Structure-reactivity relation for leaving-group variation in the acylation of α-chymotrypsin by N-acyl derivatives of "specific" amino acids. The data are from Table I. The solid line was calculated from Eq. (40) with the rate constants as defined in the figure [and given in Eqs. (41)–(45)].

downward (so that rate constants extrapolated from another nearby part of the curve would be larger than those observed) must signal a change in rate-determining step. In essence, a new and unexpected barrier has appeared so that extrapolation overestimates the rate. Curvature *upward*, such that extrapolated rate constants are smaller than those observed, indicates by contrast the change to a new reaction pathway ("change in mechanism"). Here a new and unexpected *channel* has appeared so that extrapolation underestimates the rate.

Beginning at the left of Fig. 8, substrates 1–4, we have rate constants just below 10^8 M^{-1} sec^{-1} and a weak structure-reactivity dependence ($m \sim 0$). This is as expected for rate-limiting binding of substrate to enzyme. Phillipp and Bender[73] have suggested that such substrates as *p*-nitrophenyl esters indeed have binding as the rate-limiting step for acylation. A similar proposal for δ-chymotrypsin was advanced by Renard and Fersht.[67]

As the electron-donating power of the substituents in the phenyl ring of these esters is increased, the rate falls off, showing that binding is probably being approached as a limiting rate but that another subsequent process is also influencing the rate as well. The logical candidate is nucleophilic attack.

If a line is drawn from around point 4 (the *p*-methoxyphenyl ester) to the cluster of points 5–9 (methyl and ethyl esters), its slope $m = -0.4$. Since (Table II) this is just the chemical expectation for nucleophilic attack, this supports the view that the falloff in rate from $k = 10^8$ M^{-1} sec^{-1} with leaving-group $pK_a \leq 10$ to around 10^5 M^{-1} sec^{-1} at $pK_a \sim 0$ results from a change in rate-determining step (downward break in the structure-reactivity plot) from binding to nucleophilic attack on the carbonyl group.

Now if the line of slope -0.4 is continued beyond the cluster of points 5–9 into the region of $pK_a > 0$, all observed rate constants fall below this extrapolated line. This is the sign of a new change in rate-determining step, to a transition state with substantially more negative charge on the leaving group (more negative value of m). As Fig. 8 shows, a straight line can be passed among points 10–14 which has a slope $m = -1.5$. The readiest thought is that the new rate-limiting step is now leaving-group expulsion from the tetrahedral adduct. These leaving groups (nitroanilines) are "good" in being able to stabilize negative charge and thus fall into the class expected to depart with solvation catalysis, as explained above. Table II suggests that $m \sim 0.6$ to -1.0 for such processes. Although the value of -1.5 developed in Fig. 8 is larger, two things should be noticed. First, a smaller value could easily be accommodated by the scattered points in this region of the plot. Second, the value of m, representing as it does the transition-state charge development on the leaving group, may be sensitive to the exact character and geometry of the acid–base catalytic machinery. Some deviation of the enzymic observations from the simplest chemical expectations might therefore not be surprising. Thus we consider the structure-reactivity break at around $pK_a = 0$ to arise from a change in rate-determining step: At lower pK_a, formation of the tetrahedral intermediate (nucleophilic attack) governs the rate, while decomposition of the tetrahedral intermediate (leaving-group expulsion with solvation catalysis) dominates at high pK_a.

Beyond about point 14, a further nonlinearity is encountered. Extrapolation of the line for leaving-group expulsion with solvation catalysis ($m = -1.5$) into the region of $pK_a > 2$ fails to predict the observations. Instead, points 15–23 show two anomalies: First, they all fall *above* the extrapolated line, and, second, they all show about the same rate constant ($10\ M^{-1}\ sec^{-1}$). Thus an *upward* break in the plot, indicative of the incursion of a new channel of reaction, is seen, and the new channel is such that $m \sim 0$. Table II shows this to be in exact accord with chemical expectation. These "poorer" leaving groups (nonnitro anilines and ammonia) prefer to bear positive rather than negative charge. The catalytic mode thus changes from solvation catalysis to proton-transfer catalysis. The transition-state charge on the leaving group is now similar to that in the acyl reactant (probably slightly positive), so $m \sim 0$.

The final limb of the plot is generated by a downward break at around $pK_a = 8$, with the extension given a slope of $m = -0.7$ to conform with chemical expectation for rate-limiting product release from the acyl enzyme. This break is only weakly indicated by the single point 24 (methylamine leaving group) in Fig. 8 and ought not to have been hypothesized on this basis alone. Another piece of evidence, not readily added to Fig. 8, however, strongly suggests that product release determines the rate for substituted-amine leaving groups. Inward and Jencks[74] found that various amines reacted with furoyl chymotrypsin at a single, common rate, and Zeeberg and Caplow[75] showed this also to be true for the specific acyl enzyme, N-acetyltyrosylchymotrypsin. These observations indicate that the transition state for aminolysis of the acyl enzyme by these amines has a leaving-group moiety essentially identical in electronic (and probably steric) character to the free amine. Since this transition state should be the same as the one for acylation of the enzyme, by the principle of microscopic reversibility, a plot of $\log k_E$ vs. pK_a should have $m = -0.7$ for alkylamine leaving groups.

Thus the structure-reactivity data for acylation of chymotrypsin by "specific" substrates indicate that by alteration of leaving-group pK one can drive the rate-limiting transition state along this cascade:

binding ($pK < -10$)

↓

nucleophilic attack
($pK \sim -9$ to -1)

↓

leaving-group departure   leaving-group departure
with solvation catalysis ⟶ with proton-transfer catalysis
($pK \sim 0$ to 1)   ($pK \sim 2$ to 6)

↓

product release
($pK > 10$)

The variation of leaving-group reactivity brings into view transition states for each of the two "physical steps," binding and product release, and three different component transition states for the "chemical step," namely, those for formation of the tetrahedral intermediate by nucleophilic attack and for its decomposition with two modes of protolytic assistance, solvation catalysis and proton-transfer catalysis. It is interesting to see only one mode of nucleophilic attack but two modes of leaving-group expulsion. This occurs because there is only a single nucleophilic structure (the serine hydroxyl) which presumably experiences assistance by solvation catalysis, falling as it does in the class of groups better able to bear negative than positive charge. The leaving-group structures, in contrast, are varied in the experiment, and one passes through both catalytic modes.

For a quantitative version of the arguments, we can return to Eq. (32b), which expresses k_E in terms of individual rate constants. We write k_E as the reciprocal in

$$k_E^{-1} = k_s^{-1} + (k_s k_2 / k_{-s})^{-1} + (k_s k_2 k_{-P1} / k_{-s} k_{-2})^{-1} \tag{37}$$

Recalling that k_2 is the overall rate constant for the "chemical step" with enzyme–substrate complex as initial reference state and noting that k_s/k_{-s} is the equilibrium constant for binding and k_2/k_{-2} the equilibrium constant for the "chemical step," we can write

$$k_E^{-1} = k_s^{-1} + k_c^{-1} + k_p^{-1} \tag{38}$$

Here the initial reference state for each term is now free enzyme–free substrate, and k_c is the overall "chemical" rate constant; the two "physical" rate constants are k_s for binding and k_p for product release. We also know now that k_c ought to be split into component rate constants for nucleophilic attack (k_N) and for leaving-group expulsion with solvation catalysis (k_L^{sc}) and with proton-transfer catalysis (k_L^{pt}). Maintaining free enzyme–free substrate as initial reference state for each of these rate constants, we have

$$k_c^{-1} = k_N^{-1} + (k_L^{sc} + k_L^{pt})^{-1} \tag{39}$$

or

$$k_E^{-1} = k_s^{-1} + k_N^{-1} + (k_L^{sc} + k_L^{pt})^{-1} + k_p^{-1} \tag{40}$$

The solid line of Fig. 8 is simply a logarithmic plot of Eq. (40) with the definitions of k_s, k_N, k_L^{sc}, k_L^{pt}, and k_p given by

$$k_s = 10^8 \ M^{-1} \sec^{-1} \tag{41}$$

$$k_N = (10^{5 \cdot 0.4 pK}) \ M^{-1} \sec^{-1} \tag{42}$$

$$k_L^{sc} = (10^{4.2 \cdot 1.5 pK}) \ M^{-1} \sec^{-1} \tag{43}$$

$$k_L^{pt} = 10 \ M^{-1} \sec^{-1} \tag{44}$$

$$k_p = (10^{6.4 \cdot 0.7 pK}) \ M^{-1} \sec^{-1} \tag{45}$$

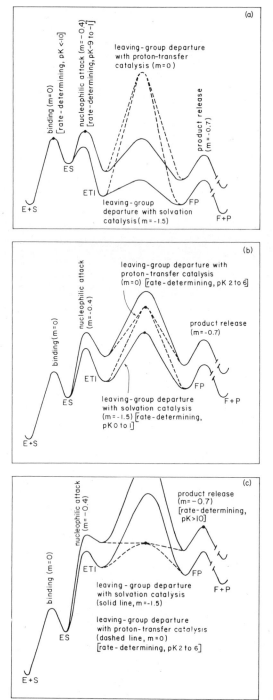

Fig. 9. Free-energy vs. reaction-progress diagrams to explain the changes in rate-determining step and reaction path induced by leaving-group variation in the acylation of chymotrypsin, as detected by the relations of Fig. 8.

The matter may also be presented by means of free-energy diagrams, as shown in Fig. 9. Figure 9(a) portrays the situation at $pK_a < 0$. The lowest-free-energy plot is, of course, that for the most reactive leaving group. For it, every transition state after that for binding to generate ES is of lower energy, so that binding limits the rate. The higher plot is for a less reactive leaving group: All states have risen in free energy by an amount determined by the quantity of negative charge in their leaving-group structures, as compared to the initial reference state. This amount of free-energy increase is zero for two transition states: that for binding and that for leaving-group departure with proton-transfer catalysis. Note that the latter transition state is not traversed in this pK region; instead, the lower-energy solvation–catalysis path is taken. The second plot is such that the nucleophilic-attack transition state has now surpassed the binding transition state. The transition between these curves thus describes the leftmost break in the plot of Fig. 8.

In Fig. 9(b), for the pK region 0–6, we see the free energies of all states (with the same two exceptions as before) at higher values. The lowest curve is one where the transition state for leaving-group departure with solvation catalysis has become rate limiting in place of the nucleophilic-attack transition state (because $m = -1.5$, the former is rising in free energy faster than the latter for which $m = -0.4$). In fact, because the substituent effect is the largest of any on the reaction path for the solvation–catalysis transition state, it would ordinarily remain rate determining at all higher pK. *However*, a *parallel* path with a lower value of m (indeed $m = 0$) exists, in the form of the proton-transfer-catalyzed leaving-group expulsion. At around pK 2, this becomes the more traversed path and solvation catalysis an unimportant high-energy route. Thus the transition from Fig. 9(a) to Fig. 9(b) represents the second break in Fig. 8, and the transition from the lower to the upper curve in Fig. 9(b) corresponds to the upward break, to proton-transfer catalysis.

Finally, Fig. 9(c) shows the situation as the pK rises beyond 6. For the lower curve, leaving-group expulsion with proton-transfer catalysis is still rate determining. As the leaving group becomes still less reactive, the proton-transfer catalysis transition state remains constant in free energy and at last it is exceeded in free energy by the product-release transition state, which has a greater electron density in the leaving group. Product release then determines the rate.

2.4. Isotope-Effect Studies

It will be useful to take brief consideration of heavy-atom isotope-effect and solvent isotope-effect measurements which have been made on chymotrypsin acylation and their relation to the hypotheses about transition states just developed from structure-reactivity data. Table IV contains a summary of the measurements to be discussed. As explained in Chapters 4, 5, 6, and 7 in this volume, it is from isotope-effect sources that we may expect the most definitive information on transition-state structures.

The first three entries in Table IV are oxygen isotope effects for ethyl and methyl esters. Their magnitudes are most readily appreciated in terms of an estimate by Klinman[80] of the maximum kinetic isotope effect k_{16}/k_{18} for fission of a C—O bond (1.068) and an excellent study by Kirsch and his group of O^{18} effects in the methyl formate ($HCOO*CH_3$) system.[81] Hydrazinolysis of methyl formate at pH 7.85 gives $k_{16}/k_{18} = 1.062$, thus showing that the leaving-group C—O bond is breaking and indeed is nearly completely broken in the transition state. This confirms Klinman's theoretical estimate and establishes an expectation for a transition-state structure with a very broken C—O bond. The hydrazinolysis at pH 10, where other evidence suggests that N to O proton transfer in a tetrahedral adduct is rate limiting, gives $k_{16}/k_{18} = 1.005$. This is then the expectation for a transition state strongly resembling the adduct but with C—O bond fission not occurring. For reactions in which formation of the adduct is probably rate limiting, effects larger than 1.005 but much smaller than 1.062 were found. Thus hydrolysis by hydroxide ion gave 1.009, and succinate-catalyzed hydrolysis gave 1.012. The most reasonable explanation of the fact that the latter two values exceed 1.005 is that the leaving-group oxygen may participate to some degree in the reaction-coordinate motion for nucleophilic attack on the carbonyl function. This will make a small normal contribution to the isotope effect, almost surely considerably less than the square root of the C^{12}—O^{16}/C^{12}—O^{18} reduced-mass ratio, which has a value of 1.025. Taking this as an upper limit, the expected range of leaving-group O^{18} isotope effects for rate-determining nucleophilic attack is between 1.000 and 1.03 ($= 1.025 \times 1.005$). The expected range for leaving-group expulsion is 1.005 to 1.068 if protolytic assistance to C—O bond fission does not "run ahead" of the fission itself. Since protonation of the labeled oxygen may produce an inverse isotope effect ($K_{16}/K_{18} = 1/1.023$),[82] protolytic assistance may reduce the effects. However, this effect should be less important with oxygen leaving groups, which prefer to bear negative charge in the transition state, than with less electronegative leaving groups.

These expectations are, of course, those for "pure" cases in which only one transition state determines the rate. If more than one transition state is involved, the observed effects will refer to a virtual transition-state structure (see Chapter 2) and will then be weighted-average effects. For example, if a nucleophilic-attack reaction state (with, say, $k_{16}/k_{18} = 1.02$) and a leaving-group expulsion transition state (with, say, $k_{16}/k_{18} = 1.06$) determined the rate equally—both of the same free energy—then the observed isotope effect would be $[(1.02 + 1.06)/2] \sim 1.04$.

There are two relevant characteristics of the three oxygen isotope effects for acylation of chymotrypsin listed in Table IV: First, they are all in the range expected for either nucleophilic attack or for leaving-group expulsion with a relatively adduct-like transition state; second, they vary substantially with what would have been considered remote or relatively unimportant changes (pH change from 6.8 to 8, change of N-acyl substituent—even ethoxyl to methoxyl leaving group seems not a very great change).

This variation definitely signals a sensitivity of transition-state structure

Table IV
Isotope Effects in the Acylation of α-Chymotrypsin

Substrate	Isotope effect	Remarks	Reference
1. Ac-Trp-O*CH$_2$CH$_3$	$k_E^{16}/k_E^{18} = 1.0180 \pm 0.0007$	pH 6.8, 25°	76
2. CH$_3$OCO-Trp-O*CH$_2$CH$_3$	$k_E^{16}/k_E^{18} = 1.0117 \pm 0.0004$	pH 6.8, 25°	76
3. Ac-Trp-O*CH$_3$	$k_E^{16}/k_E^{18} = 1.007$	pH 8, 25°	Chapter 7, this volume
4. Ac-Trp-OCH$_2$CH$_3$	$k_E^{H_2O}/k_E^{D_2O} \sim 1.8$	pH 8, 25°	77
5. Ac-Trp-N*H$_2$	$k_E^{14}/k_E^{15} = 1.0062 \pm 0.0004$	pH 6.73, 25°	78
6. Ac-Trp-N*H$_2$	$k_E^{14}/k_E^{15} = 1.0100 \pm 0.0010$	pH 8.00, 25°	78
7. Ac-Trp-N*H$_2$	$k_E^{14}/k_E^{15} = 1.0064 \pm 0.0006$	pH 9.43, 25°	78
8. Ac-Trp-NH$_2$	$V^{H_2O}/V^{D_2O} = 1.9$	pH 8, 25°	79

to the variables involved. Sawyer and Kirsch[76] noted one possible origin for the *N*-acyl change: The carbomethoxy compound may form an oxazoline intermediate. Another postulate is required for the ethoxyl–methoxyl change. Perhaps the active-site binding of the transition state is so sensitive to alkyl structure that the change from ethoxyl to methoxyl leaving group actually alters the transition-state structure and thus the isotope effect from 1.018 to 1.007. Alternatively, the combination of alkoxyl change and pH-induced minor alterations in enzyme structure might give rise to the shift. These hypotheses are not so unlikely as they may seem because the variation among the isotope effects, while large, may reflect structure changes which do not require much energy. The value of 1.018, a 1.8% effect, is 26% of the maximum of 1.068 (6.8%), while 1.012 is 18% and 1.007 is 10%. If these correspond to bond order changes from 0.74 (= 1.00–0.26) to 0.82 to 0.90, then the change in length of the C—O bond needed to cover the range (estimated from Pauling's rule) is $\Delta r \sim 0.3 \ln(0.90/0.74) \sim 0.06$ Å. For a weak C—O force constant of about 4 mdyn/Å, this change in length would require only about 2 kcal/mol of energy.

Another possibility is suggested by the fact that the alkoxyl leaving groups fall at a breakpoint on the structure-reactivity plot of Fig. 8. That plot suggests that these leaving groups are at the turning point between rate-determining nucleophilic attack (transition state **10**) and rate-determining leaving-group expulsion with solvation catalysis (transition state **11**):

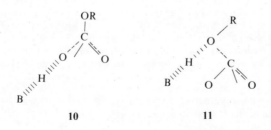

10   **11**

It is reasonable that the break occurs at these leaving groups because—other things being equal—the similarity of the serine hydroxymethyl (partially bonded in **10**) and the alkoxyl leaving group (partially bonded in **11**) would tend to make **10** and **11** of similar free energy. However, the relative energies of these two transition states might be sensitive to such variables as pH or N-acyl substituent because, as Fig. 10 emphasizes, their stereochemical disposition in the active site is rather different. Thus pH changes which affect ionizable groups even rather distant from the active site or binding of different N-acyl groups might modulate the relative contributions of the two transition states to determining the rate, through small alterations in the enzyme's framework structure.

Such small modulations of the contributions of these two structures to the virtual transition-state structure can, however, drastically affect the observed O^{18} isotope effect, which is expected to be rather different for the two structures. This can be seen by way of an example. Let us imagine that, in the case of the methoxyl substrate at pH 8, **10** determines the rate to 95% and **11** to 5% (i.e., leaving-group expulsion is 20 times faster than nucleophilic attack). Then the observed effect of 1.007 is a weighted average of the effect for **10** (for which we take $k_{16}/k_{18} \sim 1.006$) and for **11** (for which we take $k_{16}/k_{18} \sim 1.026$). The value for **10** corresponds to a very adduct-like transition state with a very slight reaction-coordinate contribution (since the isotopic

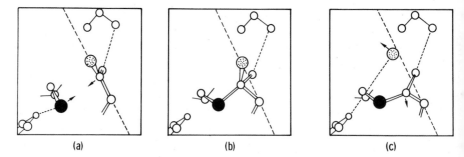

| (a) | (b) | (c) |

Fig. 10. Stereochemical model for structural alterations along the reaction path for the acylation of chymotrypsin, as developed by Blow and his collaborators.[50] Part (a) shows the acyl substrate in place in the active site, its leaving group a speckled sphere. The dashed line traversing the whole diagram indicates roughly the plane of the initial carbonyl function; it is shown for comparison in the later diagrams. The serine hydroxyl, a large black sphere, has swung into position for nucleophilic attack and is shown by the dashed line as hydrogen-bonded to the imidazole of His 57. Arrows indicate the motion of the carbonyl carbon and serine oxygen toward each other in the course of nucleophilic attack. A second dashed line portrays a hydrogen bond of the NH of Gly-193 to the carbonyl oxygen; the α-carbon and carbonyl carbon of Gly 193 are also shown as reference points. In part (b), nucleophilic attack is complete and the tetrahedral intermediate is seen. Part (c) shows the completion of the leaving-group expulsion step, a hydrogen bond now between His-57 and the leaving group. The transition states for the two steps should have structures intermediate between (a) and (b) for nucleophilic attack and between (b) and (c) for leaving-group expulsion. These sit in the enzymic environment with sufficient difference that small alterations in enzyme character could differentially affect their free energies.

bond is not breaking), while the value for **11** likewise corresponds to a quite adduct-like transition state but now with a large reaction-coordinate contribution (because here the isotopic bond is breaking). Keeping the same isotope effects, we can now calculate that contributions of 70% for **10** and 30% for **11** will generate the effects of 1.012 observed for the ethoxyl–carbomethyl compound at pH 6.8 and that contributions of 40% for **10** and 60% for **11** will generate 1.018, seen for the ethoxyl–acetyl compound at pH 6.8.

These changes, involving an alteration of the relative rates of nucleophilic attack and leaving-group expulsion from 20-fold in one direction to 1.5-fold in the other, do not demand much energy: A variation of 2 kcal/mol in the relative energies of **10** and **11** will reproduce the whole range of observations on the model just used. On the other hand, if we hold to a hypothesis that the isotope variations arise from structural changes of a single transition state, a very similar energy requirement of around 2 kcal/mol can also be estimated, as we saw above.

The two hypotheses cannot, therefore, be readily distinguished on energetics. However, the structure-reactivity plot (which breaks just at these leaving groups) militates with at least mild force in favor of the view that environmental perturbations, with these leaving groups, can shift the balance of contribution to a virtual transition-state structure of the two distinct contributing transition states **10** and **11** and thus alter the isotope effects.

Transition-state structures **10** and **11** show a proton bridge linking both the attacking nucleophilic oxygen in **10** and the departing leaving-group oxygen in **11** to the enzymic acid–base catalytic machinery, summarized as "B." One of the original evidences for this view was the depression of the rate in deuterium oxide solvent, as exemplified by entry 4 in Table IV.

This experiment is one of a series reported in a classic paper of Bender et al.,[77] which laid down all the fundamental methodology for the study of enzymic solvent isotope effects. Figure 11 portrays their data for k_E in H_2O and D_2O, as a function of pH (or pD), for the chymotryptic hydrolysis of N-acetyltryptophan ethyl ester. These, as explained before, should be data for acylation, although k_{ES} has deacylation as the rate-determining step with this substrate. The data are rather rough and only an approximate fit of the two data sets to a bell-shaped pH–rate profile has been attempted. This attempt yields $\Delta pK \sim 0.3$ for the acidic limb and $\Delta pK \sim 0.4$ for the basic limb of the profile, while $\Delta pK = 0.5 \pm 0.2$ is expected (Chapter 6). Considering the dispersion of the data, this is probably an acceptable result.

The striking fact is not that the rate is depressed in D_2O—doubtless it is, and many other cases exemplify the depression more convincingly—but that the depression is so small. The ratio of calculated maximum k_E's, from the approximate profiles of Fig. 11, is only 1.8. No reasonable fit of the data is likely to produce a ratio which is substantially larger.

This ratio needs to be considered in the light of several pieces of information from sources other than direct studies of chymotrypsin but maintaining in mind the observation in crystallographic research of the charge-relay chain

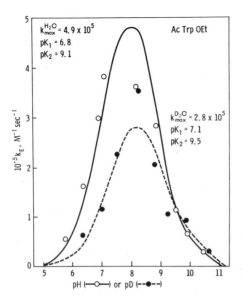

Fig. 11. The pH and pD dependence of the rate constant k_E for acylation of α-chymotrypsin in H_2O and D_2O, calculated from the data of Reference 77.

as the acid–base machinery of chymotrypsin.[51] As explained above, the apparent utility of this chain depends on its ability to move at least two protons (and conceivably three, if that of Ser-214 were to be included) as nucleophilic attack by Ser-195 occurs (the chain serving for general base catalysis) or as leaving-group expulsion occurs (the chain serving for general acid catalysis).

The points to be considered bear on the question of the nature of any proton bridge linking the enzyme to the substrate-derived portion of the transition state and they are the following: isotope-effect expectations for such a bridge if proton transfer is occurring and if hydrogen bonding is occurring in the transition state; evidence on hydrogen bridges in model systems subject to general catalysis; evidence from theory on the importance of proton transfer in this and similar processes.

First, if the bridge in question were a simple proton-transfer bridge, with the isotopically substituted proton possessing a substantial amplitude in the reaction coordinate, the expected kinetic isotope effects are quite large.[83] Such bridges appear to link the cores of E2 elimination transition states (in which heavy-atom reorganization is at work) to the general bases, accelerating such reactions, and large rate ratios for H and D substrates are common.[84] A value of around 7 is often considered a rough maximum and can be calculated from the loss of the isotope zero-point energy difference in a C—H bond of vibration frequency 3000 cm^{-1}. A similar estimate for an O—H bond of frequency 3600 cm^{-1} yields a ratio of about 11. The observation of 1.8 is well below this limit (in the usual logarithmic way it is around 25% of the limit), which suggests that if proton transfer is a component of the reaction coordinate, the amplitude of the proton is rather small ("the proton is asymmetrically located" — see Chapter 4).

Second, model systems in which general catalysis is occurring rarely show large isotopic rate ratios, although there has been insufficient examination of those where such a ratio is expected on other grounds to be very large. One case, in which a contribution of over sevenfold was dissected out for a bridging proton, was cited above (*p*-methoxy-*N*-methyltrifluoroacetanilide methanolysis, Table III). However, when general catalysis of C—O bond formation or fission is under observation, the bridges usually[8] produce isotope effects around 2. This has been interpreted[29,30,39] as hydrogen bonding (solvation catalysis) rather than proton-transfer catalysis, and the isotopic fractionation factors are then in accord with expectation for the unusual hydrogen bonds seen by Kreevoy and his group.[38] The chymotrypsin effect of 1.8 is in accord with this mechanism if only one of the charge-relay protons contributes or if two of them contribute quite small factors each ($\sim \sqrt{1.8} = 1.3$). The latter possibility would suggest less effective interaction than in general catalysis in model systems.

Third, there is the theoretical evidence. Scheiner *et al.*[85] have employed a quantum-mechanical technique (PRDDO, "partial retention of diatomic differential overlap") to examine the reaction pathway for the nucleophilic-attack step in a chymotrypsin model (cf. Chapter 3). Their main conclusion (for our purposes here) was that the two proton transfers along the charge-relay chain (which their model forced to occur in concert) are not coupled to the heavy-atom reorganization processes but occur in a separate stage of reaction. Their calculated energies suggest indeed that the motion of the protons is rate determining with heavy-atom reorganization subsequent, rapid, and possibly with no barrier. This mechanism is mildly inconsistent with the smallness of the deuterium solvent isotope effect (entry 4, Table IV) and strongly inconsistent with the oxygen isotope effects, which ought to be negligible for rate-determining proton shifts in the charge-relay chain, prior to nucleophilic attack. Thus, if the conclusion that the heavy-atom reorganization and proton-transfer processes are uncoupled is valid, the energetics must actually be such that heavy-atom motion is the chief component of the reaction coordinate in the highest-energy transition state. If this process is catalyzed by reasonably strong hydrogen bonding, then both oxygen and deuterium isotope effects are accounted for.

Another theoretical contribution of interest is a study[86] of the effects of geometrical structure and chemical constitution on the motion of protons in chains of hydrogen bonds, like the charge-relay chain. Here it was found that the degree to which the motions of two such protons in a chain are coupled to each other is very dependent on the overall length of the chain but is rather insensitive to chemical constitution. Short lengths favor coupled motion, while longer ones lead to uncoupling. Indeed, the calculations suggested that variations in chain length of tenths of an Ångstrom unit could couple or uncouple such motions. This may also help to explain how only one proton in the chain forms an effective transition-state bond, at least with these rather abbreviated substrates. It is known that the enzyme structure is altered when oligomeric substrates or inhibitors bind to it,[87] and these alterations may lead to a coupling of the charge-relay chain with longer peptide substrates but to uncoupled behavior with "minimal" substrates such as these, thus yielding quite small solvent isotope effects. This point is briefly discussed again below.

Our conclusions from the first four entries of Table IV are therefore these: Both oxygen and solvent deuterium isotope effects are consistent with either a nucleophilic-attack transition state with weak solvation catalysis or a leaving-group expulsion transition state with a similar catalysis; the variability of the oxygen effects suggest that substrate-induced and pH-induced small alterations in enzyme structure may affect the balance between these, the substrates lying at the breakpoint between the two transition states on the structure-reactivity plot.

Let us now examine entries 5–8 in Table IV, all of which relate to a common substrate, N-acetyltryptophanamide. The three entries 5–7 are nitrogen isotope effects which—except in magnitude—are entirely analogous to the oxygen isotope effects just considered. Klinman's estimate[80] for maximum C—N bond-breaking isotope effect k_{14}/k_{15} is 1.044. This value (calculated for a C—N single bond) should doubtless be increased for amide starting materials because of the partially double bond to nitrogen in amides; taking a C—N frequency of 1600 in place of Klinman's 1134 yields a value of 1.062. Again protonation should produce an inverse isotope effect, which for ammonium ion* is $K_{14}/K_{15} = 1/1.039$. Since we have no careful model study like that of the Kirsch group on methyl formate, there is no calibration point for formation of a tetrahedral adduct. The best we can do offhand is to assume the effect is about $1.062 \div 1.044 \sim 1.017$ (for conversion of an amide partially double bond to a C—N single bond). For the anticipated mechanistic sequence with amide substrates we thus have the following limiting estimates for k_E^{16}/k_E^{18} (if the transition state exactly resembles the structure shown):

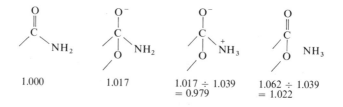

		$1.017 \div 1.039$	$1.062 \div 1.039$
1.000	1.017	$= 0.979$	$= 1.022$

---

* Thode et al.[88] report values which lead to an effect of 1.024 after correction of the ammonia fractionation factor to its aqueous phase value. This has been redetermined as 1.039 by M. H. O'Leary ["Studies of Enzyme Reaction Mechanisms by Means of Heavy-Atom Isotope Effects," *Isotope Effects on Enzyme-Catalyzed Reactions* (W. W. Cleland, M. H. O'Leary, and D. B. Northrop, eds.), pp. 233–251, University Park Press, Baltimore (1977)] and A. P. Young.

Thus a nucleophilic-attack transition state should produce normal N^{15} isotope effects of 1.000–1.017 (say, up to about 1.5–2%), the transition state for a protonation step could bring these down as far as about 2% *inverse*, a transition state for C—N bond cleavage then will produce effects between about 2% inverse and 2% normal, and finally a transition state for the product-release step should give effects around 2% normal. Note that, to the rather small extent to which one can rely on estimates like these, the transition state for leaving-group expulsion with prior protonation would lead to *inverse* N^{15} effects for transition-state structures with up to about 50% C—N fission, to no isotope effect at all at that point, and to normal effects for transition-state structures with greater degrees of C—N fission.

It is notable that in terms of the structure-reactivity plot of Fig. 8, this compound also falls at a breakpoint, between rate-limiting leaving-group expulsion and rate-limiting product release. Furthermore, the isotope effect is unusually responsive to pH alterations, an effect interpreted by O'Leary and Kluetz as a signal of a pH-dependent rate-determining step. If we imagine, purely to exemplify the argument, that at pH 8 both transition states (for leaving-group expulsion and for product release) equally determine the rate, that the C—N bond is about one-half broken in the former (so that $k_{14}/k_{15} \sim 1.00$), and that product release gives $k_{14}/k_{15} \sim 1.02$, then we generate the observed effect of 1.01 at this pH. The smaller effects at both higher and lower pH indicate leaving-group expulsion to govern the rate to a greater degree under *both* circumstances. Perhaps alterations of enzyme charge character away from the situation at pH 8 in either direction lead to a faster release of amine product.

Fersht and his group have also argued that a change in rate-determining step accounts for certain features of the pH–rate profile and of acyl-enzyme-trapping experiments in chymotryptic hydrolysis of formylphenylalanine formylhydrazide and formylphenylalanine semicarbazide.[89] These substrates have leaving groups with pK_a's similar to chloroanilines (compounds 15 and 16 in Fig. 8) and thus should also be at a breakpoint on the structure-reactivity plot. However, the break is not for a change in rate-determining step but for a shift in pathway from leaving-group expulsion with solvation catalysis to leaving-group expulsion with proton-transfer catalysis. Perhaps a differential pH sensitivity of these two pathways is the explanation of the trapping and pH–rate experiments with the hydrazide substrates.

The last entry in Table IV is a solvent isotope effect for hydrolysis of N-acetyltryptophanamide. Its magnitude is entirely similar to that seen with the ester substrates, and the same remarks with regard to the charge-relay chain could be made here. However, some factors are different. If, indeed, the two processes of leaving-group expulsion and product release both determine the rate at pH 8, one does not expect similar solvent isotope effects for them (as was the case with transition states 3 and 4). In such a situation the observed value of 1.9 would be a weighted average of a presumably larger effect for protolytically assisted leaving-group expulsion and a smaller (perhaps unit)

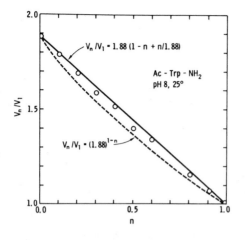

Fig. 12. Proton inventory plot for the acylation of α-chymotrypsin by N-acetyltryptophanamide. The solid line corresponds to one-proton catalysis and the dashed line to "infinite-proton" catalysis, i.e., to a large number of contributions from different sites, each of very small magnitude.

effect for product release. An attempt to illuminate the situation further is represented by the proton inventory experiment of Fig. 12.* If the entire isotope effect were a result of a single proton's changing state, then the upper straight line of Fig. 12 should describe the data. At the other extreme of possibilities, one could hypothesize that the entire effect arises from a very large number of very small isotope effects, multiplying together to generate the observed value of 1.9. If that were so, the expected dependence is the logarithmic one shown as the dashed line (see Chapter 6). Clearly, the truth is in between. In fact, a polynomial regression fit of $V_n(n)$ shows the linear and quadratic terms significant at the 95% level but no other terms at the 90% level or above. The best quadratic fit of the data gives two transition-state fractionation factors, $\phi_1 = 0.88 \pm 0.07$ and $\phi_2 = 0.59 \pm 0.04$. These correspond to isotope-effect contributions $(k_H/k_D)_1 = 1.14 \pm 0.09$ and $(k_H/k_D)_2 = 1.69 \pm 0.11$. There are thus several models consistent with the solvent isotope effects:

1. Two protons contribute to the isotope effect in a single transition state which determines the rate, one generating a modest effect of 1.7 and the other a quite small effect of 1.1.
2. One proton contributes substantially in a single transition state, while a number of others make small contributions.
3. Two or more protons (say, those of the charge-relay chain) contribute in a transition state for leaving-group expulsion, while a second transition state, which has no isotope effect (say, that for product release), also participates in determining the rate; the proton inventory—a weighted average—is then close to linear.

---

* The velocities were determined for substrate concentrations near K_m so that the effect is a weighted average of the effects on k_E and k_{ES}. For amide substrates, the transition states are believed to be the same for the two enzymic rate constants.

To summarize the isotope effects for the amide substrate, they are consistent with either leaving-group expulsion or product release or some mixture of the two (most likely) as rate limiting, but a simple proton-transfer reaction coordinate, or any form of highly coupled charge-relay action in a single transition state, is not indicated.* The situation may differ for large, more "natural" substrates, as discussed in Chapter 6.

2.5. Chemical Predestination and Transition-State Plasticity

Let us now step back from the details of chymotryptic transition states and ask whether these highly speculative interpretations of structure and reactivity, pH effects, and isotope effects provide any general message about enzyme catalytic power. The data appear to have shown us some five different transition states in the course of the acylation stage alone of chymotrypsin action. One general inquiry which might be addressed is this: Are these chemical or biological transition states? That is, has chymotrypsin (and, by inference, other enzymes) been forced by a refractory character in transition states to accept the "prebiotic" (perhaps better, "abiotic") transition-state structures of simple chemical examples of nucleophilic attack, leaving-group expulsion, etc.? Or, on the contrary, has the evolution of enzymes entailed the invention of new transition states, unknown in nonliving nature, which enzymes employ in achieving their impressive catalytic feats? The former view—if affirmed— would imply a chemical predestination which imposes itself upon biological nature and requires, in the case of enzyme catalysis, that biological evolution simply develop stabilizing protein envelopes about preexisting "chemical" transition states. The latter view expresses the idea that transition states are sufficiently plastic as to permit rather severe alterations in their structure through interaction with enzymes, so that the development of enzyme catalytic power during the course of evolutionary history would be attended by a simultaneous alteration of the transition states to new "biological" forms.

Although one can detect in contemporary discussions a good deal of sympathy for—or even dogmatic assertions of the necessity of—the "biological" viewpoint, it is worth calling attention to the fascinating volume, *Biochemical Predestination*, of Kenyon and Steinman[90] in which the following proposition is advanced:

> throughout the range of biological order . . . there appears to be an inherent tendency toward the type of organization which we observe in living cells.

* There is, however, evidence for a coupled charge-relay chain for oligopeptide substrates [R. L. Schowen, "Solvent Isotope Effects on Enzymic Reactions," in *Isotope Effects on Enzyme-Catalyzed Reactions* (W. W. Cleland, M. H. O'Leary, and D. B. Northrop, eds.), pp. 64–99, University Park Press, Baltimore (1977); M. W. Hunkapillar, M. D. Forgac, and J. H. Richards, "Mechanism of Action of Serine Proteases: Tetrahedral Intermediate and Concerted Proton Transfer," *Biochemistry*, Vol. 15, pp. 5581–5588 (1976)].

Carrying their idea one step further, it may well be that not only are biological structure and function predestined by the properties of those biomonomers produced on the primitive earth but that the occurrence of just these monomers —and their properties—was *chemically* predestined by the composition of the universe at the instant of its formation.

The most notable feature of the entire discussion of the foregoing sections has been an extraordinary agreement between *chemical expectation*, as deduced from model experiments, and *biological observations*, as exemplified by the results of chymotrypsin experiments. Admittedly, there is a rather heavy element of circularity in this conclusion, inasmuch as our interpretation of the biological observations depended on the chemical expectations and the agreement between the two. However, insofar as one can currently see, the picture is such as to favor the view that chemical transition states, within the context of those forces which enzyme structure can exert,[91] are structurally implastic and that the structures of these transition states may well have provided an invariant of biological history, to which the evolutionary process has had to adapt in a creative way.

3. ACETYLCHOLINESTERASE [92,93]

3.1. Comparison with Serine Proteases [93]

The task of acetylcholinesterase (EC 3.1.1.7) is to catalyze the hydrolysis of acetylcholine,

$$(CH_3)_3\overset{+}{N}CH_2CH_2OCOCH_3 \xrightarrow[H_2O]{} (CH_3)_3\overset{+}{N}CH_2CH_2OH + CH_3CO_2^- \qquad (46)$$

a substance responsible for transmission of information across nervous synapses and neuromuscular junctions. It is frequently thought of as an advanced form of serine protease, its assignment upgraded from the routine digestive function to a critical role in nerve action, with a corresponding increase in structural complexity. While the serine proteases consist typically of a single globular subunit of molecular weight around 25,000, the form of acetylcholinesterase most commonly used in mechanistic study is a tetramer [a dimer of dimers, the latter linked by disulfide bonds; thus $(\alpha_2)_2$] with each monomeric subunit having a molecular weight around 76,000. Each subunit possesses an active site. Preparations of the enzyme, observed electron microscopically, exhibit the fundamental tetramer with an appended "tail" and aggregates of two and three tetramers, each aggregate with a "tail."

The enzyme contains a serine which is acylated and deacylated in the course of its action. The pH dependence of the rate is similar to that seen for serine proteases, with an inflection at pH 5–7, presumably arising from an active-site histidine. The latter has been considered a possible component of a charge-relay chain like that of the serine proteases. The active site, in addition to the nucleophilic serine and some form of acid–base catalytic machinery, seems to contain a binding site for the ammonium pole of acetylcholine (rather than the hydrophobic pocket of chymotrypsin).

3.2. Transition States for Acylation

As in the serine proteases, the kinetic parameter k_E ($\equiv k_{cat}/K_m$) can be used to isolate the rate-determining transition state(s) in the reaction sequence leading up to first-product release from the acyl enzyme (the product in question being choline). A notable fact about k_E for acetylcholinesterase is that its absolute value[93] for acetylcholine as substrate is very close to 10^8 M^{-1} sec^{-1}, only about tenfold slower than the maximum expected for diffusion of a small substrate toward acetylcholinesterase (see Table II). This in itself suggests that this diffusion and/or some reorganization of the enzyme may be rate limiting in acetylcholine hydrolysis.

Such a hypothesis and arguments in its favor have been developed by Rosenberry.[94] First, there is practically no depression of k_E for acetylcholine hydrolysis in deuterium oxide. If acid–base-catalyzed nucleophilic attack or leaving-group expulsion were rate limiting, such a depression would be expected. Indeed, for k_{ES} which is determined in major part by deacetylation,[92] the solvent isotope effect $k_{H_2O}/k_{D_2O} \sim 2.3$. Second, various substrates produce differing values of apparent pK_a in the pH–k_E profile. There are several possible explanations for such an observation, but the other data just cited suggested that a conformation change was rate limiting to a greater or lesser degree with the different substrates. Since this implies that the rate of conformation change is substrate dependent, the hypothesis is of an *induced-fit conformation change*.

The information in Tables V, VI, and VII lends support to this hypothesis. These tables are concerned with secondary deuterium isotope effects in acetyl-cholinesterase action. Tables V and VI help to establish chemical expectations for possible transition states in acylation, and Table VII presents the enzymic results.

Table V contains data on the β-secondary deuterium isotope effect for reactions in which nucleophilic attack at carbonyl is occurring. The simplest interpretation of such effects (see Chapter 5) is that carbonyl addition reduces hyperconjugation from the β-CH bonds into the carbonyl π orbital. This tightens the CH bonds, producing an inverse isotope effect. On the other hand, effects which *increase* hyperconjugation, such as electrophilic interaction or desolvation of the carbonyl group, produce normal isotope effects. If it is assumed that all β-CH bonds contribute about equally, then a "per D" effect can be calculated as the Nth root of the observed effect (for N deuteriums) and expressed as a percentage effect (third column of the table). The equilibrium hydration of a ketone (entry 1) produces an effect of $-4.6\%/D$, which should be near the maximum for a kinetic isotope effect resulting from loss of hyperconjugation, corresponding to a transition state that exactly resembles a

Table V

Secondary Deuterium Isotope Effects for Some Processes Involving Nucleophilic and Electrophilic Interaction at Carbonyl

Reaction (L = H or D)	k_H/k_D, observed	$100[(k_H/k_D)^{1/N} - 1]$ ($\%$/D)	Reference
1. Equilibrium hydration of $(ClCL_2)_2CO$, $15°-46°$	0.83 ± 0.04	$-4.6 \pm 1.3\%$	95
2. Deacetylation of CL_3CO-α-chymotrypsin, $25°$	0.94 ± 0.01	$-2.0 \pm 0.4\%$	96
3. Acidic hydrolysis of CL_3COOCH_3, $0°-42°$	0.93 ± 0.02	$-2.4 \pm 0.7\%$	95
4. Basic hydrolysis of CL_3COOCH_3 or $CL_3COOCH_2CH_3$, $25°$	0.90 ± 0.02	$-3.5 \pm 0.7\%$	95
5. Basic hydrolysis of $CL_3COOC_6H_5$, $25°$	0.98 ± 0.01	$-0.7 \pm 0.3\%$	96
6. Basic hydrolysis of $CL_3COOC_6H_4NO_2$-p, $25°$	0.97 ± 0.01	$-1.0 \pm 0.3\%$	96
7. Equilibrium protonation of $CL_3COC_6H_5$, $25°$	1.29 ± 0.08	$+8.9 \pm 2.7\%$	97
8. Equilibrium extraction into chlorocyclohexane from water of $CL_3CONHC_6H_4NO_2$-p, $25°$, and of $CL_3COOC_6H_4NO_2$-p from water into cyclohexane, $25°$	1.06 ± 0.03; 1.15 ± 0.04	$+2.0 \pm 1.0\%$ $+4.8 \pm 1.2\%$	98
9. Basic hydrolysis of CL_3COOCH_3, $50°$, or $CL_3COOC_2H_5$, $65°$	1.03 ± 0.10; 1.15 ± 0.09	$+1.0 \pm 3.3\%$; $+4.8 \pm 3.0\%$	95, 99

tetrahedral adduct. As expected, many rate processes with quasi-tetrahedral transition states produce β-D effects which are between the limits of 0 and -4.6%. The variation in these effects doubtless reflects variation in the structures of the transition states. The effects shown include the deacetylation of chymotrypsin acyl enzyme, acidic hydrolysis of alkyl acetates, basic hydrolysis of aryl acetates, and basic hydrolysis of alkyl acetates near $25°$. Entries 7–9

Table VI[95]

Secondary Deuterium Isotope Effects for Nonenzymic, Basic Hydrolysis of Acetylcholine at $25°$

Substrate	k^H/k^D, observed	$100[(k^H/k^D)^{1/H} - 1]$ ($\%$/D)
$CL_3COOCH_2CH_2N(CH_3)_3^+$	0.96 ± 0.01	$-1.4 \pm 0.3\%$
$CH_3COOCL_2CH_2N(CH_3)_3^+$	1.01 ± 0.01	$0.5 \pm 0.5\%$
$CH_3COOCH_2CL_2N(CH_3)_3^+$	1.01 ± 0.01	$0.5 \pm 0.5\%$
$CH_3COOCH_2CH_2N(CL_3)_3^+$	0.994 ± 0.003	$-0.07 \pm 0.03\%$

of Table V show how electrophilic interactions or decreased nucleophilic interaction at carbonyl can weaken the β-CH bonds through an increase in hyperconjugation, thus producing normal isotope effects. Protonation of the carbonyl oxygen, placing a large positive charge on carbonyl carbon, generates an effect of about $+9\%/D$. Extraction of carbonyl compounds from water, where nucleophilic solvation is occurring, into organic solvents, where no such solvation is anticipated, produces effects of $2-5\%/D$. At temperatures of $50°-65°$, the basis hydrolysis of alkyl acetates seems to show a change in rate-determining step to one in which (presumably) nucleophilic interaction at carbonyl is reduced, giving rise to effects of a similar magnitude.

Table VI focuses on the nonenzymic, basic hydrolysis of acetylcholine itself and includes both β-D acyl effects and effects of substitution in the cholin moiety. The β-D effect is -1.4%, between the values of about 1% seen for aryl acetates and -2 to -3% seen for alkyl acetates. Since choline is an alcohol of leaving-group capacity intermediate between alkanols and phenols, this result is quite reasonable. Substitution of deuterium at points in the choline moiety leads to effects which are—at these levels of precision—essentially negligible. This is again as expected because in the transition state for non-enzymic hydrolysis, the isotopic binding at these sites will be unchanged from the reactant-state situation.

These effects thus establish the expectations for the β-D effect in an acetylcholinesterase acylation transition state in which nucleophilic inter-actions predominate: an inverse effect of about -1 to $-4\%/D$. The effects in the choline moiety would not necessarily be negligible for such an enzyme acylation transition state, because enzymic interaction with the choline is expected. If, for example, the torsional motion about the ethylene group were converted to an isotopically sensitive wagging of the methylenes by tight bind-ing of the ester function into the catalytic site and the ammonium pole to the "anionic site," a reasonable inverse isotope effect would arise. A frequency of 1000 cm^{-1} with an isotopic frequency ratio of 1.36 would generate an effect $k_H/k_D \sim 1.9/D$. In fact, the effect would probably be considerably smaller because of lower isotopic sensitivity but should not be negligible. If, as is currently thought, the methyl groups of the ammonium head are deposited in a hydrophobic environment,[100] effects like those observed by Tanaka and Thornton are expected.[101] They found that protium materials were extracted into the hydrophobic medium μ-Bondapak-C$_{18}$ more favorably than deuterium materials by factors of $+0.2$ to $+0.8\%/D$. If similar factors were at work in an acylation transition state with acetylcholinesterase, deuterium substitution at the nine hydrogenic positions of the trimethylammonium group of acetyl-choline should lead to observed isotope effects $k_E^H/k_E^D = 1.02$ to 1.07.

The actual situation is as portrayed in Table VII. First, the β-D acyl effect is within experimental error of zero, and although the error is uncom-fortably large, the indication is that little or no nucleophilic-attack contribu-tion is present. Second, the effects for deuteration on the ethylene unit are very small, suggesting no substantial transition-state restriction to torsion by bind-

Table VII

Secondary Deuterium Isotope Effects for the Acylation of
Acetylcholinesterase at 25°, pH 7.5

Substrate	k_E^H/k_E^D, observed	$100[(k_E^H/k_E^D)^{1/N} - 1]$, (%/D)	Reference
$CL_3COOCH_2CH_2N(CH_3)_3^+$	0.99 ± 0.03	$-0.3 \pm 1\%$	95
$CH_3COOCL_2CH_2N(CH_3)_3^+$	1.00 ± 0.07	$0.0 \pm 4\%$	95
$CH_3COOCH_2CL_2N(CH_3)_3^+$	1.01 ± 0.03	$0.5 \pm 1\%$	95
$CH_3COOCH_2CH_2N(CL_3)_3^+$	1.05 ± 0.07	$0.5 \pm 1\%$	95
$CL_3COOC_6H_4NO_2\text{-}p$, H_2O	0.99 ± 0.02	$-0.3 \pm 0.7\%$	96
$CL_3COOC_6H_4NO_2\text{-}p$, D_2O	1.00 ± 0.01	$0.0 \pm 0.3\%$	96
$CL_3COOCH_3{}^a$	1.10 ± 0.06	$+3.2 \pm 1.9\%$	95

^{a} This effect was measured at pH 6.5.

ing at the two ends. Third, the effect in the ammonium methyl groups is within experimental error of zero, although it is consistent with some hydrophobic interaction at this center. A simple model for the transition state is then one for a conformation change of acetylcholinesterase, perhaps being induced by interaction of the ammonium head with its binding site.

Table VII also contains two β-D effects for *p*-nitrophenyl acetate as substrate, one measured in protium oxide and one in deuterium oxide (the solvent isotope effect is considered below). The errors here are slightly better, and the indication is strong that the transition state contains a trigonal, not a quasi-tetrahedral, acyl function, with neither nucleophilic or electrophilic interactions at carbonyl. The most likely model is thus that a conformation change is wholly or largely rate limiting, in both H_2O and D_2O, for *p*-nitrophenyl acetate also. This is in spite of the fact that k_E is on the order of only $10^5\ M^{-1}$ sec^{-1} for this substrate, slower by a 1000-fold than for acetylcholine. This supports the pH-dependency evidence in favor of a substrate-controlled rate of conformation change.[94]

Finally, methyl acetate, for which k_E is about $10^3\ M^{-1}\ sec^{-1}$, exhibits and isotope effect of $+3.2\%/D$. This is in the direction and of the magnitude expected for either electrophilic interaction at the carbonyl (for example, by the proton of an enzymic catalytic function) or desolvation of the carbonyl preparatory to nucleophilic attack. Such effects are suggestive of the incursion of the "chemical steps" as rate determining, although detailed analysis will require a better understanding of the model reactions. Whether this incursion has been caused solely by the change in substrate structure or also by the pH change is unknown.

In advancing the induced-fit hypothesis, Rosenberry cited the negligible solvent isotope effect on k_E for acetylcholine and larger normal effects for slower substrates where acid–base-catalyzed steps might be coming into play.[94] He found for *p*-nitrophenyl acetate at pH 8.5, 0.1-*M* NaCl, 0.05-*M* Tris · HCl buffer, a value of $k_E^{H_2O}/k_E^{D_2O} = 1.93 \pm 0.07$. We find with less pure commercial enzyme, pH 7.5, 0.03-M Tris · HCl buffer, $k_{H_2O}/k_{D_2O} = 1.53 \pm 0.02$.

A proton inventory study of the solvent isotope effect on k_E in acetyl-cholinesterase-catalyzed hydrolysis of p-nitrophenyl acetate is shown in Fig. 13. The shape of this curve is "bulging upward," which indicates in the simplest cases (see Chapter 6) either an inverse isotope-effect contribution or a change in rate-determining step. Both models are roughly fitted to the data (nothing should be attributed to the differences in fit—either model, by a small adjustment of fractionation factors, can be brought into equally good agreement with the data). One model assumes a reactant-state fractionation factor around 0.7, such as that envisioned for a cation-binding site of formyl tetrahydrofolate synthetase.[102] A transition-state isotope effect of 2.2 then brings prediction into rough agreement with observation. In the second model, it is imagined that the change from H_2O to D_2O causes a shift in the rate-limiting balance of two serial steps, one with an isotope effect of 3.3 (ascribed to two protons with equal contributions of 1.8) and one with no isotope effect. The two steps proceed at about equal rates in D_2O, while one is about 3.5-fold faster than the other in H_2O.

The importance of the β-D measurement for p-nitrophenyl acetate in D_2O (Table VII) arises from this second model. If the rate-limiting balance of two steps, one with no solvent isotope effect and one with a substantial solvent isotope effect, is affected by the deuteration of the solvent, the absence of a β-D effect shows that nucleophilic attack is not occurring in either. More simply put, although p-nitrophenyl acetate does show a solvent isotope effect, the "chemical" step(s) with quasi-tetrahedral transition states do not seem to be rate determining. The β-D effect for methyl acetate is consistent with a

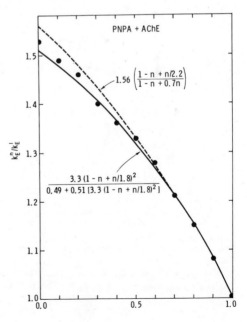

Fig. 13. Proton inventory plot for the acylation of acetylcholinesterase by p-nitrophenyl acetate. The dashed line is for a model with one transition-state fractionation factor of 2.2^{-1} (expected for a catalytic bridge) and one reactant-state fractionation factor of 0.7. The solid line is for a model in which two transition states in series contribute to a virtual transition state; the solvent isotope effects are different for the two, so that the balance of their contributions alters with atom fraction of deuterium (plotted on the abscissa). One transition state has two isotope-effect contributions of 1.8 each for a net effect of 3.3, the other transition state has no isotope effect. In D_2O, the two transition states contribute about equally (49% and 51%) to the virtual hybrid.

$$1.56 \left(\frac{1 - n + n/2.2}{1 - n + 0.7n} \right)$$

$$\frac{3.3 \, (1 - n + n/1.8)^2}{0.49 + 0.51 \, [3.3 \, (1 - n + n/1.8)^2]}$$

"chemical" transition state. For this substrate, Rosenberry[94] has estimated $k_E^{H_2O}/k_E^{D_2O} \sim 2$. Therefore, it appears likely that three situations can be discerned:

1. Acetylcholine as substrate: $k_E \sim 10^8 \ M^{-1} \ sec^{-1}$; no solvent isotope effect; no β-D isotope effect.
2. p-Nitrophenyl acetate as substrate: $k_E \sim 10^5 \ M^{-1} \ sec^{-1}$; $k_E^{H_2O}/k_E^{D_2O} \sim 1.5–1.9$; no β-D isotope effect in either H_2O or D_2O.
3. Methyl acetate as substrate: $k_E \sim 10^3 \ M^{-1} \ sec^{-1}$; $k_E^{H_2O}/k_E^{D_2O} \sim 2$; β-D isotope effect $+3.2\%/D$.

A mechanistic scheme consistent with these results is shown in Fig. 14. The absence of a β-D effect for both acetylcholine and p-nitrophenyl acetate indicates a rate-limiting conformation change for both. The absence of a solvent isotope effect for acetylcholine and the presence of one for p-nitrophenyl acetate makes it probable that these are *different* conformation changes. The two might be in series or in parallel. We have chosen the serial alternative, imagining that the induced fit is generated in two stages: one without a solvent isotope effect in which the general active-site structure is assembled (labeled R-I for "first reorganization" in Fig. 14) and a second (R-II) with a substantial solvent isotope effect, in which the acid–base machinery is prepared for catalysis. If the rate of R-II is more sensitive to substrate structure than the rate of R-I, then R-I could be rate determining for acetylcholine and R-II

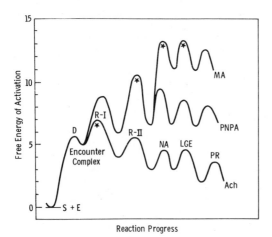

Fig. 14. Free-energy vs. reaction-progress diagram for the acylation of acetylcholinesterase by acetylcholine (ACh), p-nitrophenyl acetate (PNPA), and methyl acetate (MA). The barriers to the acylation process are taken to be those for diffusion (D) to form an encounter complex, a first conformational reorganization of the enzyme in the presence of substrate (R-I) which is imagined to assemble the substrate-binding structure of the active site, a second substrate-induced conformational reorganization of the enzyme (R-II) which assembles the acid–base catalytic machinery and involves a solvent isotope effect, nucleophilic attack (NA), leaving-group expulsion (LGE), and first-product release from the acyl enzyme (PR). Acetylcholine ($k_E = 10^8$, no CD_3 effect, no solvent isotope effect) is shown as having R-I rate determining. p-Nitrophenyl acetate ($k_E = 10^5$, no CD_3 effect, $k_{H_2O}/k_{D_2O} = 1.5–1.9$) has R-II rate determining. Methyl acetate ($k_E = 10^3$, $k_{CH_3}/k_{CD_3} = 1.1$, $k_{H_2O}/k_{D_2O} = 2$) has NA or LGE or, possibly, both as rate determining.

for *p*-nitrophenyl acetate. Finally, the protolytically catalyzed "chemical" steps of nucleophilic attack and/or leaving-group expulsion become rate limiting with methyl acetate. This substrate then produces both a β-D isotope effect and a solvent isotope effect.

3.3. Transition-State Binding and Catalysis

The conclusion, from Rosenberry's work and the considerations adduced just above, that an induced reorganization process is rate limiting for acylation of acetylcholinesterase by acetylcholine provides an opportunity to consider the question of transition-state stabilization and choice of standard reaction when a "physical" step limits the rate. In Chapter 2, Section 2.4, the acetylation of acetylcholinesterase was treated in terms of three standard reactions: basic, neutral, and acidic hydrolysis of acetylcholine. These three transition states, for which structural hypotheses are given in Fig. 2 of Chapter 2, were shown to bind to acetylcholinesterase to generate the rate-limiting enzymic transition state with a free-energy release of 10.6 kcal/mol (basic), 13.9 kcal/mol (neutral), and 17.3 kcal/mol (acidic). We are now in a position to make a structural proposal for the enzymic transition state and to speculate about the components of distortion energy and vertical binding energy that go into the empirical binding energies just mentioned.

The data given in the last section indicate that, at most, some interaction of the ammonium pole of acetylcholine with the enzyme is occurring in the rate-limiting transition state. A structural hypothesis is, therefore, formulated as

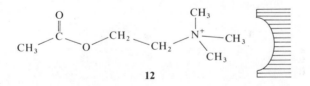

12

We can consider the binding equilibrium for a single standard reaction, say the basic hydrolysis. That equilibrium is then given by

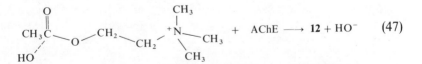

$$+ \; AChE \longrightarrow 12 + HO^- \qquad (47)$$

for which $\Delta G_b^* = -10.6$ kcal/mol. The empirical binding energy in this formulation has components which can be directly measured to a greater extent than is usual. The distortion energy ΔG_D, for example, will have the following contributions:

1. Bonding changes in the acetylcholine core:

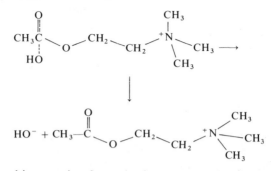

We know this quantity, from the free energy of activation of the standard reaction, to be $\Delta G = -33.2$ kcal/mol.

2. Changes in enzyme from "resting" structure to that of the conformation-change (R-I) transition state. We might estimate this by assuming that the conformation change R-I is tenfold faster than R-II in the case of *p*-nitrophenyl acetate as substrate and that both R-I and R-II in this case are "uninduced," thus simulating development of a poised structure. This leads to an estimate of $\Delta G = +9$ kcal/mol.

3. Assuming an interaction of the ammonium head and the anionic site in **12**, with both partially desolvated, the energy of partial desolvation of both structures will contribute (call this unknown quantity ΔG_{desol}).

These changes should generate the poised structures which will then interact to liberate the vertical binding energy of the transition state, ΔG_{vb}^* (see Chapter 2). Since $\Delta G_b^* = \Delta G_D + \Delta G_{vb}^*$,

$$\Delta G_{vb}^* \simeq +33.2 - 9 - \Delta G_{desol} - 10.6$$

$$\Delta G_{vb}^* \simeq +13.6 - \Delta G_{desol}$$

In this particular case, the free energy of vertical binding of the transition state may actually be positive—it depends on the degree of desolvation and desolvated ion–ion interaction present in the R-I transition state. If this is negligible ($\Delta G_{desol} \sim 0$), then $\Delta G_{vb}^* \sim +13.6$ kcal/mol. Then also, $\Delta G_D = -33.2 + 9 = -24.2$ kcal/mol. Thus the distortion energy would be negative. The empirical binding energy is then -10.6, the sum of these, and the enzyme derives its catalytic power from conversion of the partially bonded transition state for base catalysis into an essentially unperturbed reactant molecule, present at the core of the rate-limiting enzymic transition state, that for the R-I conformation change.

This analysis, although it gives a formally correct account of the source of the catalytic power of acetylcholinesterase, leaves a bad taste with most mechanisms scientists. A moment's thought shows that the enzyme has "really" achieved such a stable transition state as that for the R-I step by so greatly stabilizing a series of "chemical" transition states (for nucleophilic attack and for leaving-group expulsion) that they are no longer kinetically significant. This leaves behind the R-I transition state to limit the rate.

This means then that the chemically interesting questions now revolve around transition states not directly accessible with the physiological substrate acetylcholine. To expose them and to study them, one must employ unnatural (and much more slowly reacting) substrates such as methyl acetate. Here the "chemical" transition states are probably in view, although k_E is only 10^{-5} that for acetylcholine. What one must do is to study the nature of these transition states and then try to understand what factors have increased their energies and thus how to extrapolate to the situation in the similar inaccessible transition states with acetylcholine. This is not a simple assignment. Here is still another effort in which one can hope that theoretical chemistry will provide much support.

4. CATECHOL-O-METHYLTRANSFERASE[17,103]

4.1. General Mechanistic Features

Catechol-O-methyltransferase (COMT; EC 2.1.1.6) is found in various organisms and tissues, including the brain and liver,[104] where it serves such functions as inactivation of catecholamines and the detoxification of catechols of external origin. It catalyzes the transfer of the methyl group of S-adenosylmethionine (AdoMet) to one or both of the oxygens,

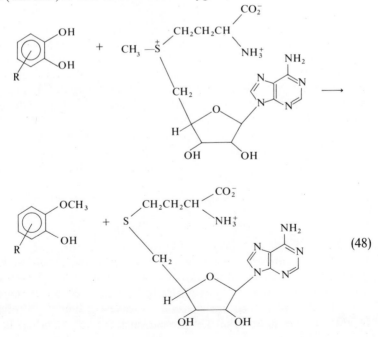

$$(48)$$

of catechol substrates. The enzyme is a single-subunit protein of molecular weight 24,000 and requires Mg^{2+} for activity. The active site contains two sulfhydryl groups, derivatization of which destroys activity and interferes with binding of the substrates[105-107] As yet, no direct mechanistic role has been assigned to the sulfhydryls. The active site is quite specific for the AdoMet structure, the L-amino acid unit, both sugar hydroxyls, the adenine amino group, and the proper chirality at the sulfonium center all being prerequisites for an active methyl donor.[108] The catechol unit is required for acceptor activity, but various side chains will be accepted.[109,110] Charged side chains, as in the natural catecholamine substrates, lead to a predominance of *m*-methylated product (typical meta/para ratios range from 2 to 20, with 6–7 being typical). Nonpolar side chains increase the proportion of *p*-methylation, the ratio becoming unity for 4-ethylcatechol. This had led to the proposal that an active-site hydrophobic pocket, which excludes the charged side chain, thus forces substrates with such structures into an orientation relative to the methyl donor which favors *m*-methylation. Binding then becomes random with nonpolar side chains, which are not excluded from the hydrophobic region, and both *m*- and *p*-products are formed. The ratio is independent of the degree of purification of the enzyme, showing that a single enzyme is catalyzing both reactions.

Various attempts have been made to deduce whether a methylated enzyme is involved as a reaction intermediate. Initial-velocity kinetic studies and inhibition studies are in principle capable of demonstrating whether such an intermediate is present as a free species,[43] but the current reports are conflicting.[111-115] In any case, such approaches can never test for the intermediacy of a methylated enzyme to which both acceptor and demethylated donor are simultaneously bound, such as would be formed by a transitory donation of methyl from donor to enzyme and then immediate transfer from enzyme to prebound acceptor. The best evidence to date is that of Floss and his collaborators,[45] who showed inversion of configuration at chiral methyl in a plant enzyme system. Such inversion is expected for direct donor–acceptor transfer, while retention is expected for double displacement (donor–enzyme, enzyme–acceptor) whether the methylated enzyme were ever a free species or not. If this finding can be extended to COMT, the indication is that both AdoMet and catechol are bound together and that direct transmethylation then occurs.

4.2. Chemical Expectations

The experimental probes of transition-state structure to be used for COMT action are the α-deuterium secondary kinetic isotope effect for CH_3-AdoMet vs. CD_3-AdoMet, and the ^{13}C primary kinetic isotope effect for $^{12}CH_3-$AdoMet vs. $^{13}CH_3-$AdoMet (cf. Chapters 5 and 7). Later, we shall also consider solvent isotope effects in the investigation of protolytic catalysis by the enzyme. If we take the mechanism to be the Random Bi Bi process described above, then a minimal version (for which we can explore possible values of the α-D and ^{13}C isotope effects) is that of

$$E + CatOH + AdoMet \xrightleftharpoons[\text{steps}]{\text{binding}} E^{\text{CatOH}}_{\text{AdoMet}} \qquad (49)$$

$$E^{\text{CatOH}}_{\text{AdoMet}} \xrightleftharpoons[\text{displacement}]{\text{nucleophilic}} E^{\text{CatOCH}_3}_{\text{AdoHcy}} \qquad (50)$$

$$E^{\text{CatOCH}_3}_{\text{AdoHcy}} \xrightleftharpoons[\text{steps}]{\text{product-release}} E + CatOCH_3 + AdoHcy \qquad (51)$$

Here CatOH stands for the catechol acceptor and AdoHcy for the demethylated donor *S*-adenosylhomocysteine.

The strategy of approach is to make rather detailed predictions of the isotope effects to be expected for each possible transition-state structure. We can begin with the two sets of "physical" steps, for binding and for product release, and then proceed to a thorough consideration of the "chemical" or nucleophilic displacement step. If some stage of the binding process is rate determining, and if the nature of the binding is such that the methyl-group vibration frequencies in the AdoMet are unaltered from their free-solution values, then isotope effects (both α-D and ^{13}C) on k_E should be negligible. The prediction is thus a value 1.000 for both effects. For measurements conducted on k_{ES}, some conformation change of the enzyme preceding nucleophilic attack could be rate limiting. Again, unless direct interactions with the methyl group alter its vibration frequencies, no isotope effects should be seen. The conclusion is, therefore, that unit isotope effects, both α-D and ^{13}C, should be observed if events in advance of the nucleophilic displacement determine the rate.

If product release, of AdoHcy or of the methylated catechol or of both, should determine the rate, then α-D and ^{13}C isotope effects should reflect *completed* transfer of the methyl group: i.e., they should be equilibrium isotope effects for transfer of a methyl group from a sulfonium center to an oxygen center. While no direct measure of such an effect is available, a rough estimate can be made through the calculated D and ^{13}C fractionation factors of Hartshorn and Shiner.[116] We assume that a fluorine can simulate the oxygen center (Cordes' rule: "fluorine = oxygen") and that a chlorine simulates a sulfur, neglecting the effect of the sulfonium charge and substituents. Clearly the estimate will be very rough. The relevant data from Hartshorn and Shiner are the following fractionation factors relative to acetylene:

$$CH_3F: \quad 1.465 \,(\alpha\text{-D}); \quad 1.0259 \,(^{13}C)$$

$$CH_3Cl: \quad 1.405 \,(\alpha\text{-D}); \quad 1.0058 \,(^{13}C)$$

The factors are for preference of the heavier isotope, so the estimates of equilibrium isotope effects are thus

$$K_{3H}/K_{3D} \sim (1.405/1.465)^3 = 0.88$$

$$K_{12}/K_{13} \sim (1.0058/1.0259) = 0.98$$

The prediction is that if product-release steps are rate limiting, both the α-D and ^{13}C isotope effects should be inverse, the former by (say) 10–15% and the latter by about 2%.

Now we consider the nucleophilic displacement step. The transition state for this process can be characterized by two principal reaction variables (see Chapter 3 and earlier sections of this chapter), the Pauling bond order of the C—O bond, n_{CO}, and the Pauling bond order of the C—S bond, n_{CS} (structure **13**). Other structural variables can be related to these two so that an MAR diagram (n_{CO} vs. n_{CS}) can express a complete set of reasonable hypothetical

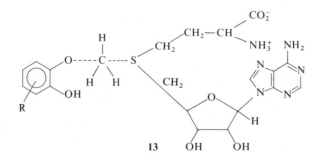

13

structures for the nucleophilic displacement transition state. Protolytic bridges from the oxygens to enzyme functional groups will be considered below in connection with solvent isotope effects. Transition-state structures fall into about four categories of pairwise values of (n_{CO}, n_{CS}), and rough statements can be made for each category about expected α-D and ^{13}C effects. The principles are these:

1. The ^{13}C effects are essentially primary carbon isotope effects for a carbon-transfer reaction. Just as with hydrogen isotope effects for hydrogen transfer (Chapter 4), the ^{13}C effect should be normal and largest for "symmetrical" transition states ($n_{CO} \sim n_{CS}$) and smaller for "unsymmetrical" ones ($n_{CS} > n_{CO}$ or $n_{CO} > n_{CS}$).[117]
2. The α-D effects arise chiefly from changes in the bending potentials of the isotopic hydrogens.[118] If the entering and leaving groups are both close in the transition state (n_{CS}, n_{CO} large), then the bending force constants will be larger than in the reactant and $k_D > k_H$; if the entering and leaving groups are very distant (n_{CS}, n_{CO} small), the bending force constants will be smaller than in the reactant and $k_H > k_D$.

The predictions for the various classes of nucleophilic displacement transition state are then

1. n_{CO} small, n_{CS} large ("reactant-like"): α-D near 1.0, ^{13}C near 1.0.
2. n_{CS} small, n_{CO} large ("product-like"): α-D near equilibrium effect (~ 0.9), ^{13}C near 1.0.

3. n_{CS} small, n_{CO} small ("loose" or "carbonium-like"): α-D normal, ^{13}C normal.

4. n_{CS} large, n_{CO} large ("tight" or "near-pentavalent"): α-D inverse, ^{13}C normal.

It will be desirable to make these predictions more quantitative, and in the next section that is done.

4.3. Isotope-Effect Predictions for Carbon and Hydrogen

As emphasized Chapters 4 through 6 in this volume, a knowledge of re-actant-state and transition-state vibration frequencies permits the calculation, by the Bigeleisen equation, of the kinetic isotope effect. Wolfsberg and Stern first worked out computer programs from which isotope-effect predictions could readily be generated from models consisting of geometrical and force constant information.[119] This procedure is by now highly systematic, and a particularly good "user-oriented" program, BEBOVIB, has been written by L. B. Sims and his group and made available with very full documentation from the Quantum Chemistry Program Exchange.* A version of this program has been used to make quantitative predictions of the expected ^{13}C and α-D effects for various nucleophilic displacement transition-state models related to COMT action.[120]

The model employed is shown in Fig. 15. The principal reaction variables were taken as the Pauling bond orders, n_{CO} and n_{CS}, and related to distances by Pauling's equation, $n = \exp[(r_1 - r_n)/0.3]$ (distances in angstroms). The geometries and force constants of the model reactant and product, $CH_3SH_2^+$ and $CH_3OH_2^+$, were inferred from known values for related compounds. Transition-state properties which were expected to have values intermediate between those of reactant and product were interpolated, assuming that they had progressed by a fraction $[n_{CO}/(n_{CO} + n_{CS})]$. Force constants for the CO and CS partial bonds were taken as simply proportional to the bond orders. The HCS and HCO bending force constants were assumed to change in pro-portion to the CO and CS bond orders.

The calculated isotope effects are portrayed on MAR diagrams in Figs. 16 and 17. The values are for 37°, the temperature at which experiments were conducted. As is apparent, the general expectations developed above are ful-filled in the quantitative treatment. The program now becomes to measure the α-D effect and to trace out on Fig. 16 an area of structures consistent with that effect (including its experimental error), to measure the ^{13}C effect and to trace out a corresponding area on Fig. 17, and then to superimpose these and thus to find that range of structures simultaneously consistent with both the α-D and ^{13}C effects.

---

* Dr. R. W. Counts, Quantum Chemistry Program Exchange, Department of Chemistry, Indiana University, Bloomington, Ind. 47401.

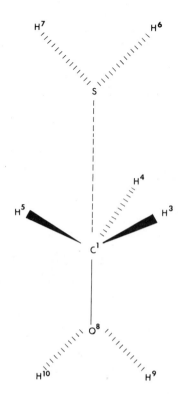

Fig. 15. Model employed for the vibrational analysis estimation of deuterium and ^{13}C kinetic isotope effects for various transition-state structures in methyl transfer.

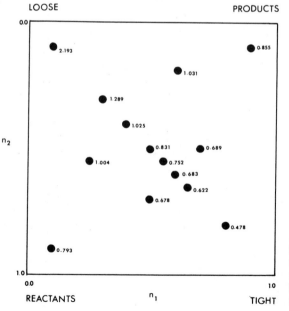

Fig. 16. Predicted isotope effects, k_{CH_3}/k_{CD_3}, as a function of transition-state structure (n_1 = carbon–oxygen bond order, n_2 = carbon–sulfur bond order) for methyl transfer.

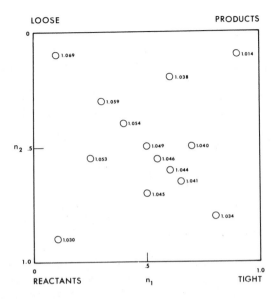

LOOSE   PRODUCTS

Fig. 17. Predicted kinetic isotope effects, k_{12}/k_{13}, as a function of transition-state structure (n_1 = carbon–oxygen bond order, n_2 = carbon–sulfur bond order) for methyl transfer.

REACTANTS n_1 TIGHT

4.4. Isotope-Effect Measurements for Carbon and Hydrogen[120,121]

The α-D and ^{13}C isotope effects were obtained by direct rate measurement on the COMT system, using 2,4-dihydroxyacetophenone,

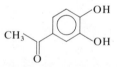

as substrate. The enzyme was obtained from rat liver. It was necessary to take some care in rate measurement in order to assure effects of meaningful precision. The assay was spectrophotometric, and the method shown in Fig. 18 was employed to obtain several hundred good kinetic points during each run. Labeled and unlabeled AdoMet runs were conducted in alternation and isotope effects calculated from the ratios of adjacent rates, a procedure which minimizes the influence of variations in enzyme activity. It was also necessary to have the labeled (CD_3 and $^{13}CH_3$) AdoMet samples with the proper chirality at the sulfonium center. This was accomplished by preparation of the properly labeled methionines, which are achiral at sulfur, by chemical synthesis and adenosylation of these by the yeast *Saccharomyces cervisiae*.[122]

The isotope effects which were measured in this way at saturating concentration of AdoMet were in essential agreement with those from a Michaelis–Menten treatment of data for independent studies of the labeled and unlabeled cofactors, although small deviations from Michaelis–Menten kinetics complicate the picture. Insofar as can be seen, the effects on k_E and k_{ES} are the same.

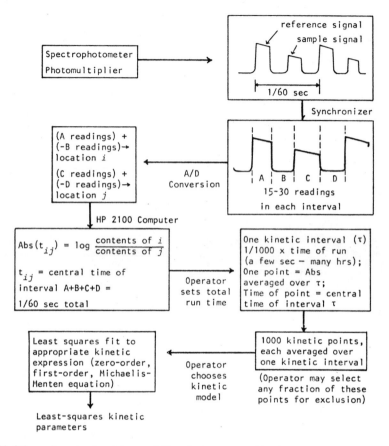

Fig. 18. Scheme for extensive data acquisition during enzyme kinetics measurements in order to generate data of unusual precision (Dr. Wesley White).

The values found were

$$k_{CH_3}/k_{CD_3} = 0.83 \pm 0.05$$

$$k_{CH_3}^{12}/k_{CH_3}^{13} = 1.09 \pm 0.04$$

Both observations serve to exclude events in advance of nucleophilic displacement as rate limiting. While the α-D effect is consistent with the expected equilibrium effect and thus would agree with a product-release transition state, the large normal ^{13}C effect excludes this possibility. The nucleophilic displacement transition state must limit the rate, therefore, and the next step is to trace out, on the superimposed MAR's from Figs. 16 and 17, that space of structures which is simultaneously consistent with both carbon and hydrogen isotope effects.

Fig. 19. Diagram showing the ranges of transition-state structures consistent with the measured deuterium and carbon kinetic isotope effects for methyl transfer catalyzed by catechol-O-methyltransferase. The slanted lozenge includes the structures consistent with deuterium effects in the measured range 0.83 ± 0.05. The expanse at the upper left includes structures for which k_{12}/k_{13} is calculated to be larger than 1.05 and thus consistent with the measured range of 1.09 ± 0.04. The shaded area contains those structures (essentially those with both bond orders about 0.5) that are simultaneously in agreement with both hydrogen and carbon kinetic isotope effects.

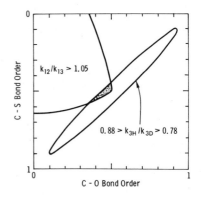

4.5. Transition-State Structure

Figure 19 shows an MAR diagram on which have been plotted two spaces. The elliptical region corresponds to those transition-state structures which, from the calculations represented in Fig. 16, are consistent with α-D effects in the observed range of 0.83 ± 0.05, i.e., between 0.78 and 0.88. The second region, "northwest" in the diagram, is derived from the calculations from the ^{13}C effect in Fig. 17. The measured ^{13}C effect is 1.09 ± 0.04, corresponding to effects between 1.05 and 1.13. No effects as large as 1.13 were calculated, so the space shown includes all effects larger than 1.05.

As is apparent, the two regions overlap only in a small area where $n_{CS} \sim 0.45$–0.60, $n_{CO} \sim 0.35$–0.5. Considering the simplicity of the model calculations, the conservative conclusion is that the structure of the transition state is approximated by that for $n_{CS} \sim \frac{1}{2}$, $n_{CO} \sim \frac{1}{2}$. From Pauling's rule, this corresponds to an extension of each bond by about 0.2 Å from its equilibrium length. An approximate representation of the enzymic transition-state structure is given in Fig. 20.

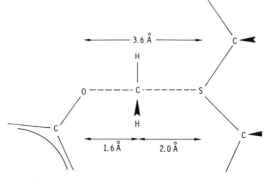

Fig. 20. Approximately scaled representation of methyl-transfer transition state structure with both bond orders about 0.5, for COMT-catalyzed methylation of a catechol substrate by S-adenosylmethionine.

4.6. Bridges to the Enzyme: Solvent Isotope Effects

As noted in Section 1.3, general acid–base catalysis of alkyl transfer is not a very common phenomenon. Furthermore, in those cases where it is observed, the isotope effects for the bridging proton seem considerably smaller than those seen in acyl transfer. This suggests that the hydrogen-bonding stabilization of the alkyl-transfer transition state may be less effective than that of acyl-transfer transition states. It is thus of some special interest to know whether COMT employs general catalysis to stabilize the methyl-transfer transition state.

Preliminary measurements[120] of the solvent isotope effect with 2,4-dihydroxyacetophenone as substrate give a rate ratio for H_2O vs. 68% D_2O of 1.04 ± 0.12 and with norepinephrine as substrate a ratio for H_2O vs. 85% D_2O of 1.24 ± 0.18. These ratios are consistent with either the presence or absence of protolytic bridges to the enzyme structure, and resolution of the problem must await further study.

5. SUMMARY

Structure-reactivity investigations and isotope-effect studies of the acylation of α-chymotrypsin by specific substrates reveal transition states all along the reaction pathway, from binding of the substrate to the enzyme through nucleophilic attack and leaving-group expulsion both with solvation catalysis and proton-transfer catalysis to release of first product from acyl enzyme. The properties of the "chemical" transition states do not appear to differ substantially from those for similar nonenzymic reactions, suggesting that transition states may not be very plastic within the constraints on the exertion of force which enzymes suffer. Their character may be chemically predetermined, and biological evolution may have experienced these structures as historical invariants.

Isotope-effect data for the acylation of acetylcholinesterase support the view that the natural substrate induces a conformation change which determines the rate. This fact does not preclude a formal analysis of catalytic power in terms of transition-state stabilization. Poorer substrates lead to exposure of the "chemical" transition states here also.

Carbon primary and hydrogen secondary isotope effects, taken together, determine a transition-state structure for the nucleophilic displacement step of the action of catechol-O-methyltransferase in which both partial bonds to the transferring methyl group are about half-bonds, extended by about 0.2 Å from their stable lengths.

REFERENCES

1. J. F. Kirsch, in: *Advances in Linear Free Energy Relationships* (N. B. Chapman and J. Shorter, eds.), pp. 369–400, Plenum Press, New York (1972).
2. W. P. Jencks, Structure-reactivity correlations and general acid–base catalysis in enzymic transacylation reactions, *Cold Spring Harbor Symp. Quant. Biol.* **36**, 1–11 (1972).
3. K. T. Douglas, A basis for biological phosphate and sulfate transfers—transition state properties of transfer substrates, *Prog. Bioorg. Chem.* **4**, 193–238 (1976).
4. C. K. Ingold, *Structure and Mechanism in Organic Chemistry*, 2nd ed., Cornell University Press, Ithaca, N.Y. (1969), Chap. 7.
5. J. Hine, *Physical Organic Chemistry*, 2nd ed., McGraw-Hill, New York (1962), Chaps. 6, 7.
6. C. A. Bunton, *Nucleophilic Substitution at a Saturated Carbon Atom*, Elsevier, Amsterdam (1963).
7. A. Streitwieser, *Solvolytic Displacement Reactions*, McGraw-Hill, New York (1962).
8. S. L. Johnson, General base and nucleophilic catalysis of ester hydrolysis and related reactions, *Adv. Phys. Org. Chem.* **5**, 237–330 (1967).
9. W. P. Jencks, *Catalysis in Chemistry and Enzymology*, McGraw-Hill, New York (1969).
10. M. L. Bender, *Mechanisms of Homogeneous Catalysis from Protons to Proteins*, Wiley-Interscience, New York (1971).
11. A. J. Kirby, in: *Comprehensive Chemical Kinetics* (C. H. Bamford and C. F. H. Tipper, eds.), Vol. 10, 57–207, Elsevier, Amsterdam (1972).
12. R. J. E. Talbot, in: *Comprehensive Chemical Kinetics* (C. H. Bamford and C. F. H. Tipper, eds.), Vol. 10, 209–293, Elsevier, Amsterdam (1972).
13. W. P. Jencks, General acid–base catalysis of complex reactions in water, *Chem. Rev.* **72**, 705–718 (1972).
14. W. P. Jencks, Enforced general acid–base catalysis of complex reactions and its limitations, *Acc. Chem. Res.* **9**, 425–432 (1976).
15. W. P. Jencks and J. M. Sayer, Structure and mechanism in complex general acid–base catalyzed reactions, *Faraday Symp. Chem. Soc.* **10**, 41–49 (1975).
16. P. D. Boyer, ed., *The Enzymes*, Vols. 3, 4, 5, 7, 8, 9, 3rd ed., Academic Press, New York, (1971–1973).
17. S. H. Mudd, in: *Metabolic Conjugation and Metabolic Hydrolysis* (W. H. Fishman, ed.), Vol. 3, pp. 297–350, Academic Press, New York (1973).
18. G. M. Maggiora and R. E. Christoffersen, Chapter 3 in this volume.
19. T. C. Bruice, Some pertinent aspects of mechanism as determined with small molecules, *Annu. Rev. Biochem.* **45**, 331–373 (1976).
20. F. G. Bordwell, P. F. Wiley, and T. G. Mecca, Mode of solvent participation in solvolysis reactions at a tertiary carbon atom, *J. Am. Chem. Soc.* **97**, 132–136 (1975).
21. D. J. McLennan, A case for the concerted S_N2 mechanism of nucleophilic aliphatic substitution, *Acc. Chem. Res.* **9**, 281–287 (1976).
22. W. E. Buddenbaum and V. J. Shiner, Jr., ^{13}C kinetic isotope effects and reaction coordinate motions in transition states for S_N2 displacement reactions, *Can. J. Chem.* **54**, 1146–1161 (1976).
23. (a) S. Seltzer and A. Zavitsas, Correlation of isotope effects in substitution reactions with nucleophilicities. Secondary α-deuterium isotope effect in the iodide-131-exchange of methyl-d_3 iodide, *Can. J. Chem.* **45**, 2023–2031 (1967). (b) C. M. Won and A. V. Willi, Kinetic deuterium isotope effects in the reactions of methyl iodide with azide and acetate ions in aqueous solution, *J. Phys. Chem.* **76**, 427–432 (1972).
24. K. C. Westaway, An unusually large α-secondary deuterium kinetic isotope effect, *Tetrahedron Lett.* 4229–4232 (1975).
25. E. K. Thornton and E. R. Thornton, Chapter 1 in this volume.
26. E. M. Arnett, Quantitative comparisons of weak organic bases, *Prog. Phys. Org. Chem.* **1**, 223–403 (1963).

27. J. P. Fox and W. P. Jencks, General acid and general base catalysis of the methoxyaminolysis of 1-acetyl-1,2,4-triazole, *J. Am. Chem. Soc.* **96**, 1436–1449 (1974).

28. J. L. Kurz, Transition states as acids and bases, *Acc. Chem. Res.* **5**, 1–9 (1972).

29. R. L. Schowen, Mechanistic deductions from solvent isotope effects, *Prog. Phys. Org. Chem.* **9**, 275–332 (1972).

30. C. G. Swain, D. A. Kuhn, and R. L. Schowen, Effect of structural changes in reactants on the position of hydrogen-bonding hydrogens and solvating molecules in transition states. The mechanism of tetrahydrofuran formation from 4-chlorobutanol, *J. Am. Chem. Soc.* **87**, 1553–1561 (1965).

31. T. H. Cromartie and C. G. Swain, Chlorine kinetic isotope effects in the cyclization of chloroalcohols, *J. Am. Chem. Soc.* **97**, 232–233 (1975).

32. J. K. Coward, R. Lok, and O. Takagi, General base catalysis in nucleophilic attack at sp^3 carbon of methylase model compounds, *J. Am. Chem. Soc.* **98**, 1057–1059 (1976).

33. I. Mihel, J. O. Knipe, J. K. Coward, and R. L. Schowen, to be published.

34. M. Choi and E. R. Thornton, A kinetic study of the hydrolysis of substituted *N*-benzoyl-imidazoles and *N*-benzoyl-N^1-methylimidazolium Ions in light and heavy water. Hydrogen bridging without rate-determining proton transfer as a model for enzymic charge-relay, *J. Am. Chem. Soc.* **96**, 1428–1436 (1974).

35. E. K. Thornton and E. R. Thornton, Chapter 1, and G. M. Maggiora and R. E. Christoffersen, Chapter 3 in this volume.

36. P. Schuster, G. Zundel, and C. Sandorfy, eds., *The Hydrogen Bond, Recent Developments in Theory and Experiments*, North-Holland, Amsterdam (1976).

37. R. A. More O'Ferrall, The fractionation of hydrogen and deuterium isotopes in solutions of sodium methoxide, *Chem. Commun.* **1969**, 114–115.

38. M. M. Kreevoy, T.-M. Liang, and K.-C. Chang, The structures and isotopic fractionation factors of complexes, AHA–, *J. Am. Chem. Soc.* **99**, 5207–5209 (1977).

39. S. S. Minor and R. L. Schowen, One-proton solvation bridge in intramolecular carboxylate catalysis of ester hydrolysis, *J. Am. Chem. Soc.* **95**, 2279–2281 (1973).

40. N. Gravitz and W. P. Jencks, Mechanism of the hydrolysis of a phthalinidium cation. Direct observation and trapping of the tetrahedral intermediate and the effect of strong acid on rate and equilibrium constants of the reversible reaction, *J. Am. Chem. Soc.* **96**, 489–499 (1974).

41. N. Gravitz and W. P. Jencks, The mechanism of formation and breakdown of amine tetrahedral addition compounds of phthalinidium cation. The relative leaving-group abilities of amines and alkoxide ions, *J. Am. Chem. Soc.* **96**, 499–506 (1974).

42. N. Gravitz and W. P. Jencks, Mechanism of general acid–base catalysis of the breakdown and formation of tetrahedral addition compounds from alcohols and a phthalinidium cation. Dependence of Brønsted slopes on alcohol acidity, *J. Am. Chem. Soc.* **96**, 507–522 (1974).

43. W. W. Cleland, in: *The Enzymes*, 3rd ed. (P. D. Boyer, ed.), Vol. 2, pp. 1–65, Academic Press, New York (1970).

44. P. D. Boyer, ed., *The Enzymes*, Vol. 3, 3rd ed., Academic Press, New York (1971).

45. L. Mascaro, Jr., R. Hörhammer, S. Eisenstein, L. K. Sellers, K. Mascaro, and H. G. Floss, Synthesis of methionine carrying a chiral methyl group and its use in determining the steric course of the enzymatic C-methylation of indolepyruvate during indolmycin biosynthesis, *J. Am. Chem. Soc.* **99**, 273–274 (1977).

46. Structure and function of proteins at the three-dimensional level, *Cold Spring Harbor Symp. Quant. Biol.* **36**, (1972).

47. H. Fritz, H. Tschesche, L. J. Greene, and E. Truscheit, eds., *Proteinase Inhibitors, Proceedings of the 2nd International Research Conference*, Springer-Verlag New York, Inc., New York (1974).

48. E. Reich, D. B. Rifkin, and E. Shaw, eds., *Proteases and Biological Control*, Cold Spring Harbor Laboratory, Cold Spring Harbor, N.Y. (1975).

49. T. H. Fife, Physical organic model systems and the problem of enzymatic catalysis, *Adv. Phys. Org. Chem.* **11**, 1–122 (1975).

50. D. M. Blow, Structure and mechanism of chymotrypsin, *Acc. Chem. Res.* **9**, 145–152 (1976).

51. D. M. Blow, J. J. Birktoft, and B. S. Hartley, Role of a buried acid group in the mechanism of action of chymotrypsin, *Nature (London)* **221**, 337–340 (1969).

52. A. R. Fersht, D. M. Blow, and J. Fastrez, Leaving-group specificity in the chymotrypsin-catalyzed hydrolysis of peptides. A stereochemical interpretation, *Biochemistry* **12**, 2035–2041 (1973).

53. R. E. Williams and M. L. Bender, Substituent effects on the chymotrypsin-catalyzed hydrolysis of specific ester substrates, *Can. J. Biochem.* **49**, 210–217 (1971).

54. S. A. Bizzozero, W. K. Baumann, and H. Dutler, Kinetic investigation of the α-chymotrypsin-catalyzed hydrolysis of peptide-ester substrates. The relationship between the structure of the peptide moiety and reactivity, *Eur. J. Biochem.* **58**, 167–176 (1975).

55. F. E. Brot and M. L. Bender, Use of the specificity constant of α-chymotrypsin, *J. Am. Chem. Soc.* **91**, 7187–7191 (1969).

56. B. Zerner and M. L. Bender, Acyl-enzyme intermediates in the α-chymotrypsin catalyzed hydrolysis of "specific" substrates. The relative rates of hydrolysis of ethyl, methyl and *p*-nitrophenyl esters of *N*-acetyl-L-tryptophan, *J. Am. Chem. Soc.* **85**, 356–358 (1963).

57. H. F. Bundy and C. L. Moore, Chymotrypsin-catalyzed hydrolysis of *m*-, *p*- and *o*-nitroanilides of *N*-benzoyl-L-tyrosine, *Biochemistry* **5**, 808–811 (1966).

58. M. Philipp, R. M. Pollack, and M. L. Bender, Influences of leaving-group electronic effect on α-chymotrypsin: Catalytic constants of specific substrates, *Proc. Nat. Acad. Sci. USA* **70**, 517–520 (1973).

59. T. Inagami, S. S. York, and A. Patchornik, An electrophilic mechanism in the chymotrypsin-catalyzed hydrolysis of anilide substrates, *J. Am. Chem. Soc.* **87**, 126–127 (1965).

60. C.-A. Bauer, R. C. Thompson, and E. R. Blout, The active centers of *Strephomyces griseus* protease 3 and α-chymotrypsin: Enzyme–substrate interactions remote from the scissile bond, *Biochemistry* **15**, 1291–1295 (1976).

61. R. J. Foster and C. Niemann, Reevaluation of kinetic constants of previously investigated specific substrates of α-chymotrypsin, *J. Am. Chem. Soc.* **77**, 1886–1892 (1955).

62. G. Hein and C. Niemann, An interpretation of the kinetic behavior of model substrates of α-chymotrypsin, *Proc. Nat. Acad. Sci. USA* **47**, 1341–1355 (1961).

63. G. D. Giles and C. F. Wells, Acid–base equilibria involving oxygen-containing molecules in dilute aqueous solution, *Nature (London)* **201**, 606–607 (1969).

64. G. D. Fasman, ed., *Handbook of Biochemistry and Molecular Biology*, Vol. 1, 3rd ed., CRC Press, Cleveland (1976).

65. M. Eigen and G. G. Hammes, Elementary steps in enzyme reactions, *Adv. Enzymol. Relat. Subj. Biochem.* **25**, 1–38 (1963).

66. S. H. Smallcombe, B. Ault, and J. H. Richards, Magnetic resonance studies of protein-small molecule interactions. Dynamics of binding between *N*-acetyl-D-tryptophan and α-chymotrypsin, *J. Am. Chem. Soc.* **94**, 4585–4590 (1972).

67. M. Renard and A. R. Fersht, Anomalous pH dependence of k_{cat}/K_m in enzyme reactions. Rate constants for the associations of chymotrypsin with substrates, *Biochemistry* **12**, 4713–4718 (1973).

68. C. G. Mitton, R. L. Schowen, M. Gresser, and J. Shapley, Isotope exchange in the basic methanolysis of aryl esters. Molecular interpretation of free energies, enthalpies and entropies of activation, *J. Am. Chem. Soc.* **91**, 2036–2044 (1969).

69. L. D. Kershner and R. L. Schowen, Proton transfer and heavy-atom reorganization in amide hydrolysis. Valence–isomeric transition states, *J. Am. Chem. Soc.* **93**, 2014–2024 (1971).

70. R. L. Schowen, C. R. Hopper, and C. M. Bazikian, Substituent effects and solvent isotope effects in the basic methanolysis of amides, *J. Am. Chem. Soc.* **94**, 3095–3097 (1972).

71. C. R. Hopper, R. L. Schowen, K. S. Venkatasubban, and H. Jayaraman, Proton inventories of transition states for solvation catalysis and proton-transfer catalysis. Decomposition of the tetrahedral intermediate in amide methanolysis, *J. Am. Chem. Soc.* **95**, 3280–3283 (1973).

72. V. Gold and S. Grist, Deuterium solvent isotope effects on reactions involving the aqueous hydroxide ion, *J. Chem. Soc. Perkin Trans. 2* **1972**, 89–95.
73. M. Phillipp and M. L. Bender, Is binding the rate-limiting step in acylation of α-chymotrypsin by specific substrates?, *Nature (London) New Biol.* **241**, 44 (1973).
74. P. W. Inward and W. P. Jencks, The reactivity of nucleophilic reagents with furoyl-chymotrypsin, *J. Biol. Chem.* **240**, 1986–1996 (1965).
75. B. Zeeberg and M. Caplow, Transition-state charge distribution in reactions of acetyl-tyrosyl-chymotrypsin intermediate, *J. Biol. Chem.* **248**, 5887–5891 (1973).
76. C. B. Sawyer and J. F. Kirsch, Kinetic isotope effects for the chymotrypsin catalyzed hydrolysis of ethoxyl-^{18}O labeled specific ester substrates, *J. Am. Chem. Soc.* **97**, 1963–1964 (1975).
77. M. L. Bender, G. E. Clement, F. J. Kézdy, and H. d'A. Heck, The correlation of the pH(pD) dependence and the stepwise mechanism of α-chymotrypsin-catalyzed reactions, *J. Am. Chem. Soc.* **86**, 3680–3690 (1964).
78. M. H. O'Leary and M. D. Kluetz, Nitrogen isotope effects on the chymotrypsin-catalyzed hydrolysis of N-acetyl-L-tryptophanamide, *J. Am. Chem. Soc.* **94**, 3585–3589 (1972).
79. K. S. Venkatasubban, Proton bridges in enzymic and nonenzymic anide solvolysis, Ph.D. thesis, University of Kansas, Lawrence (1974).
80. J. P. Klinman, Kinetic isotope effects in enzymology, *Adv. Enzymol. Relat. Areas Mol. Biol.* in press.
81. C. B. Sawyer and J. F. Kirsch, Kinetic isotope effects for reactions of methyl formate-methoxyl-^{18}O, *J. Am. Chem. Soc.* **95**, 7375–7381 (1973).
82. E. R. Thornton, Solvent isotope effects in H_2O^{16} and H_2O^{18}, *J. Am. Chem. Soc.* **84**, 2474–2475 (1962).
83. R. A. More O'Ferrall, in: *Proton-Transfer Reactions* (E. Caldin and V. Gold, eds.), Chapman & Hall, London (1975).
84. W. H. Saunders, Jr., Distinguishing between concerted and nonconcerted eliminations, *Acc. Chem. Res.* **9**, 19–25 (1976).
85. S. Scheiner, D. A. Kleier, and W. N. Lipscomb, Molecular orbital studies of enzyme activity: I: Charge relay system and tetrahedral intermediate in acylation of serine proteases, *Proc. Nat. Acad. Sci. USA* **72**, 2606–2610 (1975).
86. R. D. Gandour, G. M. Maggiora, and R. L. Schowen, Coupling of proton motions in catalytic activated complexes. Model potential-energy surfaces for hydrogen-bond chains, *J. Am. Chem. Soc.* **96**, 6967–6979 (1974).
87. R. Huber, D. Kukla, W. Bode, P. Schwager, K. Bartels, J. Deisenhofer, and W. Steigemann, Structure of the complex formed by bovine trypsin and bovine pancreatic trypsin inhibitor. II. Crystallographic refinement at 1.9 Å resolution, *J. Mol. Biol.* **89**, 73–101 (1974).
88. H. G. Thode, R. L. Graham, and J. H. Ziegler, A mass spectrometer and the measurement of isotope exchange factors, *Can. J. Res. Sect. B* **23**, 40–47 (1945).
89. (a) A. R. Fersht and M. Renard, pH dependence of chymotrypsin catalysis. Appendix: Substrate binding to dimeric α-chymotrypsin studied by X-ray diffraction and the equilibrium method, *Biochemistry* **13**, 1416–1426 (1974). (b) A. R. Fersht and Y. Requena, Mechanism of the α-chymotrypsin-catalyzed hydrolysis of amides. pH dependence of k_c K_m. Kinetic detection of an intermediate, *J. Am. Chem. Soc.* **93**, 7079–7087 (1971).
90. D. H. Kenyon and G. Steinman, *Biochemical Predestination*, McGraw-Hill, New York (1969).
91. A. R. Fersht, Catalysis, binding and enzyme–substrate complementarity, *Proc. R. Soc. London Ser. B* **187**, 397–407 (1974).
92. H. C. Froede and I. B. Wilson, in: *The Enzymes*, 3rd ed. (P. D. Boyer, ed.), Vol. 5, pp. 87–114, Academic Press, New York (1971).
93. T. L. Rosenberry, Acetylcholinesterase, *Adv. Enzymol. Relat. Areas Mol. Biol.* **43**, 159–171 (1975).
94. T. L. Rosenberry, Catalysis by acetylcholinesterase: Evidence that the rate-limiting step with certain substrates precedes general acid–base catalysis, *Proc. Nat. Acad. Sci. USA* **72**, 3834–3838 (1975).

95. J. L. Hogg, Transition-state structures for catalysis by serine hydrolases and for related organic reactions, Ph.D. thesis, University of Kansas, Lawrence (1974).

96. J. P. Elrod, A comparative mechanistic study of a set of serine hydrolases using the proton inventory technique and beta-deuterium probe, Ph.D. thesis, University of Kansas, Lawrence (1975).

97. E. M. Arnett, T. Cohen, A. A. Bothner-By, R. D. Bushick, and G. Sowinski, A large beta-deuterium isotope effect in the protonation of acetophenone, *Chem. Ind. (London)* **1961**, 473–474.

98. D. M. Quinn, Approaches to transition-state structure for various enzyme-catalyzed acyl transfers, Ph.D. thesis, University of Kansas, Lawrence (1977).

99. E. A. Halevi and Z. Margolin, Temperature dependence of the secondary isotope effect on aqueous alkaline ester hydrolysis, *Proc. Chem. Soc. (London)* **1964**, 174.

100. G. M. Steinberg, M. L. Mednick, J. Maddox, R. Rice, and J. Cramer, A hydrophobic binding site in acetylcholinesterase, *J. Med. Chem.* **18**, 1056–1061 (1975).

101. N. Tanaka and E. R. Thornton, Isotope effects in hydrophobic binding measured by high-pressure liquid chromatography, *J. Am. Chem. Soc.* **98**, 1617–1619 (1976).

102. J. A. K. Harmony, R. H. Himes, and R. L. Schowen, The monovalent cation-induced association of formyltetrahydrofolate synthetase subunits: A solvent isotope effect, *Biochemistry* **14**, 5379–5386 (1975).

103. G. L. Cantoni, Biological methylation: Selected aspects, *Annu. Rev. Biochem.* **44**, 335–451 (1975).

104. G. D. Rock, J. H. Tong, and A. D. D'Iorio, A comparison of brain and liver catechol-*O*-methyltransferases, *Can. J. Biochem.* **48**, 1326–1331 (1970).

105. R. T. Borchardt and D. Thakker, Affinity labeling of catechol-*O*-methyltransferase with *N*-iodoacetyl-3,5-dimethoxy-4-hydroxyphenylethylamine, *Biochem. Biophys. Res. Commun.* **54**, 1233–1239 (1973).

106. R. T. Borchardt and D. Thakker, Catechol-*O*-methyltransferase. 6. Affinity labeling with *N*-haloacetyl-3,5-dimethoxy-4-hydroxyphenylalkylamines, *J. Med. Chem.* **18**, 152–158 (1975).

107. R. T. Borchardt and D. R. Thakker, Affinity labeling of catechol-*O*-methyltransferase by *N*-haloacetyl derivatives of 3,4-dimethoxy-4-hydroxyphenylethylamine. Kinetics of inactivation, *Biochemistry* **14**, 4543–4551 (1975).

108. Y. S. Wu, Synthesis and biological activity of analogs of *S*-adenosylmethionine and *S*-adenosylhomocysteine, Ph.D. thesis, University of Kansas, Lawrence (1975).

109. C. R. Creveling, N. Dalgard, H. Shimizu, and J. W. Daly, Catechol-*O*-methyltransferase. III. *m*- and *p*-*O*-methylation of catecholamines and their metabolites, *Mol. Pharmacol.* **6**, 691–696 (1970).

110. C. R. Creveling, N. Morris, H. Shimizu, H. H. Ong, and J. Daly, Catechol-*O*-methyltransferase. IV. Factors affecting *m*- and *p*-methylation of substituted catechols, *Mol. Pharmacol.* **8**, 398–409 (1972).

111. L. Flohe and K.-P. Schwabe, Kinetics of purified catechol-*O*-methyltransferase, *Biochim. Biophys. Acta* **220**, 469–476 (1970).

112. J. K. Coward, E. P. Slisz, and F. Y.-H. Wu, Kinetic studies on catechol-*O*-methyltransferase. Product inhibition and the nature of the catechol binding site, *Biochemistry* **12**, 2291–2296 (1973).

113. R. T. Borchardt, Catechol-*O*-methyltransferase. 1. Kinetics of tropolone inhibition, *J. Med. Chem.* **16**, 377–382 (1973).

114. R. T. Borchardt, Catechol-*O*-methyltransferase. 2. *In vitro* inhibition by substituted 8-hydroxyquinolines, *J. Med. Chem.* **16**, 382–387 (1973).

115. R. T. Borchardt, Catechol-*O*-methyltransferase. 4. *In vitro* inhibition by 3-hydroxy-4-pyrones, 3-hydroxy-2-pyindones and 3-hydroxy-4-pyridones, *J. Med. Chem.* **16**, 581–583 (1973).

116. S. R. Hartshorn and V. J. Shiner, Jr., Calculation of H/D, $^{12}C/^{13}C$ and $^{12}C/^{14}C$ fractionation factors from valence force fields derived for a series of simple organic molecules, *J. Am. Chem. Soc.* **94**, 9002–9012 (1972).

117. L. B. Sims, A. Fry, L. T. Netherton, J. C. Wilson, K. D. Reppond, and S. W. Crook, Variations of heavy-atom kinetic isotope effects in S_N2 displacement reactions, *J. Am. Chem. Soc.* **94**, 1364–1365 (1972).

118. V. J. Shiner, Jr., in: *Isotope Effects in Chemical Reactions* (C. J. Collins and N. S. Bowman, eds.), pp. 90–159, Van Nostrand Reinhold, New York (1970).

119. M. J. Stern and M. Wolfsberg, Simplified procedure for the theoretical calculation of isotope effects involving large molecules, *J. Chem. Phys.* **15**, 4105–4124 (1966).

120. M. F. Hegazi, Transition state structures for: I. Catechol-*O*-methyltransferase trans-methylation. II. Water and acetate catalyzed hydrolyses of ethyl *o*-nitrophenyl oxalate, Ph.D. thesis, University of Kansas, Lawrence (1976).

121. M. F. Hegazi, R. T. Borchardt, and R. L. Showen, S_N2-Like transition state for methyl transfer catalyzed by catechol-*O*-methyltransferase, *J. Am. Chem. Soc.* **98**, 3048–3049 (1976).

122. M. F. Hegazi, R. T. Borchardt, S. Osaki, and R. L. Schowen, in: *New or Improved Syntheses, Methods or Techniques in Nucleic Acid Chemistry* (L. B. Townsend, ed.), Wiley, New York (1976).

Transition States for Hydrolysis of Acetals, Ketals, Glycosides, and Glycosylamines

E. H. Cordes and H. G. Bull

1. INTRODUCTION

The point of this review is to define, within limits of current understanding, transition-state structures for enzymic hydrolysis of glycosyl compounds. Transition-state structures for corresponding nonenzymic reactions will be developed to the extent that such information is useful in consideration of the enzymic reactions.

We shall consider the mechanism of action of two subclasses of enzymes which catalyze hydrolysis of glycosides: (1) the glycoside hydrolases (EC 3.2.1) and (2) the nucleosidases (EC 3.2.2). The former group of enzymes includes amylases, cellulases, hyaluronidases, neuraminidases, chitinases, and debranching enzymes, among others; the latter group includes enzymes which catalyze the hydrolysis of both purine and pyrimidine nucleosides as well as that of nicotinamide adenine dinucleotide (NAD). The total number of enzymes which have been investigated that fall into these categories is substantial. No attempt is made here to be comprehensive. Quite the contrary, the number of glycoside-hydrolyzing enzymes which have been subjected to close mechanistic scrutiny is small, and we shall focus attention on them. Hopefully, mechanisms defined for these enzymes may apply in part also to those for enzymes less well studied at this time. Moreover, we shall not attempt to discuss any single enzyme comprehensively: Those lines of evidence which pertain directly to the definition of transition-state structure will be developed; physical

E. H. Cordes and H. G. Bull • Department of Chemistry, Indiana University, Bloomington, Indiana.

properties, biological function, comparative enzymology, and related matters will be considered only insofar as is necessary or helpful to the understanding of mechanism.

There are relatively few experimental approaches to transition-state structure for enzyme-catalyzed reactions:

1. Study of steady-state kinetics and inhibition patterns for enzymic reactions sometimes makes possible the assignment of rate-determining step for the overall process, a crucial piece of information for transition-state-structure definition. Moreover, knowledge of rate-determining step is usually required for proper interpretation of isotope and substituent effects, as is developed below. Analysis of initial velocity and inhibition patterns to understand pathway and mechanism has been reviewed in detail.[1-3]

2. Analysis of transient-state enzyme kinetic data may yield information pertinent to existence of intermediates and the nature of rate-determining step. With exception of important work on hen egg white lysozyme, rather little transient-state kinetic work has been done with the enzymes under consideration here.

3. Measurement of isotope effects for enzymic reactions is a productive approach to transition-state structure. Useful information has been obtained from measurements of both primary, including solvent and heavy-atom, and secondary isotope effects. The chapters in this volume by J. P. Klinman, J. L. Hogg, K. B. J. Schowen, and M. H. O'Leary treat isotope effects in enzyme mechanism studies in detail, and we shall use this information as background for our considerations.

4. The sensitivity of rates of enzymic reactions to the nature of polar substituents in the substrate may offer important information concerning transition-state structure. This topic has been reviewed in detail recently by Kirsch.[4]

5. The detailed knowledge of enzyme structure resulting from application of X-ray diffraction techniques has provided a great deal of significant information concerning transition-state structures. A particularly beautiful example is provided by the work of Huber and associates on the structure of the trypsin inhibitor–trypsin complex.[5] Among enzymes considered in this review, only lysozyme has been subjected to detailed crystallographic analysis.

6. The concept of transition-state analogs obviously bears directly on the question of transition-state structure. If the philosophy of these measurements—that the enzyme binds the transition state more tightly than it binds the substrates or products—is correct, careful design, construction, and testing of transition-state analogs should reveal useful information concerning geometry in the transition state.

7. Finally, much of what one knows about transition-state structures for nonenzymic reactions may be directly applicable to corresponding enzymic reactions. We begin our considerations at this point.

2. NONENZYMIC HYDROLYSIS OF ACETALS, KETALS, AND ORTHOESTERS

Mechanism and catalysis for nonenzymic hydrolysis of acetals and related substrates has been reviewed in detail in 1967[6] and 1974.[7] In addition, a briefer account is available,[8] as is a review directed specifically to understanding acetal hydrolysis as it relates to the mechanism of action of lysozyme.[9] Consequently we shall simply summarize the principal findings that pertain to identification of transition-state structure for these reactions.

2.1. Reaction Pathway

Hydrolysis of acetals, ketals, orthoesters, and related substrates is described by

$$
\begin{array}{c}
R_1 \quad\!\! O{-}R \\
\diagdown C \diagup \\
\diagup \quad \diagdown \\
R_2 \quad\!\! O{-}R
\end{array}
\;+\; H_2O \;\rightleftharpoons\;
\begin{array}{c}
R_1 \\
\diagdown \\
\quad C{=}O + 2\,ROH \\
\diagup \\
R_2
\end{array}
\tag{1}
$$

The complete reaction requires the rupture of two carbon–oxygen bonds, the addition of a molecule of water to the substrate, and several proton-transfer reactions. It follows that the hydrolytic reaction must involve formation of a number of intermediates along the reaction pathway. This simple realization immediately raises several questions concerning reaction mechanisms, such as the following: What is the structure of the intermediates involved? Which step of the reaction sequence is rate determining? Superimposed on each of the above questions are those concerned with catalysis. It is well established that hydrolysis of most acetals, ketals, and orthoesters occurs with specific acid catalysis and that hydrolysis of certain of these substrates occurs with general acid catalysis as well. What is the mechanism of such catalysis? What is the role of proton-transfer reactions in determining overall reaction rates? How do these findings relate to the observation of enzymic catalysis for substrates of related structure?

Certain aspects of the hydrolysis of acetals, ketals, and orthoesters appear so well established as to not require development here. Important among them is the site of bond cleavage for these reactions. In principle, carbon–oxygen bond cleavage might occur so as to yield either a carboxonium-ion or alkyl carbonium-ion intermediate:

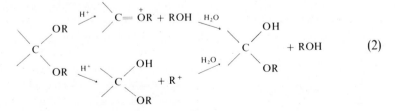

$$\tag{2}$$

On the basis of the relative stabilities of the two intermediates, one might expect the reaction to occur with cleavage of that bond leading to formation of the carboxonium ion. This expectation has been fully borne out. The earliest convincing evidence for this point of view is the important work of Lucas and his associates on the hydrolysis of acetals derived from optically active alcohols. For example, hydrolysis of the D-(+)-2-octanol acetal of acetaldehyde in dilute aqueous phosphoric acid yields 2-octanol having the same optical rotation as the original alcohol from which the acetal was synthesized.[10,11] This finding excludes formation of the alkyl carbonium ion, in which case substantial or complete racemization of the alcohol would be expected, and an A2 reaction involving nucleophilic attack of solvent on the alcohol, in which case optical inversion of the alcohol would be expected. Similarly, the formal, acetal, and carbonate derived from D-(−)-2,3-butanediol and the acetal derived from D-(+)-2-butanol undergo acid-catalyzed hydrolysis with complete retention of configuration at the carbinol carbon of the alcohol.[12]

Bourns *et al.* corroborated the above conclusion in an isotope tracer study of acetal formation and hydrolysis.[13] The condensation of benzaldehyde and *n*-butyraldehyde, enriched in ^{18}O, with *n*-butyl and allyl alcohols yielded acetals of normal isotopic abundance and ^{18}O-enriched water:

$$R-C{\overset{^{18}O}{\underset{H}{}}} + 2\,R'OH \xrightleftharpoons{H^+} R-C{\overset{OR'}{\underset{OR'}{}}}-H + H_2{}^{18}O \qquad (3)$$

In a like fashion, hydrolysis of benzaldehyde di-*n*-butyl acetal and *n*-butyraldehyde di-*n*-butyl acetal in ^{18}O-enriched water yielded alcohols of normal isotopic content [the reverse of Eq. (3)]. Thus, these reactions clearly proceed with carbonyl carbon–oxygen bond cleavage (or formation).

These results should not be interpreted to indicate that cleavage of the alcohol carbon–oxygen bond, by either a unimolecular or bimolecular process, never occurs in these reactions. But they do serve to indicate that such a reaction pathway will be restricted to substrates of unusual structure or will contribute in only a minor way to the overall hydrolytic process for substrates of ordinary structure.

A second matter which seems to be generally agreed upon is the non-involvement of water as a nucleophilic reagent in the transition state for hydrolysis of acetals, ketals, and orthoesters. This point has been explicitly assumed in the formulation of Eq. (2), which indicates carbonium-ion pathways only. Several lines of evidence, including volumes of activation, entropies of activation, structure-reactivity correlations, solvent isotope effects, and the like, suffice to indicate the involvement of carbonium ions in the transition state for hydrolysis of most of the substrates in this class.[6] Again, this is not to suggest that A2 reactions never occur for the hydrolysis of these substrates; in fact, we shall devote some effort to those cases in which such a route may be important. Moreover, we shall have the occasion to review much of the information which has led to the conclusion that water is not involved as a nucleophile in the transition state for these reactions. But our discussion will be in the context of the assumption that this is the case. For extensive documentation of the point, the reader may refer to an earlier review.[6]

2.2. Rate-Determining Step

The considerations just cited permit us to formulate the hydrolysis of acetals, ketals, and orthoesters according to the multistep pathway

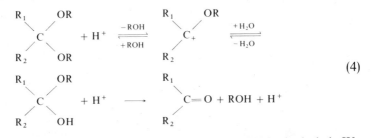

$$(4)$$

in which proton-transfer reactions have not been explicitly included. We shall consider the importance of proton-transfer reactions for hydrolysis of these substrates below. At the moment, let us consider which of the steps of Eq. (4) is the rate-determining step. In solutions containing little alcohol, rate-determining reaction of the carbonium ion with solvent is extremely unlikely since this requires that alcohol react with the carbonium ion, regenerating starting material, more rapidly than water reacts with the carbonium ion, yielding products. Since the rate constants for reaction of alcohol and water are almost certainly about the same, the rate for the latter reaction must be greater than that for the former. A similar argument suggests that the tetrahedral-intermediate decomposition, the final step, is not rate determining. Since the overall equilibrium constant for interconversion of substrate and tetrahedral intermediate should be about unity, the latter would be present in much greater concentration than the former were equilibrium established, owing to the high concentration of water relative to alcohol. Since the rate constants for decomposition of these species should be about equal, the rate for intermediate decomposition should be greater than the corresponding quantity for the substrate. These conclusions are fully corroborated in kinetic studies of ketal and orthoester hydrolysis conducted in the presence of deuterated alcohols.

A study of the kinetics and product composition for the hydrolysis of methyl ketals and methyl orthoesters in methanol-d_4-deuterium oxide mixtures,

$$\begin{array}{c}\diagup\\ \diagdown\end{array}C\begin{array}{c}\diagup OCH_3\\ \diagdown OCD_3\end{array} + D^+ \rightleftharpoons \begin{array}{c}\diagup\\ \diagdown\end{array}C^+{-}OCD_3 + CH_3OD \qquad (5)$$

$$\begin{array}{c}\diagup\\ \diagdown\end{array}C\begin{array}{c}\diagup OCD_3\\ \diagdown OCD_3\end{array} + D^+ \rightleftharpoons \begin{array}{c}\diagup\\ \diagdown\end{array}C^+{-}OCD_3 + CD_3OD$$

$$\begin{array}{c}\diagup\\ \diagdown\end{array}C^+{-}OCH_3 + D_2O \longrightarrow \begin{array}{c}\diagup\\ \diagdown\end{array}C{=}O + CH_3OD + D^+$$

$$\begin{array}{c}\diagup\\ \diagdown\end{array}C^+{-}OCD_3 + D_2O \longrightarrow \begin{array}{c}\diagup\\ \diagdown\end{array}C{=}O + CD_3OD + D^+$$

employing proton magnetic-resonance spectroscopy as an analytical tool, has provided a simple and straightforward experimental distinction between several of the possible rate-determining steps for these reactions.[14] The experimental quantities determined by this method in the study of, for example, methyl orthobenzoate hydrolysis, include the first-order rate constants for the disappearance of the methoxy protons of the orthoester, $k_{orthoester}$; for the appearance of the methyl protons of methanol, k_{MeOH}; and for the appearance of the methoxy protons of the carboxylic ester, k_{ester}. Since the proton resonance singlets for each of these groups are well separated, the rate constants can be determined simultaneously. In addition, the ratio of integrated proton intensities at infinite time of the products to some internal, time-independent standard provides a quantitative measure of product composition for many substrates.

Both the product composition and the relative magnitudes of the various rate constants are functions of the nature of the rate-determining step. If carbonium-ion formation were rapid and reversible (i.e., carbonium-ion formation not rate determining), the methoxy groups of the starting material would be rapidly exchanged for deuteriomethoxy groups through reaction of the carbonium ion with solvent deuteriomethanol. The orthoester would be converted more slowly to carboxylic ester product. Thus, $k_{orthoester}$ and k_{MeOH} would be considerably larger than k_{ester}. Furthermore, little or no carboxylic ester product containing methoxy protons would be formed since virtually all of the orthoester would have been converted into the corresponding deuterated material in the preequilibrium exchange reactions. In contrast, if carbonium-ion formation were rate determining, methanol would not be exchanged for deuteriomethanol in a preequilibrium reaction; hence, $k_{orthoester}$, k_{MeOH}, and k_{ester} would be nearly identical. In addition, only protio carboxylic ester would be produced as reaction product.

Studies of this type have been performed employing 2,2-dimethoxypropane, 6,6,6-trimethoxyhexanonitrile, methyl orthobenzoate, and methyl orthocarbonate as substrates.[14] The course of the hydrolysis of methyl orthobenzoate as a function of time is indicated in Fig. 1. As may be judged qualitatively from this figure, the rate of disappearance of the methoxy protons of the orthoester is comparable to the rate of appearance of the corresponding protons of methyl benzoate. Furthermore, a rather substantial amount of methyl benzoate, as opposed to deuteriomethyl benzoate, is formed in the

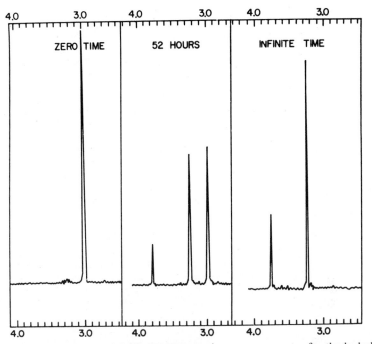

Fig. 1. Initial, intermediate, and final proton magnetic-resonance spectra for the hydrolysis of methyl orthobenzoate in an equimolar mixture of deuterium oxide and methanol-d_4. Methoxy protons of orthoester appear at 3.0 ppm, those of methyl benzoate at 3.8 ppm, and those of methanol at 3.25 ppm. (Reproduced from Reference 14 with permission of the copyright owner, the American Chemical Society.)

reaction as judged from the intensity of the appropriate signal in the infinite-time spectrum. These results are just those predicted on the basis of rate-determining carbonium-ion formation.

To this point, we have established that the transition state for hydrolysis of acetals, ketals, and orthoesters involves carbon–oxygen bond cleavage, which may be accompanied by proton transfer, leading to formation of a carbonium ion derived from the carbonyl moiety of the substrate. In an effort to define the transition-state structure more precisely, let us review several aspects of these reactions including structure-reactivity relationships, secondary deuterium isotope effects, and solvent deuterium isotope effects. These considerations will lay the foundation for a consideration of catalytic mechanisms subsequently.

2.3. Structure-Reactivity Relationships

A great deal of information has been collected relating to structure-reactivity correlations for hydrolysis of acetals, ketals, and orthoesters. Representative data are summarized in Table I. Salient features include the following.

Table I

Linear Free-Energy Relationships for Hydrolysis of Acetals, Glycosides,
and Related Substrates[a]

Substrates	Catalysts	ρ	Ref.
	Acid	−1.16	b
	Acid	−3.3	c
	Acid	−2.0[d]	e
	Acid Base	−0.006 2.8	f f
	Acid Base β-Glucosidase	−0.66 2.5 1.0	g g h
	Acid	−0.11	i
β-(substituted phenyl)-di-N- acetylchitobiosides	Lysozyme	1.23	j

[a] For an extensive collection of data, see E. H. Cordes and H. G. Bull, *Chem. Rev.*, **74**, 581 (1974).
[b] H. G. Bull, K. Koehler, T. C. Pletcher, J. J. Ortiz, and E. H. Cordes, *J. Am. Chem. Soc.* **93**, 3002 (1971).
[c] R. B. Dunlap, G. A. Ghanim, and E. H. Cordes, *J. Phys. Chem.* **73**, 1898 (1969).
[d] The linear free-energy relationship is $\log(k/k_0) = \rho[\sigma + 0.5(\sigma^+ - \sigma)]$.
[e] T. H. Fife and L. K. Jao, *J. Org. Chem.* **30**, 1492 (1965).
[f] A. N. Hall, S. Hollingshead, and H. N. Rydon, *J. Chem. Soc.* **1961**, 4290.
[g] A. N. Hall, S. Hollingshead, and H. N. Rydon, *Biochem. J.* **57**, 1 (1954).
[h] R. L. Nath and H. N. Rydon, *Biochem. J.* **57**, 1 (1954).
[i] D. Piszkiewicz and T. C. Bruice, *J. Am. Chem. Soc.* **90**, 2156 (1968).
[j] G. Lowe, G. Sheppard, M. L. Sinnott, and A. Williams, *Biochem. J.* **104**, 893 (1967).

First, second-order rate constants for specific-acid-catalyzed hydrolysis of these substrates are quite sensitive to the nature of polar substituents in the carbonyl moiety of the substrate; without exception, rate constants increase markedly with increasing electron donation from the polar substituent.

This is, of course, precisely the expected result on the basis of a transition state which has partial carbonium-ion character: The developing carbonium ion will be stabilized relative to the substrate by those substitutents which donate electrons. These data argue strongly against reaction mechanisms involving nucleophilic attack by water in the transition state and those involving rate-determining substrate protonation, for which much smaller effects of polar substituents would be expected.

Hydrolysis of orthobenzoates in water is less sensitive to the nature of polar substituents than is the hydrolysis of acetals of benzaldehyde in the same solvent. This observation provides mild support for the idea that the transition state for acetal hydrolysis may have more carbonium-ion character than that for orthoester hydrolysis, a point to which we shall return later.

Second, for hydrolysis of glycosides and acetals, rate constants are not markedly sensitive to the nature of polar substituents in the leaving group. In most cases, small negative values of ρ are obtained. This is reasonably interpreted to reflect the opposing effects of polar substituents on substrate protonation and carbon–oxygen bond cleavage. Electron donation from a polar substituent, for example, would tend to increase the extent of protonation but retard the rate of decomposition of the protonated substrate. The observation of negative values of ρ suggests that the former effect is somewhat the more important of the two. Alternatively, the effects of polar substituents can be viewed simply as stabilization of the departing leaving group, which bears a partial positive charge in the transition state, relative to the uncharged ground state by electron donation.

Third, linear free-energy relationships for general-acid-catalyzed hydrolysis of acetals, ketals, and orthoesters have been investigated in a few cases. Of considerable interest is the observation that general-acid-catalyzed hydrolysis of 2-(4-substituted phenoxy)tetrahydropyrans and benzaldehyde methyl 3-substituted phenyl acetals is characterized by positive values of ρ even though the specific-acid-catalyzed hydrolysis of the former compounds at least has a negative value of ρ. This result has been interpreted to suggest that carbon–oxygen bond cleavage is particularly important relative to substrate protonation for the general-acid-catalyzed reaction.[15,16]

The structure-reactivity correlations just described indicate that the individual reactions may differ in two important respects: timing of proton transfer relative to cleavage of carbon–oxygen bonds (specific acid vs. general acid catalysis) and extent of cleavage of carbon–oxygen bonds in the transition state.

2.4. Secondary Deuterium Isotope Effects

α-Deuterium isotope effects for the hydrolysis of acetals and orthoformates have been probed.[17] Isotope effects of this type have been developed into relatively reliable indicators of transition-state structure for a variety of reactions. For reactions in which hybridization at carbon changes from sp^3

to sp^2,

$$(6)$$

k_H/k_D is a measure of the net change between the initial and transition state of the bending force constants for CHX. Since this bending mode is lost to a variable degree depending on hybridization at carbon in the transition state, isotope effects will vary from unity to a maximal value, near 1.23 when $X = \text{}^-OR$,[18,19] with increasing cleavage of the C—X bond in the transition state. Thus, for those reactions in which the transition state is substrate-like, values of k_H/k_D near unity will be obtained, and for those having carbonium-ion-like transition states, values near 1.23 will be observed. Interpolation between these extremes will provide information concerning transition-state structure for intermediate cases.

α-Deuterium isotope effects for the hydrolysis of propionaldehyde diethyl acetal, three substituted benzaldehyde diethyl acetals, and ethyl orthoformate are collected in Table II.[17]

The secondary deuterium isotope effect for the hydrolysis of substituted benzaldehyde diethyl acetals increases with increasing electron-withdrawing power of the polar substituent. These data indicate that the extent of C—O bond cleavage in the transition state increases markedly as the stability of the intermediate carbonium-ion diminishes. Thus, the transition state for hydrolysis of the p-methoxy derivative has little carbonium-ion character, while that for hydrolysis of the p-nitro derivative has a good deal of such charac-ter. The change in transition-state structure as a function of substrate reactivity for hydrolysis of the benzaldehyde acetals is perhaps the largest yet observed among carbonium-ion processes; the tendency of the transition state to in-creasingly resemble the product carbonium ion as substrate reactivity decreases (corresponding to a decrease in carbonium-ion stability) is in qualitative agreement with the predictions of Leffler,[20] Hammond,[21] and Thornton.[22]

Table II
Secondary Deuterium Isotope Effects for the Rates
of Hydrolysis of Benzaldehyde Diethyl Acetals,
Propionaldehyde Diethyl Acetal, and Ethyl
Orthoformate[a]

Substrate	$k_H/k_{\alpha D}$
p-Nitrobenzaldehyde diethyl acetal	1.15 ± 0.01
Benzaldehyde diethyl acetal	1.09 ± 0.01
p-Methoxybenzaldehyde diethyl acetal	1.04 ± 0.01
Propionaldehyde diethyl acetal	1.17 ± 0.01
Ethyl orthoformate	1.05 ± 0.01

[a] Reference 17.

Comparison of the isotope effects for the hydrolysis of ethyl orthoformate and propionaldehyde diethyl acetal with those for hydrolysis of the benzaldehyde acetals reveals, in light of the above discussion, that the transition state for the former substrate has little and that that for the latter substrate has much carbonium-ion character. These conclusions are in accord with our expectations in light of the transition-state-structure–carbonium-ion stability correlation developed above.

Generalizing, it seems fair to conclude that hydrolysis of orthocarbonates and orthoesters is characterized by reactant-like transition states, while hydrolysis of acetals and ketals derived from aliphatic substrates is characterized by carbonium-ion-like transition states. Acetals and ketals derived from aromatic substrates may be expected, as is the case directly studied here, to occupy either of the above categories, or an intermediate one, depending on the nature of the polar substituents (one may note that, for example, *p*-methoxybenzaldehyde diethyl acetal is a "phenyligous" orthoformate). Note that these considerations are fully consistent with conclusions based on structure-reactivity correlations. Specifically, hydrolysis of acetals and ketals is generally more sensitive to the nature of polar substituents than is that for orthoesters (Table I). Thus, the extent of C—O bond cleavage deduced on the

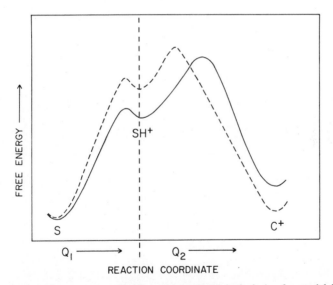

Fig. 2. Diagrammatic free-energy profiles for acid-catalyzed hydrolysis of a model ketal (solid line) and orthoester (dashed line). The reaction coordinate is broken into two parts: Q_1 refers to formation of the protonated substrate, SH^+, and Q_2 refers to cleavage of the carbon–oxygen bond of SH^+ to yield a carbonium ion, C^+. Note that the orthoester is less basic than the ketal and is less reactive than the ketal. The degree of covalent bond cleavage in the transition state is greater for decomposition of the conjugate acid of the ketal than for the orthoester. (Reproduced from Reference 7 with permission of the copyright owner, the American Chemical Society.)

basis of isotope effects parallels the susceptibility of these reactions to effects of polar substituents.

These considerations are summarized graphically in Fig. 2.

2.5. General Acid Catalysis

General acid catalysis for orthoester hydrolysis was first detected by Brønsted and Wynne-Jones in 1929.[23] Since that time, such catalysis has been searched for with a wide variety of acetals, ketals, and orthoesters and has been observed in several such cases. Work in this area has been reviewed in some detail.[8]

Attention here is focused on ascertaining those factors which cause hydrolysis of these substrates to become subject to general acid catalysis and to defining the mechanism of this catalysis. These are matters of great general interest in physical organic chemistry. Jencks has formulated a rule (the "libido rule") which attempts to identify those cases in which proton transfer is concerted with making or breaking of covalent bonds[24]:

> Concerted general acid–base catalysis of complex reactions in aqueous solution can occur only (a) at sites that undergo a large change in pK in the course of the reaction, and (b) when this change in pK converts an unfavorable to a favorable proton transfer with respect to the catalyst; i.e., the pK of the catalyst is intermediate between the initial and final pK values of the substrate site.

Hydrolysis of the compounds under consideration here certainly meets these criteria. For example, general acid catalysis of orthoester hydrolysis involves proton transfer from the hydrated proton, p$K_a = -1.74$, to a site having an initial pK near -8 and a final one near 17. The rule suggests that we might expect the probability of observing general acid catalysis to increase with decreasing substrate basicity, which has the effect of increasing the difference between the basicities of the site at which catalysis occurs between the initial and final states. Moreover, we might anticipate that general acid catalysis would be increasingly easy to observe with increasing stability of the carboxonium-ion intermediate. Both of these expectations are in fact borne out.

On the basis of a large number of studies, reviewed in detail in Reference 7, one can conclude the following regarding general acid catalysis for acetals and related substrates.

First, hydrolysis of orthoesters derived from aliphatic alcohols is subject to general acid catalysis; the corresponding Brønsted α values are large, 0.8–1.0. Second, hydrolysis of diethyl phenyl orthoformate, an orthoester with a phenolic leaving group, is subject to general acid catalysis with a Brønsted exponent near 0.5.[25] Third, hydrolysis of acetals and ketals derived from ordinary carbonyl compounds and aliphatic alcohols is not subject to general acid catalysis. Fourth, hydrolysis of acetals and ketals which have a phenolic or other very weakly basic leaving group or which form particularly stable carbonium ions is subject to general acid catalysis.[8] Values of the Brønsted exponent vary from 0.75 to 0.50. Fifth, hydrolysis of benzaldehyde *t*-butyl acetals is subject to general acid catalysis.[26]

In attempting to rationalize these observations, the reaction-coordinate contour diagrams first employed by More O'Ferrall and applied by Jencks to related reactions[24] will prove useful. In Fig. 3, a two-dimensional representation of the three-dimensional reaction surface for specific-acid-catalyzed hydrolysis of an acetal or related substrate is provided. The progress of proton transfer is shown along the vertical axis and progress of C—O bond cleavage along the horizontal axis. As this contour diagram is constructed, the preferred pathway involves protonation of the substrate followed by cleavage of the C—O bond, i.e., the classical A1 mechanism. Note that a variety of additional mechanisms for acetal hydrolysis can be visualized through making simple changes in this diagram. For example, if we suppose that the substrate is less basic, making formation of the conjugate acid less favorable, and the carbonium ion more stable, we could arrive at a mechanism in which C—O bond cleavage occurs prior to protonation. In Fig. 3, this would involve moving first along the horizontal axis and then vertically, rather than vice versa. Kinetically, this would correspond to a pH-independent hydrolysis reaction. Note that several such cases have been identified[7]; these all involve either very weakly basic substrates or formation of particularly favorable carbonium ions or both, as we anticipated.

One can imagine an intermediate case between the two extremes thus far

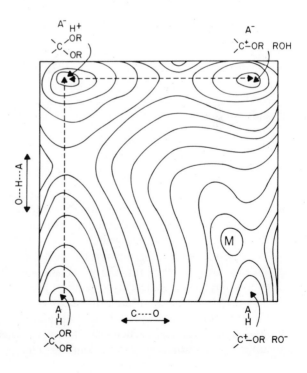

Fig. 3. Contour diagram illustrating the pathway for specific-acid-catalyzed hydrolysis of an acetal or related substrate. *M* denotes an energy maximum. (Reproduced from Reference 7 with permission of the copyright owner, the American Chemical Society.)

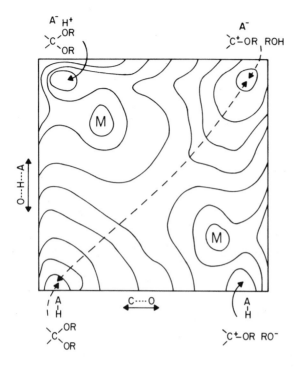

Fig. 4. Contour diagram for general-acid-catalyzed hydrolysis of an acetal or related substrate involving concerted substrate protonation and carbon–oxygen bond cleavage. M denotes an energy maximum. (Reproduced from Reference 7 with permission of the copyright owner, the American Chemical Society.)

considered; it is shown diagrammatically in Fig. 4. As the substrate becomes less basic and the carboxonium ion more stable, one forms a lowest-energy pathway in which proton transfer and covalent bond cleavage are concerted. Again, these are precisely the conditions under which general acid catalysis is observed for acetal, ketal, and orthoester hydrolysis. In most cases, values of the Brønsted exponent for general acid catalysis of these reactions are large, strongly suggesting that the proton is largely transferred from catalyst to substrate in the transition state.

2.6. Intramolecular Facilitation of Acetal Hydrolysis

The complete tertiary structure of hen egg white lysozyme reveals that two carboxyl groups, Asp-52 and Glu-35, are located in the immediate vicinity of the glycosidic C—O bond cleaved in the course of the catalytic process (see below). This observation has generated considerable speculation concerning the possible role of these groups in the enzymic reaction. The mechanism of lysozyme-catalyzed reactions and its relationship to mechanisms for acetal hydrolysis have been reviewed in detail.[9] Since bond-changing reactions in enzymic processes occur within a substrate–enzyme complex, these considerations have focused attention on intramolecular facilitation of the hydrolysis of glycosides and acetals, particularly those in which one or more carboxyl functions are positioned so as to permit participation in the reaction.

SCHEME II

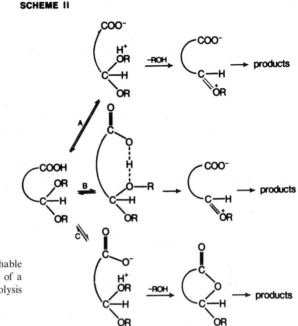

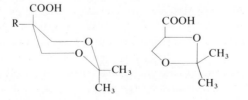

Fig. 5. Kinetically indistinguishable mechanisms for participation of a carboxyl function in the hydrolysis of acetals.

At the outset, we may note that pH-independent hydrolysis of an acetal or related compound bearing a carboxyl group may reflect three kinetically indistinguishable mechanisms: specific-acid-catalyzed hydrolysis of the carboxylate form of the substrate (which may or may not involve facilitation through electrostatic stabilization of the developing carbonium ion), intramolecular general acid catalysis by the carboxylic acid group, and nucleophilic attack by the carboxylate on the protonated substrate. These possibilities are depicted in Fig. 5. Since kinetics will not serve to distinguish these possibilities, alternative methods must be employed for those cases in which evidence for intramolecular participation is observed.

In 1967, Bruice and Piszkiewicz carefully reviewed previous claims for intramolecular participation in glycoside and acetal hydrolysis and found little firm evidence establishing such participation.[27] Moreover, these workers carefully examined the kinetics of hydrolysis of 1,3-dioxanes and 1,3-dioxolanes substituted with carboxyl functions (1 and 2) and found no evidence for intramolecular facilitation in either case.[27] A subsequent investigation by the same workers failed to reveal evidence for such facilitation for glycoside hydrolysis.[28]

Subsequently, Capon and co-workers did observe abnormally rapid hydrolysis for 2-carboxyphenyl-β-D-glucoside and O-carboxyphenyl methyl acetals, indicating some form of participation.[29] On the basis of the solvent isotope effect, these workers suggested that facilitation involved general acid catalysis by the carboxylic acid function (pathway B, Fig. 5).

Dunn and Bruice, in an extensive series of investigations, have confirmed the intramolecular participation of the carboxyl group in hydrolysis of o-carboxyphenyl alkyl acetals.[30] However, these workers attribute the facilitated hydrolysis to electrostatic stabilization by the carboxylate ion (pathway A, Fig. 5) rather than to general acid catalysis. Rate increases of 350- to 1000-fold have been observed.

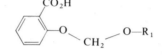

An alternative approach to study of intramolecular carboxyl group participation in acetal hydrolysis has been taken by Fife and Anderson, who have elected to search for such participation in substrates for which inter-molecular general acid catalysis has been observed for the unsubstituted compounds. Hydrolysis of 2-(o-carboxyphenoxy)tetrahydropyran and ben-zaldehyde methyl (o-carboxyphenyl)acetal proceeds with pronounced carboxyl

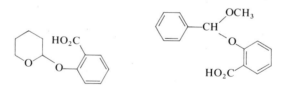

group participation.[31] Rate increases over the corresponding unsubstituted compounds in the range of 10^4–10^6, depending on substrate structure and solvent, are observed. In contrast to cases previously discussed, it is reasonable to attribute these rate increases to intramolecular general acid catalysis since the unsubstituted compounds exhibit intermolecular general acid catalysis. More-over, the magnitude of calculated rate constants and the magnitude of the observed facilitation appear too large to be accounted for in terms of electro-static facilitation. Note that a concentration of at least 580 M formic acid would be required to achieve a rate of hydrolysis of unsubstituted tetrahydropyran phenol acetal equal to that observed for the o-carboxyl-substituted one.[31] Clearly, intramolecular general acid catalysis must involve some reaction facilitation that does not derive solely from approximation of substrate with acid catalyst.

A particularly striking rate increase for acetal hydrolysis deriving from intramolecular carboxyl-group participation is provided by benzaldehyde disalicyl acetal. Hydrolysis of this compound exhibits a bell-shaped dependence of rate on pH, as does the hydrolysis of glycosides by lysozyme, with a rate maximum near pH 6 in 50% dioxane.[32] At the rate maximum, this compound

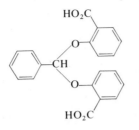

is 2.7×10^9 times more reactive than the corresponding dimethyl ester, which may reflect both intramolecular general acid catalysis and electrostatic facilitation by a carboxylate function.[32]

Electrostatic facilitation of acetal hydrolysis by micelle-forming ionic surfactants is also well established.[7]

3. HYDROLYSIS OF GLYCOSIDES

Our discussion of the hydrolysis of glycosides begins with a brief account of how the nonenzymic reaction differs from that for simple acetals and other structurally related substrates. Subsequently, mechanisms for the reactions catalyzed by lysozyme, β-galactosidase, α-glucosidase, and β-glucuronidase are developed.

3.1. Nonenzymic Hydrolysis of Glycosides

Structurally, glycosides are a subclass of acetals. Consequently, it is not surprising that the mechanism of glycoside hydrolysis is closely related to that presented above for acetals and related compounds. However, the fact that one of the two alkoxy groups characteristic of acetals is incorporated into the cyclic structure of a sugar molecule has its own peculiar implications for reactivity and pathway. The subject of glycoside hydrolysis has been reviewed in detail.[33]

There are two peculiar structural features of glycosides that tend to distinguish the associated hydrolytic reactions from those for simpler acetals. First, one of the oxygen atoms attached to the glycosidic carbon atom is part of the cyclic sugar moiety. This implies that carbonium-ion formation must involve an associated substrate conformation change. Specifically, the most stable conformation for most pyranose glycosides is the chair structure. But maximal stabilization of the corresponding carbonium ion by the unshared pair of electrons on the ring oxygen requires that the ring adopt a half-chair structure:

$$\text{(7)}$$

This is less stable than the chair form by a few kilocalories per mole and, therefore, has the effect of making glycosides less reactive than simpler acetals in unimolecular acid-catalyzed hydrolysis reactions. Second, the sugar moiety bears a number of hydroxyl (or amino or amide) functions on the ring. The inductive electron-withdrawing power of these groups tends to destabilize carbonium ions formed in hydrolytic reactions. The factor too will tend to reduce the reactivity of glycosides compared to acetals. Moreover, these groups may participate as intramolecular nucleophilic reagents in the expulsion of the leaving alkoxy function. Taken together, these considerations suggest that (1) acid-catalyzed unimolecular hydrolysis of glycosides should be much slower than the corresponding reaction for hydrolysis of simple acetals, and (2) hydrolysis pathways in which participation of nucleophilic reagents occurs should be more common for hydrolysis of glycosides than for that for acetals. Both of these expectations are borne out experimentally.

Acid-catalyzed glycoside hydrolysis is several orders of magnitude slower than that for hydrolysis of simple acetals. Thus, hydrolysis of, for example, benzaldehyde diethyl acetal is rapid at pH 3 at 25°. In contrast, hydrolysis of methyl glucoside is a slow reaction in boiling $1N$ HCl. A variety of lines of evidence establish that acid-catalyzed hydrolysis of glycosides occurs via a unimolecular pathway similar to that for simple acetals.[33] Of specific note is the magnitude of the alpha secondary deuterium isotope effect (Table III).

In addition to the acid-catalyzed hydrolysis, glycosides frequently show a base-catalyzed reaction as well. This may involve a direct nucleophilic attack of hydroxide ion from the solvent or, more frequently, the intramolecular participation of the hydroxyl or acetamide function at C2 of the substrate.[33]

Since acid-catalyzed glycoside hydrolysis is so slow, it has been difficult to search for general acid catalysis for this reaction. In view of the considerations developed earlier in this review, it is not likely that the nonenzymic reaction is highly susceptible to such catalysis. Little evidence for electrostatic facilitation of glycoside hydrolysis has been accumulated, but there is no reason to expect that this should be less important than for the case of acetal hydrolysis (*vide supra*).

3.2. Lysozyme

Among those enzymes which catalyze the hydrolysis of glycosidic bonds, the reaction mechanism is best understood for lysozyme. The natural substrate for this enzyme is bacterial wall polysaccharides; it has maximal enzymic activity with substrates containing six or more residues of N-acetylglucosamine or of this sugar alternating with N-acetylmuramic acid. The structure of the

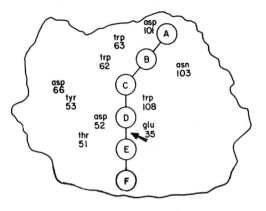

Fig. 6. Schematic representation of the active site of hen egg white lysozyme. The individual units of a hexasaccharide substrate are designated A–F. Hydrolysis occurs between units D and E as indicated by the arrow. A number of amino acid residues which are involved in either substrate binding or catalysis of bond-changing reactions are indicated.

enzyme and the mechanism of the catalytic process have been reviewed in detail.[9,34]

Determination of the structure of hen egg white lysozyme through crystallographic analysis reveals that six sugar units of a polysaccharide substrate can be accommodated within a cleft in the enzyme surface at sites which are designated A through F.[35] This is shown schematically in Fig. 6. Cleavage of the substrate occurs between the D and E sites, as shown by the arrow in the figure. Three carboxyl functions make contact with bound substrate: Glu-35, Asp-52, and Asp-101. The former two were suggested to play essential roles in catalysis, Glu-35 as a general acid catalyst and Asp-52 as a negative site for stabilization of the developing carbonium ion in the transition state.[35] Support for this point of view has come from the pH-rate profile, chemical modification experiments, and determination of values of pK_a for the groups involved.[9,34,36,37]

Of particular note is the concept of distortion of ring D of the substrate toward the half-chair conformation, i.e., toward the transition-state geometry. This suggestion was initially made by the Phillips group based on model building studies which revealed that optimal binding of a hexameric substrate required such a distortion.[35] Subsequently, several lines of investigation have provided experimental support for this idea. In a series of substrate-binding experiments, it has been demonstrated that filling site D on the enzyme is an energy-requiring process. In fact, filling site D requires more energy than is gained by filling sites E and F.[38] The consequence is that substrates preferentially bind at the top of the cleft, in sites A, B, and C. Specifically, *N*-acetylglucosamine binds at site C, the dimer of this sugar at site BC, and the trimer at site ABC.[38] The tendency of substrates (or potential substrates) to avoid binding at site D is a reflection of the positive free-energy change associated with such binding, a possible consequence of the distortion of ring D of the substrate. This also raises the question of the formation of nonproductive complexes between enzyme and substrate, a matter considered below.

Catalysis by distortion is also supported by results of a study of the binding of the lactone analog of tetra-N-acetylglucosamine to lysozyme.[39] This substrate is a conformational analog for the carbonium ion of tetra-N-acetylglucosamine:

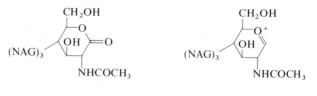

lactone analog of tetra-NAG   Carbonium ion from tetra-NAG

It has been estimated that the affinity of the lysozyme D site for the half-chair conformer of N-acetylglucosamine is 6000-fold greater than its affinity for the chair conformer.[39] This result is consistent with and support for a carbonium-ion-like transition state. X-ray diffraction studies of the lysozyme–tetrasaccharide lactone complex have revealed that the lactone bound at site D actually adopts a conformation which is closer to a boat or sofa than to a half-chair.[40] This observation is supportive of the concept of catalysis by substrate distortion but suggests that the original suggestion about substrate conformation at site D may not be precisely correct.

The strongest evidence for a carbonium-like transition state for lysozyme-catalyzed reactions comes from a series of measurements of alpha secondary deuterium isotope effects by Raftery and his co-workers.[41-43] The pertinent data are collected in Table III. All substrates for lysozyme examined reveal isotope effects significantly larger than unity, strongly suggesting that the transition state possesses considerable carbonium-ion character. The isotope effects for the enzymic reactions are similar to those for acid-catalyzed glycoside hydrolysis, suggesting similar transition-state geometries. Comparison of the observed isotope effects with the maximal value expected, about 1.23, for complete formation of the carbonium ion is consistent with a transition-state geometry in which the hybridization at the reaction site is about midway between sp^2 and sp^3. This conclusion is based on the assumption that the isotope effect is a maximal one, that is, that the reaction involving C—O bond cleavage

Table III

Alpha Secondary Deuterium Isotope Effects for Hydrolysis of Glycosides

Substrate	Catalyst	k_H/k_D	Ref.
Phenyl-β-D-glucopyranoside	Acid	1.13	41
Phenyl-β-D-glucopyranoside	Methoxide	1.03	41
Pehnyl-β-D-glucopyranoside	Lysozyme	1.11	41, 42
Phenyl-β-D-glucopyranoside	β-Glucosidase	1.01	42
Chitotriose	Lysozyme	1.14	43

is solely rate determining. To the extent that other steps, for which no change in hybridization at the reaction center is involved, contribute to the overall rate, the observed isotope effect will be less than that for the bond-changing reaction itself. Thus, the experimental isotope effect may tend to underestimate the carbonium-ion character of the transition state. However, it appears that k_{cat} for lysozyme-catalyzed hydrolysis of chitohexose and, particularly, chitotriose reflects largely the bond-breaking reaction itself.[44] Thus, the secondary deuterium isotope effects observed for lysozyme-catalyzed reactions are less ambiguous in interpretation than are those for several other enzymic reactions. Finally, note that the isotope effects are for V/K_m since they were measured by a double-label competition technique. It is not likely that there exists a detectable secondary deuterium isotope effect on K_m, even for substrates for which k_{cat} is solely rate determining, as a consequence of the likely distortion of the substrate in the binding process, as discussed above. If the substrate undergoes significant distortion toward carbonium-ion geometry, even without C—O bond stretching, this should generate a small but not insignificant secondary deuterium isotope effect. Measurement of such effects for binding of substrates to enzymes would be technically challenging but of great interest.

Structure-reactivity correlations for lysozyme-catalyzed hydrolysis of four aryl β-di-N-acetylchitobiosides yield a value of ρ of 1.2[45] (Table I).

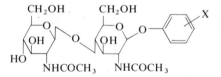

This value refers solely to V_{max}; K_m was found to be independent of substituent, evidence that bond cleavage is solely rate determining for lysozyme-catalyzed hydrolysis of these substrates (which are, incidentally, about 10^6 less reactive than the best of the lysozyme substrates). The observed value of ρ is considerably more positive than is that for acid-catalyzed hydrolysis of glycosides[7,28,33,46]; less positive than that for alkaline hydrolysis, which proceeds via direct nucleophilic attack; and about the same as that for general-acid-catalyzed hydrolysis of 2-(4-substituted phenoxy)tetrahydropyrans[47] and general-acid-catalyzed hydrolysis of 3-substituted phenyl acetals.[48] Thus, this result, taken by itself, provides evidence both for formation of a carbonium-ion intermediate and for general acid catalysis in the lysozyme-catalyzed hydrolysis of these substrates.

However, the structure-reactivity correlations for lysozyme-catalyzed reactions are not so simple as implied thus far. Work subsequent to that just described, in which electron-donating substituents were included in the series, yields the biphasic Hammett plot shown in Fig. 7.[49] Substrates possessing electron-withdrawing substituents yield a value of ρ of 1.2, as noted above, but those with electron-donating substituents yield a value of ρ of -3.0. Assuming that this Hammett plot is mechanistically meaningful, it requires that

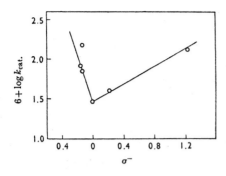

Fig. 7. Biphasic Hammett plot obtained for the lysozyme-catalyzed hydrolysis of aryl β-di-N-acetylchitobiosides. (Reproduced from Reference 49 with permission of the copyright owner, the Biochemical Society.)

two different mechanisms obtain for the two classes of substrate. Consistent with this interpretation is that fact that appreciable solvent deuterium isotope effects, $k_{cat}^{H}/k_{cat}^{D} = 1.7–2.0$, are observed only for those substrates possessing electron-withdrawing substituents. This result is consistent with general acid catalysis for hydrolysis of these substrates only.[4,49] Both the strongly negative value of ρ and the absence of an appreciable solvent deuterium isotope effect for the other substrates suggest that these reactions proceed via classical A1 reaction pathways.

Of particular interest for the understanding of the mechanism of lysozyme-catalyzed reactions is a series of studies in which data from rapid reaction studies have been correlated with those from steady-state kinetic measurements and thermochemical studies to yield a detailed description of the nature and thermodynamics of the reaction pathway for hexasaccharide hydrolysis.[44,49] The pathway is

$$EU_2^\alpha \underset{k_{32}^\alpha}{\overset{k_{23}^\alpha}{\rightleftharpoons}} EU_1^\alpha \overset{K_{12}^\alpha}{\rightleftharpoons} E + S \overset{K_{12}^\beta}{\rightleftharpoons} ES_1^\beta \underset{k_{32}^\beta}{\overset{k_{23}^\beta}{\rightleftharpoons}} ES_2^\beta \underset{k_{43}^\gamma}{\overset{k_{34}^\gamma}{\rightleftharpoons}} ES^\gamma \overset{k_{cat}}{\longrightarrow} E + P \qquad (8)$$

$$ES^*$$

EU$_2^\alpha$ and EU$_1^\alpha$ are nonproductive complexes, ES$_1^\beta$ and ES$_2^\beta$ are productive complexes, and ES$^\gamma$ is the form that participates in the bond-changing reaction, k_{cat}. The free-energy profile associated with the pathway of Eq. (8) is provided in Fig. 8. The free energies of all enzyme–substrate complexes fall in a narrow range 5.5–7.0 kcal below the free energy of free enzyme and substrate. The free energies for the transition states for their interconversions, as well as for k_{cat}, all lie in a narrow range 9–12 kcal above free enzyme and substrate. The enzyme forms EU$_2^\alpha$, ES$_2^\beta$, and ES$^\gamma$ are present in comparable concentrations, and, hence, all contribute to experimental equilibrium and steady-state rate constants.

Of particular note is the fact that rate and equilibrium constants for the preequilibrium steps are nearly the same for tetra-, penta-, and hexa-N-acetylglucosamine despite the fact that there exist large differences in the overall rates of lysozyme-catalyzed hydrolysis of these substrates. Thus,

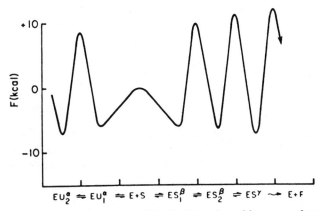

Fig. 8. Reaction profile for the mechanism of Eq. (8). The values of free energy for transition and equilibrium states were determined from kinetic, thermodynamic, and activation parameters for the individual steps in the mechanism. (Reproduced from Reference 50 with permission of the copyright owner, the American Society of Biological Chemists.)

the rate difference must reflect differences in the transition states for k_{cat}. This has been interpreted to suggest that the bond-changing reaction in ES^γ leading to the carbonium-ion intermediate is concerted with a rearrangement of this complex leading to full interaction of the substrate with the DEF site.[50]

3.3. β-Galactosidase

β-Galactosidase catalyzes the transfer of D-galactose from a variety of β-galactosides to a variety of acceptors, including water. Galactosyl transfer occurs with retention of anomeric configuration. This enzyme occurs widely distributed among living organisms; the enzyme from *E. coli* has been particularly thoroughly studied. This enzyme is a tetramer, molecular weight 500,000. β-Galactosidases have been the subject of a careful review.[51]

Although lactose is the physiological substrate for β-galactosidase and a careful study of the hydrolase and transgalactosidase activities of the *E. coli* enzyme with this substrate has been accomplished,[52] the great bulk of mechanistic work has utilized substrates whose hydrolysis is more convenient to follow. 2-Nitrophenyl-β-D-galactoside has been a frequent choice.

The pathway for the β-galactosidase-catalyzed reactions appears complicated. There is good evidence to suggest the formation of at least two intermediates: an enzyme-bound galactosyl carbonium ion and a covalent galactosyl enzyme.

Evidence for formation of a galactosyl carbonium ion derives largely from measurements of alpha secondary deuterium isotope effects. The initial measurements were made with a series of aryl-2',3',4',6'-tetra-O-acetyl-β-D-galactopyranosides[53]:

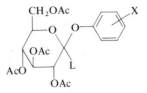

Neither K_m nor k_{cat} for these substrates shows a simple dependence on aglycone acidity. This observation suggests that different steps may be rate determining (or that different mechanisms may operate) depending on the nature of the leaving group. This conclusion is borne out by variation in the magnitude of the alpha secondary deuterium isotope effects from 1.25 for the most reactive substrate, the 2,4-dinitrophenyl galactoside, to 1.00 for the least reactive substrate, the 4-bromophenyl galactoside. The former value is near the upper limit expected for formation of a carbonium ion from a substrate of this type and strongly suggests that a carbonium ion is formed in the transition state. The value of unity may reflect either (1) a transition to a rate-determining step in which the hybridization at C'-1 of the substrate does not change or (2) incursion of a mechanism involving participation of a nucleophilic reagent in the transition state. The former explanation is preferred, and rate-determining enzyme conformation change has been advanced as a plausible explanation for the observed isotope-effect results.[53]

These results have been thrown into clearer focus by the interesting observation that β-galactosidase catalyzes the hydrolysis of β-D-galactosyl-pyridinium ions.[54] Such substrates are of particular note since the leaving group bears a full positive charge, and, consequently, hydrolysis cannot be susceptible to general acid catalysis through partial protonation of the leaving group. A closely related example, enzyme-catalyzed hydrolysis of NAD, is considered below. The demonstration that β-galactosidase is active with these substrates does not indicate that general acid catalysis is unimportant for catalysis of hydrolysis of other substrates, but it does place limits on the extent to which such catalysis can contribute to the total catalytic effect. The pyridinium-ion substrates are, depending on the nature of the leaving group, two or three orders of magnitude less reactive in β-galactosidase-catalyzed reactions than is 2-nitrophenyl-β-D-galactoside.

Unlike the behavior of substrates bearing phenolic leaving groups, log k_{cat} for β-galactosidase-catalyzed hydrolysis of the galactosyl pyridinium ions is linearly dependent on the pK_a of the parent amine.[55,56] This suggests that a single step, the cleavage of the C—N bond of the substrate, may be rate determining for all of the pyridinium ions. Alpha secondary deuterium isotope effects for β-galactosidase-catalyzed hydrolysis of the pyridinium ion and 4-bromoisoquinolinium ion are 1.14 ± 0.04 and 1.19 ± 0.05, respectively, confirming the formation of carbonium ions in the rate-determining step. The magnitude of these isotope effects suggests that the transition-state geometry is about midway between tetrahedral and trigonal. In addition, the alpha secondary deuterium isotope effect for enzymic hydrolysis of β-D-galacto-pyranosyl azide is 1.10 ± 0.03, also requiring carbonium-ion formation in the transition state.[55] For this substrate, it has been possible to demonstrate that general acid catalysis of the departure of the azide group cannot account for more than a rate acceleration of a factor of 70.

The concept of enzyme-bound carbonium ions as intermediates ln *E. coli* β-galactosidase-catalyzed reactions also receives support from the observation that 5-galactonolactone, a transition-state analog for the carbonium ion, is an effective inhibitor of this enzyme.[57]

Data thus far cited strongly suggest that the enzyme-catalyzed reactions involve formation of an enzyme-bound carbonium ion as an intermediate on the reaction pathway. Formation or decomposition of this species may be rate determining for the pyridinium-ion substrates and for aryl galactosides having acidic leaving groups. Carbonium-ion formation is preceded on the reaction pathway by a conformation change of an enzyme–substrate intermediate, and this step is rate determining for aryl galactosides bearing less acidic leaving groups. A remaining question deals with the fate of the carbonium ion once it is formed.

Several lines of evidence strongly suggest that the predominant fate of enzyme-bound galactosyl carbonium ions is collapse to form a covalent β-D-galactopyranosylenzyme.[55] This suggestion is consistent with the alpha deuterium isotope effects just discussed and is supported by the following additional observations.

First, the reaction of 2-nitrophenylgalactoside with *E. coli* β-galactosidase yields an initial rapid burst of 2-nitrophenolate ion, strongly suggesting the formation of a covalent enzyme–substrate intermediate.[58] Second, treatment of the enzyme with galactal leads to the formation of a covalent enzyme–galactal compound.[59] This too is good support for occurrence of a corresponding intermediate along the reaction pathway.

All of the observations indicated above can be successfully rationalized by the mechanism suggested by Sinnott and Souchard[53] and provided in Fig. 9.

3.4. β-Glucosidase

β-Glucosidase catalyzes the hydrolysis of aryl, alkyl, and oligosaccharide β-glucosides.[60] The enzyme has low specificity for the aglycone and is not absolutely specific for the glucose moiety; for example, sweet almond β-glucosidase catalyzes the hydrolysis of β-galactosides and β-fucosides as well as β-glucosides.[61] In contrast to the behavior of lysozyme, it exhibits maximal activity with aryl glucosides, and in contrast to both lysozyme and β-galactosidase, only water serves as glycosyl group acceptor.[60] The transfer of the glucosyl moiety from glucosides to water occurs with retention of anomeric configuration.[62]

A central question of enzyme mechanism that seems reasonably clear for

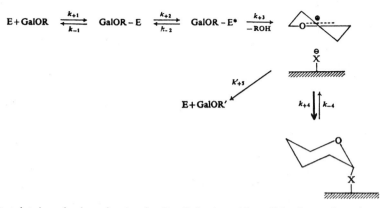

Fig. 9. Postulated mechanism of action for *E. coli* β-galactosidase. Gal refers to β-D-galactopy-ranosyl and k_2 represents a conformation change. (Reproduced from Reference 53 with permission of the copyright owner, the Biochemical Society.)

the cases of lysozyme and β-galactosidase, the occurrence of an enzyme-bound carbonium-ion intermediate along the reaction pathway, is less clear in the case of β-glucosidase. Both structure-reactivity correlations and alpha secondary deuterium isotope effects yield results which are mechanistically ambiguous.

In an unusually thorough study, Nath and Rydon examined the kinetics of β-glucosidase-catalyzed hydrolysis of 21 substituted phenyl β-glucosides.[63] Both values of K_m and V_{max} are well correlated with substituent constants: ρ for K_m is near -1.0 and that for V_{max} is near 1.0 (Table I). The value for V_{max} is close to that for lysozyme-catalyzed hydrolysis of related substrates (Table I) and suggests a similar reaction mechanism: general-acid-catalyzed cleavage of the C—O bond leading to formation of a carbonium ion in the transition state. However, this value of ρ is also consistent with nucleophilic displacement of phenol in the transition state by either a solvent-derived species or some group at the enzyme active site. Note in particular that the value for V_{max}/K_m, the second-order rate constant at low substrate concentrations, is near 2.0. This value suggests development of considerable negative charge on the leaving oxygen atom and is somewhat difficult to rationalize on the basis of general acid catalysis for phenol departure.

The substantial negative value of ρ for K_m for β-glucosidase-catalyzed reactions is not easy to interpret. It has been suggested that K_m may be a dissociation constant for the Michaelis–Menten complex and that the value of ρ_{K_m} of -1.0 reflects electronic effects on hydrogen-bonding interactions between the enzyme and the glucosyl moiety.[63] However, it seems likely that such electronic effects would not be so pronounced, and it is more satisfying to assume that the rate of cleavage of the C—O bond is reflected, in part, in the dependence of K_m on the nature of polar substituents.

The alpha secondary deuterium isotope for β-glucosidase-catalyzed hydrolysis of phenyl-β-D-glucopyranoside is 1.01.[42] Although this value was originally considered to reflect nucleophilic displacement of the leaving phenol in the transition state, it was recognized that an alternative interpretation is a rate-determining step in which C—O bond cleavage is not involved. In fact, it has been suggested that this isotope effect reflects a rate-determining conformation change of the enzyme–substrate complex, a situation analogous to that which obtains for certain substrates, as noted above, with β-galactosidase.[55] At this point, it is not possible to distinguish between these alternatives.

5-Gluconolactone, a transition-state analog that resembles the stereochemical and dipolar[64] features of a glucosyl carbonium-ion intermediate, is an effective inhibitor of β-glucosidase.[65-67] Moreover, nojirimycin (5-deoxy-5-amino-glucopyranose), also considered to be a carbonium ion-like transition-state analog, is an even more potent inhibitor of this enzyme.[67] In contrast, D-glucal[67] and cyclic hexitols including inositols,[66] transition-state analogs for possible tetrahedral intermediates, are bound only weakly by β-glucosidase. Thus the weight of evidence from transition-state analogs suggests the existence of an enzyme-bound carbonium-ion intermediate.

It has recently been demonstrated that an inositol epoxide, conduritol B epoxide, is an irreversible active-site labeling reagent for β-glucosidase.[66] There is chemical evidence to suggest that the alkylated enzyme results from attack of an enzyme carboxylate group on the epoxide. Consistent with this finding is the observation that β-glucosylamine is a very potent inhibitor of yeast β-glucosidase, a finding interpreted to suggest that this enzyme has an anionic group at the active site.[68] These findings tentatively suggest that there may be a significant relationship between the mechanism of β-glucosidase- and lysozyme-catalyzed reactions. This lends very modest support to the concept of a carbonium-ion pathway for the former, as in the case of the latter, enzyme.

4. HYDROLYSIS OF GLYCOSYLAMINES

Glycosylamines are a group of substrates structurally related to the glycosides. The only difference is that an amine rather than an alcohol is the leaving group in the hydrolysis reactions. However, this simple change causes, as is indicated below, a number of complications in understanding the associated reaction mechanisms. The routes for both enzymic and nonenzymic hydrolysis of glycosylamines are not so well understood as are those for hydrolysis of glycosides.

Glycosylamines of biochemical interest include the nucleosides and nucleotides. In addition to considering the mechanism of the nonenzymic hydrolysis of these compounds, we shall briefly consider what little is known about transition-state structures for those enzymes which catalyze the hydrolysis of these substrates.

4.1. Nonenzymic Hydrolysis of Nucleosides

The pathway for hydrolysis of simple glycosylamines, such as glucosylaniline, has been established to involve formation of a Schiff base as an intermediate[69]:

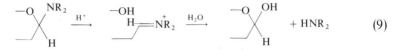

$$\qquad (9)$$

That is, the initial bond-changing reaction is rupture of the carbon–oxygen bond, not the carbon–nitrogen one. Several years ago, it was proposed that nucleosides, which have structurally more complex leaving groups, hydrolyze by the same route.[70-72] However, a number of lines of evidence now strongly suggest that the rate-determining step for the hydrolysis of purine nucleosides at least,[73-78] and probably the pyrimidine nucleosides as well,[79-81] occurs with rate-determining unimolecular dissociation of the mono- and dications of the substrates with rupture of the carbon–nitrogen bond and formation of a carboxonium ion:

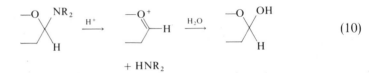

$$\qquad (10)$$

Evidence supporting this contention includes the following: (1) Values of entropy of activation are near zero or positive, suggestive of a unimolecular reaction pathway[74-77]; (2) hydrolysis frequently occurs via the dication, an observation which is easier to rationalize in terms of C—N cleavage than in terms of C—O bond rupture[73,74,76,77]; (3) 2-deoxyribonucleosides are significantly more reactive than the ribonucleosides,[69,73,76,77] the observation expected on the basis of the polar effect of the 2′-hydroxyl group on the stability of a developing carboxonium ion in the transition state; (4) the hydrolysis of 1,7-dimethylguanosine reveals a marked pH-independent reaction,[73] which is consistent with C—N bond cleavage but difficult to account for in terms of C—O bond cleavage since no driving force exists for the latter reaction; (5) there is no anomerization concomitant with hydrolysis for purine nucleosides as would be expected if C—O bond cleavage were more rapid than C—N bond cleavage[69]; and (6) electron-releasing polar substituents in the purine moiety decrease the rate of nucleoside hydrolysis as expected for C—N bond cleavage in the transition state.[75,76] The conclusion of rate-determining unimolecular cleavage of the C—N bond for purine nucleoside hydrolysis has been corroborated by determination of the kinetic alpha secondary deuterium isotope effects for these reactions.

The pH-rate profile for nonenzymic hydrolysis of inosine establishes that inosine hydrolyzes through both the monocation and dication:

$$S + H^+ \rightleftharpoons SH^+ \longrightarrow products$$

$$\big\updownarrow$$

$$SH_2^{2+} \longrightarrow products \tag{11}$$

Consequently, the alpha secondary deuterium isotope effect for this reaction was measured at both pH 1.0, where hydrolysis via the monocation is the predominant pathway, and at pH 0, where hydrolysis occurs largely via the dication. The results are collected in Table IV. Note that, in both cases, the reaction reveals a large normal isotope effect, establishing that the reactions involve carboxonium-ion formation in the transition state, as was suggested by earlier results summarized above. In addition, the magnitude of the isotope effect strongly suggests a quite late transition state in which the C-1 of the ribose moiety is largely trigonal. A similar result was obtained for adenosine hydrolysis (Table IV).

This information provides a sensibly secure knowledge of the pathway of purine nucleoside hydrolysis and of the transition-state structure in terms of extent of C—N bond cleavage. Consequently, the essential features of the reaction mechanism, particularly as they relate to the provision of the necessary background information for understanding of corresponding enzymic reaction mechanisms, are within reach.

The nonenzymic hydrolysis of the *N*-glycosidic linkage of the nicotinamide nucleotide coenzymes is much less well studied but, at the same time, offers less challenging mechanistic questions. The fact that the leaving nicotinamide moiety bears a full positive charge in the substrate essentially eliminates the possibility of initial C—O bond cleavage [Eq. (9)]; the essential mechanistic question, then, is how breakage of the C—N bond occurs: with or without the participation of solvent as a nucleophilic reagent. Structurally, the site of bond cleavage in NAD and related compounds strongly resembles that in protonated acetals:

It has been established that acid-catalyzed hydrolysis of acetals occurs with unimolecular decomposition to yield the carboxonium ion[7]; by analogy, one would expect NAD to hydrolyze by a pH-independent pathway to generate the corresponding carboxonium ion. In fact, it has been established that NAD does hydrolyze by a pH-independent route, as well as a base-catalyzed one.[82] Alpha secondary deuterium isotope effects for the pH-independent hydrolysis of both NAD and NMN establish that these reactions are, in fact, unimolecular and involve carboxonium-ion formation (Table V).

Table IV
Alpha Secondary Deuterium Isotope Effects for
Hydrolysis of Inosine and Adenosine[a]

Reaction	Substrate	k_H/k_D
Hydrolysis, pH 1.0, 25°	Inosine	1.17 ± 0.01
Hydrolysis, pH 1.0, 50°	Inosine	1.18 ± 0.01
Hydrolysis, pH 0.0, 25°	Inosine	1.23 ± 0.01
Hydrolysis, pH 0.0, 50°	Inosine	1.22 ± 0.07
Hydrolysis, pH 1.0, 25°	Adenosine	1.235 ± 0.015

[a] R. Romero, R. Stein, H. G. Bull and E. H. Cordes, unpublished results.

Table V
Alpha Secondary Deuterium Isotope Effects for
Hydrolysis of NAD and NMN[a]

	Isotope Effect, k_H/k_D	
Reaction	NAD	NMN
pH-Independent nonenzymic hydrolysis	1.11 ± 0.01	1.11 ± 0.01
Pig brain NAD glycohydrolase	0.98 ± 0.02	1.13 ± 0.01
N. crassa NAD glycohydrolase	1.00 ± 0.01	1.10 ± 0.01
E. coli NMN glycohydrolase	—	0.99 ± 0.02

[a] H. G. Bull, J. Ferraz, and E. H. Cordes, unpublished results.

4.2. NAD Glycohydrolases

These enzymes form a heterogeneous group (EC 3.2.2.5) which are found widely distributed in mammalian tissues and are present in certain microbiological sources as well. They catalyze the cleavage of nicotinamide from NAD and, sometimes, from NMN and nicotinamide riboside:

$$\text{NAD} \xrightarrow{\text{NAD glycohydrolase}} \text{nicotinamide} + \text{ADP–ribose} \qquad (12)$$

NAD glycohydrolases, together with nucleotide pyrophosphatases, provide the only known mechanism for the degradation of this coenzyme and may play an important role in metabolic regulation.[83]

The NAD glycohydrolases pose an interesting question of enzymic reaction mechanism similar to that indicated above for β-galactosidase-catalyzed hydrolysis of glycosyl pyridinium ions. If we suppose that the enzymic route is similar to the nonenzymic one in the sense that unimolecular decomposition with generation of the carboxonium ion is involved, many of the catalytic processes commonly considered to be employed by enzymes cannot function. The absence of a nucleophilic reagent in the transition state eliminates both nucleophilic and general base catalysis; the presence of a full positive charge on the leaving group provides little possibility for general acid catalysis. Approximation and orientation effects cannot be important for unimolecular reactions. Finally, the fact that both substrate and transition state bear similar charges, although the distribution is significantly different, argues against a large role for electrostatic facilitation. Nevertheless, one must account for the fact that the enzymic hydrolysis of NAD is at least 1 million times faster than the nonenzymic reaction under the same conditions. The possibility of some form of catalysis by distortion is strongly suggested.

NAD glycohydrolases of mammalian tissues occur in both the nucleus and cytoplasm. The nuclear enzymes catalyze the formation of adenosine-diphosphate–ribose polymers, a reaction qualitatively distinct from that catalyzed by the cytoplasmic enzymes [Eq. (12)]. The cytoplasmic enzymes are usually membrane bound and form part of the endoplasmic reticulum; the bull semen enzyme is a prominent exception, being soluble. Several of these enzymes have been purified to apparent homogeneity including the enzymes from pig brain,[84] calf spleen,[85] and bull semen.[86,87] These enzymes are of relatively low molecular weight and apparently contain neither metal ions nor other cofactors with the exception of the bull semen enzyme, which contains zinc.

Two of the microbiological NAD glycohydrolases have been carefully purified: those from *Neurospora crassa* and *Bacillus subtilis*.[88-90] These enzymes are soluble and of low molecular weight. Of particular interest is the fact that both contain large amounts of carbohydrate of unknown structure which is covalently associated. The *N. crassa* enzyme contains 74% by weight carbohydrate; the corresponding figure for the *B. subtilis* enzyme is 54%. These observations both raise the question of the role of the carbohydrate moiety in the catalytic process.

Despite the availability of pure enzymes and the existence of the interesting mechanistic questions raised above, these enzymes have been rather little studied. Only the bull semen enzyme has received systematic study with respect to the kinetics and nature of the substrate-binding site.[91-95] The kinetics of the pig brain enzyme have been probed to some extent,[84,96] particularly with reference to enzyme inactivation,[97-99] and a nice kinetic study of the calf spleen enzyme has recently been completed.[85]

In an effort to establish the basic feature of the enzymic reaction mechanism—whether the reaction occurs via a carbonium-ion intermediate—the kinetic alpha secondary deuterium Isotope effects were measured for (1) pH-independent hydrolysis of NAD, (2) pig brain NAD glycohydrolase-catalyzed hydrolysis of NAD, and (3) *N. crassa* NAD glycohydrolase-catalyzed hydrolysis of NAD (Table V). The enzymic hydrolysis of NAD did not yield an isotope effect detectably different from unity in either case. This observation is consistent with one of three interpretations: (1) Substrate binding by the NAD glycohydrolases is irreversible; (2) carbon–nitrogen bond cleavage is not involved in the rate-determining step; or (3) carbon–nitrogen bond cleavage occurs with participation of a nucleophilic reagent. An attempt was made to distinguish among these alternatives by using a poorer substrate, in the hope that C—N bond cleavage might be slowed sufficiently with respect to other processes involved so as to become partially or completely rate determining. Since both enzymes utilize NMN as substrate, though less effectively than NAD, that nucleotide seemed a good choice for pursuit of this point.

The kinetic alpha secondary deuterium isotope effect for pH-independent nonenzymic hydrolysis of NMN was determined and is included in Table V. The value is identical, within experimental error, to that obtained for the corresponding reaction of NAD, as expected. This established that NMN, too, hydrolyzes via unimolecular decomposition to yield a carboxonium ion. The isotope effect for NMN hydrolysis catalyzed by both NAD glycohydrolases was determined (Table V). In contrast to the case for NAD hydrolysis, values significantly greater than unity were obtained in each case. This observation established that NAD glycohydrolase-catalyzed hydrolysis of NMN proceeds via rate-determining (or partially rate-determining) unimolecular cleavage of the C—N bond to yield a carboxonium-ion intermediate. Moreover, it very strongly suggests that the route of NAD hydrolysis catalyzed by the same enzymes also involves a carboxonium-ion route but that C—N bond formation is not the rate-determining step.

To pursue this point a little further, the alpha secondary deuterium isotope effect for hydrolysis of NMN catalyzed by a crude NMN glycohydrolase present in E. coli was measured. This enzyme is unique in that NMN, rather than NAD, is the preferred substrate. As indicated in Table V, an isotope effect of unity was obtained, suggesting that hydrolysis of NMN by the NMN glycohydrolase also does not occur with C—N bond cleavage in the rate-determining step, analogous to NAD hydrolysis catalyzed by the NAD glycohydrolases.

4.3. Purine Nucleosidases

Among this group of enzymes, only one example, AMP nucleosidase (EC 3.2.2.4), has been purified extensively and carefully characterized:

$$AMP + H_2O \longrightarrow adenine + ribose\text{-}5\text{-}phosphate \qquad (13)$$

The enzyme from Azotobacter vinelandii has been crystallized and shown to contain six identical subunits comprising a total molecular weight of 320,000.[100] The basic kinetic properties of the enzyme have been established.[101-103] Of particular note is the observation that formycin-5-phosphate is a potent inhibitor of this enzyme and a possible transition-state analog.[104] Since the crystal structure of this compound is known, this provides insight into transition-state geometry for the glycone portion of the substrate.

Other than the enzyme described above, purine nucleosidases (EC 3.2.2.1) have not been well characterized.

REFERENCES

1. (a) W. W. Cleland, The kinetics of enzyme-catalyzed reactions with two or more substrates or products. I. Nomenclature and rate equations, *Biochim. Biophys. Acta* **67**, 104–137 (1963). (b) W. W. Cleland, The kinetics of enzyme-catalyzed reactions with two or more substrates or products. II. Inhibition: Nomenclature and theory, *Biochim. Biophys. Acta* **67**, 173–187 (1963). (c) W. W. Cleland, The kinetics of enzyme-catalyzed reactions with two or more substrates or products. III. Prediction of initial velocity and inhibition patterns by inspection, *Biochim. Biophys. Acta* **67**, 188–196 (1963).
2. I. H. Segel, *Enzyme Kinetics,* Wiley, New York (1975).
3. W. W. Cleland, in: *The Enzymes* (P. D. Boyer, ed.), Vol. II, pp. 1–65, Academic Press, New York (1970).
4. J. F. Kirsch, in: *Advances in Linear Free Energy Relationships* Plenum Press, New York (1972).
5. A. Rühlmann, D. Kukla, P. Schwager, K. Bartels, and R. Huber, Structure of the complex formed by bovine trypsin and bovine pancreatic trypsin inhibitor, *J. Mol. Biol.* **77**, 417–436 (1973).
6. E. H. Cordes, Mechanism and catalysis for the hydrolysis of acetals, ketals, and ortho esters, *Prog. Phys. Org. Chem.* **4**, 1–44 (1967).
7. E. H. Cordes and H. G. Bull, Mechanism and catalysis for hydrolysis of acetals, ketals, and ortho esters, *Chem. Rev.* **74**, 581–603 (1974).
8. T. H. Fife, General acid catalysis of acetal, ketal, and ortho ester hydrolysis, *Acc. Chem. Res.* **5**, 264–272 (1972).
9. B. M. Dunn and T. C. Bruice, Physical organic models for the mechanism of lysozyme action, *Adv. Enzymol.* **37**, 1–60 (1973).
10. J. M. O'Gorman and H. J. Lucas, Hydrolysis of the acetal of D(+)-2-octanol, *J. Am. Chem. Soc.* **72**, 5489–5490 (1950).
11. H. K. Garner and H. J. Lucas, Preparation and hydrolysis of some acetals and esters of D(−)-2,3-butanediol, *J. Am. Chem. Soc.* **72**, 5497–5501 (1952).
12. E. R. Alexander, H. M. Busch, and G. L. Webster, Preparation and hydrolysis of optically active 2-butyl acetal, *J. Am. Chem. Soc.* **74**, 3173 (1952).
13. F. Stasiuk, W. A. Sheppard, and A. N. Bourns, An oxygen-18 study of acetal formation and hydrolysis, *Can. J. Chem.* **34**, 123–127 (1956).
14. A. M. Wenthe and E. H. Cordes, Concerning the mechanisms of acid-catalyzed hydrolysis of ketals, ortho esters, and orthocarbonates, *J. Am. Chem. Soc.* **87**, 3173–3180 (1965).
15. T. H. Fife and L. H. Brod, General acid catalysis and the pH-independent hydrolysis of 2-(*p*-nitrophenoxy) tetrahydropyran, *J. Am. Chem. Soc.* **92**, 1681–1684 (1970).
16. E. Anderson and B. Capon, Intermolecular general-acid catalysis in acetal hydrolysis, *J. Chem. Soc. B* **1969**, 1033–1037.
17. H. G. Bull, K. Koehler, T. C. Pletcher, J. J. Ortiz, and E. H. Cordes, Effects of α-deuterium substitution, polar substituents, temperature, and salts on the kinetics of hydrolysis of acetals and ortho esters, *J. Am. Chem. Soc.* **93**, 3002–3011 (1971).
18. V. J. Shiner, M. W. Rapp, E. A. Halevi, and M. Wolfsberg, Solvolytic α-deuterium effects for different leaving groups, *J. Am. Chem. Soc.* **90**, 7171–7172 (1968).
19. A. Streitwieser, Jr., and G. A. Dafforn, Secondary deuterium isotope effects in trifluoro-acetolysis of isopropyl *p*-toluenesulfonate, *Tetrahedron Lett.* **16**, 1263–1266 (1969).
20. J. E. Leffler, Parameters for the description of transition states, *Science* **117**, 340–341 (1953).
21. G. S. Hammond, A correlation of reaction rates, *J. Am. Chem. Soc.* **77**, 334–338 (1955).
22. E. R. Thornton, A simple theory for predicting the effects of substituent changes on transition-state geometry, *J. Am. Chem. Soc.* **89**, 2915–2927 (1967).

23. J. N. Brønsted and W. F. K. Wynne-Jones, Acid catalysis in hydrolytic reactions, *Trans. Faraday Soc.* **25**, 59–76 (1929).

24. W. P. Jencks, General acid–base catalysis of complex reactions in water, *Chem. Rev.* **72**, 705–718 (1972).

25. E. Anderson and T. H. Fife, General acid catalysis of ortho ester hydrolysis, *J. Org. Chem.* **37**, 1993–1996 (1972).

26. E. Anderson and T. H. Fife, General acid catalysis of acetal hydrolysis. The hydrolysis of substituted benzaldehyde di-tert-butyl acetals, *J. Am. Chem. Soc.* **93**, 1701–1704 (1971).

27. T. C. Bruice and Dennis Piszkiewicz, A search for carboxyl-group catalysis in ketal hydrolysis, *J. Am. Chem. Soc.* **89**, 3568–3576 (1967).

28. D. Piszkiewicz and T. C. Bruice, Glycoside hydrolysis. II. Intramolecular carboxyl and acetamido group catalysis in β-glycoside hydrolysis, *J. Am. Chem. Soc.* **90**, 2156–2163 (1968).

29. B. Capon, M. C. Smith, E. Anderson, R. H. Dahm, and G. H. Sankey, Intramolecular catalysis in the hydrolysis of glycosides and acetals, *J. Chem. Soc. B* **1969**, 1038–1047.

30. (a) B. M. Dunn and T. C. Bruice, Steric and electronic effects on the neighboring general acid catalyzed hydrolysis of methyl phenyl acetals of formaldehyde, *J. Am. Chem. Soc.* **92**, 2410–2416 (1970). (b) B. M. Dunn and T. C. Bruice, Further investigation of the neighboring carboxyl group. Catalysis of hydrolysis of methyl phenyl acetals of formaldehyde. Electrostatic and solvent effects, *J. Am. Chem. Soc.* **92**, 6589–6594 (1970). (c) B. M. Dunn and T. C. Bruice, Electrostatic catalysis. IV. Intramolecular carboxyl group electrostatic facilitation of the A-1-catalyzed hydrolysis of alkyl phenyl acetals of formaldehyde. The influence of oxocarbonium ion stability, *J. Am. Chem. Soc.* **93**, 5725–5731 (1971).

31. T. H. Fife and E. Anderson, Intramolecular carboxyl group participation in acetal hydrolysis, *J. Am. Chem. Soc.* **93**, 6610–6614 (1971).

32. E. Anderson and T. H. Fife, Carboxy-group participation in acetal hydrolysis. The hydrolysis of benzaldehyde disalicyl acetal, *Chem. Commun.* **1971**, 1470–1471.

33. B. Capon, Mechanism in carbohydrate chemistry, *Chem. Rev.* **69**, 407–498 (1969).

34. T. Imoto, L. N. Johnson, A. C. T. North, D. C. Phillips, and J. A. Rupley, in: *The Enzymes* (P. D. Boyer, ed.), Vol. VII pp. 665–868, Academic Press, New York (1970).

35. C. C. F. Blake, L. N. Johnson, G. A. Mair, A. C. T. North, D. C. Phillips, and V. R. Sarma, Crystallographic studies of the activity of hen egg-white lysozyme, *Proc. Roy. Soc. London Ser. B* **167**, 378–388 (1967).

36. S. Kuramitsu, K. Ikeda, K. Hamaguchi, H. Fujio, T. Amano, S. Miwa, and T. Nishina, Ionization constants of Glu 35 and Asp 52 in hen, turkey, and human lysozymes, *J. Biochem. (Tokyo)* **76**, 671–683 (1974).

37. S. K. Banerjee and J. A. Rupley, Turkey egg white lysozyme, free energy, enthalpy, and steady state kinetics of reaction with N-acetylglucosamine oligosaccharides, *J. Biol. Chem.* **250**, 8267–8274 (1975).

38. J. A. Rupley, L. Butler, M. Gerring, F. J. Hartdegen, and R. Pecoraro, Studies on the enzymic activity of lysozyme, III. The binding of saccharides, *Proc. Nat. Acad. Sci. U.S.A.* **57**, 1088–1095 (1967).

39. I. I. Secemski, S. S. Lehrer, and G. E. Lienhard, A transition state analog for lysozyme, *J. Biol. Chem.* **247**, 4740–4748 (1972).

40. L. O. Ford, L. N. Johnson, P. A. Machin, D. C. Phillips, and R. Tjian, Crystal structure of a lysozyme–tetrasaccharide lactone complex, *J. Mol. Biol.* **88**, 349–371 (1974).

41. F. W. Dahlquist, T. Rand-Meir, and M. A. Raftery, Demonstration of carbonium ion intermediate during lysozyme catalysis, *Proc. Nat. Acad. Sci. U.S.A.* **61**, 1194–1198 (1968).

42. F. W. Dahlquist, T. Rand-Meir, and M. A. Raftery, Application of secondary α-deuterium kinetic isotope effects to studies of enzyme catalysis. Glucoside hydrolysis by lysozyme and β-glucosidase, *Biochemistry* **8**, 4214–4221 (1969).

43. L. E. H. Smith, L. H. Mohr, and M. A. Raftery, Mechanism for lysozyme-catalyzed hydrolysis, *J. Am. Chem. Soc.* **95**, 7497–7500 (1973).

44. E. Holler, J. A. Rupley, and G. P. Hess, Productive and nonproductive lysozyme–chitosaccharide complexes. Kinetic investigations, *Biochemistry* **14**, 2377–2385 (1975).

45. G. Lowe, G. Sheppard, M. L. Sinnott, and A. Williams, Lysozyme-catalysed hydrolysis of some β-aryl di-*N*-acetylchitobiosides, *Biochem. J.* **104**, 893–899 (1967).
46. A. N. Hall, S. Hollingshead, and H. N. Rydon, The acid and alkaline hydrolysis of some substituted phenyl α-glucosides, *J. Chem. Soc.* pp. 4290–4295 (1961).
47. T. H. Fife and L. K. Jao, General acid catalysis of acetal hydrolysis. The hydrolysis of 2-aryloxytetrahydropyrans, *J. Am. Chem. Soc.* **90**, 4081–4085 (1968).
48. E. Anderson and B. Capon, Intermolecular general acid catalysis in acetal hydrolysis, *J. Chem. Soc. B* pp. 1033–1037 (1969).
49. C. S. Tsai, J. Y. Tang, and S. C. Subbarao, Substituent effect on lysozyme-catalysed hydrolysis of some β-aryl di-*N*-acetylchitobiosides, *Biochem. J.* **114**, 529–534 (1969).
50. S. K. Banerjee, E. Holler, G. P. Hess, and J. A. Rupley, Reaction of *N*-acetylglucosamine oligosaccharides with lysozyme. Temperature, pH, and solvent deuterium isotope effects; equilibrium, steady state, and pre-steady state measurements, *J. Biol. Chem.* **250**, 4355–4367 (1975).
51. K. Wallenfels and R. Weil, in: *The Enzymes* (P. D. Boyer, ed.), Vol. VII, pp. 617–663, Academic Press, New York (1972).
52. R. E. Huber, G. Kurz, and K. Wallenfels, A quantitation of the factors which affect the hydrolase and transgalactosylase activities of β-galactosidase (*E. coli*) on lactose, *Biochemistry* **15**, 1994–2001 (1976).
53. M. L. Sinnott and I. J. L. Souchard, The mechanism of action of β-galactosidase. Effect of aglycone nature and α-deuterium substitution on the hydrolysis of aryl galactosides, *Biochem. J.* **133**, 89–98 (1973).
54. M. L. Sinnott, β-Galactosidase-catalysed hydrolysis of the β-D-galactopyranosylpyridinium cation, *J. Chem. Soc. Chem. Commun.* **1973**, 535–536.
55. M. L. Sinnott and S. G. Withers, The β-galactosidase-catalysed hydrolysis of β-D-galactopyranosyl pyridinium salts, *Biochem. J.* **143**, 751–762 (1974).
56. M. L. Sinnott, O. M. Viratelle, and S. G. Withers, pH- and magnesium ion-dependence of the hydrolyses of β-D-galactopyranosyl pyridinium salts catalysed by *Escherichia coli* β-galactosidase, *Biochem. Soc. Trans.* **3**, 1006–1009 (1975).
57. J. Conchie, A. J. Hay, I. Strachan, and G. A. Levvy, Inhibition of glycosidases by aldonolactones of corresponding configuration, *Biochem. J.* **102**, 929–941 (1967).
58. A. L. Fink and K. J. Angelides, β-Galactosidase catalyzed hydrolysis of *O*-nitrophenyl-β-D-galactoside at subzero temperatures. Evidence for a galactosyl-enzyme intermediate, *Biochem. Biophys. Res. Commun.* **64**, 701–708 (1975).
59. D. F. Wentworth and R. Wolfenden, Slow binding of D-galactal, a "reversible" inhibitor of bacterial β-galactosidase, *Biochemistry* **13**, 4715–4720 (1974).
60. T. E. Barman, *Enzyme Handbook*, Vol. 2, Springer-Verlag New York, Inc., New York (1969), pp. 578–779.
61. W. W. Pigman, Specificity, classification, and mechanism of action of glvcosidases. *Adv. Enzymol.* **4**, 41–74 (1944).
62. G. Legler, Labelling of the active centre of a β-glucosidase, *Biochim. Biophys. Acta* **131**, 728–729 (1968).
63. R. L. Nath and H. N. Rydon, The influence of structure on the hydrolysis of substituted phenyl β-D-glucoside by emulsin, *Biochem. J.* **57**, 1–10 (1954).
64. G. A. Levvy and S. M. Snaith, The inhibition of glycosidases by aldonolactones, *Adv. Enzymol.* **36**, 151–181 (1972).
65. J. Conchie, A. L. Gelman, and G. A. Levvy, Inhibition of glycosidases by aldonolactones of corresponding configuration. The C-4- and C-6-specificity of β-glucosidase and β-galactosidase, *Biochem. J.* **103**, 609–615 (1967).
66. G. Legler and F. Witassek, Anzahl der aktiven Zentren der β-Glucosidasen A und B aus dem Süssmandel-emulsin durch Fluoreszenzmessungen, *Hoppe-Seyler's Z. Physiol. Chem.* **355**, 617–625 (1974).
67. E. T. Reese and F. W. Parrish, Nojirimycin and D-glucono-1,5-lactone as inhibitors of carbohydrases, *Carbohydr. Res.* **18**, 381–388 (1971).

68. H. L. Lai and B. Axelrod, 1-Aminoglycosides, a new class of specific inhibitors of glycosidases, *Biochem. Biophys. Res. Commun.* **54**, 463–468 (1973).

69. B. Capon, Mechanism in carbohydrate chemistry, *Chem. Rev.* **69**, 407–498 (1969).

70. C. A. Dekker, Nucleic acids. Selected topics related to their enzymology and chemistry, *Annu. Rev. Biochem.* **29**, 453–474 (1960).

71. F. Micheel and A. Heesing, Über die stabilität der *N*-glykoside, insbesondere der guanidinglykoside und der nucleoside, *Chem. Ber.* **94**, 1814–1824 (1961).

72. G. W. Kenner, in: *The Chemistry and Biology of Purines*, (G. E. W. Wolstenholme and C. M. O'Connor, eds.), pp. 312–313, Little, Brown, Boston (1957).

73. J. A. Zoltewicz, D. F. Clark, T. W. Sharpless, and G. Grahe, Kinetics and mechanism of the acid-catalyzed hydrolysis of some purine nucleosides, *J. Am. Chem. Soc.* **92**, 1741–1750 (1970).

74. J. A. Zoltewicz and D. F. Clark, Kinetics and mechanism of the hydrolysis of guanosine and 7-methylguanosine nucleosides in perchloric acid, *J. Org. Chem.* **37**, 1193–1197 (1972).

75. R. P. Panzica, R. J. Rousseau, R. K. Robins, and L. B. Townsend, A study on the relative stability and a quantitative approach to the reaction mechanism of the acid-catalyzed hydrolysis of certain 7- and 9-β-D-ribofuranosylpurines, *J. Am. Chem. Soc.* **94**, 4708–4714 (1972).

76. E. R. Garrett and P. J. Mehta, Solvolysis of adenine nucleosides. I. Effects of sugars and adenine substituents on acid solvolyses, *J. Am. Chem. Soc.* **94**, 8532–8541 (1972).

77. L. Hevesi, E. Wolfson-Davidson, J. B. Nagy, O. B. Nagy, and A. Bruylants, Contribution to the mechanism of the acid-catalyzed hydrolysis of purine nucleosides, *J. Am. Chem. Soc.* **94**, 4715–4720 (1972).

78. E. R. Garrett, Kinetics of the hydrolytic degradation of a nucleoside, the antibiotic psicofuranine, *J. Am. Chem. Soc.* **82**, 827–832 (1960).

79. R. Shapiro and S. Kang, Uncatalyzed hydrolysis of deoxyuridine, thymidine, and 5-bromodeoxyuridine, *Biochemistry* **8**, 1806–1810 (1969).

80. R. Shapiro and M. Danzig, Acidic hydrolysis of deoxycytidine and deoxyuridine derivatives. The general mechanism of deoxyribonucleoside hydrolysis, *Biochemistry* **11**, 23–29 (1972).

81. J. Cadet and R. Teoule, Nucleic acid hydrolysis. I. Isomerization and anomerization of pyrimidic deoxyribonucleosides in an acidic medium, *J. Am. Chem. Soc.* **96**, 6517–6519 (1974).

82. B. M. Anderson and C. D. Anderson, The effect of buffers on nicotinamide adenine dinucleotide hydrolysis, *J. Biol. Chem.* **238**, 1475–1478 (1963).

83. N. Kaplan, in: *Current Aspects of Biochemical Energetics*, (N. O. Kaplan and E. P. Kennedy, eds.), pp. 447–458, Academic Press, New York (1966).

84. N. I. Swislocki and N. O. Kaplan, Purification and characterization of diphosphopyridine nucleosidase from pig brain, *J. Biol. Chem.* **242**, 1083–1088 (1967).

85. F. Schuber and P. Travo, Calf-spleen nicotinamide-adenine dinucleotide glycohydrolase. Solubilization purification and properties of the enzyme, *Eur. J. Biochem.* **65**, 247–255 (1976).

86. J. H. Yuan and B. M. Anderson, Bull semen nicotinamide adenine dinucleotide nucleosidase, *J. Biol. Chem.* **246**, 2111–2115 (1971).

87. J. H. Yuan, L. B. Barnett, and B. M. Anderson, Bull semen nicotinamide adenine dinucleotide nucleosidase. II. Physical and chemical studies, *J. Biol. Chem.* **247**, 511–514 (1972).

88. J. Everse and N. O. Kaplan, Characteristics of microbial diphosphopyridine nucleosidases containing exceptionally large amounts of polysaccharides, *J. Biol. Chem.* **243**, 6072–6075 (1968).

89. J. Everse, K. E. Everse, and N. O. Kaplan, The pyridine nucleosidases from *Bacillus subtilis* and *Neurospora crassa*. Isolation and structural properties, *Arch. Biochem. Biophys.* **169**, 702–713 (1975).

90. J. Everse, J. B. Griffin, and N. O. Kaplan, The pyridine nucleosidase from *Bacillus subtilis*. Kinetic properties and enzyme-inhibitor interactions, *Arch. Biochem. Biophys.* **169**, 714–723 (1975).

91. B. M. Anderson, C. J. Ciotti, and N. O. Kaplan, Chemical properties of 3-substituted pyridine analogues of diphosphopyridine nucleotide, *J. Biol. Chem.* **234** 1219–1225 (1959).

92. C. Zervos, R. Apitz, A Stafford, and E. H. Cordes, Kinetic properties of bull semen NAD glycohydrolase, *Biochim. Biophys. Acta* **220**, 636–638 (1970).
93. J. H. Yuan and B. M. Anderson, Bull semen nicotinamide adenine dinucleotide nucleosidase. III. Properties of the substrate binding site, *J. Biol. Chem.* **247**, 515–520 (1972).
94. J. H. Yuan and B. M. Anderson, Bull semen nicotinamide adenine dinucleotide nucleosidase. V. Kinetic studies, *J. Biol. Chem.* **248**, 417–421 (1973).
95. J. H. Yuan and B. M. Anderson, Bull semen nicotinamide adenine dinucleotide nucleosidase. IV. Nonpolar interactions of inhibitors with the substrate binding site, *Arch. Biochem. Biophys.* **149**, 419–424 (1972).
96. N. I. Swislocki, M. I. Kalish, F. I. Chasalow, and N. O. Kaplan, Solubilization and comparative properties of some mammalian diphosphopyridine nucleosidases, *J. Biol. Chem.* **242**, 1089–1094 (1967).
97. R. Apitz, K. Mickelson, K. Shriver, and E. H. Cordes, Some properties of reactions catalyzed by pig brain NAD glycohydrolase, *Arch. Biochem. Biophys.* **143**, 359–364 (1971).
98. E. Cayama, R. Apitz-Castro, and E. H. Cordes, Substrate-dependent, thiol-dependent inactivation of pig brain nicotinamide adenine dinucleotide glycohydrolase, *J. Biol. Chem.* **248**, 6479–6483 (1973).
99. (a) S. Green and A. Dobrjansky, pH-Dependent inactivation of nicotinamide-adenine dinucleotide glycohydrolase by its substrate, oxidized nicotinamide-adenine dinucleotide, *Biochemistry* **10**, 2496–2500 (1971). (b) S. Green and A. Dobrjansky, Inactivation of nicotinamide-adenine dinucleotide glycohydrolase from livers of different mammalian species by nicotinamide-adenine dinucleotide, *Biochemistry* **10**, 4533–4538 (1971).
100. V. L. Schramm and L. I. Hochstein, Purification, crystallization, and subunit structure of allosteric adenosine 5-monophosphate nucleosidase, *Biochemistry* **11**, 2777–2783 (1972).
101. V. L. Schramm, Kinetic properties of allosteric adenosine monophosphate nucleosidase from *Azotobacter vinelandii*, *J. Biol. Chem.* **249**, 1729–1736 (1974).
102. V. L. Schramm and L. I. Hochstein, Stabilization of allosteric monophosphate nucleosidase by inorganic salts, substrate, and essential activator, *Biochemistry* **10**, 3411–3417 (1971).
103. V. L. Schramm and J. F. Morrison, Studies on the allosteric modification of nucleoside diphosphatase activity by magnesium nucleoside triphosphates and inosine diphosphate, *Biochemistry* **10**, 2272–2277 (1971).
104. V. L. Schramm, P. Freidenreich, and F. Fatabene, The inhibition of AMP nucleosidase by formycin-5-phosphate, a possible transition state analog, *Fed. Proc. Fed. Am. Soc. Exp. Biol.* **35**, 1706 (1976).

12

Decarboxylations of β-Keto Acids and Related Compounds

Ralph M. Pollack

1. INTRODUCTION

The transfer of a proton from one atom to another is one of the most fundamental processes in chemistry and biochemistry. The detailed mechanism of this reaction has been given much attention both for proton transfers to and from carbon and the more electronegative oxygen and nitrogen.[1] An important class of proton transfers involves those reactions in which another bond is made or broken in addition to the proton transfer itself. Two extreme conditions may occur. The two steps of the reaction may be concerted (all bond breaking and making simultaneous) or stepwise (formation or cleavage of one bond leading to a true intermediate, followed by further bond making and/or breaking). We shall use the definition of a "true intermediate" proposed by Bauer,[2] that is, any species with a lifetime of greater than one molecular vibration: in other words a species with restoring forces for all of its vibrational motions.

A particularly intriguing set of reactions which may occur either by stepwise or concerted pathways is the decarboxylation of β-keto acids and related compounds (**1a–c**):

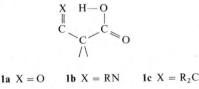

1a X = O   **1b** X = RN   **1c** X = R₂C

Ralph M. Pollack • Laboratory for Chemical Dynamics, Department of Chemistry, University of Maryland Baltimore County, Baltimore, Maryland.

Each of these reactions may be envisioned to proceed through either a concerted mechanism,

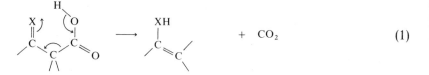

$$+ \quad CO_2 \qquad (1)$$

or a stepwise mechanism,

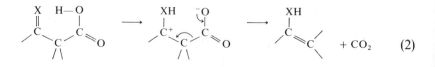

$$+ CO_2 \qquad (2)$$

These systems provide an excellent opportunity to investigate the nature of cyclic decompositions involving proton transfer. In this chapter we shall discuss the currently available evidence for the transition state structures for decarboxylations of β-keto acids (**1a**), Schiff bases of β-keto acids (**1b**), and β,γ-unsaturated acids (**1c**). This evidence will be interpreted in terms of the nature of the reaction pathway, i.e., whether it is concerted or stepwise. For reactions which are concerted we shall also be concerned with the exact nature of the proton transfer. Is the proton in a potential well in the transition state, or is it actually undergoing translation? In addition, the evidence for the transition-state structures for enzymic β-keto acid decarboxylation by acetoacetate decarboxylase will be reviewed and discussed.

2. DECARBOXYLATION OF β-KETO ACIDS

2.1. General Mechanism

Decarboxylation of β-keto acids (e.g., acetoacetic acid, **2**) occurs spontaneously in solution to give carbon dioxide and a ketone:

$$\underset{\textbf{2}}{CH_3\overset{\overset{\textstyle O}{\|}}{C}-CH_2-COOH} \longrightarrow CH_3\overset{\overset{\textstyle O}{\|}}{C}CH_3 + CO_2 \qquad (3)$$

Both the free acid and the anion are generally labile with respect to decarboxylation, although the undissociated acids usually are more reactive. A similar decarboxylation also occurs in the case of malonic acid and related 1,1-dicarboxylic acids:

$$HOOC-CH_2-COOH \longrightarrow HOOC-CH_3 + CO_2 \qquad (4)$$

The basic mechanistic features of the decarboxylation have been delineated by several workers over the last 50 years. In 1929, Pedersen[3] proposed that the reactive form of the free keto acids is the ketone and that the corresponding enol is unreactive toward decarboxylation. This conclusion was based on the fact that undissociated dimethylacetoacetic acid, which cannot enolize, is 4.5-fold more reactive than acetoacetic acid itself.[3] Other nonenolizable compounds which undergo facile decarboxylation include α-bromocamphorcarboxylic acid,[4] dialkylmalonic acids,[5] and α,α-dimethylbenzoylacetic acid.[6] From the near identity of reaction rates for the halogenation and the decarboxylation of dimethylacetoacetic acid, Pedersen[7] concluded that decarboxylation of β-keto acids leads directly to the enol, which in the presence of bromine or iodine is rapidly halogenated:

$$\underset{slow}{CH_3\overset{O}{\overset{\|}{C}}C(CH_3)_2CO_2H \longrightarrow} CH_3\overset{OH}{\overset{|}{C}}=C(CH_3)_2 \underset{fast}{\overset{X_2}{\longrightarrow}} CH_3\overset{O}{\overset{\|}{C}}-C(CH_3)_2X \qquad (5)$$

This suggestion is supported by the resistance of bridgehead β-keto acids to decarboxylation. For example, 4-methylbicyclo[3.3.1]non-3-en-9-one-1-carboxylic acid (3) and its saturated analog (4) do not decarboxylate when heated at 250° for 30 min[8]:

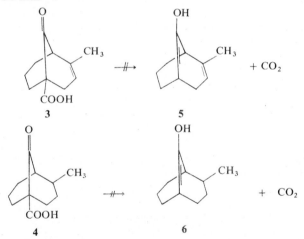

3   **5**

4   **6**

This result is a consequence of the difficulty in forming a highly strained double bond in the product enols (5 and 6) as predicted by Bredt's rule.[9] Bicyclo[3.3.1]-nonan-2-one-1-carboxylic acid (7), which can form a less strained enol (8), decarboxylates readily at 145°[10]:

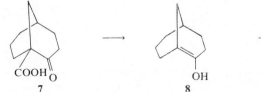

7   **8** $+ \quad CO_2$ (6)

Several transition states have been proposed to account for these observations. They differ primarily in terms of the relative timing of hydrogen transfer and carbon–carbon bond cleavage. Pedersen[11] initially proposed the intermediacy of the dipolar form of the acid (9), which breaks down to products,

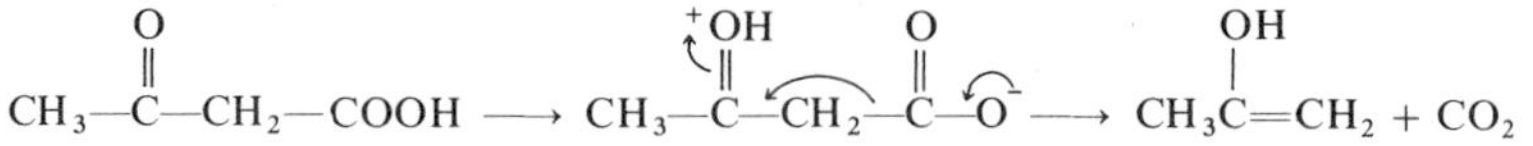

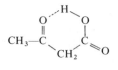

$$CH_3-\overset{O}{\underset{\|}{C}}-CH_2-COOH \longrightarrow CH_3-\overset{+OH}{\underset{\|}{C}}-CH_2-\overset{O}{\underset{\|}{C}}-O \longrightarrow CH_3C=CH_2 + CO_2 \qquad (7)$$

9

but he did not specify whether proton transfer or carbon–carbon bond cleavage is rate determining. Westheimer[12] subsequently suggested, on the basis of solvent effects, that the decarboxylation occurs through the intermediacy of a cyclic form (**10**):

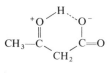

10

However, as Hine[13] has pointed out, electrostatic interactions as well as hydrogen-bonding considerations should give the dipolar ion a structure (**11**) very similar to **10**:

11

There are several questions which might be asked concerning the detailed nature of this transition state: (1) What is the relative timing of the bond-making and bond-breaking processes? Is the hydrogen transferred prior to, subsequent to, or concurrent with the carbon–carbon bond cleavage? (2) What is the magnitude of the charge separation in the transition state? (3) How does the structure of the transition state vary with changes in reactants and/or conditions? Much evidence is available which allows a description of the transition state for this process in some detail. The decarboxylation of the undissociated acids will be considered first, followed by the reaction of the carboxylic acid anions.

2.2. Nature of the Transition State for the Free Acids

2.2.1. Medium Effects

In 1941, Westheimer and Jones[12] showed that the rate of carbon dioxide evolution from dimethylacetoacetic acid is independent of the polarity of the medium for water, 25%, 50%, and 75% methanol and 50% dioxane as solvents. They interpreted this result to mean that the reaction does not occur via the dipolar ion. This formulation has been criticized by Hine,[13] who noted that even though the concentration of the dipolar ion would decrease in the more nonaqueous solvents, its rate of decomposition to products should increase since this process involves charge destruction. It would appear, however, that neither of these two views is strictly correct. According to transition-state theory,[14] a change in the rate of reaction induced by a solvent modification only gives information about the relative energies of the reactant(s) and the transition state. The lack of rate dependence of the decarboxylation on solvent polarity means simply that the polarities of the transition state and the ground state are very similar. Since the ground state has only slight charge separation, the transition state must be essentially nonpolar. The solvent effect does not, however, give any definitive information about the relative timing of hydrogen transfer and carbon–carbon bond cleavage.

Other investigations have also shown that solvent effects on β-keto acid decarboxylations are small and somewhat variable. Benzoylacetic acid decarboxylates slightly more rapidly in mixtures of isopropanol–water and acetonitrile–water than in pure water.[15] In dioxane–water mixtures, benzoylacetic acid shows a minor variation in rate with solvent, with a slight rate maximum at 50% dioxane water.[16] α,α,-Dimethylbenzoylacetic acid, on the other hand, reacts about threefold slower in 50% dioxane–water than in completely aqueous solution.[17] A rate maximum has also been observed in mixtures of organic solvents and water for acetone dicarboxylic acid[18] and phenylmalonic acid.[19]

The rates of decarboxylation of acetoacetic acid[20] and benzoylacetic acid[21] have been examined in concentrated mineral acid solutions. Increased concentrations of acid inhibit these reactions, although the effect is not large. Since inorganic salts also act as inhibitors and the effect is similar to that observed with acids, the acid inhibition has been ascribed to a simple salt effect.[20]

2.2.2. Substituent Effects

The above conclusion that the transition state is only slightly polar is supported by kinetic studies of the decarboxylation of benzoylacetic acids. For a series of ring-substituted benzoylacetic acids (**12**) in aqueous solution,

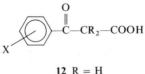

12 R = H
13 R = CH$_3$

Straub and Bender[15] have found that, within experimental error, the rates of decarboxylation are independent of substituent. Similarly, Hay and Tate[16]

observed that p-methoxybenzoylacetic acid reacts only slightly faster than p-nitrobenzoylacetic acid (ca. 1.5-fold) in 50% dioxane. In contrast, Logue et $al.$[6] have found a tendency for electron-withdrawing substituents to accelerate the decomposition of α,α-dimethylbenzoylacetic acids (13) in aqueous hydrochloric acid, but this effect is quite small ($\rho = +0.3$). These results confirm the conclusions reached earlier on the basis of solvent effects that the reaction occurs through a transition state with little charge separation.

In contrast to the lack of substituent dependence observed in aqueous solutions and 50% dioxane, it has been reported[22] that substituent effects on the decarboxylation of 12 in benzene are substantial ($\rho = -1.0$), suggesting that the transition state is polar in benzene and nonpolar in water. It has recently been shown,[6] however, that benzoylacetic acid is extensively enolized in benzene. Consequently, these rate constants do not refer solely to the decarboxylation process, and conclusions regarding the nature of the transition state in benzene are unreliable.

2.2.3. Isotope Effects

The independence of the decarboxylation rate of β-keto acids on both ring substituent and solvent suggests that a cyclic transition state is involved, with carbon–carbon bond cleavage synchronous with hydrogen transfer (14):

$$\tag{8}$$

The most direct method for probing the synchrony of this process is the use of isotope effects. If a bond to the isotopic atom is being broken in the rate-determining step, an isotope effect should be observed. In 1960, Hodnett and Rowton[23] reported carbon-14 isotope effects for the decarboxylation of 2-benzoylpropionic acid (15). They found an isotope effect for substitution at the carboxyl carbon ($k^{12}/k_*^{14} = 1.074$) and the α-carbon ($k^{12}/k_\alpha^{14} = 1.051$). No effect was observed for isotopic substitution at C_β ($k^{12}/k_\beta^{14} = 1.000$). These results indicate that the rate-determining step in the decarboxylation involves cleavage of the C_*—C_α bond, with little or no change in the bonding at C_β. Isotope effects for ^{13}C and ^{14}C substitution at the carboxyl group have also been found for oxaloacetic acid[24,25] and malonic acid[26-31] decarboxylations, although the exact magnitude of the effects in malonic acid is somewhat uncertain.[32]

Initial investigations[22] of hydrogen isotope effects with ring-substituted benzoylacetic acids in benzene gave somewhat unusual results. In this system isotope effects for decarboxylation were reported to vary from 0.8 to 2.8 (k_H/k_D), suggesting that hydrogen transfer is part of the reaction coordinate in

Table I

Hydrogen Isotope Effects
in the Decarboxylation of p-Substituted
α,α-Dimethylbenzoylacetic Acids

X	k_H/k_D
p-CH$_3$O	1.14 ± 0.02
H	1.20 ± 0.02
p-Cl	1.20 ± 0.02
p-NO$_2$	1.41 ± 0.04

some systems but not in others or that the degree of hydrogen transfer varies markedly with ring substituents. However, as mentioned previously, these compounds enolize extensively in benzene, so that conclusions drawn from this work are unreliable. A reinvestigation of hydrogen isotope effects using dimethylbenzoylacetic acids, which are incapable of enolizing, in water has shown the hydrogen isotope effects to be uniformly small (Table I).[6] These low isotope effects suggest that either the hydrogen is not undergoing translation in the rate-determining step or that the transition state is either very early or very late with regard to hydrogen transfer.[33] Let us first consider the possibility of a very asymmetric position for the hydrogen in a concerted mechanism. From the relative rates of decarboxylation of the free α,α-dimethylbenzoylacetic acids and their anions,[17] $pK_a^\ddagger$ values for the transition states of the decomposition of the free acids may be calculated.[34] These values correspond to virtual equilibrium constants for the dissociation of the proton in the transition state for decarboxylation of the acids,

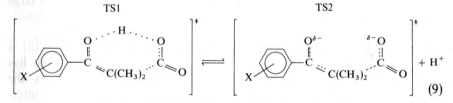

$$\tag{9}$$

and are given in Table II along with the pK_a's for the dimethylbenzoylacetic acids themselves. The transition state for the decomposition of the acid (TS1) has a $pK_a^\ddagger$ which is close to that of a normal carboxylic acid for all substituents. This result may be used to rule out a structure for TS1 with the hydrogen almost fully transferred, i.e., a late transition state such as 16,

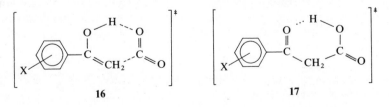

16   17

Table II

pK$_a$ Values for the Dissociation of
α,α-Dimethylbenzoylacetic Acids and the
Transition States for Their Decarboxylations

Substituent	pK$_a{}^a$	pK$_a^{\ddagger}$
p-CH$_3$O	3.43	4.78
H	3.40	4.44
p-Cl	3.38	4.02
p-NO$_2$	3.24	3.16

^{a} pK$_a$ values for the corresponding benzoylacetic acids are used as estimates.

where both the hydrogen transfer and carbon–carbon bond cleavage are virtually complete. A structure of this type should show a pK$_a$ very similar to the enol of the corresponding acetophenone, which for acetophenone itself has been estimated to be about 11.[35] It is clear that TS1 is a much stronger acid than would be predicted on the basis of a late transition state. On the other hand, TS1 has a pK$_a$ quite similar to the reactant acids themselves, and one is tempted to assign a structure to TS1 in which there is very little breakage of the initial C—COOH bond (17). A look, however, at the substituent dependence of pK$_a^{\ddagger}$ shows that this structure, too, must be incorrect, since there is a large substituent dependence of pK$_a^{\ddagger}$ ($\rho = 1.7$), although the pK$_a$'s of benzoylacetic acids are virtually independent of substituent ($\rho = +0.2$). Since neither a transition state which has very little hydrogen transfer nor one which has almost complete hydrogen transfer is consistent with the pK$_a^{\ddagger}$ values, the conclusion must be drawn that at the transition state the hydrogen is intermediate between the two oxygens.

It is also possible to use the transition-state pK$_a^{\ddagger}$ values as evidence against the zwitterionic mechanism. If this were the reaction pathway, then the transition state should have a structure intermediate between that of the zwitterion and the enol. Making the reasonable assumption that the extent of reaction can be estimated from the pK$_a^{\ddagger}$ of the transition state relative to that of the reactant and products, it is possible to estimate what the charge separation would have to be if this were the mechanism. Using a pK$_a$ value of -6.4 for protonated acetophenone[36,37] as a model for the pK$_a$ of the zwitterion and a pK$_a$ of 11 for the enol of acetophenone, it can be calculated[34] that a pK$_a^{\ddagger}$ of 4.4 would correspond to a structure about 60% along the reaction coordinate from zwitterion to enol. Consequently, if the reaction were proceeding via rate-determining cleavage of the zwitterion, the transition state should have substantial charge separation (about 0.4 charge units).* A charge separation

---

* Acetophenone has been used as a model for the zwitterion even though internal hydrogen bonding almost surely exists in the zwitterion. Although this would lower the acidity of the zwitterion substantially, incorporation of a correction factor would only increase the resemblance of the transition state to the zwitterion.

of this magnitude would seem to be ruled out by both the solvent and substituent effects.

At this point, it is possible to characterize the transition state as a concerted one with approximately equal carbon–carbon bond cleavage and hydrogen transfer (by solvent effects and substituent effects) which is neither very early nor very late (by p$K_a^{\ddagger}$ data). However, the uniformly small solvent hydrogen isotope effects observed in the decarboxylation of α,α-dimethyl-benzoylacetic acids in water suggest that the proton is not undergoing translation in the reaction coordinate. These isotope effects may be interpreted in terms of the model proposed by Swain et al.[38,39] in which proton transfer between electronegative atoms accompanying heavy-atom reorganization is not part of the reaction-coordinate motion. Consequently, primary hydrogen isotope effects in these transfers need not be observed. The transition state can then be represented as **18**, with the wavy lines representing a stable potential well and the dotted lines translation*:

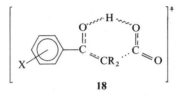

18

A more detailed analysis of the reaction-coordinate geometry for this process may be obtained by examining the implications of the lack of coupling of hydrogen transfer and carbon–carbon bond cleavage, following the ideas of Schowen[39] and Thornton.[41] The most nucleophilic electrons at the carbonyl group should be the *n* electrons, rather than the π electrons which which are still partially bound to the carbonyl carbon in the transition state. Consequently, the newly forming O—H bond must be in the same plane as the original C—C=O system (**19**):

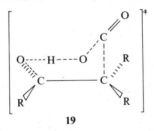

19

The carbon–carbon bond which is breaking, on the other hand, must be perpendicular to this plane to allow overlap of the incipient *p* orbital with the *p* orbital of the carbonyl carbon. The transition state, therefore, must have five atoms coplanar with only the leaving carbon above the plane. This arrange-

---

* An exception to this model for the transition state is found in the report that $k_{H_2O}/k_{D_2O} = 3.1$ for the decarboxylation of 1-ethyloxalacetate.[40] However, this work could not be reproduced.[6]

ment nicely explains the low isotope effect because coupling of the asymmetric stretch leading to C—C bond cleavage is with a bending mode of the O—H—O system rather than a stretching mode. Consequently, during carbon–carbon bond cleavage, both the symmetric and antisymmetric stretching vibrations are stable and retain their zero-point energy.

A possible flaw in a six-membered ring transition state of this type which has been pointed out by Hine and Sachs[42] is that it would be impossible for the O—H—O system to attain linearity in a ring this small. While it is no doubt true that great strain would be necessary to meet all of the stereoelectronic constraints in a transition state such as **19**, a relaxation of the requirement for strict linearity in the hydrogen bond should allow a transition state similar to this one. Theoretical calculations have shown[43] that although hydrogen bonds do show a preference for linearity, substantial deviations can be tolerated without much expenditure of energy. Furthermore, it is not necessary to invoke the commonly accepted bond angle of 120° for the $\diagup$C=O $\cdots$ H hydrogen bond.[44] Molecular models show that a narrower angle in the transition state increases the linearity of the O—H—O bond. Gandour[45] has suggested that proton-transfer reactions might involve the intermediacy of a water molecule through which proton transfer could take place. This would give an eight-membered ring in this case, allowing a linear arrangement of donor–proton–acceptor. However, a water molecule is probably not involved in the decomposition of β-keto acids since α,α-dimethylacetoacetic acid decarboxylates slightly faster in benzene than in water.[22]

In summary, it appears that a six-membered cyclic transition state, such as **19**, is involved in β-keto acid decarboxylations. This transition state has little or no charge separation, with both carbon–carbon bond cleavage and hydrogen transfer occurring to about the same extent. The hydrogen is probably bound to the n electrons, and its motion is not coupled with heavy-atom reorganization.

2.3. Nature of the Transition State for the Anions

The decomposition of the carboxylate anions of β-keto acids has not been studied as extensively as the corresponding reaction of the undissociated acids. However, some information does exist which may be used to describe the transition state for this process. The mechanism for monoanion decarboxylation is generally accepted as being a simple carbon–carbon bond cleavage to give carbon dioxide and the enolate ion:[46]

$$R_1-\underset{R_2}{\overset{O}{\underset{}{\parallel}}}\!\!\!\overset{}{\underset{R_3}{C}}-\overset{O}{\underset{}{\overset{\parallel}{C}}}-O^- \longrightarrow \underset{R_1}{\overset{O^-}{\diagdown}}C=C\underset{R_2}{\overset{R_3}{\diagup}} + CO_2 \tag{10}$$

A description of the nature of the transition state of this reaction can be gleaned from an analysis of the observed substituent effects. For the decomposition of the anions of ring-substituted benzoylacetic acids in water, Straub and Bender[15] found a value of $\rho = +1.42$. A similar substituent dependence exists for these compounds in aqueous dioxane ($\rho = +1.7$),[16] as well as for the decomposition of the anions of α,α-dimethylbenzoylacetic acids in water ($\rho = +1.7$).[17]

An appropriate reaction which may be used as a model for the anionic decomposition is the ionization of ring-substituted acetophenones:

$$X\text{—}\underset{}{\bigcirc}\text{—}\overset{O}{\overset{\|}{C}}\text{—}CH_3 \rightleftharpoons X\text{—}\underset{}{\bigcirc}\text{—}\overset{O^-}{\overset{|}{C}}\diagdown_{CH_2} + H^+ \qquad (11)$$

A comparison of the ρ values for the decarboxylation reactions with the ρ for equilibrium ionization of acetophenones should give an estimate of the extent of charge formation in the incipient enol at the transition state for the decarboxylation. Unfortunately, ρ for the equilibrium in Eq. (11) is unknown in aqueous solution. However, the substituent dependence for the *rate* of ionization of acetophenones in aqueous hydroxide ion solutions is known ($\rho = +1.0$).[47] In addition, the magnitude of the primary tritium isotope effects for this reaction ($k_H/k_T = 12$–18, equivalent to $k_H/k_D = 5.6$–7.4)[47,48] suggests that the transition state for deprotonation is nearly symmetrical.[49,50] This conclusion is supported by the near equivalence of the pK_a's of water and acetophenone.* Since ρ for decarboxylation of benzoylacetic acids ($+1.4$ to $+1.7$) is greater than 1.0, it is likely that the transition state for this reaction is one in which the carbon–carbon bond is almost completely broken, (i.e., product-like rather than reactant-like).

It will be noted that the substituent effect on the stability of the benzoylacetate ions has been ignored in the above analysis. However, this effect should be small, as evidenced by the small dependence of the pK_a of these acids on substituent ($\rho = +0.2$).[15] In any case, inclusion of this effect would only strengthen the argument for a late transition state.

2.4. Amine Catalysis

Primary amine catalysis of the decarboxylation of β-keto acids is of wide interest as a model system for acetoacetate decarboxylase since enzymic decarboxylation of acetoacetic acid is known to proceed through an imine intermediate.[51,52] The primary amine-catalyzed reaction also proceeds through an imine intermediate,[12,53-55] shown for acetoacetic acid in

$$RNH_2 + CH_3\overset{O}{\overset{\|}{C}}CH_2\overset{O}{\overset{\|}{C}}OH \longrightarrow CH_3\overset{RN}{\overset{\|}{C}}CH_2COOH \longrightarrow CH_3\overset{RNH}{\overset{|}{C}}{=}CH_2 + CO_2$$

$$\downarrow H_2O \qquad (12)$$

$$RNH_2 + CH_3\overset{O}{\overset{\|}{C}}CH_3$$

---

* Loudon[35] has estimated the carbon pK_a of acetophenone in water to be about 16.

Evidence for this intermediate comes from the fact that primary amines are much better catalysts than secondary amines, while tertiary amines are inactive.[56] In addition, Taguchi and Westheimer[57] have shown that the Schiff bases **20** and **21** decarboxylate about 10^5- to 10^7-fold more rapidly than the corresponding β-keto acids in decalin and dioxane:

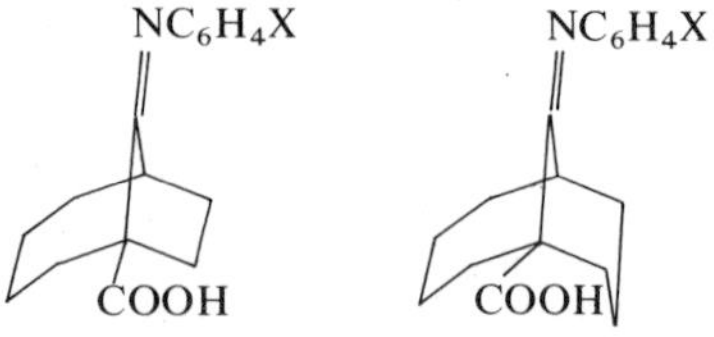

20a X = OCH_3 **21a** X = OCH_3
20b X = H **21b** X = H
20c X = Cl **21c** X = Cl

It has been suggested[58] that the aniline-catalyzed decarboxylation of oxaloacetic acid may proceed through the carbinolamine rather than the Schiff base in water:

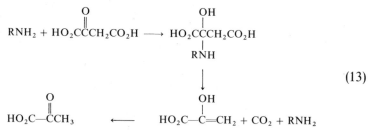

$$RNH_2 + HO_2C\overset{O}{\overset{\|}{C}}CH_2CO_2H \longrightarrow HO_2C\underset{RNH}{\underset{|}{\overset{OH}{\overset{|}{C}}}}CH_2CO_2H$$

$$\downarrow \qquad (13)$$

$$HO_2C{-}\overset{O}{\overset{\|}{C}}CH_3 \longleftarrow HO_2C{-}\overset{OH}{\overset{|}{C}}{=}CH_2 + CO_2 + RNH_2$$

However, there seems to be no compelling argument in favor of this suggestion, and in view of the fact that imines of β-keto acids are labile toward decarboxylation, an imine intermediate appears more reasonable.

Very little information is available about the nature of the transition state for decarboxylation of β-iminoacids in water. The major drawback to obtaining the requisite data is the fact that the rate constant for hydrolysis of the imine is usually similar in magnitude to the rate constant for decarboxylation,[54,55,59] making extraction of the rate constant for the decarboxylation difficult.[53] Furthermore, rate constants for isolable intermediates such as **20** and **21** cannot be determined in water since they hydrolyze to the corresponding ketone, rather than decarboxylate.

One may speculate with reasonable assurance, however, that these reactions are two-step processes, involving preequilibrium proton transfer followed by carbon–carbon bond cleavage:

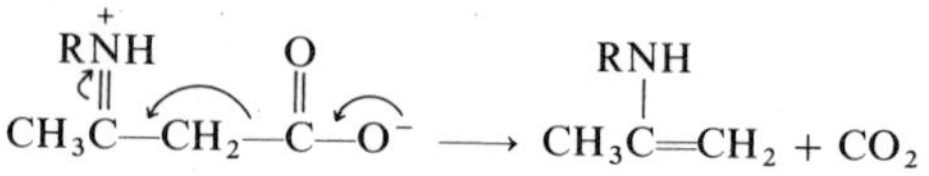

(14)

22

Since pK_a values for imines are generally somewhat higher than for carboxylic acids,[60-62] the imine intermediate will no doubt exist as the zwitterion (22) in aqueous solutions. Decarboxylation can then occur simply by heavy-atom reorganization with no need for proton transfer to take place. A similar situation exists in the thermal decarboxylation of 2-pyridylacetic acid (23):

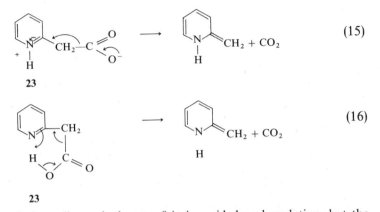

This reaction is formally equivalent to β-iminoacid decarboxylation, but the pyridine ring lends sufficient stability to the imine bond so that hydrolysis does not occur. On the basis of the similarity of the rates of decarboxylation of 23 and the corresponding 4-substituted pyridylacetic acid, Stermitz and Huang[63] postulated that the decomposition occurs through the zwitterion [Eq. (15)] rather than a cyclic transition state [Eq. (16)]. Since the cyclic mechanism is unavailable to 4-pyridylacetic acid, this conclusion is undoubtedly correct. Taylor[64,65] has found that this mechanism is general for a variety of heterocyclic acetic acids which formally resemble imines.

Although it is difficult to investigate the nature of the transition state for these reactions in aqueous solution, it has been possible to obtain some insight into the process in nonaqueous solvents. Taguchi and Westheimer[57] examined the decarboxylation of 20 and 21 in decalin and dioxane and found that the rate of decarboxylation depends only sightly on the substituent in the aniline portion of the molecule ($\rho = -0.5 \pm 0.2$), suggesting a relatively nonpolar transition state. In addition, these reactions show a negligible deuterium isotope effect ($k_H/k_D = 1.0$). These features, coupled with the fact that the imino acid is probably present in the uncharged form rather than the zwitterion, suggest that the transition state is cyclic with the proton in a stable potential well, similar to what is postulated for the free β-keto acids.

The difference in the transition states for imino acid decarboxylation in water and in nonpolar solvents can be rationalized in the following way. According to the "libido" rule postulated by Jencks,[66] concerted reactions are only expected to occur when a stepwise pathway would lead to a very unstable intermediate. In the case of decarboxylation of β-imino acids, a stepwise mechanism would involve the formation of the zwitterion 22:

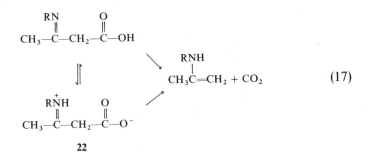

$$\text{(17)}$$

In aqueous solution this intermediate is more stable than the neutral form and, consequently, the stepwise process is favored. In nonpolar solvents, however, the zwitterion becomes extremely unstable. A concerted pathway then allows reaction to occur without the intervention of this high-energy intermediate.

3. DECARBOXYLATIONS OF β,γ-UNSATURATED ACIDS

3.1. General Mechanism

The thermal (ca. 250°) decomposition of β,γ-unsaturated carboxylic acids has been postulated[67] to proceed through a mechanism which is formally identical to the one for β-keto acids:

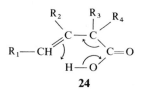

24

a. $R_1 = C_6H_5, R_2 = H, R_3 = R_4 = CH_3$ **d.** $R_1 = R_3 = R_4 = CH_3, R_2 = H$
b. $R_1 = R_2 = H, R_3 = R_4 = CH_3$ **e.** $R_1 = H, R_2 = C_6H_5, R_3 = R_4 = CH_3$
c. $R_1 = R_2 = R_3 = R_4 = H$ **f.** $R_1 = H, R_2 = R_3 = R_4 = CH_3$

As would be predicted for a mechanism of this type, the reaction is unimolecular[68] and has a negative entropy of activation.[68-70] Where stereoisomerism is possible, the more thermodynamically favored *trans* olefins are formed; however, the product is invariably the terminal olefin even when $R_1 = H$ and the internal olefins are more stable, in accordance with a concerted mechanism.[71] It has also been shown that when the carboxyl group is deuterated, the deuterium is found on the γ-carbon of the product, suggesting intramolecular hydrogen transfer.[71] Preequilibrium hydrogen transfer can be ruled out by the observation that there is no deuterium exchange with the C-4 hydrogen prior to decarboxylation.[72]

3.2. Nature of the Transition State

3.2.1. Isotope Effects

Both carbon and hydrogen isotope effects have been examined for the gas phase decarboxylation of 2,2-dimethyl-4-phenylbut-3-enoic acid (**24a**) at about 280°C.[72] For carboxyl-labeled ^{14}C, an isotope effect of 1.035 (k^{12}/k^{14}) was found, clearly implicating carbon–carbon bond cleavage in the rate-limiting transition state. A hydrogen isotope effect of $k_H/k_D = 2.8$ was observed for the same compound labeled at the carboxyl hydrogen. These results demand that both hydrogen transfer and carbon–carbon bond cleavage be occurring in the transition state. Hydrogen isotope effects (k_H/k_D) of 1.8 to 3.0 (187°C) were also found for the decarboxylation of a series of ring-substituted 2,2-dimethyl-3-phenylbut-3-enoic acids (**24e**).[73] Extrapolation of these values to 50° gives isotope effects of 2.4 to 4.8 for this system, in agreement with those expected for rate-limiting hydrogen transfer.

Although the transition state for the decarboxylation of β,γ-unsaturated acids closely resembles that for the decarboxylation of β-keto acids, the difference in the hydrogen isotope effects in the two systems suggests a subtle difference in their structures. Whereas the hydrogen isotope effects in the β-keto acid series ($k_H/k_D \simeq 1.3$) indicate that the hydrogen is in a potential well in the rate-limiting transition state (*vide supra*), the corresponding isotope effects in the β,γ-unsaturated acids ($k_H/k_D \simeq 2.4$–4.8) show that translation of the hydrogen is occurring. That is, hydrogen transfer and carbon–carbon bond cleavage are coupled when hydrogen transfer is to carbon but uncoupled when hydrogen transfer is to oxygen. This difference may be rationalized by noting that carbon has no nonbonding electrons, and consequently hydrogen transfer must be to the π electrons of the double bond (**25**):

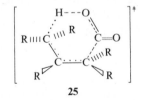

25

Facile coupling of the asymmetric stretch leading to C—C bond cleavage with the asymmetric stretch of the O—H—O system is then possible, leading to the observed hydrogen isotope effects.[39] In contrast, the availability of *n* electrons on oxygen leads to the lack of coupling observed with the benzoylacetic acids. An analogous distinction has previously been drawn for eliminations to form olefins vs. eliminations to form carbonyl groups.[39]

3.2.2. Substituent Effects

As would be expected for a concerted reaction, alkyl substitution on the α-carbon increases the rate of the reaction. The effect of alkyl substitution on the rate was examined by Bigley and May,[74] who found that 2,2-dimethylbut-3-enoic acid (24b) decarboxylates 5.6-fold faster than the parent acid but-3-enoic acid (24c). Since in the transition state a double bond is migrating from the β,γ position to the α,β position and more highly substituted double bonds are more stable, alkyl substitution at the α-carbon should increase the rate of decarboxylation. Similarly, alkyl substitution at the γ-carbon is expected to retard the rate, and this effect is seen in the rate decrease of 6.7-fold for decarboxylation of 2,2-dimethyl-pent-3-enoic acid (24d) compared to 2,2-dimethylbut-3-enoic acid (24b).[74] The effect of substitution at the α-carbon is very similar to what is observed for the β-keto acid series; dimethylacetoacetic acid decomposes 4.5-fold faster than acetoacetic acid itself at 18°C.[3] However, α,α-dimethylbenzoylacetic acid decarboxylates slightly slower than benzoylacetic acid itself in water.[6,15] Possibly steric interactions between the methyl groups and the orthohydrogens of the phenyl ring in the transition state lower the rate of decarboxylation for the dimethyl derivative.

Substituent effects in the β position are also quite revealing. Both phenyl substitution and methyl substitution at this position cause large rate enhancements for decarboxylation. For example 2,2-dimethyl-3-phenylbut-3-enoic acid (24e) decarboxylates 55 times faster than 2,2-dimethyl-but-3-enoic acid itself (24b),[70,74] and 2,2,3-trimenthybut-3-enoic acid (24f) reacts 30-fold more rapidly than the corresponding 2,2-dimethyl derivative (24b). As Bigley and May[74] observed, one would initially expect little or no effect from substitution at the β-carbon since it should stabilize both the newly forming double bond and the disappearing one approximately equally. Although these authors initially attributed this rate enhancement to distortion of the double bond, it is equally logical that a partial positive charge is being produced at the β-carbon and the phenyl and methyl groups act to stabilize this charge. To distinguish between these two interpretations Bigley and Thurman[73,75] examined ring-substituted 2,2-dimethyl-3-phenylbut-3-enoic acids. They found a good Hammett correlation of the rates, with σ^+ giving a ρ value of -1.2. From this result they estimated that a partial positive charge of about 0.2 unit is formed at the β-carbon in the transition state. Therefore, although the reaction is concerted with simultaneous transfer of the hydrogen and carbon–carbon bond cleavage, the extent of hydrogen transfer must be somewhat greater than the extent of carbon–carbon bond cleavage at the transition state.

In this context, it is interesting to reexamine the question of charge separation in the benzoylacetic acid series. Both the decarboxylation of benzoylacetic acids[15] and α,α-dimethylbenzoylacetic acids[6] show ρ values near zero, suggesting little charge buildup in the transition state. In each case, however, there is some charge separation in the starting ketones. Since substituent effects are a function of the *difference* between the polarity of the reactants and the transition state, it is evident that a ρ of near zero for the decarboxylation of benzoylacetic acids implies a transition state with charge separation comparable to the initial keto acids.

Both the decarboxylations of β,γ-unsaturated acids and of β-keto acids then have transition states which are virtually identical in that they are both cyclic with hydrogen transfer slightly more advanced than carbon–carbon bond cleavage. It appears that these transition states differ only in whether the hydrogen is transferred to the n electrons of the carbonyl group (β-keto acids) or the π electrons of the methylene (β,γ-unsaturated acids). In the former, the hydrogen is in a stable potential well (not undergoing translation) at the transition state, whereas in the β,γ-unsaturated acids proton transfer is coupled to carbon–carbon bond cleavage.

4. ACETOACETATE DECARBOXYLASE

4.1. General Mechanism

The enzymic decarboxylation of acetoacetic acid by acetoacetate decarboxylase has been the subject of many recent investigations,[76] and it is now possible to specify the nature of the transition state for this process in some detail. The mechanism has been shown to involve the formation of a Schiff base intermediate from the ε-amino group of a lysine residue of the enzyme and the keto group of the substrate. This intermediate then decomposes, presumably via the zwitterion, to carbon dioxide and the enamine of acetone. Subsequent hydrolysis of the enamine produces acetone and regenerates the active enzyme:

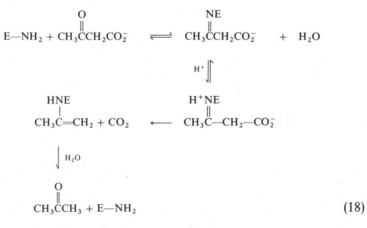

$$\tag{18}$$

This mechanism, which is similar to the corresponding reaction catalyzed by primary amines, was initially postulated on the basis of complete exchange of labeled oxygen from solvent water into the carbonyl group of the substrate during enzymic decarboxylation.[77] Later, it was demonstrated that the enzyme is irreversibly inactivated by borohydride in the presence of acetoacetate by reduction of either the intermediate Schiff base or enamine.[51] Subsequent hydrolysis of the reduced intermediate produced ε-N-isopropyllysine, implicating a lysine residue at the enzyme active site.[52]

One of the most intriguing results to come out of the work on this enzyme is the finding that the active site amine group has an abnormally low pK_a. By an analysis of the pH-rate profile for acylation of this moiety with 2,4-dinitrophenyl propionate, Schmidt and Westheimer[78] concluded that the amino group has a pK_a of 5.9, about 4 units less than the value of a normal lysine ε-amino group. The assignment of this pK_a to the essential lysine is supported by the fact that enzyme inhibited with acetopyruvate or acetic anhydride is inactive toward acylation by 2,4-dinitrophenyl propionate, showing that it is the active-site amino group which reacts with dinitrophenyl propionate. Furthermore, acylated enzyme is completely unreactive toward decarboxylation of acetoacetic acid.

The pK_a of the essential amino group was also estimated by using 5-nitrosalicylaldehyde as a reporter group.[79,80] Reduction of the aldehyde onto the active site with sodium borohydride,

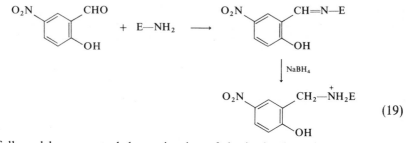

(19)

was followed by a spectral determination of the ionization constants of both the phenolic hydroxyl group and the ammonium ion. These pK_a values were found to be 2.4 and 6.0, respectively, substantially lower than the corresponding values for the model compound N-methyl-2-hydroxy-5-nitrobenzylamine (5.9 and 10.7, respectively). The pK_a of 6, assigned to the ammonium ion, suggested that the pK_a of the active-site amino group of the native enzyme is about 6, in agreement with the kinetic results obtained with 2,4-dinitrophenyl propionate. Since the acidity of the phenolic moiety was substantially increased in the modified enzyme relative to N-methyl-2-hydroxy-5-nitrobenzylamine, it was argued[80] that the cause of the abnormal lysine pK_a is the presence of one or more positive charges at the enzyme active site. An alternative explanation based on a hydrophobic active site was discarded since it would predict a decrease in the phenol acidity rather than the increase actually observed.

The reduced basicity of the essential lysine should have a mechanistic advantage since, as Kokesh and Westheimer point out,[80] it is necessary for this group to be unprotonated in order to react with the carbonyl of the substrate to form a Schiff base. An enzyme with an unperturbed lysine residue having a pK_a of about 11 would have only a small fraction of its essential amine groups free for nucleophilic attack at a pH near neutrality, and Schiff base formation would be expected to be slow. A group with a pK_a of near 6, on the other hand, will be largely deprotonated under these conditions and able to react with the carbonyl.

4.2. Nature of the Transition State

An assessment of the detailed nature of the transition state for acetoacetate decarboxylase is complicated by the fact that there is no single rate-limiting step. O'Leary and Baughn[59] have shown by the use of carbon isotope effects that the decarboxylation step is only partially rate limiting in both the enzymic reaction and the model reaction catalyzed by aminoacetonitrile (pK_a 5.3). A comparison of these two reactions shows that catalysis by acetoacetate de-decarboxylase is about 10^3-fold more efficient than catalysis by aminoacetonitrile.[54] Since both reactions have the decarboxylation step only partially rate limiting, it is apparent that both formation and decarboxylation of the enzymic Schiff base are enzyme catalyzed, and the magnitude of the enzymic reaction cannot be accounted for solely on the basis of formation of a Schiff base. This conclusion is supported by the finding that the rate of decarboxylation of the imine from aminoacetonitrile and acetoacetic acid is 100-fold slower than the rate of the acetoacetate decarboxylase-catalyzed reaction.[55] Clearly, there are factors which are involved in stabilization of the transition states for both Schiff base formation and decarboxylation which are not operative in the model system.

Although no studies appear to have been made on the mechanism of enzymic Schiff base formation with this enzyme, it is possible to speculate on the factors which may be involved from results on simple systems. It has been found that both Schiff base formation and hydrolysis are subject to general acid–base catalysis in aqueous solution,[81] and it seems likely that this mechanism is operating with the enzyme as well. Particularly interesting in this regard are the findings of intramolecular general catalysis of imine formation by amine groups[82,83] and of imine hydrolysis by carboxylate ions.[84] Furthermore, it has been shown that general catalysis by a carboxylate ion is markedly enhanced in nonpolar solvents.[85] The combination of a general base catalyst in a hydrophobic environment at the enzymic active site is an attractive possibility to explain the enhanced rate of Schiff base formation with this enzyme.

The manner in which the enzyme catalyzes the decarboxylation of the intermediate Schiff base is still unclear. One suggestion, advanced by O'Leary and Baughn,[59] is that a reduced polarity at the active site, relative to water, could accelerate the rate of decomposition of the zwitterion. A similar effect has been demonstrated[86] for the decarboxylation of 2-(1-carboxy-1-hydroxy-ethyl)-3,4-dimethylthiazolium chloride (26):

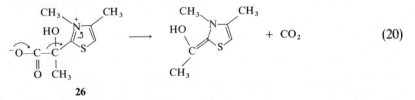

$$\text{26}$$

$$(20)$$

However, a rate increase of this type would be expected only if the most stable form of the enzyme Schiff base were the zwitterion (27) and not the anion (28) or the uncharged form (**29**):

$$\overset{\overset{+}{H}NE}{\underset{\|}{CH_3-C-CH_2-CO_2^-}} \rightleftharpoons \overset{NE}{\underset{\|}{CH_3-C-CH_2-CO_2^-}} \rightleftharpoons \overset{NE}{\underset{\|}{CH_3C-CH_2-CO_2H}}$$

$$\quad\quad\quad 27 \quad\quad\quad\quad\quad\quad\quad 28 \quad\quad\quad\quad\quad\quad\quad 29$$

However, the fact that pK_a's of Schiff bases are generally about 3 units lower than the corresponding amine from which they are formed,[60,87,88] coupled with a pK_a of 6 for the active site lysine, indicates that the predominant form of the enzymic Schiff base at neutral pH is either **28** or **29**. At this point, one might speculate that since it is the zwitterionic form which is active toward decarboxylation, it is unlikely that this enzyme would have evolved in a manner not favoring this form. One way in which the zwitterion could be stabilized relative to structures **28** and **29** would be if formation of the Schiff base removes the factor causing the lowered amine-group basicity. If this were the case, then the intermediate Schiff base would have a pK_a of about 8 and consequently would exist primarily as the zwitterion, enabling facile decarboxylation to occur.

One method by which the perturbation affecting the lysine residue might be altered is a conformation change upon Schiff base formation. The positively charged group(s) responsible for destabilizing the protonated nitrogen could be removed somewhat from the active center, thereby attenuating the effect. Another possibility is that upon binding of acetoacetate the substrate carboxyl group shields the lysine nitrogen from the perturbing positive charge(s). This is a distinct possibility if this positively charged group is the same one which is responsible for the binding of the carboxylate portion of the substrate. Decarboxylation would then lead to the enzymic enamine of acetone in which the perturbation is regenerated, lowering the nitrogen pK_a and allowing facile hydrolysis back to acetone. It would appear that the most advantageous situation for the enzyme would be one in which the nitrogen has a pK_a near neutrality in all forms, free amine, Schiff base, and enamine. In the free-amine form, of course, a pK_a of about 6 would allow almost all of the amine to be unprotonated at neutral pH but yet maintain as much nucleophilicity as possible. In the Schiff base, a pK_a of about 8 would be optimum since this would stabilize the zwitterion form (**27**) over **28** or **29** but the nitrogen would not be so basic as to retard the decarboxylation step. In the product enamine the situation is similar to the free enzyme; the nitrogen must be unprotonated, but it also must accept a positive charge readily so that protonation of the carbon–carbon double bond is not inhibited. Again a pK_a of about 6 appears to be optimum.

If the substrate carboxylate is truly altering the lysine pK_a, one might speculate further as to the effect of binding of acetoacetate on the amine nucleophilicity. The nucleophilicity (and basicity) of this group should take a sudden jump upon binding of the substrate, increasing markedly both the rates of attack on the carbonyl to form the Schiff base and the rate of protonation to form an inactive immonium ion. If nucleophilic attack is faster than proton transfer from solvent, then Schiff base formation would be quite rapid indeed. The enzyme would have the best of all possible worlds, a highly nucleophilic amine residue existing in an unprotonated form at neutral pH. To assess the validity of this hypothesis it is necessary to estimate the rate of protonation of the amine residue at the active site and compare it with the turnover number of the enzyme (300 sec^{-1}).[54] By using a pK_a of 10.7 for the lysine amino group and 15.7 for water, it can readily be calculated[1] that at pH 7 the rate constant for protonation of the amine by hydronium ion would be about 10^4 sec^{-1} and for protonation by water about 10^7 sec^{-1} if diffusion processes are as rapid at the enzyme active site as they are in strictly aqueous solution. However, it is unlikely that proton transfer from water at the enzyme active site is this rapid. Since the active site is thought to be somewhat hydrophobic,[76] the extensive hydrogen-bonding lattice of water probably does not exist there. Generation of a hydroxide ion in the absence of hydrogen bonding would be expected to be quite unfavorable based on the enormous increase in basicity of hydroxide in aqueous dimethylsulfoxide solutions.[62] Consequently, the acidity of a water molecule at the active site should be much lower, both kinetically and thermodynamically, than a water molecule in the bulk solution, and the calculated rate constant may be too large by several powers of 10. Furthermore, protonation by water would involve the generation of a hydroxide ion at the active site near the substrate carboxylate ion, which is likely to be energetically unfavorable due to electrostatic repulsion. A similar explanation has been advanced to account for the failure of borohydride to trap the Schiff base from the enzyme and acetonylphosphonate[89] and the relatively slow reduction of the Schiff base from acetonylsulfonate.[76]

Although it can be argued that protonation of the essential amine should be slower at the enzyme active site than it would be in bulk solution, it is difficult to assess the magnitude of these effects. Little is known about water structure and proton transfer under these conditions. A decision concerning the relative rates of Schiff base formation and ammonium-ion formation must await further study.

A factor which may be responsible for an enhanced rate of decarboxylation on the enzyme surface is a conformational restriction of the substrate.[55] The conformation of the Schiff base which is reactive toward decarboxylation should be the one in which the bond between the leaving group carboxylate and the methylene carbon is perpendicular to the plane of the imine (**30**):

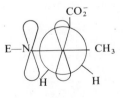

30

This conformation should be the most labile since the incipient p orbital formed from carbon–carbon bond cleavage is parallel to the p orbitals of the imine carbon and nitrogen, allowing facile overlap to occur in the transition state. If the enzyme were to lock the substrate into this conformation, then decarboxylation should be enhanced.

Evidence that this conformational restriction does, in fact, operate at the enzyme active site has been presented by Kluger and Nakaoka.[89] They found that, although acetoacetate decarboxylase catalyzes the exchange of protons in acetone and 2-butanone[90,91] as well as a variety of other inhibitors,[89] there is no enzyme-catalyzed hydrogen exchange with the competitive inhibitor acetonylphosphonate (31):

$$
\begin{array}{ccc}
\overset{\displaystyle O}{\underset{\displaystyle \|}{CH_3-C-CH_3}} + ENH_2 & \rightleftharpoons \overset{\displaystyle \overset{+}{H\overset{}{N}E}}{\underset{\displaystyle \|}{CH_3-C-CH_3}} & \rightleftharpoons \overset{\displaystyle H\overset{}{N}E}{\underset{\displaystyle |}{CH_3C=CH_2}} \quad (21)
\end{array}
$$

$$
\underset{\displaystyle \textbf{31}}{\overset{\displaystyle O}{\underset{\displaystyle \|}{CH_3CCH_2PO_3H^-}}} + ENH_2 \longrightarrow \text{no exchange} \qquad (22)
$$

Since acetonylphosphonate is most likely bound to the enzyme in a Schiff base linkage, it was postulated[89] that the lack of exchange of protons at the 2 position is due to the inability of the carbon–hydrogen bonds to assume the correct orientation for proton abstraction. A reasonable explanation for this phenomenon is that acetonylphosphonate binds similarly to acetoacetate, placing the carbon–phosphorus bond parallel to the π system of the imine. However, carbon–hydrogen bond cleavage to form the enamine requires the carbon–phosphorus bond to be in the same plane as the C—C—N system, which would necessitate loss of much of the binding energy of the phosphonate group. The fact that exchange does not occur suggests that there is a marked orientational requirement for binding of the phosphonate group and, by implication, of the carboxylate group of acetoacetate.

REFERENCES

1. R. P. Bell, *The Proton in Chemistry*, 2nd ed., Cornell University Press, Ithaca, N.Y. (1973).
2. S. H. Bauer, Operational criteria for concerted bond breaking in gas-phase molecular elimination reactions, *J. Am. Chem. Soc.* **91**, 3688–3689 (1969).
3. K. J. Pedersen, The ketonic decomposition of beta-keto carboxylic acids, *J. Am. Chem. Soc.* **51**, 2098–2107 (1929).
4. W. Pastanagoff, Über die kinetik der katalytischen zersetzung der bromkamphokarbonsaure, *Z. Phys. Chem. (Leipzig)* **112**, 448–460 (1924).
5. A. L. Bernoulli, H. Jakubowics, Zerfallsgeschwindigkeit mono- und disubstituierter Malonsäuren, *Helv. Chim. Acta* **4**, 1018–1029 (1921).
6. M. W. Logue, R. M. Pollack, and V. P. Vitullo, The nature of the transition state for the decarboxylation of beta-keto acids, *J. Am. Chem. Soc.* **97**, 6868–6869 (1975).
7. K. J. Pedersen, Dimethylacetoacetic acid. Hydrolysis of the ethyl ester. Ketonic decomposition

reaction with iodine and bromine. Dissociation constant, *J. Am. Chem. Soc.* **58**, 240–246 (1936).

8. V. Prelog, P. Barman, and M. Zimmerman, Zur Kenntnis des Kohlenstoffringes. Weitere Untersuchungen über die Gültigkeitsgrenzen der Bredt'schen Regel. Eine Variante der Robinson'schen Synthese von cyclischen ungesättigten Ketonen, *Helv. Chim. Acta* **32**, 1284–1296 (1949).

9. G. Kobrich, Bredt compounds and the Bredt rule, *Angew. Chem. Int. Ed. Engl.* **12**, 464–473 (1973).

10. J. P. Ferris and N. C. Miller, The decarboxylation of β-keto acids. II. An investigation of the Bredt rule in bicyclo [3 · 2 · 1] octane systems, *J. Am. Chem. Soc.* **88**, 3522–3527 (1966).

11. K. J. Pedersen, The decomposition of α-nitrocarboxylic acids with some remarks on the decomposition of β-ketocarboxylic acids, *J. Phys. Chem.* **38**, 559 (1934).

12. F. H. Westheimer and W. H. Jones, The effect of solvent on some reaction rates, *J. Am. Chem. Soc.* **63**, 3283–3286 (1941).

13. J. Hine, *Physical Organic Chemistry*, 2nd ed., McGraw-Hill, New York (1962), p. 305.

14. L. P. Hammett, *Physical Organic Chemistry*, 2nd ed., McGraw-Hill, New York (1970), Chap. 5.

15. T. S. Straub and M. L. Bender, Cycloamylases as enzyme models. The decarboxylation of benzoylacetic acids, *J. Am. Chem. Soc.* **94**, 8881–8888 (1972).

16. R. W. Hay and K. R. Tate, The kinetics of decarboxylation of benzoylacetic acid and its *p*-methoxy and *p*-nitro derivatives in dioxane–water mixtures, *Aust. J. Chem.* **23**, 1407–1413 (1970).

17. M. W. Logue, R. M. Pollack, and V. P. Vitullo, unpublished observations.

18. E. O. Wiig, Carbon dioxide cleavage from acetone dicarboxylic acid, *J. Phys. Chem.* **32**, 961–989 (1928).

19. G. A. Hall, Jr., and E. S. Hanrahan, Kinetics of the decarboxylation of phenylmalonic acid, *J. Chem. Phys.* **69**, 2402–2406 (1965).

20. K. J. Pedersen, The hydrolysis of ethyl acetoacetate and the decarboxylation of acetoacetic acid in strongly acid solution, *Acta Chem. Scand.* **15**, 1718–1722 (1961).

21. D. S. Noyce and Sr. M. A. Matesich, The decarboxylation of benzoylacetic acids in aqueous sulfuric acid, *J. Org. Chem.* **32**, 3243–3244 (1967).

22. C. G. Swain, R. F. W. Bader, R. M. Esteve, Jr., and R. N. Griffin, Use of substituent effects on isotope effects to distinguish between proton and hydride transfers. Part II. Mechanisms of decarboxylation of β-keto acids in benzene, *J. Am. Chem. Soc.* **83**, 1951–1955 (1961).

23. E. M. Hodnett and R. L. Rowton, C[14]-isotope effects in the decarboxylation of 2-benzoylpropionic acid, *Radioisotopes Phys. Sci. Ind. Proc. Conf. Use, Copenhagen 1960* **3**, 225–233 (1962).

24. A. Wood, Carbon isotope effects in the decarboxylation of oxaloacetic acid, *Trans. Faraday Soc.* **60**, 1263–1267 (1964).

25. A. Wood, Carbon isotope effects in the decarboxylation of oxaloacetic acid, *Trans. Faraday Soc.* **62**, 1231–1235 (1966).

26. J. Bigeleisen and L. Friedman, C[13] isotope effect in the decarboxylation of malonic acid, *J. Chem. Phys.* **17**, 998–999 (1949).

27. P. E. Yankwich and M. Calvin, An effect of isotopic mass on the rate of a reaction involving the carbon–carbon bond, *J. Chem. Phys.* **17**, 109–110 (1949).

28. J. G. Lindsay, A. N. Bourns, and H. G. Thode, C[13] isotope effect in the decarboxylation of normal malonic acid, *Can. J. Chem.* **29**, 192–200 (1951).

29. P. E. Yankwich and A. L. Promisolow, Intramolecular carbon isotope effect in the decarboxylation of liquid malonic acid near the melting point, *J. Am. Chem. Soc.* **76**, 4648–4651 (1954).

30. A. Roe and M. Hellmann, Determination of an isotope effect in the decarboxylation of malonic-1-C[14] acid, *J. Chem. Phys.* **19**, 660 (1951).

31. P. E. Yankwich, A. L. Promisolow, and R. F. Nystrom, C[14] and C[13] intramolecular isotope effects in the decarboxylation of liquid malonic acid at 140.5°, *J. Am. Chem. Soc.* **76**, 5893–5895 (1954).

32. A. Fry, in: *Isotope Effects in Chemical Reactions* (C. J. Collins and N. S. Bowman, eds.), p. 364–414, Van Nostrand Reinhold, New York (1970).

33. F. H. Westheimer, The magnitude of the primary kinetic isotope effect for compounds of hydrogen and deuterium, *Chem. Rev.* **61**, 265–273 (1961).

34. J. Kurz, Transition state characterization for catalyzed reactions, *J. Am. Chem. Soc.* **85**, 987–991 (1963).

35. G. M. Loudon, Aminolysis of α-acetoxystyrenes. The pK_a of acetophenones in aqueous solution, *J. Am. Chem. Soc.* **98**, 3591–3597 (1976).

36. E. M. Arnett, Quantitative comparisons of weak organic bases, *Prog. Phys. Org. Chem.* **1**, 223–405 (1963). The pK_a value was corrected to the H_0 scale as evaluated by Jorgenson and Hartter.[37]

37. M. J. Jorgenson and D. R. Hartter, A critical re-evaluation of the Hammett acidity function at moderate and high acid concentrations of sulfuric acid. New H_0 values based solely on a set of primary aniline indicators, *J. Am. Chem. Soc.* **85**, 878–883 (1963).

38. C. G. Swain, D. A. Kuhn, and R. L. Schowen, Effect of structural changes in reactants on the position of hydrogen-bonding hydrogens and solvating molecules in transition states. The mechanism of tetrahydrofuran formation from 4-chlorobutanol, *J. Am. Chem. Soc.* **87**, 1553–1561 (1965).

39. R. L. Schowen, Mechanistic deductions from solvent isotope effects, *Prog. Phys. Org. Chem.* **9**, 275–332 (1972).

40. C. S. Tsai, Y. T. Lin, and E. E. Sharkawi, Mechanism of the decarboxylation of monoethyl oxalacetate, *J. Org. Chem.* **37**, 85–87 (1972).

41. E. R. Thornton, A simple theory for predicting the effects of substituent changes on transition state geometry, *J. Am. Chem. Soc.* **89**, 2915–2927 (1967).

42. J. Hine and W. H. Sachs, Possible bifunctional catalysis by 2-dimethylaminoethylamine in the dealdolization of diacetone alcohol, *J. Org. Chem.* **39**, 1937–1944 (1974).

43. J. Del Bene and J. A. Pople, Theory of molecular interactions. I. Molecular orbital studies of water polymers using a minimal Slater-type basis, *J. Chem. Phys.* **52**, 4858–4866 (1970).

44. P. A. Kollman, A theory of hydrogen bond directionality, *J. Am. Chem. Soc.* **94**, 1837–1842 (1972).

45. R. D. Gandour, Structural requirements for intramolecular proton transfers, *Tetrahedron Lett.* **1974**, 295–298.

46. R. W. Hay and M. A. Bond, Kinetics of the decarboxylation of acetoacetic acid, *Aust. J. Chem.* **20**, 1823–1828 (1967).

47. J. R. Jones, R. E. Marks, and S. C. Subbarao, Kinetic isotope effects. Part 2. Rates of abstraction of hydrogen and tritium from acetophenone and some para- and meta-substituted acetophenones in alkaline media, *Trans. Faraday Soc.* **63**, 111–123 (1967).

48. C. G. Swain, E. C. Stivers, J. F. Reuwer, Jr., and L. J. Schaad, Use of hydrogen isotope effects to identify the attacking nucleophile in the enolization of ketones catalyzed by acidic acid, *J. Am. Chem. Soc.* **80**, 5885–5893 (1958).

49. J. E. Dixon and T. C. Bruice, Dependence of the primary isotope effect (k^H/k^D) on base strength for the primary amine catalyzed ionization of nitroethane, *J. Am. Chem. Soc.* **92**, 905–909 (1970).

50. R. P. Bell and D. M. Goodall, Kinetic hydrogen isotope effects in the ionization of some nitroparaffins, *Proc. R. Soc. London Ser. A* **294**, 273–296 (1966).

51. I. Fridovich and F. H. Westheimer, On the mechanism of the enzymatic decarboxylation of acetoacetate. II, *J. Am. Chem. Soc.* **84**, 3208–3209 (1962).

52. S. Warren, B. Zerner, and F. H. Westheimer, Acetoacetate decarboxylase. Identification of lysine at the active site, *Biochemistry* **5**, 817–822 (1966).

53. K. J. Pedersen, Amine catalysis in the decarboxylation of oxalacetic acid, *Acta Chem. Scand.* **8**, 710–722 (1954).

54. J. P. Guthrie and F. H. Westheimer, Cyanomethylamine as a model for acetoacetate decarboxylase, *Fed. Proc. Fed. Am. Soc. Exp. Biol.* **26**, 562 (1967).

55. J. P. Guthrie and F. Jordan, Amine-catalyzed decarboxylation of acetoacetic acid. The rate constant for decarboxylation of a β-imino acid, *J. Am. Chem. Soc.* **94**, 9136–9141 (1972).

56. B. R. Brown, The mechanism of thermal decarboxylation, *Q. Rev., Chem. Soc.* **1951**, 131–146.

57. K. Taguchi and F. H. Westheimer, Decarboxylation of Schiff bases, *J. Am. Chem. Soc.* **95**, 7413–7423 (1973).

58. R. W. Hay, The aniline catalyzed decarboxylation of oxaloacetic acid, *Aust. J. Chem.* **18**, 337–351 (1965).

59. M. H. O'Leary and R. L. Baughn, Acetoacetate decarboxylase. Identification of the rate-determining step in the primary amine catalyzed reaction and in the enzymic reaction, *J. Am. Chem. Soc.* **94**, 626–630 (1972).

60. J. Hine, B. C. Menon, J. H. Jensen, and J. Mulders, Catalysis of α-hydrogen exchange. II. Isobutyraldehyde 2-d exchange via *n*-methyliminium ion formation, *J. Am. Chem. Soc.* **88**, 3367–3373 (1966).

61. M. L. Bender and A. Williams, Ketimine intermediates in amine-catalyzed enolization of acetone, *J. Am. Chem. Soc.* **88**, 2502–2508 (1966).

62. J. Hine, J. C. Craig, J. Underwood II, and F. A. Via, Kinetics and mechanism of the hydrolysis of *N*-isobutylidenemethylamine in aqueous solution, *J. Am. Chem. Soc.* **92**, 5194–5199 (1970).

63. F. R. Stermitz and W. H. Huang, Thermal and photodecarboxylation of 2-, 3-, and 4-pyridyl-acetic acid, *J. Am. Chem. Soc.* **93**, 3427–3431 (1971).

64. P. J. Taylor, The decarboxylation of some heterocyclic acetic acids, *J. Chem. Soc. Perkin Trans. 2, 1972*, 1077–1086.

65. R. G. Button and P. J. Taylor, The decarboxylation of some heterocyclic acetic acids. Part II. Direct and indirect evidence for the zwitterionic mechanisms, *J. Chem. Soc. Perkin Trans 2* **1973**, 557–567.

66. W. P. Jencks, General acid–base catalysis of complex reactions in water, *Chem. Rev.* **72**, 705–718 (1972).

67. R. T. Arnold, O. C. Elmer, and R. M. Dodson, Thermal decarboxylation of unsaturated acids, *J. Am. Chem. Soc.* **72**, 4359–4361 (1950).

68. G. G. Smith and S. E. Blau, Decarboxylation, I. Kinetic study of the vapor phase thermal decarboxylation of 3-butenoic acid, *J. Phys. Chem.* **68**, 1231 (1964).

69. D. B. Bigley and J. C. Thurman, Studies in decarboxylation. Part II. Kinetic evidence for the mechanism of thermal decarboxylation of β, γ-unsaturated acids, *J. Chem. Soc.* **1965**, 6202–6205.

70. B. D. Bigley and J. C. Thurman, Studies in decarboxylation. Part III. The thermal decarboxylation of 2,2-dimethyl-3-phenylbut-3-enoic acid, *J. Chem. Soc. B* **1966**, 1076–1077.

71. D. B. Bigley, Studies in decarboxylation. Part I. The mechanism of decarboxylation of unsaturated acids, *J. Chem. Soc. B* **1964**, 3894–3899.

72. D. B. Bigley and J. C. Thurman, Studies in decarboxylation. Part V. Kinetic isotope effects in the gas-phase thermal decarboxylation of 2,2-dimethyl-4-phenylbut-3-enoic acid, *J. Chem. Soc B* **1967**, 941–943.

73. D. B. Bigley and J. C. Thurman, Studies in decarboxylation. Part VI. A comparison of the transition states for the decarboxylation of β-keto and β,γ-unsaturated acids, *J. Chem. Soc. B* **1968**, 436–440.

74. D. B. Bigley and R. W. May, Studies in decarboxylation. Part IV. The effect of alkyl substituents on the rate of gas-phase decarboxylation of some β,γ-unsaturated acids, *J. Chem. Soc.* **1967**, 557–561.

75. D. B. Bigley and J. C. Thurman, On the transition state for decarboxylation of β-keto acids and β,γ-unsaturated acids, *Tetrahedron Lett.* **1967**, 2377–2380.

76. I. Fridovich, Acetoacetate decarboxylase, *Enzymes* **6**, 255–271 (1972).

77. G. A. Hamilton and F. H. Westheimer, On the mechanism of the enzymatic decarboxylation of acetoacetate, *J. Am. Chem. Soc.* **81**, 6332–6333 (1959).

78. D. E. Schmidt and F. H. Westheimer, p*K* of the lysine amino group at the active site of acetoacetate decarboxylase, *Biochemistry* **10**, 1249–1253 (1971).

79. P. A. Frey, F. C. Kokesh, and F. H. Westheimer, A reporter group at the active site of acetoacetate decarboxylase. I. Ionization constant of the amino group, *J. Am. Chem. Soc.* **93**, 7266–7269 (1971).

80. F. C. Kokesh and F. H. Westheimer, A. reporter group at the active site of acetoacetate de-

carboxylase. II. Ionization constant of the amino group, *J. Am. Chem. Soc.* **93**, 7270–7274 (1971).

81. W. P. Jencks, *Catalysis in Chemistry and Enzymology,* McGraw-Hill, New York (1969), pp. 490ff.

82. J. Hine, M. S. Cholod, and W. K. Chess, Jr., Kinetics of the formation of imines from acetone and primary amines. Evidence for internal acid-catalyzed dehydration of certain intermediate carbinolamines, *J. Am. Chem. Soc.* **95**, 4270–4276 (1973).

83. J. Hine and W. S. Li, Internal catalysis in imine formation from acetone and acetone-d_6 and conformationally constrained derivatives of *N,N*-dimethyl-1,3-propanediamine, *J. Org. Chem.* **40**, 2622–2626 (1975).

84. R. M. Pollack and R. H. Kayser, unpublished observations.

85. R. M. Pollack and M. Brault, Synergism of the effect of solvent and of general base catalysis in the hydrolysis of a Schiff base, *J. Am. Chem. Soc.* **98**, 247–248 (1976).

86. J. Crosby, R. Stone, and G. E. Lienhard, Mechanisms of thiamine-catalyzed reactions. Decarboxylation of 2-(1-carboxy-1-hydroxyethyl)-3,4-dimethylthiazolium chloride, *J. Am. Chem. Soc.* **92**, 2891–2900 (1970).

87. M. L. Bender and A. Williams, Ketimine intermediates in amine-catalyzed enolization of acetone, *J. Am. Chem. Soc.* **88**, 2502–2508 (1966).

88. C. H. Rochester, *Acidity Functions,* Academic Press, New York (1970), Chap. 7.

89. R. Kluger and K. Nakaoka, Inhibition of acetoacetate decarboxylase by ketophosphonates. Structural and dynamic probes of the active site, *Biochemistry* **13**, 910–914 (1974).

90. W. Tagaki and F. H. Westheimer, Acetoacetate decarboxylase. Catalysis of hydrogen-deuterium exchange in acetone, *Biochemistry* **7**, 901–905 (1968).

91. G. Hammons, F. H. Westheimer, K. Nakaoka, and R. Kluger, Proton-exchange reactions of acetone and butanone. Resolution of steps in catalysis by acetoacetate decarboxylase, *J. Am. Chem. Soc.* **97**, 1568–1671 (1975).

13

The Mechanism of Phosphoryl Transfer

S. J. Benkovic and K. J. Schray

1. INTRODUCTION

Our intent in this chapter is to present an abbreviated but inclusive description of the probable transition states involved in phosphoryl-transfer reactions as elucidated by physical organic studies. A number of critical, extensive reviews on particular aspects of this topic have recently appeared.[1-9] Probable modes of catalysis will be delineated particularly in relation to their anticipated involvement in biological phosphoryl transfer.

2. THE DISSOCIATIVE MECHANISM

The dissociative monomeric metaphosphate mechanism postulated for the hydrolysis of monoester monoanions derived from alcohols and phenols, thiols, and amines is supported by a variety of data including (1) the general observation of P—O, P—S, or P—N bond cleavage,[10-16] (2) entropies of activation close to zero in contrast to biomolecular or associative solvolysis where entropies are usually more negative by 20 eu,[17] and (3) the partitioning of products between methanol and water to yield a product ratio that approximates the molar composition of the mixed solvent or favors methyl phosphate formation.[18,19] Recently evidence has been presented for the formation in the gas phase of monomeric methyl metaphosphate by pyrolysis of 2-butenyl-phostonate and its trapping by N-methylaniline to yield N-methyl-N-phenyl-phosphoramidate.[20] Linear free-energy relationships between the logarithmic rates of hydrolysis of the monoanions and the dissociation constants of the

S. J. Benkovic • Department of Chemistry, The Pennsylvania State University, University Park, Pensylvania. K. J. Schray • Department of Chemistry, Lehigh University, Bethlehem, Pennsylvania.

corresponding leaving group[11,14,15,21] or the last dissociating proton of the phosphate[22] exhibit a small β_{lg} dependence in accord with the departure of a neutral ligand, suggesting that the hydrolysis proceeds through

$$
\begin{array}{c}
\quad\quad\quad\; O \\
\quad\quad\quad\; \| \\
\overset{\delta+}{RX}\cdots\cdots P\!=\!\!O^- \\
H\quad\; | \\
\quad\quad\quad O^{\delta-}
\end{array}
\qquad X = O,\, S,\, NH
\qquad (1)
$$

with negligible assistance by attacking solvent. Proton transfer apparently proceeds directly or through intervening solvent molecules and may become rate determining for O monoesters with $pK_a < 7$[21] and for S monoesters throughout the pK_a range of 3–12.[14] Evidence supporting this suggestion is derived from the observation of deuterium solvent isotope effects (k_{H_2O}/k_{D_2O}) of 1.4–1.8 in the regions of interest for both O and S esters as well as the more rapid hydrolysis (ca. 10^2) of acetyl phosphate monoanion than predicted by the above correlations. The latter may hydrolyze via an internal proton transfer through a cyclic six-membered ring, not requiring solvent participation:

$$(2)$$

In the case of phosphoramidates it appears likely that proton transfer to the amine leaving group is essentially complete. For phosphoramidates comprised of an amine whose $pK_a < 7.2$, the observed rate constant for monoanion hydrolysis is relatively constant, indicative of a cancellation of effects between zwitterion formation and amine expulsion.[15] Phosphoramidates with an amine whose $pK_a > 7.2$ and which exist predominantly as the zwitterion, exhibit a $\beta_{lg} \simeq -1$, in accord with nearly complete P—N bond cleavage in the transition state. No deuterium solvent isotope is observed.[23]

An inquiry of interest is the rate of decomposition of the reactive zwitterionic species in several representative cases. A sample calculation, based on an estimate of the zwitterion concentration, predicts the first-order rate coefficient to be 10^7 min^{-1} (39°) for phenyl phosphate monoanion.[21] For N-phosphoryl N'-methylimidazolium ion and phosphoramidate, this value is 10^{-3}–10^{-4} min^{-1} (39°),[15] and for phenyl thiophosphate monoanion (assuming a $pK_a - 10$ for $R\overset{+}{S}HPO_3^{2-}$), this term (35°) is greater than diffusion controlled.[14] This fact supports the previous contention that the proton transfer may be rate determining in S-phosphate monoanion hydrolysis. The calculation is slightly sensitive to leaving-group pK_a; thus, the above should be viewed as median values. In the absence of other effects these coefficients furnish a preliminary estimate as to the efficiency of enzymic catalysis provided the latter merely optimized the zwitterion concentration without changing the intrinsic pK_a of the departing group. The overall order of hydrolytic reactivity ($S \geq N >$ O) for phosphate esters, which extends at least for the phosphoramidates and O phosphate monoesters to other nucleophiles, offers a tentative rationale for the intermediacy of S and N esters as kinetically competent intermediates in enzyme-mediated processes.

The metaphosphate mechanism also applies to the hydrolysis of phosphate monoester dianions or polyphosphates possessing excellent leaving groups, e.g., carboxylate,[12] phosphate monoanion,[24,25] phenylpropyl phosphate monoanion,[26,27] and 2,4- and 2,6-dinitrophenolate,[21,28] and is supported by similar experimental criteria. Dianion hydrolysis is highly sensitive to the pK_a of the leaving group, $\beta_{lg} = -1.2$, experimentally identical to that for the equilibrium or complete transfer of the phosphoryl moiety.[15] This result is consistent with a dissociative transition state,

$$
\text{RO}^{\delta-} \cdots\cdots\;\; \overset{\displaystyle\text{O}}{\underset{\;\;\delta^-\text{O}\overset{}{\diagup}\quad \overset{}{\diagdown}\text{O}^-}{\text{P}}}
\tag{3}
$$

where P—O bond cleavage is well advanced. This description is further supported by the observation of a substantial O^{18} isotope effect in the hydrolysis of 2,4-dinitrophenyl phosphate in which the ester oxygen has been enriched with the heavy isotope[29] and a negative deviation from the linear free-energy relationship for those esters whose leaving group is less capable of stabilizing the developing negative charge by resonance, e.g., phosphate.[2]

The formation of metaphosphate-related species is also indicated by the results of kinetic investigations on the hydrolyses of phosphordiamidic and phosphorothiodiamidic halides and esters[30-33] where "metaphosphorimidates" apparently are produced. Structure-reactivity correlations are suggestive of an E1cB mechanism with P—O bond cleavage only slightly advanced in the transition state for both the P=O and P=S systems as illustrated in

$$\tag{4}$$

for an N,N'-diphenylphosphorodiamidate ester. The presence of a planar intermediate species was also inferred by the formation of racemic hydrolysis products upon solvolysis of N-cyclohexylphosphoramidothioic chloride.[34] Additional evidence for the intermediacy of "metaphosphorimidates" has been obtained by generating this species from an insoluble, polymer-bound precursor and trapping it on a second solid phase suspended in the same solution but physically precluded from direct reaction.[35] Although dimers or oligomers of the "metaphosphorimidate" have not been unequivocally excluded as the actual phosphorylating agent, their formation would logically be derived from the monomeric intermediate.

The spectrum of known dissociative mechanisms in displacement at phosphorus also includes the S_N1 solvolysis of di-t-butylphosphinyl chloride,[36]

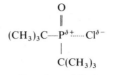

$$(CH_3)_3C—P^{\delta+}\cdots Cl^{\delta-} \qquad (5)$$

and the A1 hydrolysis of N-p-nitrophenyldiphenylphosphinamide,[37]

$$C_6H_5—P^{\delta+}\cdots NH_2C_6H_4NO_2 \qquad (6)$$

both of which lead to a phosphinylium ion. In the former it is apparent that large steric effects are necessary for the observation of the S_N1 mechanism, since the rate of hydrolysis of the corresponding isopropyl derivative by an associative mechanism is nearly a million times faster than the t-butyl halide. The difficulty in generating the phosphinylium ion from the halide is probably a matter of bond energies; the P—Cl bond has been assigned an approximate bond energy of 76 kcal/mol from measurements of PCl_3, and this deficit would not be compensated for by an increase in P—O π-bond energy.[38,39] This barrier may be circumvented with phosphinamides, which possess excellent leaving groups, permitting observation of an A1 mechanism for systems which generally proceed via acid-catalyzed associative mechanisms.[40,41] Although the latter classes of esters generally do not serve as substrates for known phosphohydrolases, their mode of hydrolysis serves to emphasize the importance of acid catalysis in P—N bond cleavage and defines the limits of the dissociative mechanism. In summation there is little justification for extending the concept of pentacovalent intermediates to the hydrolytic reactions of phosphate monoesters.

3. THE ASSOCIATIVE MECHANISM

Nucleophilic addition–elimination reactions to tetracoordinate phosphorus involving the intermediacy of a pentacovalent or phosphorane species comprise the associative mechanism. To appreciate the possible complexities attendant with such a process, the structural characteristics of phosphoranes and their ability to undergo intramolecular ligand exchange must first be considered.

Phosphoranes generally exhibit trigonal-bipyramidal geometry.[42-44] The structure of a typical oxyphosphorane, the phenanthrenequinonetri-

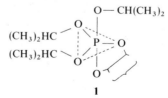

1

isopropyl phosphite adduct (**1**),[45] has the following pertinent features: (1) the phosphorus atom lies within a triangle defined by three bonding equatorial ligands which form the basal plane of the trigonal bipyramid, and their bonds subtend an angle of ca. 120°; (2) the remaining two apical ligands are situated above and below the basal plane, and their bonds subtend an angle of ca. 180°; and (3) an O—P—O ring bond angle of ca. 90° with the five-membered ring spanning one apical and equatorial position. Feature (3) applies to adducts which possess a five-membered ring and is probably a general phenomenon.[46] Within **1** are, however, extensive steric interactions which occur owing to several short nonbonded distances; i.e., the apical oxygens are within 2.7 Å of the first carbon of the equatorial isopropyl groups. Nonetheless, cyclic oxyphosphoranes are generally more stable than acyclic ones owing to the constraints imposed by the five-membered ring which minimize additional steric repulsions.[46] Stable oxyphosphoranes with five monodentate ligands consequently are relatively fewer in number. However, the pentaethoxy-, pentamethoxy-, and pentaphenoxyphosphoranes have been reported with the pentacovalency of phosphorus based on the positive chemical shifts in their ^{31}P NMR spectra relative to 85% H_3PO_4.[47-49] Although trigonal-bipyramidal geometry has not been unequivocally established for these compounds, temperature-dependent ^{1}H and ^{31}P NMR spectroscopy of derivatives where alkyl or phenyl groups replace one or more alkoxy substituents is consistent with this type of geometry. In addition, an extensive number of orbital calculations point to the trigonal-bipyramidal structure as the more stable one.[50]

Distortion toward the alternate stable geometry, the tetragonal pyramid structure, has been found in spirophosphoranes, for example, the 1,3,2-dioxa-

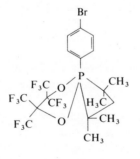

2

phospholan (**2**) formed from hexafluoroacetone and 2,2,4,4-tetramethyl-1-*p*-bromophenyl phosphetane.[51] The bond angles from the apical *p*-bromophenyl carbon to the four equatorial ligands range from 100–105°, whereas the smaller O—P—C bond angles are ca. 90° in contrast to the 120° in a trigonal-bipyramidal molecule. Apparently the particular combination of both a four- and five-membered ring in this case permits square pyramidal geometry to become the more stable.

The geometry adopted by a given phosphorane obviously should be greatly affected by the nature of the ligands.* Both calculation and observation support the argument that in the trigonal-bipyramidal structure the more electronegative substituents will preferentially occupy the apical positions.[42,50] In terms of a molecular orbital picture the relative accumulation of electron density at apical rather than equatorial positions in the trigonal-bipyramid geometry—a tendency further enforced by the introduction of $3d$ orbitals on phosphorus[52]—favors bonding of electronegative substituents at these two positions. This situation is reversed in the square pyramid where the basal positions are relatively higher in electron density.[50] It follows that π acceptors will prefer apical sites in the trigonal bipyramid and π donors equatorial positions, whereas the converse is predicted for the square pyramid. Finally, both calculations and observation indicate that apical bonds are generally longer and less stable than equatorial ones.[44] The separation between trigonal-bipyramidal and tetragonal-pyramidal geometries in pentacovalent phosphorus compounds as implied above is not rigid; in fact the latter serves as either a transition state or metastable intermediate for the intramolecular ligand exchange found with phosphoranes[53]—a process often referred to as Berry pseudorotation.[54]

With the latter mechanism, pairwise exchange of apical and equatorial ligands takes place in a concerted fashion by way of a tetragonal pyramidal transition state:

$$\text{(7)}$$

The pivotal point, an equatorial ligand—arbitrarily 3—remains equatorial in the process and occupies the apex position in the transition state. Proceeding from left to right, the two apical groups designated 1 and 2 may be viewed as undergoing a motion which results in closing the original apical–apical bond angle from 180° to 120°, whereas the remaining equatorial groups 4 and 5 appear to open the initial 120° bond angle to 180°. The result is a new trigonal bipyramid in which the pivotal group has remained in an equatorial position, but the other four groups have exchanged their positions. An alternate permutationally indistinguishable process for intramolecular ligand exchange termed the "turnstile" mechanism[55] can be pictured in terms of the following model:

$$\text{(8)}$$

* The crystal structure of $(C_6H_5O)_5P$ reveals that phosphorus is at the center of a nearly perfect trigonal bipyramid. X-ray analysis of a 5,5-spirobicyclic homophosphorane (five oxygen ligands), i.e., the spirobicyclic pentaoxy phosphorane derived from catechol and phenol, indicates a 15°-turnstile configuration.[156] The latter configuration has been regarded as a mix of tetragonal-pyramidal and trigonal-bipyramidal geometry.[157]

In the course of turnstile rotation the following sequence of internal movements takes place: (1) The two equatorial ligands move toward each other in the equatorial plane, closing the original 120° angle to 90°; (2) one apical ligand and one equatorial ligand move ca. 9° from the vertical and horizontal plane, respectively; (3) an internal rotation of ligand pairs 1, 3 and 2, 4, 5 with the central phosphorus atom as pivot occurs; and (4) the original bond angles of a pentacovalent trigonal bipyramid are restored. The sequence depicted above has the same result as a Berry pseudorotation process with ligand 4 as pivot; both processes result in simultaneous exchange of the two apical ligands with two equatorial ligands. The turnstile rotation does not, however, involve a tetragonal-pyramidal intermediate but one of lower symmetry. Furthermore, there are four distinct turnstile rotation processes, i.e., where ligand 4 remains in the equatorial plane, for every Berry pseudorotation process. Although in principle there are alternate modes of rearrangement, other pathways must involve improbable planar five-coordinate intermediates and are not serious possibilities.[56] A choice between the pseudorotation and turnstile mechanisms is problematic; various calculations on the fluorophoranes support the former route for ligand exchange in PF_5 and PF_3H_2,[57] whereas data obtained from NMR studies on polycyclic oxyphosphoranes can be more readily rationalized in terms of the turnstile pathway.[58] In this chapter the more widely adopted pseudorotation mechanism will be employed since the stereochemical consequences, which are identical, rather than the means of exchange are important in the processes discussed here.

The preference rules for pseudorotation, i.e., which of the ligand permutations is favored kinetically and thermodynamically, are dictated by the relative stability of the intervening tetragonal-pyramidal and resulting trigonal-bipyramidal species. Consequently an exchange of an equatorial ligand with an apical position will be thermodynamically and kinetically preferred in the Berry mechanism for electronegative atoms or π-electron acceptors, since in the transition the ligand will occupy a basal position in the square-pyramidal transition state. Apicophilicity orders have been established experimentally by measuring the activation energy for pseudorotation process in a variety of spirophosphoranes and by determining the stereochemical nature of the products formed upon alkaline hydrolysis of a series of phosphetanium salts.[59-62] These results suggest the following order of apicophilicities: aryl, vinyl < alkyl < dimethylamino < phenoxy, phenylthio, benzoyl $\simeq$ alkoxy, alkythio $\simeq$ chloro < H. The observed order is not readily explicable in terms of electronegativity arguments but underscores the importance of π-donor ability overcompensating for electronegativity. In addition the apical position preference of substituents with low-lying unfilled orbitals as in the case of sulfur appears to be important. However, it is apparent that this order is sensitive to the nature of the other substituents on phosphorus and is a crude index at best. Pentacovalent species containing bidentate ligands are also subject to preference rules. Five- or four-membered rings generally will span equatorial–apical positions but not equatorial–equatorial positions since the distortion of the phosphorus-ring ligand bond angle from 90° to 120° will be accompanied by a significant increase in ring strain. Six-membered rings, however, can often be situated diequatorially. Apical–apical spanning is precluded by steric requirements in four-, five-, and six-membered ring systems.

Restricting our choice of compounds to those with biological relevance, nucleophilic attacks on cyclic phosphate and phosphonates remain among the best examples of the associative mechanism. The acid- and base-catalyzed hydrolyses of cyclic five-membered esters of phosphoric acid proceed 10^6–10^8 times more rapidly than those of the corresponding acyclic analogs.[5] The acid-catalyzed hydrolysis of ethylene hydrogen phosphate is accompanied by almost equally rapid oxygen exchange into the unreacted substrate.[63] Likewise hydrolysis of methyl ethylene phosphate is accompanied by rapid cleavage of the exocyclic methoxy group, a process that corresponds to the ^{18}O exchange noted with the parent compound.[64] A considerable portion of this increased reactivity resides in the strain of the five-membered ring based on evidence provided by measurements of the heats of saponification of the acyclic and five-membered cyclic compounds[65,66] and supported by calculation of the strain energy.[67] Relief of strain in the phosphate ring on passing from the ground state with a ring angle at phosphorus of 99° to a trigonal-bipyramidal transition state (ring angle, 90°) would account for about 10^4 of the increased reactivity. It should be noted that the observed acceleration may be attributed to σ-bond strain imposed by the ring or differences in π energy between cyclic and acyclic esters owing to distortion of the π bonding in the former.[68,69] Greater weakening of the P—O bond by additional antiperiplanar lone pairs on the remaining oxygen atoms in the cyclic vs. acyclic esters also may be involved.[70] Regardless of the actual cause, it is obvious that a factor of 10^2 remains unaccountable. This figure, then, may represent a measure of the lower free energy of a hydrolysis pathway proceeding through a metastable pentacovalent intermediate with a lifetime of $> 10^{-13}$ sec, relative to one in which only the transition state has pentacoordinate character.

Ring strain *per se* does not account for the observations of equally rapid ^{18}O exchange or methoxyl cleavage which proceed with conservation of the ring structure. Consequently an associative mechanism,

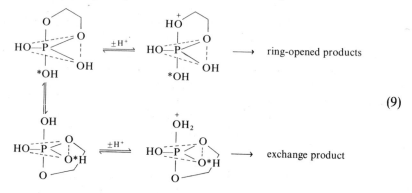

(9)

involving pentacovalent intermediates able to interconvert through pseudoro-tation is favored as illustrated for ethylene phosphate hydrolysis and exchange. That apical attack and apical departure are the preferential modes of bond making and bond breaking,[71] at least when the ligands are relatively electro-negative, follows from several independent lines of argument, namely, (1) as noted above, apical bonds are longer and weaker than equatorial ones; (2) if apical attack is preferred, then for symmetry reasons apical departure is pre-ferred to the same extent; and (3) the hydrolysis of **3** occurs almost exclusively with ring opening,[72]

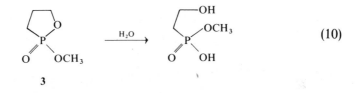

$$(10)$$

3

whereas equatorial attack by water and departure by methoxyl involving a pentacovalent species (**4**) is not observed. Species **4** in principle could be formed

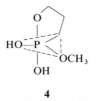

4

by apical attack of water and decompose with equatorial loss of the methoxyl group, which again is contrary to experiment. Argument (2) is an extended application of the principle of microscopic reversibility[73] and in a restrictive sense only accurately applies to the ^{18}O exchange reaction of ethylene phos-phate.

The importance of a pseudorotation process in producing the ^{18}O ex-change into the unreacted ester is supported by analogous studies on the pH dependence of the product distribution in the hydrolysis of methyl ethylene phosphate.[74] At lower pH the fraction of exocyclic (methoxyl) cleavage falls toward zero, consistent with proton trapping of **5** competing with the pseudo-rotation of **5** to **6**. Berry pseudorotation of **7** to **8** without a proton switch is an unfavorable process owing to positioning of a positively charged substituent at an equatorial position. This example also stresses the need for neutral species for allowed pseudorotations; similar reasoning shows that dianionic but not monoanionic species are restricted in their available pseudorotation modes. The lack of methoxyl group loss from **3** apparently stems from the unfavorable pseudorotations necessary to place this group in an apical position—one mode forces the five-membered ring diequatorially, and the second positions a carbon atom apically. It should be emphasized that the preference rules are not inviolate but merely describe the lowest free energy for the process. Numerous examples are known where these barriers are penetrated.[8]

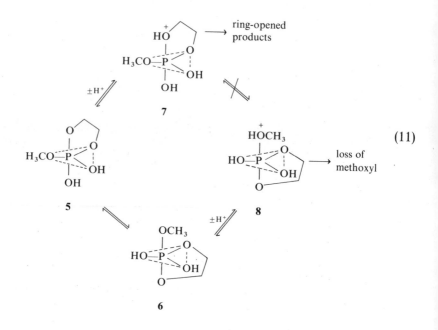

(11)

The above examples also suffice to demonstrate the effects of pseudorotations of pentacoordinate intermediates on the stereochemical course of the reaction. Retention of configuration at phosphorus may be conceived as proceeding through a pathway involving apical attack, pseudorotation of the incipient leaving group from an equatorial position, and apical departure. Topological representations dealing with multiple pseudorotation processes have been dealt with elsewhere.[8]

4. NUCLEOPHILIC REACTIONS AT ACYCLIC PHOSPHORUS

The β parameters, obtained from structure-reactivity correlations of reactivity as a function of the pK_a of the attacking nucleophile or leaving group, may be utilized as approximate indices of the fraction of charge transferred to the nucleophile or leaving group in the transition state and correspondingly may reflect the degree of bond formation or breakage.[75] Since one is correlating the kinetics of a phosphoryl transfer reaction with the equilibrium for a proton-transfer process, i.e., pK_a, it is necessary prior to analysis to recalibrate in terms of equilibrium phosphoryl transfer. The β values for complete transfer of $[PO_3^{2-}]$ between the donor phosphoramidate and a series of pyridines is 1.2.[76-78] Likewise, the equilibrium transfer of an O-phosphate diester moiety between phenolic oxygens is characterized by β of 1.2.[79] This value, which should be independent of the chemical nature of the donor, will be employed in the ensuing discussion and should be a reasonable approximation for all classes of esters. The moieties $[PO_3^{2-}]$ or $[(RO)_2PO_2^-]$ approximate the electropositive character of a proton, for which β by definition is unity, rather than an acyl group where $\beta = 1.6–1.7$.[75] For the general case the unit charge transferred to or lost by the nucleophile may be estimated by $\beta_n/1.2$, with a concomitant change in charge of $\beta_{lg}/1.2$ on the departing group.

Studies of the reactions of nucleophiles at the phosphorus atom of monoester dianions are characterized by their marked lack of sensitivity to the basicity of the nucleophile.[80-82] The logarithms of the second-order rate coefficients for the attack of various amines on a series of 2-nitro-4-substituted-phenyl phosphate dianions have been correlated as a function of both amine and leaving-group pK_a. The slopes, β, of these linear plots are tabulated in Table I. Since there is no doubt that a bimolecular displacement reaction on phosphorus is involved—isolation of the anticipated phosphoramidate product, entropy of activation $- 20$ eu, etc.—it is remarkable that the rates of the substitution reactions remain constant, independent of the basicity of the pyridines employed with the dianion of 2,4-dinitrophenyl phosphate; i.e., $\beta_n = 0$.[82] The reaction obviously is a limiting case and may be described in terms of a specific molecular interaction more akin to solvation of the developing electrophilic center, i.e., pyridine replacing water, rather than the mechanism of displacement encountered with sp^3 hybridized carbon. As the pK_a of the leaving group is increased, β for the series of nucleophiles increases only slightly, 0–0.1. The description of the above reaction that emerges features a loose, uncoupled transition state; i.e., the degree of bond cleavage is not directly proportional to bond formation. A similar situation obtains for nucleophilic attack on

Table I

A Comparison of β Values for the Incoming and Leaving Groups for Phosphate Esters

Type of reaction	β nucleophile	Ref.	β leaving group	Ref.
Phosphoramidates plus amines	0.2	76, 77	−0.9 to −1.1	77
Phosphate monoester dianions plus amines	0–0.1	81	−1.2	82
Phosphate diester monoanions plus amines	0.3–0.4	83	−0.9 to −1.0	83
Phosphonate diester monoanions plus amines	0.35	84	—	
Phosphate triester plus oxyanions	0.3–0.6	85	−0.3 to −0.6	85
Phosphoro- and phosphonohalides plus hydroxamate	0.8–0.9	86, 87	—	

phosphoramidate monoanions[76,77] and presumably extends to all other substrates whose hydrolysis may be described in terms of the dissociative mechanism:

$$X = NH_2^-, O^-$$

(12)

As the degree of esterification or protonation is progressively increased,
the β parameters gradually increase. For example, the logarithms of the rate
coefficients for the reaction of the same series of pyridines with the monoanion
of 2,4-dinitrophenyl phosphate or 2,4-dinitrophenyl methyl phosphate are a
linear function of nucleophile pK_a with a β_n of 0.3–0.4.[83] As for the above case
of the dianionic species, the rate of nucleophilic attack by amines on phosphorus
is still extremely dependent on the pK_a of the leaving group and rapidly becomes less significant as the pK_a increases. Consequently, the transition state
for these reactions still closely resembles structure (12), although it is beginning
to assume some characteristics of a displacement or associative process. A
similar value for β_n is observed in the reactions of amines with an unhindered
phosphonate diester.[84]

In the reactions of triesters, one now encounters approximately equal
values of β both for nucleophile and leaving group.[84] For a series of pyridines,
nucleophilic attack on the 2,4-dinitrophenyl ester of **9** defines a $\beta_n = 0.6$. Other

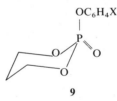

9

nucleophiles, including, for example, phosphate dianion, acetate, and trifluoro-
ethoxide, exhibit their own linear free-energy relationships. The β_{lg} changes
from -0.6 to -0.9 for phosphate dianion and acetate, respectively, whereas
β_n for the oxyanions increases from 0.3 to 0.5 with increasing pK_a of the leaving
group. Higher values of β_n are found in the nucleophilic reactions of hydroxam-
ates with phosphoryl halides.[86,87] Consequently the unimolecular dissociative aspect has disappeared, since the driving force for metaphosphate expulsion
derived from excess electron density has been blocked by esterification or
protonation. Consequently the transition state resembles that encountered in
associative mechanisms, i.e., a pentacovalent phosphorus species, consistent
with β_{lg} equal to or less than β_n. In accordance with the characteristics of such
species, the most plausible formulation of the transition state is a trigonal bipyramid with entering nucleophile and departing group occupying apical
positions.

The stereochemical consequences of this transition-state model, namely inversion of configuration at phosphorus, has been demonstrated for nucleophilic displacement reactions by hydroxide and methoxide on **10** and **11**,

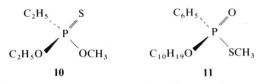

10   **11**

respectively,[88,89] citing but two examples. The employment of optically active acyclic dialkoxyphosphonium salts, such as **12**, which should undergo

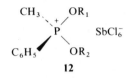

12

nucleophilic displacement by hydroxide through a phosphorane, have revealed, however, that the products of the reaction are markedly dependent on the nature of the alkoxy ligands.[90-94] In the case of $R_1 = CH_3$ and $R_2 = C_2H_5$, the predominant reaction (ca. 90%) is direct displacement of alkoxide resulting from attack of hydroxide in the face of the tetrahedral phosphonium salt opposite the alkoxy ligand. In the case of $R_1 = CH_3$ and $R_2 = C_{10}H_{19}$ (menthyl), the predominant pathway (ca. 55%) is the formation of **13**, which apparently results from the addition–pseudorotation–elimination sequence in

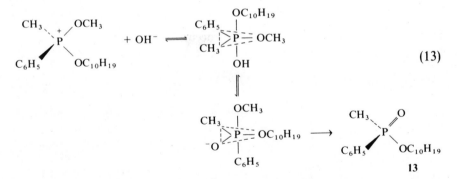

(13)

with initial attack opposite the menthyl ligand (ca. 76% of the reaction). Attack by hydroxide opposite the methoxy ligand (ca. 24%) leads to direct loss of this ligand with no detectable pseudorotation process. Replacement of an alkoxy with an alkythio ligand as in **14** revealed that alkaline hydrolysis proceeds

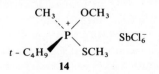

14

mainly with loss of thiol and formation of the product of retained configuration,[93] which again may be interpreted in terms of the addition–pseudorotation–elimination sequence. These examples can be extended to include nucleophilic substitution at phosphorus in (1) S-alkyl alkylphosphonothioates (**15**) and S-alkylphosphorothioates (**16**), where displacement of the thiol ligand by methoxide proceeds with inversion and retention of configuration, respec-

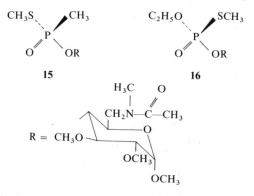

15   **16**

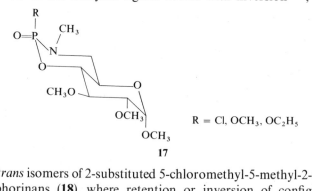

tively[94,95]; in (2) the diastereomers of tetrahydro-1,3,2-oxazophosphorines (**17**), where loss of the exocyclic ligand occurs with inversion[96]; and in (3)

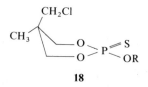

$R = Cl, OCH_3, OC_2H_5$

17

the *cis* and *trans* isomers of 2-substituted 5-chloromethyl-5-methyl-2-thio-1,3,2-dioxaphosphorinans (**18**), where retention or inversion of configuration is

18

found dependent on the nature of the nucleophile and the leaving group, as well as added salts and solvent.[97,98] Collectively these results suggest the probable presence of a metastable intermediate rather than simply a transition-state species in cases where retention of configuration is observed provided an equatorial–equatorial displacement is discounted. However, differences in free energy are evidently marginal between the various structures for the initial phosphoranes, the pseudorotation modes, and the allowed permutational isomers that ensue,[99] so that generalizations regarding the stereochemical outcome of nucleophilic attack on triesters must be applied with caution particularly when oxygen- and sulfur-containing ligands are present.

Kinetic criteria also have been applied in order to distinguish between a process involving direct displacement through a colinear transition state and one involving a pentacoordinate intermediate with significant lifetime. A com-

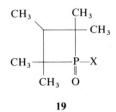

19

parison of the hydrolysis rate of the strained ring systems (**19**) and the acyclic

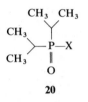

20

analogs (**20**) revealed rate retardations when $X = -Cl$ and $-N^+H(CH_3)_2$— **19** hydrolyzes more slowly than **20**—but a rate acceleration when $X = -OCH_3$.[100,101] A plausible interpretation attributes the latter acceleration to relief of strain upon formation of the trigonal-bipyramidal intermediate, whereas rate retardation occurs in its absence because of the need to expand the C—P—C ring angle in the displacement transition state. Moreover, the kinetics of the alkaline hydrolysis of the acyclic ester, methyl diisopropyl-phosphinate, are characterized by an unusual induction period, interpreted as the accumulation of a low concentration of the intermediate. However, there is no evidence in any of these systems for incorporation of O^{18} from solvent into unreacted ester, despite the fact that breakdown of the intermediate apparently is rate determining for the alkaline hydrolysis of the phosphinate esters. The absence of exchange is consistent with the above preference rules for pseudorotation. The stricter steric requirements imposed by the addition of nucleophiles to tetrahedral phosphorus relative to trigonal acyl carbon often results in general base catalysis rather than nucleophilic attack as found in the reactions of amines with bis(p-nitrophenyl) methylphosphonate[102] and aryl diphenylphosphinates[103] despite the presence of a good leaving group.

It is useful to compare the reaction rate coefficients for several standard nucleophiles with the various classes of esters in order to visualize the effects of changes in pK_a, state of ionization, and the chemical identity of the nucleo-philic hetero atom. For the oxyanions, acetate and hydroxide, the ratio of rate coefficients for reaction with the negatively charged diester and neutral triester favors the latter by an average factor of about 2000.[83] For attack by amines, pyridine, and n-butylamine, the same ratio varies over a range of only 2 to 100. The latter is probably the result of the increased electropositive character of phosphorus upon esterification.[104] Assuming that a similar effect may be ascribed to the oxyanions, a retarding factor of ca. 100 then may be attributed to unfavorable electrostatic repulsion per unit negative charge. A second point of interest is the higher reactivity of acetate or fluoride relative to pyridine at the triester level despite the lower pK_a. This is generally observed as a consequence of the greater strength of P—O or P—F vs. P—N bonds being reflected in this transition state owing to the increased importance of bond formation. This factor also undoubtedly influences whether a direct displacement process or one involving a pentacovalent species where the P=O bond energy is lost will be observed. A final important feature is that nucleophilic reactivity does not decrease monotonically from triester to monoester; the rate of reaction of pyridine, for example, with 2,4-dinitrophenyl phosphate dianion is greater than with the corresponding triesters, which is an additional manifestation of the highly uncoupled, metaphosphate character of the transition state of the former.

Replacement of a phosphoryl oxygen atom with sulfur generates a class of phosphate esters possessing a P=S center. Such reagents have found increasing use as a means for probing enzyme-catalyzed phosphoryl-transfer processes. The susceptibility of triesters containing the P=S center to nucleophilic displacement by hydroxide is considerably lessened, i.e., $k_S/k_O \simeq 0.03$, owing to decreased polarization of phosphorus upon substitution of the less electronegative sulfur.[105] For monoester hydrolysis where a metaphosphate mechanism presumably is operative, the inverse order is observed.[106] This order is not inviolate, whereas the rates of hydrolysis of the trimethyl esters in aqueous alkali decrease in the sequence P=O > P=S; in aprotic solvents the relative order is reversed.[107] The deduction of an enzyme mechanism based on determination of the k_S/k_O ratio therefore is subject to several pitfalls, not the least of which is differences in substrate binding owing to the considerable difference between P=S and P=O in interatomic distance and van der Waals radii.[108] They are, however, quite useful in stereochemical studies, since a chiral phosphate ester is more accessible in this series, as discussed in Section 6.2.

5. MECHANISM FOR CATALYSIS OF PHOSPHORYL TRANSFER

5.1. Intramolecular Acid–Base and Nucleophilic Catalysis

Models for enzymic catalysis of phosphoryl transfer have been found principally in phosphates with suitably placed intramolecular catalytic residues. The majority of productive studies have utilized the carboxyl group and have produced excellent examples of the two principal catalytic mechanisms: (1) general acid catalysis of leaving-group expulsion and (2) nucleophilic attack on phosphorus involving as a prerequisite formation of five- or six-membered rings.

5.1.1. General Acid Catalysis

Catalysis of phosphoryl transfer (principally hydrolysis) from monoprotonated monoesters remain via the metaphosphate mechanism, as manifested in the hydrolysis of salicyl and acetyl phosphate.[12,109] The location of the proton is the principal question and has been examined by hydrolyzing a series of phenyl-substituted salicyl phosphates. This allows determination of the linear free-energy correlation for the rate coefficient dependency on the pK_a of the carboxyl moiety and the phenolic leaving group.[110] These results suggest that P—O cleavage is well advanced and that proton transfer is far from complete:

 (14)

This conclusion is somewhat surprising in view of mechanism (1), which features preequilibrium transfer so that the hydrolysis of salicyl phosphate must involve considerable alteration in the transition-state structure. It has important implications for enzymic catalysis, however, showing that the free energy of the normal zwitterionic intermediate can be lowered or the species bypassed by proper placement of an acidic residue. An estimate of the efficiency of the catalysis depends on the reference standard selected. If viewed as a dianionic species, then the rate of hydrolysis is some 10^{10} greater than predicted, and if as a monoanion, ca. 10^2 faster than anticipated. The effectiveness of general acid catalysis—expressed as the ratio of the rates of hydrolysis of the *o*- and *p*-isomers—is ca. 200 for salicyl phosphate. The manifestation of catalysis is markedly dependent on the system employed; the hydrolyses of the dianions derived from phosphoenolpyruvate[111] and 2-carboxyphenyl phosphoramidate[19] are not catalyzed by the intramolecular carboxyl function. This suggests that a steric constraint disfavors a catalytic mode,

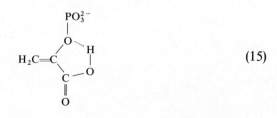 (15)

which avoids formation of a zwitterionic species and that the intermediacy of this species also cannot be circumvented in phosphoramidate hydrolysis owing to the necessity for expelling a neutral leaving group:

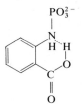

$$\tag{16}$$

5.1.2. Nucleophilic Catalysis

Appropriate di- and triesters undergo intramolecular nucleophilic attack resulting in catalysis of both endocylic and exocyclic ligand exchange since most examples involved internal cyclizations forming five- or six-membered rings. With carboxyl moieties, examples of participation by both carboxyl and carboxylate have been observed.

An example of diester, carboxyl-group catalysis is the hydrolysis of monobenzyl phosphoenolpyruvate, which proceeds via loss of benzyl alcohol (90%)—despite the fact that it is the poorer leaving group—and minor amounts (10%) of monobenzyl phosphate.[112] The pH-rate profile implicates the carboxyl group as being catalytically active. In the presence of hydroxylamine the course of the reaction is unchanged, although pyruvate oxime hydroxamate is produced. The proposed scheme,

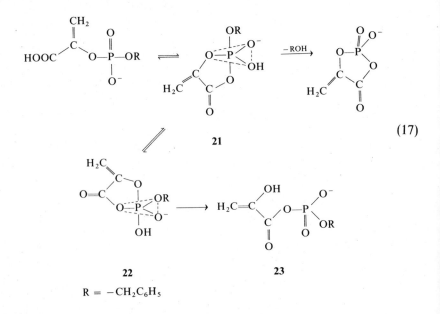

$$\tag{17}$$

incorporates the pentacovalent species **21** and **22** in order to rationalize the formation of **23**. Carboxylate catalysis is encountered in the hydrolysis of diesters of salicyl phosphate.[113]

Similar catalysis can be observed in triesters. Both carboxyl and carboxylate catalysis is observed in these systems.[79,114,115] The composition of the products, endo- or exocyclic displacement, depends exclusively on the basicity of the leaving group in accord with the fact that the necessary pseudorotations are allowed, and, moreover, the free-energy barrier separating them is small relative to decomposition. The hydrolysis of **24** yields mainly the product of endocyclic displacement (96%),

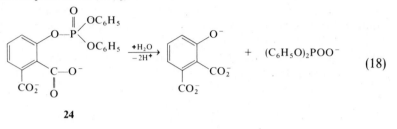

$$+ \quad (C_6H_5O)_2POO^- \tag{18}$$

24

whereas the hydrolysis of **25** leads to that arising from exocyclic displacement (98%),

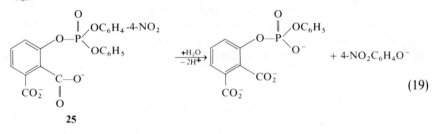

$$+ 4\text{-}NO_2C_6H_4O^- \tag{19}$$

25

The ratio of endo- to exocyclic products is unity for an exocyclic leaving group $pK_a \approx 8.5$, which is within estimates of the basicity of the salicylate oxygen. Linear free-energy relationships involving the rate coefficients for exocyclic and endocyclic displacement as a function of the pK_a of the exocyclic leaving group exhibit a high sensitivity, $\beta_{lg} = -1.4 \pm 0.2$ and $\beta_n = -0.3$, respectively. The former implicates a transition state for exocyclic displacement with considerable P—O bond fission and conforms to the supposed ionic character of the apical bond. The response of the endocyclic rate to changes in exocyclic substitution of phosphorus is consistent with the previous hypothesis that the transition state in triester nucleophilic displacement reactions is coupled and furthermore implies that a small percentage of leaving group sensitivity resides in bond formation. The above examples have been selected as illustrative; others have been reported involving carboxyl,[116-118] hydroxyl,[119] and amide[120] functions.

Several generalizations may be drawn from these and other examples. (1) Catalyses by carboxyl and/or carboxylate are important in P—O bond cleavage reactions. However, no consistent rationale has been developed concerning the differing efficacies of carboxyl vs. carboxylate in the nucleophilic cyclizations. Catalysis by the carboxyl function is thus readily envisioned throughout the biological pH range. (2) The ratio of k^{intra}/k^{inter}, an index of effective molarity, is $\simeq$ unity for triesters and $\simeq 40\,M$ for the diester monoanions. At concentrations of acetate equivalent to that for the o-carboxy esters, the rates of the intramolecular reactions are 10^4–10^6 more rapid than their bimolecular counterparts. (3) Most examples of pentacovalent phosphorus formation involve cyclic systems which contain five-membered rings. However, in nucleophilic catalysis five- and six-membered ring formation appears to be roughly comparable in catalytic effectiveness in many but not all cases.[116] Although it may appear surprising that (1) ring size is not manifest in these reactions and (2) cyclization is a kinetically favorable process when the reverse process is so favorable, these findings are in accord with the relief of ring strain that is achieved in attainment of pentacovalency at phosphorus. Since there are a number of difficulties associated with the dissection of kinetic rate coefficients in intramolecular systems, the contribution of a stable pentacovalent species to the overall rate cannot be readily assessed,[121] although it is probably not greater than 10^2.

Catalysis of phosphoryl transfer by nucleophilic addition to intermediate phosphoranes appears plausible in principle. The phosphorylation of R^2OH by the phosphorane **26** in the presence of imidazole has been speculated to proceed via the hexacoordinate species, yielding **27**, which is subsequently hydrolyzed to the $R^1OPO(OR^2)OH$ diester[122]:

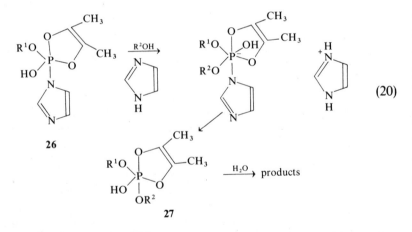

$$(20)$$

However, such a process should have severe steric constraints, and it remains to be conclusively demonstrated that imidazole is not acting in a nonnucleophilic catalytic role.

5.2. Catalysis by Metal Ions

The widespread occurrence of divalent metal ions as cofactors in phosphoryl transfer reactions has continued to encourage the search for models of such catalysis. The metal ion may be postulated to act catalytically by one or more of several mechanisms: (1) as a template for binding and orienting substrates and enzymic catalytic groups, i.e., an entropic rate enhancement; (2) charge neutralization to enhance approach of a negatively charged nucleophile to anionic phosphorus; (3) chelation of the leaving group to facilitate its loss via either of the above mechanisms; (4) complexation with a pentacovalent intermediate to control the stereochemical course or increase the rate of the reaction; and (5) a hydroxyl carrier to promote hydrolysis. Several representative model systems will illustrate these points. It should be clear that separation of some of these effects is difficult, if not impossible.

The Zn^{2+}-catalyzed phosphoryl transfer between phosphorylimidazolium ion and pyridine-2-carbaldoxime and the virtual reverse reaction is believed to proceed through a species represented in[123,124]

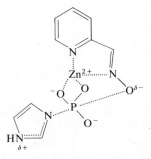

$$(21)$$

It is obvious that the Zn^{2+} can be viewed as a template for bringing the reacting species together. It, however, also serves to neutralize the charge repulsion between the anions of the aldoxime and the phosphoryl moiety in the expulsion of the imidazole. The reverse reaction, i.e., phosphoryl transfer to nucleophiles from pyridine-2-carbaldoximyl phosphate, is, as expected, also Zn^{2+}-catalyzed and involves basically the same transition state characterized by high metaphosphate character. Catalysis in this direction is usually expressed in terms of leaving-group activation, i.e., reduction of the leaving-group pK_a. Thus the distinction between modes (2) and (3) in this case is principally semantic, since both involve stabilization of the same transition state.

Numerous other examples exist of leaving-group chelation in order to facilitate metaphosphate formation, all possessing a bidentate ligand order to produce productive complexation.[125-127] Rate enhancements of 10^5–10^6 for structure (21) and for

$$(22)$$

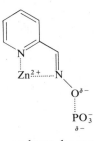

(23)

are directly proportional to the resultant decrease in leaving-group pK_a owing to the metal ion–oxyanion complex and fit predictions based on the structure-reactivity correlation for the hydrolysis of phosphate monoester dianions. The leaving-group dependence for phosphoryl transfer from dianions is greater than for monoanions; consequently the former are more susceptible to this mode of catalysis than the latter species.

A number of attempts have been made to demonstrate metal-ion catalysis of phosphoryl transfer from ATP and similar polyphosphate chains,[128,129] but large rate enhancements have been absent. Since the incipient oxyanion leaving group does not lie within the coordination sphere of the metal ion, it follows that complexation must be very strong and may not be attainable in purely aqueous solutions. However, in largely nonaqueous solution, e.g., 70% dimethyl sulfoxide, significant catalysis of phosphoryl transfer from ATP by Mg^{2+} occurs.[130,131] The hydrolysis reaction has the additional requirements of a dicarboxylic acid and arsenate, while transfer to inorganic phosphate requires only Mg^{2+} and a dicarboxylic acid. Although the system is clearly quite complex, the results suggest a nucleophilic displacement sequence with an arsenatophosphate intermediate in the hydrolysis reaction.

Several examples postulated to involve metal-ion interaction with covalent phosphorus intermediates have been reported.[116,132-134] In lactic acid O-phenyl phosphate hydrolysis, which is thought to proceed through a cyclic acyl phosphate pentacovalent intermediate, catalysis by Mg^{2+} and other divalent metal ions is observed. The 10^2–10^3 rate enhancement is consistent with stabilization of the intermediate and/or a complexation of the leaving group in the rate-determining breakdown of the pentacovalent structure **28**.

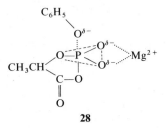

28

Other examples show that nitrogen may be involved as an equatorial ligand and that catalysis occurs with both five- and six-membered rings.

Bidentate coordination of the tetravalent ester to the metal ion resulting in catalysis of exocyclic ligand loss, possibly owing to stabilization of the ensuing pentacovalent intermediate, has been invoked in the catalysis of methyl phosphate hydrolysis by triethylenetetramine Co^{3+}. The acceleration in rate is about 100-fold relative to monoanionic methyl phosphate.[135] No catalysis is observed with Co^{3+} complexes where bidentate coordination is precluded. The crucial question is whether the chemical identity of the reactive species is **29** or **30**. If **29**, then catalysis may be interpreted in terms of hydroxyl attack

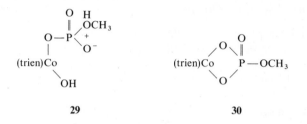

on a monoester monoanion and would be a combined example of modes (1) and (5). If **30**, then the rate of hydrolysis is ca. 10^6 faster than trimethyl phosphate, with the metal ion inducing strain comparable to that observed for ethylene phosphate.

In summary, although a considerable extension of our understanding of the mechanisms and transition-state structures is desirable, significant catalysis by metal ions is feasible for either dissociative or associative phosphoryl-transfer mechanisms. It should be emphasized that several of the rather arbitrarily defined modes of metal-ion catalysis may operate in concert. Clearly the best illustrated mechanism is enhancement of the leaving group by complexation, which may occur in the associative or dissociative processes. The template effect in which the metal ion may serve as a bidentate binding site on an enzyme also is readily envisioned. Little evidence has been accumulated at present quantitating the contribution to catalysis by metal-ion charge neutralization of the anionic phosphoryl moiety in displacements by anionic nucleophiles.

6. MECHANISMS AND PROBES OF ENZYMIC PHOSPHORYL TRANSFER

The number of enzymes catalyzing phosphoryl transfer is, of course, very large. In numerous cases our understanding of the mechanism has advanced far enough to cautiously attempt to crudely outline the probable mechanism of these transfers. In this section we shall summarize the information derived from model studies which is pertinent to the enzyme reaction and shall consider the methods applicable to elucidating the transfer mechanisms. Our starting point is to consider what is reasonable and effective in the light of the known model chemistry. Then we may attempt to define methods which will be informative in evaluating the mode of enzymic catalysis and the associated mechanism of transfer. The vast majority of biological esters are monoesters, and in the biological pH ranges their nonenzymic phosphoryl transfer almost invariably involves metaphosphate or species with considerable metaphosphate character. However, protonation of the phosphoryl oxygens (or possibly tight metal-ion chelation) effectively converts these to di- and triesters. Thus both mechanisms must be considered in enzymic transformations.

6.1. General Acid Catalysis

General acid catalysis or metal-ion chelation of the leaving group should be enzymically important regardless of the mechanism of phosphorylation, i.e., in the generation of metaphosphate or in the bimolecular displacement proceeding through a transition state with pentacovalent structure or a true intermediate. That leaving-group stability is important in mono- and dianion hydrolysis of monoesters via the metaphosphate mechanism is unmistakable, owing to the large values for β_{lg}. This likewise is true for transfer to nucleophiles other than water. While monoesters with excellent leaving groups, e.g., ATP^{4-}, acetyl phosphate, hydrolyze readily via expulsion of the conjugate base of the leaving group, it is clear that a suitably positioned proton would further enhance these rates. The dianion of glucose-6-phosphate, like salicyl phosphate, is hydrolytically labile because of the availability of a proton from the neighboring hydroxyl group which is effective although only weakly acidic $(pK_a \simeq 11)$.[136,137]

We may attempt to evaluate the rate increase which might be achieved enzymically via proton donation to monoesters. If we consider the decomposition of the zwitterionic species formed by protonation of the leaving group, the rate coefficient for P—O fission of $RX^+HPO_3^{2-}$ ranges from 10^{-3} to 10^9 min^{-1} for various leaving groups, as noted previously. It is, however, difficult to envision a simple process outside microsolvent effects for increasing the concentration of the zwitterionic species of a monoester or the unsymmetrical di- or trianionic polyphosphate tautomer by increasing the basicity of the incipient departing ligand without a compensating decrease in its reactivity. A similar argument views the attendant difficulties of increasing the former's concentration by increasing the acidity of the enzymic functional group above that of hydronium ion. Indeed, actual preequilibrium protonation of the leaving ligand probably is not necessary, as seen above for intramolecular systems. If one analogously assumes that the interaction between an acidic enzyme functional group and the leaving ligand generates an assemblage whose pK_a for proton dissociation is 4–5 pK_a units greater than the free ligand, then rate acceleration of ca. 10^4 may be attained for oxygen esters based on the data for salicyl phosphate. Perturbations of pK_a of this order of magnitude have been observed.[138] Di- and triesters undergoing bimolecular displacement likewise exhibit a large dependence on the leaving-group pK_a. General acid catalysis in intramolecular systems is usually accompanied by nucleophilic participation so that it is reasonable that a suitably positioned proton donor would facilitate ligand expulsion in these systems as well.

Tight chelation of the leaving group by divalent metal ions may be viewed similarly and is plausible for either mechanism. For dianionic hydrolysis and transfer to other nucleophiles from both mono- and dianionic monoesters this dependency is ca. tenfold per pK unit, illustrating the high catalytic potential of this mode. The required tight leaving-group chelation is difficult to obtain in aqueous solution but may be greatly enhanced in the enzyme active-site environment.

6.2. Pseudocyclization

There are a limited number of biological esters with nucleophiles positioned intramolecularly so as to be involved in cyclic pentacovalent phosphate formation. However, a mechanism involving a pentacovalent intermediate need not be limited to substrates with suitably juxtaposed functional groups. Various mechanisms of constraint may be envisioned whereby the geometry of the pentacovalent phosphorus may be forced on the substrate: (1) The attacking nucleophile may be positioned so as to mimic its being a member of a five-membered ring with either the leaving group and/or a phosphoryl oxygen, and (2) the phosphate oxygens and/or leaving group may be forced into such a configuration. Both of these would be expected to facilitate nucleophile attack if they closely approximate a five-membered ring. This same process can, of course, occur for a phosphoryl-enzyme intermediate. The coupling between enzyme, donor, and acceptor may be seen as a derivative of the strain theory of catalysis.[139] In mode (2) "strain" is manifest in a higher potential energy reactant ground state; in mode (1) the product after ligand loss is of higher potential energy—in relationship to the unbound donor–acceptor. It is clear, however, that the stabilization of the pentacovalent species arises from a minimization of unfavorable nonbonded interactions so that modes (1) and (2) represent limiting means for attaining its intermediacy relative to merely a transition-state lifetime. These mechanisms require several binding groups at the active site or possibly a metal ion to accomplish the pseudocyclization. Moreover, the rather rigid stereochemical requirements for the apical addition and departure of ligands from pentacovalent phosphorus dictate specific orientations for donor–acceptor binding in the absence of pseudorotations.

6.3. General Base Catalysis and Environmental Effects

There is no requirement for general base catalysis in the metaphosphate mechanism. The reaction of the nucleophile with metaphosphate is rapid—not rate limiting—and solvent could readily absorb the excess proton if one is liberated. For the associative process, however, substantial catalysis by suitably positioned basic residues might be obtained in those cases where rate-determining decomposition of the pentacovalent species occurs. In the case of amine attack, for example, general base catalysis would serve to trap the initially formed intermediate by proton abstraction, preventing its return to starting materials[140]:

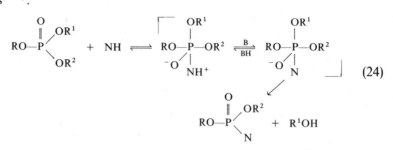

$$(24)$$

The process may take place in a stepwise or concerted fashion.

Since dissociative decomposition of either monoester mono- or dianions must proceed through transition states involving charge separation, large acceleratory effects may be anticipated for these reactions conducted in less polar solvents. At present it is difficult to assess the magnitude of possible catalysis caused by substrate location in hydrophobic regions of low polarity at the active site. Estimates derived from experiments in mixed solvent are not applicable owing to specific solvent effects, but micellar studies indicate increased rate factors of at least 100-fold and possibly larger.[121,141,142] These effects should be attenuated in the associative mechanism. One may therefore reasonably argue that combinations of these modes of catalysis with the loss of translational and rotational entropies upon binding of substrates by the enzyme[121] should suffice to attain the catalytic advantage found in a comparison of enzymic-catalyzed phosphoryl transfer to model reactions.[2]

6.4. Experimental Approaches

Several approaches have been used to distinguish between the dissociative and associative mechanisms in enzyme-catalyzed phosphoryl transfer. We shall consider a specific example of each approach, the information yielded, and the limitation of the method.

Usher et al.[143-145] have examined the stereochemistry of both the cyclization and ring opening steps of the ribonuclease reaction to obtain evidence for possible pentacovalent intermediates. Of the two plausible directions of nucleophilic addition to phosphorus—the "in-line" mechanism of nucleophilic approach 180° from the leaving atom (equivalent to apical attack) and the "adjacent" mechanism of approach 90° from the leaving atom (equivalent to equatorial attack)—the latter requires the existence of a pentacovalent intermediate which must pseudorotate in order to expel the departing ligand from an apical position. The former does not distinguish between the existence of a pentacovalent species and simple bimolecular displacement. The two mechanisms yield products of predictable and different stereochemistry provided a chiral phosphorus center is employed. The stereochemistry resulting from ribonuclease-catalyzed cyclization and ring opening was probed by use of the separated diastereomers of uridine-2′,3′-cyclic phosphorothioate (**31**). In the

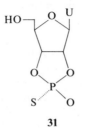

31

experimental protocol the chirality at phosphorus upon ring opening was maintained by employing H_2O^{18} in the medium. Recyclization was achieved by a chemical process of known stereochemistry, permitting determination of the stereochemistry of the ribonuclease-catalyzed ring opening upon analysis of the O^{18} content and establishing the identity of the diastereomer. The resulting stereochemical inversion found in each step rules out the possibility of an adjacent mechanism.

Although direct evidence may be obtained for the adjacent mechanism, if stereochemical retention of configuration is observed, the finding of inversion does not distinguish between a simple bimolecular displacement and the intermediacy of a pentacovalent species. A second experimental limitation is obviously the requirement for separated diasteriomeric forms of substrate.

Phosphate diesters hydrolyze considerably slower than monoesters since the dissociative mechanism, i.e., expulsion of metaphosphate, is inhibited by esterification. Several diester derivatives of normal monoester substrates have been examined for enzymic hydrolytic activity.[146] We may reason that if the enzymic mechanism is of the dissociative type the diesters will not function effectively (or do so very poorly) as substrates. If a bimolecular or associative process is operative, diesterification should not prevent enzymic catalysis of the transfer. The limitations of this probe lie in the obvious question of the effect of the substitution of $-O(H)$ by alkyl, alkoxy, or halo groups on the interaction of substrate and enzyme. Clearly utilization of a diesterified substrate by enzyme would strongly argue for the operation of an associative mechanism. An absence of activity, however, may be due to poor or nonproductive binding rather than a dissociative process.

Examples of this approach include the testing of γ-alkoxy derivatives of GTP as substrates in the ribosomal elongation factor G-dependent reaction[147] and of p-nitrophenyl esters of phosphonic acids as substrates for wheat germ phosphatase.[148] The GTP analogs were found not to be hydrolyzed by the associated GTPase activity, but were strong competitive inhibitors. They also were effective in the nucleotide-dependent binding of elongation factor G to ribosomes. Thus these compounds appear to interact with the active site similarly to GTP but are not hydrolyzed, suggesting a metaphosphate mechanism. On the other hand the phosphonate esters where (O^-) is replaced by an alkyl or aryl function are hydrolyzed at rates comparable to the monoesters by the wheat germ phosphatase. This finding is most simply interpreted on the basis of an associative mechanism, in this case an active site histidine is the nucleophilic species.

A choice between mechanistic alternatives has been based on a substitution of sulfur for oxygen to form a phosphorothioate.[149] This modification should reduce the rate of the bimolecular process but increase the rate of a metaphosphate-type mechanism. For reasons noted earlier, i.e., rate differences caused by changes in binding or microenvironmental effects, such rate changes between the two systems might not be diagnostic.

A correlation between ^{31}P nuclear magnetic-resonance chemical shift and the oxygen–phosphorus–oxygen bond angle has been described for phosphate esters.[150] Since a five-membered cyclic phosphate or any phosphate which might be constrained to mimic cyclization has a distinctive O—P—O bond angle, the ^{31}P chemical shift is potentially diagnostic for strain activation in the enzyme–phosphate complex. The limitations of this method are principally in the still-tentative nature of this correlation and whether any unusual effects on ^{31}P chemical shift may occur within the enzyme active site invalidating the correlation. An unusual chemical shift downfield from inorganic phosphate has been observed in alkaline phosphatase and assigned to the phosphoserine enzyme intermediate.[151] On the basis of the above correlation a strained O—P—O bond angle has been postulated. Thus a rate enhancement of phosphoryl transfer induced by strain and proceeding through an associative mechanism is plausible.

As has been reviewed extensively elsewhere,[152] nuclear magnetic resonance may also give considerable information on the interaction of a paramagnetic metal ion and a phosphate substrate in the presence of enzyme. Metal-ion binding to the phosphoryl oxygen(s) may be indicative of an associative mechanism, whereas chelation of the leaving group would be compatible with either mechanism. The principal problem with this type of information is whether the structure measured is relevant to the transition states or to a single structure or is a composite of several pre-transition-state species.

One of the probes used in nonenzymic phosphate hydrolysis to gather evidence for a pentacovalent intermediate is incorporation of more than a single mole of $H_2^{18}O$ in the unreacted ester. Such incorporation shows the existence of an intermediate which contains ^{18}O, pseudorotates, and returns to starting material with the loss of $H_2^{16}O$. This criterion has also been applied to enzyme-catalyzed reactions.[153] However, although in model reactions a pentacovalent species is the only realistic intermediate, in enzyme-mediated exchanges we may readily envision other "intermediate" complexes. For example, the mitochondrial ATP synthetase incorporates ^{18}O into ATP. While this in principle could occur through a pentacovalent intermediate, it may also result from reversal of the ATP cleavage reaction, prior to the release of the ^{18}O-containing inorganic phosphate from the enzyme.[154,155] The incorporation of ^{18}O requires that the bound inorganic phosphate possess rotational freedom. In solution such a reversal would be thermodynamically unfavorable, but there is growing evidence that the phosphate cleavage equilibrium can approach unity in the enzyme active site.[121,154] Thus enzyme-catalyzed reversal may be a reasonable explanation of ^{18}O incorporation. It is also necessary to point out that failure to observe ^{18}O exchange does not rule out a pentacovalent species as water exchange may be excluded by the enzyme.

In summary, while some of the above methods may yield definitive mechanistic information, it appears that more often the results will be ambiguous or circumstantial. In those cases, as usual we shall have to rely on the weight of evidence gleaned from a number of experiments in order to deduce reliable mechanistic conclusions.

REFERENCES

1. T. C. Bruice and S. J. Benkovic, *Bioorganic Mechanisms*, Vol. 2, Benjamin, Reading, Mass. (1966), pp. 1–176.
2. S. J. Benkovic and K. J. Schray, in: *The Enzymes* (P. D. Boyer, ed.), Vol. 8, pp. 201–238, Academic Press, New York (1973).
3. A. J. Kirby and S. G. Warren, *The Organic Chemistry of Phosphorus*, Elsevier, Amsterdam (1967).
4. S. J. Benkovic, in: *Comprehensive Chemical Kinetics* (C. H. Bamford and C. F. Tipper, eds.), Vol. 10, pp. 1–51, American Elsevier, New York (1972).
5. F. H. Westheimer, Pseudorotation in the hydrolysis of phosphate esters, *Acc. Chem. Res.* **1**, 70–78 (1968).
6. W. E. McEwen, in: *Topics in Phosphorus Chemistry* (M. Grayson and E. J. Griffith, eds.), Vol. 2, pp. 1–41, Wiley, New York (1965).
7. R. F. Hudson and C. Brown, Reactivity of heterocyclic phosphorus compounds, *Acc. Chem. Res.* **5**, 204–211 (1972).
8. K. Mislow, Role of pseudorotation in the stereochemistry of nucleophilic displacement reactions, *Acc. Chem. Res.* **3**, 321–331 (1970).
9. P. Gillespie, F. Ramirez, I. Ugi, and D. Marquarding, Displacement reactions of phosphorus (V) compounds and their pentacovalent intermediates, *Angew. Chem. Int. Ed. Engl.* **12**(2), 91–119 (1973).
10. C. A. Bunton, D. R. Llewellyn, K. G. Oldham, and C. A. Vernon, Reactions of organic phosphates. I. Hydrolysis of methyl dihydrogen phosphate, *J. Chem. Soc.* **1958**, 3574–3587.
11. C. A. Bunton, E. J. Fendler, E. Humeres, and K. Yang, The hydrolysis of some monophenyl phosphates, *J. Org. Chem.* **32**, 2806–2811 (1967).
12. G. Di Sabato and W. P. Jencks, Mechanism and catalysis of reactions of acyl phosphates. II. Hydrolysis, *J. Am. Chem. Soc.* **83**, 4400–4405 (1961).
13. D. C. Dittmer and O. B. Ramsey, Reactivity of thiophosphates. I. Hydrolysis of phosphorothioic acid, *J. Org. Chem.* **28**, 1268–1272 (1963).
14. S. Milstien and T. H. Fife, The hydrolysis of S-aryl phosphorothioates, *J. Am. Chem. Soc.* **89**, 5820–5826 (1967).
15. S. J. Benkovic and E. J. Sampson, Structure-reactivity correlation for hydrolysis of phosphoramidate monoanions, *J. Am. Chem. Soc.* **93**, 4009–4016 (1971).
16. J. D. Chanley and E. Feageson, Hydrolysis of phosphoramides, *J. Am. Chem. Soc.* **80**, 2686–2691 (1958).
17. L. L. Schaleger and F. A. Long, in: *Advances in Physical Organic Chemistry* (V. Gold, ed.), Vol. 1, pp. 1–33, Academic Press, New York (1963).

18. J. D. Chanley and E. Feageson, A study on the hydrolysis of phosphoramides. II. Solvolysis of phosphoramidic acid and comparison with phosphate esters, *J. Am. Chem. Soc.* **85**, 1181–1190 (1963).
19. S. J. Benkovic and P. A. Benkovic, Hydrolytic mechanisms of phosphoramidates of aromatic amino acids, *J. Am. Chem. Soc.* **89**, 4714–4722 (1967).
20. C. H. Clapp and F. H. Westheimer, Monomeric methyl metaphosphate, *J. Am. Chem. Soc.* **96**, 6710–6714 (1974).
21. A. J. Kirby and A. G. Varvoglis, The reactivity of phosphate esters. Monoester hydrolysis, *J. Am. Chem. Soc.* **89**, 414–423 (1967).
22. Y. Murakami and J. Sunamoto, Solvolysis of organic phosphates. Part IX. Structure-reactivity correlations for the hydrolysis of organic orthophosphate monoesters, *J. Chem. Soc. Perkin Trans. 2* **1973**, 1235–1241.
23. G. W. Allen and P. Haake, Hydrolysis of phosphoroguanidines. A model system for phosphorylation by phosphorocreatine, *J. Am. Chem. Soc.* **95**, 8080–8087 (1973).
24. C. A. Bunton and H. Chaimovich, The acid-catalyzed hydrolysis of pyrophosphoric acid, *Inorg. Chem.* **4**, 1763–1766 (1965).
25. D. Samuel and B. Silver, The existence of the $P'P'$-diethyl pyrophosphate ion, *J. Chem. Soc.* **1961**, 4321–4324.
26. D. L. Miller and F. H. Westheimer, The hydrolysis of γ-phenylpropyl di- and triphosphates, *J. Am. Chem. Soc.* **88**, 1507–1511 (1966).
27. D. L. Miller and T. Ukena, P_1P_1-Diethyl pyrophosphate, *J. Am. Chem. Soc.* **91**, 3050–3053 (1969).
28. C. A. Bunton, E. J. Fendler, and J. H. Fendler, The hydrolysis of dinitrophenyl phosphates, *J. Am. Chem. Soc.* **89**, 1221–1230 (1967).
29. D. G. Gorenstein, Oxygen-18 isotope effect in the hydrolysis of 2,4-dinitrophenyl phosphate. A monomeric metaphosphate mechanism, *J. Am. Chem. Soc.* **94**, 2523–2525 (1972).
30. F. H. Westheimer, Studies of the solvolysis of some phosphate esters, *Chem. Soc. Spec. Publ.* **8**, 1–15 (1957).
31. P. S. Traylor and F. H. Westheimer, Mechanisms in the hydrolysis of phosphorodiamidic chlorides, *J. Am. Chem. Soc.* **87**, 553–559 (1965).
32. A. Williams and K. T. Douglas, E1cB Mechanisms. II. Base hydrolysis of substituted phenyl phosphorodiamidate, *J. Chem. Soc. Perkin Trans. 2* **1972**, 1454–1459.
33. A Williams and K. T. Douglas, E1cB Mechanisms. Part IV. Base hydrolysis of substituted phenyl phosphoro- and phosphorothio-diamidate, *J. Chem. Soc. Perkin Trans. 2* **1973**, 318–324.
34. A. F. Gerrard and N. K. Hamer, Evidence for a planar intermediate in alkaline solvolysis of methyl *N*-cyclohexylphosphoramiothioic chlorine, *J. Chem. Soc. B* **1968**, 539–543.
35. J. Rebek and F. Gavina, The three-phase test for reaction intermediates, metaphosphates, *J. Am. Chem. Soc.* **97**, 1591–1592 (1975).
36. P. Haake and P. S. Ossip, S_N1 mechanism in displacement at phosphorus. Solvolysis of phosphinyl chlorides, *J. Am. Chem. Soc.* **93**, 6924–6930 (1971).
37. P. Haake and D. A. Tyssee, Dissociative displacement at phosphorus by unimolecular cleavage of a P—N bond. Entropy of activation as a criterion of mechanism, *Tetrahedron Lett.* **40**, 3513–3516 (1970).
38. S. B. Hartley, W. S. Holmes, J. K. Jacques, M. F. Mole, and J. C. McCoupbrey, Thermochemical properties of phosphorus compounds, *Q. Rev. Chem. Soc.* **17**, 204–223 (1963).
39. K. A. R. Mitchell, The use of outer *d* orbitals in bonding, *Chem. Rev.* **69**, 157–178 (1969).
40. G. Capozzi and P. Haake, Acid-dependent associative and dissociative mechanisms of displacement at phosphorus, *J. Am. Chem. Soc.* **94**, 3249–3450 (1972).
41. P. Haake and T. Koizumi, Hydrolysis of phosphinamides and the nature of the P—N bond, *Tetrahedron Lett.* **55**, 4845–4848 (1970).
42. E. L. Muetterties and R. A. Schuunn, Pentacoordination, *Q. Rev. Chem. Soc.* **20**(2), 245–299 (1966).
43. R. Schmutzler, Chemistry and stereochemistry of fluorophosphoranes, *Angew. Chem.* **77**, 530–541 (1965).

44. R. Luckenbach, *Dynamic Stereochemistry of Pentaco-ordinated Phosphorus and Related Elements*, Thieme, Stuttgart (1973).
45. W. C. Hamilton, S. J. LaPlaca, F. Ramirez, and C. P. Smith, Crystal and molecular structures of pentacoordinated Group V compounds. I. 2,2,2-Triisopropoxy-4,5-(2′,2″-biphenyleno)-1,3,2-dioxaphospholene, *J. Am. Chem. Soc.* **89**, 2268–2272 (1967).
46. F. Ramirez, Oxyphosphoranes, *Acc. Chem. Res.* **1**, 168–174 (1968).
47. D. B. Denney and H. M. Relles, Pentaethoxyphosphorus, *J. Am. Chem. Soc.* **86**, 3897–3898 (1964).
48. D. B. Denney and S. T. D. Gough, Pentaalkoxyphosphoranes, *J. Am. Chem. Soc.* **87**, 138–139 (1965).
49. F. Ramirez, A. J. Bigler, and C. P. Smith, Pentaphenoxyphosphorane, *J. Am. Chem. Soc.* **90**, 3507–3511 (1968).
50. R. Hoffmann, J. M. Howell, and E. L. Muetterties, Molecular orbital theory of pentacoordinate phosphorus, *J. Am. Chem. Soc.* **94**, 3047–3058 (1972).
51. J. A. Howard, D. R. Russell, and S. Trippet, Square pyramidal phosphorus. X-ray analysis of the 1,3,2-dioxaphospholanes from hexafluoroacetone and phosphetans, *Chem. Commun.* **1973**, 856–857.
52. P. Gillespie, P. Hoffman, H. Klusacek, D. Marquarding, S. Pfohl, F. Ramirez, E. A. Tsolis, and I. Ugi, Non-rigid Molecular skeletons—Berry pseudorotation and turnstile rotation, *Angew. Chem. Int. Ed. Engl.* **10**(10), 687 (1971).
53. G. M. Whitesides and H. L. Mitchell, Pseudorotation in tetrafluorodimethylaminophosphorane, *J. Am. Chem. Soc.* **91**, 5384–5386 (1969).
54. R. S. Berry, Correlation of rates of intramolecular tunneling processes, with application to some Group V compounds, *J. Chem. Phys.* **32**, 933–938 (1960).
55. I. Ugi, D. Marquarding, H. Klusacek, P. Gillespie, and F. Ramirez, Berry pseudorotation and turnstile rotation, *Acc. Chem. Res.* **4**, 288–296 (1971).
56. D. Britton and J. D. Dunitz, Isomerization of pentacoordinated molecules, *J. Am. Chem. Soc.* **97**, 3836–3837 (1975).
57. A. Strich and A. Veillard, Electronic structure of phosphorus pentafluoride and polytopal rearrangement in phosphoranes, *J. Am. Chem. Soc.* **95**, 5574–5581 (1973).
58. F. Ramirez and I. Ugi, in: *Advances in Physical Organic Chemistry* (V. Gold, ed.), Vol. 9, pp. 25–126, Academic Press, New York (1971).
59. R. K. Oram and S. Trippett, Reactions of 1-substituted 2,2,3,4,4-pentamethylphosphetanes with hexafluoroacetone and the fluorine-19 nuclear magnetic resonance spectra of the resulting 1,3,2-dioxaphospholanes, *J. Chem. Soc. Perkin Trans. 1* **12**, 1300–1310 (1973).
60. S. Trippett and P. J. Whittle, Apicophilicity of the benzoyl group in five-coordinate phosphoranes, *J. Chem. Soc. Perkin Trans. 1* **13**, 1220–1222 (1975).
61. S. Bone, S. Trippett, and P. J. Whittle, Apicophilicity of thio substituents in trigonal bipyramidal phosphoranes, *J. Chem. Soc. Perkin Trans. 1* **18**, 2125–2132 (1974).
62. K. E. DeBruin, A. G. Padilla, and M. T. Campbell, Alkaline hydrolysis of 1-×-1-Alkoxy-2,2,3,4,4-pentamethylphosphetanium salts. Unusual order of ligand kinetic axiophilicities, *J. Am. Chem. Soc.* **95**, 4681–4687 (1973).
63. P. Haake and F. H. Westheimer, Hydrolysis and exchange in esters of phosphoric acid, *J. Am. Chem. Soc.* **83**, 1102–1109 (1961).
64. F. Covitz and F. H. Westheimer, The hydrolysis of methyl ethylene phosphate: Steric hindrance in general base catalysis, *J. Am. Chem. Soc.* **85**, 1773–1777 (1963).
65. J. R. Cox, R. E. Wall, and F. H. Westheimer, Thermochemical demonstration of strain in a cyclic phosphate, *Chem. Ind. (London)* **1959**, 929.
66. E. T. Kaiser, M. Panar, and F. H. Westheimer, The hydrolysis of some cyclic esters of sulfuric acid, *J. Am. Chem. Soc.* **85**, 602–607 (1963).
67. D. A. Usher, E. A. Dennis, and F. H. Westheimer, Calculation of the bond angles and conformation of methyl ethylene phosphate and related compounds, *J. Am. Chem. Soc.* **87**, 2320–2321 (1965).
68. M. G. Newton, J. R. Cox, and J. A. Bertrand, The crystal and molecular structure of methyl pinacol phosphate, *J. Am. Chem. Soc.* **88**, 1503–1506 (1966).

69. D. B. Boyd, Mechanism of hydrolysis of cyclic phosphate esters, *J. Am. Chem. Soc.* **91**, 1200–1205 (1969).

70. J. M. Lehn and G. Wipff, Stereoelectric effects in phosphoric acid and phosphate esters, *Chem. Commun.* **1975**, 800–802.

71. S. I. Miller, in: *Advances in Physical Organic Chemistry* (V. Gold, ed.), Vol. 6, pp. 185–332, Academic Press, New York (1968).

72. E. A. Dennis and F. H. Westheimer, The rates of hydrolysis of esters of cyclic phosphonic acids, *J. Am. Chem. Soc.* **88**, 3431–3432 (1966).

73. R. L. Burwell, Jr., and R. G. Pearson, The principle of microscopic reversibility, *J. Phys. Chem.* **70**, 300–304 (1966).

74. R. Kluger, F. Covitz, E. Dennis, L. D. Williams, and F. H. Westheimer, pH-Product and pH-rate profiles for the hydrolysis of methyl ethylene phosphate. Rate-limiting pseudorotation, *J. Am. Chem. Soc.* **91**, 6066–6072 (1969).

75. W. P. Jencks and M. Gilchrist, Nonlinear structure-reactivity correlations. The reactivity of nucleophilic reagents toward esters, *J. Am. Chem. Soc.* **90**, 2622–2637 (1968).

76. W. P. Jencks and M. Gilchrist, Electrophilic catalysis. The hydrolysis of phosphoramidates, *J. Am. Chem. Soc.* **86**, 1410–1417 (1964).

77. W. P. Jencks and M. Gilchrist, Reaction of nucleophilic reagents with phosphoramidate, *J. Am. Chem. Soc.* **87**, 3199–3209 (1965).

78. G. W. Jameson and J. M. Lawlor, Aminolysis of *N*-phosphorylated pyridines, *J. Chem. Soc. B* **1970**, 53–57.

79. R. H. Bromilow, S. A. Khan, and A. J. Kirby, Intramolecular catalysis of phosphate triester hydrolysis–nucleophilic catalysis by neighboring carboxyl group of hydrolysis of dialkyl 2-carboxyphenol phosphate, *J. Chem. Soc. B* **1971**, 1091–1097.

80. G. Di Sabato and W. P. Jencks, Mechanism and catalysis of reactions of acyl phosphates. I. Nucleophilic reactions, *J. Am. Chem. Soc.* **83**, 4393–4400 (1961).

81. A. J. Kirby and W. P. Jencks, The reactivity of nucleophilic reagents toward the *p*-nitrophenyl phosphate dianion, *J. Am. Chem. Soc.* **87**, 3209–3216 (1967).

82. A. J. Kirby and A. G. Varvoglis, The reactivity of phosphate esters: Reaction of monoesters with nucleophiles. Nucleophilicity independent of basicity in a biomolecular substitution reaction, *J. Chem. Soc. B* **1968**, 135–141.

83. A. J. Kirby and M. Younas, Reactivity of phosphate esters. Reactions of diesters with nucleophiles, *J. Chem. Soc. B* **1970**, 1165–1172.

84. H. J. Brass, J. O. Edwards, and M. J. Biallas, Reactions of phosphoric acid esters with nucleophiles. III. Reactivity of amines toward *p*-nitrophenyl methyl phosphonate, *J. Am. Chem. Soc.* **92**, 4675–4681 (1970).

85. S. A. Khan and A. J. Kirby, Reactivity of phosphate esters. Multiple structure reactivity correlations for the reaction of triesters with nucleophiles, *J. Chem. Soc. B* **1970**, 1172–1182.

86. R. Swidler, R. E. Plapinger, and G. M. Stein, The kinetics of the reaction of isopropyl methylphosphonofluoridate (sarin) with substituted benzohydroxamic acids, *J. Am. Chem. Soc.* **81**, 3271–3274 (1959).

87. A. L. Green, G. L. Sainsbury, B. Saville, and M. Stansfield, The reactivity of some active nucleophilic reagents with organophosphorus anticholinesterases, *J. Chem. Soc.* **1958**, 1583–1587.

88. J. Michalski, M. Mikolajczyk, and J. Omelanczuk, Chemical evidence for Walden inversion at the thiophosphoryl center based on the Pishschimuka reaction. Synthesis of optically active *N,N*-diethylphosphonamidic chloride, *Tetrahedron Lett.* **32**, 3565–3568 (1968).

89. W. B. Farnham, K. Mislow, N. Mandel, and J. Donohue, Stereochemistry of methanolysis of menthyl *S*-methyl phenylphosphonothioate, *Chem. Commun.* **3**, 120–121 (1972).

90. K. E. DeBruin and J. R. Petersen, Steric and electronic effects on the stereochemistry of the alkaline hydrolysis of acyclic dialkoxyphosphonium salts. Pseudorotation of intermediates in phosphorus ester reactions, *J. Org. Chem.* **37**, 2272–2278 (1972).

91. K. E. DeBruin and D. M. Johnson, Relative energetics of modes for phosphorane formation and decomposition in nucleophilic displacement reactions at a cyclic phosphorus. Alkaline hydrolysis of alkoxy(alkylthio)phosphonium salts. *J. Am. Chem. Soc.* **95**, 4675–4681 (1973).

92. K. E. DeBruin and K. Mislow, Stereochemistry of the alkaline hydrolysis of dialkoxyphosphonium salts. *J. Am. Chem. Soc.* **91**, 7393–7397 (1969).
93. N. J. De'ath, K. Ellis, D. J. H. Smith, and S. Trippett, Alkaline hydrolysis of alkoxy(methylthio)phosphonium salts with retention of configuration at phosphorus, *Chem. Commun.* **1971**, 714.
94. T. D. Inch, G. J. Lewis, R. G. Wilkinson, and P. Watts, Differences in mechanisms of nucleophilic substitution at phosphorus in *S*-alkyl alkylphosphinothioates and *S*-alkyl phosphorothioates, *Chem. Commun.* **13**, 500–501 (1975).
95. J. M. Harrison, T. D. Inch, and G. J. Lewis, Use of carbohydrate derivatives for studies of phosphorus stereochemistry. 5. Preparation of some reactions of tetrahydro-1,3,2-oxazaphosphorine-2-ones and tetrahydro-1,3,2-oxazaphosphorine-2-thiones, *J. Chem. Soc. Perkin Trans. 1* **1975**, 1892–1902.
96. J. M. Harrison, T. D. Inch, and G. L. Lewis, Use of carbohydrate derivatives for studies of phosphorus stereochemistry. 4. Ring opening of 1,3,2-dioxaphosphorinan-2-ones and related compounds with Grignard reagents and with sodium methoxide, *J. Chem. Soc. Perkin Trans. 1* **1974**, 1058–1068.
97. W. S. Wadsworth, Jr., S. Larsen, and H. L. Horten, Nucleophilic substitution at phosphorus, *J. Org. Chem.* **38**, 256–263 (1973).
98. W. S. Wadsworth, Jr., and Y. Tsay, Nucleophilic substitution at phosphorus. Phosphorothioates, *J. Org. Chem.* **39**, 984–989 (1974).
99. K. E. DeBruin and D. M. Johnson, Preferential mode for nucleophilic attack by methoxide ion on *O,S*-dimethyl phenylphosphonothiolate. A contrasting behavior to reactions on analogous phosphonium salts, *J. Am. Chem. Soc.* **95**, 7921–7923 (1973).
100. R. D. Cook, C. E. Diebert, W. Schwarz, P. C. Turley, and P. Haake, Mechanism of nucleophilic displacement at phosphorus in the alkaline hydrolysis of phosphinate esters, *J. Am. Chem. Soc.* **95**, 8088–8096 (1973).
101. T. Koizumi and P. Haake, Acid-catalyzed and alkaline hydrolysis of phosphinamides. The lability of phosphorus–nitrogen bonds in acid and the mechanisms of reaction, *J. Am. Chem. Soc.* **95**, 8073–8079 (1973).
102. H. J. Brass and M. L. Bender, Reactions of general bases and nucleophiles with bis(*p*-nitrophenyl)methylphosphonate, *J. Am. Chem. Soc.* **94**, 7421–7428 (1972).
103. A. Williams and R. A. Naylor, Hydrolysis of phosphonic esters. General-base catalysis by imidazole, *J. Chem. Soc. B* **1971**, 1967–1972.
104. R. L. Collin, The electronic structure of phosphate esters, *J. Am. Chem. Soc.* **88**, 3281–3287 (1966).
105. J. A. A. Ketelaar, H. R. Gersmann, and K. Koopmans, The rate of hydrolysis of some *p*-nitrophenol esters of ortho-phosphoric and thio-phosphoric acids, *Recl. Trav. Chim. Pays-Bas* **71**, 1253–1258 (1952).
106. R. Breslow and I. Katz, Relative reactivities of *p*-nitrophenyl phosphate and phosphorothioate toward alkaline phosphatase and in aqueous hydrolysis, *J. Am. Chem. Soc.* **90**, 7376–7377 (1968).
107. V. E. Bel'skii, M. M. Bezzubova, M. V. Efremova, and I. Nuretdinov, Kinetics of the alkaline hydrolysis of some organoselenophosphorus compounds, *Zh. Obshch. Khim.* **43**, 1255–1257 (1973).
108. W. Saenger and F. Eckstein, Stereochemistry of a substrate for pancreatic ribonuclease. Crystal and molecular structure of the triethylammonium salt of uridine 2′,3′-*O,O*-cyclophosphorothioate, *J. Am. Chem. Soc.* **92**, 4712–4718 (1970).
109. M. L. Bender and J. M. Lawlor, Isotopic and kinetic studies of the mechanism of hydrolysis of salicyl phosphate. Intramolecular general acid catalysis, *J. Am. Chem. Soc.* **85**, 3010–3017 (1963).
110. R. H. Bromilow and A. J. Kirby, Intramolecular general acid catalysis of phosphate monoester hydrolysis. Hydrolysis of salicyl phosphate, *J. Chem. Soc. Perkin Trans. 2* **1972**, 149–155.
111. S. J. Benkovic and K. J. Schray, Kinetics and mechanisms of phosphoenolpyruvate hydrolysis, *Biochemistry* **7**, 4090–4096 (1968).

112. K. J. Schray and S. J. Benkovic, Mechanisms of hydrolysis of phosphate ester derivatives of phosphoenolpyruvic acid, *J. Am. Chem. Soc.* **93**, 2522–2529 (1971).

113. S. A. Khan, A. J. Kirby, M. Wakselman, D. P. Horning, and J. M. Lawlor, Intramolecular catalysis of phosphate diester hydrolysis. Nucleophilic catalysis by the neighboring carboxy group of the hydrolysis of aryl 2-carboxyphenyl phosphates, *J. Chem. Soc. B* **1970**, 1182–1187.

114. J. Steffens, E. Sampson, I. Siewers, and S. J. Benkovic, Effects of divalent metal ions on the intramolecular nucleophilic catalysis of phosphate diester hydrolysis, *J. Am. Chem. Soc.* **95**, 936–938 (1973).

115. R. H. Bromilow, S. A. Khan, and A. J. Kirby, Intramolecular catalysis of phosphate triester hydrolysis. Nucleophilic catalysis by the neighboring carboxy group of the hydrolysis of diaryl 2-carboxyphenyl phosphates, *J. Chem. Soc. Perkin Trans. 2* **1972**, 911–918.

116. J. J. Steffens, I. J. Siewers, and S. J. Benkovic, Catalysis of phosphoryl group transfer. The role of divalent metal ions in the hydrolysis of lactic acid *O*-phenyl phosphate and salicylic acid *O*-aryl phosphates, *Biochemistry* **14**, 2431–2440 (1975).

117. G. M. Blackburn and M. J. Brown, The mechanism of hydrolysis of diethyl 2-carboxylphenyl-phosphates, *J. Am. Chem. Soc.* **96**, 6492–6498 (1974).

118. S. S. Simons, Jr., Carboxyl-assisted hydrolysis. Synthesis and hydrolysis of diphenyl *cis*-2-(3-carboxy) norbornyl phosphates, *J. Am. Chem. Soc.* **96**, 6492–6498 (1974).

119. D. A. Usher, D. I. Richardson, Jr., and D. G. Oakenfull, Models of ribonuclease action. II. Specific acid, specific base, and neutral pathways for hydrolysis of a nucleotide diester analog, *J. Am. Chem. Soc.* **92**, 4699–4712 (1970).

120. R. Kluger and J. L. W. Chen, Phosphorylation of amides. Evidence for participation in catalysis, *J. Am. Chem. Soc.* **95**, 2362–2364 (1973).

121. W. P. Jencks, in: *Advances in Enzymology* (A. Meister, ed.), Vol. 43, pp. 219–410, Wiley-Interscience, New York (1976).

122. F. Ramirez, J. F. Marecek, and H. Okazaki, One-flask synthesis of unsymmetrical phosphodiesters. Selective amine catalysis of phosphorylation of primary vs. secondary alcohols, *J. Am. Chem. Soc.* **97**, 7181–7182 (1975).

123. G. J. Lloyd and B. S. Cooperman, Nucleophilic attack by zinc(II)–pyridine-2-carboxaldoxime anion on phosphorylimidazole. A model for enzymic phosphate transfer, *J. Am. Chem. Soc.* **93**, 4883–4889 (1971).

124. C. Hsu and B. S. Cooperman, personal communication.

125. S. J. Benkovic and L. K. Dunikoski, Jr., Unusual rate enhancement in metal ion catalysis of phosphate transfer, *J. Am. Chem. Soc.* **93**, 1526–1527 (1971).

126. Y. Murakami and M. Takagi, Solvolysis of organic phosphates. I. Pyridylmethyl phosphates, *J. Am. Chem. Soc.* **91**, 5130–5135 (1969).

127. Y. Murakami and J. Sunamoto, Solvolysis of organic phosphates. IV. 3-Pyridyl and 8-quinolyl phosphates as effected by the presence of metal ions, *Bull Chem. Soc. Jpn.* **44**, 1827–1834 (1971).

128. B. Cooperman, A model for the role of metal ions in the enzyme-catalyzed hydrolysis of polyphosphates, *Biochemistry* **8**, 5005–5010 (1969).

129. M. Tetas and J. M. Lowenstein, The effect of bivalent metal ions on the hydrolysis of adenosine di- and triphosphate, *Biochemistry* **2**, 350–357 (1963).

130. N. Nelson and E. Racker, Phosphate transfer from adenosine triphosphate in a model system, *Biochemistry* **12**, 563–566 (1973).

131. A. Lewis, N. Nelson, and E. Racker, Laser Raman spectroscopy as a mechanistic probe of the phosphate transfer from adenosine triphosphate in a model system, *Biochemistry* **14**, 1532–1535 (1975).

132. E. J. Sampson, J. Fedor, P. A. Benkovic, and S. J. Benkovic, Intramolecular and divalent metal ion catalysis. Hydrolytic mechanism of *O*-phenyl *N*-(glycyl)phosphoramidate, *J. Org. Chem.* **38**, 1301–1306 (1973).

133. A. S. Mildvan, in: *The Enzymes* (P. D. Boyer, ed.), Vol. II, pp. 445–536, Academic Press, New York (1970).

134. H. Ikenaga and Y. Inoue, Metal(II) ion catalyzed transphosphorylation of four homodinucleotides and five pairs of dinucleotide sequence isomers, *Biochemistry* **13**, 577–582 (1974).

135. F. J. Farrell, W. A. Kjellstrom, and T. G. Spiro, Metal-ion activation of phosphate transfer by bidentate coordination, *Science* **164**, 320–321 (1969).
136. C. A. Bunton and H. Chaimovich, The hydrolysis of glucose 6-phosphate, *J. Am. Chem. Soc.* **88**, 4082–4089 (1966).
137. C. Degani and H. Halmann, Solvolysis of phosphoric acid esters. Hydrolysis of glucose 6-phosphate. Kinetic and tracer studies, *J. Am. Chem. Soc.* **88**, 4075–4181 (1966).
138. P. A. Frey, F. C. Kokesh, and F. H. Westheimer, Reporter group at the active site of acetoacetate decarboxylase. I. Ionization constant of the nitrophenol, *J. Am. Chem. Soc.* **93**, 7266–7269 (1971).
139. W. P. Jencks, *Catalysis in Chemistry and Enzymology*, McGraw-Hill, New York (1969).
140. S. J. Benkovic and R. Lazarus, unpublished results.
141. G. J. Buist, C. A. Bunton, L. B. Kobinson, G. L. Sepulveda, and M. Stam, Micellar effects upon the hydrolysis of bis(2,4-di-nitrophenyl)hydrogen phosphate, *J. Am. Chem. Soc.* **92**, 4072–4078 (1970).
142. C. A. Bunton, E. J. Fendler, G. L. Sepulveda, and K. Yang, Micellar-catalyzed hydrolysis of nitrophenyl phosphates, *J. Am. Chem. Soc.* **90**, 5512–5518 (1968).
143. D. A. Usher, Mechanism of ribonuclease action, *Proc. Nat. Acad. Sci. USA* **62**, 661–667 (1969).
144. D. A. Usher, D. I. Richardson, Jr., and F. Eckstein, Absolute stereochemistry of the second step of ribonuclease action, *Nature (London)* **228**, 663–665 (1970).
145. D. A. Usher, E. S. Erenrich, and F. Eckstein, Geometry of the first step in the action of ribonuclease A, *Proc. Nat. Acad. Sci. USA* **69**, 115–118 (1972).
146. R. G. Yount, in: *Advances in Enzymology* (A. Meister, ed), Vol. 43, pp. 1–56, Wiley-Interscience, New York (1975).
147. F. Eckstein, W. Bruns, and A. Parmeggiani, Synthesis of guanosine 5′-di- and triphosphate derivatives with modified terminal phosphates: Effect on ribosome-elongation factor G-dependent reactions, *Biochemistry* **14**, 5225–5232 (1975).
148. M. E. Hickey, P. P. Waymack, and R. L. Van Etten, pH-Dependent leaving group effects on hydrolysis reaction of phosphate and phosphonate esters catalyzed by wheat germ acid phosphatase, *Arch. Biochem. Biophys.* **172**, 439–448 (1976).
149. J. F. Chlebowski and J. E. Coleman, Mechanism of hydrolysis of *O*-phosphorothioates and inorganic thiophosphate by *Escherichia coli* alkaline phosphatase, *J. Biol. Chem.* **249**, 7192–7202 (1974).
150. D. G. Gorenstein, Dependence of ^{31}P chemical shifts on oxygen–phosphorus–oxygen bond angles in phosphate esters, *J. Am. Chem. Soc.* **97**, 898–900 (1975).
151. J. F. Chlebowski, I. M. Armtage, P. P. Tusa, and J. E. Coleman, ^{31}P NMR of phosphate and phosphonate complexes of metalloalkaline phosphatase, *J. Biol. Chem.* **251**, 1207–1216 (1976).
152. A. S. Mildvan, in: *Annual Review of Biochemistry* (E. E. Snell, ed.), Vol. 43, pp. 357–399, Annual Reviews, Inc., Palo Alto, Calif. (1974).
153. J. H. Young, E. F. Korman, and J. Mclick, On the mechanism of ATP synthesis in oxidative phosphorylation: A review, *Bioorg. Chem.* **3**, 1–15 (1974).
154. C. R. Bagshaw, D. R. Trentham, R. G. Wolcott, and P. D. Boyer, Oxygen exchange in the γ-phosphoryl group of protein-bound ATP during Mg^{2+}-dependent adenosine triphosphatase activity of myosin, *Proc. Nat. Acad. Sci. USA* **72**, 2592–2596 (1975).
155. P. D. Boyer, Energy transduction and proton translocation by adenosine triphosphates, *FEBS Lett.* **50**, 91–94 (1975).
156. R. Sarma, F. Ramirez, B. McKeever, J. F. Marecek, and S. Lee, Crystal and molecular structure of pentaphenoxyphosphorane, $(C_6H_5O)_5P$. The configuration of acyclic, monocyclic, and spirobicyclic pentaoxyphosphoranes, *J. Am. Chem. Soc.* **98**, 581–587 (1976).
157. R. R. Holmes, Conformational preferences of pentacoordinate spirocyclic phosphorus compounds, *J. Am. Chem. Soc.* **96**, 4143–4149 (1974).

14

Intramolecular Reactions and the Relevance of Models

Richard D. Gandour

1. INTRODUCTION

In the past few years, there has been a growing interest in chemical models of enzyme action. In this large body of research, two distinct types of models emerge. The first type is mimetic models; i.e., the reactions model specific enzymes. This area will not be covered in this chapter, but the reader is referred to an excellent review by Fife[1] which covers this approach for three enzymic reactions. The second type is nonmimetic models; i.e., the reactions or interactions model a specific feature of the general process of enzyme catalysis. Among nonmimetic models are two further subdivisions. One subdivision encompasses catalysis by complexation, thus modeling the binding of the substrate to the protein. This subdivision includes reactions catalyzed by micelles, cyclodextrins, hydrophobic interactions in aqueous solution, charge-transfer complexes, and polar associations in apolar solvents. The other subdivision encompasses catalysis by functional groups, thus modeling catalysis by functional groups on side chains of the peptide backbone in the enzyme–substrate complex. This topic is our major concern in this chapter.

In this chapter the focus is primarily on intramolecular reactions, i.e., reactions in which the reacting groups are bonded to the same molecular framework. The basic premise is that reacting groups juxtaposed on the same molecule serve as a valid model for a reaction in an enzyme–substrate complex. In this regard, intramolecular reactions of both the mimetic and nonmimetic type have been studied. The importance of the intramolecular mimetic models is manifested in the direct correlation between the chemical model and the

Richard D. Gandour • Department of Chemistry, Louisiana State University, Baton Rouge, Louisiana.

enzymic reaction. The significance of the nonmimetic models is more general. Thus, for example, principles governing acyl transfer developed from model reactions would apply to the appropriate chemical and enzymic reactions. If an enzymic reaction appeared to violate one of these principles, then some special feature of the enzyme's action has been unmasked.

In this chapter we shall cover two aspects of intramolecular nonmimetic reactions. The first is the role of intramolecular reactions in the development of theories of the sources of enzymic catalytic power. The second is the use of intramolecular reactions in the development of methods for the quantitative resolution of transition-state structures.

Exploration of the origins of catalytic power is the most controversial biochemical area involving intramolecular nonmimetic reactions. Rate accelerations observed in a series of mechanistically similar intramolecular reactions are frequently used to support various theories explaining why enzymic reactions are so much faster than their bimolecular counterparts. It is with this topic of catalytic power that the discussion of the relevance of model studies with intramolecular reactions begins.

2. ORIGINS OF CATALYTIC POWER

The desire to assign quantitative estimates to various factors (proximity, orientation, restriction of rotation, etc.) responsible for the rate enhancements observed in enzymic reactions compared to bimolecular analogs has generated an interest in the study of rate accelerations in model systems. Initial attempts to quantify these various factors originated with Koshland.[2,3] Since that time there has occurred a proliferation of theories on the importance of individual factors in the total rate acceleration. Bruice[4] suggested that much of enzymic catalytic power was derived from the formation of a productive enzyme–substrate complex, which froze out translational and rotational entropy and properly oriented the reacting groups. Page and Jencks[5] quantified this idea entirely in terms of entropy loss. Page[6] further discussed in great detail a number of individual factors important in the energetics of neighboring group participation. Jencks[7] recently proposed the "Circe effect" as a major contributor to the catalytic power of enzymes. His proposal holds that the binding of the substrate to the enzyme, accompanied by an entropy loss, can lead to a substantial rate acceleration.

Other factors have been suggested for the origins of catalytic power of enzymes. These are too numerous to mention individually and have been summarized elsewhere.[7-9] Recently, the importance of vibrational factors in enzyme reactions has been discussed.[10-12]

One of the more widely held beliefs is that the entire catalytic power of an enzyme lies in its ability to stabilize the transition state(s) of a reaction. Three decades ago, Pauling suggested this same idea in terms of a complementary relationship of the enzyme's active site to the activated complex in preference to the reactant.[13] Thus, the transition state is stabilized more than the reactant state and catalysis occurs. Wolfenden[14] and Lienhard[15] have popularized this theory through the synthesis of transition-state analogs, which are potent inhibitors of the particular enzyme of interest. Critics of this theory have suggested that the predicted binding constants for "perfect" transition-state analogs are overestimated because the proponents of the transition-state-analog approach have neglected the entropy loss in the binding of the substrate.[7,16]

In this chapter, the focus is on the role that intramolecular reactions played in the development of the theories of stereopopulation control,[17] orbital steering,[18] and proximity effects.[19,20]

2.1. Stereopopulation Control

Cohen and his co-workers[21-23] have provided a rather complete set of data on rate accelerations in their "trimethyl lock" models for three separate reactions. They examined acid-catalyzed lactonization of *o*-hydroxhydrocin-namic acids[21] (**1**), cyclization of 3-(*o*-hydroxyphenyl)-1-propyl mesylates[22] (**2**), and cyclic anhydride formation in homophthalic acids[23] (**3**). These reactions are shown here, and relative rates for each series are given in Table I.

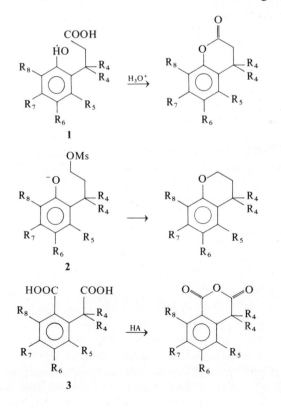

Table I
Effect of Methyl Substitution on Relative Rates of Cyclization

R_4	R_5	R_6	R_7	R_8	1	1^a	2	2^b	3	3^c
H	H	H	H	H	1	1	1	1	1	1
H	H	H	H	CH_3					10	
H	H	H	CH_3	CH_3			3.5			
H	CH_3	H	CH_3	CH_3	6.8		7			
H	CH_3	H	H	H		1^d		1.6^d	52	1
H	CH_3	H	H	CH_3					50	
CH_3	H	H	H	H	4,440	4,440	3,100	3,100	82,000	82,000
CH_3	H	H	H	CH_3	16,700					
CH_3	H	H	CH_3	CH_3				10,464		
CH_3	CH_3	H	CH_3	H	8×10^{10}					
CH_3	CH_3	H	H	CH_3	3×10^{11}					
CH_3	CH_3	H	CH_3	CH_3	3×10^{11}		9×10^5			
CH_3	CH_3	CH_3	CH_3	CH_3					8×10^5	
CH_3	CH_3	H	H	H		5×10^{10d}		2×10^{5d}		$16,000^d$

[a] Corrected for steric and electronic effects; see Reference 21.
[b] Corrected for steric and electronic effects; see Reference 22.
[c] Corrected for steric effects by this author.
[d] Extrapolated value not actually studied or measured.

Table I reveals the tremendous rate enhancements observed in these systems when both R_4 and R_5 are methyl groups. The rate acceleration of 10^{11} observed for the lactonization reaction is especially noteworthy since enhancements in rate of this magnitude begin to approach enzymic catalytic power. Milstien and Cohen[17] attribute this rate acceleration factor to an interlocking of methyl groups, which produces a conformational freezing of the side chain into the most productive rotamer for reaction. This phenomenon of greatly increasing the population of one conformation, ideally the most productive conformer for reaction, has been termed "stereopopulation control." Previous estimates for the value of rate acceleration derived from conformational restriction were placed at approximately 10^3.[24] Milstien and Cohen[17] believe that this value may have been "grossly underestimated." They do not necessarily attribute all of their observed rate enhancement to conformational restriction but do suggest that this restriction is the primary factor.

It is quite interesting to compare the three reactions listed in Table I. The intramolecular nucleophilic substitution (2) and the cyclic anhydride formation (3) show rate enhancements on the order of 10^5 and 10^4, respectively, when conformationally restricted by the "trimethyl lock." These values perhaps represent the typical rate enhancement factors for conformational restriction; the lactonization data set would then represent an atypical case. A possible reason for this difference might be orientational factors. (It has been postulated that an optimal orientation would lead to a rate enhancement of 10^4,[18] and this coupled with the 10^5 factor from conformational freezing would give a net factor of 10^9, close to the observed value.) A suggestion would be that an optimal orientation for reaction has been achieved for lactonization but not for nucleophilic substitution or anhydride formation. Another possibility is that orientational factors are less important and that the rate enhancement factor for stereopopulation control is much greater than 10^5. This would imply that reaction barriers have been introduced in the nucleophilic substitution and anhydride formation reactions which lower the overall rate enhancement. These barriers might arise from steric and/or orientational factors.

One of the more striking results is the sensitivity of anhydride formation (3) to methyl substitution for R_4 and R_5 separately. Introduction of a methyl group for R_4 in 3 results in a larger rate acceleration than the same substitution in 1 and 2 by factors of 18 and 26, respectively. Introduction of a methyl group for R_5 in 3 produces a larger rate acceleration than the same substitution in 1 and 2 by approximately a factor of 7 for both. However, when both R_4 and R_5 are substituted by methyl groups, 3 shows the slowest relative acceleration of all three reactions. Hillery and Cohen[23] explain this by suggesting that conformational freezing has already taken place in 3 by just the substitution of methyl for R_4. Thus, addition of methyl for R_5 should not significantly increase the rate.

Criticisms of Cohen's proposals have focused mainly on the model systems used in developing his theory. This is especially true of the lactonization reaction. The X-ray crystal structures of the lactone and the primary alcohol derived from reduction of 1 with $R_4 = R_5 = R_7 = R_8 = CH_3$ have been determined.[25] The crystallographic data revealed that significant deviations occurred in the bond angles of the benzene ring from the idealized values of $120°$ in order to accommodate the methyl groups at R_4 and R_5. Karle and Karle concluded that these angular deviations were large enough to permit rotation of the side chain bearing the *gem*-dimethyl groups, thus questioning the idea that a "lock" existed.[25] Bruice concluded from these data that relief of ground-state angle strain was the driving force for reaction.[20]

In a recent study, the rates of lactonization of compounds 4 and 5 were measured.[26] In these structures there was no question of conformational restriction. The observed rate accelerations when compared to 1 with R_4–R_8 =

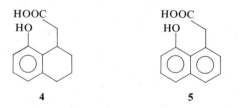

4　　　　　　5

H were 150-fold for 4 and 21,000-fold for 5. It was concluded that the maximum value for rate enhancement derived from conformational freezing is on the order of 10^3–10^4 and that some other factor must be responsible for the additional acceleration observed for **1** with $R_4 = R_5 = R_7 = R_8 = CH_3$. The authors suggested that this additional rate enhancement arose from steric factors and supported this with the observation of a secondary deuterium isotope effect of 1.09 for **1** with $R_4 = R_5 = R_7 = CH_3$ vs. **1** with $R_4 = CD_3$ and $R_5 = R_7 = CH_3$. This isotope effect was attributed to steric relief of ground-state interactions and is one of the largest observed of the remote secondary deuterium type.[27] This result clearly indicated the existence of relief of buttressing as a driving force in the lactonization reaction, but just how much this steric relief contributed to the rate enhancement was still not certain.

Winans and Wilcox estimated the importance of relief of steric strain in the lactonization of **1** with $R_4 = R_5 = CH_3$ by an empirical force field model calculation.[28] They concluded that a factor of 10^7 could be accounted for by relief of conventional strain. They also concluded that conformational restriction would only contribute on the order of 10^4 to the rate enhancement. Thus, the entire rate acceleration of 10^{11} was accounted for by the two factors. Page also suggested this explanation from a calculation using a nonbonded potential function,[6] which gave a minimum factor of 10^4 for steric relief in these cyclizations.

Hillery and Cohen[23] concluded that relief of strain is not the major driving force for cyclization from their study on cyclic anhydride formation in **3**. Examination of Table I reveals a rate enhancement of only a factor of 10 when comparing **3** with $R_4 = R_5–R_8 = CH_3$ to **3** with $R_4 = CH_3$. This acceleration factor is quite modest when compared to the rate enhancements brought about by similar substitutions in **1** and **2**, 10^7 and 10^2, respectively. Comparisons of the corrected rate constants in these model systems point out even more clearly the effect of introducing a methyl for R_5 when $R_4 = CH_3$. When corrections are made for steric factors from the introduction of methyl for R_5 and R_8 (a factor of 50) then a decrease in rate of five-fold is calculated for the introduction of methyl for R_5 when $R_4 = CH_3$ in **3**. The corresponding effects for corrected relative rates in **1** and **2** are increases in rate of 10^7-fold and 10^2-fold, respectively.

A detailed study of lactonization in coumarinic acids (**6**) by Hershfield and Schmir[29] bears further information on this point. The rate constant for acid-catalyzed lactonization of **6** with $R_4–R_8 = H$ is identical to the similar

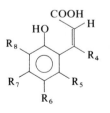

6

process in **1** with $R_4 = CH_3$. In the lactonization of various methyl derivatives of **6** there is only a small change in rate (less than an order of magnitude in either direction) except for **6** with $R_4 = R_5 = R_7 = CH_3$, which shows a *decrease* of 10^3. An interpretation of these results is that the carboxyl side chain in **6** with R_4–R_8 = H has restricted rotation due to two factors: overlap of the pi electrons of the double bond with the aromatic ring and intramolecular hydrogen bonding of the carboxyl with the phenolic OH. Thus, the maximal rate acceleration from conformational restriction would already be achieved in the unsubstituted compound. This rate enhancement is 10^4 when compared to **1** with R_4–R_8 = H. Introduction of methyl groups for R_4 and R_5 disrupts by a buttressing effect the coplanarity of the double bond and the benzene ring. This disruption of coplanarity produces a less favorable arrangement of reacting groups and creates a steric barrier to lactonization with a consequent reduction in rate.

In summary, intramolecular reactions in model systems in which reacting groups are conformationally restricted show sizable rate enhancements. In the reactions discussed the typical rate enhancement factor for conformational restriction appears to be in the range of 10^3–10^5.[23,24,26,29] The lactonization of **1** is the only reaction with a larger rate enhancement factor, 10^{11}.[21] The enormous rate acceleration in this reaction probably arises from additional factors of which one is undoubtedly relief of steric strain.[6,20,26,28]

2.2. Orbital Steering

Koshland has outlined a theory for explaining the catalytic efficiency of enzymes.[30,31] His suggestion is that enzymic catalytic power arises from "orbital steering,"[18] i.e., the directing of reacting atoms along selected pathways determined by the angular components of the reacting orbitals. The rate enhancement factor due to orbital steering is predicted to be as large as 10^4 per reacting atom with an orbital alignment preference.[18]

Dafforn and Koshland have provided a theoretical basis for orientational factors as large as 10^4.[32,33] They utilized two calculational methods. The collision theory approach considered the magnitude of contributions from two sources: bending vibrations and the distribution of approach orientations.[32] The other approach, using transition-state theory, considered contributions from rotational entropy, bending vibrations, and stretching vibrations.[33]

Storm and Koshland provided a set of relative rate data on lactonizations[34,35] and thiolactonizations[35] in some structurally related aliphatic molecules. Selections from these data are shown in Table II. Corrected relative rates are derived by making adjustments for proximity and strain. These corrected rate enhancements should represent contributions solely from orientational factors.

Examination of the relative rates for lactonization of **7**, **8**, and **9** reveals a steady increase in rate enhancement. This was explained by the individual

Table II
Effect of Structure on Lactonization Rates

		X = O		X = S	
		Relative rates	Corrected[a] relative rates	Relative rates	Corrected[b] relative rates
$CH_3COOH + CH_3CH_2$—XH		1	1	1	1
(structure **7**: COOH / XH)		80	413	384	2,020
(structure **8**: COOH, CH_2—XH)		6,620	1,660	90	5
(structure **9**: XH, COOH)		1,030,000	18,700	821,000	15,000
(structure **10**: R_2, OH, COOH)	$R_2 = H$	873			
	$R_2 = CH_3$	98,700			
(structure **11** / **12**: H_3C, COOH, OH)		276			
(structure **13** / **14**: R_6, COOH, HO)	$R_6 = H$	9,200			
	$R_6 = CH_3$	480,000			

[a] See Ref. 34.   [b] See Ref. 35.   [c] Structures incorrectly assigned by Storm and Koshland; see Refs. 36 and 37.

molecular framework's ability to direct the approach of the reacting atoms.[18] Thiolactonization for **7**, **8**, and **9** shows a marked change in relative rates by comparison with lactonization in these compounds. The different orbital structure and bond lengths for a sulfhydryl group would alter the orientation between the reacting groups.[35]

Relative rates of lactonization are sensitive also to subtle changes in structure. Comparison of lactonization rates between **9**, a bicyclo[2.2.1] molecule, and **10** with $R_2 = H$, a bicyclo[2.2.2] compound, reveals that the former reacts 1200-fold faster than the latter.[35] An opposite result is observed, however, when comparing **8** to its bicyclo[2.2.2] analog (not shown). In this case the latter compound lactonizes 20-fold faster than **8**.[38] Storm and Koshland suggested that these effects were due to small changes in orientation of reacting centers.[35]

Methyl substitution on the carbon attached to the carboxyl group affects the rate of lactonization. A 100-fold rate acceleration is observed when comparing **10** with $R_2 = CH_3$ to **10** with $R_2 = H$. Similarly, **14** with $R_6 = CH_3$ lactonizes 50-fold faster than **14** with $R_6 = H$. Structures represented by **14** were incorrectly assigned[36,37] as **13**. Another incorrect assignment of structure, **11** for **12**, suggested an anomalous effect for methyl substitution. Comparison of lactonization rates of **12**, thought to be **11**, and **9** shows a 4000-fold decrease in rate. Prior to the report of the incorrect structure assignment, Koshland emphasized that this anomalous effect points to the large sensitivity in rate on small changes in orientation.[31] With this anomalous effect resolved, the data set can be interpreted in more conventional chemical language; i.e., addition of a methyl group to carbon in a cyclizing chain leads to an increase in cyclization rate.

The rate increase is paralleled by an increase in the equilibrium constant for lactone (ester) formation. If the log of rates[34] of esterification of acetic acid and lactonization of **7**, 4-hydroxypentanoic acid (not shown), **8**, and **9** are plotted against the log of their respective equilibrium constants[34] for lactone (ester) formation, it is seen that all the reactions except **9** correlate to a straight line with a slope of approximately 0.9. The slope suggests that 90% of the equilibrium driving force is present in the transition state. A similar treatment of the data with the thio analogs[35] reveals a slope of about 0.7, with **9** deviating from this line also. Deviation of **9** from both these correlations is about 2 log units above the line. This deviation might be a measure of the orientation effect, a factor of 100, or an indication that the equilibrium constant is too low due to a strain effect not present in the transition state. It is proposed then that in these model reactions a dynamic orientation effect of the magnitude suggested by Koshland has not been demonstrated since equilibrium effects can adequately account for the rate enhancements.

Capon[39] has labeled orbital steering an "unnecessary concept." His criticisms were directed toward the correction factors used by Koshland for proximity and strain. He offered that loss of internal rotational entropy should be included in the proximity term and that strain factors were underestimated.

DeTar[40] computed the amount of rate acceleration expected from steric effects for the lactonization of 9 by means of molecular mechanics. He concluded that the steric effects gave a good account of the entire rate acceleration observed for 9 with reference to 7. From a CNDO/2 study of the rotational conformation of the hydroxyl group in 7,[41] a shallow potential energy well was computed for the angular requirement for alignment of reacting orbitals. It was concluded from this study that small changes in orientation would not have a large effect on the rate of lactone formation.* An EHT study of the addition of H_2O to CH_2O probed the change in activation energy due to variation in orientation of reactants.[42] These calculations predicted relative rates for 7, 8, and 9, based solely on orientation factors, to be at least an order of magnitude lower than those suggested by Storm and Koshland.[18]

Hershfield and Schmir[43] demonstrated in the thiolactonization of 9 that a change in rate-determining step occurred at low pH. They doubted that the same rate-determining step was being observed in the other thiolactonizations studied.[35] They concluded that comparisons of relative rates among reactions proceeding by different rate-limiting steps should be viewed with apprehension. Comparisons of lactonization–thiolactonization ratios for compounds 7, 8, and 9 determined by experiment were much greater than the ratios determined from the overlap integrals computed by the CNDO/2 method.[44] It was concluded from this comparison that the experimentally determined ratios could not have resulted from changes in differential orbital overlap.

Certain calculations, however, support the orbital steering hypothesis. Hoare suggested that the energy required to form a misaligned transition state in a solvent cage indicated that precise orbital alignment was important.[45] He suggested that two factors were important for the higher activation energy for a misaligned pair of molecules. These were bond strain energy in a transition state where a bent bond was being formed and energy required for displacement of unproductive complexes in a solvent cage. A CNDO/2 study on the mechanism of α-chymotrypsin, indicated that orbital alignment was quite important.[46] In contrast to the calculation on compound 7,[41] a rather deep potential well was computed for displacement away from the perpendicular approach of the serine oxygen to the substrate carboxyl carbon.

The biggest controversy regarding orbital steering has centered on the magnitude of the proximity correction. Bruice[19,47] argued that experimentally observed intramolecular–intermolecular rate ratios were much larger than 55. Page and Jencks[5] suggested, based on calculations for loss of entropy in a cyclopentadiene dimerization, that the proximity factor is approximately 10^5.

---

* From a recent theoretical study of the addition of CH_3OH to HCOOH utilizing the PRDDO procedure, a conclusion was drawn that orientational requirements for this reaction were minimal.[89] This calculation strongly contrasted with a previous statistical-mechanical computation on the same reaction.[33]

Dafforn and Koshland[33] answered these criticisms by disagreeing with the interpretation of the experimental rate ratios and by providing a different model, the combination of two bromine atoms, for calculation of entropy loss. They calculated an entropy loss for the formation of Br_2 corresponding to a factor of 55. Page[48] pointed out the deficiencies of this model, which were that the product, Br_2, contained internal entropic contributions not present in the reactants. Dafforn and Koshland[49] rebutted with the suggestion that this internal rotational entropy must be included. Jencks and Page[50] reaffirmed their description of entropy loss from rotational and translational restriction as the best description for rate acceleration in intramolecular reactions. They agreed that descriptive terms such as "orbital steering" may be similar to translational and rotational restriction when carefully defined but that unless the loss of translational and overall rotational entropy is taken into account, estimation of the proximity factor will yield unrealistic values.

2.3. Propinquity

Bruice[20] has labeled the propinquity effect "the commonsense phenomenon." Certainly, he and his co-workers have presented an impressive array of data over the past 20 years supporting this idea. Bruice[4,19,20,51] has periodically summarized this work and the work of others. Consequently, a detailed discussion of these results would be superfluous.

The propinquity effect (sometimes referred to as the proximity effect) is the rate acceleration achieved by bringing reacting groups together on the same molecule. Bruice and Pandit[52] suggested that the rate acceleration was due to a decrease in translational entropy on the formation of the transition state. Jencks and Page[5] supported this idea with a statistical-mechanics calculation of the total entropy loss for bringing two atoms together. Their calculations revealed that the loss of translational entropy was on the order of 25–30 eu, corresponding to a factor of 10^5–10^6 in rate acceleration. This factor corresponded nicely to the observed rate enhancement seen for succinate aryl half-ester hydrolysis over nucleophilically carboxylate-catalyzed hydrolysis of aryl acetates. It was also noted that a similar enhancement was observed for acid–anhydride equilibrium constants when comparing succinic vs. acetic acid.

DeLisi and Crothers[53] have presented an alternative method for estimating the contribution of proximity to rate accelerations. Since the reaction rate is proportional to the probability that the reacting groups have the correct spatial and directional positioning for reaction, they reasoned that this probability could be obtained from a distance distribution function. They offered that spatial constraints can lead to accelerations for five-membered cyclizations on the order of 10^7 in rigid systems and 10^3 in flexible systems such as succinate aryl half-ester closure to succinic anhydride. The values correspond closely to those previously observed.[52]

3. MODEL TRANSITION-STATE STRUCTURES

In this author's opinion, the primary importance of intramolecular reactions is that they permit the study of a reaction where there is a defined geometry between the reacting groups. Because of this definition of geometry, development of transition-state structures is much easier. The dynamics are simplified also, since the translational aspects of the reaction have been converted into vibrational and rotational motions.

The study of intramolecular reactions has contributed to an understanding of chemical catalysis. In the previous sections, the emphasis was on rate accelerations and relationships with enzymic catalysis. In the case of transition-state-structure elucidation, rate inhibitions are just as important as rate accelerations. Mechanistic changes are often observed when comparing intramolecular reactions with their intermolecular analogs. Catalytic groups of similar acid–base strength acting as protolytic catalysts in bimolecular reactions can be converted to nucleophilic catalysts in intramolecular reactions. Development of transition-state structures will help to explain the kinetic effects and mechanistic changes.

In this section, no attempt to be comprehensive will be made. A number of excellent reviews have appeared recently on intramolecular catalysis.[54-59] The intent here is to concentrate on two well-characterized intramolecularly catalyzed reactions. Utilizing data from structure-reactivity studies and kinetic isotope effects, a detailed description of transition-state structure will be presented for each of these reactions.

3.1. Protolytic Catalysis of Ester Hydrolysis by Carboxylate

Kirby and Fersht[54] have reviewed thoroughly the experimental work that led them to conclude that hydrolysis of O-acetylsalicylic acid in neutral pH media occurred with intramolecular protolytic catalysis by an ionized

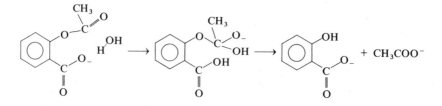

carboxyl group. It was also proposed that formation of a tetrahedral intermediate was rate limiting. This mechanism was further solidified by the observation of Minor and Schowen[60] that the rate constant for the intramolecular catalyzed hydrolysis of O-dichloroacetyl salicyclic acid at neutral pH was a linear function of mole fraction of deuterium in mixtures of H_2O and D_2O. This observation suggested that only one proton was contributing to the solvent isotope effect. This result strongly corroborated the water-bridge transition state suggested by Fersht and Kirby.[61]

The exact position of the proton in the transition state is uncertain. The rho value for the carboxyl group as determined from a set of ring-substituted derivatives is -0.52.[62] This would suggest that transfer of a proton is 52% completed, since the expected rho value for complete transfer is -1. However, this does not seem to agree with the magnitude of the observed kinetic solvent isotope effect. It may be that the proton is never completely transferred and that catalysis occurs through the formation of a strong hydrogen bond.[60]

Another question arises here. To what extent is the oxygen of the water molecule bonded to the carbon of the acetate carbonyl? The rho value for the phenoxy leaving group is 0.96.[62] This would suggest considerable charge buildup on the phenoxy oxygen, but what percentage this represents for nucleophilic addition is uncertain. Recently, the β-secondary deuterium kinetic isotope effect was determined for O-acetylsalicyclic acid vs. O-acetyl-d_3-salicyclic acid.[63] The inverse kinetic isotope effect of 0.956 ± 0.010 can be interpreted in terms of a significant addition of the oxygen of the bridging water to the carbonyl carbon of the acetate group.

A model calculation of the β-deuterium isotope effect for the addition of hydroxide ion to acetaldehyde found that the magnitude of the isotope effect was a linear function of bond order.[64] For complete addition, the expected value is 0.85. It has recently been shown for carbonyl additions that the distance of the approaching nucleophile was related to the bond order by Pauling's equation.[65] Combining these observations suggests a bond order of 0.29 $[(1 - 0.956)/(1 - 0.85)]$, which translates into an oxygen–carbon distance of 1.79 Å, using a C—O bond distance of 1.41 in Pauling's equation.[84] These ideas coupled with the previous data on the position of the proton suggest a transition state structure such as **15**.

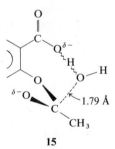

15

Umeyama[66] has suggested from considerations of crystal structure data a transition-state structure with an O—O distance of 2.9 Å for the hydrogen bond and a forming O—C bond of 2.0 Å. He performed a CNDO/2 calculation with all geometry fixed except for the transferring proton and computed an energy barrier of 37 kcal/mol. His barrier-height calculations, although too large, did predict correctly the increase in rate for substitution of methanol and ethanol for a water molecule in the bridge.[67]

He further suggested that the water molecule in the bridge was a monomer, based on a calculation that hydrogen bonding to the static hydrogen of the water bridge would produce a barrier lower in energy than that predicted for ethanol as a bridge. The monomeric water bridge appears to be borne out by the observation that there is no change in rate constant in mixtures of dioxane and water up to 60% dioxane.[68] Furthermore, there was no difference in rate in switching from 50% DMSO to 80% DMSO (although a tenfold increase was observed in comparison to water).[68]

Kirby and Lloyd examined the rates of hydrolysis of hydroxy acyl derivatives of salicylic acid, **16**.[69] In neutral pH, they suggested the hydroxy group replaced the water molecule as the bridge. They reported rate enhancements of 85, 15, and 21 for **16** with $m = 2$ and $R = H$, $m = 2$ and $R = CH_3$, and

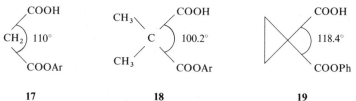

16

$m = 3$ and $R = H$, respectively. The solvent isotope effects were 2.28, 2.48, and 2.52, respectively, and compared favorably with water as a bridge, which showed a 2.20 solvent isotope effect.[61] These intramolecular alcohol bridges were only slightly rate accelerating compared to their bimolecular counterparts.[67] The reason that the rate accelerations were modest was attributed to the entropy loss in the restriction of the alcohol side chain.[69]

In summary, aspirin solvolysis serves as the prototype transition-state structure for tetrahedral-intermediate formation from protolytic catalysis by a carboxylate group. Evidence, thus far, indicates that addition to the carbonyl carbon is occurring. This bond formation may be coupled with a proton transfer or a preformed catalytic hydrogen bridge.

Just recently, Kirby and Lloyd demonstrated that intramolecular catalysis by carboxylate occurred in monoaryl malonates.[70] They observed the expected pH profile for a series of aryl hydrogen malonates (**17**), aryl hydrogen dimethylmalonates (**18**), and phenyl hydrogen cyclopropane-1,1-dicarboxylate (**19**). Their study to examine the effect of changes in geometry on catalysis revealed that there was only a modest dependence on the angle between the reacting groups. The ionized forms of **19**, **17** with Ar = phenyl, and **18** with

Ar = phenyl underwent hydrolysis with relative rates of 5.2, 2.2, and 1, respectively.

Based largely on the solvent isotope-effect criterion and other similarities (activation parameters, Hammet rho, and rate enhancement in methanol and ethanol) with aspirin hydrolysis, a protolytic mechanism was proposed. They suggested a transition-state structure such as **20**. This author has suggested,

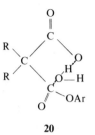

20

based on model-building studies, that transition state structures such as **20** are not likely.[71] The most likely arrangement for proton transfer is a linear alignment of donor–proton–acceptor, which cannot be achieved in a six-membered cyclic transition state without serious distortions of bond angles and bond lengths. To reconcile the data of the malonate systems with this proposal, this author suggests that the transfer of the proton does not take place and that only a "catalytic hydrogen bond" is formed and/or that electrostatic stabilization of a zwitterion, formed from nucleophilic attack of a water molecule, has occurred.

3.2. Nucleophilic Catalysis of Ester Hydrolysis by Carboxylate

Nucleophilic catalysis of a carboxylate group in bimolecular ester hydrolysis is limited to esters with extremely good leaving groups.[72] The mechanism of acetate-catalyzed hydrolysis of a series of aryl acetates depended on the pK_a of the conjugate acid of the leaving aryloxide ion. Protolytic catalysis was the exclusive mechanism for those esters with leaving groups of pK_a equal to 8.40 or larger. Nucleophilic catalysis was observed as the predominant mechanism for those esters with leaving groups of pK_a equal to 7.14 or smaller. It was further suggested that the rate-limiting step was the breakdown of the initially formed tetrahedral intermediate for those esters whose leaving group pK_a was less than 7.14 but more than 4.76, the pK_a of acetate. For intermolecular carboxylate-catalyzed ester hydrolysis, it was concluded[72] that a mechanism change occurred when ΔpK_a [pK_a(leaving group) − pK_a(carboxylate)] was greater than 3.

For intramolecular nucleophilic catalysis by carboxylate ΔpK_a can be as large as 13.[73] The reason for this enhanced nucleophilicity undoubtedly lies in the favorable equilibrium for cyclic anhydride formation (see below). At one time, it was thought the enhanced nucleophilicity was due to desolvation of the carboxylate group. It was felt that the carboxyl group and the reacting ester group were sufficiently crowded together to exclude solvent molecules.[9] The idea, although undoubtedly correct, was rendered unimportant by the observation that the intermolecular–intramolecular rate ratio for acetate-catalyzed ester hydrolysis vs. succinate intramolecularly catalyzed ester hydrolysis was identical in water and in 1-M water in dimethyl sulfoxide.[74]

Structure-reactivity studies focusing on the leaving group have been performed for a number of intramolecular carboxylate-catalyzed ester hydrolyses.[73,75,76] These studies revealed that the rate of hydrolysis was highly dependent on the pK_a of leaving group. Plots of the log of the rate constant for carboxylate-catalyzed hydrolysis vs. the pK_a of the leaving group yielded the β values shown in Table III. From these data it was concluded that the rate-

Table III
Summary of Nucleophilic Carboxylate-Catalyzed Ester Hydrolyses

System	pK_a range of leaving groups	β_{lg}	$K_{eq}{}^a$
AcO^- 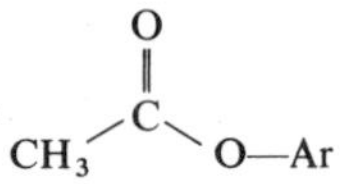 + $CH_3-C(=O)-O-Ar$	3.71–5.42	−0.6	3×10^{-12} [b]
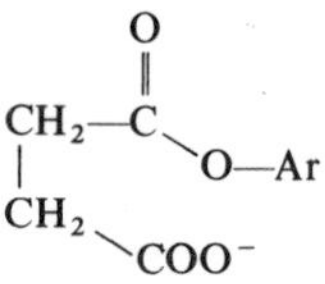 $CH_2-C(=O)-O-Ar$, CH_2-COO^-	8.47–10.20	−1.14	1×10^{-6} [c]
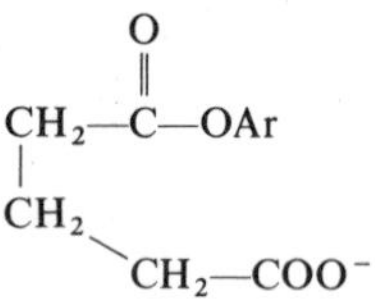 $CH_2-C(=O)-OAr$, $CH_2-CH_2-COO^-$	8.47–10.20	−1.17	1×10^{-9} [d]
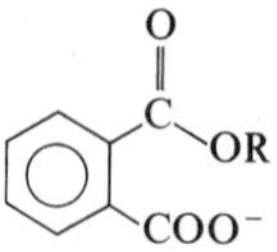 (benzene) $C(=O)-OR$, COO^-	9.98–13.55	−1.47	5×10^{-3} [e]
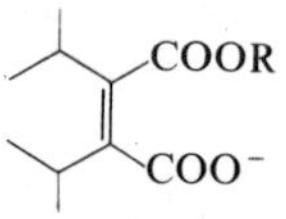 $=C-COOR$, $=C-COO^-$	13.55–16.57	−1.43	6×10^0 [f]

[a] Acid–anhydride equilibrium constant at 25°C.   [b] Reference 77.   [c] Extrapolated value; Reference 78.
[d] Estimated by comparison with succinic anhydride in CH_3COOH; Reference 79.   [e] Estimated; Reference 80.
[f] Estimated from value at 60°C; Reference 81.

limiting transition-state structure involved the breakdown of the initially formed tetrahedral intermediate. In this transition-state structure, a substantial amount of charge had developed on the leaving group.

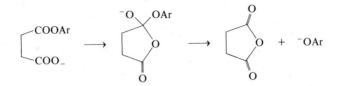

A more quantitative way of looking at the transition-state structure is to compare the value of β to the maximum possible value of β for acyl transfer, 1.7.[82] For the succinate hydrolyses, then, the transition-state structure would have 67% cleavage of the bond to the leaving group, or a bond order for the breaking bond of 0.33 (ideally, this method should be restricted to the one-step reactions[83]). Substitution of this value into Pauling's equation,[84] using a C—O single bond distance of 1.41, gives the breaking C—O bond distance of 1.75 Å in the transition-state structure **21**.

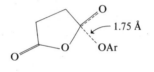

21

The β-secondary deuterium kinetic isotope effect was measured for the hydrolysis of mono-*p*-bromophenyl succinate and succinate-d_4.[85] A normal isotope effect $k^H/k^D = 1.03$, was found, which is contrary to expected result for an acyl transfer proceeding via a tetrahedral intermediate (see Chapter 5). Observation of a *normal* isotope effect implies that in the transition state the succinyl hydrogens (deuterium atoms) are in a "looser" vibrational force field than in the reactant; i.e., a relief of steric interactions has taken place in going from the reactant state to the transition state. The relief of ground-state strain was suggested as the driving force behind rate accelerations observed when the succinyl backbone was substituted with alkyl groups but was believed to be absent in the unsubstituted compound.[19] Furthermore, it has been suggested that there is an increase in steric interactions between the hydrogens when the ring is formed.[7,20] This suggestion is contrary to the results. It is concluded, then, that the equilibrium driving force for ring closure is a dominant factor in these reactions.

β values, the sensitivity of the rate of hydrolysis to the pK_a of the leaving group, for a series of intramolecular ester hydrolyses are shown in Table III. The magnitude of the β values is directly proportional to the ΔpK_a values of the leaving groups; i.e., the larger the ΔpK_a, the greater the value of β. This observation supports the idea of a nonlinear relationship between pK_a

and $\log k$, an idea suggested for proton-transfer reactions.[86] Breakdown of the intermediate, the rate-limiting step, is mostly dependent on the leaving group.

The first step, formation of the intermediate, is largely dependent on the base strength of the attacking carboxylate and the equilibrium constant for anhydride formation. Support for the latter point is found in good linear correlation for a plot of $\log k$ of hydrolysis of a series of alkylated succinate catechol half-esters vs. $\log K$ for anhydride formation from the free acid.[87] The slope of this plot was 0.53, suggesting that 53% of the equilibrium driving force was present in the transition state.

These observations are strongly suggestive that the rates for all nucleophilic carboxylate-catalyzed ester hydrolyses can be correlated from a knowledge of two factors. One factor is the equilibrium constant for anhydride formation from free acid, K_{eq}, and the second factor is ΔpK_a, $\log(K_a^N/K_a^{LG})$. The dependence of the observed rate constant on these two factors is illustrated by the reaction-progress diagram in Fig. 1. This diagram represents the anhydride formation or the rate-limiting part of the mechanism. The suggestion is that there are two major contributors to the energy of activation. One contribution is proportional to the driving force for anhydride formation (A), and the other is proportional to the difference in basicity between the nucleophile and the leaving group (B).

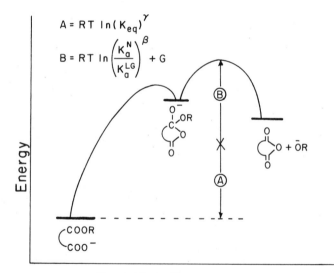

Reaction Progress

Fig. 1. Reaction-progress diagram of anhydride formation resulting from intramolecular nucleophilic attack of carboxylate at ester carbonyl. A represents the contribution to the activation energy arising from driving force for ring closure. B represents the contribution from the Brønsted relationship plus a constant, G.

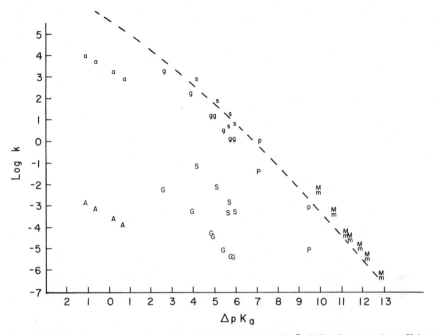

Fig. 2. Plot log of rate constant for carboxylate catalysis vs. ΔpK_a [pK_a(leaving group) $-$ pK_a(carboxylate)]. Uppercase letters represent data from the literature standardized to 25°: A, acetate[72]; G, glutarate[75]; S, succinate[75]; P, phthalate[76]; M, diisopropylmaleate.[73] Lowercase letters represent these standardized rate constants corrected by 0.53 log K_{eq} (see Table III).

Figure 2 presents a test of this idea, by plotting ΔpK_a vs. log k for some known examples of nucleophilic intramolecular catalysis by carboxylate in the hydrolysis of esters. The uppercase letters are the standardized observed rate constants, and the lowercase letters are rate constants corrected for equilibrium of anhydride formation, δ log K_{eq}, with $\delta = 0.53$. The corrected values, except for the acetate-catalyzed cases, show a nice curvilinear correlation in support of a nonlinear Bronsted relationship. The deviation of the acetate-catalyzed cases of approximately 2 log units is in agreement, perhaps by coincidence, with the proposed magnitude of the proximity factor for the comparison of intermolecular and intramolecular reactions.[5,52] Although a quantitative relationship cannot be derived at this time, the data do indicate a definite trend in support of the idea that correlations can be made for reactions of this type.

4. SUMMARY

Intramolecular catalytic reactions have been used extensively as simple models to test the importance of certain factors responsible for large rate accelerations observed for enzyme catalysis. A modest controversy has surrounded this area of research. This controversy has centered on the exact magnitude assigned to the various effects proposed. The "orbital steering" proposal has lost some of its attractiveness due to criticisms of the models studied. These criticisms are misassignments of structures, changes in rate-determining step, and the observation that the rate enhancements in the models studied can be almost completely accounted for in equilibrium terms. The "stereopopulation control" proposal has also come under heavy criticism. These criticisms have mainly been that the enormous rate acceleration observed for lactonization is largely due to steric strain in the model system used. Other criticism has focused on the suggestion that the catalytic boast for stereopopulation control arises from an increase in population of reactive but unfavorable ground-state conformations.[88] The proximity-effect factor of approximately 10^3 proposed over a decade ago seems to have garnered the most support theoretically and experimentally. However, there is still considerable rate enhancement beyond this to be satisfactorily accounted for by any intramolecular model reaction system.

The more recent work with intramolecular reactions has focused on the reaction itself. In particular, the reactions are being studied with the emphasis of uncovering the factors responsible for the rate accelerations observed for the intramolecular reaction over its intermolecular analog. Intramolecular reactions appear to provide the perfect frame of reference for elucidation of detailed transition-state structures. Although the study of intramolecular reactions began as enzyme models, the full potential of these reactions as models for all chemical processes is just now being realized.

REFERENCES

1. T. H. Fife, Physical organic model systems and the problem of enzymatic catalysis, *Adv. Phys. Org. Chem.* **11**, 1–122 (1975).
2. D. E. Koshland, Jr., Molecular geometry in enzyme action, *J. Cell. Comp. Physiol.* **47**, Supp. 1, 217–234 (1956).
3. D. E. Koshland, Jr., The comparison of non-enzymic and enzymic reaction velocities, *J. Theor. Biol.* **2**, 75–86 (1962).
4. T. C. Bruice, Intramolecular catalysis and the mechanism of chymotrypsin actions, *Brookhaven Symp. Biol.* **15**, 52–84 (1962).
5. M. I. Page and W. P. Jencks, Entropic contributions to rate accelerations in enzymic and intramolecular reactions and the chelate effect, *Proc. Nat. Acad. Sci. USA* **68**, 1678–1683 (1971).
6. M. I. Page, Energetics of neighboring group participation, *Chem. Soc. Rev.* **2**, 295–323 (1973).
7. W. P. Jencks, Binding energy, specificity, and enzymic catalysis: The Circe effect, *Adv. Enzymol. Relat. Areas Mol. Biol.* **43**, 219–410 (1975).
8. R. Lumry, in: *The Enzymes,* 2nd ed. (P. D. Boyer, ed.), Vol. 1, pp. 157–231, Academic Press, New York (1959).
9. W. P. Jencks, *Catalysis in Chemistry and Enzymology,* McGraw-Hill, New York (1969).
10. L. Schafer, S. J. Cyvin, and J. Brunwall, Zur Verschiebung von Schwingungsfrequenzen durch kinematische Kopplung, *Tetrahedron* **27**, 6177–6179 (1971).

11. R. A. Firestone and B. G. Christensen, Vibrational activation I. A source for the catalytic power of enzymes, *Tetrahedron Lett.* **1973**, 389–392.
12. D. B. Cook and J. McKenna, A contribution to the theory of enzyme catalysis. The potential importance of vibrational activation entropy, *J. Chem. Soc. Perkin Trans. 2* **1974**, 1223–1225.
13. L. Pauling, Molecular architecture and biological reactions, *Chem. Eng. News* **24**, 1375–1377 (1946).
14. R. Wolfenden, Analog approaches to the structure of the transition state in enzyme reactions, *Acc. Chem. Res.* **5**, 10–18 (1972).
15. G. E. Lienhard, Enzymatic catalysis and transition-state theory, *Science* **180**, 149–154 (1973).
16. K. Schray and J. P. Klinman, The magnitude of enzyme transition state analog binding constants, *Biochem. Biophys. Res. Commun.* **57**, 641–648 (1974).
17. S. Milstien and L. A. Cohen, Rate accelerations by stereopopulation control: Models for enzyme action, *Proc. Nat. Acad. Sci. USA* **67**, 1143–1147 (1970).
18. D. R. Storm and D. E. Koshland, Jr., A source for the special catalytic power of enzymes: Orbital steering, *Proc. Nat. Acad. Sci. USA* **66**, 445–452 (1970).
19. T. C. Bruice, in: *The Enzymes,* 3rd ed. (P. D. Boyer, ed.), Vol. 2, pp. 217–279, Academic Press, New York (1970).
20. T. C. Bruice, Some pertinent aspects of mechanism as determined with small molecules, *Annu. Rev. Biochem.* **45**, 331–373 (1976).
21. S. Milstien and L. A. Cohen, Stereopopulation control. I. Rate enhancement in the lactonizations of *o*-hydroxyhydrocinnamic acids, *J. Am. Chem. Soc.* **94**, 9158–9165 (1972).
22. R. T. Borchardt and L. A. Cohen, Stereopopulation control. II. Rate enhancement of intramolecular nucleophilic displacement, *J. Am. Chem. Soc.* **94**, 9166–9174 (1972).
23. P. S. Hillery and L. A. Cohen, Stereopopulation control in the formation of cyclic anhydrides. Demonstrations of rate and equilibrium enhancement, general catalysis and duality of mechanism (manuscript in preparation).
24. T. C. Bruice and U. K. Pandit, The effect of geminal substitution, ring size and rotamer distribution on the intramolecular nucleophilic catalysis of the hydrolysis of monophenyl esters of dibasic acids and the solvolysis of the intermediate anhydrides, *J. Am. Chem. Soc.* **82**, 5858–5865 (1960).
25. J. M. Karle and I. L. Karle, Correlation of reaction rate acceleration with rotational restriction. Crystal-structure analysis of compounds with a trialkyl lock, *J. Am. Chem. Soc.* **94**, 9182–9189 (1972).
26. C. Danforth, A. W. Nicholson, J. C. James, and G. M. Loudon, Steric acceleration of lactonization reactions: an analysis of "stereopopulation control," *J. Am. Chem. Soc.* **98**, 4275–4281 (1976).
27. J. L. Fry and R. C. Badger, Evidence for a remote secondary kinetic deuterium isotope effect arising from a sterically congested ground state, *J. Am. Chem. Soc.* **97**, 6276–6277 (1975).
28. R. E. Winans and C. F. Wilcox, A comparison of stereopopulation control with conventional steric effects in lactonization of hydrocoumarinic acids, *J. Am. Chem. Soc.* **98**, 4281–4285 (1976).
29. R. Hershfield and G. L. Schmir, Lactonization of ring-substituted coumarinic acids. Structural effects on the partitioning of tetrahedral intermediates in esterification, *J. Am. Chem. Soc.* **95**, 7359–7369 (1973); Lactonization of coumarinic acids. Kinetic evidence for three species of the tetrahedral intermediate, *J. Am. Chem. Soc.* **95**, 8032–8040 (1973).
30. D. E. Koshland, Jr., K. W. Carraway, G. A. Dafforn, J. D. Gass, and D. R. Storm, The importance of orientation factors in enzymatic reactions, *Cold Spring Harbor Symp. Quant. Biol.* **36**, 13–20 (1971).
31. D. E. Koshland, Jr., The catalytic power of enzymes, *Proc. Robert A. Welch Found. Conf. Chem. Res.* **15**, 53–91 (1972).
32. G. A. Dafforn and D. E. Koshland, Jr., The sensitivity of intramolecular reactions to the orientation of reacting atoms, *Bioorg. Chem.* **1**, 129–139 (1971).
33. A. Dafforn and D. E. Koshland, Jr., Theoretical aspects of orbital steering, *Proc. Nat. Acad. Sci. USA* **68**, 2463–2467 (1971).

34. D. R. Storm and D. E. Koshland, Jr., An indication of the magnitude of orientation factors in esterification, *J. Am. Chem. Soc.* **94**, 5805–5814 (1972).

35. D. R. Storm and D. E. Koshland, Jr., Effect of small changes in orientation on reaction rate *J. Am. Chem. Soc.* **94**, 5815–5825 (1972).

36. R. M. Moriarity and T. Adams, A criticism of the use of certain bridged bicyclic hydroxy-carboxylic acids as model compounds for the concept of orbital steering, *J. Am. Chem. Soc.* **95**, 4070–4071 (1973).

37. T. Adams and R. M. Moriarity, Acid-catalyzed lactonization of *exo*- and *endo*-bicyclo[2.2.2] oct-5-ene-carboxylic acids. Structural clarifications, *J. Am. Chem. Soc.* **95**, 4071–4073 (1973).

38. D. R. Storm, R. Tjian, and D. E. Koshland, Jr., Rate acceleration by alteration in the orientation of reacting atoms. Comparisons of lactonizations in bicyclo[2,2,2] and bicyclo[2,2,1] ring structures, *Chem. Commun.* **1971**, 854–855.

39. B. Capon, Orbital steering: An unnecessary concept, *J. Chem. Soc. B* **1971**, 1207.

40. D. F. DeTar, Calculation of steric effects in reactions, *J. Am. Chem. Soc.* **96**, 1254–1255 (1974); Quantitative predictions of steric acceleration, *J. Am. Chem. Soc.* **96**, 1255–1256 (1974).

41. T. C. Bruice, A. Brown, and D. O. Harris, On the concept of orbital steering in catalytic reactions, *Proc. Nat. Acad. Sci. USA* **68**, 658–661 (1971).

42. C. E. Kim and L. L. Ingraham, The question of orbital steering in the reaction, $H_2O + CH_2O$, *Biochim. Biophys. Acta* **297**, 220–228 (1970).

43. R. Hershfield and G. L. Schmir, Mechanism of acid-catalyzed thiolactonization. Kinetic evidence for tetrahedral intermediates, *J. Am. Chem. Soc.* **94**, 6788–6793 (1972).

44. G. N. J. Port and W. G. Richards, Orbital steering and the catalytic power enzymes, *Nature (London)* **231**, 321–322 (1971).

45. D. G. Hoare, Significance of molecular alignment and orbital steering in mechanisms for enzymatic catalysis, *Nature (London)* **236**, 437 (1972).

46. H. Umeyama, A. Imamura, C. Nagata, and M. Hanano, Molecular orbital study on the enzymic reaction mechanism of α-chymotrypsin, *J. Theor. Biol.* **41**, 485–502 (1973).

47. T. C. Bruice, Views on approximation, orbital steering, and enzymatic and model reactions, *Cold Spring Harbor Symp. Quant. Biol.* **36**, 21–27 (1971).

48. M. I. Page, Entropic rate accelerations and orbital steering, *Biochem. Biophys. Res. Commun.* **49**, 940–944 (1972).

49. A. Dafforn and D. E. Koshland, Jr., Proximity, entropy, and orbital steering, *Biochem. Biophys. Res. Commun.* **52**, 779–785 (1973).

50. W. P. Jencks and M. I. Page, "Orbital steering," entropy, and rate accelerations, *Biochem. Biophys. Res. Commun.* **57**, 887–892 (1974).

51. T. C. Bruice and S. J. Benkovic, *Bioorganic Mechanisms*, Vol. 1, Benjamin, Reading, Mass. (1966).

52. T. C. Bruice and U. K. Pandit, Intramolecular models depicting the kinetic importance of "fit" in enzymatic catalysis, *Proc. Nat. Acad. Sci. USA* **46**, 402–404 (1960).

53. C. DeLisi and D. M. Crothers, The contribution of proximity and orientation to catalytic reaction rates, *Biopolymers* **12**, 1689–1704 (1973).

54. A. J. Kirby and A. R. Fersht, Intramolecular catalysis, *Prog. Bioorg. Chem.* **1**, 1–82 (1971).

55. B. Capon, Intramolecular catalysis, *Essays Chem.* **3**, 127–156 (1972).

56. M. L. Bender, *Mechanisms of Homogeneous Catalysis from Protons to Proteins*, Wiley-Interscience, New York (1971), Chap. 9.

57. R. D. Gandour and R. L. Schowen, Intramolecular catalysis in medicinal chemistry, *Annu. Rep. Med. Chem.* **7**, 279–288 (1972).

58. B. Capon, in: *Proton-Transfer Reactions* (E. F. Caldin and V. Gold, eds.), pp. 339–384, Chapman & Hall, London (1975).

59. M. Balakrishnan, G. V. Rao, and N. Venkatassubramian, Neighboring group participation in ester hydrolysis, *J. Sci. Ind. Res.* **33**, 641–651 (1974).

60. S. S. Minor and R. L. Schowen, One proton solvation bridge in intramolecular catalysis of ester hydrolysis, *J. Am. Chem. Soc.* **95**, 2279–2281 (1973).

61. A. R. Fersht and A. J. Kirby, The hydrolysis of aspirin. Intramolecular general base catalysis of ester hydrolysis, *J. Am. Chem. Soc.* **89**, 4857–4863 (1967).
62. A. R. Fersht and A. J. Kirby, Structure and mechanism in intramolecular catalysis. The hydrolysis of substituted aspirins, *J. Am. Chem. Soc.* **89**, 4853–4857 (1967).
63. R. D. Gandour, C. Olomon, and M. Sneller, unpublished results.
64. John L. Hogg, Transition state structures for catalysis by serine hydrolases and for related organic reactions, Ph.D. thesis, University of Kansas, Lawrence (1974).
65. H. B. Bürgi, J. M. Kehn, and G. Wipff, An *ab initio* study of nucleophilic addition to a carbonyl group, *J. Am. Chem. Soc.* **96**, 1956–1957 (1974).
66. H. Umeyama, A molecular orbital study on the solvolysis of aspirin derivatives and acyl-α-chymotrypsin, *Chem. Pharm. Bull.* **22**, 2518–2529 (1974).
67. E. R. Garrett, The kinetics of solvolysis of acyl esters of salicylic acid, *J. Am. Chem. Soc.* **79**, 3401–3408 (1957).
68. G. V. Rao, Solvent effects on intramolecular catalysis, *Indian J. Chem.* **13**, 608–609 (1975).
69. A. J. Kirby and G. J. Lloyd, Intramolecular general base catalysis of intramolecular nucleophilic catalysis of ester hydrolysis, *J. Chem. Soc. Perkin Trans. 2* **1974**, 637–642.
70. A. J. Kirby and G. T. Lloyd, Structure and efficiency in intramolecular and enzymic catalysis: Intramolecular general base catalysis. Hydrolysis of monoaryl malonates, *J. Chem. Soc. Perkin Trans. 2* **1976**, 1753–1761.
71. R. D. Gandour, Structural requirements for intramolecular proton transfers, *Tetrahedron Lett.* **1974**, 295–299.
72. V. Gold, D. G. Oakenfull, and T. Riley, The acetate-catalyzed hydrolysis of aryl acetates, *J. Chem. Soc. B* **1968**, 515–519.
73. M. F. Aldersley, A. J. Kirby, and P. W. Lancaster, Intramolecular displacement of alkoxide ions by the ionized carboxy group: Hydrolysis of alkyl hydrogen dialkylmaleates, *J. Chem. Soc. Perkin Trans. 2* **1974**, 1504–1510.
74. T. C. Bruice and A. Turner, Solvation and approximation. Solvent effects on the bimolecular and intramolecular nucleophilic attack of carboxyl anion on phenyl esters, *J. Am. Chem. Soc.* **92**, 3422–3428 (1970).
75. E. Gaetjens and H. Morawetz, Intramolecular carboxylate attack on ester groups. The hydrolysis of substituted phenyl acid succinates and phenyl acid glutarates, *J. Am. Chem. Soc.* **82**, 5328–5335 (1960).
76. J. W. Thanassi and T. C. Bruice, Neighboring carboxyl group participation in the hydrolysis of monoesters of phthalic acid. The dependence of mechanisms on leaving group tendencies, *J. Am. Chem. Soc.* **88**, 747–752 (1966).
77. W. P. Jencks, F. Barley, R. Barnett, and M. Gilchrist, The free energy of hydrolysis of acetic anhydride, *J. Am. Chem. Soc.* **88**, 4464–4467 (1966).
78. T. Higuchi, L. Eberson, and J. D. McRae, Acid anhydride–free acid equilibria in water in some substituted succinic acid systems and their interaction with aniline, *J. Am. Chem. Soc.* **89**, 3001–3004 (1967).
79. M. J. Haddadin, T. Higuchi, and V. Stella, Solvolytic reactions of cylic anhydrides in anhydrons acetic acid, *J. Pharm. Sci.* **64**, 1759 (1975).
80. M. D. Hawkins, Hydrolysis of phthalic and 3,6-dimethylphthalic anhydrides, *J. Chem. Soc. Perkin Trans. 2* **1975**, 282–284.
81. L. Eberson and H. Welinder, Studies on cyclic anhydrides. III. Equilibrium constants for the acid–anhydride equilibrium in aqueous solutions of certain vicinal diacids, *J. Am. Chem. Soc.* **93**, 5821–5826 (1971).
82. J. Gerstein and W. P. Jencks, Equilibria and rates for acetyl transfer among substituted phenyl acetates, acetylimidazole, *O*-acylhydroxamic acids and thiol esters, *J. Am. Chem. Soc.* **86**, 4655–4663 (1964).
83. R. Fuchs and E. S. Lewis, in: *Techniques of Chemistry* (E. S. Lewis, ed.), Vol. 6, p. 812, Wiley, New York (1974).
84. L. Pauling, *The Nature of the Chemical Bond*, 3rd ed., Cornell University Press, Ithaca, N.Y. (1960), p. 239.

85. R. D. Gandour, V. J. Stella, M. Coyne, R. L. Schowen, and E. A. Icaza, Secondary isotope effects in intramolecular catalysis. Mono-*p*-bromophenyl succinate hydrolysis, *J. Org. Chem.* **43**, in press.

86. J. R. Murdoch, Rate-equilibria relationships and proton-transfer reactions, *J. Am. Chem. Soc.* **94**, 4410–4418 (1972).

87. L. Eberson and L. A. Svensson, Studies on catechol esters. Part III. Hydrolysis of *O*-hydroxylphenyl acid succinates; competing intramolecular nucleophilic and general base catalysis, *Acta Chem. Scand.* **26**, 2631–2641 (1972).

88. D. C. Best, G. M. Underwood, and C. A. Kingsbury, On conformation–reactivity correlations, *J. Org. Chem.* **40**, 1984–1987 (1975).

89. S. Scheiner, W. N. Lipscomb, and D. A. Kleier, Molecular orbital studies of enzyme activity. 2. Nucleophilic attack on carbonyl systems with comments on orbital steering, *J. Am. Chem. Soc.* **98**, 4770–4777 (1976).

PART **IV**

Applications of Biochemical Transition-State Information

It is remarkable that a field of study as young as the investigation of biochemical transition states has already seen so vigorous an application of its concepts and results to other problems. In fact, the early applications seem to have had considerable impact in stimulating further fundamental investigations.

This is particularly true in the field of inhibitor design. Wolfenden, whose 1969 paper in *Nature* (Reference 5 of Chapter 15) has been the single most influential publication in the field, discusses enzymic transition-state structures as targets for inhibitor synthesis, provides examples, and shows how such inhibitors can be employed in elucidating the sources of enzyme catalysis. The power of such feedback for both theoretical and practical developments is apparent.

Chapter 16, by Coward, considers the application of enzymic transition-state information to the area of drug design in two different modes. One is the synthesis of transition-state analog drugs, which should be ultrapotent enzyme inhibitors. The other involves the use of transition-state information to find molecules which will themselves experience a strong transition-state interaction with the enzyme, thus undergoing catalyzed reaction, but will yield in place of a stable product a highly reactive, irreversible inhibitor. These "suicide substrates" not only hold great promise in medicine and biological research but may explain many natural phenomena ill-understood until now.

15

Transition-State Affinity as a Basis for the Design of Enzyme Inhibitors

Richard Wolfenden

To bring about the large rates of reaction which occur in the presence of enzymes, the interaction between reacting substrates and enzymes must be strong and intimate. The fact that substrates form complexes with enzymes has been known since the time of Henri[1] and Michaelis and Menten[2] and has been exploited in the design of inhibitors that resemble substrates. Enzyme–substrate affinity is far surpassed by the affinity which is developed transiently during catalysis, as was recognized by Pauling[3] before much was known about enzyme mechanisms. This enhanced affinity can be exploited to produce more effective inhibitors, designed to resemble the substrate during its transformation to form products.[4-8] These inhibitors, usually designated "transition-state analogs,"[4] are bound very much more tightly than substrates. As of mid-1975, some 60 possible examples were known, and several potential transition-state analogs had also been found in nature.[9,10]

1. FREE ENERGY SURFACE OF ENZYME REACTIONS

An advantage of absolute reaction-rate theory, especially in treating such complex processes as enzyme catalysis, is that it focuses our attention on a structure rather than a process. Product formation may occur by a variety of paths, of which those involving the lowest free energy of activation are favored. The major transition state is of special mechanistic interest, since it is usually the only high-energy intermediate about which structural inferences can be drawn with any confidence. Indeed, no pathway which connects a transition

Richard Wolfenden • Department of Biochemistry, University of North Carolina, Chapel Hill, North Carolina.

state with a ground state, of equal molecularity and atomic composition, can be ruled out using kinetic information of the kinds that are usually available.

An uncatalyzed reaction may proceed through several transition states falling on roughly isoenergetic pathways across a surface of high potential energy. Catalysis, as was recognized by Polanyi,[11] lowers the activation energy of a reaction, and a strongly catalyzed reaction can be represented as a deep cleft in the potential energy surface for the uncatalyzed reaction. In enzyme reactions the energy of activation is lowered by an amount usually well in excess of 12 kcal, suggesting that an enzyme reaction will usually proceed through a narrow defile with a transition state of fairly discrete structure. The very depth of this cleft raises serious experimental problems. The environment of the reacting substrate in an enzyme reaction is highly anisotropic and so different from its environment in free solution that the observed effects of conventional variables of the kind often used in kinetic studies (ionic strength, solvent polarity, temperature, pressure, substituent effects) can seldom be used with confidence in attempting to distinguish between alternative mechanisms. Effects of these variables may also be obscured, wholly or in part, by events such as product release, that may dominate the kinetic behavior of an enzyme reaction but have no counterpart in nonenzymic reactions. As these limitations of kinetic approaches have been recognized, students of enzyme mechanisms have come to place greater reliance on additional information which can be gleaned from the three-dimensional structure of enzyme molecules. The use of transition-state analogs provides a stochastic way of testing hypotheses about mechanisms and is a possible basis for relating kinetic and structural approaches to mechanism.

2. TRANSITION-STATE AFFINITY

The principle of transition-state affinity is illustrated in Fig. 1 by a machine for facilitating the transport of cannonballs over a wall. The military need for this device is open to question, but it could serve as a primitive catalyst.

When it is said that a catalyst lowers the activation energy of a reaction, it is implied that interaction between the catalyst and the reacting substrate is stronger in the transition state than it is in the initial complex which is formed at the instant of encounter. This is shown for an enzyme reaction in Fig. 2, where ΔF_S is the free energy of formation of the initial complex formed by bimolecular collision of enzyme and substrate and ΔF_{TX} represents the free energy of attraction between the enzyme and the altered substrate in the transition state. The difference between these free energies is the same as the difference between free energies of activation for the nonenzymic reaction, $\Delta F_N^{\ddagger}$, and for the enzymic reaction, $\Delta F_{ES}^{\ddagger}$. The binding enhancement that occurs in the transition state (relative to the encounter complex) is thus identical to the observed rate enhancement when the rate of reaction of the enzyme–substrate complex is compared with the rate of reaction in the absence of enzyme. This follows from any rate theory which assumes that rate constants

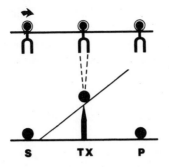

Fig. 1. Without doing physical work, the rolling magnet reduces the energy of activation for transferring the cannonball from point S to point P. The catalytic function of the magnet depends on its greater attraction for the cannonball at the point of closest approach, TX, than at points S and P. Note that the catalyst is equally effective for reaction in either direction. If the cannonball is not heavy enough, it will remain suspended at the midposition, as in the complex formed between an enzyme and a stable inhibitor that resembles an activated intermediate in substrate transformation.

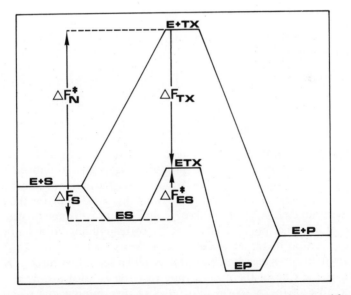

Fig. 2. Free-energy changes for an uncatalyzed reaction in water (upper pathway) and for the same reaction catalyzed by an enzyme (lower pathway). It is assumed that the altered substrate's structure in the transition state, TX, is identical in both reactions.

include an equilibrium constant for activation and a common rate constant for decomposition of all activated complexes.

For an enzyme reaction, this principle can be derived in two ways. The first of these does not explicitly take into consideration the existence of a Michaelis complex, and so avoids certain problems associated with non-productive binding and the interpretation of K_m and V_{max} values. The second-order rate constant for a one-substrate reaction,

$$E + S \longrightarrow E + P$$

is equal to k_{cat}/K_m, where k_{cat} is the turnover number and K_m is the Michaelis constant. The rate of the reaction in the absence of catalyst, but under otherwise identical conditions, is described by the first-order rate constant k_n:

$$S \xrightarrow{k_n} P$$

Equilibrium constants for activation in these two reactions lie in the same ratio as their rate constants:

$$K_E^{\ddagger}/K_N^{\ddagger} = (k_{cat}/K_m)/k_n$$

This ratio, which normally has dimensions of molarity, is equal to the equilibrium constant for formal dissociation of the enzyme–substrate complex *in the transition state* to yield the free enzyme and TX, the altered substrate in free solution:

$$(1)$$

The same relationship can be expressed in more familiar terms by taking the Michaelis complex ES into explicit consideration:[5]

$$(2)$$

Scheme (2) differs from scheme (1) in that $K_{ES}^{\ddagger}$ describes a first-order process, the decomposition of ES to form products. K_S is the equilibrium constant for decomposition of the ES complex to free enzyme and free substrate.

A typical enzyme reaction might exhibit $K_m = 10^{-3}$ M for the substrate, a turnover number (k_{cat}) of 100 sec^{-1}, and a rate constant in water, in the absence of enzyme, of 2.2×10^{-8} sec^{-1}, corresponding to a half-time of 1 year. Application of the above equations shows that $K_{TX} = 2.2 \times 10^{-13}$ M. There are many enzymes which exhibit much larger values for k_{cat}/K_m, and in most cases the rate of the nonenzymic reaction is known only as an upper limit, because of the practical difficulties of determining low reaction rates. K_{TX} may therefore often be much lower than this.

3. ASSUMPTIONS AND BOUNDARY CONDITIONS

Chemical reactions in general appear to follow the principle of absolute reaction rates, in the sense that rates are usually found to vary with temperature, pressure, and ionic strength in the manner expected if equilibria of activation were to behave in a manner similar to ordinary chemical equilibria. Whether the various detailed assumptions of the Polanyi–Eyring theory apply to enzyme reactions is not easy to establish. Provided only that activated complexes decompose at the same rate, K_{TX} can be calculated from the ratio of rate constants for the catalyzed and uncatalyzed reactions. Failure of the assumption that all activated complexes decompose at the same rate seems most likely to occur in very fast reactions during which there is insufficient time for equilibrium of activation to be established.[12] In that event, the equilibrium constant for activation would be underestimated. This is presumably more likely for a rapid enzymic reaction than for its slow nonenzymic counterpart and would result in a tendency to overestimate K_{TX}, i.e., to underestimate enzyme–substrate affinity in the transition state.

The derivation which we have considered could be applied to any catalytic phenomenon in any phase or medium and was developed independently for acid-catalyzed reactions by Kurz.[13] It involves the assumption that *similar* transition states are involved in the reactions being compared. This assumption is almost certain to be incorrect at some level of detail, for it is hardly likely that the substrate portion of the transition state will be unaffected by its interaction with the catalyst. Two kinds of exceptions appear probable: The transition state may be reached at a different point on the reaction coordinate for the catalyzed and uncatalyzed reactions, or the two transition states may be quite unrelated in structure. Both kinds of exception seem likely to lead to an underestimate of transition-state affinity.

If the transition state is reached earlier or later during bond making or breaking in the substrate for the enzyme reaction than for the nonenzymic reaction, then the reduction in free energy of activation which is observed, ΔF_{obs}, is actually less than ΔF_{TX}, the stabilization which the enzyme brings about, by tight binding, for the substrate structure which corresponds to the transition state for the nonenzymic reaction. The transition state for the nonenzymic reaction has become stabilized to such a degree that the rate is now limited by another shifted transition state with a different structure. In an extreme case, the release of a fully formed product may be rate determining (Fig. 3). The structure of the product of the nonenzymic reaction is then the same as the structure of the altered substrate in the transition state for the enzyme reaction. Evidently the chemically altered substrate, as it would occur in the transition state for the uncatalyzed reaction, must be bound even more tightly than the rate comparison would suggest (Fig. 3).[7,8] Analogous considerations apply if the transition state is reached *before* any change has occurred in the substrate.[10]

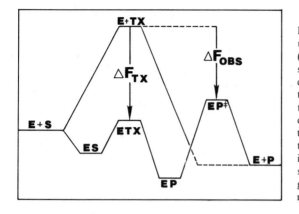

Fig. 3. Free-energy changes for an uncatalyzed reaction in water (upper pathway) and for the same reaction catalyzed by an enzyme (lower pathway). In this case the rate-determining transition state (EP‡) is reached during product release. TX, the transition state in substrate transformation, is assumed to be identical in both reactions and is stabilized to an extent, ΔF_{TX}, greater than the comparison of rates suggests, ΔF_{obs}.

In another exceptional case, there may be such a radical difference in mechanism that there is no obvious relationship between the structures of the altered substrate in the transition state for the enzymic and the nonenzymic reactions. TX for the enzymic reaction has a structure which may be supposed to be capable of an independent existence, at least in principle (TX, Fig. 4). If TX_{obs} for the nonenzymic pathway actually differs from this structure, it presumably has a lower energy, since the nonenzymic pathway will tend to follow that pathway with the lowest available free energy of activation. The enzyme may then be said to stabilize, through an affinity (ΔF_{TX}) even higher than that which the rate ratio would seem to require (ΔF_{obs}), an intermediate that is of higher energy than the actual transition state for the nonenzymic reaction (Fig. 4).[5]

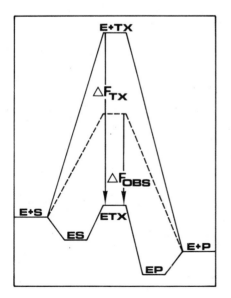

Fig. 4. Free-energy changes for an uncatalyzed reaction in water (dashed line) and for a reaction catalyzed by an enzyme, proceeding through a transition state of different structure (lower pathway). TX, the transition state in enzymic transformation of the substrate, is bound even more tightly than would be suggested by the observed rate ratio, because TX in water is less stable than the structure that corresponds to the transition state for the major reaction pathway of the uncatalyzed reaction in water.

Exceptions of these kinds probably constitute the majority of cases, since it appears improbable that any reaction would reach a transition state at just the same point (defined in terms of the lengths of bonds being formed and broken) in free solution and at the active site of an enzyme. Scheme (1) thus provides only a minimal notion of the magnitude of enzyme–substrate affinity which is likely to be required at some point during the reaction in order to explain the rate enhancement that is observed.[7,10]

4. DOES HIGH TRANSITION-STATE AFFINITY ENSURE EFFICIENT CATALYSIS?

On the basis of the previous discussion, one might be inclined to suppose that high affinity for the altered substrate, in the transition state for its transformation to product, is the sole property required of an efficient catalyst. There appear, however, to be several additional requirements.

Unless k_{cat} substantially exceeds the rate constant for the nonenzymic reaction, no catalysis can be expected under any conditions. It would be possible in principle to achieve a high value of k_{cat}/K_m (and thus a high transition-state affinity) simply by lowering the value of K_m. It is immediately evident that changes of this kind increase reaction rates only at low substrate concentrations; at higher substrate concentrations than K_m, the rate is constant and unaffected by changing substrate concentrations—a situation that may be disadvantageous for physiological control. In the absurd extreme, it is possible to imagine a "catalyst" for which k_{cat}/K_m is at the limit imposed by diffusion control, but for which $[E]k_{cat}$, where $[E]$ is the concentration of enzyme, is much less than the rate constant for the nonenzymic reaction. This "catalyst," for which K_m would have to be very low indeed, would in fact exhibit negligible activity as a catalyst. The important "trick" which an enzyme must perform is to bring about a large *increase* in affinity for the substrate in passing from the ES complex to the transition-state complex ETX. Some possible structural requirements of this process are discussed in Section 13.

From a physiological standpoint, it is reasonable to suppose that k_{cat} need only be as large (relative to k_n) as is required to allow the enzyme to maintain kinetic control (rate sensitive to substrate concentration) at physiological concentrations of substrate.[14,15a] For example, if the physiological level of substrate is 1 mM, then a K_m of 10 mM and a k_{cat} of 100 sec^{-1} would allow as effective control of reaction rate (by changing substrate concentration) as would a K_m of 1M and a k_{cat} of 1000 sec^{-1}. The first of these arrangements requires less change in affinity in passing from ES to ETX and might arise more readily during evolution for this reason.

Many enzymes act on specific substrates at rates that approach the limits imposed by diffusion (see Section 13). It is sometimes suggested that this represents a state of evolutionary perfection, on which no further improvement is possible.[15b] Such arguments assume existing physiological substrate concentrations as a given quantity. It may be more reasonable to suppose that physiological substrate concentrations have "evolved" to fit the control requirements and structural limitations of enzymes as they presently exist, at least to some extent. Adopting this broader viewpoint, it is conceivable that future development may result in the evolution of organisms with higher concentrations of metabolites than those in present-day organisms, and enzymes with higher values for V_{max} and K_m. This might allow miniaturization of the machinery required for routine metabolic transformations, and could presumably continue indefinitely.

5. MODEL REACTIONS AND THE ANALYSIS OF TRANSITION-STATE AFFINITY

One of the more useful ways of attempting to understand enzyme action is through the preparation and study of model catalysts. In the final extreme, one might be able to make a synthetic catalyst equivalent to the enzyme itself in structure and activity. If we postulate that the altered substrate in the transition state has a particular structure, we can suppose that it is bound very effectively to the enzyme by a number of forces. Some of these may be primarily ionic in nature and some may be covalent, and there are presumably a variety of intermediate descriptions. One could in principle eliminate these interactions one by one by modifying the enzyme. Each decrement in binding energy (expressed in cratic, or mole fraction, concentration units) would be matched by an increment in free energy of activation, until the last bond was broken and one would be left with the reaction in water (Fig. 5). Any of the catalyzed paths in Fig. 5 could be compared with the normal enzyme reaction path and their *relative* values of K_{TX} calculated. As better model catalysts are developed, it will be of interest to examine the affinity of these catalysts (and of partially denatured enzymes) for transition-state analogs as an aid to analyzing the forces responsible for catalysis.

In comparing enzymic and nonenzymic reactions, it seems likely that major contributions arise from the effects of solvation by water. There are indications that solvation shells, surrounding different parts of the same molecule, may interact with each other in such a way as to produce substantial nonadditive changes in entropy and volume of reaction with apparently minor variations in structure.[16,17a] These observations suggest that similar anomalies are likely to arise in equilibria of activation, especially in the many cases where reactants and products resemble each other more closely than they resemble the transition state.

Effects of solvation presumably apply with equal force to the rate enhancement produced by a catalyst and to the affinity with which it binds a transition-state analog. That these effects may be very large is evident from the marked differences in free energy of solvation between different organic

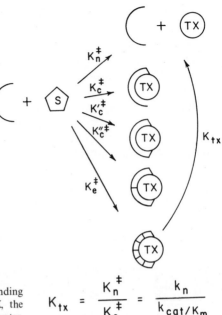

Fig. 5. Incremental effects of individual binding determinants in catalytic stabilization of TX, the transition state for the uncatalyzed reaction in water.

$$K_{tx} = \frac{K_n^{\ddagger}}{K_e^{\ddagger}} = \frac{k_n}{k_{cat}/K_m}$$

functional groups in water (Table I). Such effects provide, for example, the entire driving force (some 7 kcal) for the ammonolysis of ethyl acetate aqueous solution, a reaction which is endergonic in the vapor phase but strongly exergonic in water. If, during the hydrolysis of an amide, this substrate were withdrawn into a cavity of dielectric constant 1, then a "transition-state analog," rendered hydrophobic simply by replacing the amide function by an ethylene residue, would be expected to be bound more than eight orders of magnitude more tightly than the substrate due to solvation effects alone.[17b]

Table I
Absolute Distribution Coefficients
for Transfer from Dilute Aqueous
Solution to the Vapor Phase at 25°C[a]

Ethane	22
Ethylene	9.6
Acetylene	1.1
Dimethyl ether	4.1×10^{-2}
Ethyl acetate	5.4×10^{-3}
Acetone	1.3×10^{-3}
Ethylamine	4.1×10^{-4}
Ethanol	2.1×10^{-4}
Acetic acid	1.1×10^{-5}
Acetamide	7.6×10^{-8}

[a] Equilibria in moles per liter in vapor divided by moles per liter in dilute aqueous solution.[17b]

Against the background of the vapor phase, water itself may act as a catalyst. The use of the reaction in water, for comparison with the rate of the catalyzed reaction in the estimation of transition-state affinity, is necessary if this affinity is to be related to the observed binding of inhibitory analogs in aqueous solution. Since it is improbable that the mechanisms in water and on the enzyme are ever identical in detail, this procedure provides (for reasons discussed earlier) only a minimal estimate of the binding affinity expected of an ideal transition-state analog. Exact analysis of solvation effects will ultimately require much new information about reaction thermodynamics and equilibria in the vapor phase (in which intermolecular contacts are absent).

6. THE QUEST FOR TRANSITION-STATE ANALOGS

If the altered substrate, during substrate transformation, is bound as tightly as the above considerations suggest, then it should be possible (at least in principle) to design an unreactive analog with similar properties. A number of potent inhibitors have been designed on this premise. There has also been a growing recognition that the term "transition-state analog" can be understood too literally and that an "ideal" analog may be difficult to recognize in practice. If one had made a perfect analog of the reacting substrate in the transition state, one could not be sure that the observed affinity of the analog for an enzyme was high enough to account for the observed rate enhancement, because K_{TX} can only be determined as an upper limit. Another practical difficulty in recognizing a transition-state analog is that a compound which possessed the required properties (such as an appropriately situated electrostatic charge) in exaggerated form (several charges, for example) might be bound even more tightly than K_{TX} would require.[18] The method is thus mainly qualitative, useful for determining or exploiting differences between the substrate in the ground state and in high-energy forms and for obtaining information about the binding properties of the active site.

The ideal transition-state analog may be equivalent to a Platonic "form," possible to imitate but never to attain. The attempt may, however, yield indications of mechanism which are independent of other kinetic and structural evidence. During synthetic work it may be possible to determine whether compounds possess worthwhile activity by testing them at low states of purity, and at this stage contaminants may be encountered which are of considerable interest. For example, benzylsuccinate, a potent inhibitor of carboxypeptidase A,[19] was encountered during efforts to synthesize an inhibitor of quite different structure.

7. MULTISUBSTRATE ANALOGS AND COOPERATIVITY

Scheme (3) describes a reaction involving two or more substrates. Here the nonenzymic reaction is described by a second-order rate constant, k_2, and the enzymic reaction is described by a third-order rate constant equivalent to the turnover number k_{cat}, divided by the product of the K_m values for the two substrates. K_{TX}, the dissociation constant of the altered substrates in the transition state, is equivalent to the quotient of these rate constants and is expressed in the same units as the K_m values, normally as moles per liter.

$$E + R + S \underset{K_N^{\ddagger}}{\overset{K_E^{\ddagger}}{\rightleftharpoons}} \begin{array}{c} ERS^{\ddagger} \\ \Big\downarrow K_{TX} \\ E + RS^{\ddagger} \end{array} \searrow \nearrow E + P + Q \tag{3}$$

$$K_{TX} = \frac{K_N^{\ddagger}}{K_E^{\ddagger}} = \frac{k_n}{k_{cat}/K_{m_R} \cdot K_{m_S}}$$

As in scheme (1), the actual units in which activities or concentrations are expressed are unimportant, provided that consistency is maintained.

Tight binding of a multisubstrate analog does not require that the substrates, as they would occur in the configuration reflected by this analog, be electronically activated. K_{TX}, the dissociation constant of the substrates reacting with each other in the transition state, is not directly comparable to the product of the dissociation constant of either substrate, since this ratio depends on the convention used for expressing concentrations. This reflects a more fundamental difficulty which arises in attempting to analyze the various factors, including approximation, orientation, and internal entropy reduction, which an enzyme may bring to bear in catalyzing multisubstrate reactions. In a careful review of this problem, Jencks[18] suggests that it is not possible to analyze these contributions on the basis of presently available information but that the contribution from entropic factors alone (quite aside from chemical activation) may approach factors as large as 10^8 in an extreme case for a two-substrate reaction. It is therefore *not* possible, at present, to distinguish in real cases between analogs whose affinity depends on a resemblance to chemically activated intermediates in a multisubstrate reaction and analogs whose affinity is based on purely entropic factors (i.e., on a resemblance to an intermediate stage in the normal reaction in which substrates have undergone a reduction in translational, orientational, or internal entropy without having undergone any substantial redistribution of electrons). Multisubstrate analogs of various kinds may be useful for obtaining experimental evidence about the contributions that an enzyme can make to catalysis simply by binding reactants in an appropriate arrangement at the active site.

Efforts to design multisubstrate analogs for pyridoxal-containing enzymes, amino-acid-activating enzymes, aspartyl transcarbamylase, carboxypeptidase A, adenylate kinase, and a number of pyridine-linked dehydrogenases have been rewarded with considerable success. Multisubstrate analogs hold particular promise as antimetabolites for biosynthetic reactions in which coenzymes are involved, and some of them are discussed by Coward.[20]

8. EXAMINING POTENTIAL TRANSITION-STATE ANALOGS

Once an effective inhibitor has been discovered, several questions arise. Is the inhibitor bound for the reasons which served as a basis for its design? What is the form of the bound inhibitor? Does structural reorganization of the active site accompany binding? What components of the active site contribute to the exceptional affinity observed (i.e., what are the "catalytic residues")?

Answers to these questions can be found, at least in principle, by structural methods which are presently available. Structures of enzyme–inhibitor complexes can be compared with the native enzyme or with ES complexes which may be isolable under special conditions (for example, at low temperature). Strong interactions that occur in analog complexes should serve to immobilize particularly the groups directly involved in the catalytic process.

In what follows, no attempt will be made to review the large number of possible transition-state analogs which have been studied. Instead, we shall consider several cases that exemplify the problems which may be encountered.

9. PROTEASES, ALDEHYDES, AND THE PANCREATIC TRYPSIN INHIBITOR

It has been known for some time that proteases such as chymotrypsin and papain catalyze substrate breakdown through the intervention of acyl enzymes.[21,22] These are formed as a result of attack by an active-site nucleophile on the substrate's peptide bond, with displacement of the amino group. The resulting esters are later hydrolyzed to yield free enzyme and the carboxylic acid product. Precedent in organic chemistry suggests that the formation and breakdown of these esters might be expected to proceed through tetrahedral intermediates, formed by addition of the active-site nucleophile to the carbonyl group of the peptide bond.[23]

Based on this supposition, it seemed worthwhile to examine the possibility that aldehydes, capable of forming similar adducts spontaneously (especially with thiols[24] which correspond to the active-site nucleophile of papain), might serve as inhibitors of proteases (Fig. 6). Some proteases in particular exhibit considerable specificity for the acyl substituent of the peptide bond which is cleaved, and related aldehydes might thus be bound with some specificity. When aldehydes of this kind were examined as inhibitors of papain, they were found to serve as potent competitive inhibitors,[25] with affinities varying in proportion to second-order rate constants observed for enzymic hydrolysis of the corresponding esters. Similar observations have been described for elastase,[26] L-asparaginase,[27] and a bacterial amidase.[28]

It remains to be demonstrated whether aldehydes are in fact bound in the

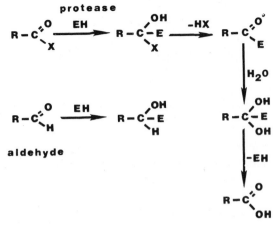

Fig. 6. Intermediates in substrate hydrolysis by proteases which act through double displacement mechanisms. The putative aldehyde adduct resembles sp^3 hybridized intermediates in formation and hydrolysis of the acyl-enzyme intermediate.

manner anticipated. There are, in fact, indications that aldehyde adducts are no more stabilized (relative to model compounds), for example, than acyl enzymes, and it is therefore questionable whether aldehydes should be regarded as transition-state analogs. Their efficient binding does suggest that tetrahedral intermediates may be on the catalytic pathway, as suggested by kinetic studies, but the transition state is as likely to resemble the acyl enzyme as an aldehyde adduct.[25] Perhaps the most convincing indications that aldehyde adducts bear some resemblance to intermediates in catalysis is the observation, in the case of elastase, that distant substituents which, in substrates, affect V_{max} have similar effects on the binding of aldehydes.[29] It has also been shown that changing pH affects aldehyde binding in the same way that it affects V_{max} for the action of papain on substrates.[30]

Aldehydes might be bound as hydrates or as free aldehydes unaltered by nucleophilic addition. Both aldehydes and nitriles, which are held considerably less tightly, could be bound unaltered at a hydrophobic cavity which might tend to exclude or strain more bulky substrates, thus contributing to catalysis. This appears difficult to reconcile with the very weak binding observed with analogs in which the —CHO or —CN group of the inhibitor is replaced by —C_2H_5, —CH_3, or simply —H, which are equally or less bulky.[31] An alternative possibility, that aldehydes are bound as hydrates,[28] might be explained by a resemblance between the *gem*-diol and an adduct formed in *direct* water attack on the substrate. This is rendered unlikely in the case of papain by the total failure of the alcohol, in which —CH_2OH replaces the putative —$CH(OH)_2$, to serve as an inhibitor.[30] (*Note added in proof*: By comparing the secondary isotope effect of deuterium substitution in aldehydes on equilibrium addition of various nucleophiles, with the corresponding isotope effect on K_i for aldehyde inhibitors[31] it has been possible to ascertain that these inhibitors are bound as thiohemiacetals by papain.)

Structural studies of inhibitory complexes of boronic acid derivatives, related to substrates of proteases, have progressed much further. As anticipated by Koehler and Lienhard,[32a] covalent reaction with an active-site nucleophile leads to a tetrahedral adduct, as has been shown by crystallographic studies of subtilisin BPN[32b] and laser Raman studies of chymotrypsin.[33]

It is of interest to inquire how an intermediate in uncatalyzed hydrolysis of a peptide in water might be stabilized through covalent interaction with a catalyst. The transition state for uncatalyzed peptide hydrolysis in water might resemble a tetrahedral intermediate formed by covalent hydration of the peptide bond. This species could be bound at an active site by replacing the —OH group of the tetrahedral intermediate with an active-site nucleophile (a roughly isoenergetic process, if the nucleophile were oxygen). If "distant" interactions of specificity determinants acted to stabilize such a covalent bond (by tending to force a roughly tetrahedral arrangement in which it was included), the complex might be considerably stabler than would be a simple alcohol adduct to the carbonyl group. Formation of carbonyl adducts of esters and acids has been shown by Guthrie[34] to be highly unfavorable. It is thus of special interest that bovine pancreatic trypsin inhibitor is a polypeptide that appears to form a "tetrahedral" adduct with trypsin which, although it has this unfavorable kind of bond, is so stabilized by distant interactions that the complex is very tight indeed.[35] Not surprisingly, the site on the inhibitor at which addition and pyramidalization occur adopts a similar structure in the complex with anhydrotrypsin, where no addition occurs.[36] This suggests that distant interactions do in fact "force" a quasi-tetrahedral structure on the crucial bond whether or not it has undergone addition. The trypsin inhibitor may, in this very limited sense, be considered a kind of transition-state analog, contributing properties (in the complex) similar to those of an aldehyde.

A serious theoretical problem in using this approach on double-displacement mechanisms is that there is really no useful basis for calculating the affinity expected of an inhibitor of this kind in an absolute sense.[7,8,10,18] Just as there is no uncatalyzed reaction in water which is truly analogous to the transformations involved in covalent catalysis, the corresponding inhibitors become transition-state analogs only after combining covalently with the enzyme.

10. TRIOSEPHOSPHATE ISOMERASE, GLYCONIC ACIDS, AND THE PHOSPHORYL SUBSTITUENT EFFECT

Of a variety of compounds related to substrates of triosephosphate isomerase, the distinctive affinity exhibited by 2-phosphoglycolic acid is understandable on the basis of its stereochemical similarity to an ene-diol or ene-diolate intermediate.[37] The hydroxamic acid of phosphoglycolate is also an extremely effective inhibitor of triosephosphate isomerase.[38] Figure 7 shows the structures. Similar observations have been made for phosphoglucose isomerase[39] and, more recently, for phosphoribose isomerase (W. W. Woodruff and R. Wolfenden, in preparation).

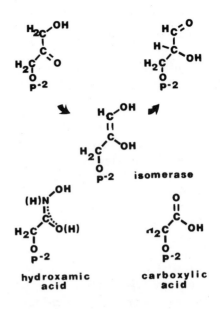

Fig. 7. Inhibitors of triosephosphate isomerase, which may resemble an ene-diol intermediate in substrate transformation.

An interesting problem is raised by the possible role of substrate phosphoryl groups in catalysis by triosephosphate isomerase. The limiting rate of reaction (V_{max}) falls off at low pH values, consistent with protonation of an essential base with a pK_a of about 6.1, whereas K_m is virtually unaffected by changing pH in the neutral range.[37,40,41] Similar behavior, exhibited by the kinetics of inactivation of the enzyme by the alkylating agent glycidol phosphate, suggested the presence of a catalytic group on the enzyme with pK_a about 6.[42a] However, chloroacetol sulfate, with no pK_a in the neutral range, inactivates the enzyme at a rate which is pH independent in this range (the competitive inhibitor sulfate likewise exhibits a K_i value which is pH independent).[41] Chloroacetol sulfate alkylates a glutamyl residue at the active site, also alkylated by glucidol phosphate. The kinetics of inactivation of the yeast enzyme by chloroacetol sulfate suggest a pK for the active site glutamyl residue in the neighborhood of 4. It appears unnecessary to postulate the invol ement of an ionizing of the enzyme to explain the pH dependence of V_{max}, which can simply be attributed to substrate ionization. Only dianionic forms of substrates appear to combine productively with the enzyme.

Situations in which V_{max}, but not K_m, is sensitive to an experimental variable provide an excellent opportunity for testing potential transition-state analogs. Theory predicts that transition-state analogs (unlike substrates or substrate analogs) should exhibit an affinity which varies in a way which can be predicted from the variation of V_{max} [scheme (2)]. This was found to be true, at least approximately, for the affinity of isomerase for 2-phosphoglycolate and for the corresponding hydroxamate, each appearing to be bound as a dianion.[41] In the case of 2-phosphoglycolate, the major dianion in solution possesses one charge on the carboxyl group and one charge on the phosphoryl group. The dianionic species with two charges on the phosphoryl group, which may be the actual inhibitor, is relatively rare, constituting no more than 1% of total dianion in solution (based on the pK_a of the corresponding carboxymethyl ester).[37] The bound form of phosphoglycolate remains to be determined by exact structural studies. (*Note added in proof*: Examination of the bound inhibitor, by phosphorus and carbon magnetic resonance spectroscopy, shows that it appears to be bound as a *tri*anion, corresponding to an ene-diolate intermediate with two negative charges on the phosphoryl group.[42b] It follows that the enzyme itself must bind a proton as it takes up the trianion of 2-phosphoglycolate, and this would be understandable if the catalytic glutamyl residue were protonated in an analog of the normal catalytic process.)

Active-site conformation changes, provided they are not strongly endergonic, may promote catalysis by making possible strong binding contacts between the enzyme and the reacting substrate which were not present in the "open" ES complex formed at initial encounter.[43] In this particular case, the unit cell has been observed to undergo a reversible contraction along the major axis when the transition-state analog 2-phosphoglycolate is introduced.[44] The nature of this structural change remains to be elucidated, but it seems likely that the active site possesses some flexibility, i.e., that its native structure does not correspond to a deep well in a potential energy surface. One may therefore speculate that the dianionic phosphoryl group and the ene-diol protion of the substrate (or its analog, an undissociated carboxylic acid) may be bound *cooperatively* in a tightly enclosed transition-state complex. For maximal binding interaction to occur in the transition-state complex (and in the transition-state analog complex) it is apparently required that the phosphoryl group be doubly charged.

11. ADENOSINE DEAMINASE, PTERIDINE HYDRATION, AND THE RIBOSE SUBSTITUENT EFFECT

Hydrolytic deaminases, which act on adenosine and related compounds, also catalyze the hydrolytic removal of a wide variety of leaving groups other than the normal product, ammonia.[45] The fact that leaving groups as disparate as ammonia and chloride are hydrolyzed from purine ribonucleoside at similar limiting rates (despite a vast difference in the energy of their bonds to carbon) provided an early indication that the rate of reaction was probably determined, at least in part, by some step other than cleavage of the scissile bond of the substrate.[46,47] It was at first believed that this step might be hydrolysis of a common intermediate, formed as a result of leaving-group displacement by enzyme, but it was then found that the rate of reaction was hardly affected by the substitution of deuterium oxide for solvent water.[48] This suggested

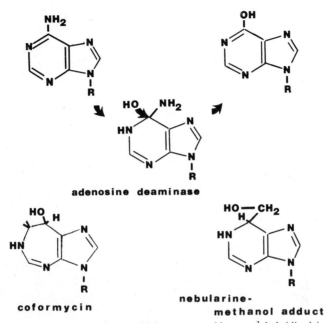

adenosine deaminase

coformycin

nebularine-
methanol adduct

Fig. 8. Inhibitors of adenosine deaminase, which may resemble an sp^3 hybridized intermediate in displacement of ammonia by water.

that the transition state might be reached very early during reaction, and a mechanism involving direct water attack on the substrate was considered a possibility (Fig. 8).

A route to a very approximate analog of a tetrahedral intermediate, formed during direct water attack on the substrate adenosine, was provided by the photoaddition of methanol to purine ribonucleoside. The adduct proved to be a strong inhibitor of deaminase activity.[49] Pteridine, earlier shown by Albert to be covalently hydrated, was a good inhibitor of adenosine deaminase, consistent with the activity of the enzyme on 4-aminopteridine as a substrate.[50] When an attempt was made to compare pteridine and its covalent hydrate as inhibitors of the enzyme, the enzyme proved unexpectedly to be a *catalyst* of the reversible hydration reaction.[51] The hydratase activity of the enzyme provided strong support for a mechanism involving direct water attack on substrates for hydrolysis (rather than a double displacement mechanism), especially when it was found that the rates of pteridine hydration and of 4-aminopteridine hydrolysis, both catalyzed by the enzyme, were of comparable magnitude. The hydration reaction was somewhat the more rapid of the two, consistent with the hypothesis that the hydration reaction is an analog of a partial reaction in the normal mechanism of catalysis.[52] Not unexpectedly, the hydration reaction was also found to be stereoselective, and the hydrate was a more potent inhibitor than pteridine itself.

Confirmation of this mechanism was provided by the recent discovery of the structure of coformycin,[53] an antibiotic long known to inhibit adenosine deaminase very strongly. This inhibitor, more tightly bound than the methanol photoadduct previously synthesized (which also served as a synthetic intermediate in the proof of structure of coformycin), resembles the hypothetical tetrahedral intermediate even more closely.

Although these results are self-consistent, none of these compounds can be considered transition-state analogs in the simplest sense, since the previously mentioned studies in deuterium oxide suggest that the transition state for the enzymic reaction may be reached before any chemical transformation of the substrate has taken place. This may well be an early conformation change within the ES complex.

Possibly related to this is the odd behavior of ribose as a substrate substituent. Adenosine deaminase catalyzes deamination of adenosine at a limiting rate 70,000 times faster than the limiting rate at which it deaminates adenine, whereas K_m values for the two compounds differ by a factor of only 3.[50] Recent results show that the two compounds do not differ appreciably in susceptibility to nucleophilic attack in model systems, so that this very large ribosyl substituent effect is purely enzymic in origin. The effect of ribose in enhancing V_{max} is reflected by the affinity which the enzyme exhibits for the analog obtained by photoaddition of methanol to purine, the resulting aglycone being vastly inferior to the methanol photoadduct of purine ribonucleoside as an inhibitor.[54] Whatever the origin of the ribosyl substituent effect on catalysis, it is also manifest in the binding of this potential transition-state analog.

12. CATALYTIC EFFECTS OF NONREACTING PORTIONS OF THE SUBSTRATE

In each of the cases we have considered (and even more strikingly in results of the elegant studies of Ray and Long[55] on phosphoglucomutase) there is evidence that cooperative binding interactions, present in the transition state for enzyme-catalyzed reactions, include groups on the substrate which might not have been expected to participate directly in the catalytic process. Any binding determinant on the substrate is normally expected to enhance k_{cat}/K_m. But when a group such as the ribose of adenosine or the dianionic phosphoryl group of glyceraldehyde 3-phosphate exerts an effect specifically on k_{cat}, its interaction with the enzyme is manifest *only* in the transition state. If any attractive interaction between this group and the enzyme is present in the ES complex, it is so compensated by distortion that no net effect on the value of K_m is observed. Alternatively, one may suppose that the group remains solvated in the encounter complex and only "swings into place" with generation of the transition-state complex. In the special case where the accessory binding determinant is not actually connected to the reacting substrate, as in the experiments described by Ray and Long, one is led to conclude that the determinant plays an allosteric role, adjusting the active site in such a way that it is better able to accommodate the transition state in bond making and breaking.

Although these possibilities have been recognized and clearly described,[4] it is usually difficult or impossible to conceive of a way of distinguishing between them experimentally. Indeed the question is often moot, because one cannot be assured that isolable intermediates represent steps on the major pathway, and kinetic measurements can tell us only about the reactants and the transition state.

13. CONFLICTING REQUIREMENTS OF CATALYTIC EFFICIENCY AND TRANSITION-STATE AFFINITY

Despite these difficulties, it may be possible occasionally to draw some limiting conclusions about the structures of reactants in enzyme reactions and to decide whether or not the substrate (or the enzyme) is distorted in the complex which is formed in the initial encounter event.

All schemes for expressing transition-state affinity allow for the possibility that the enzyme reacts with the substrate in a rare or strained configuration (possibly even one approaching the transition state in structure), which is selected from a much larger population of normal substrate molecules. The "observed" equilibrium constant for substrate binding then includes two terms, one for formation of the rare species from the major species and a second for binding of the rare species by the enzyme. This possible mechanism for enzyme action suffers from one severe disadvantage: It does not allow for high values of k_{cat}/K_m. This kinetic parameter, a measure of overall catalytic effectiveness, cannot exceed the rate of enzyme–substrate encounter in solution. If only a small portion of the total substrate in solution is available for reaction, then k_{cat}/K_m (expressed in terms of bulk concentrations of substrate and enzyme) cannot approach the limit imposed by the total rate of enzyme–substrate encounter in solution. Thus, to be catalytically effective, enzymes must combine with substrates in their ground-state structures or at least in structures which are reasonably populous. There are in fact many enzyme reactions which proceed with very large second-order rate constants; indeed there appear to be at least five cases where k_{cat}/K_m exceeds $10^9 \ M^{-1} \ sec^{-1}$.[43] In these cases, one may infer that mass transfer occurs largely as a result of combination of reactants E and S in forms which are not far removed from their ground-state structures. Rare species do not appear to be obligatory reactants, although they may sometimes participate. Since this is true for both enzyme and substrate, it seems that the active site is effectively open to substrate access in the ground state most of the time (Fig. 9). Otherwise a rare species, with an open active site, would be one of the true reactants, and the observed second-order rate constant might be expected to reflect its scarcity. These tentative inferences need not apply to less efficient enzymes. The "open"

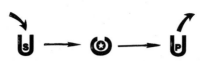

Fig. 9. Possible resolution of the conflicting requirements of rapid substrate access and high transition-state affinity.

encounter complex, which seems to be required by these almost diffusion-limited rates, could presumably be formed at a point on the enzyme distant from the catalytic site to which the substrate would later proceed by diffusion.

There is no reason why the apparent "openness" of the encounter complex should be maintained in the transition state. Both enzyme and substrate presumably tend toward structural adaptation, which is both the cause and effect of catalysis. The larger molecule (the enzyme) may tend to adapt itself to the smaller molecule by surrounding it as completely as possible in such a way as to optimize structural complementarity (Fig. 9). If the enzyme possesses some flexibility of active-site residues, the advantage of bringing new binding determinants to bear may outweigh the energetic cost of distorting the native site from its original configuration (Fig. 10). This general mechanism differs from older versions of the induced-fit theory[54] in that it involves an explicit contribution to catalytic efficiency from structural reorganization of the active site. The induced-fit theory was intended to explain certain examples of enzyme specificity, particularly the nonreactivity of smaller homologs of good substrates. The active configuration, into which the enzyme is drawn in the ES complex according to the induced-fit theory, is energetically less favorable than the inactive configuration of the free enzyme. This would presumably

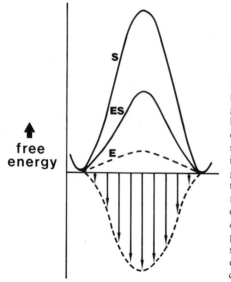

free energy

Fig. 10. A hypothetical free-energy profile, for reaction of an enzyme–substrate complex to give an enzyme–product complex, is indicated by ES. Profiles are also shown for the separated enzyme (E) and substrate (S), undergoing similar transformations as independent species in free solution. The enzyme is shown as undergoing a thermodynamically unfavorable distortion during catalysis, in curve E. This is more than counterbalanced by an increase in enzyme–substrate affinity, indicated by *vertical arrows*, which distortion of the enzyme makes possible through new binding contacts (as suggested in Fig. 9). The sum of these two effects is equivalent to the difference between curves S and ES.

result in a loss of catalytic efficiency, as compared with the efficiency of another catalyst that might be fixed in an active configuration.[4,18] In the present mechanism, enzyme–substrate contacts change in concerted fashion, maintaining optimal structural complementarity throughout the catalytic process, except at the point of substrate entry and product egress. In the ideal case, a more efficient catalyst may not be conceivable.

14. SUMMARY

Transition-state affinity, estimated by comparing the rates of catalyzed and uncatalyzed reaction, provides a minimal idea of the enzyme–substrate affinity which must be generated at some point during the catalyzed reaction. This affinity, always very large, is partly reflected by the affinity of enzymes for stable inhibitors which can form complexes resembling activated intermediates in substrate transformation.

The design of transition-state analogs rests on the assumption that the structural theory of chemistry is correct, that enzyme reactions tend to follow ordinary chemical principles, that theories of absolute reaction rates are reasonable (at least in a qualitative sense), that reactions proceed through intermediates whose structures can be inferred from indirect kinetic evidence, and that catalysis requires stabilization of these intermediates. When this approach is successful, it provides independent evidence that these assumptions are correct, as well as tending to confirm the *specific* mechanism on which its design was based.

Factors which contribute to the binding of transition-state analogs, and to the efficiency of the catalytic process itself, are matters of continuing interest. These factors can be evaluated through kinetic and structural studies of enzyme complexes with transition-state analogs. In due course these studies can be expected to reveal the form of the bound inhibitor and to suggest the nature of structural reorganizations of the active site that may accompany substrate transformation.

REFERENCES

1. V. Henri, *Lois Generales De L'Action Des Diastases,* Hermann, Paris (1903).
2. H. L. Segal, in: *The Enzymes* (P. D. Boyer, H. Lardy, and K. Myrbäck, eds.), Vol. 1, pp. 1–48, Academic Press, New York (1959).
3. L. Pauling, Molecular architecture and biological reactions, *Chem. Eng. News* **24**, 1375–1377 (1946).
4. W. P. Jencks, in: *Current Aspects of Biochemical Energetics* (N. O. Kaplan and E. P. Kennedy, eds.), pp. 273–298, Academic Press, New York (1966).
5. R. Wolfenden, Transition state analogues for enzyme catalysis, *Nature (London)* **223** 704–705 (1969).

6. G. E. Lienhard, I. I. Secemski, K. A. Koehler, and R. N. Lindquist, Enzymatic catalysis and the transition state theory of reaction rates: Transition state analogs, *Cold Spring Harbor Symp. Quant. Biol.* **36**, 45–51 (1971).
7. R. Wolfenden, Analog approaches to the transition state of enzyme reactions, *Acc. Chem. Res.* **5**, 10–18 (1972).
8. G. E. Lienhard, Enzymatic catalysis and transition state theory, *Science* **180**, 149–154 (1973).
9. R. N. Lindquist, in: *Drug Design* (E. J. Ariens, ed.), Vol. 5, pp. 24–80, Academic Press, New York (1975).
10. R. Wolfenden, Reversible analog inhibitors and enzyme catalysis, *Annu. Rev. Biophys. Bioeng.* **5**, 271–306 (1976).
11. G-M. Schwab, *Katalyse, vom Standpunkt der chemischen Kinetik,* Springer-Verlag, Berlin (1931).
12. L. P. Hammett, *Physical Organic Chemistry,* McGraw-Hill, New York (1970).
13. J. L. Kurz, Transition state characterization for catalyzed reactions, *J. Am. Chem. Soc.* **85**, 987–991 (1963).
14. W. W. Cleland, Enzyme kinetics, *Annu. Rev. Biochem.* **36**, 77–112 (1967).
15. (a) A. R. Fersht, Catalysis, binding and enzyme–substrate complementarity, *Proc. Roy. Soc. London Ser. B* **187**, 397–407 (1974). (b) W. J. Albery, and J. R. Knowles, Evolution of enzyme function and the development of catalytic efficiency, *Biochemistry* **15**, 5631–5640 (1976).
16. W. Kauzmann, A. Bodansky, and J. Rasper, Volume changes in protein reactions, *J. Am. Chem. Soc.* **84**, 1777–1788 (1962).
17. (a) C. A. Lewis, Jr., and R. Wolfenden, Influence of pressure on the equilibrium of hydration of aliphatic aldehydes, *J. Am. Chem. Soc.* **95**, 6686–6688 (1973). (b) R. Wolfenden, Free energies of hydration and hydrolysis of gaseous acetamide, *J. Am. Chem. Soc.* **98**, 1987 (1976).
18. W. P. Jencks, in: *Advances in Enzymology* (A. Meister, ed.), Vol. 43, pp. 220–410, Academic Press, New York (1975).
19. L. D. Byers and R. Wolfenden, A potent reversible inhibitor of carboxypeptidase A, *J. Biol. Chem.* **247**, 606–608 (1972).
20. J. K. Coward, Chapter 16 in this volume.
21. G. Lowe and A. Williams, Direct evidence for an acylated thiol as an intermediate in papain- and ficin-catalysed hydrolysis of esters, *Proc. Chem. Soc. London* **1964**, 140–141.
22. L. J. Brubacher and M. L. Bender, The preparation and properties of *trans*-cinnamoyl-papain, *J. Am. Chem. Soc.* **88**, 5871–5880 (1966).
23. M. Caplow, Chymotrypsin catalysis, evidence for a new intermediate, *J. Am. Chem. Soc.* **91**, 3639–3645 (1969).
24. G. Lienhard and W. P. Jencks, Thiol addition to the carbonyl group. Equilibria and kinetics, *J. Am. Chem. Soc.* **88**, 3982 (1966).
25. J. O. Westerik and R. Wolfenden, Aldehydes as inhibitors of papain, *J. Biol. Chem.* **247**, 8195–8197 (1972).
26. R. C. Thompson, Use of peptide aldehydes to generate transition-state analogs of elastase, *Biochemistry* **12**, 47–51 (1973).
27. J. O. Westerik and R. Wolfenden, Aspartic-β-semialdehyde: A potent inhibitor of *E. coli* L-asparaginase, *J. Biol. Chem.* **249**, 6351–6353 (1974).
28. J. D. Findlater and B. A. Orsi, Transition-state analogs of an aliphatic amidase, *FEBS Lett.* **35**, 109–111 (1973).
29. R. C. Thompson, Binding of peptides to elastase: Implications for the mechanism of substrate hydrolysis, *Biochemistry* **13**, 5495–5501 (1974).
30. J. O. Westerik, Aldehyde inhibitors of papain and L-asparaginase, Ph.D. thesis, University of North Carolina, Chapel Hill (1974).
31. (a) C. A. Lewis, Jr., and R. Wolfenden, Antiproteolytic aldehydes and ketones: Substituent and secondary deuterium isotope effects on equilibrium addition of water and other nucleophiles, *Biochemistry* **16**, 4886–4889 (1977). (b) C. A. Lewis, Jr., and R. Wolfenden, Thiohemiacetal formation at the active site of papain, *Biochemistry* **16**, 4890–4895 (1977).

32. (a) K. A. Koehler and G. E. Lienhard, 2-Phenylethaneboronic acid, a possible transition-state analog for chymotrypsin, *Biochemistry* **10**, 2477–2483 (1971). (b) D. A. Mathews, R. A. Alden, J. J. Birktoft, S. T. Freer, and J. Kraut, X-ray crystallographic study of boronic acid adducts with subtilisin BPN' (novo), *J. Biol. Chem.* **250**, 7120–7126 (1975).

33. G. P. Hess, D. Seybert, A. Lewis, J. Spoonhower, and R. Cookingham, Tetrahedral inter-mediate in a specific α-chymotrypsin inhibitor complex detected by laser Raman spectroscopy, *Science* **189**, 384–386 (1975).

34. J. P. Guthrie, Hydration of carboxylic acids and esters. Evaluation of the free energy change for addition of water to acetic and formic acids and their methyl esters, *J. Am. Chem. Soc.* **95**, 6999–7003 (1973).

35. R. Huber, D. Kukla, W. Bode, P. Schwager, K. Bartels, J. Deisenhofer, and W. Steigemann, Structure of the complex formed by bovine trypsin and bovine pancreatic trypsin inhibitor, *J. Mol. Biol.* **89**, 73–101 (1974).

36. R. Huber, W. Bode, D. Kukla, and U. Kohl, Structure of the anhydro-trypsin–inhibitor complex, *Biophys. Str. Mech.* **1**, 189–201 (1975).

37. R. Wolfenden, Binding of substrate and transition state analogs to triosephosphate isomerase, *Biochemistry* **9**, 3404–3407 (1970).

38. K. D. Collins, An activated intermediate analog, *J. Biol. Chem.* **249**, 136–142 (1974).

39. J. M. Chirgwin, and E. A. Noltmann, 5-Phosphoarabonate as a potential "transition state analogue" for phosphoglucose isomerase, *Fed. Proc., Fed. Am. Soc. Exp. Biol.* **32**, 667 (1973).

40. B. Plaut and J. R. Knowles, pH-dependence of the triose phosphate isomerase reaction, *Biochem. J.* **129**, 311–320 (1972).

41. F. C. Hartman, G. M. LaMoraglia, Y. Tomozawa, and R. Wolfenden, The influence of pH on the interaction of inhibitors with triosephosphate isomerase and determination of the pK_a of the active site carboxyl group, *Biochemistry* **14**, 5274–5279 (1975).

42. (a) S. G. Waley, The pK of the carboxyl group at the active site of triosephosphate isomerase, *Biochem. J.* **126**, 255–256 (1972). (b) I. D. Campbell, P. A. Kiener, S. G. Waley and R. Wolfen-den, Triosephosphate isomerase: Interactions with ligands studied by nuclear magnetic resonance, *Biochem. Soc. Trans.* **5**, 750–752 (1977).

43. R. Wolfenden, Enzyme catalysis: Conflicting requirements of substrate access and transition state affinity, *Mol. Cell. Biochem.* **3**, 207–211 (1974).

44. L. N. Johnson and R. Wolfenden, Changes in absorption spectrum and crystal structure of triosephosphate isomerase brought about by 2-phosphoglycollate, a potential transition state analogue, *J. Mol. Biol.* **47**, 93–100 (1970).

45. J. G. Cory and R. J. Suhadolnik, Structural requirements of nucleosides for binding by adenosine, *Biochemistry* **4**, 1729–1732 (1965).

46. R. Wolfenden, Enzymatic hydrolysis of 6-substituents on purine ribosides, *J. Am. Chem. Soc.* **88**, 3157–3158 (1966).

47. H. Bär and G. I. Drummond, On the mechanism of adenosine deaminase action, *Biochem. Biophys. Res. Commun.* **24**, 584 (1966).

48. R. Wolfenden, On the rate-determining step in the action of adenosine deaminase, *Biochemistry* **8**, 2409–2412 (1969).

49. B. E. Evans and R. Wolfenden, A potential transition state analog for adenosine deaminase, *J. Am. Chem. Soc.* **92**, 4751–4752 (1970).

50. R. Wolfenden, J. Kaufman, and J. B. Macon, Ring-modified substrates of adenosine de-aminase, *Biochemistry* **8**, 2412–2415 (1969).

51. B. E. Evans and R. Wolfenden, Hydratase activity of a hydrolase, *J. Am. Chem. Soc.* **94**, 5902–5903 (1972).

52. B. E. Evans and R. Wolfenden, Catalysis of the covalent hydration of pteridine by adenosine deaminase, *Biochemistry* **12**, 392–398 (1973).

53. (a) H. Nakamura, G. Koyama, Y. Iitaka, M. Ohno, N. Yagisawa, S. Kondo, K. Maeda, and H. Umezawa, Structure of coformycin, an unusual nucleoside of microbial origin, *J. Am. Chem. Soc.* **96**, 4327 (1974). (b) P. K. Woo, H. W. Dion, S. M. Lange, L. F. Dahl, and L. J. Durham, A novel adenosine and ara-A deaminase inhibitor, *J. Heterocycl. Chem.* **11**, 641–643 (1974).

54. D. F. Wentworth, G. Mitchell, and R. Wolfenden, Influence of substituent ribose on transition state stabilization by adenosine deaminase *Biochemistry* (in press).

55. W. J. Ray, Jr., and J. W. Long, Thermodynamics and mechanism of the PO_3-transfer process in the phosphoglucomutase reaction, *Biochemistry* (in press).

16

Transition-State Theory and Reaction Mechanism in Drug Action and Drug Design

James K. Coward

1. INTRODUCTION

Rational development of potent and specific new drugs requires a detailed knowledge of the chemical mechanism of action for available pharmacologic agents and/or the enzymes which catalyze a drug-sensitive reaction. Numerous drugs now in the pharmacopeia have been discovered largely by serendipity and by testing thousands of analogs of known effective agents. However, the emergence of bioorganic chemistry has provided a mechanistic foundation on which medicinal chemists and pharmacologists can construct a detailed understanding of the mode of action of drugs currently available and can improve the design of new agents. In previous chapters of this book, others have presented detailed descriptions of the determination of transition-state structures and their role in biochemical reactions. In this chapter, the application of these concepts to understanding the mode of drug action and to the design of new drugs will be considered. The use of transition-state analogs and multisubstrate adducts as potent and specific chemotherapeutic agents (the theoretical aspects of which were described by Wolfenden in Chapter 15) will be presented as the logical extension of transition-state theory into pharmacology. The preceding chapters have dealt with phenomena for which there is a considerable amount of literature available. In this final chapter, we shall discuss an approach to the study of drug action and drug design about which there is very little in the literature. It is hoped that this brief presentation will stimulate others to attempt to bridge the gap between bioorganic chemistry and pharmacology.

James K. Coward • Department of Pharmacology, Yale University School of Medicine, New Haven, Connecticut.

2. REACTION MECHANISM AND THE ACTION OF DRUGS

Of the hundreds of drugs currently used in clinical practice,[1] the chemical mechanism of drug action is known in only a few cases. A major reason for this is that a large number of drugs act at ill-defined receptors[2] where the mode of action of the natural agonist, let alone the (ant)agonistic drug, is not well understood. As progress is made in the isolation and characterization of receptors,[3] a more detailed understanding of their mechanism of action will become available. In contrast, many of the myriad of antimetabolites which have been isolated and/or synthesized[1,4] act at clearly defined loci, either because they are incorporated into growing macromolecules[5] or because they act on a

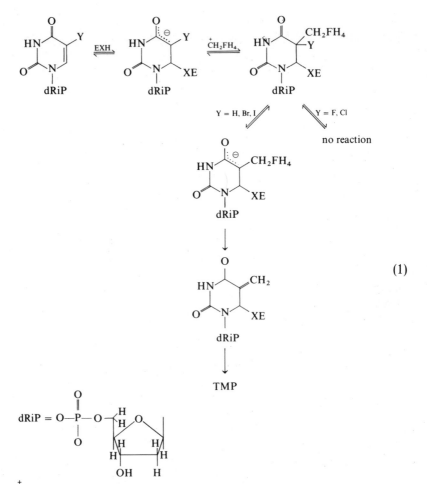

$$(1)$$

$\overset{+}{C}H_2FH_4$ = 5,10-Methylene tetrahydrofolate (protonated) specific enzyme. However, examination of the literature reveals a paucity of data available which is relevant to the *chemical* mechanism of antimetabolite action. An exception to this general statement is to be found in the work of Santi and Heidelberger and their co-workers[6-11] on the mechanism of action of 5-fluorouracil (5-FU) and the corresponding deoxyribotide (F-dUMP). This drug is known to be a very potent inhibitor of the enzyme thymidylate synthetase (EC 2.1.1.45),[1] and these workers have shown that F-dUMP can act as a "quasi-substrate"[7] where the drug is attached to the enzyme to form a reversible covalent drug–enzyme complex. Based on mechanistic studies using model compounds,[10] Santi and co-workers have postulated the reaction sequence shown in scheme (1) to explain the inhibition of thymidylate synthetase by F-dUMP. In contrast to F-dUMP, the substrate d-UMP readily loses a proton from the covalent ES complex enroute to product (TMP) formation. Confirmation of these mechanistic proposals comes from the recent demonstration of thymidylate synthetase-catalyzed TMP formation from 5-halogenated pyrimidine nucleotides in which the halogen can be abstracted with partial positive charge, e.g., 5-BrdUMP and 5-IdUMP.[11] Some uncertainty remains as to the exact nature of the bond between F-dUMP and the tetrahydrofolate cofactor[7,9] as well as to the identity of the amino acid (EXH) which is involved in the initial nucleophilic attack at C-6 of (F)-dUMP.[9,12,13] However, the overall mechanism of action of F-dUMP at the enzyme level is quite well understood in chemical terms.

In contrast to the mechanistic data available relative to the action of F-dUMP, relatively few data are available on the chemical mechanism by which methotrexate (MTX), a widely used antineoplastic drug,[1] acts. Although it has been known for some time that MTX is a very potent inhibitor of the enzyme dihydrofolate reductase (EC 1.5.1.3) with $K_i = 3 \times 10^{-11} M$ ($K_m^{FH_2} = 1.3 \times 10^{-6} M$),[14,15] very little is known about the chemical basis for this potent inhibition. A recent study has indicated that the pK_a of MTX is increased by at least 3 pH units when bound to the enzyme, thus suggesting the presence of an anionic binding site on the enzyme.[16] Since inhibitors with such low K_i values are rare, even with transition-state analogs (cf. Chapter 15), it would be of great value to accumulate more data on the chemical basis of this strong interaction between MTX and dihydrofolate reductase.

3. TRANSITION-STATE ANALOGS AND MULTISUBSTRATE ADDUCTS

Wolfenden's resurrection of Pauling's ideas on the tight binding of transition-state structures to enzymes[17] has had a most stimulating effect. This can be appreciated by the recent growth in the literature on transition-state analogs as potent enzyme inhibitors.[18] In this section, we shall discuss specifically the development of a potent new inhibitor of uridine biosynthesis, *N*-(phosphono-acetyl)-L-aspartate (PALA),[19-21] and the implications of this approach for the design of new drugs in general. Considering the mechanism of action of aspartate transcarbamylase (ATCase, EC 2.1.3.2), the first enzyme involved in the *de novo* synthesis of pyrimidines,[22] Collins and Stark synthesized PALA, the structure of which is shown together with the proposed tetrahedral intermediate in Fig. 1. Using *E. coli* ATCase, these workers showed the K_i for PALA to be 2.7×10^{-8} M, compared to $K_m = 2.7 \times 10^{-5}$ M for carbamyl phosphate and 1.7×10^{-2} M for L-aspartate. Similar data have been obtained for PALA inhibition of ATCase from mouse spleen[23] and hamster kidney cells.[20] In addition, a structurally related compound in which the amide–NH moiety of PALA is replaced by —CH_2— (DIKEP) is also a potent inhibitor of ATCase, with K_i ca. 10 times that for PALA.[20] Thus, PALA and DIKEP are potent competitive inhibitors of ATCase, with K_i values considerably lower than either K_m value and in the case of PALA ca 5×10^{-2} times the product of the K_m's ($K_m^{CAP} \times K_m^{ASP} = 4.6 \times 10^{-7}$ M). Although some questions have been raised concerning the differences in observed vs. theoretical K_i values in this study[19] and others[24] (cf. Chapter 15), there is no question that inhibitors of this type, rationally designed from mechanistic considerations, are extremely potent and specific.[18]

It would be expected that an inhibitor of aspartate transcarbamylase would have a considerable effect on pyrimidine biosynthesis in whole cells. This has

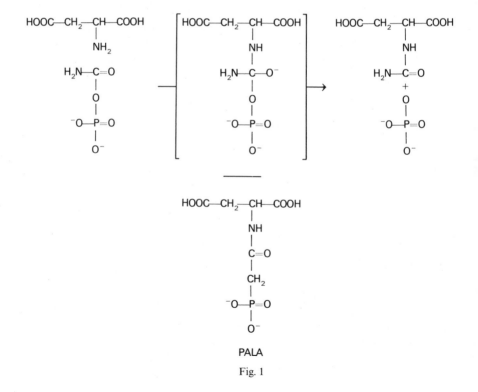

PALA

Fig. 1

been shown to be the case using PALA in several cultured cell lines.[20] In these studies it was shown that at concentrations of 1 and $2 \times 10^{-4} M$ PALA completely inhibited the growth of virus-transformed hamster cells (C13/5V) and mouse cells (SV 3T3), respectively. The drug apparently penetrated the cell membrane and was not extensively metabolized under the conditions employed. These statements are based on experiments where cellular extracts were added to the standard ATCase assay mixture. Extracts from cells which had been treated with PALA, appropriately diluted to give a final PALA assay concentration equal to I_{50}, gave about 50% inhibition of ATCase compared to control extracts. In addition, ion-exchange chromatography of the cellular extract demonstrated that the radioactive isotope was associated with unchanged PALA. These findings are important because they indicate that the charged phosphonate, PALA, can penetrate the membranes of these cells and that the phosphonate moiety of PALA appears resistant to metabolism in these cells.

Similar experiments have shown that the drug is not appreciably metabolized in whole animals.[21] Measurements of plasma levels of PALA, together with tissue concentration in spleen, liver, and jejunum, have shown that the drug rapidly passes from the plasma to give peak tissue concentrations within 2 to 4 hr after subcutaneous administration.[21] Similarly, measurements of ATCase activity in tissue extracts showed that the spleen enzyme was ca. 80% inhibited within 1 hr of oral or subcutaneous administration (the jejunum enzyme was comparably inhibited only after 24 hr), and this inhibition was maintained for periods up to 72 hr. Finally, an experiment was designed to establish the effect of PALA on *de novo* pyrimidine biosynthesis. It was not possible to measure directly *de novo* synthesis *in vivo*, and it was difficult to establish that the potent inhibition of ATCase by PALA rendered this enzyme rate limiting in the *de novo* pathway. In their experiment, however, the authors were able to demonstrate that PALA completely inhibits the isoproterenol-stimulated incorporation of thymidine into DNA.[21]

The ability of a multisubstrate adduct such as PALA to exert its inhibitory effect on the appropriate biochemical reaction in intact cells and whole animals is very encouraging. However, it should be noted that although PALA apparently penetrates the cell membranes,[20,21] the closely related substrate, carbamyl phosphate, does not gain entry into cells.[21] Certainly, a more complete understanding of the transport process would be of enormous value in designing multisubstrate adducts for use *in vivo*. The inability of N-(phosphopyridoxyl)ornithine, a multisubstrate adduct inhibitor of ornithine decarboxylase,[25] to affect the activity of the enzyme in drug-treated H-35 cells or to affect the biosynthesis of polyamines suggests that the drug is not readily transported. The use of tritiated inhibitor in cell culture revealed that the cells took up less than 1% of the administered drug, suggesting that the cells are impermeable to this compound.[26] In contrast are the results obtained using methionyladenylate as an inhibitor of the biosynthesis of f-Met t-RNA.[27] The fact that protein synthesis and virus focus formation were blocked in cells treated with this compound led the authors to conclude that the cell was permeable to this type of molecule. However, only nonradioactive inhibitor was used, and it was not clearly established that the charged drug was able to penetrate the cell membrane.

A major advantage to be expected in the use of multisubstrate adducts as enzyme inhibitors, in addition to their potency as analogs of reaction intermediates, is a high degree of specificity for a particular enzyme found in the complex cellular milieu. Unfortunately, data on this point in the literature are limited. For example, what are the effects of PALA on other enzymes which use either aspartate or carbamyl phosphate as a substrate? In the case of multisubstrate adducts which inhibit pyridoxal(phosphate)-dependent enzymes, some data of this type are available. Heller et al.[25] found that N-(phosphopyridoxyl)ornithine was a potent inhibitor of ornithine decarboxylase but had no effect on S-adenosylmethionine decarboxylase. Since these two enzymes are involved in the synthesis of the two precursors of spermidine, namely putrescine and decarboxylated S-adenosylmethionine, specificity of the type observed aids considerably in the interpretation of drug effects on polyamine synthesis in more complex systems such as cultured cells or whole animals. Turano et al.[28] have also observed a similar specificity in the action of several N-(phosphopyridoxyl)amino acids on tyrosine aminotransferase. N-(Phosphopyridoxyl)tyrosine is a much more potent inhibitor of this enzyme ($K_i = 10^{-9}$ M) than noncognate amino acid adducts; e.g., K_i for N-(phosphopyridoxyl)phenylalanine $= 10^{-7}$ M. It should be noted that the specificity of this type of compound for the appropriate enzyme is not absolute. Thus, N-(phosphopyridoxal)ornithine is also an excellent inhibitor of lysine decarboxylase, so that the extra methylene in lysine seems not to be critical for inhibitory activity of the adduct, but it is critical for substrate activity.[25] Similar results which demonstrate a lack of absolute specificity with this type of drug have been obtained with other N-(phosphopyridoxyl)amino acids such as glutamate vs. aspartate[29] and phenylalanine vs. tyrosine.[30]

The lack of specificity observed when studying analogs of substrates or products which participate in a number of enzyme-catalyzed reactions is a major problem in the design of new drugs. Thus, the potent inhibition of a number of S-adenosylmethionine (SAM)-dependent reactions, by one of the products, S-adenosylhomocysteine (SAH), has led to the synthesis of several analogs of SAH which have been tested as inhibitors for most of the SAM-dependent methylases.[31] The analogs which have proven to be good inhibitors are generally nonspecific within this class of enzymes,[32] although recent studies have demonstrated that this type of inhibitor can block the methylation of t-RNA[33] or catecholamines[34] in specially selected cell lines. It would appear, however, that to inhibit methylation, or any other biochemical reaction involving two substrates, the use of a multisubstrate adduct would greatly increase the chances of achieving the desired specificity of drug action.

4. DRUG METABOLISM—INACTIVATION AND LETHAL SYNTHESIS

Given the myriad of drug metabolism pathways operative in various tissues,[2] the mechanisms of metabolic reactions are of considerable importance in the design of new drugs. Thus, phosphonates, such as PALA[19-21] and several analogs of glycolytic intermediates,[35] are much more stable to hydrolysis than the corresponding acyl and alkyl phosphates. Similarly, phosphonate analogs (2) of 5'-aminoacyladenylates (1) should be more stable than the corresponding phosphates (3) in terms of hydrolytic breakdown catalyzed by phosphoric

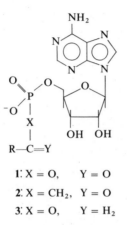

1: X = O,   Y = O

2: X = CH$_2$, Y = O

3: X = O,   Y = H$_2$

diester hydrolases (EC 3.1.4). Unfortunately, only one example of an analog of type 2 is to be found in the literature. Although the compound [2, R = C$_6$H$_5$CH$_2$CH(NH$_2$)—] is reported to be an inhibitor of phenylalanyl t-RNA synthetase,[36] no detailed biochemical data are reported. Stability of type 3 analogs has not been investigated with radiolabeled drug,[27] although compounds of this type are known to be potent inhibitors of f-Met t-RNA synthetase[27] and asparagine synthetase.[37]

Another example which illustrates the use of mechanistic information to design stable drugs is associated with the metabolism of nucleosides. Adenosine derivatives, such as SAH, are primarily metabolized by deamination at the 6 position and/or cleavage of the purine–ribose bond.[38] A methylase inhibitor, S-tubercidinylhomocysteine (STH),[32] was designed to resist metabolism via those two routes due to the presence of the tubercidin (7-deazaadenosine) moiety. It is known that tubercidin is not a substrate for adenosine deaminase and that tubercidin is stable to mild acid hydrolysis since preequilibrium protonation of the purine ring at N^7 cannot occur.[5] However, another pathway recently demonstrated in rat liver for SAH,[39] namely cleavage of the 5'-thio-ether bond, apparently predominates in the metabolism of SAH in cultured cells.[33,34] A nonthioether analog of SAH was synthesized in order to eliminate this route of metabolic inactivation of the drug, but this compound proved to be a poor methylase inhibitor in cell-free systems.[40] A related compound, the 7-deaza analog of 5′-methylthioadenosine (MTA), 5′-methylthiotubercidin (MTT), has been shown to inhibit the hydrolysis of MTA by crude rat prostate homogenates, and MTT is not itself a substrate for this enzyme-catalyzed reaction.[41] Thus, a new class of adenosine analogs can be considered if resistance to hydrolytic cleavage of the purine–ribose bond is desired, as, for example, in the study of polyamine biosynthesis, where the 2 moles of MTA produced per mole of spermine are rapidly metabolized by hydrolysis.

The role of drug metabolism in producing the active form of a drug *in vivo* has been recognized for many years.[4] The classic experiments with the azo dye, Prontosil, led to the discovery that *p*-aminobenzenesulfonamide is, in fact, the active antibacterial agent, resulting from reductive cleavage of the azo moiety of the Prontosil.[4] More recently, the conversion of dihydroxyphenylalanine (L-DOPA) to dopamine in the brain has led to a restorative treatment of Parkinson's disease, presumably due to the increased amount of dopamine available at the partially degenerated dopaminergic neurons.[1] The development of L-DOPA provides an excellent example of the use of "pro-drugs"[4] to deliver an active form of the pharmacologic agent to the desired site of action. The active agent in this case, dopamine, is a primary amine and thus positively charged at neutral pH. It does not penetrate the brain, but the amino acid precursor, L-DOPA, is readily transported. Having crossed the blood-brain barrier, the pro-drug, L-DOPA, is converted to the active form dopamine, and the desired effect of increasing the concentration of dopamine in the brain is accomplished.[1] Most of the examples of pro-drugs to be found in the literature involve esters, amides, etc., of the active drug.[42] These pro-drugs are presumably converted to the active form by nonspecific hydrolytic enzymes, although little experimental data are available to confirm this. A more selective conversion to the active form is to be found in the design of antiinflammatory agents[43] based on the observation that connective tissue has a high hyaluronidase (EC 3.2.1.35 and .36) activity. To avoid the undesired effects associated with steroid therapy, a steroid glycoside was synthesized in which 2-deoxy-2-acetamidoglucose was coupled to the exocyclic primary hydroyl group of prednisolone. The pro-drug was not hydrolyzed in plasma but was rapidly converted to the active drug either by purified hyaluronidase or by fluid from the inflamed joint. In experimental animals one of the most undesired effects of steroid therapy, ulcerogenicity, was less than 5% of that observed with the parent drug prednisolone.[43]

More recently, Higuchi and co-workers[44,45] have used the pro-drug approach to develop a new antidote for diisopropyl fluorophosphate (DFP) poisoning. One of the earliest examples of drug development based on a detailed chemical understanding of the enzyme reaction involved was in the development of DFP antidotes.[1] The mechanism of action of acetylcholinesterase (AChE) is now well established,[46] and its inactivation by DFP led to an understanding of how various structurally related nerve gases act. This is outlined in scheme (2) and is readily recognized as the prototype "serine esterase." The antidote, 2-pyridinealdoxime methiodide (2-PAM), was designed to reactivate AChE by first binding to the anionic site and then dis-

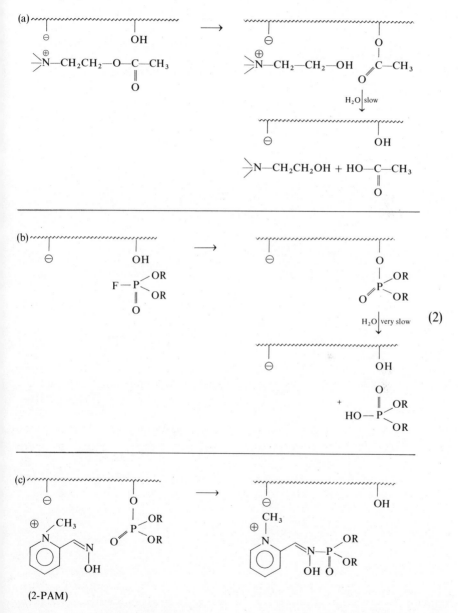

(2)

$$R = -CH(CH_3)_2$$

placing the enzyme-bound phosphate ester via nucleophilic attack of the oxime moiety.[2] However, 2-PAM is a quaternary pyridinium derivative and as such only slowly penetrates the brain. Higuchi and co-workers synthesized Pro-2-PAM, which is much more readily taken up by the brain and more rapidly reactivates AChE in rat brains.[45] Pro-2-PAM also contains a quaternary nitrogen as synthesized, but the low pK_a of 6.32 would indicate that an acidic hydrogen of the pro-drug is removed in neutral media to produce a noncharged molecule, which penetrates the brain where it is oxidized to the active form, 2-PAM. This is outlined in scheme (3):

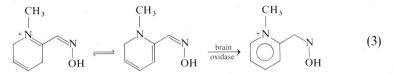

(3)

    In the examples of lethal synthesis discussed thus far, an inactive form of the drug (pro-drug) is converted to the active form via an enzyme-catalyzed reaction not normally involved in the metabolism of the active drug. In contrast are the class of pro-drugs called "suicide substrates" or "k_{cat} inhibitors."[47,48] This class of agents has been extensively reviewed,[48-50] and in this section the discussion will center on a classic and well-documented example, namely the irreversible inhibition of β-hydroxydecanoyl thioester dehydrase by $\Delta^{3,4}$-decenoyl N-acetyl cysteamine, as depicted in scheme (4).[51]

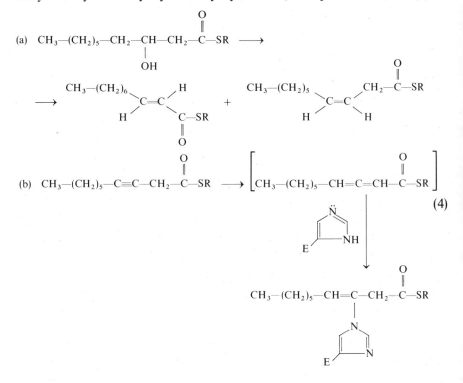

(4)

The acetylenic compound acts as a substrate for the enzyme; however, the allene intermediate which is generated is irreversibly alkylated by a proximate histidine residue, thus irreversibly inactivating the enzyme. Parallel structural variations in the acetylenic inhibitor vs. the β-hydroxy ester substrates result in parallel effects on inhibitory and substrate activity, respectively.[52] This lends strong support to the hypothesis that the inhibitor acts first as a substrate to generate a reactive intermediate at the enzyme active site. In *E. coli*, the acetylenic compound is a potent inhibitor of fatty acid biosynthesis,[53] thus indicating a high degree of specificity in a complex cellular milieu. Inhibitors of the type just discussed are an exciting new addition to the growing list of pro-drugs which require a more complete understanding of underlying enzyme mechanisms than heretofore utilized in drug design. Incorporating mechanistic information of the type considered in this and earlier chapters should enable persons interested in drug research to more accurately understand the mode of drug action and to design more effective agents.

ACKNOWLEDGMENT

Research carried out in the author's laboratory was supported by grants from the U.S. Public Health Service, #MH-18,038 and #CA-10,748.

REFERENCES

1. L. S. Goodman and A. Gilman, eds., *The Pharmacological Basis of Therapeutics*, 5th ed., Macmillan, New York (1975).
2. A. Goldstein, L. Aronow, and S. M. Kalman, eds., *Principles of Drug Action*, 2nd ed., Wiley, New York (1974).
3. H. P. Rang, ed., *Drug Receptors*, University Park Press, Baltimore (1973).
4. A. Albert, *Selective Toxicity; the Physico-Chemical Basis of Therapy*, 5th ed., Chapman & Hall, London (1973).
5. R. J. Suhadolnik, *Nucleoside Antibiotics*, Wiley-Interscience, New York (1970).
6. D. V. Santi and Charles S. McHenry, 5-Fluoro-2'-deoxyuridylate: Covalent complex with thymidylate synthetase, *Proc. Nat. Acad. Sci. USA* **69**, 1855–1857 (1972).
7. D. V. Santi, C. S. McHenry, and H. Sommer, Mechanism of interaction of thymidylate synthetase with 5-fluorodeoxyuridylate, *Biochemistry* **13**, 471–481 (1974).
8. R. J. Langenback, P. V. Danenberg, and C. Heidelberger, Thymidylate synthetase: Mechanism of inhibition by 5-fluoro-2'-deoxyuridylate, *Biochem. Biophys. Res. Commun.* **48**, 1565–1571 (1972).
9. P. V. Danenberg, R. J. Langenback, and C. Heidelberger, Structures of reversible and irreversible complexes of thymidylate synthetase and fluorinated pyrimidine nucleotides, *Biochemistry* **13**, 926–933 (1974).
10. A. L. Pogolotti, Jr., and D. V. Santi, Model studies of the thymidylate synthetase reaction. Nucleophilic displacement of 5-*p*-nitrophenoxymethyluracils, *Biochemistry* **13**, 456–466 (1974), and references therein.
11. Y. Wataya and D. V. Santi, Thymidylate synthetase catalyzed dehalogenation of 5-bromo- and 5-iodo-2'-deoxyuridylate, *Biochem. Biophys. Res. Commun.* **67**, 818–823 (1975).
12. T. I. Kalman, Glutathione-catalyzed hydrogen isotope exchange of position 5 of uridine. A model for enzymic carbon alkylation reactions of pyrimidines, *Biochemistry* **10**, 2567–2573

(1971); Inhibition of thymidylate synthetase by showdomycin and its 5'-phosphate, *Biochem. Biophys. Res. Commun.* **49**, 1007–1013 (1972).

13. (a) C. S. McHenry and D. V. Santi, A sulfhydryl group is not the covalent catalyst in the thymidylate synthetase reaction, *Biochem. Biophys. Res. Commun.* **57**, 204–208 (1974). (b) H. Sommer and D. V. Santi, Purification and amino acid analysis of an active site peptide from thymidylate synthetase containing covalently bound 5-fluoro-2'-deoxyuridylate and methylenetetrahydrofolate, *Biochem. Biophys. Res. Commun.* **57**, 689–695 (1974). (c) R. L. Bellisario, G. F. Maley, J. H. Galivan, and F. Maley, Amino acid sequence at the FdUMP binding site of thymidylate synthetase, *Proc. Natl. Acad. Sci. U.S.A.* **73**, 1848–1852 (1952).

14. W. C. Werkheiser, Specific binding of 4-amino folic acid analogues by folic acid reductase, *J. Biol. Chem.* **236**, 888–893 (1961).

15. J. R. Bertino, B. A. Booth, A. L. Bieber, A. Cashmore, and A. C. Sartorelli, Studies on the inhibition of dihydrofolate reductase by the folate antagonists, *J. Biol. Chem.* **239**, 479–485 (1964).

16. M. Poe, N. J. Greenfield, J. M. Hirshfield, and K. Hoogsteen, Dihydrofolate reductase from a methotrexate-resistant strain of *Escherichia coli*: Binding of several folates and pteridines as monitored by ultraviolet difference spectroscopy, *Cancer Biochem. Biophys.* **1**, 7–11 (1974).

17. R. Wolfenden, Analog approaches to the structure of the transition state in enzyme reactions, *Acc. Chem. Res.* **5**, 10–18 (1972).

18. R. Wolfenden, Analog inhibitors and enzyme catalysis, *Annu. Rev. Biophys. Bioeng.* **5**, 271–305 (1976).

19. K. D. Collins and G. R. Stark, Aspartate transcarbamylase. Interaction with the transition state analogue *N*-(phosphonacetyl)-L-aspartate, *J. Biol. Chem.* **246**, 6599–6605 (1971).

20. E. A. Swyryd, S. A. Seaver, and G. R. Stark, *N*-(Phosphonacetyl)-L-aspartate, a potent transition state analog inhibitor of aspartate transcarbamylase, blocks proliferation of mammalian cells in culture, *J. Biol. Chem.* **249**, 6945–6950 (1974).

21. T. Yoshida, G. R. Stark, and N. J. Hoogenraad, Inhibition by *N*-(phosphonacetyl)-L-aspartate of aspartate transcarbamylase activity and drug-induced cell proliferation in mice, *J. Biol. Chem.* **249**, 6951–6955 (1974).

22. G. R. Jacobson and G. R. Stark, in: *The Enzymes*, 3rd ed. (P. D. Boyer, ed.), Vol. IX, pp. 226–308, Academic Press, New York (1973).

23. N. J. Hoogenraad, Reaction mechanism of aspartate transcarbamylase from mouse spleen, *Arch. Biochem. Biophys.* **161**, 76–82 (1974).

24. K. Schray and J. P. Klinman, The magnitude of enzyme transition state analog binding constants, *Biochem. Biophys. Res. Commun.* **57**, 641–648 (1974).

25. J. S. Heller, E. S. Canellakis, D. L. Bussolotti, and J. K. Coward, Stable multisubstrate adducts as enzyme inhibitors. Potent inhibition of ornithine decarboxylase by *N*-(*S'*-phosphopyridoxyl)ornithine, *Biochim. Biophys. Acta* **403**, 197–207 (1975).

26. J. S. Heller, N. C. Motola, J. K. Coward, and E. S. Canellakis, unpublished results.

27. M. Robert-Gero, F. Lawrence, and P. Vigier, Inhibition by methioninyl adenylate of focus formation by rous sarcoma virus, *Cancer Res.* **35**, 3571–3576 (1975), and references therein,

28. C. Borri Voltattorni, A. Orlacchio, A. Giartosio, F. Conti, and C. Turano, The binding of coenzymes and analogues to the substrate–coenzyme complex to tyrosine aminotransferase, *Eur. J. Biochem.* **53**, 151–160 (1975).

29. E. S. Severin, N. N. Gulyaev, E. N. Khurs, and R. M. Khomutov, The synthesis and properties of phosphopyridoxyl amino acids, *Biochem. Biophys. Res. Commun.* **35**, 318–323 (1969).

30. C. Borri Voltattorni, A. Minelli, and C. Turano, *Boll. Soc. Ital. Biol. Sper.* **47**, 700–702 (1971).

31. R. T. Borchardt, in: *The Biochemistry of S-Adenosylmethionine* (F. Salvatore and E. Borek, eds.), Columbia University Press, New York in press.

32. J. K. Coward, D. L. Bussolotti, and C.-D. Chang, Analogs of *S*-adenosylhomocysteine as potential inhibitors of biological transmethylation. Inhibition of several methylases by *S*-tubercidinylhomocysteine, *J. Med. Chem.* **17**, 1286–1289 (1974).

33. C.-D. Chang and J. K. Coward, Effect of *S*-adenosylhomocysteine and *S*-tubercidinylhomo-

cysteine on transfer ribonucleic acid methylation in phytohemagglutinin-stimulated lymphocytes, *Mol. Pharmacol.* **11**, 701–707 (1975).

34. R. J. Michelot, N. Lesko, R. W. Stout, and J. K. Coward, Effect of S-adenosylhomocysteine and S-tubercidinylhomocysteine on catecholamine methylation in neuroblastoma cells, *Mol. Pharmacol.* **13**, 368–373 (1977).

35. P-J. Cheng, W. D. Nunn, R. J. Tyhach, S. L. Goldstein, R. Engel, and B. E. Trapp, Investigations concerning the mode of action of 3,4-dihydroxybutyl-1-phosphonate on *Escherichia coli*. *In vitro* examination of enzymes involved in glycerol 3-phosphate metabolism, *J. Biol. Chem.* **250**, 1633–1639 (1975).

36. G. Goring and F. Cramer, Synthese von Inhibitoren für die Phenylalanyl-tRNA-Synthetase: Methylen-Analoge des Phenylalanyl-adenylats, *Chem. Ber.* **106**, 2460–2467 (1973).

37. J. Uren and P. K. Chang, personal communication.

38. J. Frank Henderson and A. R. P. Paterson, *Nucleotide Metabolism,* Academic Press, New York (1973).

39. R. D. Walker and J. A. Duerre, S-Adenosylhomocysteine metabolism in various species, *Can. J. Biochem.* **53**, 312–319 (1975).

40. C.-D. Chang and J. K. Coward, Analogues of S-adenosylhomocysteine as potential inhibitors of biological transmethylation. Synthesis of analogues with modifications at the 5′-thioether linkage, *J. Med. Chem.* **19**, 684–691 (1976).

41. J. K. Coward, N. C. Motola, and J. D. Moyer, Polyamine biosynthesis in rat prostate. Substrate and inhibitor properties of 7-deaza analogues of decarboxylated S-adenosylmethionine and 5′-methylthioadenosine, *J. Med. Chem.* **20**, 500–505 (1977).

42. A. A. Sinkula, The prodrug approach in drug design, *Annu. Rep. Med. Chem.* **10**, 306–316 (1975).

43. R. Hirschmann, R. G. Strachan, P. Buchschacher, L. H. Sarrett, S. L. Steelman, and R. Silber, An approach to an improved antiinflammatory steroid. The synthesis of 11β,17-dihydroxy-3,20-dione-1,4-pregnadien-21-yl 2-adetamido-2-deoxy-β-D-glucopyranoside, *J. Am. Chem. Soc.* **86**, 3903–3904 (1964).

44. N. Bodor, E. Shek, and T. Higuchi, Delivery of a quaternary pyridinium salt across the blood-brain barrier by its dihydropyridine derivative, *Science* **190**, 155–156 (1975).

45. (a) N. Bodor, E. Shek, and T. Higuchi, Improved delivery through biological membranes. 1. Synthesis and properties of 1-methyl-1,6-dihydropyridine-2-carbaldoxime, a pro-drug of N-methylpyridinium-2-carbaldoxime chloride, *J. Med. Chem.* **19**, 102–107 (1976). (b) E. Shek, T. Higuchi, and N. Bodor, Improved delivery through biological membranes. 2. Distribution, excretion, and metabolism of N-methyl-1,6-dihydropyridine-2-carbaldoxime hydrochloride, a pro-drug of N-methylpyridinium-2-carbaldoxime chloride, *J. Med. Chem.* **19**, 108–112 (1976). (c) E. Shek, T. Higuchi, and N. Bodor, Improved delivery through biological membranes. 3. Delivery of N-methylpyridinium-2-carbaldoxime chloride through the blood-brain barrier in its dihydropyridine pro-drug form, *J. Med. Chem.* **19**, 113–117 (1976).

46. T. L. Rosenberry, Acetylcholinesterase, *Adv. Enzymol.* **43**, 103–218 (1975).

47. R. H. Abeles and A. L. Maycock, Suicide enzyme inactivators, *Acc. Chem. Res.* **9**, 313–319 (1976).

48. R. R. Rando, Chemistry and enzymology of k_{cat} inhibitors, *Science* **185**, 320–324 (1974).

49. R. R. Rando, Mechanisms of action of naturally occurring irreversible enzyme inhibitors, *Acc. Chem. Res.* **8**, 281–288 (1975).

50. R. R. Rando, Commentary on the mechanism of action of antibiotics which act as irreversible enzyme inhibitors, *Biochem. Pharmacol.* **24**, 1153–1160 (1975).

51. K. Endo, G. M. Helmkamp, and K. Block, Mode of inhibition of β-hydroxydecanoyl thioester dehydrase by 3-decynoyl-N-acetylcysteamine, *J. Biol. Chem.* **245**, 4293–4296 (1970).

52. G. M. Helmkamp, R. R. Rando, D. J. H. Brock, and K. Block, β-Hydroxydecanoyl thioester dehydrase. Specificity of substrates and acetylenic inhibitors, *J. Biol. Chem.* **243**, 3229–3231 (1968).

53. L. R. Kass, The antibacterial activity of 3-decynoyl-N-acetylcysteamine, *J. Biol. Chem.* **243**, 3223–3228 (1968).

Author Index

Italic numbers indicate pages where complete reference citations are given.

Subject Index